Springer Collected Works in Mathematics

More information about this series at http://www.springer.com/series/11104

JACK KIEFER
1961

Jack Carl Kiefer

Collected Papers II

Statistical Inference and Probability (1964–1984)

Editors
Lawrence D. Brown
Ingram Olkin
Jerome Sacks
Henry P. Wynn

Reprint of the 1986 Edition

Springer

Author
Jack Carl Kiefer (1924–1981)
University of California
Berkeley, CA
USA

Editors
Lawrence D. Brown
University of Pennsylvania
Philadelphia, PA
USA

Ingram Olkin
Stanford University
Stanford, CA
USA

Jerome Sacks
Duke University
Durham, NC
USA

Henry P. Wynn
The London School of Economics
London
UK

Published with the cooperation of the Institute of Mathematical Statistics

ISSN 2194-9875
Springer Collected Works in Mathematics
ISBN 978-1-4939-3499-7 (Softcover)
978-1-4613-8507-3 (Hardcover)

Library of Congress Control Number: 2012954381

Springer New York Heidelberg Dordrecht London

Printed on acid-free paper

Springer Science+Business Media LLC New York is part of Springer Science+Business Media
(www.springer.com)

Jack Carl Kiefer
Collected Papers II

Statistical Inference and Probability
(1964–1984)

Published with the co-operation of the
Institute of Mathematical Statistics
and edited by

Lawrence D. Brown
Ingram Olkin
Jerome Sacks
Henry P. Wynn

Springer-Verlag
New York Berlin Heidelberg Tokyo

Lawrence D. Brown
Department of Mathematics
Cornell University
Ithaca, NY 14853
U.S.A.

Ingram Olkin
Department of Statistics
Stanford University
Stanford, CA 94305
U.S.A.

Jerome Sacks
Department of Mathematics
University of Illinois
Urbana, IL 61801
U.S.A.

Henry P. Wynn
Imperial College
London, England

AMS Subject Classification: 62-XX

Library of Congress Cataloging in Publication Data
Kiefer, Jack, 1924–1981
Jack Carl Kiefer collected papers.
"Published with the co-operation of the Institute of Mathematical Statistics."
Bibliography: p.
Contents: 1. Statistical inference and probability, 1951–1963—2. Statistical inference and probability, 1964–1984—3. Design of experiments.
1. Mathematical statistics—Collected works.
I. Brown, Lawrence D. II. Institute of Mathematical Statistics. III. Title.
QA276.A12K54 1984 519.5 84-10598

Softcover reprint of the hardcover 1st edition 1985

ISBN-13: 978-1-4613-8507-3 e-ISBN-13: 978-1-4613-8505-9
DOI: 10.1007/978-1-4613-8505-9

9 8 7 6 5 4 3 2 1

Preface

Jack Kiefer's sudden and unexpected death in August, 1981, stunned his family, friends, and colleagues. Memorial services in Cincinnati, Ohio, Berkeley, California, and Ithaca, New York, shortly after his death, brought forth tributes from so many who shared in his life. But it was only with the passing of time that those who were close to him or to his work were able to begin assessing Jack's impact as a person and intellect.

About one year after his death, an expression of what Jack meant to all of us took place at the 1982 annual meeting of the Institute of Mathematical Statistics and the American Statistical Association. Jack had been intimately involved in the affairs of the IMS as a Fellow since 1957, as a member of the Council, as President in 1970, as Wald lecturer in 1962, and as a frequent author in its journals. It was doubly fitting that the site of this meeting was Cincinnati, the place of his birth and residence of his mother, other family, and friends. Three lectures were presented there at a Memorial Session—by Jerry Sacks dealing with Jack's personal life, by Larry Brown dealing with Jack's contributions in statistics and probability, and by Henry Wynn dealing with Jack's contributions to the design of experiments. These three papers, together with Jack's bibliography, were published in the *Annals of Statistics* and are included as an introduction to these volumes. For those who knew Jack, they serve us as a reminder of the impact he had and how dearly he was held. For those who didn't know Jack, they provide some insight into his character and the development of his research efforts.

Jack's career was centered first at Cornell University and later at the University of California in Berkeley. Jack's presence as an influential faculty member for 25 years at Cornell was only superseded by his magnetic affect on the statistical community at large, drawing any number of visitors and associates from nearby and far. Although at Berkeley only a short time, by the time of his death Jack had become a focus of research activity among students, faculty, and visitors. Cornell and Berkeley each paid tribute to Jack by holding symposia in his honor, shared by Jack Wolfowitz at Cornell and Jerzy Neyman at Berkeley.

The symposia helped express a sense of loss of the person and a sense of honor and respect for his achievements. Above all, they showed how current and vital his work was even after his death. For it was Jack's scientific work that heavily affected the profession at large. His collected papers, numbering over 100 and comprising more than 1600 pages, exhibit an amazing range of interests. They cover sequential analysis, nonparametric analysis, decision theory, multivariate analysis, inventory theory, stochastic processes, and design of experiments. Brought together, these papers are a testament to a pre-eminent scientific career.

Though these papers stand by themselves, and so many of the papers contain introductions or comments reflecting Jack's acute sense of where matters stood at the time they were written, we thought it useful to provide some overview of the works, how they related to research at the time, and their impact on subsequent work. Therefore, we asked a number of people to provide commentary on some of these papers. The response was overwhelming in that virtually everyone responded with thoughtful and perceptive comments. Though our intention was to help readers and scholars to see a large picture, by being selective we run the risk of perhaps neglecting certain issues, threads, and pertinent facts. There are many who were not asked but could easily have given equally responsive comments. We know that they will understand that no slight was intended on their appreciation of Jack's works.

Many of Kiefer's papers cover a wide spectrum of areas for which there is no neat rubric to describe them. For lack of a well-defined title we label these Statistical Inference and Probability (Volumes I and II). A second group of papers can very naturally be characterized as Design of Experiments, and these comprise Volume III.

A few papers, especially the early ones, could be labeled either as Design or as Statistical Inference and Probability. Because of the length of the Design Volume, we have placed these in Volumes I and II.

The Bibliography is the basic source of referencing, and Kiefer's papers are referred to by [·]. Other references follow standard citation style.

Since many of the commentaries are on groups of papers, it was difficult to assign a particular commentary to a single paper. The commentary section for papers on Statistical Inference and Probability is located at the end of Volume II. The commentary section for Design is at the end of Volume III. Some duplication of commentary was unavoidable; these serve to offer different perspectives on a particular paper or issue. Some papers which have no direct comment are discussed in the introductory articles by Brown and Wynn.

To the commentators and to all those, including Jack's family, who helped in the task of assembling and preparing these volumes for publication, we wish to express our deepest thanks.

January, 1984

Lawrence Brown
Ingram Olkin
Jerome Sacks
Henry Wynn

Contents

*Papers with asterisks have commentaries (see pages vii–xiii).

Bibliography of Jack Kiefer

A. Papers (Published)

[1] Almost subminimax and biased minimax procedures, (with P. Frank). *Ann. Math. Statist.* **22** (1951), 465–468. [MR 13 (1952) 143, Zbl 43 (1952) 346]. (I)

[2] The inventory problem: I. Case of known distributions of demand, (with A. Dvoretzky and J. Wolfowitz). *Econometrica* **20** (1952), 187–222. [MR 13 (1952) 856, Zbl 46 (1953) 376]. (I)

[3] The inventory problem: II. Case of unknown distributions of demand, (with A. Dvoretzky and J. Wolfowitz). *Econometrica* **20** (1952), 450–466. [MR 14 (1953) 301, Zbl 48 (1953) 371]. (I)

[4] Stochastic estimation of the maximum of a regression function, (with J. Wolfowitz). *Ann. Math. Statist.* **23** (1952), 462–466. [MR 14 (1953) 299, Zbl 49 (1954) 366]. (I)

[5] Sequential minimax estimation for the rectangular distribution with unknown range. *Ann. Math. Statist.* **23** (1952), 586–593. [MR 14 (1953) 487, Zbl 48 (1953) 121]. (I)

[6] On minimum variance estimators. *Ann. Math. Statist.* **23** (1952), 627–629. [MR 15 (1954) 241, Zbl 48 (1953) 120]. (I)

[7] On Wald's complete class theorems. *Ann. Math. Statist.* **24** (1953), 70–75. [MR 14 (1953) 998, Zbl 50 (1954) 140]. (I)

[7a] Correction of a proof. *Ann. Math. Statist.* **24** (1953), 680. (I)

[8] Sequential minimax search for a maximum. *Proc. Amer. Math. Soc.* **4** (1953), 502–506. [MR 14 (1953) 1103, Zbl 50 (1954) 357]. (I)

[9] Sequential decision problems for processes with continuous time parameter. Testing hypotheses, (with A. Dvoretzky and J. Wolfowitz). *Ann. Math. Statist.* **24** (1953), 254–264. [MR 14 (1953) 997, 1279, Zbl 50 (1954) 148]. (I)

[9a] Corrections to "Sequential decision problems for processes with continuous time parameter testing hypotheses". *Ann. Math. Statist.* **30** (1959), 1265. (I)

[10] Sequential decision problems for processes with continuous time parameter. Problems of estimation, (with A. Dvoretzky and J. Wolfowitz). *Ann. Math. Statist.* **24** (1953), 403–415. [MR (1954) 242, Zbl 51 (1954) 366]. (I)

[11] On the optimal character of the (s, S) policy in inventory theory, (with A. Dvoretzky and J. Wolfowitz). *Econometrica* **21** (1953), 586–596. [MR 15 (1954) 333, Zbl 53 (1956) 279]. (I)

[12] On the theory of queues with many servers, (with J. Wolfowitz). *Trans. Amer. Math. Soc.* **78** (1955), 1–18. [MR 16 (1955) 601, Zbl 64 (1956) 133]. (I)

[13] On tests of normality and other tests of goodness of fit based on distance methods, (with M. Kac and J. Wolfowitz). *Ann. Math. Statist.* **26** (1955), 189–211. [MR 17 (1956) 55, Zbl 66 (1956–1957) 123]. (I)

[14] On the characteristics of the general queueing process, with applications to random walk, (with J. Wolfowitz). *Ann. Math. Statist.* **27** (1956), 147–161. [MR 17 (1956) 980, Zbl 70 (1957) 366]. (I)

[15] Asymptotic minimax character of the sample distribution function and of the classical multinomial estimator, (with A. Dvoretzky and J. Wolfowitz). *Ann. Math. Statist.* **27** (1956), 642–669. [MR 18 (1957) 772, Zbl 73 (1959–60) 146]. (I)

[16] Consistency of the maximum likelihood estimator in the presence of infinitely many incidental parameters, (with J. Wolfowitz). *Ann. Math. Statist.* **27** (1956), 887–906. [MR 19 (1958) 189, Zbl 73 (1959–60) 147]. (I)

[17] Sequential tests of hypotheses about the mean occurrence time of a continuous parameter Poisson process, (with J. Wolfowitz). *Naval Res. Logist. Quart.* **3** (1956), 205–219. [MR 18 (1957) 833]. (I)

[18] Some properties of generalized sequential probability ratio tests, (with L. Weiss). *Ann. Math. Statist.* **28** (1957), 57–74. [MR 19 (1958) 333, Zbl 79 (1959) 354]. (I)

[19] Invariance, minimax sequential estimation, and continuous time processes. *Ann. Math. Statist.* **28** (1957), 573–601. [MR 19 (1958) 1097, Zbl 80 (1959) 130]. (I)

[20] Optimum sequential search and approximation methods under minimum regularity assumptions. *J. Soc. Indust. Appl. Math.* **5** (1957), 105–136. [MR 19 (1958) 1097, Zbl 81 (1959) 385]. (I)

[21] On the deviations of the empiric distribution function of vector chance variables, (with J. Wolfowitz). *Trans. Amer. Math. Soc.* **87** (1958), 173–186. [MR 20 (1959) #5519, Zbl 88 (1961) 113]. (I)

[22] On the nonrandomized optimality and randomized nonoptimality of symmetrical designs. *Ann. Math. Statist.* **29** (1958), 675–699. [MR 20 (1959) #4910, Zbl 92 (1962) 361]. Corrections (and comments), unpublished. (III)

[23] Optimum designs in regression problems, (with J. Wolfowitz). *Ann. Math. Statist.* **30** (1959), 271–294. [MR 21 (1960) #3079, Zbl 90 (1961) 114]. (III)

[24] *K*-sample analogues of the Kolmogorov–Smirnov and Carmér-V. Mises tests. *Ann. Math. Statist.* **30** (1959), 420–447. [MR 21 (1960) #1668, Zbl 134 (1967) 367]. (I)

[25] Asymptotic minimax character of the sample distribution function for vector chance variables, (with J. Wolfowitz). *Ann. Math. Statist.* **30** (1959), 463–489. [MR 21 (1960) #6642, Zbl 93 (1962) 156]. (I)

[26] Optimum experimental designs. *J. Roy. Statist. Soc., Ser. B* **21** (1959), 272–319. [MR 22 (1961) #4101, Zbl 108 (1964) 153]. (III)

[27] A functional equation technique for obtaining Wiener process probabilities associated with theorems of Kolmogorov–Smirnov type. *Proc. Cambridge Philos. Soc.* **55** (1959), 328–332. [MR 22 (1961) #8557, Zbl 96 (1962) 334]. (I)

[28] Optimum experimental designs V, with applications to systematic and rotatable

designs. *Proc. 4th Berkeley Sympos. Math. Statist. and Prob.* **1** (1960), 381–405, Univ. California Press, Berkeley, Calif. [MR 24 (1962) #A3765, Zbl 134 (1967) 366]. (III)

[29] The equivalence of two extremum problems, (with J. Wolfowitz). *Canad. J. Math.* **12** (1960), 363–366. [MR 22 (1961) #8616, Zbl 93 (1962) 156]. (III)

[30] Distribution free tests of independence based on the sample distribution function, (with J. R. Blum and M. Rosenblatt). *Ann. Math. Statist.* **32** (1961), 485–498. [MR 23 (1962) #A2989, Zbl 139 (1968) 363]. (I)

[31] Optimum designs in regression problems, II. *Ann. Math. Statist.* **32** (1961), 298–325. [MR 23 (1962) #A735, Zbl 99 (1963) 135]. (III)

[32] On large deviations of the empiric d.f. of vector chance variables and a law of the iterated logarithm. *Pacific J. Math.* **11** (1961), 649–660. [MR 24 (1962) #A1732, Zbl 119 (1966) 349]. (I)

[33] Two more criteria equivalent to D-optimality of designs. *Ann. Math. Statist.* **33** (1962), 792–796. [MR 25 (1963) #701, Zbl 116 (1965) 113]. (III)

[34] An extremum result. *Canad. J. Math.* **14** (1962), 597–601. [MR 26 (1963) #1968, Zbl 134 (1967) 369]. (III)

[35] Channels with arbitrarily varying channel probability functions, (with J. Wolfowitz). *Information and Control* **5** (1962), 44–54. [MR 24 (1962) #B2506, Zbl 107 (1964) 345]. (I)

[36] Minimax character of Hotelling's T^2 test in the simplest case, (with N. Giri and C. Stein). *Ann. Math. Statist.* **34** (1963), 1524–1535. [MR 27 (1964) #6331, Zbl 202 (1971) 495]. (I)

[37] Asymptotically optimum sequential inference and design, (with J. Sacks). *Ann. Math. Statist.* **34** (1963), 705–750. [MR 27 (1964) #893, Zbl 255 (1973) 62063]. (I)

[38] Local and asymptotic minimax properties of multivariate tests, (with N. Giri). *Ann. Math. Statist.* **35** (1964), 21–35. [MR 28 (1964) #2605, Zbl 133 (1967) 418]. (II)

[39] Minimax character of the R^2-test in the simplest case, (with N. Giri). *Ann. Math. Statist.* **35** (1964), 1475–1490. [MR 29 (1965) #6579, Zbl 137 (1967) 368]. (II)

[40] Optimum extrapolation and interpolation designs, I, (with J. Wolfowitz). *Ann. Inst. Statist. Math.* **16** (1964), 79–108. [MR 31 (1966) #2806, Zbl 137 (1967) 131]. (III)

[41] Optimum extrapolation and interpolation designs, II, (with J. Wolfowitz). *Ann. Inst. Statist. Math.* **16** (1964), 295–303. [MR 31 (1966) #2806, Zbl 137 (1967) 131]. (III)

[42] Admissible Bayes character of T^2-, R^2-, and other fully invariant tests for classical multivariate normal problems, (with R. Schwartz). *Ann. Math. Statist.* **36** (1965), 747–770. [MR 30 (1965) #5430; 50 (1975) #11567, Zbl 137 (1967) 36; 249 (1973) 62058]. (II)

[42a] Correction to "Admissible Bayes character of T^2-, R^2-, and other fully invariant tests for classical multivariate normal problems". *Ann. Math. Statist.* **43** (1972), 1742. (II)

[43] On a problem connected with the Vandermonde determinant, (with J. Wolfowitz). *Proc. Amer. Math. Soc.* **16** (1965), 1092–1095. [MR 32 (1966) #115, Zbl 142 (1968) 269]. (III)

[44] On a theorem of Hoel and Levine on extrapolation designs, (with J. Wolfowitz). *Ann. Math. Statist.* **36** (1965), 1627–1655. [MR 32 (1966) #3230, Zbl 138 (1967) 140]. (III)

[45] Multivariate optimality results. *Multivariate Analysis. Proceedings of an International Symposium* (ed. by P. R. Krishnaiah), (1966), 255–274, Academic Press, New York. [MR 37 (1969) #2372, Zbl 218 (1972) 448]. (II)

[46] Optimum multivariate designs, (with R. H. Farrell and A. Walbran). *Proc. Fifth Berkeley Sympos. Math. Statist. and Probability* **1** (1967), 113–138. [MR 35 (1968) #5099, Zbl 193 (1970) 171]. (III)

[47] On Bahadur's representation of sample quantiles. *Ann. Math. Statist.* **38** (1967), 1323–1342. [MR 36 (1968) #933, Zbl 158 (1960) 370]. (II)

[48] Statistical inference, (panel discussion with G. A. Barnard, L. M. LeCam, and L. J. Salvage). *The Future of Statistics* (Proceedings of a Conference on the Future of Statistics held at the University of Wisconsin, Madison, Wisconsin, June 1967) (ed. by D. G. Watts), (1968), 139–160, Academic Press, New York. (II)

[49] Statistical inference. *The Mathematical Sciences*, (1969), 60–71, The M.I.T. Press, Cambridge. [Reprinted in *Math. Spectrum*. **3** (1970–71), 1–11]. (II)

[50] On the deviations in the Skorokhod–Strassen approximation scheme. *Z. Wahrsch. Verw. Gebiete*. **13** (1969), 321–332. [MR 41 (1971) #1117, Erratum 41, p. 1965, Zbl 176 (1969) 482]. (II)

[51] Old and new methods for studying order statistics and sample quantiles. *Nonparametric Techniques in Statistical Inference* (ed. by M. L. Puri), (1970), 349–357, Cambridge University Press, London. [MR 44 (1972) #3442]. (II)

[52] Deviations between the sample quantile process and the sample df. *Nonparametric Techniques in Statistical Inference* (ed. by M. L. Puri), (1970), 299–319, Cambridge University Press, London. [MR 43 (1972) #2808]. (II)

[53] Optimum experimental designs. *Actes du Congrès International des Mathématiciens, Nice*, **3** (1970), 249–254. [MR 54 (1977) #8993, Zbl 237 (1972) 62050]. (III)

[54] Iterated logarithm analogues for sample quantiles when $p_n \downarrow 0$. *Proc. Sixth Berkeley Symp. on Mathematical Statistics and Probability* **1** (1970), 227–244, Univ. California Press, Berkeley. [MR 53 (1977) #6696, Zbl 264 (1974) 62815]. (II)

[55] The role of symmetry and approximation in exact design optimality. *Statistical Decision Theory and Related Topics*. (*Proc. Symp.*) (ed. by S. S. Gupta and J. Yackel), (1971), 109–118, Academic Press, New York. [MR 50 (1975) #3447, Zbl 274 (1974) 62050]. (III)

[56] Skorohod embedding of multivariate rv's, and the sample df. *Z. Wahrsch. Verw. Gebiete*. **24** (1972), 1–35. [MR 49 (1975) #6382, Zbl 267 (1974) 60034]. (II)

[57] Optimum designs for fitting biased multiresponse surfaces. *Multivariate Analysis — III. Proceedings of the Third International Symposium on Multivariate Analysis* (ed. by P. R. Krishnaiah), (1973), 287–297, Academic Press, New York. [MR 51 (1976) #2188, Zbl 291 (1975) 62093]. (III)

[58] General equivalence theory for optimum designs (approximate theory). *Ann. Statist.* **2** (1974), 849–879. [MR 50 (1975) #8856, Zbl (1975) 62092]. (III)

[59] Discussion on the paper: Planning experiments for discriminating between models, by A. C. Atkinson and D. R. Cox. *J. Roy. Statist. Soc. Ser. B* **36** (1974), 345–346. (III)

[60] Balanced block designs and generalized Youden designs, I. Construction (patchwork). *Ann. Statist.* **3** (1975), 109–118. [MR 51 (1976) #4578, Zbl 305 (1976) 62052]. (III)

[61] Construction and optimality of generalized Youden designs. *A Survey of Statistical Design and Linear Models* (ed. by J. N. Srivastava), (1975), 333–353, North-Holland Pub. Co, Amsterdam. [MR 52 (1976) #15877, Zbl 313 (1975) 62057]. (III)

[62] Review of the paper: Bayesian analysis of generic relations in Agaricales, by R. E. Machol and R. Singer. *Mycologia* **67** (1975), 203–205. (II)

[63] Optimal design: Variation in structure and performance under change of criterion. *Biometrika* **62** (1975), 277–288. [MR 52 (1976) #2064, Zbl 321 (1976) 62086]. (III)

[64] Optimal designs for large degree polynomial regression, (with W. J. Studden). *Ann. Statist.* **4** (1976), 1113–1123. [MR 54 (1977) #11676, Zbl 357 (1978) 62051]. (III)

[65] Asymptotically minimax estimation of concave and convex distribution functions, (with J. Wolfowitz). *Z. Wahrsch. Verw. Gebiete.* **34** (1976), 73–85. [MR 53 (1977) #1829, Zbl 354 (1978) 62035]. (II)

[66] Large sample comparison of tests and empirical Bayes Procedures, (with D. S. Moore). *On the History of Statistics and Probability: Proc. of a Symp. on the American Mathematical Heritage* (ed. by D. B. Owen), (1976), 347–365, M. Dekker, New York. (II)

[67] Admissibility of conditional confidence procedures. *Ann. Statist.* **4** (1976), 836–865. [MR 55 (1978) #11454, Zbl 353 (1975) 62008]. (II)

[68] Asymptotically minimax estimation of concave and convex distribution functions. II, (with J. Wolfowitz). *Statistical Decision Theory and Related Topics, II* (ed. by S. S. Gupta and D. S. Moore), (1977), 193–211, Academic Press, New York. [MR 56 (1978) #1572, Zbl 418 (1980) 62031]. (II)

[69] Conditional confidence statements and confidence estimators. *J. Amer. Statist. Assoc.* **72** (1977), 789–827. [MR 58 (1979) #24638, Zbl 375 (1978) 62023]. (II)

[70] The ideas of conditional confidence in the simplest setting, (with C. Brownie). *Comm. Statist.—Theor. Methods.* **A6**(8) (1977), 691–751. [MR 56 (1978) #3993, Zbl 392 (1979) 62002]. (II)

[71] Conditional confidence and estimated confidence in multidecision problems (with applications to selection and ranking). *Multivariate Analysis—IV. Proceedings of the Fourth International Symposium on Multivariate Analysis* (ed. by P. R. Krishnaiah), (1977), 143–158, North-Holland, Amsterdam. [MR 58 (1979) #18810, Zbl 381 (1979) 62009]. (II)

[72] Comparison of rotatable designs for regression on balls, I (Quadratic), (with Z. Galil). *J. Statist. Plann. Inference* **1** (1977), 27–40. [MR 58 (1979) #24769, Zbl 394 (1979) 62058]. (III)

[73] The foundations of statistics—are there any? *Synthese* **36** (1977), 161–176. [MR 58 (1979) #31488, Zbl 375 (1978) 60005]. (II)

[74] Comparison of design for quadratic regression on cubes, (with Z. Galil). *J. Statist. Plann. Inference* **1** (1977), 121–132. [MR 58 (1979) #24770, Zbl 381 (1979) 62062]. (III)

[75] Comparison of simplex designs for quadratic mixture models, (with Z. Galil). *Technometrics* **19** (1977), 445–453. [MR 57 (1978) #17972, Zbl 372 (1978) 62058]. (III)

[76] Comparison of Box–Draper and D-optimum designs for experiments with mixtures, (with Z. Galil). *Technometrics* **19** (1977), 441–444. [MR 58 (1979) #3233, Zbl 369–389 (1978) 62087]. (III)

[77] Asymptotic approach to familes of design problems. *Comm. Statist.—Theory Methods* **A7** (1978), 1347–1362. [MR 82g (1982) 62106, Zbl 389 (1979) 62058]. (III)

[78] A Diophantine problem in optimum design theory. *Utilitas Math.* **14** (1978), 81–98. [MR 80b (1980) 62091, Zbl 391 (1979) 62055]. (III)

[79] Comment on paper: Pseudorandom number assignment in statistically designed simulation and distribution sampling experiments, by L. W. Schruben and B. H. Margolin. *J. Amer. Statist. Assoc.* **73** (1978), 523–524. (III)

[80] Extrapolation designs and Φ_p-optimum designs for cubic regression on the *q*-ball, (with Z. Galil). *J. Statist. Plann. Inference* **3** (1979), 27–38. [MR 81a (1981) 62074, Zbl 412 (1980) 62055]. (III)

[81] Sequential statistical methods. *Studies in Probability Theory* (ed. by M. Rosenblatt), (1978), 1–23, Studies in Math., 18, Math. Assoc. Amer., Washington, D.C. [MR 80m (1980) 62077, Zbl 412 (1980) 62056]. (II)

[82] Comments on taxonomy, independence, and mathematical models (with reference to a methodology of Machol and Singer). *Mycologia* **71** (1979), 343–378. (II)

[83] Optimal design theory in relation to combinatorial design. *Ann. Discrete Math.* **6** (1980), 225–241. [MR 82a (1982) 62107, Zbl 463 (1982) 62066]. (III)

[84] Designs for extrapolation when bias is present. *Multivariate Analysis — V. Proc. Fifth International Symp.* (ed. by P. R. Krishnaiah), (1980), 79–93, North-Holland Pub. Co., Amsterdam. [MR 81i (1981) 62130, Zbl 458 (1982) 62063]. (III)

[85] *D*-optimum weighing designs, (with Z. Galil). *Ann. Statist.* **8** (1980), 1293–1306. [MR 82g (1982) 62104, Zbl 466 (1982) 62066]. (III)

[86] Time- and space-saving computer methods, related to Mitchell's DETMAX, for finding D-optimum designs, (with Z. Galil). *Technometrics* **22** (1980), 301–313. [MR 81j (1981) 62147, Zbl 459 (1982) 62060]. (III)

[87] Optimum weighing designs, (with Z. Galil). *Recent Developments in Statistical Inference and Data Analysis. Proceedings of the International Conference in Statistics in Tokyo* (ed. by K. Matusita), (1980), 183–189, North-Holland Pub. Co. Amsterdam. [MR 82a (1982) 62108, Zbl 462 (1982) 62059]. (III)

[88] Optimum balanced block and Latin square designs for correlated observations, (with H. P. Wynn). *Ann. Statist.* **9** (1981), 737–757. [MR 82h (1982) 62122]. (III)

[89] The interplay of optimality and combinatorics in experimental design. *Canad. J. Statist.* **9** (1981), 1–10. [MR 82m (1982) 62167]. (III)

[90] Relationships of optimality for individual factors of a design, (with J. Eccleston). *J. Statist. Plann. Inference* **5** (1981), 213–219. [MR 83c (1983) 62115, Zbl 481 (1982) 62059]. (III)

[91] Optimum rates for non-parametric density and regression estimates, under order restrictions. *Statistics and Probability, Essays in Honor of C. R. Rao* (ed. by G. Kallianpur, P. R. Krishnaiah, J. K. Ghosh), (1982), 419–428, North-Holland Pub. Co., Amsterdam. (II)

[92] On the characterization of D-optimum weighing designs for $n \equiv 3 \pmod 4$, (with Z. Galil). *Statistical Decision Theory and Related Topics III* (ed. By S. S. Gupta and J. O. Berger), **1** (1982), 1–35, Academic Press, New York. (III)

[93] Eight lectures on mathematical statistics. (Chinese) *Advances in Mathematics* (*Beijing*) (Shuxue Jin Zhan) **10** (1981), 94–130. (II)

[94] Conditional inference. *Encyclopedia of Statistical Sciences*, Volume 2 (ed. By S. Kotz, N. L. Johnson, and C. B. Read), (1982), 103–109, Wiley-Interscience. (II)

[95] Construction methods for *D*-optimum weighing designs when $n \equiv 3 \pmod 4$, (with Z. Galil). *Ann. Statist.* **10** (1982), 502–510. [MR 83i (1983) 62139]. (III)

[96] Autocorrelation-robust design of experiments, (with H. P. Wynn). *Scientific Inference, Data Analysis, and Robustness* (ed. by T. Leonard and C. -F. Wu), (1983), 279–299, Academic Press, New York. (III)

[97] Comparison of designs equivalent under one or two criteria, (with Z. Galil). *J.*

Statist. Plann. Inference **8** (1983), 103–116. (III)

[98] Optimum and minimax exact treatment designs for one-dimensional autoregressive error processes, (with H. P. Wynn). *Ann. Statist.* **12** (1984), 414–450. (III)

B. Book Reviews

[99] Review of *The Advanced Theory of Statistics*, Volume 2, "Inference and Relationship," M. G. Kendall and A. Stuart. *Ann. Math. Statist.* **35** (1964), 1371–1380. (II)

[100] Review of *The Savory Wild Mushroom*, M. McKenney. (Second edition, revised and enlarged by D. E. Stuntz). *Quarterly Review Biology* **47** (1972), 342–343. (II)

[101] Review of *A Field Guide to Western Mushrooms*, A. H. Smith. *Quarterly Review Biology* **52** (1977), 91. (II)

C. Books, Edited Volumes, Lecture Notes

[102] *Sequential identification and ranking procedures, with special reference to Koopman–Darmois populations*, (with R. E. Bechhofer and M. Sobel), Statistical Research Monographs, Vol. 3, 1968, The University of Chicago Press, Chicago, Illinois. [MR 39 (1970) #6445, Zbl 208 (1971) 446].

[103] Jacob Wolfowitz, selected papers (ed. by J. Kiefer, with the assistance of U. Augustin and L. Weiss). 1980, Springer-Verlag, New York. [MR 83d (1983) 01080, Zbl 447 (1981) 62001].

[104] Lectures on design theory, (1974). Mimeograph Series #397, Department of Statistics, Purdue University.

[105] Contributions to the theory of games and statistical decision functions. Doctoral dissertation, 1952, Columbia University.

[106] Notes on decision theory (1953), (notes recorded by J. Sacks).

D. Papers (Unpublished)

[107] Note on asymptotic efficiency of M.L. estimators in nonparametric problems, (with J. Wolfowitz), (circa 1960). (II)

[108] Mathematics 371 Final Examination (with D. Kiefer), (1972). (II)

[109] Lecture notes on statistical inference, (1973), to be published by Springer-Verlag, 1984.

[110] *D*-optimality of the GYD for $v \geq 6$, (1974). (III)

[111] Optimality criteria for designs, (1975). (III)

JACK KIEFER
Fall, 1974

Reprinted from THE ANNALS OF MATHEMATICAL STATISTICS
Vol. 35, No. 1, March, 1964

LOCAL AND ASYMPTOTIC MINIMAX PROPERTIES OF MULTIVARIATE TESTS

BY N. GIRI[1] AND J. KIEFER[2]

Cornell and Stanford Universities

0. Summary. This paper contains details of the results announced in the abstract by the authors (1962). Techniques are developed for proving local minimax and "type D" properties and asymptotic (that is, far in distance from the null hypothesis) minimax properties in complex testing problems where exact minimax results seem difficult to obtain. The techniques are illustrated in the settings where Hotelling's T^2 test and the test based on the squared sample multiple correlation coefficient R^2 are customarily employed.

1. Introduction. In almost all of the standard hypothesis testing problems of multivariate analysis—in particular, in the normal ones—no meaningful nonasymptotic (in the sample size) optimum properties are known, either for the classical tests or for any other tests. The property of being a best invariant test under a group G of transformations which leave the problem invariant, which is possessed by some of these tests, is often unsatisfactory because the Hunt-Stein theorem is not valid; for example, this is the case if G is the real linear group of nonsingular $p \times p$ matrices where $p \geqq 2$. The only satisfactory properties known to us at this writing are the admissibility of Hotelling's T^2 test, proved by Stein (1956), and the minimax character in a few special cases of Hotelling's test, proved recently by the authors and Stein (1963), and of the test based on the multiple correlation coefficient, proved recently by the authors (1963).

The proof of local or asymptotic (far in distance from the null hypothesis) properties, for which we herein develop simple techniques, serves two purposes. Firstly, there is the obvious point of demonstrating such properties for their own sake. But well known and valid doubts have been raised as to the extent of meaningfulness of such properties. Secondly, then, and in our opinion more important, local or asymptotic properties can give an indication of what to look for in the way of genuine minimax or admissibility properties of certain procedures, even though the latter do not follow from the local or asymptotic properties. For example, if S_1 and S_2 are independent $p \times p$ central Wishart matrices ($p \geqq 2$) with expectations Σ and $\delta\Sigma$ per degree of freedom, and if it is desired to test $H_0:\delta = 1$ against $H_1:\delta = \lambda + 1$ (specified) > 1 or $H_1':\delta > 1$, then Stein showed (see Lehmann (1959), pp. 231, 338) that the best invariant test of level $\alpha(0 < \alpha < 1)$ under the real linear group G operating as $(S_1, S_2, \delta, \Sigma) \rightarrow (gS_1g', gS_2g', \delta, g\Sigma g')$, which is also the likelihood ratio test, is inadmissible and is not

Received 14 May 1963.

[1] Research supported by ONR Contract No. Nonr-401(03).

[2] Fellow of the John Simon Guggenheim Memorial Foundation; research supported in part by ONR Contract No. Nonr-266(04) (NR 047-005).

minimax for H_1. It was possible to obtain this result without much calculation because the best invariant procedure under the group G_T of nonsingular lower triangular matrices, which is transitive on $\{\Sigma\}$ and for which the Hunt-Stein theorem is valid, is not invariant under the real linear group. But in other examples such an expeditious demonstration may not be available; we note, therefore, that the result in the present example is already indicated in the local theory (as $\lambda \to 0$), which may thus be expected to indicate the direction of such results in other cases where (unlike the present case) the nonlocal theory is much more difficult. Such examples are in fact found in the T^2 and R^2 tests which are mentioned above and are treated in the paper: the local optimality of these tests, which will be seen in Section 2 not to be very difficult, was proved at a time when the T^2 and R^2 tests were not known to be minimax; such a simple local result, or its negation, lends credence to the genuine minimax property, or its negation, and thus indicates the direction of proof or disproof which seems most promising; although the R^2 and T^2 genuine minimax properties are known in only a few cases, one's belief is strengthened in the validity of these properties in general. When, as in the examples of Section 2, the local or asymptotic property is possessed by only one of the G_T-invariant procedures, it is of course more indicative than when, as in the example of Section 4, many tests share the property.

In the negative direction, the local or asymptotic theory not only indicates, but says something definite about the genuine minimax result: In the setting of Section 2 (resp., Section 4), if a level α test ϕ^* of $H_0: \delta = 0$ against $H_1: \delta = \lambda$ (specified) > 0 maximizes the minimum power under H_1 for *every* $\lambda > 0$, then it must clearly be locally minimax as $\lambda \to 0$ (resp., asymptotically minimax as $\lambda \to \infty$). Thus, the failure of ϕ^* to possess the local or asymptotic minimax property proves that it is not minimax (uniformly) for every λ.

One can refine the local and asymptotic notions we consider to versions obtained by including one or more error terms. For example, in Section 2 the local optimality property (2.4) can be refined to ask that a level α critical region which satisfies (2.4) with $\inf_\eta P_{\lambda,\eta}\{R\} = \alpha + C_1\lambda + o(\lambda)$ as $\lambda \to 0$, also satisfies

$$\lim_{\lambda\to 0} \frac{\inf_\eta P_{\lambda,\eta}\{R\} - \alpha - C_1\lambda}{\sup_{\phi\varepsilon Q_\alpha} \inf_\eta P_{\lambda,\eta}\{\phi \text{ rejects } H_0\} - \alpha - C_1\lambda} = 1.$$

This involves more calculations; typically, in the setting (2.3) of Section 2, two moments of the a priori distribution $\xi_{1,\lambda}$, rather than just one, become important. As further refinements are invoked, more moments are brought in. In the T^2 and R^2 cases mentioned earlier wherein genuine minimax results have been obtained, the limits of the corresponding a priori distributions as $\lambda \to 0$ are positive Lebesgue densities $f(\eta)$ whose moments are determined successively by further refinements. Thus, in those cases, the first moments of f coincide with those of $\xi_{1,\lambda}$ in Examples 1 and 2 of Section 2. A similar result for the first moment holds in Example 1 of Section 4 as $\lambda \to \infty$.

The local theory developed in Section 2 uses a slightly refined version of the well known result that Bayes procedures with constant risk are minimax. Such Bayesian techniques can be used also in the classical Neyman-Pearson local theory; for example, see Kiefer (1959), p. 280, for an application to regions of type C.

In Section 3 a variant of Isaacson's type D region is discussed. Somewhat surprisingly, the T^2 and R^2 tests are not of type D among G_T-invariant tests (except, of course in the lowest dimension). This fact can be interpreted, roughly, in terms of the classical tests having constant (maximin) power on a family of ellipsoids in a reduced parameter space, while certain other tests with other ellipsoids as local contours of the power function yield a greater Gaussian curvature for the power function. Unfortunately, the action of G_T in such problems destroys symmetry in the coordinates and makes it more difficult to achieve good intuition.

It is hoped that further, more difficult multivariate examples will be treated elsewhere.

The reader is referred to Lehmann (1959) and to Anderson (1958) for the standard nomenclature which we shall use.

2. Locally minimax tests. Let X be a space with associated σ-field which, along with the other obvious measurability considerations, we will not mention in what follows. For each point (δ, η) in the parameter set Ω (where $\delta \geqq 0$), suppose that $p(\cdot; \delta, \eta)$ is a probability density function on X with respect to some σ-finite measure μ. (The range of η may depend on δ.) For fixed α, $0 < \alpha < 1$, we shall be interested in testing, at level α, the hypothesis $H_0: \delta = 0$ against the alternative $H_1: \delta = \lambda$, where λ is a specified positive value, and in giving a sufficient condition for a test to be approximately minimax in the sense of (2.4) below. This is a local theory, in the sense that $p(x; \lambda, \eta)$ is close to $p(x; 0, \eta)$ when λ is small. Thus, obviously, every test of level α would be locally minimax in the sense of the trivial criterion obtained by not subtracting α in the numerator and denominator of (2.4). As indicated in the introduction, our proof of (2.4) as it stands consists merely of considering local power behavior with sufficient accuracy to obtain an approximate version of the classical result that a Bayes procedure with constant risk is minimax. A result of the type obtained can be proved under various possible sets of conditions, of which we use a form convenient in many applications, listing possible generalizations and simplifications as remarks.

Throughout this section, such expressions as $o(1)$, $o(h(\lambda))$, etc., are to be interpreted as $\lambda \to 0$.

For each fixed α, $0 < \alpha < 1$, we shall consider critical regions of the form $R = \{x: U(x) > C_\alpha\}$ where U is bounded and positive and has a continuous d.f. for each (δ, η), equicontinuous in (δ, η) for $\delta <$ some δ_0, and where

$$(2.1) \qquad P_{0,\eta}\{R\} = \alpha, \qquad P_{\lambda,\eta}\{R\} = \alpha + h(\lambda) + q(\lambda, \eta),$$

where $q(\lambda, \eta) = o(h(\lambda))$ uniformly in η, with $h(\lambda) > 0$ for $\lambda > 0$ and $h(\lambda) = o(1)$.

We shall also be concerned with probability measures $\xi_{0,\lambda}$ and $\xi_{1,\lambda}$ on the sets $\delta = 0$ and $\delta = \lambda$, respectively, for which

$$\int p(x; \lambda, \eta)\xi_{1,\lambda}(d\eta) \Big/ \int p(x; 0, \eta)\xi_{0,\lambda}(d\eta) = 1 + h(\lambda)[g(\lambda) + r(\lambda)U(x)] + B(x, \lambda) \tag{2.2}$$

where $0 < c_1 < r(\lambda) < c_2 < \infty$ for λ sufficiently small, and where $g(\lambda) = O(1)$ and $B(x, \lambda) = o(h(\lambda))$ uniformly in x. It is clear from the form of (2.1) and (2.2) that, reparametrizing, there is no loss of generality in letting $h(\delta) = \delta$, but we retain the stated forms for use in applications.

REMARKS.

1. In many applications the set $\{\delta = 0\}$ is a single point. Also, the set $\{\delta = \lambda\}$ is often a convex finite-dimensional Euclidean set wherein each component η_i is $O(h(\lambda))$; in this case, if $p(x; \lambda, \eta)/p(x; 0, \eta)$ is of the form

$$1 + h(\lambda)U(x) + \sum_{i,j=1}^{k} S_i(x)a_{ij}(\lambda)\eta_j + B(x, \lambda, \eta) \tag{2.3}$$

with S_i and a_{ij} bounded and with $\sup_{x,\eta} B(x, \lambda, \eta) = o(h(\lambda))$, and if there exists any $\xi'_{1,\lambda}$ satisfying (2.2), then the degenerate $\xi''_{1,\lambda}$ which assigns measure 1 to the mean of $\xi'_{1,\lambda}$ also satisfies (2.2). Both of these simplifications occur in Examples 1 and 2 below.

2. Another simplification occurs if the $\xi_{i,\lambda}$ can be chosen to be independent of λ, as is the case in Examples 1 and 2. The assumptions on B and U can then be weakened. (See Remark 3.)

3. One can weaken the assumptions on U and B (and, similarly, on the S_i and a_{ij} of Remark 1), which are used only in order to verify (2.8) and (2.9) below. For example, the assumption on $B(x, \lambda)$ can be weakened to $P_{\lambda,\eta}\{|B(x,\lambda)| < \epsilon h(\lambda)\} \to 0$ as $\lambda \to 0$, uniformly in η for each $\epsilon > 0$. If the $\xi_{i,\lambda}$'s are independent of λ, the uniformity of this last condition is unnecessary. The boundedness of U and the equicontinuity of its distribution can be weakened similarly.

4. The following modifications are also trivial to introduce: consideration of critical regions of a more complicated form than $\{U > C_\alpha\}$; consideration of randomized tests rather than critical regions R; modification (in the absence of continuity of the power function) of the equality signs in (2.1) to $\leqq$ and $\geqq$ signs, respectively, with equality on the support of the $\xi_{i,\lambda}$. The conclusion of Lemma 1 clearly holds even if Q_α is modified to include every family $\{\phi_\lambda\}$ of tests of level $\alpha + o(h(\lambda))$. One can similarly consider optimality of a family $\{U_\lambda\}$ rather than of a single U, by replacing R by $R_\lambda = \{x: U_\lambda(x) > C_{\alpha,\lambda}\}$, where $P_{0,\eta}\{R_\lambda\} = \alpha - q_\lambda(\eta)$ with $q_\lambda(\eta) = o(h(\lambda))$.

LEMMA 1. *If U satisfies (2.1) and if for sufficiently small λ there exist $\xi_{0,\lambda}$ and $\xi_{1,\lambda}$ satisfying (2.2), then U is locally minimax of level α for testing $H_0: \delta = 0$ against $\delta = \lambda$ as $\lambda \to 0$; that is,*

$$\lim_{\lambda\to 0} \frac{\inf_\eta P_{\lambda,\eta}\{R\} - \alpha}{\sup_{\phi_\lambda \epsilon Q_\alpha} \inf_\eta P_{\lambda,\eta}\{\phi_\lambda \text{ rejects } H_0\} - \alpha} = 1, \tag{2.4}$$

where Q_α is the class of tests of level α.

PROOF. Write

$$(2.5)\qquad \tau_\lambda = 1/\{2 + h(\lambda)[g(\lambda) + C_\alpha r(\lambda)]\},$$

so that

$$(2.6)\qquad (1 - \tau_\lambda)/\tau_\lambda = 1 + h(\lambda)[g(\lambda) + C_\alpha r(\lambda)].$$

A Bayes critical region relative to the a priori distribution $\xi_\lambda = (1 - \tau_\lambda)\xi_{0,\lambda} + \tau_\lambda \xi_{1,\lambda}$ (for 0-1 losses) is, by (2.2) and (2.6),

$$(2.7)\qquad B_\lambda = \{x : U(x) + B(x, \lambda)/r(\lambda)h(\lambda) > C_\alpha\}.$$

Write

$$P^*_{0,\lambda}\{A\} = \int P_{0,\eta}\{A\}\xi_{0,\lambda}(d\eta) \quad\text{and}\quad P^*_{1,\lambda}\{A\} = \int P_{\lambda,\eta}\{A\}\xi_{1,\lambda}(d\eta).$$

Let $V_\lambda = R - B_\lambda$ and $W_\lambda = B_\lambda - R$. Using the fact that $\sup_x |B(x, \lambda)/h(\lambda)| = o(1)$ and our continuity assumption on the d.f. of U, we have

$$(2.8)\qquad P^*_{0,\lambda}\{V_\lambda + W_\lambda\} = o(1).$$

Also, for $U_\lambda = V_\lambda$ or W_λ,

$$(2.9)\qquad P^*_{1,\lambda}\{U_\lambda\} = P^*_{0,\lambda}\{U_\lambda\}[1 + O(h(\lambda))].$$

Write $r^*_\lambda(A) = (1 - \tau_\lambda)P^*_{0,\lambda}\{A\} + \tau_\lambda(1 - P^*_{1,\lambda}\{A\})$. From (2.5), (2.8), and (2.9), the integrated Bayes risk relative to ξ_λ is then

$$\begin{aligned}(2.10)\qquad r^*_\lambda(B_\lambda) &= r^*_\lambda(R) + (1 - \tau_\lambda)(P^*_{0,\lambda}\{W_\lambda\} - P^*_{0,\lambda}\{V_\lambda\}) \\ &\qquad + \tau_\lambda(P^*_{1,\lambda}\{V_\lambda\} - P^*_{1,\lambda}\{W_\lambda\}) \\ &= r^*_\lambda(R) + (1 - 2\tau_\lambda)(P^*_{0,\lambda}\{W_\lambda\} - P^*_{0,\lambda}\{V_\lambda\}) \\ &\qquad + P^*_{0,\lambda}\{V_\lambda + W_\lambda\}O(h(\lambda)) \\ &= r^*_\lambda(R) + o(h(\lambda)).\end{aligned}$$

If (2.4) were false we could, by (2.1), find a family of tests $\{\phi_\lambda\}$ of level α such that ϕ_λ has power function $\alpha + g(\lambda, \eta)$ on the set $\delta = \lambda$, with

$$\limsup_{\lambda\to 0}\, [\inf_\eta g(\lambda, \eta) - h(\lambda)]/h(\lambda) > 0.$$

The integrated risk r'_λ of ϕ_λ with respect to ξ_λ would then satisfy

$$\limsup_{\lambda\to 0}\, (r^*_\lambda(R) - r'_\lambda)/h(\lambda) > 0,$$

contradicting (2.10).

EXAMPLE 1. (Hotelling's T^2 test). Let $X_1, \cdots, X_N$ be independently and identically distributed normal p-vectors, each with mean vector ξ and nonsingular covariance matrix Σ. Write $N\bar{X} = \sum_1^N X_i$ and $S = \sum_1^N (X_i - \bar{X})(X_i - \bar{X})'$. Let $\delta > 0$ be specified. For testing the hypothesis $H_0: \xi = 0$ against $H_1: N\xi'\Sigma^{-1}\xi = \delta$ at significance level α $(0 < \alpha < 1)$, a commonly employed procedure is Hotelling's T^2 test, which rejects H_0 when $T^2 = N(N-1)\bar{X}'S^{-1}\bar{X} > C'$ or, equivalently, when $U = T^2/(T^2 + N - 1) > C$, where C (or C') is chosen so as to yield a test of level α. We assume $N > p$, since it is easily shown that the denominator of (2.4) is zero in the degenerate case $N \leqq p$.

In our search for a locally minimax test as $\delta \to 0$, we may restrict attention to the space of the minimal sufficient statistic $(\bar{X}, S)$. The full linear group G of $p \times p$ nonsingular matrices leaves the problem invariant, operating as $(\bar{X}, S; \xi, \Sigma) \to (g\bar{X}, gSg'; g\xi, g\Sigma g')$. However, the Hunt-Stein theorem cannot be applied for this group if $p \geqq 2$, as Stein has demonstrated in several examples. (See Stein (1955), Lehmann (1959), pp. 231 and 338, and James and Stein (1960), p. 376.) However, the theorem does apply for the smaller group G_T of nonsingular lower triangular matrices (zero above the diagonal), which is solvable. (See Kiefer (1957), Lehmann (1959), p. 345.) Thus, for each δ there is a level α test which is almost invariant (hence, in the present problem, there is such a test which is invariant; see Lehmann (1957), p. 225) under G_T and which maximizes, among *all* level α tests, the minimum power under H_1. In terms of the local point of view, the denominator in (2.4) is unchanged by the restriction to G_T-invariant tests, and for any level α test ϕ there is a G_T-invariant level α test ϕ' for which the expression $\inf_\eta P_{\lambda,\eta}\{\phi' \text{ rejects } H_0\}$ is at least as large, so that a procedure which is locally minimax among G_T-invariant level α tests is locally minimax among all level α tests.

In place of the one-dimensional maximal invariant T^2 obtained under G, one now obtains a p-dimensional maximal invariant $Z = (Z_1, \cdots, Z_p)$ defined by

$$Z_i = \bar{X}'_{[i]}(S_{[i]})^{-1}\bar{X}_{[i]}$$

where we write $C_{[i]}$ for the upper left-hand $i \times i$ submatrix of a matrix C and $b_{[i]}$ for the i-vector consisting of the first i components of a vector b. Z_i is essentially Hotelling's statistic based on the first i coordinates. (This and the other straightforward computations which follow will be found in detail in Giri, Kiefer, and Stein (1963).) We shall find it more convenient to work with the equivalent statistic $Y = (Y_1, \cdots, Y_p)'$ where

$$Y_i = NZ_i/(1 + NZ_i) - NZ_{i-1}/(1 + NZ_{i-1}) \qquad (Z_{-1} = 0).$$

It is easily seen that $Y_i \geqq 0$, $\sum_1^p Y_i \leqq 1$, and $\sum_1^p Y_i = U = T^2/(N - 1 + T^2)$.

A corresponding maximal invariant $\Delta = (\delta_1, \cdots, \delta_p)$ in the parameter space of (μ, Σ) under G_T when H_1 is true is easily seen to be given by

$$\delta_i = N\xi'_{[i]}(\Sigma_{[i]})^{-1}\xi_{[i]} - N\xi'_{[i-1]}(\Sigma_{[i-1]})^{-1}\xi_{[i-1]} \qquad (\delta_1 = N\xi_1^2/\Sigma_{11}).$$

Here $\delta_i \geqq 0$ and $\sum_1^p \delta_i = \delta$. The nuisance parameter in this reduced setup is $\eta = (\eta_1, \cdots, \eta_p)$ where $\eta_i = \delta_i/\delta \geqq 0$, $\sum_1^p \eta_i = 1$. The corresponding maximal invariant under H_0 takes on the single value $0 = (0, \cdots, 0)$; we may for convenience also write $\eta = 0$ in that case.

A straightforward but slightly tedious computation yields for the Lebesgue density of Y on $H = \{y: y_i > 0, 1 \leqq i \leqq p; \sum_1^p y_i < 1\}$ the function

$$(2.11)\quad \begin{aligned} f_\Delta(y) &= \pi^{-p/2}\Gamma(N/2)\left(1 - \sum_1^p y_j\right)^{(N-p-2)/2} \Big/ \Gamma[(N-p)/2] \prod_1^p y_i^{\frac{1}{2}} \\ &\times \exp\left\{-\frac{\delta}{2} + \sum_{j=1}^p y_j \sum_{i>j} \frac{\delta_i}{2}\right\} \prod_{i=1}^p \phi((N-i+1)/2, 1/2; y_i\delta_i/2), \end{aligned}$$

where ϕ is the confluent hypergeometric function (sometimes denoted by ${}_1F_1$),

$$(2.12)\qquad \phi(a, b; x) = \sum_{j=0}^\infty [\Gamma(a+j)\Gamma(b)/\Gamma(a)\Gamma(b+j)j!]x^j.$$

We now verify the assumptions of Lemma 1 for $U = \sum_1^p Y_i$. Those just preceding (2.1) are obvious. In (2.1) we can take $h(\lambda) = b\lambda$ with b a positive constant. Of course, $P_{\lambda,\eta}\{R\}$ does not depend on η. From (2.11) and (2.12), we have

$$(2.13)\quad \frac{f_{\lambda,\eta}(y)}{f_{0,0}(y)} = 1 + \frac{\lambda}{2}\left\{-1 + \sum_{j=1}^p y_i\left[\sum_{i>j}\eta_i + (N-j+1)\eta_j\right]\right\} + B(y, \eta, \lambda),$$

where $B(y, \eta, \lambda) = o(\lambda)$ uniformly in y and η. We have the setup of Remark 1 above, and (2.2) is satisfied by letting $\xi_{0,\lambda}$ give measure one to the single point $\eta = 0$, while $\xi_{1,\lambda}$ gives measure one to the single point η^* (say) whose jth coordinate is $(N-j)^{-1}(N-j+1)^{-1}p^{-1}N(N-p)$, so that $\sum_{i>j}\eta_i^* + (N-j+1)\eta_j^* = N/p$ for all j. Applying Lemma 1, we have

THEOREM 1. *For every p, N, and α, Hotelling's T^2 test is locally minimax for testing $\delta = 0$ against $\delta = \lambda$ as $\lambda \to 0$.*

EXAMPLE 2. (The R^2 test) With $X_1, \cdots, X_N$ as in Example 1, partition Σ as

$$\Sigma = \begin{Vmatrix} \Sigma_{11} & \Sigma_{12} \\ \Sigma_{21} & \Sigma_{22} \end{Vmatrix}$$

where Σ_{22} is $(p-1) \times (p-1)$. Write $\rho^2 = \Sigma_{12}\Sigma_{22}^{-1}\Sigma_{21}/\Sigma_{11}$. It is desired to test the hypothesis $H_0: \rho^2 = 0$ that the first component is independent of the others, against the alternative $H_1: \rho^2 = \delta$, where δ is specified, $0 < \delta < 1$. It is clear that the transformations $(\xi, \Sigma, \bar{X}, S) \to (\xi + b, \Sigma, \bar{X} + b, S)$ leave the problem invariant and, along with the group G_T' considered below, generate a group which satisfies the Hunt-Stein conditions and in which these transformations form a normal subgroup; the action of these transformations is to reduce the problem to that where $\xi = 0$ (known) and $S = \sum_1^N X_iX_i'$ is sufficient, where N has been reduced by one from what it was originally. We therefore treat this latter formulation, considering $X_1, \cdots, X_N$ to have zero mean. We assume $N \geqq p \geqq 2$, the case $N < p$ now being degenerate.

We now consider the group G_T' of nonsingular lower triangular matrices whose first column contains only zeros except for the first element. It is easily seen that this group, operating as $(S; \Sigma) \to (gSg', g\Sigma g')$, leaves the problem invariant.

The development now parallels that of Example 1. A maximal invariant $R = (R_2, \cdots, R_p)$ is defined by

$$\sum_2^i R_j = S_{12[i]}(S_{22[i]})^{-1}S_{21[i]}/S_{11}, \qquad 2 \leqq i \leqq p.$$

Thus, $R_j \geqq 0$, $\sum_2^p R_j \leqq 1$, and $\sum_2^p R_j = U$ is the squared sample multiple correlation coefficient between first and other components (usually denoted by R^2). The corresponding maximal invariant in the parameter space, $\Delta = (\delta_2, \cdots, \delta_p)$, is given by

$$\sum_2^i \delta_j = \Sigma_{12[i]}(\Sigma_{22[i]})^{-1}\Sigma_{21[i]}/\Sigma_{11}, \qquad 2 \leqq i \leqq p.$$

Thus, $\delta_i \geqq 0$, $\sum_2^p \delta_i = \rho^2$, the squared population multiple correlation coefficient. We write $\eta = (\eta_2, \cdots, \eta_p)$ with $\eta_i = \delta_i/\delta$ as before, and $(\delta, \eta) = (0, 0)$ under H_0. The Lebesgue density of R when $\rho^2 = \lambda$ can be computed (see Giri and Kiefer (1963) for details) to be

$$(2.14)\ f_{\lambda,\eta}(r)$$
$$= \frac{(1-\lambda)^{\frac{1}{2}(p-1)}\left(1-\sum_2^p r_i\right)^{\frac{1}{2}(N-p-1)}}{\left[1+\sum_2^p r_i((1-\lambda)/\gamma_i - 1)\right]^{\frac{1}{2}N}\Gamma[\frac{1}{2}(N-p+1)]\pi^{\frac{1}{2}(p-1)}}$$
$$\times \prod_2^{p-1}\{r_i^{\frac{1}{2}}\gamma_i^{\frac{1}{2}}(\pi_i+1)^{\frac{1}{2}(N-i+2)}\Gamma[\tfrac{1}{2}(N-i+2)]\}^{-1}$$
$$\times \sum_{\beta_2=0}^{\infty}\cdots\sum_{\beta_j=0}^{\infty}\Gamma\left(\sum_2^p \beta_j + \tfrac{1}{2}N\right)$$
$$\times \prod_2^p \left\{\frac{\Gamma[\frac{1}{2}(N-i+2)+\beta_i]}{(2\beta_i)!}\left[\frac{4r_i(1-\lambda)/\gamma_i(1+\pi_i^{-1})}{1+\sum_2^p r_j[(1-\lambda)/\gamma_j - 1]}\right]^{\beta_i}\right\}$$

where $\gamma_i = 1 - \sum_2^i \delta_j = 1 - \lambda \sum_2^i \eta_j$, $\pi_i = \delta_i/\gamma_i = \lambda\eta_i(1 - \lambda \sum_2^i \eta_j)$. (The expression $1/(1 + \pi_i^{-1})$ means 0 if $\delta_i = 0$.) From this we obtain

$$(2.15)\quad \frac{f_{\lambda,\eta}(r)}{f_{0,0}(r)} = 1 + \frac{N\lambda}{2}\left\{-1 + \sum_{j=2}^p r_j\left[\sum_{i>j}\eta_i + (N-j+2)\eta_j\right]\right\} + B(r, \eta, \lambda)$$

where $B(r, \eta, \lambda) = o(\lambda)$ uniformly in r and η. We see that the assumptions of Lemma 1 are again satisfied for $U = \sum_2^p y_i$ (="R^2"), with $h(\lambda) = b\lambda$ again. In fact, (2.15) becomes (2.13) if we replace $N\lambda$ in (2.15) by λ, p by $p - 1$, and the index range $2 \leqq j \leqq p$ by $1 \leqq j' = j - 1 \leqq p - 1$; thus, $\xi_{1,\lambda}$ now gives measure one to the point whose jth coordinate $(2 \leqq j \leqq p)$ is $(N - j + 1)^{-1} \cdot (N - j + 2)^{-1}(p - 1)^{-1}N(N - p + 1)$. (Of course, it is no coincidence that (2.13) and (2.15) correspond: (2.11) involves ratios of noncentral to central chi-square variables, while (2.14) involves similar ratios with random noncentrality parameters; the first order terms in the expansions, which involve only expectations of these quantities, correspond to each other.) We conclude

THEOREM 2. *For every p, N, and* α, *the critical region which consists of large values of the squared sample multiple correlation coefficient* R^2 *is locally minimax for testing* $\rho^2 = 0$ *against* $\rho^2 = \lambda$ *as* $\lambda \to 0$.

3. Type *D* and *E* regions. The notion of a type D or E region is due to Isaacson (1951). Kiefer (1958) showed that the usual F-test of the univariate linear hypothesis has this property. Lehmann (1959a) showed that, in finding regions which are of type D, invariance could be invoked in the manner of the Hunt-Stein theorem; and that this could also be done for type E regions (if they exist) provided that one works with a group which operates as the identity on the nuisance parameter set (H of the next paragraph).

Suppose, for a parameter set $\Omega' = \{(\theta, \eta): \theta \varepsilon \Theta, \eta \varepsilon H\}$ with associated distributions, with Θ a Euclidean set, that every test function ϕ has a power function $\beta_\phi(\theta, \eta)$ which, for each η, is twice continuously differentiable in the components of θ at $\theta = 0$, an interior point of Θ. Let Q_α be the class of locally strictly unbiased level α tests of $H_0: \theta = 0$ against $H_1: \theta \neq 0$; our assumption on β_ϕ implies that all tests in Q_α are similar and that $\partial\beta_\phi/\partial\theta_i \mid_{\theta=0} = 0$ for ϕ in Q_α. Let $\Delta_\phi(\eta)$ be the determinant of the matrix $B_\phi(\eta)$ of second derivatives of $\beta_\phi(\theta, \eta)$ with respect to the components of θ (that is, the Gaussian curvature) at $\theta = 0$. We assume the parametrization to be such that $\Delta_{\phi'}(\eta) > 0$ for all η for at least one ϕ' in Q_α. A test ϕ^* is said to be of *type E* if $\phi^* \varepsilon Q_\alpha$ and $\Delta_{\phi^*}(\eta) = \max_{\phi \varepsilon Q_\alpha} \Delta_\phi(\eta)$ for all η. If H is a single point, ϕ^* is said to be of *type D*.

In the examples which interest us, such as those treated in the previous section, it seems doubtful that type E regions exist. (In terms of Lehmann's development, H is not left fixed by many transformations.) Without pursuing the question of when such regions exist, we introduce two possible optimality criteria, in the same spirit as the type D and E criteria, which will always be fulfilled by some test under minimum regularity assumptions: Write $\bar{\Delta}(\eta) = \max_{\phi \varepsilon Q_\alpha} \Delta_\phi(\eta)$. A test ϕ^* will be said to be of *type* D_A if $\phi \varepsilon Q_\alpha$ and

$$\max_\eta [\bar{\Delta}(\eta) - \Delta_{\phi^*}(\eta)] = \min_{\phi \varepsilon Q_\alpha} \max_\eta [\bar{\Delta}(\eta) - \Delta_\phi(\eta)]$$

and of *type* D_M if

$$\max_\eta [\bar{\Delta}(\eta)/\Delta_{\phi^*}(\eta)] = \min_{\phi \varepsilon Q_\alpha} \max_\eta [\bar{\Delta}(\eta)/\Delta_\phi(\eta)].$$

These criteria resemble stringency and regret criteria employed elsewhere in statistics; the subscripts "A" and "M" stand for "additive" and "multiplicative" regret principles. The possession of these properties is invariant under the product of any transformation on Θ (acting trivially on H) of the same general type as those for which type D regions retain their property, and an arbitrary 1-1 transformation on H (acting trivially on Θ), but, of course, not under more general transformations on Ω'. Obviously, a type E test automatically satisfies these weaker criteria.

Suppose now that a problem is invariant under a group of transformations G for which the Hunt-Stein theorem holds and which acts trivially on Θ; that is, such that $g(\theta, \eta) = (\theta, g\eta)$ for g in G, in a usual abuse of notation. If ϕg is, as usual, the test function defined by $\phi g(x) = \phi(gx)$, a trivial computation then shows that $\Delta_{\phi g}(\eta) = \Delta_\phi(g\eta)$ and hence that $\bar{\Delta}(\eta) = \bar{\Delta}(g\eta)$. Also, if ϕ is better than ϕ' in the sense of either of the above criteria, then ϕg is clearly better than $\phi' g$. All of the requirements of Lehmann's development are easily seen to be satisfied, so that we can conclude that there is an almost invariant (hence, in our examples, an invariant) test which is of type D_A or D_M. (This differs from the way in which invariance is used in the application of (b), p. 883, of Lehmann (1959a), where it is used to reduce Θ rather than H, as here.)

If, furthermore, G is transitive on H, then $\bar{\Delta}(\eta)$ is constant, as is $\Delta_\phi(\eta)$ for an invariant ϕ (which we therefore write simply as Δ_ϕ). In this case we conclude that *if ϕ^* is invariant and if ϕ^* is of type D among invariant ϕ (that is, if Δ_{ϕ^*} maximizes Δ_ϕ over all invariant ϕ), then ϕ^* is of type D_A and D_M among all ϕ.*

Our main tool for verifying optimality in these senses is a trivial one:

LEMMA 0. *Let L be a class of non-negative definite symmetric $m \times m$ matrices, and suppose J is a fixed nonsingular member of L. If* $\operatorname{tr} J^{-1}B$ *is maximized (over B in L) by $B = J$, then* $\det B$ *is also maximized by J. Conversely, if L is convex and J maximizes* $\det B$, *then* $\operatorname{tr} J^{-1}B$ *is maximized by $B = J$.*

PROOF. Write $J^{-1}B = A$. If $A = I$ maximizes $\operatorname{tr} A$, we have $(\det A)^{1/m} \leqq \operatorname{tr} A/m \leqq 1 = \det I$. Conversely, if I maximizes $\det A$, it also maximizes $\operatorname{tr} A$, since $\operatorname{tr} B > \operatorname{tr} I$ implies $\det(\alpha B + (1-\alpha)I) > 1$ for α small and positive.

Of course, the usefulness of this tool lies in the fact that the generalized Neyman-Pearson lemma allows us to maximize $\operatorname{tr} QB_\phi$ (for fixed Q) more easily than to maximize $\Delta_\phi = \det B_\phi$, among similar level α tests. We can find, for each Q, a ϕ_Q which maximizes $\operatorname{tr} QB_\phi$; a ϕ^* which maximizes Δ_ϕ is then obtained by finding a ϕ_Q for which $B_{\phi_Q} = Q^{-1}$.

In examples of the type which interest us, the reduction by invariance under a group G which is transitive on H often results in a reduced problem wherein the maximal invariant is a vector $Y = (Y_1, \cdots, Y_m)$ whose distribution depends only on $\gamma = (\gamma_1, \cdots, \gamma_m)$ where $\gamma_i = \theta_i^2$ (where $\theta = (\theta_1, \cdots, \theta_m)$), and such that the density f_γ of Y with respect to a σ-finite measure μ is of the form

$$f_\gamma(y) = f_0(y)\{1 + \sum_1^m \gamma_i(h_i + \sum_j a_{ij}y_j)\} + Q(y, \gamma) \tag{3.1}$$

where the h_i and a_{ij} are constants and $Q(y, \gamma) = o(\sum \gamma_i)$ as $\gamma \to 0$, and where we can differentiate under the integral sign to obtain, for invariant test functions ϕ of level α,

$$\begin{aligned}\beta_\phi(\theta, \eta) = \alpha\left(1 + \sum_1^m \gamma_i h_i\right) \\ + \sum_i \gamma_i \sum_j a_{ij} \int y_j \phi(y) f_0(y) \mu(dy) + o(\sum \gamma_i)\end{aligned} \tag{3.2}$$

as θ (or γ, where $\gamma_i = \theta_i^2$) $\to 0$. We shall call this the *symmetric reduced regular* (SRR) case.

In the SRR case, every invariant ϕ has a diagonal B_ϕ whose ith diagonal entry, by (3.2), is $2[h_i\alpha + E_0\{\sum_j a_{ij}Y_j\phi(Y)\}]$. By the Neyman-Pearson lemma, tr QB_ϕ is maximized over such ϕ by a ϕ^* of the form

$$\phi^*(y) = \begin{Bmatrix}1\\0\end{Bmatrix} \quad \text{if} \quad \sum_{i,j} a_{ij}q_i y_j \begin{Bmatrix}>\\<\end{Bmatrix} C \tag{3.3}$$

where C is a constant and q_i is the ith diagonal element of Q (we need only consider diagonal Q's at this point). From this and the previous remarks, we conclude

LEMMA 2. *In the* SRR *case, an invariant test ϕ^* of level α is of type D among invariant ϕ (and, hence, of type D_A and D_M among all ϕ) if and only if ϕ^* is of the form* (3.3) *with* $q_i^{-1} = \text{const } b_{\phi^* i}$, *where $b_{\phi^* i}$ is the ith diagonal element of B_{ϕ^*} (that is,* $q_i^{-1} = \text{const } [h_i\alpha + E_0\{\sum_j a_{ij}Y_j\phi^*(Y)\}]$).

EXAMPLE 1. In the setting of Example 1 of Section 2, we let $\theta = N^{\frac{1}{2}}F^{-1}\xi$ where F is the (unique) member of G_T with positive diagonal elements and such that $FF' = \Sigma$. Then Θ is Euclidean p-space, G_T operates transitively on $H = \{$positive definite symmetric $\Sigma\}$ but trivially on Θ, and we have the SRR case with Y_i as in (2.11) and $\gamma_i = \delta_i$ of (2.11). We thus have (3.2) with $h_i = -\frac{1}{2}$, $a_{ij} = 1$ (resp., 0, $N - j + 1$) if $i > j$ (resp., $i < j$, $i = j$). Hotelling's T^2 test has a power function which depends only on $\sum \gamma_i$, so that, with the above parametrization for θ, we have B_{T^2} a multiple of the identity. Also, Hotelling's critical region is of the form $\sum y_i > C$. But, when all q_i are equal, the critical region corresponding to (3.3) is of the form $\sum_j (N + p + 1 - 2j)y_j > C$, which is not Hotelling's region if $p > 1$. We conclude, perhaps somewhat surprisingly in view of Theorem 1,

THEOREM 3. *For $0 < \alpha < 1 < p < N$, Hotelling's T^2-test is not of type D among G_T-invariant tests, and hence is not of type D_A or D_M (nor of type E) among all tests.*

The actual computation of a ϕ^* of type D among G_T-invariant tests appears difficult in view of the fact that we must (by (2.11)) compute an integral of the form

$$E_0\{Y_r^h\phi(Y)\} = \int_{\{\sum c_j y_j > C\}} \frac{\pi^{-\frac{1}{2}p}\Gamma(N/2)y_r^{h-\frac{1}{2}}(1 - \sum y_j)^{\frac{1}{2}(N-p-2)}}{\Gamma[\frac{1}{2}(N-p)]\prod_{i\neq r} y_i^{\frac{1}{2}}} \prod dy_i \tag{3.4}$$

for $h = 0$ or 1 for various choices of the c_j's and C. When α is close to 0 or 1, one can carry out approximate computations, as illustrated by the discussion of the next paragraph.

As $\alpha \to 1$, one can see that the complement $\bar{R}$ of the critical region becomes a simplex with one corner at 0. When $p = 2$, if we write $\rho = 1 - \alpha$ and consider critical regions of level α of the form $by_1 + y_2 > C$ where $0 < L^{-1} \leqq b \leqq L$, L being fixed but large (this keeps $\bar{R}$ close to the origin), we obtain easily from (3.4) that $\rho = E_0(1 - \phi(Y)) = (N-2)C/2b^{\frac{1}{2}} + o(C)$ as $C \to 0$. Similarly, $E_0\{(1 - \phi(Y))Y_i\} = (N-2)C^2b^{i-\frac{3}{2}}/8 + o(C) = \rho^2 b^{i-\frac{3}{2}}/2(N-2) + o(\rho)$ as

$\rho \to 0$, while $E_0 Y_i = 1/N$. We therefore obtain, from (2.13), for the power near H_0,

$$\begin{aligned}\alpha + \frac{\rho}{2}\Big\{\gamma_1\Big[1 - \frac{N\rho + o(\rho)}{2(N-2)b^{\frac{1}{2}}}\Big]\\ + \gamma_2\Big[1 - \frac{\rho + o(\rho)}{2(N-2)b^{\frac{1}{2}}} - \frac{(N-1)\rho b^{\frac{1}{2}} + o(\rho)}{2(N-2)}\Big]\Big\} + o\left(\sum \gamma_i\right),\end{aligned} \tag{3.5}$$

where the $o(\rho)$ and $o(\sum \gamma_i)$ terms are uniform in γ and ρ, respectively. The product Δ_ϕ of the coefficients of γ_1 and γ_2 is easily seen to be maximized when $b = (N+1)/(N-1) + o(1)$, as $\rho \to 0$; with more care, one can obtain further terms in an expansion in ρ for the type D choice of b. The argument is completed by showing that $b < L^{-1}$ implies that $\bar{R}$ lies in a strip so close to the y_1-axis as to make $E_0\{(1 - \phi(Y))Y_1\}$ too large and $E_0\{(1 - \phi(Y))Y_2\}$ too small to yield a ϕ as good as that with $b = (N+1)/(N-1)$, with a similar argument if $b > L$.

When ρ is very close to 0, we see that all choices of $b > 0$ give substantially the same power, $\alpha + \rho(\gamma_1 + \gamma_2)/2 + O(\rho^2)$, so that the relative departure from being of type D, of the T^2 test or any other critical region of the form $bY_1 + Y_2 > C_\alpha$ (b fixed and positive), approaches 0 as $\alpha \to 1$. We do not know how great the departure of Δ_{T^2} from $\bar{\Delta}$ can be for arbitrary α.

One can treat similarly the case $p > 2$ and also the case $\alpha \to 0$.

EXAMPLE 2. We have already noted the correspondence between (2.13) and (2.15). Thus, for the setting of Example 2 of Section 2, we obtain by an argument like that used for Theorem 3,

THEOREM 4. *For $0 < \alpha < 1$, $p > 2$, and $N \geqq p$ or $N > p$ depending on whether or not the mean ξ is known, the critical region consisting of large values of the squared sample multiple correlation coefficient R^2 is not of type D among G_T-invariant tests, and hence is not of type D_A or D_M (nor of type E) among all tests.*

Approximate computations of G_T-invariant type D tests can be carried out when α is close to 0 or 1, as in Example 1.

4. Asymptotically minimax tests. In this section we treat the setting of Section 2 when $\lambda \to \infty$, and expressions such as $o(1)$, $o(H(\lambda))$, etc., are to be interpreted in this light. We are now interested in minimaxing a probability of error which is going to zero. Readers who are familiar with asymptotically large sample size theory (referred to in the remark below) will recall that, in that setting, it is difficult directly to compare approximations to such small probabilities for different families of tests, and one instead compares their logarithms. While our considerations are asymptotic in a sense not involving sample sizes (although some examples, such as that of testing whether the mean of a normal variate with unit variance is 0 or λ, fall equivalently into either framework), we encounter the same difficulty, which accounts for the form of (4.4).

As in Section 2, various possible sets of assumptions could be used. We choose one which differs slightly from the form used in Section 2. Modifications are remarked on, below.

Suppose then that the region $R = \{x: U(x) \geqq C_\alpha\}$ satisfies (in place of (2.1))

$$P_{0,\eta}\{R\} = \alpha, \qquad P_{\lambda,\eta}\{R\} = 1 - \exp\{-H(\lambda)[1 + o(1)]\}, \tag{4.1}$$

where $H(\lambda) \to +\infty$ with λ and the $o(1)$ term is uniform in η. Suppose, replacing (2.2), that

$$\begin{aligned}\int p(x; \lambda, \eta)\xi_{1,\lambda}(d\eta) \Big/ \int p(x; 0, \eta)\xi_{0,\lambda}(d\eta) \\ = \exp\{H(\lambda)[G(\lambda) + R(\lambda)U(x)] + B(x, \lambda)\}\end{aligned} \tag{4.2}$$

where $\sup_x |B(x, \lambda)| = o(H(\lambda))$ and $0 < c_1 < R(\lambda) < c_2 < \infty$. Our only other regularity assumption is that C_α is a point of increase from the left of the d.f. of U, when $\delta = 0$, uniformly in η; that is,

$$\inf_\eta P_{0,\eta}\{U \geqq C_\alpha - \epsilon\} > \alpha \tag{4.3}$$

for every $\epsilon > 0$.

Remark. The reader will find no difficulty in giving analogues here of the specializations and generalizations remarked upon in Section 2. One further variation, which is more relevant here (where $1 - P_{\lambda,\eta}\{R\} \to 0$ as $\lambda \to \infty$) than it would be in the setting of Section 2, is to let $\alpha \to 0$ as $\lambda \to \infty$ in such a way that both $P_{0,\eta}\{R\}$ and $1 - P_{\lambda,\eta}\{R\}$ go to zero, perhaps at different rates. One obtains asymptotic minimax results which have some formal resemblance to familiar results which are asymptotic (in the sense that the sample size $n \to \infty$) for testing between simple hypotheses, as considered by Chernoff, Bahadur, Hodges and Lehmann, and others.

Lemma 3. *If U satisfies* (4.1) *and* (4.3), *and if for sufficiently large λ there exist $\xi_{0,\lambda}$ and $\xi_{1,\lambda}$ satisfying* (4.2), *then U is asymptotically logarithmically minimax of level α for testing $H_0: \delta = 0$ against $\delta = \lambda$ so $\lambda \to \infty$; that is,*

$$\lim_{\lambda\to\infty} \frac{\inf_\eta \{-\log[1 - P_{\lambda,\eta}\{R\}]\}}{\sup_{\phi_\lambda \epsilon Q_\alpha} \inf_\eta \{-\log[1 - P_{\lambda,\eta}\{\phi_\lambda \text{ rejects } H_0\}]\}} = 1. \tag{4.4}$$

Proof. Suppose, contrary to (4.4), that there is an $\epsilon > 0$ and an unbounded sequence Γ of values λ with corresponding tests ϕ_λ in Q_α for which

$$P_{\lambda,\eta}\{R\} > 1 - \exp\{-H(\lambda)(1 + 5\epsilon)\} \tag{4.5}$$

for all η.

There are two cases, (4.6) and (4.9). If $\lambda \epsilon \Gamma$ and

$$-1 - G(\lambda) \leqq R(\lambda)C_\alpha + 2\epsilon, \tag{4.6}$$

consider the a priori distribution given by the $\xi_{i,\lambda}$ and by τ_λ satisfying

$$\tau_\lambda/(1 - \tau_\lambda) = \exp\{H(\lambda)(1 + 4\epsilon)\}. \tag{4.7}$$

The integrated risk of any Bayes procedure B_λ must satisfy

$$\begin{aligned}r_\lambda^*(B_\lambda) \leqq r_\lambda^*(\phi_\lambda) &\leqq (1 - \tau_\lambda)\alpha + \tau_\lambda \exp\{-H(\lambda)(1 + 5\epsilon)\} \\ &= (1 - \tau_\lambda)[\alpha + \exp\{-\epsilon H(\lambda)\}],\end{aligned} \tag{4.8}$$

by (4.5) and (4.7). But, according to (4.2), a Bayes critical region is

$$B_\lambda = \{x : U(x) + B(x, \lambda)/R(\lambda)H(\lambda) \geqq [-(1 + 4\epsilon) - G(\lambda)]/R(\lambda)\}.$$

Hence, if λ is so large that $\sup_x |B(x, \lambda)/H(\lambda)R(\lambda)| < \epsilon/c_2$, we have, from (4.6),

$$B_\lambda \supset \{x : U(x) > C_\alpha - \epsilon/c_2\} = B'_\lambda \quad \text{(say)}.$$

The assumption (4.3) implies that $P_{0,\eta}\{B'_\lambda\} > \alpha + \epsilon'$ with $\epsilon' > 0$, contradicting (4.8) for sufficiently large λ.

On the other hand, if $\lambda \varepsilon \Gamma$ and

$$-1 - G(\lambda) > R(\lambda)C_\alpha + 2\epsilon, \tag{4.9}$$

let

$$\tau_\lambda/(1 - \tau_\lambda) = \exp\{H(\lambda)(1 + \epsilon)\}. \tag{4.10}$$

Then, by (4.2),

$$B_\lambda = \{x : U(x) + B(x, \lambda)/R(\lambda)H(\lambda) \geqq [-(1 + \epsilon) - G(\lambda)]/R(\lambda)\}.$$

Hence, if $\sup_x |B(x, \lambda)/H(\lambda)R(\lambda)| < \epsilon/2c_2$, we conclude from (4.9) that $B_\lambda \subset R$, so that, by (4.1) and (4.10),

$$r^*(B_\lambda) > \tau_\lambda \exp\{-H(\lambda)[1 + o(1)]\} = (1 - \tau_\lambda) \exp\{H(\lambda)(\epsilon - o(1))\}. \tag{4.11}$$

But

$$r^*(B_\lambda) \leqq r^*(\phi_\lambda) \leqq (1 - \tau_\lambda)\alpha + \tau_\lambda \exp\{-H(\lambda)(1 + 5\epsilon)\}$$
$$= (1 - \tau_\lambda)[\alpha + \exp\{-4\epsilon H(\lambda)\}],$$

which contradicts (4.11) for sufficiently large λ.

EXAMPLE 1. In the setting of Example 1 of Section 2, with

$$U = T^2/(N - 1 + T^2) = \sum_1^p Y_i$$

again, (2.11) and (2.12) yield (since $\phi(a, b; x) = \exp\{x(1 + o(1))\}$ as $x \to \infty$)

$$\frac{f_{\lambda,\eta}(y)}{f_{0,0}(y)} = \exp\left\{\frac{\lambda}{2}\left[-1 + \sum_{j=1}^p y_j \sum_{i \geqq j} \eta_i\right][1 + B(y, \eta, \lambda)]\right\} \tag{4.12}$$

with $\sup_{y,\eta} |B(y, \eta, \lambda)| = o(1)$ as $\lambda \to \infty$. From this and the smoothness of $f_{0,0}$ (or from the well known form of the density of T^2) we see (for example, putting $\eta_p = 1$, the density of U being independent of η) that

$$P_{\lambda,\eta}\{U < C_\alpha\} = \exp\{\tfrac{1}{2}\lambda(C_\alpha - 1)[1 + o(1)]\} \tag{4.13}$$

as $\lambda \to \infty$; thus, (4.1) is satisfied with $H(\lambda) = (1 - C_\alpha)/2$. Next, letting $\xi_{1,\lambda}$ assign measure one to the point $\eta_1 = \cdots = \eta_{p-1} = 0$, $\eta_p = 1$, and $\xi_{0,\lambda}$ assign measure one to (0, 0), we obtain (4.2). Finally, (4.3) is trivial. We conclude, from Lemma 3,

THEOREM 5. *For every α, p, N, Hotelling's test is asymptotically minimax for testing $\delta = 0$ against $\delta = \lambda$ as $\lambda \to \infty$.*

Although no critical region of the form $\sum_1^p a_i Y_i > C$ other than Hotelling's would have been locally minimax in the considerations of Section 2, many regions of this form are asymptotically minimax (which, of course, makes Theorem 5 less of an argument in support of the use of the T^2 test):

THEOREM 6. *If $C < 1$ and $1 = b_1 \leqq b_2 \leqq \cdots \leqq b_p$, then the critical region $\{\sum_1^p b_j Y_j > C\}$ is asymptotically minimax (among tests of the same size) as $\lambda \to \infty$.*

PROOF. Since the maximum of $\sum_1^p y_j \sum_{i \geqq j} \eta_j$ subject to $\sum_1^p b_j y_j \leqq C$ is clearly achieved at $y_1 = C, y_2 = \cdots = y_p = 0$, integration of $f_{\lambda,\eta}(y)$ over a small region near that point yields (4.13) with C_α replaced by C. Since the b_j's are nondecreasing in j, it is obvious from (4.12) that $\xi_{1,\lambda}$ can be chosen to yield (4.2) with $U = \sum b_j Y_j$. Again, (4.3) is trivial.

The reader may find it interesting, in the case $p = 2$, to note geometrically what happens to (4.13) if $C > 1$, and to note the dependence of the power on η if $b_2 < b_1$.

The result of Theorem 5 is obviously related, in the underlying structure which yields it, to Stein's (1956) admissibility result, although neither implies the other. It is interesting to note also that the same departure from this structure (in the behavior as $\rho^2 \to 1$) which prevents Stein's method from proving the admissibility of the R^2 test, also prevents us from applying Lemma 3 to Example 2 of Section 2 as $\rho^2 \to 1$.

REFERENCES

ANDERSON, T. W. (1958). *Introduction to Multivariate Statistical Analysis.* Wiley, New York.

GIRI, N. and KIEFER, J. (1962). Minimax properties of Hotelling's and certain other multivariate tests (abstract). *Ann. Math. Statist.* **33** 1490–1491.

GIRI, N. and KIEFER, J. (1963). Minimax character of the R^2 test in the simplest case. To be published.

GIRI, N., KIEFER, J. and STEIN, C. (1963). Minimax character of Hotelling's T^2 test in the simplest case. *Ann. Math. Statist.* **34** 1524–1535.

ISAACSON, S. L. (1951). On the theory of unbiased tests of simple statistical hypotheses specifying the values of two or more parameters. *Ann. Math. Statist.* **22** 217–234.

JAMES, W. and STEIN, C. (1960). Estimation with quadratic loss. *Proc. Fourth Berkeley Symp. Math. Statist. Prob.* **1** 361–379.

KIEFER, J. (1957). Invariance, minimax sequential estimation, and continuous time processes. *Ann. Math. Statist.* **28** 573–601.

KIEFER, J. (1958). On the nonrandomized optimality and randomized nonoptimality of symmetrical designs. *Ann. Math. Statist.* **29** 675–699.

KIEFER, J. (1959). Optimum experimental designs. *J. Roy. Statist. Soc. Ser. B* **21** 272–319.

LEHMANN, E. L. (1959). *Testing Statistical Hypotheses.* Wiley, New York.

LEHMANN, E. L. (1959a). Optimum invariant tests. *Ann. Math. Statist.* **30** 881–884.

STEIN, C. (1955). On tests of certain hypotheses invariant under the full linear group (abstract). *Ann. Math. Statist.* **26** 769.

STEIN, C. (1956). The admissibility of Hotelling's T^2 test. *Ann. Math. Statist.* **27** 616–623.

Reprinted from THE ANNALS OF MATHEMATICAL STATISTICS
Vol. 35, No. 4, December, 1964

MINIMAX CHARACTER OF THE R^2-TEST IN THE SIMPLEST CASE

BY N. GIRI[1] AND J. KIEFER[2]

Cornell University

Summary. In the first nontrivial case, dimension $p = 3$ and sample size $N = 3$ or 4 (depending on whether or not the mean is known), it is proved that the classical level α normal test of independence of the first component from the others, based on the squared sample multiple correlation coefficient R^2, maximizes, among all level α tests, the minimum power on each of the usual contours where the R^2-test has constant power. A corollary is that the R^2-test is most stringent of level α in this case.

1. Introduction. Let $X_1, \cdots, X_N$ be independent normal p-vectors with common mean vector ξ and common nonsingular covariance matrix Σ. Write $N\bar{X} = \sum_1^N X_i$ and $S = \sum_1^N (X_i - \bar{X})(X_i - \bar{X})'$. Partition Σ and S as

$$\begin{pmatrix} \Sigma_{11} & \Sigma_{12} \\ \Sigma_{21} & \Sigma_{22} \end{pmatrix} \text{ and } \begin{pmatrix} S_{11} & S_{12} \\ S_{21} & S_{22} \end{pmatrix},$$

respectively, where Σ_{22} and S_{22} are $(p-1) \times (p-1)$. Write $\rho^2 = \Sigma_{12}\Sigma_{22}^{-1}\Sigma_{21}/\Sigma_{11}$. Let $\delta (0 < \delta < 1)$ be specified. For testing the hypothesis $H_0 : \rho^2 = 0$ against $H_1 : \rho^2 = \delta$ at significance level α, a commonly employed procedure is the test based on the squared sample multiple correlation coefficient R^2, which rejects H_0 when $R^2 = S_{12}S_{22}^{-1}S_{21}/S_{11} > C$, where C is chosen so as to yield a test of level α. Throughout this paper $0 < \alpha < 1$, so that $0 < C < 1$.

In this paper we are interested in a minimax question regarding the R^2-test, namely, whether or not that test maximizes, among all level α tests, the minimum power under H_1. We succeed in proving that, for each possible choice of δ and α, the answer is affirmative in the first nontrivial case, $p = 3, N = 4$ (or $N = 3$ for the corresponding problem where ξ is known).

Our method of proof parallels that of Giri, Kiefer, and Stein (1963) (hereafter referred to as GKS) for the corresponding T^2-test result; the steps are the same, the detailed calculations in the present case being slightly more complicated. The remarks in GKS on the indications that the result holds for general p and N (in particular, from the local results of Giri and Kiefer (1962)), but of the inadequacy of the present method, apply also here. The reader is referred to GKS for a discussion of the Hunt-Stein theorem, its validity under the group of real lower-triangular matrices and its failure under the full linear group, and for other comments. Anderson (1958) is referred to for multivariate theory and Lehmann (1959) for testing (including invariance and minimax) theory.

Received January 14, 1964.
[1] Research supported by ONR contract No. Nonr-401(03).
[2] Fellow of the John Simon Guggenheim Memorial Foundation; research supported in part by ONR contract No. Nonr-266(04) (NRO 47-005).

It is well known that among tests based on the sufficient statistic $(\bar{X}, S)$, the R^2-test is best invariant under the group G of transformations of the form $(\xi, \Sigma, \bar{X}, S) \to (A\xi + b, A\Sigma A', A\bar{X} + b, ASA')$ where A is nonsingular and $A_{12} = A'_{21} = 0$. (For $p > 2$, this does not imply our minimax result, because of the failure of the Hunt-Stein theorem.) Simaika (1941) showed that the R^2-test is uniformly most powerful among all level α tests whose power is a function only of ρ^2, a result which is also implied by stronger results of Wolfowitz (1945) as well as by the best invariant character just mentioned. When $p = 2$ we have the well known properties of the standard two-tailed test based on the sample correlation coefficient; and when $N \leqq p$ if ξ is unknown or $N \leqq p - 1$ if ξ is known, it is easy to see that the infimum over H_1 of the power of every test equals the size of the test (for example, in this case $z_p = 1$ in (2.2), so that the distribution of z does not depend on the correlation between first and last components). Hence, the case $p = 3$, $N = 4$ (or $N = 3$ if ξ is assumed known) is the simplest one to be considered.

We now outline briefly our method of proof. We may restrict attention to the space of the minimal sufficient statistic $(\bar{X}, S)$. In the next section we shall first reduce the problem to the case where ξ is known and N is reduced by one, and shall then apply the Hunt-Stein theorem for an appropriate group G_T of $p \times p$ matrices (essentially the direct sum of the $(p - 1) \times (p - 1)$ lower triangular matrices and the nonzero reals), which is solvable. (See Kiefer (1957), Lehmann (1959), p. 345.) Thus, there is a test of level α which is almost invariant (hence, in the present problem, there is such a test which is invariant; see Lehmann (1957), p. 225) under G_T and which maximizes, among *all* level α tests, the minimum power over H_1. Whereas R^2 was a maximal invariant under G, with a single distribution under each of H_0 and H_1, the maximal invariant under G_T is a $(p - 1)$ dimensional statistic $R = (R_2, \cdots, R_p)'$ with a single distribution under H_0 but with a distribution which depends continuously on a $(p - 2)$-dimensional parameter $\Delta = (\delta_2, \cdots, \delta_p)$, $\delta_i \geqq 0$, $\sum_2^p \delta_i = \delta$ (fixed), under H_1. Thus, when $N > p > 2$ (or $N \geqq p > 2$ if ξ is known), there is no UMP invariant test under G_T as there was under G. We compute the Lebesgue densities f_Δ^* and f_0^* of R, under H_0 and H_1. Because of the compactness of the reduced parameter spaces $\{0\}$ and $\Gamma = \{(\delta_2, \cdots, \delta_p): \delta_i \geqq 0, \sum_2^p \delta_i = \delta\}$ and the continuity of f_Δ^* in Δ, it follows (see Wald (1950)) that every minimax test for the reduced problem in terms of R, is Bayes. In particular, the R^2-test, $\sum_2^p R_i > C$, (where $\sum_2^p R_i$ is what is usually called R^2), which is G_T-invariant, maximizes the minimum power over H_1 if and only if there is a probability measure λ on Γ such that, for some constant K,

$$\int_\Gamma \frac{f_\Delta^*(r_2, \cdots, r_p)}{f_0^*(r_2, \cdots, r_p)} \lambda(d\Delta) >, =, < K \tag{1.1}$$

according to whether $\sum_2^p r_i >, =, < C$, except possibly for a set of measure zero. (Here C depends on the specified α, and λ and K may depend only on C and the specified value $\delta > 0$.) An examination of the integrand in (1.1) will allow us to replace (1.1) by the equivalent

$$\int_\Gamma \frac{f_\Delta^*(r_2, \cdots, r_p)}{f_0^*(r_2, \cdots, r_p)} \lambda(d\Delta) = K \quad \text{if} \quad \sum_2^p r_i = C. \tag{1.2}$$

We are able to evaluate the unique value which K must take on in order that (1.2) can be satisfied, and are then faced with the question of whether or not there exists a probability measure λ satisfying the left half of (1.2). The development thus far, which holds for general p and $N > p$, is carried out in Sections 2 and 3. In Section 4 we then obtain a λ and carry out the proof that it satisfies the left half of (1.2) in the special case $p = 3$, $N = 4$ (or $N = 3$ if ξ is known).

2. Reduction of the Problem to (1.2). Throughout this paper, we shall find it convenient to index the components of $(p - 1)$-vectors by subscripts $2, 3, \cdots, p$, with a corresponding convention for $(p - 1) \times (p - 1)$ matrices.

For testing H_0 against H_1 we need only consider test functions which depend on the statistic $(\bar{X}, S)$, sufficient for (ξ, Σ). It can easily be verified that the group H of transformations $(\xi, \Sigma, \bar{X}, S) \to (\xi + b, \Sigma, \bar{X} + b, S)$ leaves the testing problem in question invariant, that H is normal in the group G^* generated by H and the group of transformations G_T considered below, and that G_T and H (and hence G^*) satisfy the Hunt-Stein conditions; the action of the transformations in H is to reduce the problem to that where $\xi = 0$ (known) and $S = \sum_{i=1}^N X_i X_i'$ is sufficient for Σ, where N has been reduced by 1 from what it was originally. Using the standard method of reduction in steps, we can therefore treat this latter formulation, considering $X_1, X_2, \cdots, X_N$ to have zero mean. We assume also $N \geq p \geq 2$, it having been shown in Section 1 that, in the degenerate case $N < p$, the maximin value of the power equals the size. Furthermore, with this formulation, we need only consider test functions which depend on the sufficient statistic $S = \sum_{i=1}^N X_i X_i'$, the Lebesgue density of which is

$$f_\Sigma(s_{11}, s_{12}, s_{22}) = c(\det \Sigma)^{-N/2} \exp\left[-(\tfrac{1}{2}) \operatorname{tr} \Sigma^{-1} s\right] \times (\det s)^{(N-p-1)/2} \tag{2.1}$$

where

$$c^{-1} = 2^{Np/2} \pi^{p(p-1)/4} \prod_{i=1}^p \Gamma((N + 1 - i)/2).$$

We now consider the group G_T of nonsingular lower triangular matrices (zero above the main diagonal) whose first column contains only zeros except for the first element. A typical element g of G_T can be represented as $g = \begin{pmatrix} g_{11} & 0 \\ 0 & g_{22} \end{pmatrix}$ where g_{22} is $(p - 1) \times (p - 1)$ lower triangular. As we have stated earlier, it is easily seen that this group operating as $(S; \Sigma) \to (gSg'; g\Sigma g')$, leaves the problem invariant. We now compute a maximal invariant of S under the action of the group G_T in the usual fashion: If a function ϕ of S is invariant under G_T, then $\phi(S) = \phi(gSg')$ for all S and all $g \in G_T$, i.e., $\phi(S_{11}, S_{12}, S_{22}) = \phi(g_{11}S_{11}g_{11}, g_{11}S_{12}g_{22}', g_{22}S_{22}g_{22}')$. We may consider the domain of S to be symmetric positive definite matrices, which have probability one for all Σ; then there is an F in G_T

with positive diagonal elements such that $FF' = \begin{pmatrix} S_{11} & 0 \\ 0 & S_{22} \end{pmatrix}$. Putting $g = LF^{-1}$ where L is any diagonal matrix with values ± 1 in any order on the main diagonal, we see that ϕ is a function only of $L_{22}F_{22}^{-1}S_{21}L_{11}/F_{11}$ and hence, because of the freedom of choice of L, of $|F_{22}^{-1}S_{21}/F_{11}|$; or, equivalently, of the $(p-1)$-vector whose ith component $Z_i (2 \leqq i \leqq p)$ is the sum of squares of the first i components[3] of $|F_{22}^{-1}S_{21}/F_{11}|$ (whose components are indexed $2, 3, \cdots, p$). Write $b_{[i]}$ for the $(i-1)$-vector consisting of the first $i-1$ components of the $(p-1)$-vector b, and $c_{[i]}$ for the upper left hand $(i-1) \times (i-1)$ submatrix of a $(p-1) \times (p-1)$ matrix c. Then Z_i can be written as $(F_{22}^{-1}S_{21}/F_{11})'_{[i]} \cdot (F_{22}^{-1}S_{21}/F_{11})_{[i]}$. Since $(F_{22[i]})^{-1} = (F_{22}^{-1})_{[i]}$, we have, for $2 \leqq i \leqq p$,

$$Z_i = \frac{S_{12[i]}(F_{22[i]}^{-1})'(F_{22[i]})^{-1}S'_{12[i]}}{S_{11}} \tag{2.2}$$
$$= \frac{S_{12[i]}S_{22[i]}^{-1}S_{21[i]}}{S_{11}}$$

The vector $Z = (Z_2, \cdots, Z_p)'$ is thus a maximal invariant under G_T if it is invariant under G_T, and it is easily seen to be the latter. Z_i is essentially the squared sample multiple correlation computed from the first i coordinates of the X_i. Let us define a $(p-1)$-vector $R = (R_2, R_3, \cdots, R_p)'$ by

$$\sum_2^i R_j = Z_i \qquad 2 \leqq i \leqq p; \tag{2.3}$$

i.e., $R_i = Z_i - Z_{i-1}$ where we define $Z_1 = 0$. It follows trivially from above that R is also a maximal invariant under G_T. It is easily verified that $R_j \geqq 0$ for each j, $\sum_2^p R_j \leqq 1$, and of course $\sum_2^p R_j = S_{12}S_{22}^{-1}S_{21}/S_{11}$ is the squared sample multiple correlation coefficient between the first and other components (usually denoted by R^2). We shall find it more convenient to work with the equivalent statistic R instead of with Z.

A corresponding maximal invariant $\Delta = (\delta_2, \cdots, \delta_p)'$ in the parametric space of Σ under G_T when H_1 is true is given by

$$\sum_2^i \delta_j = \Sigma_{12[i]}(\Sigma_{22[i]})^{-1}\Sigma_{21[i]}/\Sigma_{11}, \qquad 2 \leqq i \leqq p. \tag{2.4}$$

It is clear that $\delta_j \geqq 0$ and $\sum_2^p \delta_j = \rho^2$, the squared population multiple correlation coefficient. The corresponding maximal invariant under H_0 takes on the single value $0 = (0, \cdots, 0)'$. It is well-known that the Lebesgue density function f_Δ^* of the maximal invariant depends only on Δ under H_1 and is a fixed f_0^* under H_0. We must now compute f_Δ^* and f_0^*. Actually we need only obtain the ratio f_Δ^*/f_0^* for use in (1.2), so we could proceed without keeping track of factors

[3] On page 1527 of GKS, Z_i should be defined similarly, instead of as the square of the ith component.

not depending on Δ in this computation. However, it is not much extra work to keep track of these factors, so we shall do so. There are several ways of computing f^*_Δ. For example, one method different from that which we shall use, but parallel to the method used by Anderson for computing the distribution of R^2, is to use the Bartlett decomposition to write

$$R_i \Big/ \left(1 - \sum_2^p R_j\right) = \left[\left(1 - \sum_2^i \delta_j\right)^{\frac{1}{2}} N_i + \delta_i^{\frac{1}{2}} \chi_{N-i+2}\right]^2 \Big/ (1 - \delta)\chi^2_{N-p+1}$$

where N_i are normal, χ_j is a chi-variable with j degrees of freedom, and all N_i and χ_j are independent, and then to integrate out on the χ_{N-i+2} and, finally, on χ^2_{N-p+1}.

We can assume $\Sigma_{11} = 1, \Sigma_{22} = I$ (the $(p-1) \times (p-1)$ identity matrix), and $\Sigma_{21} = (\delta_2^{\frac{1}{2}}, \cdots, \delta_p^{\frac{1}{2}})' = \delta^*$ (say) in (2.1), since f^*_Δ depends on Σ only through Δ. For this choice Σ^* (say) of Σ, (2.1) can be rewritten as

$$\begin{aligned} &f_{\Sigma^*}(s_{11}, s_{12}, s_{22}) \\ (2.5) \qquad &= c(1-\rho^2)^{-N/2} \exp\left[-\tfrac{1}{2} \operatorname{tr} (A_{11}s_{11} + A_{12}s_{12}' + A_{12}'s_{12} + A_{22}s_{22})\right] \\ &\quad \times (\det(s))^{(N-p-1)/2} \end{aligned}$$

where

$$\begin{aligned} A_{11} &= (\Sigma_{11} - \Sigma_{12}\Sigma_{22}^{-1}\Sigma_{21})^{-1} = (1-\rho^2)^{-1}, \\ A_{12} &= (\Sigma_{12}\Sigma_{22}^{-1}\Sigma_{21} - \Sigma_{11})^{-1}\Sigma_{12}\Sigma_{22}^{-1} = -(1-\rho^2)^{-1}\delta^{*\prime}, \\ A_{22} &= (\Sigma_{22} - \Sigma_{21}\Sigma_{11}^{-1}\Sigma_{12})^{-1} = (I - \delta^*\delta^{*\prime})^{-1}. \end{aligned}$$

Let B be the unique lower triangular $p \times p$ matrix belonging to G_T with positive diagonal elements $b_{ii}(1 \leqq i \leqq p)$ and such that $S_{22} = B_{22}B_{22}'$, $S_{11} = b_{11}^2$, and let $V = B_{22}^{-1}S_{21}$. One can easily compute the Jacobians

$$\partial S_{22}/\partial B_{22} = 2^{p-1}\prod_2^p (b_{ii})^{p+1-i}, \qquad \partial S_{21}/\partial V = \prod_2^p b_{ii},$$

and $\partial S_{11}/\partial b_{11} = 2b_{11}$, so that the joint density of b_{11}, V, and B_{22} is

$$(2.6) \qquad h_{\Sigma^*}(b_{11}, v, b_{22}) = 2^p f_{\Sigma^*}(b_{11}^2, v'b_{22}', b_{22}b_{22}')b_{11}\prod_{i=2}^p b_{ii}^{p+2-i}.$$

Putting $W = (W_2, \cdots, W_p)'$ with $W_i = |V_i|$ $(2 \leqq i \leqq p)$, and noting that the $(p-1)$-vector W can arise from any of the 2^{p-1} vectors $V = M_{22}W$ where M_{22} is a $(p-1) \times (p-1)$ diagonal matrix with diagonal entries ± 1, we can write $g = bM$ where $M = \begin{pmatrix} M_{11} & 0 \\ 0 & M_{22} \end{pmatrix}$ with $M_{11} = \pm 1$ and g ranging over *all* matrices in G_T; we obtain for the density of W, writing $\bar{g}_{ij}$ $(i \geqq j \geqq 2)$ for the components of g_{22},

$$
\begin{aligned}
h^*_{\Sigma^*}(w) &= 2^p \int f_{\Sigma^*}(g_{11}^2, w'g_{22}', g_{22}g_{22}') \prod_{i=2}^p |\bar{g}_{ii}|^{p+2-i} |g_{11}| \times \prod_{i \geqq j \geqq 2} d\bar{g}_{ij}\, dg_{11} \\
&= (1-\rho^2)^{-N/2} 2^p c \int \exp\{-[2(1-\rho^2)]^{-1} \operatorname{tr}(g_{11}^2 - \delta^* w' g_{22}' \\
&\qquad - \delta^{*\prime} g_{22} w + (1-\rho^2)(I - \delta^*\delta^{*\prime})^{-1} g_{22} g_{22}')\} \\
&\qquad \times \prod_{i=2}^p |\bar{g}_{ii}|^{N+1-i} |g_{11}|^{N-p} (1 - w'w/g_{11}^2)^{(N-p-1)/2} \prod_{i \geqq j \geqq 2} d\bar{g}_{ij}\, dg_{11}
\end{aligned} \tag{2.7}
$$

Writing $W = g_{11}U$, we obtain from (2.7) that the density of U is

$$
\begin{aligned}
h^{**}_{\Sigma^*}(u) &= (1-\rho^2)^{-N/2} 2^p c \int \exp\{-[2(1-\rho^2)]^{-1} \operatorname{tr} \\
&\quad (g_{11}^2 - g_{11}\delta^* u' g_{22}' - g_{11}\delta^{*\prime} g_{22} u + (1-\rho^2)(I - \delta^*\delta^{*\prime})^{-1} g_{22} g_{22}')\} \\
&\quad \times |g_{11}|^{N-1} \prod_{i=2}^p |\bar{g}_{ii}|^{N+1-i} (1 - u'u)^{(N-p-1)/2} \prod_{i \geqq j \geqq 2} d\bar{g}_{ij}\, dg_{11},
\end{aligned} \tag{2.8}
$$

the range of integration being from $-\infty$ to ∞ in each variable. It is easily checked that $U_j^2 = R_j (2 \leqq j \leqq p)$. Hence, the density of $R = (R_2, \cdots, R_p)'$ is given by

$$
\begin{aligned}
f^*_\Delta(r) &= \frac{(1-\rho^2)^{-N/2} 2c}{\prod_{i=2}^p r_i^{\frac{1}{2}}} \int \exp\{-[2(1-\rho^2)]^{-1} \operatorname{tr} \\
&\quad (g_{11}^2 - 2g_{11}\delta^{*\prime} g_{22} r^* + (1-\rho^2)(I - \delta^*\delta^{*\prime})^{-1} g_{22} g_{22}')\} \\
&\quad \times \left(1 - \sum_{j=2}^p r_j\right)^{(N-p-1)/2} |g_{11}|^{N-1} \prod_{i=2}^p |\bar{g}_{ii}|^{N+1-i} \prod_{i \geqq j \geqq 2} d\bar{g}_{ij}\, dg_{11}
\end{aligned} \tag{2.9}
$$

where $r^* = (r_2^{\frac{1}{2}}, \cdots, r_p^{\frac{1}{2}})'$. Let $\bar{C} = (1-\rho^2)^{-1}(I - \delta^*\delta^{*\prime})$. Since $\bar{C}$ is positive definite, there exists a lower triangular $(p-1) \times (p-1)$ matrix T with positive diagonal elements $T_{ii} (2 \leqq i \leqq p)$ such that $T\bar{C}T' = I$. Writing $h = Tg_{22}$, we obtain $\partial h / \partial g_{22} = \prod_2^p T_{ii}^{i-1}$. Let us define $\gamma_i (2 \leqq i \leqq p)$ by

$$
\gamma_i = 1 - \sum_{j=2}^i \delta_j, \qquad \gamma_1 = 1 \tag{2.10}
$$

(so that $\gamma_p = 1 - \rho^2$) and $\alpha_i (2 \leqq i \leqq p)$ by

$$
\alpha_i = [\delta_i \gamma_p / \gamma_{i-1}\gamma_i]^{\frac{1}{2}}. \tag{2.11}
$$

Writing $\alpha = (\alpha_2, \cdots, \alpha_p)'$, a simple calculation (similar to that used to obtain (2.3) of GKS) shows that $(T_{[i]}\delta^*_{[i]})'(T_{[i]}\delta^*_{[i]}) = \gamma_p(1-\gamma_i)/\gamma_i$, so that $\alpha = T\delta^*$. Since $\bar{C}\delta^* = \delta^*$ by direct computation, we obtain $\alpha = T\bar{C}\delta^* = T^{-1\prime}\delta^*$. From this and the easy computation $\det \bar{C} = (1-\rho^2)^{2-p}$, we obtain

$$f_\Delta^*(r) = 2c(1-\rho^2)^{-N(p-1)/2} \prod_2^p r_i^{-\frac{1}{2}}$$

$$\times \int \exp\{-[2(1-\rho^2)]^{-1} \operatorname{tr}(g_{11}^2 - 2g_{11}\alpha' hr^* + hh')\}$$

$$\times \left(1 - \sum_{j=2}^p r_j\right)^{(N-p-1)/2} |g_{11}|^{N-1} \prod_{i=2}^p |h_{ii}|^{N+1-i} \prod_{i \geqq j \geqq 2} dh_{ij}\, dg_{11}$$

$$= 2c(1-\rho^2)^{-N(p-1)/2} \prod_2^p r_i^{-\frac{1}{2}} \left(1 - \sum_{j=2}^p r_j\right)^{(N-p-1)/2} \tag{2.12}$$

$$\times \int \exp\{-g_{11}^2/2(1-\rho^2)\} |g_{11}|^{N-1}$$

$$\times \left\{ \int \exp\{-[2(1-\rho^2)]^{-1} \sum_{i \geqq j \geqq 2} [h_{ij}^2 - 2\alpha_i r_j^{\frac{1}{2}} h_{ij} g_{11}]\}\right.$$

$$\left. \times \prod_{i=2}^p |h_{ii}|^{N+1-i} \prod_{i \geqq j \geqq 2} dh_{ij} \right\} dg_{11},$$

the integration again being from $-\infty$ to ∞ in each variable. For $i > j$ the integration with respect to h_{ij} yields a factor $(2\pi)^{\frac{1}{2}}(1-\rho^2)^{\frac{1}{2}} \exp[\alpha_i^2 r_j g_{11}^2/2(1-\rho^2)]$. For $i = j$, we obtain a factor

$$(2\pi)^{\frac{1}{2}}(1-\rho^2)^{(N+2-i)/2} \exp[\alpha_i^2 r_i g_{11}^2/2(1-\rho^2)]$$

$$\times E(\chi_1^2(\alpha_i^2 r_i g_{11}^2/(1-\rho^2))^{(N+1-i)/2}) = [2(1-\rho^2)]^{(N-i+2)/2} \tag{2.13}$$

$$\times \Gamma((N-i+2)/2)\phi((N-i+2)/2, \tfrac{1}{2}, r_i\alpha_i^2 g_{11}^2/2(1-\rho^2)),$$

where $\chi_1^2(\beta)$ is a noncentral chi-square variable with one degree of freedom and noncentrality parameter $\beta = E\chi_1^2(\beta) - 1$ and where ϕ is the confluent hypergeometric function (sometimes denoted by ${}_1F_1$),

$$\phi(a, b; x) = \sum_{j=0}^\infty \frac{\Gamma(a+j)\Gamma(b)x^j}{\Gamma(a)\Gamma(b+j)j!}.$$

Thus for $r \varepsilon H = \{r{:}r_i \geqq 0, 2 \leqq i \leqq p; \sum_2^p r_i < 1\}$ we have (noting that the exponent of the factor $(1-\rho^2)$ vanishes)

$$f_\Delta^*(r) = (2\pi)^{(p-1)(p-2)/4}\left(1 - \sum_{j=2}^p r_j\right)^{(N-p-1)/2}$$

$$\times 2c \prod_2^p r_i^{-\frac{1}{2}} \int_{-\infty}^{\infty} \exp\left\{[-[2(1-\rho^2)]^{-1} g_{11}^2 \left(1 - \sum_{j=2}^p r_j \sum_{i>j} \alpha_i^2\right)\right\} \tag{2.14}$$

$$\times \prod_{i=2}^p [2^{(N+2-i)/2}\Gamma((N-i+2)/2)\phi((N-i+2)/2, \tfrac{1}{2}; r_i\alpha_i^2 g_{11}^2/2(1-\rho^2))]$$

$$\times |g_{11}|^{N-1}\, dg_{11}.$$

Integrating with respect to g_{11}, the density of r can be written as

$$
\begin{aligned}
f_\Delta^*(r) &= \frac{(1-\rho^2)^{N/2}\left(1-\sum_{i=1}^p r_i\right)^{(N-p-1)/2}}{\left(1+\sum_{i=2}^p r_i((1-\rho^2)/\gamma_i - 1)\right)^{N/2}\Gamma((N-p+1)/2)\pi^{(p-1)/2}} \\
&\times \frac{1}{\prod_{i=2}^p \{r_i^{\frac{1}{2}}\Gamma((N-i+2)/2)\}} \times \sum_{\beta_2=0}^\infty \cdots \sum_{\beta_p=0}^\infty \Gamma\left(\sum_{j=2}^p \beta_j + N/2\right) \\
&\times \prod_{i=2}^p \left\{\frac{\Gamma((N-i+2)/2+\beta_i)}{(2\beta_i)!} \times \left[\frac{4r_i\alpha_i^2}{1+\sum_{j=2}^p r_j((1-\rho^2)/\gamma_j - 1)}\right]^{\beta_i}\right\}.
\end{aligned} \tag{2.15}
$$

Hence

$$
\begin{aligned}
\frac{f_\Delta^*(r)}{f_0^*(r)} &= \frac{(1-\delta)^{N/2}}{\left(1+\sum_{i=2}^p r_i((1-\delta)/\gamma_i - 1)\right)^{N/2}} \\
&\times \sum_{\beta_2=0}^\infty \cdots \sum_{\beta_p=0}^\infty \frac{\Gamma\left(\sum_{j=2}^p \beta_j + N/2\right)}{\Gamma(N/2)} \\
&\times \prod_{i=2}^p \left\{\frac{\Gamma((N-i+2)/2+\beta_i)}{\Gamma((N-i+2)/2)(2\beta_i)!} \times \left[\frac{4r_i\alpha_i^2}{1+\sum_{j=2}^p r_j((1-\delta)/\gamma_j - 1)}\right]^{\beta_i}\right\}.
\end{aligned} \tag{2.16}
$$

The continuity of f_Δ^* in Δ over its compact domain $\Gamma = \{(\delta_2, \cdots, \delta_p): \delta_i \geqq 0, \sum_{i=2}^p \delta_i = \delta\}$ is evident, so we conclude that the minimax character of the critical region $\sum_{i=2}^p R_i \geqq C$ is equivalent to the existence of a probability measure λ satisfying (1.1). Clearly (1.1) implies (1.2). On the other hand, if there is a λ and a K for which (1.2) is satisfied and if $\bar{r} = (\bar{r}_2, \cdots, \bar{r}_p)$ is such that $\sum_2^p r_i = C' > C$, writing $f = f_\Delta^*/f_0^*$ and $\tilde{r} = C\bar{r}/C'$, we see at once that

$$f(\bar{r}) = f(C'\tilde{r}/C) > f(\tilde{r}) = K,$$

because of the form of f_Δ^*/f_0^* and the fact that $C'/C > 1$ and $\sum_2^p \tilde{r}_i = C$. It is to be noted that $\gamma_i^{-1}(1-\delta) - 1 = -\sum_{j>i} \delta_j/\gamma_i$ and that $\gamma_i > 0$. This and a similar argument for the case $C' < C$ show that (1.1) implies (1.2). The remaining computations of the paper are somewhat simplified by the fact that for fixed C and δ we can at this point easily compute the unique value of K for which (1.2) can possibly be satisfied.

3. Evaluation of K. Let $\hat{R} = (R_2, \cdots, R_{p-1})'$ and write $f_\Delta^*(\hat{r} \mid u)$ for the version of the conditional Lebesgue density of $\hat{R}$ given that $\sum_2^p R_i = u$, which is continuous in $\hat{r}$ and u for $r_i > 0$, $\sum_2^{p-1} r_i < u < 1$, and is 0 elsewhere; also write $f_\delta^{**}(u)$ for the Lebesgue density of $R^2 = \sum_2^p R_i$ which is continuous for $0 < u < 1$ and vanishes elsewhere (and which depends on Δ only through δ). Then (1.2) can be written as

$$\int f_\Delta^*(r \mid C)\, d\lambda(\Delta) = \left[\frac{K f_0^{**}(C)}{f_\delta^{**}(C)}\right] f_0^*(\hat{r} \mid C) \tag{3.1}$$

for $r_i > 0$ and $\sum_2^{p-1} r_i < C$. The integral of (3.1), being a probability mixture of probability densities, is itself a probability density in $\hat{r}$, as is $f_0^*(\hat{r} \mid C)$. Hence the expression in square brackets equals one. It is well known that for $0 < C < 1$

$$f_\delta^{**}(C) = \frac{(1-\delta)^{N/2}\Gamma(N/2)}{\Gamma((N-p+1)/2)\Gamma((p-1)/2)} C^{(p-3)/2}(1-C)^{(N-p-1)/2} \times F(N/2, N/2; (p-1)/2; C\delta), \tag{3.2}$$

where $F(a, b; c; x)$ is the ordinary $({}_2F_1)$ hypergeometric series, given by

$$\begin{aligned} F(a, b; c; x) &= \sum_{r=0}^{\infty} \frac{x^r \Gamma(a+r)\Gamma(b+r)\Gamma(c)}{r!\,\Gamma(a)\Gamma(b)\Gamma(c+r)} \\ &= \sum_{r=0}^{\infty} \frac{(a)_r (b)_r}{(c)_r\, r!} x^r, \end{aligned} \tag{3.3}$$

where we write $(a)_r = \Gamma(a+r)/\Gamma(a)$. (See Anderson (1958) or use (2.15).) Hence from (3.1) the value of K which satisfies (1.2) is given by

$$K = (1-\delta)^{N/2} F(N/2, N/2; (p-1)/2; C\delta). \tag{3.4}$$

Hence by (3.4) and (2.16) the condition (1.2) becomes

$$\begin{aligned} &\int_\Gamma \left[1 + \sum_2^p r_i((1-\delta)/\gamma_i - 1)\right]^{-N/2} \sum_{\beta_2=0}^{\infty} \cdots \sum_{\beta_p=0}^{\infty} \frac{\Gamma\left(\sum_2^p \beta_j + N/2\right)}{\Gamma(N/2)} \\ &\times \prod_{i=2}^{p} \left\{ \frac{\Gamma((N-i+2)/2 + \beta_i)}{\Gamma((N-i+2)/2)(2\beta_i)!} \times \left[\frac{4 r_i \alpha_i^2}{1 + \sum_2^p r_j((1-\delta)/\gamma_j - 1)}\right]^{\beta_i} \right\} d\lambda(\Delta) \\ &= F(N/2, N/2; (p-1)/2; C\delta) \end{aligned} \tag{3.5}$$

for all r with $r_i > 0$ and $\sum_2^p r_i = C$. Unlike the corresponding equation (2.8) of GKS, (3.5) does not yield an obvious conclusion regarding the dependence of λ on C and δ only through $C\delta$, although we shall obtain this conclusion in the case treated in the next paragraph.

4. The case $p = 3$, $N = 3$ (or $N = 4$ if ξ is unknown). In this case (3.5) can be written (as can be seen, for example, by writing $\phi(\frac{3}{2}, \frac{1}{2}; x) = (1 + 2x)e^x$ for $i = 2$ in (2.14), and then carrying out the integration) as

$$\int_\Gamma \sum_{n=0}^{\infty} \left\{ \frac{(1+2n)}{(1-r_2\delta)^{\frac{3}{2}}} \left(\frac{r_3\delta_3}{(1-\delta_2)(1-r_2\delta)} \right)^n + \frac{r_2\delta_2(1-\delta)(2n+1)(2n+3)}{(1-\delta_2)(1-r_2\delta)^{\frac{5}{2}}} \left(\frac{r_3\delta_3}{(1-\delta_2)(1-r_2\delta)} \right)^n \right\} d\lambda(\Delta) = F(\tfrac{3}{2}, \tfrac{3}{2}; 1; C\delta). \tag{4.1}$$

One could presumably try to solve for λ by using the theory of Meijer transforms with kernel $F(\frac{3}{2}, \frac{3}{2}; 1; x)$. We proceed instead, as in GKS for the T^2 problem, by expanding (4.1) in an appropriate power series. Write Γ_1 for the unit one-dimensional simplex $\{(\beta_1, \beta_2): \beta_i \geq 0, \sum_1^2 \beta_i = 1\}$ and make the change of variables $t_1 = r_2/(1-r_2)$, $t_1 + t_2 = (r_2 + r_3)/[1-(r_2+r_3)]$, $\eta_1 = \delta_2/(1-\delta_2)$, $\eta_1 + \eta_2 = \delta/(1-\delta) = \delta'$ (say), $C^* = C/(1-C)$, $y = t_2\delta'/(1+t_1+\delta')(1+C^*)$, and $\beta_i = \eta_i/\delta'(i = 1, 2)$. Write λ^* for a measure for β_2 on Γ_1 associated with λ in the obvious way, and denote by $\mu_i = \int_0^1 \beta_2^i \, d\lambda^*(\beta_2)$ the ith moment of λ^*. Finally, write $z = C\delta = C^*\delta'/(1+C^*)(1+\delta')$. We then obtain from (4.1)

$$(1-z)\sum_{n=0}^{\infty} y^n(2n+1)\mu_n + (z-y)\sum_{n=0}^{\infty} y^n(2n+1)(2n+3) \times (\mu_n - \mu_{n+1}) = (1-z)^{\frac{3}{2}}F(\tfrac{3}{2}, \tfrac{3}{2}; 1; z)(1-y)^{-\frac{3}{2}}. \tag{4.2}$$

Writing $B_z = (1-z)^{\frac{3}{2}}F(\frac{3}{2}, \frac{3}{2}; 1; z)$, we obtain upon equating coefficients of like powers of y on the two sides of (4.2), the following set of equations as equivalent to (4.1):

$$\begin{aligned} &\text{(a) } 1 + 2z - 3z\mu_1 = B_z \\ &\text{(b) } -(2n-1)\mu_{n-1} + (2n + z(2n+2))\mu_n - z(2n+3)\mu_{n+1} = B_z \frac{\Gamma(n+\frac{1}{2})}{\Gamma(\frac{1}{2})n!}, \quad n \geq 1. \end{aligned} \tag{4.3}$$

(Of course $\mu_0 = 1$ for λ^* to be a probability measure.) It is clear from (4.3) that λ^*, if it exists, depends on C and δ only through their product. One could now try to show that the sequence $\{\mu_i\}$ defined by $\mu_0 = 1$ and (4.3) satisfies the classical necessary and sufficient conditions for it to be a moment sequence of a probability measure on [0, 1] or, equivalently, that the Laplace transform $\sum_0^\infty \mu_j(-t)^j/j!$ is completely monotone on $[0, \infty)$, but we have been unable to proceed successfully in this way. Instead, we shall obtain, in the next paragraph, a function $m_z(x)$, which we then prove in the succeeding paragraphs below to be the Lebesque density $d\lambda^*(x)/dx$ of an absolutely continuous probability measure λ^* satisfying (4.3) (and hence (4.1)). That proof does not rely on the somewhat heuristic development of the next paragraph, but we nevertheless sketch that development to give an idea of where the $m_z(x)$ of (4.8) comes from.

The generating function $\phi(t) = \sum_{j=0}^\infty \mu_j t^j$ of the sequence $\{\mu_j\}$ satisfies a differential equation which is obtained in the usual way by multiplying (4.3) (b) by t^{n-1} and summing with respect to n from 1 to ∞:

$$2(1-t)(t-z)\phi'(t) - t^{-1}(t^2 - 2zt + z)\phi(t) = B_z(1-t)^{-\frac{1}{2}} - 1 - zt^{-1}. \tag{4.4}$$

This is solved by treatment of the corresponding homogeneous equation and by variation of parameter, to yield

$$\phi(t) = \left[\frac{t-z}{(1-t)t}\right]^{\frac{1}{2}} \int_0^t \left[\frac{B_z \tau^{\frac{1}{2}}}{2(1-\tau)(\tau-z)^{\frac{3}{2}}} - \frac{\tau^{\frac{1}{2}}}{(\tau-z)^{\frac{3}{2}}(1-\tau)^{\frac{1}{2}}} + \frac{1}{2[\tau(1-\tau)(\tau-z)]^{\frac{1}{2}}}\right] d\tau. \tag{4.5}$$

The constant of integration has been chosen to make ϕ continuous at 0 with $\phi(0) = 1$, and (4.5) defines a single-valued analytic function on the complex plane cut from 0 to z and from 1 to ∞. Now, if there did exist an absolutely continuous λ^* whose suitably regular derivative m_z satisfied

$$\int_0^1 m_z(x)\, dx/(1 - tx) = \phi(t), \tag{4.6}$$

we could obtain m_z by using the simple inversion formula

$$m_z(x) = \frac{1}{2\pi i x} \lim_{\epsilon \downarrow 0} [\phi(x^{-1} + i\epsilon) - \phi(x^{-1} - i\epsilon)]. \tag{4.7}$$

Since there is nothing in the theory of Stieltjes transforms which tells us that an m_z satisfying (4.7) does satisfy (4.6) (and hence (4.1)), we will use (4.7) only as a formal device to obtain m_z which we shall then prove, in the remaining paragraphs, satisfies (4.1). From (4.5) and (4.7) we obtain, for $0 < x < 1$,

$$\begin{aligned} m_z(x) = \frac{(1-zx)^{\frac{1}{2}}}{2\pi x^{\frac{1}{2}}(1-x)^{\frac{1}{2}}} \Bigg\{ & B_z \int_0^x \frac{du}{(1-u)(1-zu)^{\frac{1}{2}}} \\ & + \int_0^\infty \left[\frac{B_z u^{\frac{1}{2}}}{(1+u)(z+u)^{\frac{3}{2}}} + \frac{1}{[u(1+u)(z+u)]^{\frac{1}{2}}} - 2\frac{u^{\frac{1}{2}}}{(1+u)^{\frac{1}{2}}(z+u)^{\frac{3}{2}}}\right] du \Bigg\} \\ & = \frac{(1-zx)^{\frac{1}{2}}}{2\pi(x(1-x))^{\frac{1}{2}}} \{B_z Q_z(x) + c_z\} \quad \text{(say)}. \end{aligned} \tag{4.8}$$

c_z can be evaluated by making the change of variables $v = (1+u)^{-1}$ and using (4.11) below. We obtain

$$c_z = \tfrac{2}{3}B_z F(\tfrac{3}{2}, 1; \tfrac{5}{2}; 1-z) + \pi F(\tfrac{1}{2}, \tfrac{1}{2}; 1; 1-z) - \pi F(\tfrac{3}{2}, \tfrac{1}{2}; 2; 1-z)$$

and

$$Q_z(x) = 2(1-z)^{-1}[1 - (1-zx)^{-\frac{1}{2}}] + (1-z)^{-\frac{3}{2}} \log\left[\frac{(1-zx)^{\frac{1}{2}} + (1-z)^{\frac{1}{2}}}{(1-zx)^{\frac{1}{2}} - (1-z)^{\frac{1}{2}}} \cdot \frac{1-(1-z)^{\frac{1}{2}}}{1+(1-z)^{\frac{1}{2}}}\right].$$

Now to show that $d\lambda^*(x) = m_z(x)\, dx$ (with m_z defined by (4.8)) satisfies

(4.1) with λ^* a probability measure, we must show that

$$
\begin{aligned}
&\text{(a) } m_z(x) \geqq 0 \text{ for almost all } x, 0 < x < 1;\\
&\text{(b) } \int_0^1 m_z(x)\,dx = 1;\\
(4.9)\quad &\text{(c) } \mu_1 = \int_0^1 x m_z(x)\,dx \text{ satisfies (4.3)(a)};\\
&\text{(d) } \mu_n = \int_0^1 x^n m_z(x)\,dx \text{ satisfies (4.3)(b) for } n \geqq 1.
\end{aligned}
$$

Condition (4.9) (a) will follow from (4.8) and the positivity of B_z and c_z for $0 < z < 1$. The former is obvious. To prove the positivity of c_z, we first note that $F(\frac{3}{2}, \frac{3}{2}; 1; z) \geqq (1 - z)^{-2}$; this is seen by comparing the two power series, the coefficients of z^j being $[(\frac{3}{2})_j/j!]^2$ and $(j + 1)$, and the ratio of the former to the latter being $\prod_{i=1}^{j} (i + \frac{1}{2})^2/i(i + 1) \geqq 1$. We thus have $B_z \geqq 1 - z$. Substituting this lower bound into the expression for c_z and writing $u = 1 - z$, the resulting lower bound for c_z has a power series in u (convergent for $|u| < 1$) whose constant term is 0 and whose coefficient of u^j for $j \geqq 1$ is $(j + \frac{1}{2})^{-1} - \Gamma^2(j + \frac{1}{2})/\Gamma(j)\Gamma(j + 1)(j + 1)$; by the well known logarithmic convexity of the Γ-function, $\Gamma^2(j + \frac{1}{2}) < \Gamma(j)\Gamma(j + 1)$, so the coefficient of u^j for $j \geqq 1$ is $> (j + \frac{1}{2})^{-1} - (j + 1)^{-1} > 0$. Hence, $c_z > 0$ for $0 < z < 1$.

To prove (4.9) (d) we note that $m_z(x)$ defined by (4.3) satisfies the differential equation

$$
(4.10)\quad m_z'(x) + \tfrac{1}{2}m_z(x)\left[\frac{1 - 2x + zx^2}{x(1 - x)(1 - zx)}\right] = B_z/2\pi x^{\frac{1}{2}}(1 - x)^{\frac{1}{2}}(1 - zx),
$$

so that an integration by parts yields, for $n \geqq 1$,

$$
\begin{aligned}
-z(n + 2)\mu_{n+1} &+ (1 + z)(n + 1)\mu_n - n\mu_{n-1}\\
&= \int_0^1 \{-z(n + 2)x^{n+1} + (1 + z)(n + 1)x^n - nx^{n-1}\}m_z(x)\,dx\\
&= \int_0^1 x^n(1 - (1 + z)x + zx^2)m_z'(x)\,dx\\
&= \tfrac{1}{2}\left\{-\mu_{n-1} + 2\mu_n - z\mu_{n+1} + B_z\frac{\Gamma(n + \frac{1}{2})}{n!\,\Gamma(\frac{1}{2})}\right\},
\end{aligned}
$$

which is (4.3) (b).

The proofs of (4.9) (b) and (c) rely on certain identities involving hypergeometric functions. In the next paragraph we list some of the properties of hypergeometric functions which will be used in these proofs.

The material presented in this paragraph can be found in Erdélyi (1953), Chapter 2. The hypergeometric function $F(a, b; c; x)$ has the following integral representation when $\text{Re}(c) > \text{Re}(b) > 0$:

$$F(a, b; c; x) = \frac{\Gamma(c)}{\Gamma(b)\Gamma(c-b)} \int_0^1 t^{b-1}(1-t)^{c-b-1}(1-tx)^{-a}\, dt. \tag{4.11}$$

We will also use the representation

$$\log\left(\frac{1+x}{1-x}\right) = 2xF(\tfrac{1}{2}, 1; \tfrac{3}{2}; x^2) \tag{4.12}$$

and the identities

$$F(a, b; c; x) = F(b, a; c; x); \tag{4.13}$$

$$\begin{aligned}(c-a-1)F(a, b; c; x) &+ aF(a+1, b; c; x) \\ &- (c-1)F(a, b; c-1; x) = 0,\end{aligned} \tag{4.14}$$

$$\begin{aligned}&\lim_{c\to -n}[\Gamma(c)]^{-1}F(a, b; c; x) \\ &= \frac{(a)_{n+1}(b)_{n+1}}{(n+1)!}x^{n+1}F(a+n+1, b+n+1; n+2; x) \text{ for } n = 0, 1, 2, \cdots.\end{aligned} \tag{4.15}$$

$$\begin{aligned}c(1-x)F(a, b; c; x) &- cF(a-1, b; c; x) \\ &+ (c-b)xF(a, b; c+1; x) = 0;\end{aligned} \tag{4.16}$$

$$\begin{aligned}&F(\tfrac{1}{2}+\lambda, -\tfrac{1}{2}-\nu; 1+\lambda+\mu; x)F(\tfrac{1}{2}-\lambda, \tfrac{1}{2}+\nu; 1+\nu+\mu; 1-x) \\ &+ F(\tfrac{1}{2}+\lambda, \tfrac{1}{2}-\nu; 1+\lambda+\mu; x)F(-\tfrac{1}{2}-\lambda, \tfrac{1}{2}+\nu; 1+\nu+\mu; 1-x) \\ &- F(\tfrac{1}{2}+\lambda, \tfrac{1}{2}-\nu; 1+\lambda+\mu; x)F(\tfrac{1}{2}-\lambda, \tfrac{1}{2}+\nu; 1+\nu+\mu; 1-x) \\ &= \frac{\Gamma(1+\lambda+\mu)\Gamma(1+\nu+\mu)}{\Gamma(\lambda+\mu+\nu+\tfrac{3}{2})\Gamma(\tfrac{1}{2}+\mu)};\end{aligned} \tag{4.17}$$

$$F(a, b; c; x) = (1-x)^{c-a-b}F(c-a, c-b; c; x). \tag{4.18}$$

We are now in a position to prove (4.9) (b) and (c). From (4.8), using (4.11) and (4.12), we obtain

$$\begin{aligned}\int_0^1 m_z(x)\, dx &= (1-z)^{\frac{1}{2}}F(\tfrac{3}{2}, \tfrac{3}{2}; 1; z)F(-\tfrac{1}{2}, \tfrac{1}{2}; 1; z) \\ &\times [1 - F(\tfrac{1}{2}, 1; \tfrac{3}{2}; 1-z) + \tfrac{1}{3}(1-z)F(\tfrac{3}{2}, 1; \tfrac{5}{2}; 1-z)] \\ &- (1-z)^{\frac{3}{2}}F(\tfrac{3}{2}, \tfrac{3}{2}; 1; z) + (\pi/2)[F(\tfrac{1}{2}, \tfrac{1}{2}; 1; 1-z) - F(\tfrac{3}{2}, \tfrac{1}{2}; 2; 1-z)] \\ &\times F(-\tfrac{1}{2}, \tfrac{1}{2}; 1; z) \\ &+ B_z \int_0^1 \frac{(1-zx)^{\frac{1}{2}}}{2\pi(1-z)^{\frac{1}{2}}x^{\frac{1}{2}}(1-x)^{\frac{1}{2}}} \log\left(\frac{1+(1-z)^{\frac{1}{2}}(1-zx)^{-\frac{1}{2}}}{1-(1-z)^{\frac{1}{2}}(1-zx)^{-\frac{1}{2}}}\right) dx.\end{aligned} \tag{4.19}$$

The first expression in square brackets in (4.19) vanishes, as is easily seen from the power series (3.3). Using the power series for $\log(1+u^{\frac{1}{2}})/(1-u^{\frac{1}{2}})$, the integral of (4.19) can be written as

$$\frac{1}{\pi(1-z)}\sum_{n=0}^{\infty}\frac{(1-z)^n}{2n+1}\int_0^1\frac{dx}{x^{\frac{1}{2}}(1-x)^{\frac{1}{2}}(1-zx)^n}$$

$$= (1-z)^{-1} + (1-z)^{-1}\sum_{n=1}^{\infty}\frac{(1-z)^n}{2n+1}F(n, \tfrac{1}{2}; 1; z)$$

$$= (1-z)^{-1} + \sum_{n=0}^{\infty}\frac{(1-z)^n}{2n+3}F(n+1, \tfrac{1}{2}; 1; z)$$

$$(4.20)\qquad = (1-z)^{-1} + \tfrac{1}{2}\sum_{m=0}^{\infty}\frac{z^m(\frac{1}{2})_m}{(m!)^2}\sum_{n=0}^{\infty}\frac{(n+m)!\,(1-z)^n}{n!\,(n+\frac{3}{2})}$$

$$= (1-z)^{-1} + \tfrac{1}{2}\sum_{m=0}^{\infty}\frac{z^m(\frac{1}{2})_m}{(m!)^2}(1-z)^{-\frac{3}{2}}\int_0^{1-z}\frac{m!\,t^{\frac{1}{2}}}{(1-t)^{m+1}}\,dt$$

$$= (1-z)^{-1} + \tfrac{1}{2}(1-z)^{-\frac{3}{2}}\int_0^{1-z}\frac{t^{\frac{1}{2}}\,dt}{(1-t)^{\frac{1}{2}}(1-z-t)^{\frac{1}{2}}}$$

$$= (1-z)^{-1} + (\pi/4)(1-z)^{-\frac{1}{2}}F(\tfrac{1}{2}, \tfrac{3}{2}; 2; 1-z).$$

Hence, from (4.19) and (4.20) after cancellation one gets

$$(4.21)\qquad \int_0^1 m_z(x)\,dx = (\pi/2)F(-\tfrac{1}{2}, \tfrac{1}{2}; 1; z)[F(\tfrac{1}{2}, \tfrac{1}{2}; 1; 1-z) - F(\tfrac{3}{2}, \tfrac{1}{2}; 2; 1-z)] + (\pi/4)(1-z)^2F(\tfrac{3}{2}, \tfrac{3}{2}; 1; z)F(\tfrac{3}{2}, \tfrac{1}{2}; 2; 1-z).$$

Using (4.14) with $a = \frac{1}{2}$, $b = \frac{1}{2}$, $c = 2$, and (4.18) with $a = b = \frac{3}{2}$, $c = 1$, we obtain from (4.21)

$$(4.22)\qquad \int_0^1 m_z(x)\,dx = (\pi/4)\{F(-\tfrac{1}{2}, \tfrac{1}{2}; 1; z)F(\tfrac{1}{2}, \tfrac{1}{2}; 2; 1-z) + F(\tfrac{3}{2}, \tfrac{1}{2}; 2; 1-z)[F(-\tfrac{1}{2}, -\tfrac{1}{2}; 1; z) - F(-\tfrac{1}{2}, \tfrac{1}{2}; 1; z)]\}.$$

By the use of (4.14) with $a = b = -\frac{1}{2}$, $c = 1 + \epsilon$ and (4.15) with $n = 0$, $c = \epsilon \to 0$, (4.22) reduces to

$$(4.23)\qquad (\pi/4)\{F(-\tfrac{1}{2}, \tfrac{1}{2}; 1; z)F(\tfrac{1}{2}, \tfrac{1}{2}; 2; 1-z) + (z/2)F(\tfrac{3}{2}, \tfrac{1}{2}; 2; 1-z)F(\tfrac{1}{2}, \tfrac{1}{2}; 2; z)\}.$$

Now, by (4.17) with $\mu = 1$, $\lambda = -1$, $\nu = 0$, the expression (4.23) equals one if we have

$$F(\tfrac{3}{2}, \tfrac{1}{2}; 2; 1-z)[F(-\tfrac{1}{2}, \tfrac{1}{2}; 1; z) - F(-\tfrac{1}{2}, -\tfrac{1}{2}; 1; z) + (z/2)F(\tfrac{1}{2}, \tfrac{1}{2}; 2; z)] = 0.$$

The expression inside the square brackets is easily seen to be zero by using (3.3) and computing the coefficient of z^n. Thus (4.9) (b) is proved.

We now verify (4.9) (c). We proceed from (4.8) in a manner parallel to that used to obtain (4.21). The integrand of (4.19) is altered by multiplication by x, and in place of (4.20) we obtain $(1 - z)^{-1}/2 - z^{-1}/3 + [\pi/4z(1 - z)^{\frac{1}{2}}] \cdot F(-\frac{1}{2}, \frac{3}{2}; 2; 1 - z)$. The analogue of (4.21) is

$$
\begin{aligned}
\mu_1 = \int_0^1 x m_z(x)\, dx &= (\pi/4) F(-\tfrac{1}{2}, \tfrac{3}{2}; 2; z) \\
&\times [F(\tfrac{1}{2}, \tfrac{1}{2}; 1; 1 - z) - F(\tfrac{3}{2}, \tfrac{1}{2}; 2; 1 - z)] + (\pi/4)(1 - z)^2/z \\
&\times F(\tfrac{3}{2}, \tfrac{3}{2}; 1; z) F(-\tfrac{1}{2}, \tfrac{3}{2}; 2; 1 - z) - [(1 - z)^{\frac{5}{2}}/3z] F(\tfrac{3}{2}, \tfrac{3}{2}; 1; z).
\end{aligned} \tag{4.24}
$$

To verify (4.9) (c) we then have to prove the following identity (using (4.3) (a)):

$$
\begin{aligned}
(1 + 2z)/3z = (\pi/4) F(-\tfrac{1}{2}, \tfrac{3}{2}; 2; z) [F(\tfrac{1}{2}, \tfrac{1}{2}; 1; 1 - z) \\
- F(\tfrac{3}{2}, \tfrac{1}{2}; 2; 1 - z)] + (\pi/4)[(1 - z)^2/z] \\
\times F(\tfrac{3}{2}, \tfrac{3}{2}; 1; z) F(-\tfrac{1}{2}, \tfrac{3}{2}; 2; 1 - z).
\end{aligned} \tag{4.25}
$$

Using (4.18) with $c = 1$, $a = b = \frac{3}{2}$, then (4.16) with $a = \frac{1}{2}$, $b = -\frac{1}{2}$, $c = 1$, and then (4.17) with $\lambda = \mu = 0$, $\nu = 1$, (4.25) can be reduced to

$$
\begin{aligned}
(4/3\pi)(1 + 2z) &= z F(-\tfrac{1}{2}, \tfrac{3}{2}; 2; z) [F(\tfrac{1}{2}, \tfrac{1}{2}; 1; 1 - z) \\
&- F(\tfrac{3}{2}, \tfrac{1}{2}; 2; 1 - z)] + (3z/2) F(\tfrac{1}{2}, -\tfrac{1}{2}; 2; z) F(-\tfrac{1}{2}, \tfrac{3}{2}; 2; 1 - z) \\
&+ (4/3\pi)(1 - z) + (1 - z) F(\tfrac{1}{2}, \tfrac{3}{2}; 2; 1 - z) [F(\tfrac{1}{2}, -\tfrac{1}{2}; 1; z) \\
&- F(\tfrac{1}{2}, -\tfrac{3}{2}; 1; z)].
\end{aligned} \tag{4.26}
$$

Using (4.14) with $a = -\frac{3}{2}$, $b = \frac{1}{2}$, $c = 1 + \epsilon$, and then (4.15) with $n = 0$, $c = \epsilon \to 0$, the expression inside the square brackets in the last term of (4.26) can be further reduced to $\frac{1}{2} z F(-\frac{1}{2}, \frac{3}{2}; 2; z)$. Hence we are faced with the problem of establishing the identity

$$
\begin{aligned}
4/\pi &= F(-\tfrac{1}{2}, \tfrac{3}{2}; 2; z) F(\tfrac{1}{2}, \tfrac{1}{2}; 1; 1 - z) + \tfrac{3}{2} F(\tfrac{1}{2}, -\tfrac{1}{2}; 2; z) \\
&\times F(-\tfrac{1}{2}, \tfrac{3}{2}; 2; 1 - z) - [(z + 1)/2] F(\tfrac{3}{2}, -\tfrac{1}{2}, 2; z) \\
&\times F(\tfrac{1}{2}, \tfrac{3}{2}; 2; 1 - z),
\end{aligned} \tag{4.27}
$$

which finally by (4.11) with $\lambda = \nu = -1$, $\mu = 2$ reduces to

$$
\begin{aligned}
0 = F(-\tfrac{1}{2}, \tfrac{3}{2}; 2; z) [F(\tfrac{1}{2}, \tfrac{1}{2}; 1; 1 - z) + \tfrac{3}{2} F(\tfrac{3}{2}, -\tfrac{1}{2}; 2; 1 - z) \\
- \tfrac{3}{2} F(\tfrac{1}{2}, -\tfrac{1}{2}; 2; 1 - z) + ((1 - z)/2 - 1) F(\tfrac{3}{2}, \tfrac{1}{2}; 2; 1 - z)].
\end{aligned} \tag{4.28}
$$

The expression inside the square brackets in (4.28) has a power series in $1 - z$, the value of which is easily seen to be zero by computing the coefficients of various power of $1 - z$. Hence (4.9) (c) is proved.

REFERENCES

ANDERSON, T. W. (1958). *Introduction to Multivariate Statistical Analysis*. Wiley, New York.

ERDÉLYI, A., ed. (1953). *Higher Transcendental Functions*, **1**. McGraw-Hill, New York.

GIRI, N. and KIEFER, J. (1962). Minimax properties of Hotelling's and certain other multivariate tests. (abstract). *Ann. Math. Statist.* **33** 1490–1491.

GIRI, N., KIEFER, J. and STEIN, C. (1963) (referred to as (GKS)). Minimax character of Hotelling's T^2 test in the simplest case. *Ann. Math. Statist.* **34** 1524–1535.

KIEFER, J. (1957). Invariance, minimax sequential estimation, and continuous time proc esses. *Ann. Math. Statist.* **28** 573–601.

LEHMANN, E. L. (1959). *Testing Statistical Hypotheses*. Wiley, New York.

SIMAIKA, J. B. (1941). An optimum property of two statistical tests. *Biometrika* **32** 70–80.

WALD, A. (1950). *Statistical Decision Functions*. Wiley, New York.

WOLFOWITZ, J. (1945). The power of the classical tests associated with the normal distribution. *Ann. Math. Statist.* **20** 540–551.

ADMISSIBLE BAYES CHARACTER OF T^2-, R^2-, AND OTHER FULLY INVARIANT TESTS FOR CLASSICAL MULTIVARIATE NORMAL PROBLEMS

By J. Kiefer[1] and R. Schwartz[2]

Cornell University and General Electric Company

0. Summary. In a variety of standard multivariate normal testing problems, it is shown that certain procedures, often fully invariant, similar, and/or likelihood ratio, are admissible Bayes procedures. The problems include the multivariate general linear hypothesis (where some of the procedures considered were previously shown to be admissible by other methods), the testing of independence of sets of variates (where the likelihood ratio test is shown, for the first time, to be admissible), tests about only some components of the means, classification procedures (for any number of populations), Behrens-Fisher problem, tests about values of or proportionality or equality of covariance matrices, etc. A general technique is developed for obtaining certain Bayes procedures for such problems from the corresponding Bayes procedures relative to *a priori* distributions of a certain type for problems where nuisance parameter means have been deleted.

1. Notation. Before discussing the contents of this paper, we list the notation which will be used throughout.

The letters k, m, n, p, q, r, N, with or without subscripts, will denote positive integers, usually the number of rows or columns of a matrix. S, T, U, V, W, X, Y, Z, with or without subscripts or superscripts, will denote random matrices (or vectors), which in the absence of subscripts always have p rows. S and T will be square. Other Roman capital letters will denote vectors and matrices. I_q denotes the $q \times q$ identity, and 0 denotes any matrix of zeros. V will denote the entire random matrix under observation in any problem, and will always have N columns (= vector observations), independently distributed, each p-variate normal. The decomposition of a $p \times q$ matrix B into blocks will be denoted, for example, by $B = \{B_{ij}, \ 1 \leqq i \leqq k_1, \ 1 \leqq j \leqq k_2\}$, where B_{ij} is $p_i \times q_j$ with $\sum_i p_i = p$, $\sum_j q_j = q$. A decomposition into blocks of rows (resp., columns) alone will be denoted by $B = \{B_{(i)}\}$ (resp., $B = \{B_j\}$). Other decompositions will occasionally be denoted by superscripts. However, a notation like $V = (X, Y, U)$ will sometimes serve better than $V = (V_1, V_2, V_3)$ to distinguish the roles of different parts of V. Unprimed vectors will denote column vectors. B' denotes the transpose of B. The determinant of C is denoted by $|C|$, and its trace is denoted by tr C. If C is symmetric positive definite,

Received 27 August 1964.

[1] Research supported by the Office of Naval Research under contract No. Nonr-266(04) (NRO 47-005).

[2] Written, in part, while this author was a National Science Foundation Predoctoral Fellow.

$C^{\frac{1}{2}}$ will denote its unique symmetric positive definite square root. The average of the columns of any matrix X will be denoted by $\bar{X}$. The matrices S and T will be nonnegative definite symmetric, usually obtained in a problem as $S = (X - \bar{X})(X - \bar{X})'$ or $S = YY'$. Whenever such a matrix is positive definite on a set of probability one according to each θ in Ω, we shall invert it without further mention of the exceptional set.

Positive finite constants, depending on the problem but not on the parameter values, will be denoted by c or c_i. (A trivial exception to positivity and finiteness is the usage in (1.3).) The meaning of any c_i may change with the problem.

$\Omega = \{\theta\} = H_0 + H_1$ will denote the parameter space in any problem. Occasionally, to emphasize the symmetry of two hypotheses, we shall write $\Omega = H_1 + H_2$. The parameter θ will be decomposed into a collection of matrices (or vectors) ξ, ν, μ, Σ, etc., with or without subscripts; Σ without subscripts or superscripts will always be a covariance matrix which is $p \times p$ positive definite. Thus, Greek letters will be used to denote functions of θ; in addition, β, η, and γ will be reserved for other variables in terms of which it is convenient to write *a priori* densities, and of which components of θ may be functions. (See, for example, (4.1).) A Greek letter with subscript or superscript 0 (and perhaps other subscripts of superscripts) always denotes a specified value.

The use of $\sum$ to denote summation (e.g., as $\sum_i$ or $\sum_1^k$) will always be such that it cannot be mistaken for a parameter.

All probability laws of observable random variables, or functions thereof which we shall consider, will have Lebesgue densities on a Euclidean set. The Lebesgue density function of X when θ describes the underlying probability measure will be denoted by $f_X(x; \theta)$, or perhaps by $f_X(x; \beta(\theta))$ or $f_X(x; \beta)$ if this density depends on θ only through $\beta(\theta)$. We shall write $\exp a = e^a$, $\operatorname{etr} A = \exp \operatorname{tr} A$. Densities of particular interest are the multivariate normal density of a $p \times n$ matrix X of independent columns, each with nonsingular covariance matrix Σ, and with $EX = \xi$:

$$\phi_{p,n}(x; \xi, \Sigma) = c_1|\Sigma|^{-n/2} \operatorname{etr} - \tfrac{1}{2}\Sigma^{-1}(x - \xi)(x - \xi)'; \tag{1.1}$$

and, if $n \geqq p$, the central Wishart density of $W = (X - \xi)(X - \xi)'$ in this setting:

$$\psi_{p,n}(w; \Sigma) = c_2|\Sigma|^{-n/2}|w|^{(n-p-1)/2} \operatorname{etr} - \tfrac{1}{2}\Sigma^{-1}w. \tag{1.2}$$

For (1.2), the domain is $\{w_{ij}, i \geqq j: w \text{ positive definite}\}$.

A priori probability measures or positive constant multiples thereof will be denoted by Π. It is convenient to refrain from giving the explicit values of positive multiplicative constants and to require only $\Pi(\Omega) < \infty$ rather than $\Pi(\Omega) = 1$, and we shall do so. If $\Pi = \Pi_0 + \Pi_1$ with Π_i a finite measure on H_i, every Bayes critical region (for 0–1 loss function) is of the form

$$\{v: \textstyle\int f_V(v; \theta)\Pi_1(d\theta)/\int f_V(v; \theta)\Pi_0(d\theta) > c\} \cup L_c \tag{1.3}$$

for some $c(0 \leqq c \leqq \infty)$, where L_c is a measurable subset of the set obtained from the set in braces in (1.3) by replacing $>$ by $=$. In all our applications every L_c will have probability 0 for all θ in Ω, so that our Bayes procedures will be essentially unique. (An exception occurs in Corollary 3.2, where a different argument is used.) *Hence, all our Bayes procedures are admissible.*

In each example we obtain a family of tests by varying c from 0 to ∞ in (1.3). When, for example, the tests are similar, this of course yields an admissible similar test of each possible significance level.

The Π_i which arise in our examples all have Lebesgue densities on Euclidean sets, or are measures assigning all mass to a single point θ_0, or are products of these. Sometimes it will be convenient to consider Π_i or one of its factors to be a density on a Euclidean set Γ which is mapped in a given way into Ω or one of its factors. For example, it will be simpler to compute with the Lebesgue density $c|I_p + \eta\eta'|^{-m}$ on Euclidean p-space $E^p = \Gamma = \{\eta\}$, than with the induced measure on the space of positive definite $\Sigma = (I_p + \eta\eta')^{-1}$. Such Lebesgue densities (which will be integrable but not necessarily of integral one) will be denoted by $d\Pi_i(\eta)/d\eta$. The integrating Lebesgue measure in such a case will be denoted by $d\eta$.

In each example it is possible to work either with the original V, or else with a sufficient statistic. Usually the computations are such that there is no particular gain in using the reduction to the latter form.

Throughout the paper densities will be continuous on the product of sample and parameter space, both of which will be Euclidean spaces or Borel subsets thereof. *A priori* densities will be of the same character. Thus, no measurability considerations will ever be required, and they will always be omitted.

The reader is referred to the books by Roy (1957) and Anderson (1958) for descriptions of various multivariate problems and procedures, and to the book by Lehmann (1959) for general hypothesis testing theory.

2. Introduction. Admissibility of various classical statistical tests has been proved using (1) Bayes procedures, (2) exponential or other special structure of Ω, (3) invariance, and (4) local properties. (Estimation problems do not concern us here, and the techniques of this paper yield little of interest in such problems.) Some examples of (1) can be found in Lehmann and Stein (1948), Karlin (1957), Lehmann (1959), and Ellison (1962). Method (2) has been used by Birnbaum (1955), Stein (1956b), Nandi (1963), Ghosh (1964), and, more recently, by Schwartz (1964b). (It is also indicated to be the approach of Roy and Mikhail (1960), but the method is inapplicable in at least one of the cases described in their abstract, that of testing independence; this will be discussed further in the next paragraph.) Aside from the trivial case of compact groups, only the one-dimensional translation parameter case of (3) has been studied, in Lehmann and Stein (1953). The most common occurrence of (4) is with unique uniformly or locally most powerful unbiased tests in cases where unbiasedness implies similarity on that part of the boundary of H_1 which is in H_0. (See Lehmann (1959).) Uniformly most powerful tests can be regarded as a special case of (1). A result like that of Wald (1942) on the analysis of variance test can be regarded, for example, as an application of (1) to the similar tests obtained from unbiasedness considerations (4).

The use of techniques (3), (4), and (2) in standard multivariate normal problems has been limited. Best invariant procedures under the full linear group need not be minimax, let alone admissible. Most powerful unbiased tests, or analogues of Wald's theorem, fail to exist. While the exponential structure can be used to prove admissibility of Hotelling's T^2-test (Stein (1956b)) and of a class of tests of the multivariate general linear hypotheses (Schwartz (1964b)), this technique (or its generalization to certain nonexponential families) cannot be used in the problem of testing independence of sets of variates when $p \geqq 3$. This is discussed by Stein (1956b); from a slightly different viewpoint, it can be seen, even in the bivariate case, that this method fails because $f_{R^2}(r^2; \rho^2)/f_{R^2}(r^2; 0)$ (where R and ρ are the sample and population correlation coefficients) is not unbounded as $\rho \to 1$. The admissibility of the usual test in the bivariate case $p = 2$ is proved in Lehmann (1959) using (4), but this approach also fails when $p > 2$.

Our main interest is in those procedures which are invariant under all linear-affine transformations which leave the problem invariant, and which we shall call *fully invariant*. We use the Bayes technique (1) to prove admissibility of certain fully invariant tests of the general multivariate linear hypothesis and of the hypothesis of independence of sets of variates, as well as in other testing problems. The T^2- and R^2-tests are special cases. Even in the case of a test such as Hotelling's, where admissibility was proved (Stein (1956b)) by method (2), our result yields additional information on the performance of the test; for the method of (2) only insures that no other test of the same size is superior to T^2 "far" from H_0, while our Bayes result, discussed further in Section 4, reflects the behavior of T^2 closer to H_0.

Nevertheless, the Bayes technique has severe limitations. As always, it may be hard to guess the Π with respect to which a given test is Bayes, or to carry out a very explicit integration for a given Π. Moreover, many natural admissible tests cannot be proved admissible by this approach for reasons other than that of lack of integrability for minimum sample sizes which is mentioned later for certain tests. For example, Birnbaum's treatment shows that, for the problem of testing that the mean of a bivariate normal vector, with known covariance matrix, is 0, any compact convex polygonal acceptance region is admissible; but analyticity considerations show it cannot be Bayes. Thus, in problems like that of Section 4, one cannot expect tests such as the familiar one of Roy which is based on the largest characteristic root of $(XX' + YY')^{-1}XX'$, to be Bayes; but such tests can be proved admissible by technique (2). (See the references given earlier.)

Thus, we shall obtain admissibility of certain isolated but natural (and, often, well known) tests, rather than any general theorem characterizing Bayes tests. The tests obtained are often similar and unbiased, and are sometimes most powerful invariant and classical in origin (e.g., likelihood ratio tests). An admissible test of level α which is similar is of course also admissible with respect to its power function considered only on H_1, among level α tests.

There is no difficulty in constructing many noninvariant Bayes tests. For example (in a case of the setting of Section 4), if $V = (X, Y)$ with X $p \times 1$ and Y $p \times n (n \geqq p)$, with each column of V having covariance matrix Σ and with $EX = \xi$, $EY = 0$, for testing $H_0 : \xi = 0$ one easily shows that the following critical regions are Bayes: $X_{(1)} > c$ (where $X_{(i)}$ is 1×1), $X'X > c$, $X'X/\text{tr}\,(Y'Y) > c$, $\sum_1^p X_{(i)}^2/(Y'Y)_{ii} > c$, etc. The disadvantage of using any of these tests is of course that while each of them is similar for some subhypothesis of H_0, they have less satisfactory power characteristics under H_0 itself. This is why it is usually of greatest interest to find fully invariant tests. (In some examples, such as that of Section 6 (ii), it may be that the group of transformations which leaves the problem invariant is less relevant than a subgroup which leaves invariant some natural measure of distance from H_0; that is, it may be that $H_1 = \bigcup_\tau H_{1\tau}$ is invariant under a group G, but that each $H_{1\tau}$ is not.)

Such invariant tests often can be obtained, and in some cases have been obtained more than once in the literature since the first work of Jeffreys (1939) in this direction, as *formal* ("generalized") Bayes procedures with respect to invariant Π's of infinite mass. Of course, such a derivation cannot yield admissibility. In estimation problems as simple as that of estimating the mean of a standard univariate normal distribution with squared error loss, it is not hard to prove that the best invariant procedure cannot be Bayes. It was Lehmann and Stein (1948) who first showed that, in invariant testing problems, a different situation sometimes prevailed, and that best invariant procedures were sometimes genuinely Bayes for noninvariant reasons. An example of Section 7 (iii) gives a generalization of some of their univariate normal results. It will be seen that the rationale in choosing Π in other problems, such as the T^2 and R^2 generalizations of Sections 4 and 5, is somewhat different; this difference will be discussed further at the end of Section 3.

One consequence of the Bayesian character of certain invariant tests is that there is no possibility of proving an inadmissibility result for an essentially unique best invariant test where each hypothesis consists of a p-dimensional translation parameter family, analogous to the corresponding sweeping inadmissibility result of Stein (1956a) (see also Brown (1964)) in estimation problems. This is already evident in the example, covered by the results of Lehmann and Stein (1948), pp. 503–504, according to which, if $V = (V_1, \ldots, V_n)$ with $n > 1$, the V_i being independent normal p-vectors with common unknown mean and with covariance matrix $\sigma_i^2 I_p$ under H_i (with σ_i^2 specified), the essentially unique best invariant test, based on $\sum_1^n (V_i - \bar{V})'(V_i - \bar{V})$, is admissible.

It will be obvious (and will sometimes be illustrated explicitly) that in many examples there are many, and often infinitely many, linearly independent Π's relative to which a given procedure is Bayes. For example, Π_i will often assign all measure to a set where $\Sigma^{-1} = C + \eta\eta'$ where C can be taken to be an arbitrary fixed positive definite matrix (which we will usually take to be I_p) and η is a $p \times q$ random matrix; in some cases (e.g., (7.2)), even q may be varied for a fixed test. (A brief general discussion of these Π's will be found in the three paragraphs following (3.8).) The procedure is then Bayes relative to any finite (and, often, infinite) convex mixture of those Π's. This variability of the Π relative to which a given procedure is Bayes lends insight regarding the performance of the procedure. The richness of the family of Π's relative to which, for example, the T^2-test can be seen to be Bayes, may find a use in proving the minimax character of that test on a surface of constant power, and with computational ease compared with the calculation of Giri, Kiefer, and Stein (1963) in a special case. This approach has not yet succeeded.

Thus, no really novel minimax results are contained in this paper. An admissible test like (7.10), which (for appropriate c_i) has constant power on each H_i, is automatically minimax (the one-sided analogue being even simpler). Section 6(ii) gives an example where minimax properties follow from previously known results.

Regarding minimax properties, we remark that, at least locally, tests based on traces of appropriate matrices appear to be more satisfactory than those based on determinants (Section 4 and 6; see Schwartz (1964a)).

The reader will note that, in many respects, the results of Sections 4 and 5 are more satisfactory than those of some of the examples of Sections 6 and 7 (for example, 6(ii)) where the group involved is not merely a direct sum of full linear groups.

It would require too much space to list, in each setting considered herein, even a few of the tests which can be obtained by the methods of this paper. We shall therefore list a few such variants only in Sections 4 (multivariate general linear hypothesis) and 7(i); the applicability of the methods in other examples will be clear.

Moreover, there are many testing setups for normal and other exponential families which we shall omit entirely because of the space they would occupy, but in which our methods can be applied. A few of the problems we shall exclude are those concerned with hypothesized nonzero values of all or some of the canonical correlations, correlations, partial correlations, or eigenvalues of a covariance matrix; equality of such parameters of two covariance matrices; hypothesized values of certain parameters of both the mean and covariance matrix, or equality of such parameters of two mean vectors and covariance matrices; the hypotheses which arise in principal component and factor analysis.

In addition, there are multivariate analogues of many of the examples of Lehmann and Stein (1948), which we shall omit.

Our examples will be ones in which both means and covariance matrices are unknown. Where some of these are known (for example, if Σ and ν are known in Section 4), the problems are easier to solve and sometimes (as in the cases of Sections 4 and 6 if Σ is known) have well known solutions.

3. Preliminary results. We summarize here the integration results which will be used repeatedly.

From (1.1) we have, if η and z are $p \times m$ and t is $p \times p$ and positive definite,

$$\int_{E^{mp}} \operatorname{etr}\{-\tfrac{1}{2}[t\eta\eta' - 2z\eta']\}\, d\eta = c|t|^{-m/2} \operatorname{etr} \tfrac{1}{2}t^{-1}zz'. \tag{3.1}$$

If γ is a p-vector, an obvious diagonalization yields the well known relation

$$|I_p + \gamma\gamma'| = 1 + \gamma'\gamma. \tag{3.2}$$

Defining, for h real,

$$b_h = \int_{E^p} (1 + \gamma'\gamma)^{-h/2}\, d\gamma, \tag{3.3}$$

we clearly have

$$b_h < \infty \Leftrightarrow h > p. \tag{3.4}$$

With $\eta = (\eta_1, \eta_2, \ldots, \eta_m)$ $p \times m$, write $Q_j = I_p + \sum_{i=1}^{j} \eta_i\eta_i'$ and let $Q_j^{\frac{1}{2}}$ be the positive definite symmetric square root of Q_j. Using the change of variables $z_j = Q_{j-1}^{-\frac{1}{2}}\eta_j$ and (3.2)–(3.3), we have, for $j > 1$ and h real,

$$\begin{aligned} \int_{E^p} |Q_j|^{-h/2}\, d\eta_j &= \int |Q_{j-1} + \eta_j\eta_j'|^{-h/2}\, d\eta_j \\ &= |Q_{j-1}|^{(1-h)/2} \int |I_p + z_j z_j'|^{-h/2}\, dz_j \\ &= b_h |Q_{j-1}|^{(1-h)/2}. \end{aligned} \tag{3.5}$$

Hence,

$$\int_{E^{mp}} |I_p + \eta\eta'|^{-h/2}\, d\eta = \prod_{i=1}^{m} b_{h-i+1}, \tag{3.6}$$

which with (3.4) yields

$$\int_{E^{mp}} |I_p + \eta\eta'|^{-h/2}\, d\eta < \infty \Leftrightarrow h > m + p - 1. \tag{3.7}$$

We shall use the well known fact that, with η $p \times k$,

$$|I_k - \eta'(I_p + \eta\eta')^{-1}\eta| = |I_p + \eta\eta'|^{-1}. \tag{3.8}$$

This follows, for example, from a direct computation upon writing $\eta = A_p L A_k$ where A_j is orthogonal $j \times j$ and, according to whether $p \leqq k$ or $p \geqq k$, we have $L = (D_p, 0)$ or $L' = (D_k, 0)$ with D_j diagonal $j \times j$. This diagonalization also demonstrates the positive definiteness of the matrix whose determinant is on the left side in (3.8); we shall use this fact below.

A main idea in our construction of appropriate Π's is the frequent representation of covariance matrices $(I_p + \eta\eta')^{-1}$ where η is $p \times q$ for an appropriate q. Thus, we assign all measure to a set where $I_p - \Sigma$ is positive definite. The form $I_p + \eta\eta'$ and certain other forms used in the Π's were suggested by their appearance in the linear functionals used in approach (2) of Section 2 for the problem of testing the general linear hypothesis. (See Stein (1956) and Schwartz (1964a) and (1964b).)

The representation $(I_p + \eta\eta')^{-1}$ suggests a formal structure for the *a priori* densities $d\Pi_i(\eta)/d\eta$ which by (3.7) will in fact be integrable. It also yields integrals (and hence Bayes procedures) of simple functional form.

A second idea (which depends on the above-mentioned property of the set where Π is supported) is the elimination of means which are nuisance parameters by means of Lemma 3.1 below. Another, more obvious, idea, is the treatment of certain nuisance *components* of the p-vectors of means (for example, in Section 6) by letting Π assign all measure to a set where these components are independent of the others and have a single distribution under each Π_i.

We shall now formalize a technique for proving that, in some cases, a procedure Δ which is Bayes for a problem $P = \{H_0, H_1, V\}$ remains Bayes when the problem is altered to P^* by the addition of certain nuisance parameters and corresponding observables. Suppose, for a problem P, that Δ is Bayes (i.e., satisfies (3.1)) for given Π_0, Π_1, and c. Suppose P is now altered to $P^* = \{H_0^*, H_1^*, V^*\}$ as follows: $V^* = (V, U_1, U_2, \ldots, U_m)$ where U_j is $q_j \times 1$ with m and q_j arbitrary positive integers; the U_j are independent of V and each other, U_j being normal with mean vector ν_j and nonsingular covariance matrix $\Sigma^{(j)}$. The $\Sigma^{(j)}$ might be related to some of the parameters of V or to each other (for example, several might be equal). However, we assume that, for each θ in a subset of H_i to which Π_i assigns all measure, there is a corresponding set in H_i^* for which the domain of $(\nu_1, \nu_2, \ldots, \nu_m)$ is $E^{\Sigma_1^m q_i}$ and for which $\Sigma^{(j)}$, which in this instance we shall write as $\Sigma^{(i,j)}$ for clarity, can be written as

$$\Sigma^{(j)} \equiv \Sigma^{(i,j)} = (C^{(j)} + D^{(i,j)})^{-1} \tag{3.9}$$

where $C^{(j)}$ is symmetric positive definite (and does not depend on i), and $D^{(i,j)}$ is symmetric nonnegative definite. (If $\Sigma^{(j)}$ is unrelated to θ, it can of course be treated trivially by letting each Π_i^* assign all measure to any specified value of $(\Sigma^{(j)}, \nu_j)$.) If $D^{(i,j)}$ is of rank $r_{i,j}$, we can then write

$$D^{(i,j)} = \Lambda^{(i,j)}\Lambda^{(i,j)\prime} \tag{3.10}$$

where Λ is $q_j \times r_{i,j}$. (Actually, certain restrictions on the ν_j can be imposed, as will become evident in the course of the proof of Lemma 3.1.) The possible distributions of V under H_i^* are the same as under H_i. As mentioned at the end of Section 1, the H_i and H_i^* are assumed to be Euclidean Borel sets.

LEMMA 3.1. *If Δ is Bayes relative to Π for problem P, then Δ^* is Bayes relative to some Π^* for problem P^*, where*

$$\Delta^*(v, u_1, u_2, \ldots, u_m) = \Delta(v). \tag{3.11}$$

PROOF. Write $\beta = (\Sigma^{(1)}, \Sigma^{(2)}, \ldots, \Sigma^{(m)})$ and $\nu = (\nu_1, \ldots, \nu_m)$. Let the conditional *a priori* distribution of the ν_j, given θ (the parameter of V) and β, be as follows: the ν_j are conditionally independent and, under H_i^* (that is, if $\theta \varepsilon H_i$), with *a priori* probability one,

$$(\Sigma^{(i,j)})^{-1}\nu_j = \Lambda^{(i,j)}\gamma^{(i,j)} \tag{3.12}$$

where $\gamma^{(i,j)}$ is $r_{i,j} \times 1$ and is normal with mean vector 0 and covariance matrix

$$(I_{r_{i,j}} - \Lambda^{(i,j)\prime}(C^{(j)} + D^{(i,j)})^{-1}\Lambda^{(i,j)})^{-1} = (B^{(i,j)})^{-1} \quad \text{(say)}.$$

Denote this conditional distribution of ν by $\Pi_{i,\theta,\beta}^*$. The (marginal) joint distribution of θ and β under H_i^* (that is, under $c_i\Pi_i^*$) is given as follows: θ has (marginal) distribution Π_i and, given θ, the conditional distribution of β is any probability measure $\Pi_{i,\theta}$ assigning measure one to a specified value (or set of values) of β which is possible for this value of θ (that is, which is consistent with any relation which may exist between θ and β). There is no measurability difficulty in this construction, since the H_i^* are Euclidean Borel sets. We must only verify that $B^{(i,j)}$ is positive definite. Letting $C^{\frac{1}{2}}$ denote the symmetric positive definite square root of $C^{(j)}$ and $\eta = C^{-\frac{1}{2}}\Lambda^{(i,j)}$, we can write $B^{(i,j)}$ as $I_{r_{i,j}} - \eta'(I_{q_j} + \eta\eta')^{-1}\eta$ and recall the second sentence following (3.8).

From (3.8) itself we then have $|B^{(i,j)}| = |I_{q_j} + \eta\eta'|^{-1} = |C^{(j)}||\Sigma^{(i,j)}|$. Hence, if ν_j satisfies (3.12) under H_i^*, we have, omitting most superscripts in the exponential after the second expression,

$$\begin{aligned}
&f_{U_j}(u;\nu_j,\Sigma^{(i,j)})\phi_{r_{i,j}}(\gamma^{(i,j)};0,(B^{(i,j)})^{-1})\\
&= c_3|\Sigma^{(i,j)}|^{-\frac{1}{2}}|B^{(i,j)}|^{\frac{1}{2}}\operatorname{etr}-\tfrac{1}{2}\{(\Sigma^{(i,j)})^{-1}(u-\nu_j)(u-\nu_j)'\\
(3.13)\qquad &\quad + B^{(i,j)}\gamma^{(i,j)}\gamma^{(i,j)\prime}\}\\
&= c_3|C^{(j)}|^{\frac{1}{2}}\operatorname{etr}-\tfrac{1}{2}\{C^{(j)}uu'\}\operatorname{etr}-\tfrac{1}{2}\{Duu' - 2\Lambda\gamma u' + \Lambda\gamma\gamma'\Lambda'\Sigma\\
&\quad + \gamma\gamma' - \Lambda'(C+D)^{-1}\Lambda\gamma\gamma'\}\\
&= c_3|C^{(j)}|^{\frac{1}{2}}\operatorname{etr}-\tfrac{1}{2}\{C^{(j)}uu'\}\operatorname{etr}-\tfrac{1}{2}(\gamma-\Lambda'u)(\gamma-\Lambda'u)'.
\end{aligned}$$

From this we conclude that if, under H_i^*, ν_j is the function of $\gamma^{(i,j)}$ given in (3.12), then

$$(3.14)\qquad \frac{\int_{E^{r_{1,j}}} f_{U_j}(u;\nu_j,\Sigma^{(1,j)})\phi_{r_{1,j}}(\gamma^{(1,j)};0,(B^{(1,j)})^{-1})\,d\gamma^{(1,j)}}{\int_{E^{r_{0,j}}} f_{U_j}(u;\nu_j,\Sigma^{(0,j)})\phi_{r_{0,j}}(\gamma^{(0,j)};0,(B^{(0,j)})^{-1})\,d\gamma^{(0,j)}} = c_4\,.$$

Hence,

$$(3.15)\qquad \frac{\int f_{U_1,\cdots,U_m}(u_1,\cdots,u_m;\nu,\beta)\Pi^*_{1,\theta,\beta}(d\nu)}{\int f_{U_1,\cdots,U_m}(u_1,\cdots,u_n;\nu,\beta)\Pi^*_{0,\theta,\beta}(d\nu)} = c_5\,.$$

Thus, writing $\theta^* = (\theta,\nu,\beta)$, we have

$$\begin{aligned}
(3.16)\quad &\int f_{V^*}(v,u_1,\ldots,u_m;\theta^*)\Pi_1^*(d\theta^*)/\int f_{V^*}(v,u_1,\ldots,u_m;\theta^*)\Pi_0^*(d\theta^*)\\
&= c_6[\int f_V(v;\theta)\Pi_1(d\theta)/\int f_V(v,\theta)\Pi_0(d\theta)],
\end{aligned}$$

so that a proper choice of c^* in $\Pi^* = c^*\Pi_1^* + \Pi_0^*$ yields the conclusion of the lemma.

A degenerate case of the above setup occurs when V and θ are absent; that is, when, on the basis of $U_1\ U_2,\ldots,U_m$, it is desired to test some hypothesis concerning the $\Sigma^{(j)}$ (and/or a linear space of linear combinations of the ν_j which, under both H_i^*, has only 0 in common with the space spanned by $\Sigma^{(i,j)}\Lambda^{(i,j)}\gamma^{(i,j)}$). In that case let H_i^{**} be any subhypothesis of H_i^* which consists of the $(r_{i1} + \ldots + r_{im})$-dimensional Euclidean space of $(\gamma^{(i,1)},\ldots,\gamma^{(i,m)})$, the values of the $\Sigma^{(j)}$ and other linear combinations of the ν_j being specified so that (ν,β) is completely determined by the $\gamma^{(i,j)}$. Then Π_i^* (formerly $\Pi^*_{i,\theta,\beta}$) has a continuous positive Lebesgue density on each H_i^{**}, and every procedure has continuous power function on each H_i^{**}. Hence, even though the Bayes procedure in this case is not essentially unique, we conclude:

COROLLARY 3.2. *For testing between the H_i^* based on $U_1, U_2, \ldots, U_m$, for each $\alpha(0 \leqq \alpha \leqq 1)$ the randomized test which accepts H_1^* with probability α for every sample value, is an admissible Bayes procedure.*

Lehmann and Stein (1948) also gave examples where the randomized test of Corollary 3.2 is Bayes.

We remark, incidentally, that the method used by Lehmann and Stein to handle means which are nuisance parameters (other than the use of a Π_i concentrated at a single point), and which differs from that of Lemma 3.1, is (roughly) to let the means have normal *a priori* densities under H_0 (say), with means equal to those under H_1, and with variances equal to the difference between hypothesized variances under H_1 and H_0. When one or more variances are nuisance parameters, as under H_0 in Student's problem when mean and variance are both specified under H_1 (that is, the problem of Section 6(vi) below, modified to specify Σ also, under H_1), their *a priori* density again reflects the difference between variances under H_0 and H_1. We are usually unable to make use of these techniques, but the test derived in (7.8) is an exception, which uses a direct multivariate analogue of the above technique for means. Lehmann and Stein generally consider simple alternatives and thereby often obtain uniformly most powerful tests against composite alternatives; these are stronger conclusions than ours, which are often obtained in settings where no such uniformly most powerful tests exist.

4. Multivariate general linear hypothesis. In the usual formulation, $V = (V_1, \ldots, V_N)$ with cov $V_i = \Sigma$ and $EV = \xi L$, H_0 being $\xi K = \xi^{(0)}K$ (specified, and which by a translation can be taken to be 0), with L and K known matrices. This can be transformed into the canonical form wherein $V = (X, Y, U)$ with $EX = \xi(p \times r)$, $EY = O(p \times n)$, $EU = \nu(p \times h)$, where, under Ω, ξ and ν have E^{pr} and E^{ph} as their domains, and all columns of V are again independent with common unknown covariance matrix Σ; H_0 is $\xi = 0$. We treat the problem in this canonical form.

According to Lemma 3.1, results for the general case follow from those for the case $h = 0$, which we hereafter treat. In part (v) below we shall discuss the case $n < p$; until that part, we suppose $n \geqq p$.

(i) Let both Π_1 and Π_0 assign all their measure to θ's for which $\Sigma^{-1} = I_p + \eta\eta'$ for some $(p \times r)\eta$. Also, under H_1 all measure is assigned to ξ's of the form $\xi = \Sigma\eta$ (where $\Sigma^{-1} = I_p + \eta\eta'$). The Π_i's can be considered as absoutely continuous measures on the space E^{pr} of η's, and are given by

$$
\begin{aligned}
(4.1) \qquad d\Pi_1(\eta)/\,d\eta &= |I_p + \eta\eta'|^{-(r+n)/2} \operatorname{etr} \tfrac{1}{2}\{\eta'(I_p + \eta\eta')^{-1}\eta\}, \\
d\Pi_0(\eta)/\,d\eta &= |I_p + \eta\eta'|^{-(r+n)/2}.
\end{aligned}
$$

The integrability of these densities follows from (3.7) (since $n \geqq p$) and the boundedness of the nonnegative definite matrix in braces (which, according to the comment two sentences below (3.8), yields a positive definite matrix when subtracted from I_p). We then have

$$\int f_{X,Y}(x, y; \theta)\Pi_1(d\theta)/\int f_{X,Y}(x, y; \theta)\Pi_0(d\theta)$$

$$(4.2)\quad = \frac{\int |I_p + \eta\eta'|^{(r+n)/2} \operatorname{etr}\{-\frac{1}{2}(I_p + \eta\eta')(xx' + yy') + \eta x' - \frac{1}{2}(I_p + \eta\eta')^{-1}\eta\eta'\}\Pi_1(d\eta)}{\int |I_p + \eta\eta'|^{(r+n)/2} \operatorname{etr}\{-\frac{1}{2}(I_p + \eta\eta')(xx' + yy')\}\Pi_0(d\eta)}$$

$$= \frac{\operatorname{etr}\{\frac{1}{2}(xx' + yy')^{-1}xx'\} \int \operatorname{etr}\{-\frac{1}{2}(xx' + yy') \cdot(\eta - (xx' + yy')^{-1}x)(\eta - (xx' + yy')^{-1}x)'\}\, d\eta}{\int \operatorname{etr}\{-\frac{1}{2}(xx' + yy')\eta\eta'\}\, d\eta}$$

$$= \operatorname{etr}\{\tfrac{1}{2}(xx' + yy')^{-1}xx'\}.$$

(As stated in Section 1, we shall not require, here and in other examples, a discussion of the exceptional set where $XX' + YY'$ is not invertible.) Since $\operatorname{tr}(XX' + YY')^{-1}XX' = c$ with probability zero for each θ, we conclude that, for each $c \geqq 0$, the critical region

$$(4.3)\qquad \operatorname{tr}(XX' + YY')^{-1}XX' \geqq c$$

is an admissible Bayes procedure. It is fully invariant, similar, and (as a consequence of the results of Das Gupta, Anderson, and Mudholkar (1964)) unbiased.

We now give an indication of some of the many modifications in the Π_i which, as described in Section 2, still yield (4.3), and also list a few modifications which yield other tests. It will be obvious that some of these modifications can be combined. Again, Lemma 3.1 applies in all cases.

(ii) In place of I_p in (4.1) we can put any positive definite symmetric $p \times p$ matrix B and write $\Sigma^{-1} = B + \eta\eta'$, and in place of $\xi = \Sigma\eta$ we can put $\xi = b\Sigma\eta$ for any nonzero scalar b, multiplying the exponent in (4.1) by b^2. We still obtain (4.3). This means that, for each such B and b, there is a Π relative to which (4.3) is Bayes, and such that Π assigns all probability to a set of (Σ, ξ) for which Σ is smaller than B^{-1} in the sense that $B^{-1} - \Sigma$ is nonnegative definite, and for which $\operatorname{tr}\Sigma^{-1}\xi\xi' < pb^2$. Thus, the test (4.3) has good performance "near" H_0, in agreement with the local minimax character and in contrast with the "distant" goodness (obtained by method (2) of Section 2), both in Schwartz (1964a). We note that, for fixed b, the Π_1's corresponding to different B's assign all measure to disjoint sets of (ξ, Σ).

(iii) Letting $\Sigma^{-1} = I_p + \eta\eta'$ as before, suppose we now let $\xi = \Sigma\eta\beta$ under H_1, where β is $r \times r$. Let k be a fixed number satisfying $0 < k < 1$. Under H_1 the conditional density of β, given η, is $\phi_{r,r}(\beta; 0, [k^{-1}I_r - \eta'(I_p + \eta\eta')^{-1}\eta]^{-1})$. The marginal density of η under H_1 is

$$(4.4)\qquad c_7|I_p + \eta\eta'|^{-(r+n)/2}|k^{-1}I_r - \eta'(I_p + \eta\eta')^{-1}\eta|^{-r/2},$$

which is again integrable, since the second determinant of (4.4) is bounded away from zero. Π_0 is again given by the second line of (4.1). The product of $f_{X,Y}$ with the density of β and η under H_1 is then

$$(4.5)\quad c_8 \operatorname{etr}\{-\tfrac{1}{2}(xx' + yy')\} \operatorname{etr}\{-\tfrac{1}{2}(yy' + (1-k)xx')\eta\eta'\} \cdot \operatorname{etr}\{-\tfrac{1}{2}k^{-1}(\beta - kx'\eta)(\beta - kx'\eta)'\}.$$

The integration of the last factor of (4.5) with respect to β yields a constant, and comparing the integration with respect to η of the middle factor with the corresponding integration (with $k = 0$) under Π_0, we obtain the critical region

$$(4.6)\qquad |YY' + XX'|/|YY' + (1-k)XX'| \geqq c$$

as an admissible Bayes test for $0 < k < 1$. For $k = 1$, one obtains the likelihood ratio test

$$(4.7)\qquad |YY' + XX'|/|YY'| \geqq c,$$

but one change is needed in the previous derivation: the integrability of (4.4), which by (3.8) equals $c_7|I_p + \eta\eta'|^{-n/2}$ when $k = 1$, is now assured, according to (3.7), if and only if $n > p + r - 1$. Thus, only under this restriction does our method show that the test (4.7) is admissible Bayes, although the admissibility without this restriction can be proved by method (2) of Section 2. (See Schwartz (1964b).) It is not known whether or not (4.7) is Bayes when $n < p + r$, except when $r = 1$, when the treatment of part (i) applies.

(iv) Without giving any details, we list a few of the many other examples of *a priori* distributions with respect to which the Bayes procedure can be computed and is fully invariant:

(a) For $0 < k_1 < 1$ and $k_2 > 0$, modify (iii) by letting $\xi = \Sigma\eta(k_2 I_r + \beta)$ under H_1, the conditional density of the $r \times r$ matrix β, given η, being

$$g(\eta) \operatorname{etr}\{-\tfrac{1}{2}[k_1^{-1}\beta\beta' - \eta'(I_p + \eta\eta')^{-1}\eta(k_2 I_r + \beta)(k_2 I_r + \beta')]\},$$

where

$$g(\eta) = |k_1^{-1}I_r - \eta'(I_p + \eta\eta')^{-1}\eta|^{r/2} \operatorname{etr}\{\tfrac{1}{2}k_2^2\eta'(I_p + \eta\eta')^{-1}\eta \cdot [-I_r + (k_1^{-1}I_r - \eta'(I_p + \eta\eta')^{-1}\eta)^{-1}\eta'(I_p + \eta\eta')^{-1}\eta]\};$$

the marginal density of η under H_1 is now $c_8|I_p + \eta\eta'|^{-(r+n)/2}/g(\eta)$. The Bayes critical region is

$$(4.8)\quad \operatorname{etr}\{\tfrac{1}{2}k_2^2(YY' + (1-k_1)XX')^{-1}XX'\} \cdot |YY' + XX'|^{r/2}/|YY' + (1-k_1)XX'|^{r/2} \geqq c.$$

For $k_1 = 1$ we have the same modification as in (iii), so that in that case the approach only proves that (4.8) is Bayes if $n > p + r - 1$.

(b) If $r \geqq p$, alter $d\Pi_i(\eta)/d\eta\, (i = 0, 1)$ in (4.1) by multiplying it by $|\eta\eta'|^{t/2}$ where $p - r - 1 < t < n - p + 1$; these inequalities, needed for integrability of the altered $d\Pi_i(\eta)/d\eta$, come from an obvious modification of (3.7) and corresponding considerations near $|\eta\eta'| = 0$. The integrand in both the numerator and denominator of the third expression of (4.2) is multiplied by $|\eta\eta'|^{t/2}$. Using a result of Constantine (1963) (p. 1279), the resulting fully invariant critical region can be written as

(4.9) $$ {}_1F_1((r+t)/2, r/2, \tfrac{1}{2}(XX'+YY')^{-1}XX') \geqq c, $$

where ${}_1F_1$ is the hypergeometric function of matrix argument (which is a polynomial multiplied by an exponential if $t/2$ is an integer; see Herz (1955)).

(c) If (i) is modified only by putting $\xi = \Sigma\eta B$ with a corresponding change in $d\Pi_1(\eta)/d\eta$ in (4.1), where B is a fixed $r \times r$ matrix, we obtain

(4.10) $$ \operatorname{tr}(XX'+YY')^{-1}XB'BX' \geqq c, $$

which is not fully invariant (unless B is orthogonal). However, if we instead put $\xi = \Sigma\eta B\epsilon$ where B is again fixed and (4.1) is altered under H_1 by letting the $r \times r$ orthogonal matrix ϵ be uniformly distributed over the orthogonal group and independent of η, we obtain, according to James (1964) (formulas (25) and (30)), integrating first over η and then over ϵ, the fully invariant test

(4.11) $$ {}_0F_0(\tfrac{1}{2}B'B, X'(XX'+YY')^{-1}X) \geqq c, $$

where ${}_0F_0$ is the hypergeometric function of two matrix arguments.

(v) We now consider the case $n < p$. If $n < p < n + r$, the test (4.3) (for example) is a nontrivial fully invariant test, but the Bayes approach of (i) fails; the admissibility in this case is still obtained by the method of Schwartz (1964b). When $n + r \leqq p$, the only fully invariant tests are the trivial fully randomized ones (which will be seen below to be inadmissible). However, there are reasonable admissible tests which are not fully invariant. For example, let $W = AV$ for any fixed nonsingular $p \times p$ matrix A, and let $W' = (W'_{(1)}, W'_{(2)})$ be a decomposition of W with $W_{(1)}$ having n rows. We now let both Π_i assign all their measure to those θ for which $W_{(1)}$ and $W_{(2)}$ are independent and the columns of $W_{(2)}$ have any specified distribution (for example, $W_{(2)}$ can have density $\phi_{p-n,r+n}(\,.\,; 0, I_{p-n})$ under both H_i). The parameters of the distribution of $W_{(1)}$ have *a priori* densities on E^{nr} given by (4.1) with p replaced by n. The derivation proceeds as before to show that the test (4.3), with (X, Y) replaced by $W^{(1)}$, is admissible Bayes. Lemma 3.1 again extends this result to the case where $h > 0$.

We remark that, when, $n < p < n + r$ (which is only possible when $r > 1$), the test (4.3) can be shown from the results of Das Gupta, Anderson, and Mudholkar (1964) to have nontrivial mimimum power on the set $H_{1c'} = \{\operatorname{tr}\Sigma^{-1}\xi\xi' \geqq c'\}$, so that the trivial randomized test of the same size cannot be maximin on any fully invariant set contained in $H_{1c'}$ for some $c' > 0$. However, when $n + r \leqq p$ the trivial randomized test is maximin on such a set (although inadmissible if $n > 0$, as is shown by comparison with the test based on $W_{(1)}$); this follows from the fact that the maximal invariant under the group of lower triangular matrices does not depend on the last row of X (see Giri, Kiefer, and Stein (1963) for this type of computation).

The fully invariant tests for the problem of this section are well known to depend only on the nonzero latent roots t_i (say) of $(XX'+YY')^{-1}XX'$. The test (4.3), based on $\sum_i t_i$, has received much less attention than the likelihood ratio test (4.7) (based on $\prod_i (1-t_i)^{-1}$), Roy's test (based on $\max_i t_i$), or

Hotelling's T_0^2-test (based on $\sum_i t_i/(1-t_i)$). All of these tests of course reduce to Hotelling's T^2-test when $r = 1$, to the (univariate) analysis of variance F-test when $p = 1$, and to Student's two-tailed t-test when $p = r = 1$. The test (4.3) was suggested by Pillai (1955), who has studied the distribution of the statistic under H_0, and it has also been studied by Schwartz (1964a), who proved its admissibility and certain other optimum properties.

5. Independence of sets of variates. Here $V = (Y, U)$ where under Ω the columns of V are independent with common unknown nonsingular covariance matrix Σ, Y is $p \times n$ with $EY = 0$, and U is $p \times h$ with $EU = \nu$ (unknown). Let $V' = (V'_{(1)}, V'_{(2)}, \ldots, V'_{(k)})$ where $V_{(i)}$ has p_i rows and $\sum_1^k p_i = p$. Under H_0 the $V_{(i)}$ are independent, so that

$$\Sigma = \begin{pmatrix} \Sigma_{11} & 0 & 0 & \cdots & 0 \\ 0 & \Sigma_{22} & 0 & \cdots & \vdots \\ \vdots & \vdots & \vdots & \vdots & 0 \\ 0 & 0 & \cdots & 0 & \Sigma_{kk} \end{pmatrix}$$

where Σ_{ii} is $p_i \times p_i$. The problem in this form usually arises (by means of an orthogonal transformation on the right) from that of observing $V = (V_1, V_2, \ldots, V_m)$ where V_i is $p \times h$, $n = (m-1)h$, $EV_i = \nu$ for all i, and V has independent columns, each with covariance matrix Σ. In any event, we can consider the case $h = 0$ and then obtain the general result from Lemma 3.1. We shall also assume $n \geqq p$; the results when $n < p$ are parallel to those discussed for the case $n + r \leqq p$ in the previous section, in the existence of admissible tests which are better than the trivial randomized test, which are based (for example) on only some of the rows of Y, and which (like all tests) have trivial minimax properties.

We let Π assign all measure to Σ^{-1}'s of the form $I_p + \eta\eta'$ under H_1, where η is $p \times 1$, and to Σ^{-1}'s of the form

$$\begin{pmatrix} I_{p_1} + \eta_{(1)}\eta'_{(1)} & 0 & \cdots & 0 \\ 0 & I_{p_2} + \eta_{(2)}\eta'_{(2)} & \cdots & \vdots \\ \vdots & 0 & \vdots & 0 \\ 0 & \cdots & 0 & I_{p_k} + \eta_{(k)}\eta'_{(k)} \end{pmatrix}$$

under H_0, where $\eta_{(i)}$ is $p_i \times 1$. We set

$$\text{(5.1)} \qquad \begin{aligned} d\Pi_1(\eta)/d\eta &= |I_p + \eta\eta'|^{-n/2}, \\ d\Pi_0(\eta)/d\eta &= \textstyle\prod_{i=0}^k |I_{p_i} + \eta_{(i)}\eta'_{(i)}|^{-n/2}. \end{aligned}$$

According to (3.7), the densities (5.1) are integrable on E^p provided $n > p$. We obtain, under H_0,

$$\text{(5.2)} \qquad f_Y(y, \Sigma)\, d\Pi_0(\eta)/d\eta = c_1 \operatorname{etr}\{-\tfrac{1}{2}yy'\} \exp\{-\tfrac{1}{2}\textstyle\sum_i \eta'_{(i)}y_{(i)}y'_{(i)}\eta_{(i)}\},$$

and, under H_1,

$$\text{(5.3)} \qquad f_Y(y; \Sigma)\, d\Pi_1(\eta)/d\eta = c_2 \operatorname{etr}\{-\tfrac{1}{2}yy'\} \exp\{-\tfrac{1}{2}\eta'yy'\eta\}.$$

Hence, from (3.1), we obtain that, for $c \geqq 1$ and $n > p$,

$$\prod_{i=1}^{k} |Y_{(i)}Y'_{(i)}|/|YY'| \geqq c \tag{5.4}$$

is an admissible Bayes critical region.

The derivation of the test (5.4) required $n > p$, and thus that approach does not handle the "minimum sample size". In the special case $k = 2$, $p_1 = 1$, a slightly different trick, used by Lehmann and Stein (1948), will work even when $n = p$. Let Π assign all measure to Σ^{-1}'s of the form

$$I_p + \begin{pmatrix} 1 & \eta' \\ \eta & \eta\eta' \end{pmatrix} \quad \text{under} \quad H_1,$$

where η is $(p - 1) \times 1$, and to Σ^{-1}'s of the form

$$I_p + \begin{pmatrix} 1 - b & 0 \\ 0 & \eta\eta' \end{pmatrix} \quad \text{under} \quad H_0,$$

where η again is $(p - 1) \times 1$ and $0 \leqq b \leqq 1$.

We set

$$\frac{d\,\Pi_1(\eta)}{d\eta} = \left| I_p + \begin{pmatrix} 1 & \eta' \\ \eta & \eta\eta' \end{pmatrix} \right|^{-p/2},$$

$$\frac{d\,\Pi_0(\eta)}{d\eta} = \left| I_p + \begin{pmatrix} 1 - b & 0 \\ 0 & \eta\eta' \end{pmatrix} \right|^{-p/2},$$

which are integrable on E^{p-1}. Consider the particular Bayes test which rejects if

$$\int f_Y(y; \Sigma)\Pi_1(d\eta) / \int f_Y(y; \Sigma)\Pi_0(d\eta) \geqq 1. \tag{5.5}$$

Carrying out the integrations according to (3.1) yields the rejection region

$$\exp\left(\tfrac{1}{2} Y_{(1)}Y'_{(2)}(Y_{(2)}Y'_{(2)})^{-1}Y_{(2)}Y'_{(1)}\right) / \exp\left(\tfrac{1}{2} b Y_{(1)}Y'_{(1)}\right) \geqq 1. \tag{5.6}$$

Taking logarithms of both sides, we finally get the rejection region

$$Y_{(1)}Y'_{(2)}(Y_{(2)}Y'_{(2)})^{-1}Y_{(2)}Y'_{(1)} / Y_{(1)}Y'_{(1)} \geqq b, \tag{5.7}$$

which in this special case is equivalent to (5.4).

As in Section 4, an infinite-dimensional set of Π's will yield (5.4), and other fully invariant tests can be obtained. (See also the second paragraph of Section 7(i).) The test (5.4) is the likelihood ratio test. For $k = 2$, $p_1 = 1$, it is the R^2-(multiple correlation coefficient-) test, which when $p = 2$ reduces to the classical two-sided test based on the sample correlation coefficient.

The technique of Section 6(ii) below can be applied to yield tests concerning the independence of subsets of the components which involve a total of fewer than p of the components.

6. Other problems of testing means.

(i) *Generalized Behrens-Fisher problem.* Let $V = (V^{(1)}, V^{(2)}, \cdots, V^{(k)})$ where

$V^{(i)}$ is $p \times (mn_i)$. Under Ω the n_i submatrices $V_1^{(i)}, V_2^{(i)}, \cdots, V_{n_i}^{(i)}$ of $V^{(i)}$, each of size $p \times m$, are identically distributed, with $EV_t^{(i)} = \xi^{(i)}$ and with each column of $V^{(i)}$ having covariance matrix $\Sigma^{(i)}$. The problem is to test H_0: $\xi^{(1)} = \xi^{(2)} = \cdots = \xi^{(q)}$. One of the ways of treating this problem is to reduce it to that of Section 4 by considering Π's which assign all measure to a set where $a_1\Sigma^{(1)} = a_2\Sigma^{(2)} = \cdots = a_q\Sigma^{(q)}$, where the positive numbers a_i (or, equivalently, their ratios) are specified, and where also $\Sigma^{(q+1)} = \cdots = \Sigma^{(k)} = I_p$, $\xi^{(q+1)} = \cdots = \xi^{(k)} = 0$. Writing

$$
\begin{aligned}
S &= \textstyle\sum_{i=1}^q \sum_{t=1}^{n_i} a_i(V_t^{(i)} - \bar{V}^{(i)})(V_t^{(i)} - \bar{V}^{(i)})', \\
\bar{V} &= (\textstyle\sum_1^q n_i a_i)^{-1}\sum_1^q n_i a_i \bar{V}^{(i)}, \\
T &= \textstyle\sum_{i=1}^q n_i a_i(\bar{V}^{(i)} - \bar{V})(\bar{V}^{(i)} - \bar{V})', \\
U &= (\textstyle\sum_1^q n_i a_i)^{\frac{1}{2}}\bar{V},
\end{aligned} \tag{6.1}
$$

and reducing the problem to the canonical form of Section 4, we can then use all of the results of Section 4 with XX', YY', r, and n replaced, respectively, by T, S, $(q-1)m$, and $m\sum_1^q (n_i - 1)$. Thus, for example, if $m\sum_1^q (n_i - 1) \geqq p$, the critical region

$$\operatorname{tr}(S + T)^{-1}T \geqq c \tag{6.2}$$

is an admissible Bayes procedure. In the case $k = q = 2$, $m = p = 1$, we obtain Student's test if we set all a_i equal, and Welch's (simplest) test if a_i is proportional to $1/n_i(n_i - 1)$. These choices thus yield corresponding generalizations for general m and p. (However, these admissibility results have limited interest because of the lack of similarity.)

Admissible tests of equality of a subset of the components of the $\xi^{(i)}$ can be obtained, in the manner of (ii) below, by basing a test such as (6.2) only on these components.

(ii) *Tests concerning a subset of components of* ξ. If in the first paragraph of Section 4 we began with the more general form $EV = L_1\xi L_2$, with $K_1\xi K_2 = 0$ under H_0, we would have in the canonical form, in place of (X, Y), a random matrix with an expectation matrix some of whose elements (or linear combinations thereof) are 0 under Ω, and some additional ones of which are 0 under H_0. Bayes procedures for this problem can be found using modifications of the methods of Section 4.

Since the tests obtained in the most general case are less simple than those obtained in other cases, we shall for the sake of brevity mention here only the following special case: $V = (X, Y, U)$ with the assumptions of Section 4 except that $\xi' = (\xi'_{(1)}, \xi'_{(2)}, 0)$, where $\xi_{(i)}$ is $p_i \times r$ and $p_1 + p_2 \leqq p$ (here p_1 may be 0). Write $V' = (V'_{(1)}, V'_{(2)}, V'_{(3)})$, in the same way. Under H_0, $\xi_{(2)} = 0$. Here we shall give an example of a reasonable class of admissible tests which are similar but not fully invariant except when $p_3 = 0$ (i.e., under

$$X \to \begin{pmatrix} ABC \\ 0DE \\ 00F \end{pmatrix} X + \begin{pmatrix} G \\ 0 \\ 0 \end{pmatrix}$$

where $|A||D||F| \neq 0)$, but which are intuitively appealing and, as described in the next paragraph, have further justification. Let Π assign all measure to the set where the submatrices $V_{(1)}$, $V_{(2)}$, and $V_{(3)}$ are independent, and where the columns of $V_{(1)}$ and $V_{(3)}$ have zero means and identity covariance matrices. All results of Section 4 then apply, with p_2, $X_{(2)}$, $Y_{(2)}$, $\xi_{(2)}$, $\nu_{(2)}$, Σ_{22} replacing p, X, Y, ξ, ν, Σ.

We remark that any minimax or local minimax property (on a set described in terms of ξ and Σ) of a test of Section 4 also holds (in terms of $\xi_{(2)}$ and Σ_{22}) for the corresponding test based on $V_{(2)}$ in our present setting; this follows from the validity of such a minimax property on the subset of Ω where $V_{(3)}$ and $V_{(1)}$ have zero means and identity covariance matrices and are independent of $V_{(2)}$, together with the fact that the power of the test on Ω depends only on $\xi_{(2)}$ and Σ_{22}. We note that a property described in terms of $\xi_{(2)}$ and Σ_{22} (unlike one described in terms of the eigenvalues of $(\Sigma_{22} - \Sigma_{23}\Sigma_{33}^{-1}\Sigma_{32})^{-1}\xi_{(2)}\xi_{(2)}'$ is not fully invariant except when $p_3 = 0$, which is why corresponding minimax procedures which are fully invariant need not exist even in the case $p_1 = 0$, $p_2 = p_3 = 1$. The rationale behind consideration of a smaller group (with $C = 0$, $E = 0$, with respect to which appropriate tests based on $V_{(2)}$ are invariant) was discussed in Section 2. Cochran and Bliss, Stein, and Olkin and Shrikhande have considered some of the problems of this subsection.

(iii) *Testing equality of components; Scheffé tests.* As another example of the general problem outlined in the first paragraph of (ii) above, suppose $\xi' = (\xi_{(1)}', \xi_{(2)}', \xi_{(3)}', \cdots, \xi_{(k)}', \xi_{(k+1)}', 0)$ where $\xi_{(i)}$ is $p_0 \times m$ for $1 \leq i \leq k$. Under H_0, $\xi_{(1)} = \xi_{(2)} = \cdots = \xi_{(k)}$. By letting Π assign all measure to the set where the $k + 2$ submatrices of V (corresponding to the above subdivision of ξ) are independent, with the last two having 0 means and identity covariance matrices, and with $\Sigma_{11} = \Sigma_{22} = \cdots = \Sigma_{kk}$ (unknown), the treatment of this problem can be reduced to that of the Behrens-Fisher problem of (i) above. The resulting tests are of course not invariant under such transformations as $X_{(i)} \rightarrow X_{(i)} + \sum_{j=1}^{k+2} A^{(j)}X_{(j)} (1 \leq i \leq k)$, where $A^{(j)}$ is $p_0 \times p_0$ and the transformation is nonsingular; these leave the problem invariant.

Somewhat different tests can be obtained in this last problem by the following device: For simplicity, suppose $k = 2$, and dispose of the last two submatrices of V as above. (The test will again not be fully invariant.) The first two submatrices $V_{(1)}$ and $V_{(2)}$ are, however, independent with probability zero under Π; and Π assigns all measure to the set where $\Sigma_{11} = \Sigma_{22} = I_{p_0} - \Sigma_{12}$ and $\xi_{(1)} = -\xi_{(2)}$ (which last includes $\xi_{(1)} = \xi_{(2)} = 0$ under Π_0). Write $V_{(1)}^* = V_{(1)} - V_{(2)}$ and $V_{(2)}^* = V_{(1)} + V_{(2)}$. Then Π assigns all measure to the set where the parameters of $V^{*\prime} = (V_{(1)}^{*\prime}, V_{(2)}^{*\prime})$ are as follows: $\xi_{(1)}^*$ arbitrary under H_1 and 0 under H_0; $\xi_{(2)}^* = 0$, $\Sigma_{22}^* = 2I_{p_0}$, $\Sigma_{12}^* = 0$, Σ_{11}^* arbitrary positive definite under both H_i. We can now use the techniques of Section 4, just as in part (ii) above. We conclude, for example, that if $n \geq p_0$ the critical region

$$\text{(6.3)} \quad \operatorname{tr}[(Y_{(1)} - Y_{(2)})(Y_{(1)} - Y_{(2)})' + (X_{(1)} - X_{(2)})(X_{(1)} - X_{(2)})']^{-1} \cdot (X_{(1)} - X_{(2)})(X_{(1)} - X_{(2)})' \geq c$$

is an admissible similar Bayes procedure. When $p_0 = r = 1$, this is Scheffé's test for the Behrens-Fisher problem (in canonical form); however, in that context it is usually assumed that $\Sigma_{12} = 0$, which is not (and cannot be) assumed in the above development.

(iv) *Tests concerning a subset of components of* $\Sigma^{-1}\xi$. This problem, most recently considered by Giri, (1964), (1965), arises from discriminant analysis. The example of (ii) above is modified by setting $(\Sigma^{-1}\xi)' = \Gamma' = (\Gamma'_{(1)}, \Gamma'_{(2)}, 0)$, with $\Gamma_{(2)} = 0$ under H_0. This problem can be treated by modifications of the methods used above. For example, as in (ii) we can obtain tests by letting Π assign all measure to the set where $V_{(2)}$ is independent of both $V_{(1)}$ and $V_{(3)}$. As in the case of (ii), one can obtain many reasonable tests which are not fully invariant (i.e., under

$$X \to \begin{pmatrix} A00 \\ BC0 \\ DEF \end{pmatrix} X + \begin{pmatrix} G \\ 0 \\ 0 \end{pmatrix}$$

where $|A||C||F| \neq 0$).

(v) *Classification.* Suppose $V = (V^{(1)}, V^{(2)}, V^{(3)})$ where $V^{(j)} = (V_1^{(j)}, \cdots, V_{m_j}^{(j)})$, each $V_i^{(j)}$ being $p \times r$, the columns of V being independent with common unknown covariance matrix Σ, and $EV_i^{(j)} = \xi^{(j)}(p \times r)$. It is desired to test $H_1 : \xi^{(3)} = \xi^{(1)}$ against $H_2 : \xi^{(3)} = \xi^{(2)}$; that is, a sample of size m_j from population $j(j = 1, 2, 3)$ is to be used to classify "population 3" as either "population 1" or "population 2". Let $m = m_1 + m_2 + m_3$. As an example, we shall prove the admissibility of the likelihood ratio criterion, analogous to (4.7). Write $(m - 2)^{-1}r^{-1}S^{(j)}$ for the usual best unbiased estimator of Σ under H_j, and write

$$Y^{(j)} = (m_j + m_3)^{-\frac{1}{2}}(m_j\bar{V}^{(j)} + m_3\bar{V}^{(3)}),$$
$$Z^{(j)} = m_{3-j}^{\frac{1}{2}}\bar{V}^{(3-j)},$$
$$U^{(j)} = (Y^{(j)}, Z^{(j)}), EU^{(j)} = \nu^{(j)}.$$

Let $Q^{(j)}$ be any orthogonal $mr \times mr$ matrix such that $VQ^{(j)} = (W^{(j)}, U^{(j)})$ where $W^{(j)}$ is $p \times (m - 2)r$. We now consider our problem in terms of $W^{(j)}, U^{(j)}$. Under H_j the $U^{(j)}$ corresponds to the nuisance parameter $\nu^{(j)}$, which we dispose of in the manner of (3.13) with $C^{(j)} = I_p$ (again writing $\Sigma^{-1} = I_p + \eta\eta'$ with η $p \times 1$); although Lemma 3.1 does not apply directly (because the nuisance variable $U^{(j)}$ differs under the two hypotheses), we conclude from (3.13) that an admissible Bayes procedure for our problem can be obtained in the form: select H_1 or H_2 according to whether

$$\frac{\text{etr}\,\{-\frac{1}{2}u^{(1)}u^{(1)\prime}\}\int_{E^p} f_{W^{(1)}}(w^{(1)}; (I_p + \eta\eta')^{-1})\,d\Pi_1(\eta)}{\text{etr}\,\{-\frac{1}{2}u^{(2)}u^{(2)\prime}\}\int_{E^p} f_{W^{(2)}}(w^{(2)}; (I_p + \eta\eta')^{-1})\,d\Pi_2(\eta)} > \text{ or } < c. \tag{6.5}$$

Under both H_1 and H_2, we let $d\Pi_i(\eta)/d\eta = |I_p + \eta\eta'|^{-(m-2)r/2}$, which according to (3.7) is integrable provided $(m - 2)r > p$. Using (3.1) and the fact that

$W^{(j)}W^{(j)\prime} = S^{(j)}$ and $U^{(1)}U^{(1)\prime} + W^{(1)}W^{(1)\prime} = U^{(2)}U^{(2)\prime} + W^{(2)}W^{(2)\prime}$, we obtain

$$|S^{(2)}|/|S^{(1)}| > \text{ or } < c \tag{6.6}$$

as an admissible Bayes rule for classifying population 3 as being the same as 1 or 2 (respectively), provides $(m - 2)r > p$. This procedure is fully invariant, under all transformations of the form $V_i^{(j)} \to AV_i^{(j)} + B$ with A (nonsingular) and B independent of i and j. It is the likelihood ratio criterion (Anderson (1958), pp. 140–141).

This test enjoys a kind of similarity, in that it has constant power on the set where $\Sigma^{-1}(\xi^{(1)} - \xi^{(2)})(\xi^{(1)} - \xi^{(2)})'$ has specified eigenvalues; in particular, the power of the test is constant over the set where $\xi^{(1)} = \xi^{(2)}$.

Other admissible procedures can be obtained by modifications of the type considered earlier. In addition, admissible procedures can be obtained for classification into one of $k(>2)$ populations: since the left side of (6.6) is proportional to the ratio of two *a posteriori* probabilities, the analogue of (6.6) is, in an obvious notation, to choose the classification j which minimizes $c_j|S^{(j)}|$.

The test (6.6) is of course also admissible for the problem where H_1 and H_2 are enlarged so as not to assume $\Sigma^{(1)} = \Sigma^{(2)}$, H_i being $\xi^{(3)} = \xi^{(i)}$, $\Sigma^{(3)} = \Sigma^{(i)}$. However, certain additional tests which may seem more appropriate can be obtained similarly in this case. For example, a fully invariant test can be obtained by putting $\Sigma^{(i)} = (I_p + \eta_i\eta_i')^{-1}$ under H_i with η_i $p \times 1$, and with

$$d\Pi_i(\eta)/d\eta = |I_p + \eta_i\eta_i'|^{(m_i+m_3-1)r/2}|I_p + \eta_{3-i}\eta_{3-i}'|^{(m_{3-i}-1)r/2};$$

there are integrable if $(m_i - 1)r > p$ for $i = 1, 2$. Writing $T^{(i)} = V^{(i)}V^{(i)\prime} - m_i\bar{V}^{(i)}\bar{V}^{(i)\prime}$, an analysis similar to that used to obtain (6.6) yield the procedure

$$|S^{(2)} - T^{(1)}||T^{(1)}|/|S^{(1)} - T^{(2)}||T^{(2)}| > \text{ or } < c. \tag{6.7}$$

A classification problem with known covariance matrices was considered by Ellison (1962). See also Das Gupta (1965).

(vi) *Testing between two possible values of the mean.* In Section 4, suppose (in the canonical form) that H_0 is $\xi = \xi^{(00)}$, while H_1 is $\xi = \xi^{(01)}$. Letting $V^{(i)} = (X - \xi^{(i)}, Y, Z)$, we can use any of the Π_0's of Section 4 on $V^{(i)}$ under the present H_i. For example, comparing the denominator of the second expression of (4.2) for $i = 0, 1$ when x is replaced by $x - \xi^{(i)}$, we obtain

$$\operatorname{etr}\{\tfrac{1}{2}(\xi^{(01)} - \xi^{(00)})X'\}|(X - \xi^{(00)})(X - \xi^{(00)})' + YY'|^{r/2}/ |(X - \xi^{(01)})(X - \xi^{(01)})' + YY'|^{r/2} \geq c \tag{6.8}$$

as an admissible Bayes critical region. Other forms can be obtained similarly.

Modifications in these H_i can be made along the lines of problem formulations considered earlier in this section; for example, the hypotheses can specify values of only a subset of the elements of ξ, and the subset can differ under the two H_i.

(vii) *Lehmann-Stein examples.* Many of the considerations of Lehmann and Stein (1948) have obvious multivariate analogues, but of course one no longer obtains uniformly most powerful one-sided tests (as in equation (6.9) of Lehmann and Stein). For example, in the setting of Section 4 with $h = 0$, to test $H_0 : \xi = 0$ against $H_1 : \xi = \xi^{(0)}, \Sigma = \Sigma^{(0)}$ (simple alternative), write $V^* = \Sigma^{(0)-\frac{1}{2}}V$ and $\xi^* = \Sigma^{(0)-\frac{1}{2}}\xi^{(0)}$, to reduce H_1 to $H_1^* : \xi = \xi^*, \Sigma = I_p$. In this form the problem is considered on pages 509–510 of Lehmann and Stein.

7. Tests concerning the covariance matrix.

(i) *Equality of covariance matrices.* This usually arises from $V^* = (V^{*(1)}, V^{*(2)}, \cdots, V^{*(k)})$ where the columns of each $V^{*(j)}$ are identically distributed. (The reduced form obtained below also applies if several $V^{*(j)}$ are assumed to have equal $\Sigma^{(j)}$'s or, equivalently, if $V^{*(j)}$ has identically distributed $p \times d_j$ submatrices with d_j no longer necessarily 1.) After an orthogonal transformation on the right of each $V^{*(j)}$, the problem is reduced to one where, under Ω, $V = (V^{(1)}, V^{(2)}, \cdots, V^{(k)}, W)$ with $EV^{(i)} = 0, EV_1^{(i)}V_1^{(i)\prime} = \Sigma^{(i)}$ (unknown), $EW = \nu$ with all elements unknown and unrelated. Here $V^{(i)}$ is $p \times n_i$. H_0 is $\Sigma^{(1)} = \Sigma^{(2)} = \cdots = \Sigma^{(k)}$ (the common value being unspecified). As in previous sections, we may suppose W absent, since Lemma 3.1 then handles the case where W is present; in the application of Lemma 3.1 in parts (i) and (ii) of this section, a given column of W will sometimes have a covariance matrix of different form under H_0 and H_1, but the $C^{(j)}$ of (3.9) can always be taken to be a scalar multiple of I_p. Let $q_i(0 \leqq i \leqq k)$ be positive integers. Let Π_0 assign all measure to the set where $\Sigma^{(1)} = \cdots = \Sigma^{(k)} = (I_p + \eta\eta')^{-1}$ where η is $p \times q_0$, and let Π_1 assign all measure where $\Sigma^{(i)} = (I_p + \eta_i\eta_i')^{-1}$ where η_i is $p \times q_i$. Furthermore, put $N = \sum_1^k n_i$ and

$$d\Pi_0(\eta)/d\eta = |I_p + \eta\eta'|^{-N/2} \tag{7.1}$$
$$d\Pi_1(\eta)/d\eta = \prod_{i=1}^k |I_p + \eta_i\eta_i'|^{-n_i/2}.$$

According to (3.7), if $q_0 \leqq N - p$ and $q_i \leqq n_i - p$ for $1 \leqq i \leqq k$, the densities of (7.1) are integrable. In this case an integration of the type we have performed repeatedly yields, writing $S^{(i)} = V^{(i)}V^{(i)\prime}$,

$$|\textstyle\sum_{i=1}^k S^{(i)}|^{q_0} / \prod_{i=1}^k |S^{(i)}|^{q_i} \geqq c \tag{7.2}$$

as an admissible Bayes procedure, which is similar and fully invariant if $\sum_1^k q_i = q_0$. Such a test can be obtained for the simplest choice $q_i = 1$ (with $q_0 = k$) provided $n_i > p$ for $1 \leqq i \leqq k$. The likelihood ratio test (resp., Barlett's modification thereof) can be obtained in this way for some sets of values n_i, by setting $q_i = c_1(n_i + 1)$ (resp., $q_i = c_1 n_i$) and $q_0 = \sum_i q_i$, where $c_1 < 1$; the obvious choice $c_1 = 1$ does not make (7.1) integrable. This dependence on divisibility properties of the n_i can be overcome for sufficiently large n_i by using the analogue of modification (iv) (b) of Section 4, which we shall now describe.

Take all $q_i \geqq p$, and multiply $d\Pi_1(\eta)$ (resp., $d\Pi_0(\eta)$) in (7.1) by $\prod_1^k |\eta_i\eta_i'|^{t_i/2}$ (resp., by $|\eta\eta'|^{t_0/2}$) with $p - 1 < t_i + q_i < n_i - p + 1$ (resp., $p - 1 < t_0 + q_0 < N - p + 1$) for integrability. There exist such t_i provided $\min_i n_i > 2(p - 1)$, so that slightly larger sample sizes are needed than for small q_i in (7.1). We obtain the test of (7.2) with q_i replaced by $q_i + t_i$ for $0 \leqq i \leqq k$.

Setting $q_i + t_i = c_1(n_i + 1)$ (resp., $= c_1 n_i$) for $1 \leqq i \leqq k$ and $q_0 + t_0 = \sum_1^k (q_i + t_i)$, where c_1 is slightly larger than $(p-1)/\min_i n_i$ (resp., $(p-1)/\min(n_i+1)$), we obtain the likelihood ratio test (resp., Bartlett's modification), provided $\min_i n_i > 2(p-1)$. The use of this modification was pointed out to us by Professor Olkin (to whom we are also indebted for other helpful comments) in the equivalent form of replacing $|\eta\eta'|^{t/2}/|I + \eta\eta'|^{n/2}$ by the density $|\Lambda|^{(q-p+t-1)/2}/|I+\Lambda|^{n/2}$ on the positive definite matrices $\Lambda = \Sigma^{-1} - I$. This technique can also be applied in Section 5 where, however, it only exhibits a wider variety of *a priori* densities relative to which (5.4) is Bayes, and requires somewhat larger sample sizes. As we have seen in Section 4(iv)(b), the use of such densities there leads to a different test from that obtained for $t = 0$. Modifications of (7.2) and other forms of tests can be obtained by using other forms of Π, as in previous sections; in particular, EW can be treated in the manner of (iii) below instead of by means of Lemma 3.1.

As a one-sided variant of the above, suppose H_1 is altered to state that $\Sigma^{(i)} - \Sigma^{(i+1)}$ is nonnegative definite for $1 \leqq i \leqq k$, and not zero for all i. In that case $(\Sigma^{(i+1)})^{-1} - (\Sigma^{(i)})^{-1}$ is nonnegative definite, so that a possible choice of $d\Pi_1(\eta)/d\eta$ is $\prod_1^k |I_p + \sum_1^i \eta_t\eta_t'|^{-n_i/2}$ where η_t is $p \times q_t$ and $(\Sigma^{(i)})^{-1} = I_p + \sum_1^i \eta_t\eta_t'$ under Π_1. Using (3.5), a modification of the argument which led to (3.7) shows that this is integrable if $\sum_1^i n_t \geqq p + q_i$ for $1 \leqq i \leqq k$, and, with the Π_0 of (7.1), produces the admissible Bayes critical region

$$(7.3) \qquad |\textstyle\sum_{t=1}^k S^{(t)}|^{q_0} / \prod_{i=1}^k |\sum_{t=i}^k S^{(t)}|^{q_i} \geqq c,$$

which is fully invariant if $\sum_1^k q_i = q_0$. The technique of the previous paragraph can also be used to obtain (7.3) with nonintegral q_i. Among possible alternative forms is one obtained from (7.4) below.

The technique of Section 6(ii) can be applied to yield corresponding tests about a subset of the components.

(ii) *Proportionality of covariance matrices.* Suppose the setup of (i) is changed for $i = 0, 1$ to $H_i : a_{i1}\Sigma^{(1)} = a_{i2}\Sigma^{(2)} = \cdots = a_{ik}\Sigma^{(k)}$, where only the positive values a_{ij} are known. In this problem each H_i is acted upon transitively by the full linear group of nonsingular $p \times p$ matrices, as well as by the group of nonsingular lower triangular matrices. Hence, every procedure invariant under either group has constant power under each H_i, so that for each group there is a best invariant procedure of any specified size. Now, it can be checked directly, using the Neyman-Pearson lemma, that the essentially unique best triangular invariant test is not invariant under the full linear group (Lehmann (1959), p. 338 treats special cases of this fact), so that no fully invariant procedure can even be minimax, let alone admissible. Thus, we cannot hope to find any fully invariant admissible procedures for this problem. Whether the best triangular invariant procedure is admissible is unknown; we have been unsuccessful in showing that it is Bayes. Instead one can use our previous methods to find admissible procedures of simple structure which, however, are not triangular invariant.

REFERENCES

ANDERSON, T. W. (1958). *An Introduction to Multivariate Statistical Analysis*. Wiley, New York.

BIRNBAUM, A. (1955). Characterization of complete classes of tests of some multiparametric hypotheses, with applications to likelihood ratio tests. *Ann. Math. Statist.* **26** 21–36.

BROWN, L. (1964). Admissibility of translation-invariant estimators. Submitted to *Ann. Math. Stat.*

CONSTANTINE, A. G. (1963). Some non-central distribution problems of multivariate linear hypothesis. *Ann. Math. Statist.* **34** 1270–1285.

DAS GUPTA, S. (1965). Optimum classification rules. To be published in *Ann. Math. Statist.*

DAS GUPTA, S., ANDERSON, T. W., and MUDHOLKAR, G. S. (1964). Monotonicity of the power functions of some tests of the multivariate linear hypothesis. *Ann. Math. Statist.* **35** 200–205.

ELLISON, B. E. (1962). A classification problem in which information about alternative distributions is based on samples. *Ann. Math. Statist.* **35** 213–223.

GIRI, N., KIEFER, J., and STEIN, C. (1963). Minimax character of Hotelling's T^2-test in the simplest case. *Ann. Math. Statist.* **34** 1524–1535.

GIRI, N. (1964). On the likelihood ratio test of a normal multivariate testing problem. *Ann. Math. Statist.* **35** 181–189.

GIRI, N. (1965). On the likelihood ratio test of a normal multivariate testing problem, II. *Ann. Math. Statist.* **36** 1061–1065.

GHOSH, M. N. (1964). On the admissibility of some tests of Manova. *Ann. Math. Statist.* **35** 789–794.

HERZ, C. S. (1955). Bessel functions of matrix argument. *Ann. Math.* **61** 474–523.

JAMES, A. T. (1964). Distributions of matrix variates and latent roots derived from normal samples. *Ann. Math. Statist.* **35** 475–501.

JEFFREYS, H. (1939). *Theory of Probability*. Oxford Univ. Press.

KARLIN, S. (1957). Polya type distributions, II. *Ann. Math. Statist.* **28** 281–308.

LEHMANN, E. L. (1959). *Testing Statistical Hypotheses*. Wiley, New York.

LEHMANN, E. L. and STEIN, C. (1948). Most powerful tests of composite hypotheses, I. Normal distributions. *Ann. Math. Statist.* **19** 495–516.

LEHMANN, E. L. and STEIN, C. (1953). The admissibility of certain invariant statistical tests involving a translation parameter. *Ann. Math. Statist.* **24** 473–479.

NANDI, H. K. (1963). On the admissibility of a class of tests. *Calcutta Statist. Assoc. Bull.* **15** 13–18.

PILLAI, K. C. S. (1955). Some new test criteria in multivariate analysis. *Ann. Math. Statist.* **26** 117–121.

ROY, S. N. (1957). *Some Aspects of Multivariate Analysis*. Wiley, New York.

ROY, S. N. and MIKHAIL, W. F. (1960). On the admissibility of a class of tests in normal multivariate analysis (abstract). *Ann. Math. Statist.* **31** 536.

SCHWARTZ, R. (1964a). Properties of a test in Manova (abstract). *Ann. Math. Statist.* **35** 939.

SCHWARTZ, R. (1964b). Admissible invariant tests in Manova (abstract). *Ann. Math. Statist.* **35** 1398.

STEIN, C. (1956a). Inadmissibility of the usual estimator for the mean of a multivariate normal distribution. *Proc. Third Berkeley Symp. Math. Statist. Prob.* **1** 197–206. Univ. of California Press.

STEIN, C. (1956b). The admissibility of Hotelling's T^2-test. *Ann. Math. Statist.* **27** 616–623.

WALD, A. (1942). On the power function of the analysis of variance test. *Ann. Math. Statist.* **13** 434–439.

Reprinted from *Ann. Math. Statist.* **36** (1965), 747–770.

The Annals of Mathematical Statistics
1972, Vol. 43, No. 5, 1742-1743

CORRECTION NOTES

CORRECTION TO
"ADMISSIBLE BAYES CHARACTER OF T^2-, R^2-, AND OTHER FULLY INVARIANT TESTS FOR CLASSICAL MULTIVARIATE NORMAL PROBLEMS"

BY J. KIEFER AND R. SCHWARTZ
Cornell University and General Electric Company

In our paper "Admissible Bayes character of T^2-, R^2- and other fully invariant tests for classical multivariate normal problems" (*Ann. Math. Statist.* **36**, 747–770) the write-up of Lemma 3.1 is somewhat incomprehensible. We thank Tom Ferguson for pointing this out. One difficulty is that it is not made sufficiently clear that $C^{(j)}$ in (3.9) is fixed and independent of θ and *a fortiori* independent of i. The assumption is that Π, the original a priori measure for the problem without nuisance parameters, as well as any specified relationship between θ and the $\Sigma^{(i,j)}$, allow the representation (3.9) with $C^{(j)}$ fixed throughout. Similarly at the bottom of page 754, $\Pi_{i,\theta}$ must assign all measure to a set of the form (3.9) with $C^{(j)}$ constant (i.e., independent of θ and i). The reading is made easier by thinking of $C^{(j)}$ as I_p and of $D^{(i,j)}$ as $\eta\eta'$ which they usually are in the sequel.

We remark that if $r_{ij} = 0$, the representation (3.12) should be replaced by the degenerate conditional prior law which assigns probability one to any single value of γ_j (e.g., zero) under H_i^*.

The relationship of (3.14) and (3.15) to (3.16) was not made clear. From (3.13) it follows that the numerator and denominator of (3.14) and therefore of (3.15) are independent of β and θ. Hence, the numerator (resp. denominator) of (3.16) contains the numerator (resp. denominator) of (3.15) as a factor. Cancellation then yields the RHS of (3.16).

On the bottom line of page 756 I_p should read I_r.

Multivariate Optimality Results

J. KIEFER[1]

DEPARTMENT OF MATHEMATICS
CORNELL UNIVERSITY
ITHACA, NEW YORK

1. INTRODUCTION; INVARIANCE AND FORMAL BAYES PROCEDURES

In this article I shall attempt to fulfill Dr. Krishnaiah's request for me to present a survey of optimality results (i.e., mainly of admissibility and minimax results) in multivariate analysis. Hence, aside from a few observations on connections and contrasts among known results and on areas still to be explored, the paper will consist primarily of a list of known results. Even this list will be incomplete and will be accompanied by only brief commentary; it would require too much space and lead too far afield to list all the standard decision-theoretic results which, when specialized, have some bearing on usual multivariate problems. Moreover, as we shall mention, many of the most interesting unanswered optimality questions in multivariate analysis seem likely to have different answers from corresponding questions in the more-explored univariate case, or, in any event, to require different techniques of proof.

I apologize in advance for the inadvertant oversights which have undoubtedly taken place in my list of references, and will try now to indicate some of the conscious omissions which have been dictated by brevity.

In what follows we shall usually forego generality for the sake of brief and simple exposition. Mention will occasionally be made of topics related to those under discussion but not pursued here. We shall not even list sources for all such topics. No distribution theory will be considered (not even the work, quite relevant to our considerations, of Stein, Wijsman, Schwartz, on the distribution of maximal invariants). The vast topic of asymptotic (in sample size) optimality, both nonsequential and sequential, which has been worked on by many authors, will receive no attention at all; perhaps the aspect of that work which has greatest interest from the point of view of this

[1] Work supported by the Office of Naval Research under Contract Nonr-401(03).

survey is Stein's concern with asymptotic optimality questions when the dimension of the multivariate population also goes to infinity.

Uniform optimality properties (e.g., minimaxity or maximum power for a given test on each of a set of contours of alternatives) will usually not receive special mention when we describe optimality results. Power function monotonicity results, such as those of [9], and which are relevant to certain minimax statements, will not be discussed. Conditioning and fiducial inference, for which a considerable literature exists (including certain invariance considerations) will not be mentioned; nor will recent work of Stone's. Characterization of the class of *all* transformations which leave a problem invariant, begun by Lehmann and Stein [39], and work such as that of Brillinger on conditions for a Lie group to leave a problem invariant, will not be mentioned; in our invariance developments we shall take the relevant group to be given. Nor shall we consider the relationship between invariance and almost-invariance (see [35]). When inadmissibility results are mentioned, the wisdom of using various competing procedures will not be discussed.

We will usually restrict attention to parametric, often normal problems. (Nonparametric minimax and admissibility results are commonly proved by reference to a parametric subproblem.) The large literature on intuitively attractive methods for constructing procedures (ML, LR, union-intersection, information, etc.) will not be touched on. We shall deal mainly with estimation and testing results, although most of the discussion applies more generally; Lehmann's work on constructing a "good" multiple decision procedure from a collection of good tests is of interest here.

Although parts of multivariate optimality theory have nothing to do with invariance, the latter concept is involved in many of the developments which have received greatest attention. Aside from the mathematical naturalness of this, there is good practical-historical motivation: many multivariate statistical problems are invariant under large groups of transformations, and, because of intuitive appeal and computational simplicity, invariant procedures were often used in such problems; tests based on Hotelling's T^2 and the same multiple-correlation coefficient R^2 are familiar classical normal examples, while the first systematic study of (scale-location) invariance is due to Pitman [43 and 44]. Thus, attention has often centered around proving or disproving minimax or admissibility properties for these classical invariant procedures, or around finding those invariant procedures which are admissible among all procedures (rather than the full minimal complete class). The role of invariance will perhaps receive more than its due in the present survey. The other principal simplifying structure is the distribution of multivariate exponential (Koopman-Darmois) type.

Minimax and admissibility considerations will be discussed in Sections 2 and 3, respectively.

Notational simplicity is called for in an article of this nature, but the consequent absence of many definitions and explicit expressions will demand some slight background on the part of the reader. We shall use S, Ω, D, and W to denote the sample, state-of-nature, and decision spaces, and the loss function on $\Omega \times D$. (W can also be allowed to depend on the observation X without much additional difficulty, even in invariance considerations.) The loss function can usually be thought of as a simple (zero-one) in testing problems, and, typically, as convex (in the decision) or even quadratic in estimation. The risk function of a statistical procedure (δ, which last is, as usual, a function on $S \times$ subsets of D) will be written r_δ. We write $\bar{r}_\delta = \sup_F r_\delta(F)$. If π is an *a priori* probability measure on Ω, $R_\delta(\pi)$ is the *a priori* (expected) risk of δ. Measurability considerations will be omitted for brevity, but (e.g., p. 400 of [2]) pathologies can arise. In hypothesis-testing problems, β_δ will denote the function of δ, and α_δ the size; minimax testing properties usually refer to the class C_α of tests δ with size $\alpha_\delta \le \alpha$ (fixed); here $\Omega = \Omega_0 + \Omega_1$ and $\alpha_\delta = \sup_{F \in \Omega_0} \beta_\delta(F)$. A locally compact group G which leaves a problem invariant will be thought of as operating on the left on (S, Ω, D); we shall write g for an element of G, acting on S, and $\bar{g}$ and $\bar{\bar{g}}$ for the homomorphic images of g acting on Ω and D; thus, $P_\theta\{gX \in A\} = P_{\bar{g}^{\prime}\theta}\{X \in A\}$ and $W(\theta, d) = W(\bar{g}\theta, \bar{\bar{g}}d)$, and δ is invariant if $\delta(x, B) = \delta(gx, \bar{\bar{g}}B)$ for all x, B, g. The left and right invariant measures on G will be denoted by μ and ν, respectively; and the modular function, by $\Delta(g') = d\mu(gg')/d\mu(g)$. (See Halmos, "Measure Theory," or Loomis, "Abstract Harmonic Analysis," on such topics.) The class of all procedures in a problem will be denoted by C, and those invariant under a group G which leaves the problem invariant will be denoted by $C^{(G)}$. We also write $C_\alpha{}^{(G)} = C^{(G)} \cap C_\alpha$. Words such as "Bayes," "minimax," or "admissible" are meant relative to C or C_α unless further modified; for example, if G is transitive on Ω in an estimation problem, some of the main considerations of this article come down to the C-admissibility or C-minimaxity of a $C^{(G)}$-admissible, Bayes, minimax (namely, best G-invariant) procedure. A Bayes procedure relative to π will be denoted by δ_π. The observed random variable, which can be thought of as the identity function on S, will be denoted by X. When the actual state of nature has (labeled) value θ, the probability measure or density of X with respect to a given σ-finite measure λ will be denoted by P_θ or p_θ, respectively. Where "sample size n" is relevant, X is to be thought of as $(X_1, X_2, \ldots, X_n)$, where the X_i are independent, identically distributed. Where "dimension p" is mentioned, it usually refers to the range of X_1 or, in a typical modification hereafter called the "p-dimensional (or p-variate) translation- (or location-) parameter problem," to that of Z, where $X = (Z, Y)$, Ω is Euclidean p-space E^p (or $\Omega_0 = \Omega_1 = E^p$ in testing), and the P_θ law of $(Z - \theta, Y)$ is independent of θ. Although the few sequential results mentioned are somewhat general, one can think of them in the special case of independent, identically distributed X_i, with each observation costing the same amount. The univariate normal distribution with mean η and variance σ^2 is denoted by $N(\eta, \sigma^2)$.

We conclude this section with a review of the general Wald decision-theoretic results which lead up to the contents of the next two sections. Under fairly general conditions for a statistical problem (e.g., [61 and 32]), the admissible procedures and the minimax procedures (some perhaps inadmissible) are contained in the closure, in an appropriate sense, of the Bayes procedures. (Without suitable compactness of Ω and regularity of W, the Bayes procedures themselves need not suffice.) Thus, restricting ourselves for simplicity to a separable setting, we are faced with studying appropriate limits of sequences $\{\delta_{\pi_i}\}$ of Bayes procedures relative to *a priori* probability measures π_i. (For example, if

$$\bar{r}_{\delta'} = \lim_i R_{\delta_{\pi_i}}(\pi_i), \tag{1}$$

then δ' is minimax.) It is tempting to consider π_i's and positive constants k_i such that $k_i \pi_i \to \pi^*$, a nontrivial and *not necessarily finite* measure on Ω, and then to compute a *formal* Bayes procedure $\delta_\pi{}^*$ relative to this improper π^* by letting $\delta_{\pi^*}(x, \cdot)$ assign measure 1 to decisions d' for which $\int p_\theta(x) W(\theta, d')$ $\pi^*(d\theta)$ (typically finite in important examples, although $R_\delta(\pi^*)$ may be infinite for all δ) is minimized. (For example, if $\Omega = E^1$, π_i is $N(0, i)$, and $k_i = (2\pi i)^{1/2}$, then π^* is Lebesgue measure.) One could then ask: (i) Is the class of such δ_{π^*}, for a suitable class of π^* (proper and improper), complete? (ii) Which improper π^*'s yield an admissible δ_{π^*}? (iii) Can a δ' satisfying (1) be obtained as a δ_{π^*} for appropriate $\{\pi_i\}$? We are trying here to replace sequences $\{\pi_i\}$ by corresponding π^*'s.

Questions (i) and (ii) will be discussed further in Section 3, while question (iii) will be the topic of Section 2. We now indicate the role of invariance in constructing such δ_{π^*}'s.

Suppose that the problem is invariant under a group G which is large enough that the reduced problem in terms of a maximal invariant $X^{(G)}$, its range $S^{(G)}$, possible distributions $\Omega^{(G)}$, decisions D, loss function W, and of course procedures $C^{(G)}$ depending only on $X^{(G)}$, satisfy the regularity assumptions which insure that the $C^{(G)}$-Bayes procedures relative to *proper a priori* laws $\pi^{(G)}$ on $\Omega^{(G)}$ are $C^{(G)}$-complete and contain the $C^{(G)}$-minimax procedures. It is then natural (for reasons indicated later in this paragraph) to modify question (iii) to ask whether some right G-invariant measure π^* (that is, a measure for which $\pi^*(Ag) = \pi^*(A)$ for all g in G and all appropriately measurable subsets A of Ω) yields a δ_{π^*} which is minimax (and which, as a part of question (ii), is admissible). For the simplest general setting which will illustrate certain ideas, suppose S, Ω, and D are all isomorphic to G in an estimation problem, and write e for the identity element of G and $W(\theta, d) = W(e, \theta^{-1}d) \equiv w(\theta^{-1}d)$. (We have identified g, $\bar{g}$, and $\bar{\bar{g}}$ in this notation.) Let $\{p_\theta(x)\}$ be densities relative to left-invariant μ on S. Let π^* be the right-invariant measure ν on $\Omega = \{\theta\}$ (or, equivalently, and what makes this the correct choice, left-invariant measure μ on $\{\theta^{-1}\}$). Then, in minimizing the *a posteriori* risk relative to π^*, we consider

$$\begin{aligned}\int W(\theta, d')p_\theta(x)\pi^*(d\theta) &= \int w(\theta^{-1}d')p_e(\theta^{-1}x)\mu(d\theta^{-1})\\ &= \int w(\theta^{-1}d')p_e(\theta^{-1}x)\mu(d\theta^{-1}x)/\Delta(x)\\ &= \int w(tx^{-1}d')p_e(t)\mu(dt)/\Delta(x) = E_e w(Xx^{-1}d')/\Delta(x). \quad (2)\end{aligned}$$

Suppose for simplicity that $E_e w(Xg)$ is minimized uniquely by $g = g_0$. Then (2) is minimized by taking $d' = xg_0$; that is, $\delta_{\pi^*}(x, xg_0) = 1$, so that δ_{π^*} is the nonrandomized estimator Xg_0. It is easily checked that this procedure is best invariant, i.e., is the essentially unique $C^{(G)}$-admissible procedure.

(This type of computation, as well as that of the next paragraph, can be carried out in the general setting where S, Ω, D need not be isomorphic to G and where, in fact, G need not be transitive on Ω and the subgroup of G leaving a point of Ω fixed need not be trivial; see Stein [58] and Schwartz [50].)

The above computation can be used to obtain the best invariant confidence set of given "shape" and μ-measure (e.g., interval of specified length in the case of a univariate translation parameter): Let B be a fixed subset of G of finite μ-measure, and for each γ in G consider the procedure which makes the confidence statement "$\theta \in X\gamma B$." The μ-measure of the confidence set $X\gamma B$ is $\mu(B)$, and the probability of coverage is

$$P_\theta\{\theta \in X\gamma B\} = P_\theta\{\theta^{-1}X \in B^{-1}\gamma^{-1}\} = P_e\{X \in B^{-1}\gamma^{-1}\}. \quad (3)$$

This is related to our earlier discussion by interpreting decision d' as the confidence set $d'B$ with $w(\theta^{-1}d') = 0$ or 1 according to whether or not $e \in \theta^{-1}d'B$; we obtain r_δ as the probability of not covering. Thus, from the discussion following (2), we see that maximum coverage is obtained by choosing γ to maximize $E_e\{1 - w(X\gamma)\} = P_e\{X \in B^{-1}\gamma^{-1}\}$, in agreement with (3). Moreover, for any γ, dividing all members of (2) by $\int p_\theta(x)\pi^*(d\theta)$, we obtain the equality between (3) and the formal *a posteriori* probability of coverage from using $X\gamma B$, when $X = x$ and π^* is treated as the *a priori* law; this result was obtained first by Pitman [43] and in great generality by Stein [58]. Finally, if we allow B to vary subject to $\mu(B) =$ specified positive constant c (say), maximum coverage among procedures XB is obtained by maximizing

$$\int_{B^{-1}} p_e(g)\mu(dg) \quad \text{subject to} \quad \int_{B^{-1}} \nu(dg) = c,$$

which, by the Neyman-Pearson lemma and the fact that $d\mu(g)/d\nu(g) = \Delta(g)$, yields B^{-1} to be of the form

$$\{g : p_e(g)\Delta(g) > k\} \cup \text{subset of } \{g : p_e(g)\Delta(g) = k\}, \tag{4}$$

for a suitable constant k. When $X = x$, the confidence set xB corresponding to the first part of (4) is thus

$$\{\theta : p_e(\theta^{-1}x)\Delta(\theta^{-1}x) > k\}. \tag{5}$$

If G is unimodular (e.g., compact, abelian, or full linear), then (4) reduces to what has often served as an intuitive choice for the form of a confidence set; to my knowledge, in other cases it has not been made clear as a general prescription that the set (5), where the formal *a posteriori* density *with respect to* μ (given that $X = x$) is $> k$, and which thus also has a natural invariance motivation, is preferable to either of the intuitive choices $xB = \{\theta : p_e(\theta^{-1}x)\ \Delta(x) > k'\}$ where the formal *a posterior* density *with respect to* ν is large, or $xB = \{\theta : p_e(\theta^{-1}x) > k''\}$. (*Added in proof:* I have just become aware of an article by Ishii and Kudo in the 1963 *Osaka J. Math*, treating some such considerations for tolerance regions.) For an example, if G, S, and Ω are the real affine group ($\{(a, b) : -\infty < a < \infty, 0 < b < \infty\}$, with $(a, b)(a', b') = (a + a'b, bb')$), then $d\mu = da\, db/b^2$, $d\nu = da\, db/b$, and $\Delta((a, b)) = 1/b$. If $X = (U, V)$, where U is the sample mean and V is the sample standard deviation from n independent univariate $N(\theta_1, \theta_2^2)$ r.v.'s ($n \geq 2$), so that X is sufficient for $\Omega = \{(\theta_1, \theta_2)\}$, then $e = (0, 1)$ and the Lebesgue density of X is proportional to $v^{n-2}e^{-n(u^2+v^2)/2}$, so that the μ-density $f_e(u, v)$ is proportional to $v^n e^{-n(u^2+v^2)/2}$. Hence B^{-1} is of the form $\{(u, v) : v^{n-1}e^{-n(u^2+v^2)/2} > k\}$. One can object to the use of μ as the measure in the restriction $\mu(B) = c$, but to depart from it requires modification of strict invariance considerations. The role of relatively invariant measures (for which $\lambda(gA) = h(g)\lambda(A)$ for some function h, such as $b^r\, da\, db$ in the above example) should be investigated in this context; their use for π^* has been discussed by Stein [58] in connection with the Pitman-Stein result mentioned above. As we have indicated, the setting $\Omega \cong G$ was chosen for simplicity and brevity, and it is not difficult tn carry out the analogue of the development leading to (4) and (5) when Ω is no longer isomorphic to G but is instead a homogeneous space G/G', where G' (compact) leaves a point of Ω fixed.

We have spent considerable space on the above considerations because a good deal of literature, beginning with Jeffreys [23] and carried on by other authors such as Barnard, Lindley, Cornfield, Geisser (e.g., [14] or possibly the present volume), has been concerned with the construction of such formal

Bayes procedures relative to improper π^*'s; it is thus evidently of interest for further work in this direction to fit particular examples into the general picture, and to make computation-saving use of simple general results such as (2), (4), and that of Pitman-Stein. Moreover, and more important from the point of view of the present survey, the justification (or lack of it) of using such procedures, in terms of desirable risk function properties, is almost never mentioned. We now turn to the description of work in which such justifications in various problems are proved or are shown to be false.

2. MINIMAX QUESTIONS; THE HUNT-STEIN THEOREM

Minimax results have been obtained in the literature for various problems by using (a) the Bayes method indicated in Eq. (1) (including the case where all π_i are the same and thus δ' is genuinely Bayes), (b) the Cramér-Rao inequality method, and (c) the Hunt-Stein theorem.

Many authors, too numerous to list, have used (a) to obtain minimax procedures since Wald first established the Bayes-minimax connection. Most of these results are univariate. A few, beginning with Wald [61] and Wolfowitz [64], gave sequential minimax results. We shall be especially concerned in Section 3 with the results of Lehmann and Stein [37].

Method (b) is due to Hodges and Lehmann [20]; see also [18] for further examples. It actually proves that a certain estimator with *constant* risk is admissible, hence minimax. Stein [53] made a multivariate application in the problem of estimating the mean of a bivariate normal distribution with identity covariance matrix.

Method (c) is approximately twenty years old. It will be discussed in more detail below.

The advantages and disadvantages of these methods are: (a) is always applicable but may require considerable guesswork (in choosing the π_i and δ') and messy computations; (b) yields admissibility as well, but is applicable only in a few estimation problems with a special form of distribution and loss (it is not known whether the inequalities of E. Barankin and M. M. Rao can be used to extend the method, but its usefulness in multivariate problems is in any event limited, since it must yield admissibility as well); method (c) is useful only when a large group G (satisfying the hypothesis of the theorem) leaves the problem invariant, but in such cases, especially when G is transitive on Ω in an estimation problem or on each Ω_i in a testing problem, it eliminates the guesswork and computation of method (a).

Before discussing the Hunt-Stein theorem, we recall the more general question (iii) of Section 1; in particular, if $r_{\delta_{\pi^*}}$ is constant for an improper π^*, is δ_{π^*} minimax? As the various counterexamples to the conclusion of the

Hunt-Stein theorem (discussed below) show, the answer is not always affirmative. In addition, even in such a simple setting as that of a univariate translation parameter with π^* not Lebesgue measure, the following type of example (first told to me by Farrell) can occur: Let $S = \Omega = D = E^1$, let X be $N(\theta, 1)$, let $W(\theta, d) = (\theta - d)^2$, and let $d\pi^*(\theta) = e^\theta\, d\theta$; then δ_{π^*} is the estimator $X + 1$, which is even invariant (and thus has constant risk), but which is not admissible or minimax. The minimax question is a less delicate one than that (discussed in Section 3) of admissibility of δ_{π^*} [58]. (Stein also described in his 1961 Wald lectures the varying delicacy required in minimax, conditioning, and admissibility questions in terms of approximating improper π^*'s by proper *a priori* probability measures.) Nevertheless, aside from the invariant setting of the Hunt-Stein theorem, few results have been obtained on the minimax character of such δ_{π^*}. Among the few exceptions are the work of Sacks [46], Katz [27], and Farrell [11] in certain cases of a univariate translation parameter restricted to the half-line.

We now turn to the Hunt-Stein theorem. We have seen in Section 1 that, in the simple context $G = \Omega$ treated there, if $\pi^* = \nu$, then δ_{π^*} (assumed essentially unique here for simplicity) is the best invariant procedure. It has constant risk, and the question of the previous paragraph is whether it is minimax. In general, without the restriction to our simple example, we ask whether a statistical problem which is invariant under a group G has what we hereafter call the *HS* (*Hunt-Stein*) *property*: For any δ in C, there is a δ' in $C^{(G)}$ with $\bar{r}_{\delta'} \leq \bar{r}_\delta$ (or typically in testing problems, the same with C and $C^{(G)}$ replaced by C_α and $C_\alpha^{(G)}$). We have separated off the question of attainment of the minimax risk by a member of C, i.e., of existence of a minimax procedure; this can often be treated by the general decision-theoretic considerations of [61] or [32] (and which are often especially simple for $C^{(G)}$, which is all one need consider if the HS property holds); if the minimax risk is attainable, the HS property asserts that a $C^{(G)}$-minimax procedure exists and is C-minimax.

In terms of the discussion of (1), one can think of π_i as being ν restricted to an appropriate compact set C_i (that is, $\pi_i(A) = \nu(C_i \cap A)/\nu(C_i)$), where $\lim C_i = G$; this may not always be the most convenient way to approximate $\pi^* = \nu$ by a sequence $\{\pi_i\}$, but is helpful intuitively. In our simple example, or more generally when $G \cong \Omega$ (and similarly when G is transitive on Ω), if G is compact ν is finite, and we can take the normalization $\nu(G) = 1$ so that δ_ν becomes a *proper* Bayes procedure with constant risk, which is thus automatically minimax. If G is not compact, we can think of trying to approximate $\pi^* = \nu$ by a sequence $\{\pi_i\}$ (such as that just mentioned) and then (in the case where the minimax risk is attained) of trying to prove (1) with $\delta' = \delta_\nu$. Some of the proofs of the HS theorem proceed in this way.

The version of the HS theorem published by Lehmann [35] is probably close in spirit to the original unpublished one of Hunt and Stein. Another early and independent general approach is due to Peisakoff [42]. Other published proofs are due, in univariate translation parameter settings, to Girshick and Savage [18] and Blackwell and Girshick [3] (with an earlier statement but incomplete proof in Wald [59]), and, in general settings, to Kudô [31], Wesler [62], and Kiefer [28]; the last of these discusses the relationship between some of these methods of proof and the necessity of the various assumptions on S, Ω, D, W, G; under certain restrictions the HS theorem is proved here for groups (called HS groups below) with descending normal chain containing only compact or abelian factor groups (and thus including, in particular, solvable groups). Chen [6, 7] has recently given a presentation of the special cases of location-scale parameter problems, including a repetition of all measurability details outlined in the previous references and consideration of attainability of the minimax risk (for which see also two paragraphs above). Sequential problems are included in the treatment of [28]. Perhaps the best approach to proving the HS theorem (in its elimination of extraneous assumptions, minimizing of considerations such as measurability, etc.) is a streamlined version of the idea used first by Karlin [24] in the context of invariant games; the sketch which follows is of a proof due to Huber; the result is related to work on existence of almost-invariant means (von Neumann, Hewitt, etc.), which has also been noted by Lecam ([33], end of Section 4), who, based on these ideas, also obtained a proof (unpublished) related to that which we now sketch. Suppose first that G is abelian. Let δ_0 be any member of C, hereafter fixed (it suffices to consider $\bar{r}_{\delta_0} < \infty$), and let $K = \{\delta : \bar{r}_{\delta_0} \leq \bar{r}_{\delta_0}\}$. Under fairly general conditions [32], K will be compact in an appropriate topology. (In testing problems, Ω often consists of densities with respect to a fixed σ-finite measure λ, and Wald's "regular convergence" of the sequence

$$\int_A \delta_i(x, d')\lambda(dx)$$

for all A with $\lambda(A) < \infty$, which implies convergence of $\beta_{\delta_i}(F)$, is appropriate; incidentally, K would usually be replaced here and below by $K \cap C_{\alpha_{\delta_0}}$ in testing problems.) For fixed g_1 in G, define the transformation T_{g_1} on C by $(T_{g_1}\delta)(x, \Delta) = \delta(g_1 x, \bar{g}_1 \Delta)$. The problem is assumed such that for each g_1 the transformation T_{g_1} is continuous in the above topology. (If $\lambda(gA) < \infty$ whenever $\lambda(A) < \infty$, this is automatic in the testing setting just mentioned parenthetically; examples show that the HS property does not necessarily hold in other problems for HS groups without some such restriction; see [28], examples v and C.) Since T_{g_1} is a continuous linear map of convex compact K into itself, the subset K_1 of K of fixed points under T_{g_1} is nonempty (Schauder-Tychonoff), and one verifies that K_1 is compact and convex. Note that K_1 consists precisely of those δ_1 in K for which $\delta_1(g_1 x, \bar{g}_1 \Delta) = \delta_1(x, \Delta)$, i.e., for which δ_1 is invariant under g_1. Next, for fixed g_2 in G, one observes that T_{g_1} maps K_1 into itself (which is where the abelian nature of G is used) and repeats the argument to obtain for the fixed points of K_1 under T_{g_2} a nonempty convex compact set $K_2 \subset K_1$, where K_2 consists of those δ_2 in K which are invariant under both g_1 and g_2. Continuing in this way, for any finite collection $g_1, g_2, \ldots, g_m$, we obtain that $\bigcap_{i=1}^{m}\{\delta : \delta \in K,\ \delta$ invariant under $g_i\}$ is compact and nonempty. Hence by compactness, $\bigcap_{g \in G}\{\delta : \delta \in K$, δ invariant under $g\}$ is nonempty, and any of its members is a g-invariant δ' in K, proving the HS property. For G_1 a closed normal subgroup of G and $G/G_1 = G_2$ abelian, one proceeds (again as in Karlin [24]) to prove that if the HS property holds for G_1, then because of the validity of the HS theorem for abelian G_2, the result holds for G. This really amounts to invariance in steps as considered by Lehmann [35]. In summary, then, using also the validity of the HS theorem for compact groups (see the previous paragraph), one obtains that if $G = G_0 \supset G_1 \supset G_2 \ldots \supset G_h = \{\text{identity}\}$ with G_{i+1} a closed normal subgroup of G_i and G_i/G_{i+1} compact or abelian for each i (such a G herein being called HS), and if the compactness and continuity conditions on C and T_g are satisfied, then the HS property holds. Using more general fixed-point theorems (Kakutani, Bohnenblust-Karlin, Fan, etc.), one can weaken the continuity assumption.

There are many directions of generalization; to mention only one, first considered in [42], if $g\Omega \subset \Omega$ for only a suitably large subset of elements g of a noncompact G (as in the univariate translation parameter problem with $\Omega =$ half-line), the result still holds; in particular, one can consider transformation semigroups, and the earlier allusion to existence of almost invariant means is relevant.

By now it is well known from several examples of Stein [52, 35, 22][2] that the HS property is not generally valid, even in testing problems, if G is the full linear group of real nonsingular $p \times p$ matrices with $p \geq 2$. (Examples can also be constructed for other Lie groups which are not HS.) In many such examples the solvable group G_T of lower triangular $p \times p$ matrices is transitive on Ω (or on each Ω_i in testing problems) and the unique best G_T-invariant procedure is not G-invariant, so that the best G-invariant procedure cannot be minimax or even admissible. In other settings, such as that of estimating normal multiple regression coefficients [55], the use of a transitive solvable subgroup of the group G of all affine transformations leaving the problem invariant does yield the classical G-invariant procedure as minimax. In still other settings, such as the multivariate normal problems where the Hotelling T^2 test and the R^2 (squared multiple correlation coefficient) test of independence are traditionally used, and where these tests are best G-invariant, G_T is

[2] In terms of the discussion on page 263, this conclusion can also be deduced from Peisakoff's example [42] concerning the free group on two generators, Sanov's result that the latter is isomorphic to a subgroup of the full linear group for $p \geq 2$, and the fact that existence of an invariant mean for a group implies it for a subgroup.

not transitive on the Ω_i, and one is faced with using method (a) for the reduced problem in terms of a G_T-maximal invariant and (in these examples) compact parameter space. This approach was used successfully [15, 17] for only the smallest nontrivial p and n; it seems unprofitable to try for general n and p to obtain analogues of the rather messy unique least-favorable *a priori* distributions obtained in the simplest cases, although a few such cases seem approachable in the same way; there is a good possibility that the use of method (a) on the original problem, rather than on the G_T-reduced problem where much symmetry has been lost, will involve less computational complexity.

Thus, the classical minimax properties of the usual univariate normal tests (e.g., [28, 35]), as well as the results of Wald, Hsu, and Simaika (e.g., [60, 63]), do not have meaningful multivariate counterparts which can be obtained as easily. Incidentally, in contrast with these univariate ANOVA results or the minimax results of [15] and [17], the results of [48], discussed below, make it appear doubtful that any single test is minimax uniformly for a natural set of alternative contours in such settings as MANOVA.

In the absence of easily obtained genuine minimax results in multivariate testing problems, various local and asymptotic (in distance from Ω_0, not in n) results [16, 48] have been obtained, the computational complexity being reduced when one considers only the behavior of the first nonzero derivative of β_δ on the boundary Ω_0 of Ω_1 or the behavior of β_δ far from Ω_0. These generalize some of the classical Neyman-Pearson local considerations [41] and subsequent developments [21, 29, 36]. We mention as an example of the difference between univariate and multivariate problems that, whereas the classical F test both maximizes locally the Gaussian curvature of the power function and also has the well-known minimax property, the T^2 test has a local analogue of the latter but not of the former.

3. ADMISSIBILITY AND COMPLETE CLASSES

The methods which have been used to obtain admissibility results in various settings include (A) the Bayes method, of showing that a procedure is unique Bayes relative to some π, or is Bayes in a setting where all risk functions are continuous and where π gives positive measure to each open set; (B) methods which use exponential type, monotone-likelihood ratio, or similar structure; (C) Cramér-Rao inequality; (D) local properties; and (E) extensions of (A) to formal Bayes procedures relative to improper π^*'s. Method (A) has been used by many authors, of whom we mention Lehmann and Stein [37], Lehmann [35], and Ellison [10] as containing multiparameter examples. We shall also discuss below [30] and [50] in this category. The complete class result of Lehmann [34] for one-parameter exponential-type distributions, and parts of Karlin's extensive investigation of decision procedures for distributions of

Polya type (e.g., [25]) make use of the methods of (A) and (D), in the setting of (B). On the other hand, the work of Birnbaum [1], Stein [54], Nandi [40], Ghosh [19], and Schwartz [49], as well as of Roy and Mikhail [45] (where, however, the method is inapplicable in at least one of the cases treated, for reasons given at the end of this paragraph and discussed in more detail in [30]), makes use of the structure of (B) in a way particularly applicable to convex acceptance regions (in an appropriate reduced sample space) for certain testing problems. For example, suppose in the σ-finite case with λ Lebesgue measure on $S = E^k$ that Ω_0 is a single element θ_0, and that for each half-space H (with bounding hyperplane not necessarily through the origin) of S and subset N of $S - H$ with $P_{\theta_0}\{N\} > 0$, there is a sequence $\{\theta_i\}$ in Ω_1 such that $\lim_i P_{\theta_i}\{N\}/P_{\theta_i}\{H\} \to \infty$ (for which it is easy to give simple sufficient conditions in terms of p_θ). The admissibility of any convex acceptance region A is then proved by noting that if B is any other acceptance region of the same size, then we can find $H \supset A$ and $N \subset B - A$ satisfying the above description, and conclude easily that $\beta_A(\theta_i) > \beta_B(\theta_i)$ for i sufficiently large, so that no such B is better than A and thus A is admissible. Birnbaum considered the case where the p_θ are a k-parameter exponential family and Ω itself is E^k; in some of the multivariate problems considered by the other cited authors, there are restrictions on Ω in this exponential family representation, so that the existence of the sequence $\{\theta_i\}$ is more difficult to establish, and, typically, only certain convex regions A can be handled by the method. Moreover, whereas Birnbaum completely characterized the admissible procedures, in such settings as MANOVA, considered by Schwartz [49], there is a difference between the necessary and the sufficient conditions given for a procedure in $C_\alpha^{(G)}$ to be C_α-admissible, which does not seem easy to eliminate by previously used techniques. We note also that, if Ω_0 is composite, Birnbaum's method as it stands is not applicable to obtain necessary conditions for admissibility. The sufficiency (for admissibility) part of the method is inapplicable in such settings as the normal test of independence (R^2), since the required condition does not hold for any $\{\theta_i\}$, even those approaching degenerate laws ($E_{\theta_i}R^2 \to 1$).

Method (C) has already been discussed in Section 2. Examples of method (D), which goes back to the Neyman-Pearson theory, can be found in Lehmann [35]; the most common occurrence is with unique uniformly or locally most powerful unbiased tests in cases where (local) unbiasedness implies similarity on that part of the boundary of Ω_1 which is in Ω_0. Uniformly most powerful tests can be regarded as a special case of (A), whereas a result like Wald's theorem [60 and 63] can be regarded as an application of (A) to the similar tests obtained from unbiasedness considerations (D). Methods (C) and (D) have not been applied much to classical multivariate problems. In the case of (C), this is because of the limitation to estimation settings, where the desired admissibility conclusion is often false and where, even where it is true (as in [57] for the bivariate normal case), the analysis appears to be much more formidable than in the invariate case. As for (D), it seems to offer greater possibilities; what is often missing is the addition of a *uniqueness* result to local results of the type discussed in Section 2.

The questions (i) and (ii) of Section 1 have already indicated the importance of method (E), which has deservedly received continued attention in the literature for over 10 years, and especially in the last few. Most (but not all) papers have focused on question (ii) for $\pi^* = \nu$ and $G = \Omega$ in estimation or G isomorphic to each of Ω_0 and Ω_1 in testing. The first of these is the classical paper of Blyth [4] in the case $G =$ additive group of reals of estimating a univariate translation parameter for a normal or rectangular density with known scale. (Part of Farrell [11] improves Blyth's sequential results.) Blackwell [2] obtained the first results for a general collection of problems, namely, estimation for discrete univariate location parameter distributions with finite support. Stein [55] considered the univariate location-parameter problem in the absolutely continuous case with squared-error loss, and Farrell [11] considered this problem for more general loss functions, concentrating on the implication of admissibility by the condition of uniqueness of the best invariant estimator, which in turn is a consequence of strict convexity of w. (Lack of uniqueness thus provides one source of counterexamples to admissibility of best invariant procedures.) A scale parameter problem is of course reduced to a location-parameter problem by taking logarithms. Brown [5] considered this univariate location-parameter problem for general loss functions in the sequential case (which is somewhat surprising, since one cannot generally write down a best invariant procedure explicitly there). Perhaps Brown's calculations resemble Blackwell's more than Blyth's, while Stein's and Farrell's are more like Blyth's. Fox and Rubin [13] considered a specific W in detail. For this same group (additive reals), Lehmann and Stein [38] obtained admissibility results in univariate location-parameter testing problems; again, lack of uniqueness of the best invariant procedure can mean inadmissibility. The somewhat different approach of Karlin [26] will be mentioned below.

We shall not discuss here the integrability conditions on W in these references; see [5] for examples which indicate that such conditions are not completely dispensable. Nor shall we discuss the considerations there of proving that almost-admissibility of an invariant procedure δ' (nonexistence of δ with $r_\delta(\theta) \le r_{\delta'}(\theta)$ and strict inequality on a set of positive Lebesgue measure) implies admissibility. The finer nature of almost admissibility of δ' compared with minimaxity is expressed in the fact that the latter can be phrased as $\bar{r}_\delta \le \bar{r}_{\delta'} \Rightarrow \bar{r}_\delta = \bar{r}_{\delta'}$, whereas (for example, again with $G \cong \Omega \cong E^1$) the former can be phrased as $\bar{r}_\delta \le \bar{r}_{\delta'} \Rightarrow r_\delta(\theta) = \bar{r}_{\delta'}$ almost everywhere.

Blyth's approach is to find a sequence of *a priori* Lebesgue densities $\pi_i'(\theta)$ such that, for any compact subset H of Ω, $\pi_i'(\theta) \geq \varepsilon_H/i$ for $i >$ some integer q_H, with $\varepsilon_H > 0$ (the simplest choice being of the form $\pi_i'(\theta) = \pi_1'(\theta/i)/i$ with π_1' suitably smooth); and such that $\bar{r}_{\delta'} - R_{\delta_{\pi_i}}(\pi_i) = o(1/i)$. If there were a δ^* better than δ' with $\bar{r}_{\delta'} - r_{\delta^*}(\theta) > \varepsilon > 0$ on a (bounded) set of positive measure, we would then have $r_{\delta'} - R_{\delta^*}(\pi_i) > \varepsilon'/i$ for i large, with $\varepsilon' > 0$, contradicting the Bayes character of δ_{π_i}. The greater delicacy of this argument than that of Eq. (1) of Section 1 (as discussed under (C) of Section 2) in the present example is evident in the fact that there are many examples where the choice of π_i as uniform on $(-i, i)$ suffices for the minimax but not for the admissibility proof.

The two-dimensional translation-parameter estimation problem ($G \cong \Omega \cong E^2$ under vector addition) is more delicate; the choice $\pi_i'(\theta) = \pi_1'(\theta/i)/i^2$ now yields ε'/i^2 in place of ε'/i in the previous paragraph, and the $o(1/i)$ term there is also at best of order $1/i^2$, yielding no contradiction. A suitable choice of the π_i which can be used to prove admissibility is given by Stein [22] in the case of squared error. A somewhat different approach of Brown (unpublished) treats more general loss functions.

In the case of a three- or higher-dimensional translation-parameter estimation problem, inadmissibility of the best invariant estimator was shown by Stein [54] and by James and Stein [22], quadratic loss normal calculations being given in some detail. It is interesting to note that the latter reference (end of Section 2) shows the inadmissibility of classical best linear unbiased estimators even for certain nonparametric problems. Brown [5] obtains further inadmissibility results. In Stein [56], discussed already in Section 2, Stein's translation-parameter admissibility and inadmissibility results which we have mentioned are used to obtain corresponding results for certain normal multiple regression problems with n sufficiently large. Further work on high-dimensional translation-parameter estimation has been done by Baranchik, Brown, and Srivastava (none as yet published).

In all the above translation-parameter problems, the unknown parameter θ is being estimated. If a subset of the components of θ are being estimated, the situation is somewhat altered. For example, if X and θ are four-dimensional but only one component of θ is being estimated (say with squared error loss), there are examples where the unique best invariant estimator is admissible (the most trivial cases being those where the components of X are independent) and others where it is inadmissible. Blackwell [2] gave the first example of the latter, and the matter is discussed by Stein [22]; the admissible and inadmissible cases are not yet delimited.

Still referring to invariant estimation problems, one would next investigate the nonabelian noncompact groups. For example, in the case of a density depending on unknown real location and positive scale parameter, we are dealing with the real affine group. If only one of these parameters is to be estimated, we have the analogue of the situation described in the previous paragraph. There are few results here, the first being the inadmissibility result of Stein [57] for the problem of estimating the variance of a normal distribution. Inadmissibility of the best invariant estimator of a quantile other than the median in this setting was proved by Brown (unpublished), who has also considered such problems as that of [57] for other distributions. One interesting feature pointed out by Stein is that whereas inadmissibility or nonmaximality results for best invariant procedures can often be proved by showing that no $C^{(G)}$-wide-sense Bayes procedure is C-wide-sense Bayes, the classical estimator considered in [57] has the latter property but not the former!

As for estimation problems involving the full linear group or other non-HS groups, we have already seen in Section 2 that we cannot expect best invariant estimators to be minimax, let alone admissible.

We turn next to the univariate estimation problem where questions (i) and (ii) have been considered in the following cases, other than that of admissibility of best invariant estimators: (I) $\pi^* \neq \nu$ in the translation parameter problems, (II) translation parameter with truncated range, (III) extreme-value densities, and (IV) exponential families. Karlin [26] considers not only the admissibility of the best invariant estimator for the univariate translation-parameter problem (discussed earlier), but also the question of which estimators of the form cx are admissible for squared error loss in cases (III) and (IV). (In [8] linear estimators of linear functions of a multivariate normal mean are considered similarly.) The inadmissibility parts of (III) and (IV) are obtained, as in the examples we have mentioned in earlier paragraphs, by constructing better procedures. The admissibility parts of these (as well as of the location parameter considerations of [26]) are treated by explicitly representing the estimator as a δ_{π^*} and then carrying through the admissibility proof by a Fourier-analysis technique which differs from the admissibility proofs described earlier. Katz [27] and Farrell [11] considered cases of (II), while Farrell [11] also considered (I). Complete class results for the class of δ_{π^*}'s were obtained by Sacks [46, 47] for (IV) and for certain cases of (II); examples which indicate limitations on the validity of such conclusions and the additional complications of (I) over (II) are included, and multiparameter extensions are considered. In Farrell [12], limits of sequences $\{\delta_{\pi_i}\}$ are considered particularly in case (I), and the question of when these limits are δ_{π^*}'s is studied; the possibility that this is not the case and that, for example, some *a priori* probability not absorbed in π^* "escapes to ∞", is treated in detail. The general picture here appears to be quite complicated (as does the extension of some of the work of Katz-Sacks-Farrell to k-variate analogues). Related is the question now being attacked in general settings by Stein [58] of the extent to which various improper π^*'s can be sufficiently well approximated by proper π's (in the sense that there is high π-probability that the formal *a posteriori* law based on *a priori* law π^* is close in an appropriate sense to that based on π); we have mentioned earlier Stein's emphasis on this approximatability point of view. Incidentally, his necessary and sufficient condition for admissibility [51], which had not shown much applicability earlier, is invoked in this study. [*Added in proof:* In the *Fifth Berkeley Symp.* Stein carries out this development in special cases of the univariate translation-parameter problem, and the equivalence of admissibility to nonintegrability of $dv/d\pi^*$ is established under certain restrictions.] As yet no analogues of the Katz-Sacks-Farrell considerations for testing problems have been published.

The only general invariant testing result we have mentioned under (E) is that of Lehmann and Stein [38]. Presumably this can be extended to the two-dimensional translation-parameter problem, although this result has not yet appeared in print. There is evidence, however, that a much more general admissibility result for unique best invariant procedures is valid in testing problems than in estimation problems. For example, this seems to be the case for a broad class of p-variate location-parameter problems, even for $p \geq 3$, although I have been unable to obtain a satisfactory general condition; far from there being general inadmissibility results like those of [53], [22], and [5], quite the opposite may be true. The approach of [38], which is similar analytically to those we have mentioned in estimation problems, and which fails in the latter for $p \geq 3$ (where the admissibility conclusion is false), cannot be expected to be modified to work in any testing problems; a new approach seems to be called for. There is no indication yet as to whether or not the approach of [58] will satisfactorily separate the conclusions in estimation from those in testing. A further consideration is that, although some examples (e.g., [35, 52]) show that for non-HS groups best invariant tests need not be minimax, let alone admissible, there are also many examples where such procedures are minimax and admissible. This is very much unlike the estimation situation, where it seems more exceptional that a best G-invariant procedure, for G non-HS, be admissible; the estimation example of [22] appears to be more typical than are the admissible cases of [55], in this respect. One indication of this difference between estimation and testing is that, in such a simple case as that of a translation parameter (of any dimension p), it cannot be that the best invariant procedure is a genuine Bayes procedure, while this can well be the case in corresponding testing examples, where the action of G on $Đ$ is trivial. (Similar results can hold in other problems where G/J is compact, where J leaves a point of D fixed.) Thus, we return to method (A). Lehmann and Stein [37] presented the first examples of invariant tests which were admissible for the noninvariant reason of being genuinely Bayes. (They were actually concerned with proving most powerful test results which are stronger than admissibility and which we shall not discuss further here.) Kiefer and Schwartz [30] used somewhat different forms of *a priori* distributions to prove that many classical invariant tests (such as those based on T^2 and R^2) are genuinely Bayes; even in cases where admissibility had been proved earlier by method (B), the Bayes approach reflects the goodness of the power function on a different part of the parameter space. These forms of the *a priori* law are to some extent suggested by the structure of the θ_i which enter in method (B) (which, however, as we have remarked, does not actually succeed in the R^2 problem). This Bayes approach, used somewhat at random in [30], has been organized and developed by Schwartz [50] into a theory which shows for certain multivariate problems (usually exponential) that any $C^{(G)}$-Bayes procedure which satisfies an integrability condition involving the *a priori* law, is actually C-Bayes. In MANOVA, for example, the presence of this integrability condition can be exhibited in an alternative form: If, in the space of the maximal invariant (the usual set of latent roots) an acceptance region A is $C^{(G)}$-Bayes for a given n, then A is C-Bayes for slightly larger n. The method clearly has limitations; for example, the "largest root" test in MANOVA, which can be proved admissible by method (B), cannot be Bayes, on grounds of analyticity; as in [1], limits of Bayes procedures are needed. It would be interesting to extend the theory of [50] to include such considerations.

To illustrate what can happen in testing in a much simpler mechanism than that studied in [50], and which merely formalizes the structure observed in the examples of [37], suppose that $X = (Z, Y)$, $\Omega_0 \cong \Omega_1 \cong G \cong$ range (Z), and that the θ density of X with respect to σ-finite $\lambda = \gamma \times \mu$ is, in Ω_i, $g_i(y)h_i(\theta^{-1}z \mid y)$, where g_i is the marginal γ-density of the G-maximal invariant Y under Ω_i. Suppose one can find probability measures π_j on G such that $\int h_1(\theta^{-1}z \mid y)\pi_1(d\theta) = \int h_0(\theta^{-1}z \mid y)\pi_0(d\theta)$. Then the best invariant ($C^{(G)}$-Neyman-Pearson) critical region $\{y: g_1(y)/g_0(y) \geq k\}$ is clearly C-Bayes. (There are clearly many examples for each G. For a simple p-variate location parameter example from [37], let $X_1, \ldots, X_n$ be independent normal p vectors with unknown p-variate mean θ and with covariance matrix I under Ω_1 and $2I$ under Ω_0; here it suffices to concentrate π_0 on a single point and to have π_1 an appropriate multivariate normal law.) This approach can be extended to problems where G is not transitive on the Ω_i and where there is a subgroup G' which leaves a point of Ω fixed. The work of [50] can be viewed as showing in certain of these more complex settings that such π_j do indeed exist for certain multivariate (usually exponential) problems, and of characterizing these π_j's and the resulting $C_\alpha^{(G)}$ tests which are C_α-Bayes (no longer unique if G is not transitive).

REFERENCES

1. Birnbaum, A. (1955). Characterization of complete classes of tests of some multiparametric hypotheses, with applications to likelihood ratio tests. *Ann. Math. Statist.* **26** 21–36.
2. Blackwell, D. (1951). On the translation parameter problem for discrete variables. *Ann. Math. Statist.* **22** 393–399.
3. Blackwell, D. and Girshick, M. A. (1954). *Theory of Games and Statistical Decisions.* Wiley, New York.
4. Blyth, C. R. (1951). On minimax statistical decision procedures and their admissibility. *Ann. Math. Statist.* **22** 22–42.
5. Brown, L. (1964). Admissibility of translation-invariant estimators. Thesis, Cornell Univ.; published in part in *Ann. Math. Statist.*, October 1966.
6. Chen, Hsi-ju. (1964). Minimax estimates of parameter vectors of translation. *Chinese Math.* **5** 300–315.
7. Chen, Hsi-ju (1964). On minimax invariant estimation of scale and location parameters. *Scienita Sinica.* **13** 1569–1586.
8. Cohen, A. (1965). Estimates of linear combinations of the parameters in the mean vector of a multivariate distribution. *Ann. Math. Statist.* **36** 78–87. See also abstract, 1081.
9. Das Gupta, S., Anderson, T. W., and Mudholkar, G. S. (1964). Monotonicity of the power functions of some tests of the multivariate linear hypothesis. *Ann. Math. Statist.* **35** 200–205.
10. Ellison, B. E. (1962). A classification problem in which information about alternative distributions is based on samples. *Ann. Math. Statist.* **35** 213–223.
11. Farrell, R. (1964). Estimators of a location parameter in the absolutely continuous case. *Ann. Math. Statist.* **35** 949–998.
12. Farrell, R. (1965). Weak limits of sequences of Bayes procedures in estimation theory. *Proc. Fifth Berkeley Symp. Math. Statist. Prob.* To be published.
13. Fox, M. and Rubin, H. (1964). Admissibility of quantile estimates of a single location parameter. *Ann. Math. Statist.* **35** 1019–1030.
14. Geisser, S. (1965). Bayesian estimation in multivariate analysis. *Ann. Math. Statist.* **36** 150–159.
15. Giri, N., Kiefer, J., and Stein C. (1963). Minimax character of Hotelling's T^2-test in the simplest case. *Ann. Math. Statist.* **34** 1524–1535.
16. Giri, N. and Kiefer, J. (1964). Local and asymptotic minimax properties of multivariate tests. *Ann. Math. Statist.* **35** 21–35.
17. Giri, N., and Kiefer, J. (1964). Minimax character of the R^2-test in the simplest case. *Ann. Math. Statist.* **35** 1475–1490.
18. Girschick, M. A. and Savage, L. J. (1950). Bayes and minimax estimates for quadratic loss functions. *Proc. Second Berkeley Symp. Math. Statist. Prob.* 53–73.
19. Ghosh, M. N. (1964). On the admissibility of some tests of Manova. *Ann. Math. Statist.* **35** 789–794.
20. Hodges, J. L. Jr. and Lehmann, E. L. (1951). Some applications of the Cramér-Rao inequality. *Proc. Second Berkeley Symp. Math. Statist. Prob.* 13–22.
21. Isaacson, S. L. (1951). On the theory of unbiased tests of simple statistical hypotheses specifying the values of two or more parameters. *Ann. Math. Statist.* **22** 217–234.
22. James, W. and Stein, C. (1960). Estimation with quadratic loss. *Proc. Fourth Berkeley Symp. Math. Statist. Prob.* **1** 361–379.
23. Jeffreys, H. (1939). *Theory of Probability.* Oxford Univ. Press, New York.
24. Karlin, S. (1953). The theory of infinite games. *Ann. Math.* **58** 371–401.

25. KARLIN, S. (1957). Polya type distributions, II. *Ann. Math. Statist.* **28** 281–308.
26. KARLIN, S. (1958). Admissibility for estimation with quadratic loss. *Ann. Math. Statist.* **29** 406–436.
27. KATZ, M. W. (1961). Admissible and minimax estimates of parameters in truncated spaces. *Ann. Math. Statist.* **32** 136–142.
28. KIEFER, J. (1957). Invariance, minimax sequential estimation, and continuous time processes. *Ann. Math. Statist.* **28** 573–601.
29. KIEFER, J. (1958). On the nonrandomized optimality and randomized nonoptimality of symmetrical designs. *Ann. Math. Statist.* **29** 675–699.
30. KIEFER, J. and SCHWARTZ, R. (1959). Admissible Bayes character of T^2, R^2, and other fully invariant tests for classical multivariate normal problems. *Ann. Math. Statist.* **36** 747–770.
31. KUDO, H. (1955). On minimax invariant estimators of the transformation parameter. *Nat. Sci. Rep. Ochanomizu Univ.* **6** 31–73.
32. LECAM, L. (1955). An extension of Wald's theory of statistical decision functions. *Ann. Math. Statist.* **26** 69–81.
33. LECAM, L. (1964). Sufficiency and approximate sufficiency. *Ann. Math. Statist.* **35** 1419–1455.
34. LEHMANN, E. L. (1947). On families of admissible tests. *Ann. Math. Statist.* **18** 97–104.
35. LEHMANN, E. L. (1959). *Testing Statistical Hypotheses.* Wiley, New York.
36. LEHMANN, E. L. 1959). Optimum invariant tests. *Ann. Math. Statist.* **30** 881–884.
37. LEHMANN, E. L. and STEIN, C. (1948). Most powerful tests of composite hypotheses. I. Normal distributions. *Ann. Math. Statist.* **19** 495–516.
38. LEHMANN, E. L. and STEIN, C. (1953). The admissibility of certain invariant statistical tests involving a translation parameter. *Ann. Math. Statist.* **24** 473–479.
39. LEHAMNN, E. L. and STEIN, C. (1953). The totality of transformations leaving a family of normal distributions invariant (abstract). *Ann. Math. Statist.* **24** 142.
40. NANDI, H. K. (1963). On the admissibility of a class of tests. *Calcutta Statist. Assoc. Bull.* **15** 13–18.
41. NEYMAN, J. and PEARSON, E. S. (1938). Contributions to the theory of testing statistical hypotheses, III. *Statist. Res. Mem.* **2** 25–27.
42. PEISAKOFF, M. P. (1950). *Transformation Parameters.* Thesis, Princeton Univ.
43. PITMAN, E. J. G. (1939). The estimation of location and scale parameters of a continuous population of any given form. *Biometrika* **30** 391–421.
44. PITMAN, E. J. G. (1939). Tests of hypotheses concerning location and scale parameters. *Biometrika* **31** 200–215.
45. ROY, S. N. and MIKHAIL, W. F. (1960). On the admissibility of a class of tests in normal multivariate analysis (abstract). *Ann. Math. Statist.* **31** 536.
46. SACKS, J. (1960). Generalized Bayes solutions in estimation problems (abstract). *Ann. Math. Statist.* **31** 246.
47. SACKS, J. (1963). Generalized Bayes solutions in estimation problems. *Ann. Math. Statist.* **34** 751–768.
48. SCHWARTZ, R. (1964). Properties of a test in Manova (abstract). *Ann. Math. Statist.* **35** 939.
49. SCHWARTZ, R. (1964). Admissible invariant tests in Manova (abstract). *Ann. Math. Statist.* **35** 1398.
50. SCHWARTZ, R. (1966). Invariant proper Bayes procedures. This volume. Also Thesis, Cornell Univ.
51. STEIN, C. (1955). A necessary and sufficient condition for admissibility. *Ann. Math. Statist.* **26** 518–522.

52. STEIN, C. (1955). On tests of certain hypotheses invariant under the full linear group (abstract). *Ann. Math. Statist.* **26** 769.
53. STEIN, C. (1955). Inadmissibility of the usual estimator for the mean of a multivariate normal distribution. *Proc. Third Berkeley Symp. Math. Statist. Prob.* **1** 197–206.
54. STEIN, C. (1956). The admissibility of Hotelling's T^2-test. *Ann. Math. Statist.* **27** 616–623.
55. STEIN, C. (1959). The admissibility of Pitman's estimator of a single location parameter. *Ann. Math. Statist.* **30** 970–979.
56. STEIN, C. (1960). Multiple regression. *Hotelling Festschr.* Stanford Univ. Press, 424–443.
57. STEIN, C. (1964). Inadmissibility of the usual estimator for the variance of a normal distribution with unknown mean. *Ann. Inst. Statist. Math.* **16** 155–160.
58. STEIN, C. (1965). Approximation of improper prior measures by prior probability measures. *Bernoulli-Bayes-Laplace Festschr.* Springer-Verlag, New York, 217–240.
59. WALD, A. (1939). Contributions to the theory of statistical estimation and testing hypotheses. *Ann. Math. Statist.* **10** 299–326.
60. WALD, A. (1942). On the power function of the analysis of variance test. *Ann. Math. Statist.* **13** 434–439.
61. WALD, A. (1950). *Statistical Decision Functions.* Wiley, New York.
62. WESLER, O. (1959). Invariance theory and a modified minimax principle. *Ann. Math. Statist.* **30** 1–20.
63. WOLFOWITZ, J. (1949). The power of the classical tests associated with the normal distribution. *Ann. Math. Statist.* **20** 540–551.
64. WOLFOWITZ, J. (1950). Minimax estimates of the mean of a normal distribution with known variance. *Ann. Math. Statist.* **21** 218–230.

Reprinted from
Multivariate Analysis.
Proceedings of an International Symposium,
(1966), 255–274.
Academic Press, New York.

Reprinted from THE ANNALS OF MATHEMATICAL STATISTICS
Vol. 38, No. 5, October, 1967

ON BAHADUR'S REPRESENTATION OF SAMPLE QUANTILES

BY J. KIEFER[1]

Cornell University

1. Introduction and summary. Let $X_1, X_2, \cdots$ be independent and identically distributed real random variables with common df F. Suppose that $0 < p < 1$, that $F(\xi_p) = p$, that F is twice differentiable in a neighborhood of p, and that F'' is bounded in that neighborhood and $F'(\xi_p) > 0$. Let S_n be the sample df based on $(X_1, \cdots, X_n)$; i.e., $nS_n(x) =$ number of $X_i \leqq x$, $1 \leqq i \leqq n$. Let $Y_{p,n}$ be a sample p-quantile based on $(X_1, \cdots X_n)$; i.e., $S_n(Y_{p,n}-) \leqq p \leqq S_n(Y_{p,n})$; if np is an integer, so that $Y_{p,n}$ is not unique, it will be seen that any measurable definition can be used in the sequel, and for the sake of definiteness we shall take the smallest possible value. We shall write $\sigma_p = [p(1-p)]^{\frac{1}{2}}$.

Let

$$R_n(p) = Y_{p,n} - \xi_p + [S_n(\xi_p) - p]/F'(\xi_p). \tag{1.1}$$

Bahadur (1966) initiated the study of $R_n(p)$ and showed that

$$R_n(p) = O(n^{-3/4}(\log n)^{\frac{1}{2}}(\log\log n)^{\frac{1}{4}}) \tag{1.2}$$

wp 1 as $n \to \infty$. He also raised the question of finding the exact order of $R_n(p)$. In the present paper we answer this by proving

THEOREM 1. *For either choice of sign,*

$$\limsup_{n\to\infty} \pm F'(\xi_p)R_n(p)/[2^{5/4}3^{-3/4}\sigma_p^{\frac{1}{2}}n^{-3/4}(\log\log n)^{3/4}] = 1 \tag{1.3}$$

wp 1.

Later in this section we shall sketch the rationale behind this result, whose proof will occupy most of the paper.

Let τ be any positive number. For $t \in [-\tau, \tau]$, we write

$$\begin{aligned} U_n(t) &= n^{3/4}F'(\xi_{p+n^{-\frac{1}{2}}t})R_n(p + n^{-\frac{1}{2}}t), \\ U_n \equiv U_n(0) &= n^{3/4}F'(\xi_p)R_n(p), \end{aligned} \tag{1.4}$$

which by assumption on F make sense for n large. Also write

$$K_n = n^{\frac{1}{2}}(p - S_n(\xi_p)). \tag{1.5}$$

As the discussion later in this section shows, it is trivial that, for each $b > 0$, uniformly in $b^{-1} < |k_n| < b$, the conditional law of $U_n/|k_n|^{\frac{1}{2}}$, given $K_n = k_n$, is asymptotically $N(0, 1)$ as $n \to \infty$. Since K_n/σ_p is also asymptotically $N(0, 1)$, it is obvious that

$$\lim_{n\to\infty} P\{U_n \leqq u\} = 2\sigma_p^{-1} \int_0^\infty \Phi(k^{-\frac{1}{2}}u)\phi(k/\sigma_p)\, dk, \tag{1.6}$$

where Φ and ϕ are the standard $N(0, 1)$ df and density. More generally, we shall

Received 22 December 1966.

[1] Research supported by the Office of Naval Research under Contract No. NONR 401 (50). Reproduction in whole or in part is permitted for any purpose of the United States Government.

show in Section 5 that, under the same conditioning, and uniformly for k_n in a set of probability arbitrarily close to 1 according to the law of K_n for each large n, for each finite collection $t_1, t_2, \cdots, t_m$ of values in $[-\tau, \tau]$, the asymptotic joint conditional df of $\{U_n(|k_n|t_1)/|k_n|^{\frac{1}{2}}, \cdots, U_n(|k_n|t_m)/|k_n|^{\frac{1}{2}}\}$ is the same as the df of $\{J(t_1), \cdots, J(t_m)\}$, where $\{J(t), -\infty < t < \infty\}$ is a separable stationary Gaussian process with mean 0 and convariance function

$$E\{J(t_1)J(t_2)\} = \max(1 - |t_1 - t_2|, 0). \tag{1.7}$$

(The definition of $U_n(|k_n|t)/|k_n|^{\frac{1}{2}}$ can be made arbitrarily when $k_n = 0$, since $P\{K_n = 0\} \to 0$.) Let G be $N(0, \sigma_p^2)$ and independent of the J-process. Let $J^*(t) = |G|^{\frac{1}{2}}J(t/|G|)$. Our discussion above indicates a result we shall prove in Section 5.

THEOREM 2. *As $n \to \infty$ the process $\{U_n(t), |t| \leqq \tau\}$ approaches in law the process $\{J^*(t), |t| \leqq \tau\}$, whose sample paths are Lip α for all $\alpha < \frac{1}{2}$, wp 1., and* $\sup_{|t| \leqq \tau} \pm U_n(t)$ *approaches* $\sup_{|t| \leqq \tau} \pm J^*(t)$ *in law.*

We shall also prove, in Section 5:

THEOREM 3. *For either choice of sign,* (1.3) *holds if* $\pm n^{3/4}F'(\xi_p)R_n(p)$ *is replaced there by* $\sup_{|t| \leqq \tau} [\pm U_n(t)]$.

Actually, this can be strengthened by letting $\tau \to \infty$ slowly with n, as in (5.4), but this is of limited interest. (See also Remark 2 of Section 6.)

Section 6 contains other remarks on related results in the domain of Bahadur's considerations, including (Theorem 4) the observation that his technique yields the strong form of the law of the iterated logarithm for order statistics, and not just the classical form as exhibited in Equation (4) of [1], and also (Theorem 5) the validity for sample quantiles of analogues of Strassen's results [11] for the cumulative sums $nS_n(p)$. Section 2 collects most of the elementary probabilistic estimates which are needed in the proof of Theorem 1, the lower- and upper-class parts of which are contained in Sections 3 and 4.

Before proceeding further, we reduce our considerations by the following:

LEMMA 1. *If Theorem* 1, 2, *and* 3 *are valid when F is the uniform* df *on* [0, 1], *then they are valid for general F satisfying the assumptions stated at the outset of this section.*

PROOF. For F satisfying the assumptions, define independent uniformly distributed random variables $X_1^*, X_2^*, \cdots$ as follows: Let $H_1, H_2, \cdots$ be uniform on [0, 1] and independent of each other and of the original X_i. Let $X_i^* = F(X_{i-})(1 - U_i) + F(X_i)U_i$. (If F is continuous, $X_i^* = F(X_i)$.) Then the X_i^* have the desired joint uniform distribution. Moreover, if B is the hypothesized neighborhood of ξ_p in which F is twice differentiable, and if S_n^* and $Y_{p,n}^*$ are the sample df and sample p-quantile based on $(X_1^*, \cdots, X_n^*)$, then Taylor's Theorem with remainder yields for the corresponding $R_n^*(p)$, as long as $Y_{p,n} \varepsilon B$,

$$\begin{aligned} R_n^*(p) &= Y_{p,n}^* - p + S_n^*(p) - p \\ &= F(Y_{p,n}) - p + S_n(\xi_p) - p \\ &= F'(\xi_p)(Y_{p,n} - \xi_p) + F''(\xi')(Y_{p,n} - \xi_p)^2/2 + S_n(\xi_p) - p \\ &= F'(\xi_p)R_n(p) + F''(\xi')(Y_{p,n} - \xi_p)^2/2F'(\xi_p), \end{aligned} \tag{1.8}$$

where ξ' is a chance value in B. Since $Y_{p,n} \varepsilon B$ wp 1 for all large n and since $n^{\frac{1}{2}}(Y_{p,n} - \xi_p)$ in fact obeys a law of the iterated logarithm, we obtain

(1.9) $$n^{3/4}[R_n^*(p) - F'(\xi_p)R_n(p)] \to 0 \text{ wp } 1.$$

This yields the desired result regarding Theorem 1, and the results for Theorems 2 and 3 are obtained similarly; one uses the additional fact that $n^{3/4} \sup_{|t| \leqq rn^{-1/2}} (Y_{p+t,n} - \xi_{p+t})^2 \to 0$ wp 1, which is an easy consequence of the law of the iterated logarithm for the sample df (see [2]), or can be proved by the techniques of Section 5.

THROUGHOUT THIS PAPER WE HEREAFTER ASSUME F TO BE THE UNIFORM DF ON [0, 1], UNLESS EXPLICITLY STATED TO THE CONTRARY (POSSIBLY BY EXHIBITING THE SYMBOL F).

We now introduce some further notation. The complement of an event A will be denoted $\bar{A}$. In the proof of Theorem 1, γ will be a value > 1 (large in Section 3, near 1 in Section 4). For each positive integer r, we denote by n_r the greatest integer $\leqq \gamma^r$. Whenever we write $\sum_r$ or $\sum_n$ it will be understood that the summation (to ∞) begins at a large enough value that expressions like log log n_r which appear in the summand are real; a similar remark applies to other expressions. All "orders" refer to behavior as $n \to \infty$ or $r \to \infty$. We abbreviate "infinitely often" by "i.o." and "almost all n" (i.e., all natural numbers n except for a finite number) by "a.a.n". We define

(1.10) $$q_n = 2^{5/4}3^{-3/4}\sigma_p^{\frac{1}{2}}n^{-3/4}(\log\log n)^{3/4},$$
$$h_n = 2^{\frac{1}{2}}3^{-\frac{1}{2}}\sigma_p n^{-\frac{1}{2}}(\log\log n)^{\frac{1}{2}},$$

and

$$T_n(x) = S_n(x+p) - (x+p) + n^{-\frac{1}{2}}K_n,$$
(1.11) $$T_n^+(x) = \sup_{0 \leqq v \leqq x} T_n(v),$$
$$T_n^-(x) = -\inf_{0 \leqq v \leqq x} T_n(v),$$
$$T_n^*(x) = \sup_{0 \leqq v \leqq x} |T_n(v)|.$$

Note that $T_n(0) = 0$ wp 1.

We now rephrase the statement of Theorem 1 in a form that will make it simpler to explain. Suppose for the moment that the event $K_n = k_n > 0$ occurs. It is then clear that, F being uniform, if $d_n > 0$, then

(1.12) $$R_n(p) > d_n \Leftrightarrow T_n(n^{-\frac{1}{2}}k_n + d_n) < -d_n,$$
$$R_n(p) < -d_n \Leftrightarrow T_n(n^{-\frac{1}{2}}k_n - d_n) \geqq d_n,$$

except for a set of probability zero (where T_n has a jump at $n^{-\frac{1}{2}}k_n - d_n$) in the second case. (Of course, (1.12) is also valid when $d_n \leqq 0$). The event of this second line of (1.12) is illustrated in Figure 1 in the case $d_n < n^{-\frac{1}{2}}k_n$. For the case $K_n = k_n \leqq 0$, there are two analogous events, again given by (1.12). Referring to the second line of (1.12) with $d_n = \lambda q_n$, write

(1.13) $$A_n(\lambda) = \{K_n > 0 \text{ and } T_n(n^{-\frac{1}{2}}K_n - \lambda q_n) \geqq \lambda q_n\}.$$

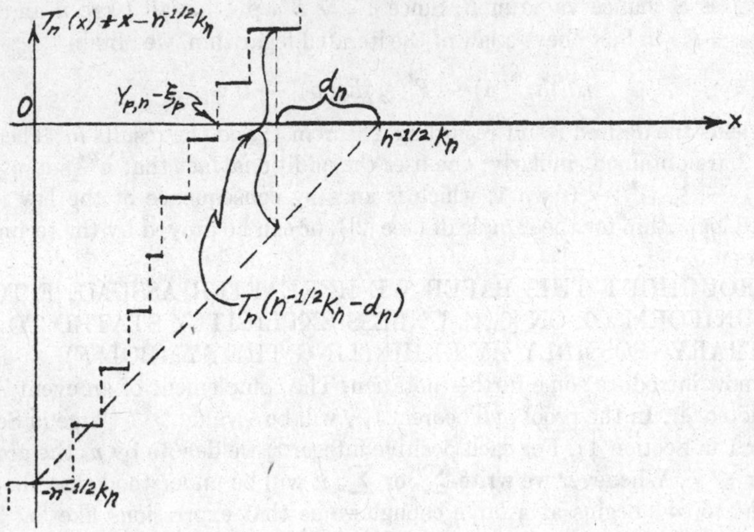

FIG. 1

Suppose we proved that

$$(1.14) \qquad P\{A_n(\lambda) \text{ i.o.}\} = 1 \text{ if } \lambda < 1, \qquad = 0 \text{ if } \lambda > 1,$$

and also proved the three analogous results for the first case of (1.12) and the two cases of $K_n \leqq 0$. This would then clearly prove Theorem 1. The cases when $K_n \leqq 0$ can be obtained by invoking the cases when $K_n > 0$ for the $(1 - p)$-tile of the random variables $1 - X_i$ and the fact that for $\lambda > 0$ the event $\{K_n = 0, T_n(\pm\lambda q_n) \pm \lambda q_n = 0 \text{ i.o.}\}$ is easily seen to have probability zero by using the Borel-Cantelli Lemma. Sections 2, 3, and 4 are devoted mainly to proving (1.14). Any modifications which are needed to prove the analogue of (1.14) for the other case of (1.12) are indicated, thus yielding Theorem 1.

As usual, for N a positive integer and $0 < \pi < 1$, we denote the binomial df by

$$(1.15) \qquad B(z, N, \pi) = \textstyle\sum_{j \leqq z} \binom{N}{j} \pi^j (1 - \pi)^{N-j},$$

and we also write

$$(1.16) \qquad B^*(\Delta, N, \pi) = B(\Delta + N\pi, N, \pi).$$

We shall frequently use the symbols Δ, N, and π to refer to quantities in the context (1.16). If $N = 0$, B and B^* assign probability one to the value zero.

We now indicate briefly the idea behind the statement (1.14). Fix $\lambda > 0$. Then $n^{-\frac{1}{2}}k_n > \lambda q_n$ clearly implies

$$(1.17) \qquad P\{A_n(\lambda) \mid K_n = k_n\} = 1 - B^*(\lambda n q_n -, n(1 - p + n^{-\frac{1}{2}}k_n), (n^{-\frac{1}{2}}k_n - \lambda q_n)/(1 - p)).$$

If $b^{-1} < k_n < b$, the Central Limit Theorem is applicable with λq_n replaced by $n^{-3/4}u$ and (with the corresponding result for $K_n < 0$) yields (1.6). In fact, if $k_n = bn^{\frac{1}{4}}h_n$ with $b > 0$, which normalization turns out to yield the crucial range of values of K_n, Lemma 2 of Section 2 implies that the normal approximation applies both to (1.17) and to the df of K_n at k_n, and yields

$$\begin{aligned} \log P\{A_n(\lambda) \mid K_n = bn^{\frac{1}{4}}h_n\} &\sim \log\,[1 - \Phi(\lambda n q_n/(bnh_n)^{\frac{1}{2}})] \\ &\sim -\tfrac{1}{2}[(4\lambda^2/3b)/\log\log n], \\ \log P\{K_n > bn^{\frac{1}{4}}h_n\} &\sim \log\,[1 - \Phi(bn^{\frac{1}{4}}h_n/\sigma_p)] \\ &\sim -\tfrac{1}{2}[(2b^2/3)\log\log n]. \end{aligned} \tag{1.18}$$

Since $\min_{b>0}[2b^2/3 + 4\lambda^2/3b] = 2\lambda^{\frac{4}{3}}$, attained at $b = \lambda^{\frac{2}{3}}$, it is not hard to show from (1.18) that

$$\log P\{A_n(\lambda)\} \sim -\lambda^{\frac{4}{3}} \log\log n, \tag{1.19}$$

so that, whatever $\lambda > 0$,

$$\begin{aligned} \textstyle\sum_r P\{A_{n_r}(\lambda)\} &= \infty \text{ if } \lambda < 1, \\ &< \infty \text{ if } \lambda > 1. \end{aligned} \tag{1.20}$$

This suggests that (1.14) can be attacked by the classical approach of using the Borel-Cantelli Lemmas for the event $\{A_{n_r}(\lambda) \text{ i.o.}\}$ when $\lambda > 1$ and for a corresponding event in terms of a sequence of independent blocks of observations (from $n_r + 1$ to n_{r+1}) when $\lambda < 1$, and by then showing that nothing much different happens for intermediate values of n.

It is this last aspect of the proof which requires some delicacy and slightly different techniques from those employed in the classical case [5], [6] or in the proof of the sample df law [2], [7]. For example, a step analogous to that of the usual approach (as in [6]) in the case of the upper class would be to show that, if $\lambda > 1$, then γ and $\lambda' > 1$ and $\epsilon > 0$ can be chosen so that, for all large r and $n_r < n \leqq n_{r+1}$,

$$P\{A_{n_{r+1}}(\lambda') \mid A_n(\lambda) \textstyle\bigcap_{n_r<j<n} \overline{A_j(\lambda)}\} > \epsilon. \tag{1.21}$$

But the intuitively obvious (1.21) seems difficult to establish in view of the complexity of the condition of (1.21) compared with the analogue in the case of partial sums or the sample df. Thus, we forego (1.21) and, roughly speaking, prove instead the superficially stronger relation

$$P\{\textstyle\bigcap_{n_r<n<n_{r+1}} \overline{A_n(\lambda)} \mid \overline{A_{n_r}(\lambda')}\} > 1 - \epsilon_r, \tag{1.22}$$

where $\sum_r \epsilon_r < \infty$. Such relations rely heavily on the persistence in n and in x of the sample df deviations $T_n(x)$, as made precise in some of the lemmas of Section 2.

2. Preliminary lemmas. Because the last argument of B^* in (1.17) is customarily small in our considerations, we shall often require the Central Limit Theorem for small tail probabilities in the context of (1.16) when $N \to \infty$,

$\pi \to 0$, $N\pi \to \infty$, and $\Delta^2/N\pi \to \infty$ at a slow enough rate:

LEMMA 2. *If* $\Delta_N > 0, \pi_N \to 0, N\pi_N \to \infty, \Delta_N^2/N\pi_N \to \infty$, *and* $\Delta_N^3/N^2\pi_N^2 \to 0$, *then, for each* $\zeta > 0$, *as* $N \to \infty$,

$$
\begin{aligned}
&B^*(-\Delta_N, N, \pi_N) \sim B^*(-\Delta_N, N, \pi_N) - B^*(-\Delta_N(1+\zeta), N, \pi_N) \\
(2.1)\qquad &\sim 1 - B^*(\Delta_N, N, \pi_N) \sim B^*(\Delta_N(1+\zeta), N, \pi_N) - B^*(\Delta_N, N, \pi_N) \\
&\sim \Phi_N(-\Delta_N/[N\pi_N(1-\pi_N)]^{\frac{1}{2}}),
\end{aligned}
$$

so that the logarithm of any of these behaves as

$$
(2.2)\qquad \log B^*(-\Delta_N, N, \pi_N) \sim -\Delta_N^2/2N\pi_N .
$$

PROOF. The standard proof in Feller [6], pp. 168–179, for the case where π_N is constant, is easily seen to apply, essentially intact, under the present assumptions. (It even yields the same conclusion with ζ replaced by ζ_N such that $N\pi_N/\Delta_N^2\zeta_N = o(1)$, but we shall not require this stronger result.) Of course, $1 - \pi_N \to 1$ is used in (2.2).

We shall also use (2.1) in the usual way to obtain estimates for the law of K_N, where $\pi_N = p$ for all N; this entails only multiplying the right side of (2.1) by $(1-p)^{-1}$, to obtain an estimate of $\log P\{N^{\frac{1}{2}}K_N > \Delta_N\}$.

The following lemmas use the sample df conditioning technique (as employed, e.g., in [7]), and make use only of Chebyshev's inequality and the fact that, given that $K_M = k > 0$, the law of $M[T_M(x) + x]$ for $x > 0$ is $B(\cdot, M(1-p) + M^{\frac{1}{2}}k, x/(1-p))$; the corresponding unconditional law is of course $B(\cdot, M, x)$. We shall not aim at sharp hypotheses or conclusions, but merely at the forms we require. Thus, all these estimates have unconditional and various conditional versions, of which we state only the forms actually used in the sequel. Also, under the conditions of Lemma 3, the analogue of (2.9) with "$>w$" replaced by "$<-w$" in both events, is easily proved in the same way; as a corollary of these two results we then have, under the same conditions,

$$
\begin{aligned}
(2.3)\qquad &P\{T_M((1-\delta)a) > (1-3\delta)w \mid T_M(a) > w, K_M = k\} \geqq \tfrac{2}{3}, \\
&P\{T_M((1-\delta)a) < -(1-3\delta)w \mid T_M(a) < -w, K_M = k\} \geqq \tfrac{2}{3},
\end{aligned}
$$

which inequalities we shall make use of; these could have been proved directly with "3δ" replaced by a smaller multiple of δ by repeating the argument of the last few lines of (2.6), and (with $\frac{2}{3}$ reduced to a smaller positive value) even with "3δ" replaced by "δ" by using a different argument. We forego such repetition and unneeded elegance. The results which are required for proving the analogue of (1.14) obtained from the first half of (1.12) are proved in almost the same way, and the form of their statements will be clear; for example, from (1.12) we see that (2.4) will be replaced by considering negative deviations over $1 \leqq z \leqq 1 + \delta$.

LEMMA 3. *Suppose* $0 < \delta \leqq \frac{1}{6}$, $a > 0$, $M\delta w \geqq 2$, *and that* $M\delta w^2/(a+w) \geqq 32$. *Then*, wp 1,

$$
(2.4)\qquad P\{\inf_{1-\delta \leqq z \leqq 1} T_M(za) > (1-3\delta)w \mid T_M(a) > w, K_M = k\} \geqq \tfrac{2}{3}.
$$

PROOF. Let Z be the largest value in $[1 - \delta, 1]$, if there is any, for which $T_M(Za-) \leqq (1 - 3\delta)w$. Given the condition of (2.4) and also that $Z = z$ and $T_M(za-) = (1 - 3\delta)s \leqq (1 - 3\delta)w$, the random function $\{T_M(x), 0 \leqq x \leqq za\}$ clearly has the same law as it does under the sole condition $T_M(za-) = (1 - 3\delta)s$, since Z is the first positive value z for which $T_M(za-) \leqq (1 - 3\delta)w$ as one decreases z from $z = 1$. Thus, since $[w - (1 - 2\delta)s/z] \geqq \delta w$,

$$\begin{aligned} &P\{T_M((1 - 2\delta)a) \\ &\quad \leqq (1 - 3\delta)w \mid T_M(a) > r, K_M = k, Z = z, T_M(za-) = (1 - 3\delta)s\} \\ &= B(M(1 - 2\delta)a + M(1 - 3\delta)w, Mza + M(1 - 3\delta)s, (1 - 2\delta)/z) \\ &= B^*(M(1 - 3\delta)[w - (1 - 2\delta)s/z], Mza + M(1 - 3\delta)s, (1 - 2\delta)/z) \\ &\geqq B^*(M(1 - 3\delta)\delta w, Mza + M(1 - 3\delta)s, (1 - 2\delta)/z) \geqq \tfrac{3}{4}, \end{aligned} \tag{2.5}$$

this last by Chebyshev's inequality using the fact that the arguments of the last B^* satisfy $\Delta/[N\pi(1 - \pi)]^{\frac{1}{2}} \geqq [M\delta w^2/8(a + w)]^{\frac{1}{2}}$. We thus have

$$\begin{aligned} &P\{\inf_{1-\delta \leqq z \leqq 1} T_M(za) \leqq (1 - \delta)w \mid T_M(a) > w, K_M = k\} \\ &= \frac{\begin{array}{l} P\{Z \text{ exists}, T_M(Za-) \leqq (1 - 3\delta)w, T_M((1 - 2\delta)a) \\ \qquad\qquad \leqq (1 - 3\delta)w \mid T_M(a) > w, K_M = k\} \end{array}}{\begin{array}{l} P\{T_M((1 - 2\delta)a) \leqq (1 - 3\delta)w \mid T_M(a) > w, K_M = k, \\ \qquad\qquad Z \text{ exists}, T_M(Za-) \leqq (1 - 3\delta)w\} \end{array}} \\ &\leqq \tfrac{4}{3} P\{T_M((1 - 2\delta)a) \leqq (1 - 3\delta)w \mid T_M(a) > w, K_M = k\} \\ &\leqq \tfrac{4}{3} \sup_{\bar{s} > w} P\{T_M((1 - 2\delta)a) \leqq (1 - 3\delta)w \mid T_M(a) = \bar{s}, K_M = k\} \\ &= \tfrac{4}{3} \sup_{\bar{s} > w} B(M(1 - 2\delta)a + M(1 - 3\delta)w, M(a + \bar{s}), (1 - 2\delta)) \\ &= \tfrac{4}{3} \sup_{\bar{s} > w} B^*(M(1 - 2\delta)(w - \bar{s}) - M\delta w, M(a + \bar{s}), (1 - 2\delta)) \leqq \tfrac{1}{3}, \end{aligned} \tag{2.6}$$

the last by Chebyshev's inequality using $(1 - 2\delta)(\bar{s} - w) + \delta w \geqq (a + \bar{s})^{\frac{1}{2}}(a + w)^{-\frac{1}{2}}w\delta/2$, so that from (1.10) the arguments of the last B^* satisfy $-\Delta/[N\pi(1 - \pi)]^{\frac{1}{2}} \geqq [M\delta w^2/8(a + w)]^{\frac{1}{2}}$. The last bound of (2.6) proves Lemma 3.

LEMMA 4. *If* $0 < \delta < 1$, $0 < a \leqq 1 - p$, $0 < w$, *then*, wp 1,

$$\begin{aligned} &P\{T_M(a) > w(1 - \delta) \mid T_M^{+}(a) > w\} \\ &\qquad \geqq \tfrac{3}{4} \quad \text{if} \quad [M/a]^{\frac{1}{2}}w(\delta - a) \geqq 2; \\ &P\{T_M(a) < -w(1 - \delta) \mid T_M^{-}(a) > w\} \\ &\qquad \geqq \tfrac{3}{4} \quad \text{if} \quad [M/a(1 + w)]^{\frac{1}{2}}w(\delta - a) \geqq 2; \\ &P\{T_M(a) > w(1 - \delta) \mid T_M^{+}(a) > w, K_M = k\} \\ &\qquad \geqq \tfrac{3}{4} \quad \text{if} \quad [M/a(1 - p)]^{\frac{1}{2}}w(\delta - a(1 - p)^{-1}) \geqq 2. \end{aligned} \tag{2.7}$$

PROOF. Let Z' be the smallest positive value for which $T_M^{+}(Z') > w$, say $T_M^{+}(Z') = s > w$. Then, for $z < a$,

$$\begin{aligned} &P\{T_M(a) > w(1 - \delta) \mid Z' = z, T_M(z) = s\} \\ &\quad = 1 - B(M[a + w(1 - \delta) - z - s], M(1 - z - s), (a - z)/(1 - z)) \end{aligned}$$

$$
\begin{aligned}
(2.8) \quad &= 1 - B^*(-Mw\delta + M[-(s-w)(1-a) + (a-z)w]/(1-z), \\
&\quad M(1-z-s), (a-z)/(1-z)) \\
&\geqq 1 - B^*(-Mw(\delta - a), M(1-z-s), (a-z)/(1-z)) \geqq \tfrac{3}{4}
\end{aligned}
$$

by Chebyshev's inequality, since the arguments of the last B^* satisfy $-\Delta/[N\pi(1-\pi)]^{\frac{1}{2}} \geqq M^{\frac{1}{2}}a^{-\frac{1}{2}}w(\delta - a)$. This yields the first line of (2.7). The other lines are proved analogously.

LEMMA 5. *Under the conditions of Lemma* 3, wp 1,

$$
\begin{aligned}
(2.9) \quad P\{\inf_{1-\delta \leqq z \leqq 1} T_M(za) > (1-3\delta)w \mid K_M = k\} \\
\geqq \tfrac{2}{3} P\{T_M(a) > w \mid K_M = k\}.
\end{aligned}
$$

Under the conditions of Lemma 4,

$$
\begin{aligned}
&P\{T_M^+(a) > w\} \leqq \tfrac{4}{3} P\{T_M(a) > w(1-\delta)\}, \\
(2.10) \quad &P\{T_M^-(a) > w\} \leqq \tfrac{4}{3} P\{T_M(a) < -w(1-\delta)\}, \\
&P\{T_M^+(a) > w \mid K_n = k\} \leqq \tfrac{4}{3} P\{T_M(a) > w(1-\delta) \mid K_n = k\} \text{ wp } 1.
\end{aligned}
$$

PROOF. (2.9) follows from (2.4) and $P\{A \mid C\} \geqq P\{A \mid BC\} P\{B \mid C\}$. (2.10) follows from (2.7) and $P\{A \mid C\} \leqq P\{B \mid C\}/P\{B \mid AC\}$ (with C = sample space in first two lines), just as used in the first three lines of (2.6).

3. Lower class result. In this section we shall prove the first half of (1.14). Throughout this section λ is fixed, $0 < \lambda < 1$, and we write $\epsilon = 1 - \lambda$. We define $S_{n_r,n}$ for $n_r < n \leqq n_{r+1}$, as the sample df based on $(X_{n_r+1}, X_{n_r+2}, \cdots, X_n)$; similarly, $K_{n_r,n} = (n - n_r)^{\frac{1}{2}}(p - S_{n_r,n}(p))$, and $T_{n_r,n}$ (with or without superscript) is then obtained by replacing S_n and K_n by $S_{n_r,n}$ and $K_{n_r,n}$ in (1.11). γ is any value which satisfies

$$
\begin{aligned}
&\gamma^{\frac{1}{2}} > 9, \\
(3.1) \quad &-\gamma^{-\frac{1}{2}} + (1 - 4\epsilon/5)(\gamma - 1)/\gamma > 1 - \epsilon, \\
&-(7/\gamma)^{\frac{1}{2}} + (1 - \epsilon/25)(\gamma - 1)/\gamma > 1 - \epsilon/20;
\end{aligned}
$$

it is clear that such a γ exists, and the somewhat redundant form of (3.1) is pointed toward its use in the sequel. We shall repeatedly use the fact that $\lim_{r\to\infty} n_{r+1}/n_r = \gamma$.

We define the following events:

$$
\begin{aligned}
L'_{r+1} &= \{1 - \epsilon/20 < K_{n_{r+1}}/n_{r+1}^{\frac{1}{2}} h_{n_{r+1}} < 1\}, \\
L''_{r+1} &= \{\inf_{1-\epsilon/10 \leqq x \leqq 1} T_{n_{r+1}}(x h_{n_{r+1}}) > q_{n_{r+1}}(1-\epsilon)\}, \\
(3.2) \quad G_r' &= \{|K_{n_r}| < 2\sigma_p(\log\log n_r)^{\frac{1}{2}}\}, \\
G_r'' &= \{T_{n_r}^-(h_{n_{r+1}}) \leqq \gamma^{\frac{1}{2}} q_{n_{r+1}}\}, \\
G'_{r,r+1} &= \{1 - \epsilon/30 < K_{n_r,n_{r+1}}/(n_{r+1} - n_r)^{\frac{1}{2}} h_{n_{r+1}} < 1 - \epsilon/60\}, \\
G''_{r,r+1} &= \{\inf_{1-\epsilon/10 \leqq x \leqq 1} T_{n_r,n_{r+1}}(x h_{n_{r+1}}) > (1 - 3\epsilon/10)(1 - \epsilon/2) q_{n_{r+1}}\}.
\end{aligned}
$$

We shall prove that the following inclusions hold for sufficiently large r (depending only on p, λ, γ):

$$L'_{r+1} \cap L''_{r+1} \subset A_{n_{r+1}}(\lambda), \tag{3.3}$$

$$G_r' \cap G'_{r,r+1} \subset L'_{r+1}, \tag{3.4}$$

$$G_r'' \cap G''_{r,r+1} \subset L''_{r+1}, \tag{3.5}$$

and we shall also prove that

$$G_r' \quad \text{occurs for a.a.r, wp 1,} \tag{3.6}$$

$$G_r'' \quad \text{occurs for a.a.r, wp 1,} \tag{3.7}$$

$$G'_{r,r+1} \cap G''_{r,r+1} \quad \text{occurs i.o., wp 1.} \tag{3.8}$$

Clearly, (1.14) follows from (3.3)–(3.8).

First, note that

$$L'_{r+1} \cap L''_{r+1} \quad \text{entails} \quad T_{n_{r+1}}(n_{r+1}^{-\frac{1}{2}} K_{n_{r+1}} - z) > \lambda q_{n_{r+1}}$$

for $0 < z < h_{n_{r+1}}\epsilon/20$; in particular, since $q_n = o(h_n)$, this holds for $z = \lambda q_{n_{r+1}}$ if r is sufficiently large. Reference to (1.12) yields (3.3).

To prove (3.4), we note that

$$n_{r+1}^{\frac{1}{2}} K_{n_{r+1}} = n_r^{\frac{1}{2}} K_{n_r} + (n_{r+1} - n_r)^{\frac{1}{2}} K_{n_r, n_{r+1}}. \tag{3.9}$$

Under G_r',

$$n_{r+1}^{-1} h_{n_{r+1}}^{-1} n_r^{\frac{1}{2}} |K_{n_r}| \sim |K_{n_r}|/\sigma_p[(2\gamma/3) \log\log n_r]^{\frac{1}{2}} < (6/\gamma)^{\frac{1}{2}}, \tag{3.10}$$

so that the left side of (3.10) is $< (7/\gamma)^{\frac{1}{2}}$ for all large r. Under $G'_{r,r+1}$,

$$\begin{aligned} n_{r+1}^{-1} h_{n_{r+1}}^{-1} (n_{r+1} - n_r)^{\frac{1}{2}} K_{n_r, n_{r+1}} &\sim [(\gamma - 1)/\gamma] K_{n_r, n_{r+1}} / (n_{r+1} - n_r)^{\frac{1}{2}} h_{n_{r+1}} \\ &= [(\gamma - 1)/\gamma][1 - (1 + \theta)\epsilon/60] \end{aligned} \tag{3.11}$$

where $0 < \theta < 1$, so that for all large r the left side is $[(\gamma - 1)/\gamma] \times [1 - (1 + \theta)\epsilon/50]$. Dividing both sides of (3.9) by $n_{r+1} h_{n_{r+1}}$ and invoking these consequences of (3.10)–(3.11) together with the third line of (3.1) (which last also implies that $(7/\gamma)^{\frac{1}{2}} + (1 - \epsilon/50)(\gamma - 1)/\gamma < 1$), we obtain (3.4).

Next, we apply

$$n_r T_{n_r} + (n_{r+1} - n_r) T_{n_r, n_{r+1}} = n_{r+1} T_{n_{r+1}} \tag{3.12}$$

to (3.2) and see that (3.5) would follow from

$$-n_r \gamma^{\frac{1}{2}} + (n_{r+1} - n_r)(1 - 3\epsilon/10)(1 - \epsilon/2) > n_{r+1}(1 - \epsilon). \tag{3.13}$$

Dividing all members of (3.13) by n_{r+1} and applying the second line of (3.1), we obtain (3.5) for all large r.

As for (3.6), it is of course a consequence of the ordinary law of the iterated logarithm for Bernoulli random variables.

To prove (3.7), we use the second line of (2.10) with $M = n_r$, $a = h_{n_{r+1}}$, $w = \gamma^{\frac{1}{2}} q_{n_{r+1}}$, and $\delta = \frac{1}{2}$, so that we have $[M/a(1 + w)]^{\frac{1}{2}} w(\delta - a) \to \infty$ as $r \to \infty$

and conclude that, for all large r,

$$(3.14)\qquad P\{\overline{G_r''}\} = P\{T_{n_r}^-(h_{n_{r+1}}) > \gamma^{\frac{1}{2}} q_{n_{r+1}}\} \leqq \tfrac{4}{3} P\{T_{n_r}(h_{n_{r+1}}) < -2^{-1}\gamma^{\frac{1}{2}} q_{n_{r+1}}\}.$$

Applying (2.2) to $B^*(-\Delta, N, \pi)$ with $N = n_r$, $\pi = h_{n_{r+1}}$, and $\Delta = 2^{-1}\gamma^{\frac{1}{2}} n_r q_{n_{r+1}}$, so that $\Delta^3/N^2\pi^2 \to 0$ and $\Delta^2/N\pi \sim \gamma^{\frac{1}{2}} 3^{-1} \log r$ (making Lemma 2 applicable) as $r \to \infty$, we obtain, for all sufficiently large r, from the first line of (3.1),

$$(3.15)\qquad \log P\{\overline{G_r''}\} < -\tfrac{4}{3} \log r.$$

Hence $\sum_r P\{\overline{G_r''}\} < \infty$, which by the Borel-Cantelli Lemma yields (3.7).

It remains to prove (3.8). Applying the classical form of (2.1) for $1 - B^*$ in the case $\pi_n \equiv p$, $N = n_{r+1} - n_r$, $\Delta_N = (1 - \epsilon/30)(n_{r+1} - n_r)h_{n_{r+1}}$, $1 + \zeta = (1 - \epsilon/60)/(1 - \epsilon/30)$, as described in the second paragraph below (2.2), we have

$$(3.16)\qquad \log P\{G'_{r,r+1}\} \sim -(3\gamma)^{-1}(\gamma - 1)(1 - \epsilon/30)^2 \log r$$

as $r \to \infty$. From (2.9) with $M = n_{r+1} - n_r$, $\delta = \epsilon/10$, $a = h_{n_{r+1}}$, $w = (1 - \epsilon/2)q_{n_{r+1}}$ (so that the conditions of Lemma 3 hold for large r), we have

$$(3.17)\qquad P\{G''_{r,r+1} \mid K_{n_r, n_{r+1}} = k\} \geqq \tfrac{2}{3} P\{T_{n_r, n_{r+1}}(h_{n_{r+1}}) > (1 - \epsilon/2)q_{n_{r+1}} \mid K_{n_r, n_{r+1}} = k\}$$

for r sufficiently large. Applying (2.2) to $1 - B^*$ for $N = (1 - p)(n_{r+1} - n_r) + (n_{r+1} - n_r)^{\frac{1}{2}} k$, $\pi_N = (1 - p)^{-1} h_{n_{r+1}}$, $\Delta_N = (n_{r+1} - n_r)(1 - \epsilon/2)q_{n_{r+1}}$ (so that the conditions of Lemma 2 are satisfied), we have, as $r \to \infty$, uniformly in k satisfying $1 - \epsilon/30 < k/(n_{r+1} - n_r)^{\frac{1}{2}} h_{n_{r+1}} < 1 - \epsilon/60$,

$$(3.18)\qquad \log P\{T_{n_r, n_{r+1}}(h_{n_{r+1}}) > (1 - \epsilon/2)q_{n_{r+1}} \mid K_{n_r, n_{r+1}} = k\} \sim -2(\gamma - 1)(3\gamma)^{-1}(1 - \epsilon/2)^2 \log r.$$

From (3.16), (3.17), and (3.18), we have for r sufficiently large,

$$(3.19)\qquad \log P\{G'_{r,r+1} \cap G''_{r,r+1}\} \geqq -(1 - \epsilon/30)^2 \log r,$$

so that $\sum_r P\{G'_{r,r+1} \cap G''_{r,r+1}\} = \infty$. The events $\{G'_{r,r+1} \cap G''_{r,r+1},\ r = 1, 2, \cdots\}$ being independent, the Borel-Cantelli Lemma yields (3.8), completing the proof of the first half of (1.14).

4. Upper class result. In this section we prove the second half of (1.14). We use the notation of the first paragraph of Section 3, except that now $\lambda = 1 + \epsilon > 1$ and (3.1) is replaced by fixing $\eta = \gamma - 1$ at any positive value satisfying

$$(4.1)\qquad \begin{aligned} &\eta < 2^{-6}3^{-2}\epsilon^2, \\ &1 + 8\eta^{\frac{1}{2}} < (.98)2^{\frac{1}{2}}3^{-\frac{1}{2}}, \\ &(1 + 10\eta^{\frac{1}{2}})(1 - \eta^{\frac{1}{2}})^{-1} < (1 + \epsilon/2), \\ &\eta < 10^{-4}. \end{aligned}$$

We now define

$$S = \{L : 2^{-1} \leqq L \leqq 2 + 4\eta^{\frac{1}{2}};\ \ L = \text{integer} \times \eta^{\frac{1}{2}} \ \text{ or}$$
$$L = 2^{-1} \ \text{ or } \ L = 2 + 4\eta^{\frac{1}{2}}\},$$
$$I_r = \{n : n_r < n \leqq n_{r+1}\},$$
$$J_r = \{|K_n| < 2\sigma_p(\log\log n_r)^{\frac{1}{2}},\ n\ \varepsilon\ I_r\},$$
$$H_r = \{|K_n - K_{n_r}| < 4\eta^{\frac{1}{2}}\sigma_p(\log\log n_r)^{\frac{1}{2}},\ n\ \varepsilon\ I_r\},$$
$$(4.2)\qquad D_r = \{(n - n_r)T_{n_r,n}^{+}(3h_{n_r}) < 2^{-1}\epsilon n_r q_{n_r},\ n\ \varepsilon\ I_r\},$$
$$E_r = \{|K_{n_r}| < 2^{-1}\sigma_p(\log\log n_r)^{\frac{1}{2}};\ \ A_n(\lambda) \ \text{ occurs for some } n\ \varepsilon\ I_r\},$$
$$B_{L,r} = \{|1 - K_{n_r}/L\sigma_p(\log\log n_r)^{\frac{1}{2}}| < \eta^{\frac{1}{2}}\} \ \text{ where } \ 2^{-1} \leqq L \leqq 2 + 4\eta^{\frac{1}{2}},$$
$$C_{L,r} = \{T_{n_r}^{+}((1 + 10\eta^{\frac{1}{2}})2^{-\frac{1}{2}}3^{\frac{1}{2}}Lh_{n_r}) > (1 + \epsilon/2)q_{n_r}\},$$
$$C'_{L,n} = \{T_n^{+}((1 + 10\eta^{\frac{1}{2}})2^{-\frac{1}{2}}3^{\frac{1}{2}}Lh_{n_r}) > \lambda q_n\} \ \text{ where } \ n\ \varepsilon\ I_r.$$

We shall prove that

(4.3) J_r occurs for a.a.r, wp 1,

(4.4) H_r occurs for a.a.r, wp 1,

(4.5) D_r occurs for a.a.r, wp 1,

(4.6) $\bar{E}_r$ occurs for a.a.r, wp 1,

(4.7) $B_{L,R} \cap C_{L,R}$ occurs only finitely often wp 1, for each fixed $L\ \varepsilon\ S$,

and we shall also prove the following hold for all sufficiently large r(depending only on p, λ, η):

$$(4.8)\qquad B_{L,r} \cap H_r \cap A_n(\lambda) \subset C'_{L,n} \quad \text{for each} \quad n\ \varepsilon\ I_r,\ \ L\ \varepsilon\ S,$$

$$(4.9)\qquad D_r \cap (\textstyle\bigcup_{n\varepsilon I_r} C'_{L,n}) \subset C_{L,r} \quad \text{for each} \quad L\ \varepsilon\ S,$$

$$(4.10)\qquad J_r \cap H_r \cap \bar{E}_r \cap A_n(\lambda) \subset \textstyle\bigcup_{L\varepsilon S} B_{L,r} \quad \text{for} \quad n\ \varepsilon\ I_r.$$

From (4.3), (4.4), (4.6), and (4.10), and from the finiteness of S, it then follows that, except on a set of probability zero, $A_n(\lambda)$ occurs i.o. only if, for some L in S, the event $B_{L,r} \cap H_r \cap (\bigcup_{n\varepsilon I_r} A_n(\lambda))$ occurs i.o. This occurrence and (4.5) imply, by (4.8)–(4.9), that $B_{L,r} \cap C_{L,r}$ occurs i.o. for some $L\ \varepsilon\ S$. Thus, (4.7) yields the desired second half of (1.14).

(4.3) is a consequence of the ordinary law of the iterated logarithm for Bernoulli variables and the fact that $\log\log n \sim \log\log n_r$ for $n\ \varepsilon\ I_r$.

To prove (4.4), let Z_i' be the indicator of the event $\{X_i \leqq p\}$, and let $Z_i = Z_i' - p$. It is well known (e.g., [6], p. 192, equation (5.7)) that there is a finite constant b_p such that, for all $\alpha_r > 0$,

$$(4.11)\qquad P\{\max_{n\varepsilon I_r} |\textstyle\sum_{n_r+1}^{n} Z_j| > \alpha_r\} \leqq b_p P\{|\textstyle\sum_{i\varepsilon I_r} Z_i| > \alpha_r\}$$

Putting $\alpha_r = 2\eta^{\frac{1}{2}}\sigma_p(n_r \log\log n_r)^{\frac{1}{2}}$ and using the classical form of (2.1) as described in the second paragraph below (2.2) with $N = n_{r+1} - n_r$, $\pi_N = p$, $\Delta_N = \alpha_r$, and the fact that $n^{-\frac{1}{2}} < n_r^{-\frac{1}{2}}$ for $n \varepsilon I_r$, we have

$$\log P\{\max_{n\varepsilon I_r} n^{-\frac{1}{2}} |\textstyle\sum_{n_r+1}^{n} Z_j| > 2\eta^{\frac{1}{2}}\sigma_p(\log\log n_r)^{\frac{1}{2}}\} \leqq -2(1 + o(1)) \log r. \tag{4.12}$$

Hence, by the Borel-Cantelli Lemma,

$$\max_{n\varepsilon I_r} n^{-\frac{1}{2}} |\textstyle\sum_{n_r+1}^{n} Z_j| \leqq 2\eta^{\frac{1}{2}}\sigma_p(\log\log n_r)^{\frac{1}{2}} \quad \text{for a.a.r, wp 1.} \tag{4.13}$$

Clearly,

$$\begin{aligned} K_{n_r} - K_n &= n^{-\frac{1}{2}} \textstyle\sum_1^n Z_i - n_r^{-\frac{1}{2}} \sum_1^{n_r} Z_i \\ &= n^{-\frac{1}{2}} \textstyle\sum_{n_r+1}^{n} Z_i + K_{n_r}[(n_r/n)^{\frac{1}{2}} - 1]. \end{aligned} \tag{4.14}$$

Since $0 < 1 - (n_r/n)^{\frac{1}{2}} < \eta$ for $n \varepsilon I_r$ and all large r, we conclude from (4.13), (4.14), the validity of (4.3) with I_r replaced by $\{n_r\}$, and the fourth line of (4.1) (which implies $\eta < \eta^{\frac{1}{2}}$) that (4.4) holds.

We turn to the proof of (4.5). If $\bar{D}_r$ occurs, let ν be the first integer $n \varepsilon I_r$ for which the inequality defining D_r fails to hold, and let Z be the smallest positive value for which

$$(\nu - n_r)T_{n_r,\nu}(3Zh_{n_r}) \geqq 2^{-1}\epsilon n_r q_{n_r}. \tag{4.15}$$

Now, for all large r, uniformly in $n \varepsilon I_r$ and $z \varepsilon (0, 1]$,

$$\begin{aligned} P\{(n_{r+1} - n)T_{n,n_{r+1}}(3zh_{n_r}) < -4^{-1}\epsilon n_r q_{n_r} \mid \bar{D}_r, \nu = n, Z = z\} \\ = B^*(-4^{-1}\epsilon n_r q_{n_r}-, (n_{r+1} - n), 3zh_{n_r}) \leqq \tfrac{1}{4}, \end{aligned} \tag{4.16}$$

by Chebyshev's inequality and the fact that the arguments of B^* in (4.16) satisfy $\Delta^2/N\pi \to \infty$ with r, uniformly. (If $n = n_{r+1}$, the probability in (4.16) is zero.) Consideration of (4.15) together with the event complementary to that of (4.16) yields, for large r,

$$P\{(n_{r+1} - n_r)T^+_{n_r,n_{r+1}}(3h_{n_r}) > 4^{-1}\epsilon n_r q_{n_r} \mid \bar{D}_r\} \geqq \tfrac{3}{4} \tag{4.17}$$

and hence, first using the familiar argument used to prove (2.10), and then using the first line of (2.10) itself with $\delta = \frac{1}{2}$,

$$\begin{aligned} \log P\{\bar{D}_r\} &\leqq \log [\tfrac{4}{3}P\{(n_{r+1} - n_r)T^+_{n_r,n_{r+1}}(3h_{n_r}) > 4^{-1}\epsilon n_r q_{n_r}\}] \\ &\leqq \log [(16/9)\{1 - B^*(8^{-1}\epsilon n_r q_{n_r}, n_{r+1} - n_r, 3h_{n_r})\}] \\ &\sim -\epsilon^2\eta^{-1}2^{-5}3^{-2} \log r, \end{aligned} \tag{4.18}$$

the last coming from an application of (2.2). Thus, (4.5) follows from the first line of (4.1) and the Borel-Cantelli Lemma for $\{\bar{D}_r\}$.

To prove (4.6), we first note that for large r the events

$$Q_r = \{T^+_{n_r}(2^{-1}(1 + 8\eta^{\frac{1}{2}})\sigma_p n_r^{-\frac{1}{2}}(\log\log n_r)^{\frac{1}{2}}) \leqq q_{n_r}\} \tag{4.19}$$

satisfy, by (2.10) with $\delta = .01$ and by (2.2),

$$\log P\{\bar{Q}_r\} < -(.98)2^{\frac{1}{2}}3^{-\frac{1}{2}}(1 + 8\eta^{\frac{1}{2}})^{-1} \log r, \tag{4.20}$$

so that the second line of (4.1) and the Borel-Cantelli Lemma for $\{\bar{Q}_r\}$ yield

$$Q_r \quad \text{occurs for a.a.r, wp 1.} \tag{4.21}$$

Note that $E_r \cap H_r$ entails $|K_n| < 2^{-1}(1 + 8\eta^{\frac{1}{2}})\sigma_p(\log\log n_r)^{\frac{1}{2}}$ for $n \varepsilon I_r$, so that by (1.12) and $n^{-\frac{1}{2}} < n_r^{-\frac{1}{2}}$ we see that $E_r \cap H_r \cap A_n(\lambda)$ entails either

$$T_n^+ (2^{-1}(1 + 8\eta^{\frac{1}{2}})\sigma_p n_r^{-\frac{1}{2}}(\log\log n_r)^{\frac{1}{2}}) > \lambda q_n \tag{4.22}$$

or else (if $0 < K_n \leqq \lambda n^{-\frac{1}{2}} q_n$)

$$T_n^+(-\lambda q_n) > \lambda q_n\,. \tag{4.23}$$

However, $Q_r \cap D_r$ entails, for $n \varepsilon I_r$,

$$T_n^+(2^{-1}(1 + 8\eta^{\frac{1}{2}})\sigma_p n_r^{-\frac{1}{2}}(\log\log n_r)^{\frac{1}{2}}) < (1 + \epsilon/2)(n_r/n)q_{n_r} < \lambda q_n\,, \tag{4.24}$$

so that (4.5) and (4.21) imply that, wp 1, the event (4.22) occurs for only finitely many n. The event (4.23), although it involves a negative argument, is even easier to handle since q_n is small compared with the argument of T_n^+ in (4.22); for example, one can consider the random variables $1 - X_i$, as described below (1.14), and use the obvious analogues of (4.5) and (4.19); we conclude that, wp 1, (4.23) also occurs for only finitely many n. This last is thus true of $E_r \cap H_r \cap A_n(\lambda)$, and by (4.4) this yields (4.6).

As for (4.7), by the classical form of (2.1) for $1 - B^*$ (described in the second paragraph below (2.2)) with $\pi_N \equiv p$, $N = n_r$, $\Delta_N = L\sigma_p(n_r \log\log n_r)^{\frac{1}{2}}[1 - \eta^{\frac{1}{2}}]$, $\zeta + 1 = [1 + \eta^{\frac{1}{2}}]/[1 - \eta^{\frac{1}{2}}]$, we have

$$\log P\{B_{L,r}\} \sim -2^{-1}L^2(1 - \eta^{\frac{1}{2}})^2 \log r. \tag{4.25}$$

On the other hand, uniformly for $|1 - k/L\sigma_p(\log\log n_r)^{\frac{1}{2}}| < \eta^{\frac{1}{2}}$, we have, from (2.2) and the third line of (2.10) with $1 - \delta = (1 + \epsilon/4)/(1 + \epsilon/2)$, for r large,

$$\begin{aligned} &\log P\{C_{L,r} \mid Kn_r = k\} \\ &\quad \leqq \log\left[\tfrac{4}{3}\{1 - B^*((1 + \epsilon/4)n_r q_{n_r}, n_r(1 - p) + n_r^{\frac{1}{2}}k,\right. \\ &\qquad \left.(1 - p)^{-1}(1 + 10\eta^{\frac{1}{2}})(\tfrac{3}{2})^{\frac{1}{2}}Lh_{n_r})\}\right] \\ &\quad \sim -L^{-1}(1 + 10\eta^{\frac{1}{2}})^{-1}(1 + \epsilon/4)^2(\tfrac{2}{3})^{\frac{1}{2}} \log r. \end{aligned} \tag{4.26}$$

Hence, writing $\bar{L} = L(1 - \eta^{\frac{1}{2}})$ and $\psi^{\frac{3}{2}} = (1 + \epsilon/2)(1 - \eta^{\frac{1}{2}})(1 + 10\eta^{\frac{1}{2}})^{-1}$, so that $\psi > 1$ by the third line of (4.1), we have, for large r,

$$\begin{aligned} -\log P\{B_{L,R} \cap C_{L,R}\}/\log r &> 2^{-1}\bar{L}^2 + (2\psi/3)^{\frac{1}{2}}\bar{L}^{-1} \\ &\geqq \min_{\alpha>0}\,[2^{-1}\alpha^2 + (2\psi/3)^{\frac{1}{2}}\alpha^{-1}] = \psi > 1. \end{aligned} \tag{4.27}$$

The Borel-Cantelli Lemma now yields (4.7).

Turning to (4.8), we see that $B_{L,r} \cap H_r$ entails, for $n \varepsilon I_r$,

$$(4.28)\qquad n^{-\frac{1}{2}}K_n < n^{-\frac{1}{2}}\sigma_p(\log\log n_r)^{\frac{1}{2}}[(1+\eta^{\frac{1}{2}})L + 4\eta^{\frac{1}{2}}] < (1+10\eta^{\frac{1}{2}})2^{-\frac{1}{2}}3^{\frac{1}{2}}Lh_{n_r},$$

since $n > n_r$ and $L \geqq \frac{1}{2}$; and similarly, it entails, for large r,

$$(4.29)\qquad n^{\frac{1}{2}}K_n - \lambda q_n > \gamma^{-1}(1-10\eta^{\frac{1}{2}})2^{-\frac{1}{2}}3^{\frac{1}{2}}Lh_{n_r} - \lambda q_n > 0$$

by the fourth line of (4.2) and the fact that $q_n = o(h_{n_r})$. Hence, (1.12), (4.28), and (4.29) yield (4.8).

As for (4.9), since $3 > (1 + 10\eta^{\frac{1}{2}})2^{-\frac{1}{2}}3^{\frac{1}{2}}L$ for $L \leqq 2 + 4\eta^{\frac{1}{2}}$ (by the fourth line of (4.1)), the occurrence of $D_r \cap C'_{L,n}$ for some $n \varepsilon I_r$ entails

$$(4.30)\qquad \begin{aligned} n_r T^+_{n_r}((1+10\eta^{\frac{1}{2}})2^{-\frac{1}{2}}3^{\frac{1}{2}}Lh_{n_r}) &> n\lambda q_n - 2^{-1}\epsilon n_r q_{n_r} \\ &> (1+\epsilon/2)n_r q_{n_r}, \end{aligned}$$

for large r, since nq_n is increasing for $n \geqq 3$. This is (4.9).

Finally, if $n \varepsilon I_r$, then $A_n(\lambda) \cap \bar{E}_{n_r} \cap H_r$ implies $K_{n_r} \geqq 2^{-1}\sigma_p(\log\log n_r)^{\frac{1}{2}}$, since $4\eta^{\frac{1}{2}} < 2^{-1}$by the last line of (4.1); on the other hand, $J_r \cap H_r$ implies $K_{n_r} < (2 + 4\eta^{\frac{1}{2}})\sigma_p(\log\log n_r)^{\frac{1}{2}}$. This proves (4.10)and completes the proof of the second half of (1.14).

5. Other proofs. We turn first to the proof of Theorem 3. The lower class result is of course implied by Theorem 1. We shall not prove the upper class result in detail, since the proof is very much like that of Section 4 for Theorem 1; instead, we shall merely indicate why only minor modifications in the latter proof are needed. As before, we assume the X_i are uniformly distributed. Denote by $K_n(p)$ and $T_n(x, p)$ the random variables defined in (1.5) and (1.11). Then, clearly,

$$(5.1)\qquad \begin{aligned} K_n(p + n^{-\frac{1}{2}}t) &= K_n(p) - n^{\frac{1}{2}}T_n(n^{-\frac{1}{2}}t, p), \\ T_n(x, p + n^{-\frac{1}{2}}t) &= T_n(x + n^{-\frac{1}{2}}t, p) - T_n(n^{-\frac{1}{2}}t, p) \end{aligned}$$

Hence, the second half of (1.12) implies that $R_n(p + n^{-\frac{1}{2}}t) < -\lambda q_n$ if and only if (except for a set of zero probability)

$$(5.2)\qquad \begin{aligned} \lambda q_n &\leqq T_n(n^{-\frac{1}{2}}K_n(p + n^{-\frac{1}{2}}t) - \lambda q_n, p + n^{-\frac{1}{2}}t) \\ &= T_n(n^{-\frac{1}{2}}t + n^{-\frac{1}{2}}K_n(p) - T_n(n^{-\frac{1}{2}}t, p) - \lambda q_n, p) - T_n(n^{-\frac{1}{2}}t, p). \end{aligned}$$

Now, the techniques used earlier can easily be employed to prove that, for $c > (2\tau)^{\frac{1}{2}}$,

$$(5.3)\qquad P\{T_n^*(n^{-\frac{1}{2}}\tau) > cn^{-3/4}(\log\log n)^{\frac{1}{2}} \text{ i.o.}\} = 0.$$

Since $n^{-3/4}(\log\log n)^{\frac{1}{2}} = o(q_n)$, the proof that, for $\lambda > 1$, wp 1 there are only finitely many n for which (5.2) occurs for some t (depending on n and the sample sequence) in $[0, \tau]$ essentially reduces to the upper class proof for Theorem 1, if one takes note of the fact that it is a *uniform* behavior of T_n over an interval (e.g., in (4.7)) which is actually proved in Section 4, and not just (1.12). In fact, it is not very difficult to see that one can even replace τ in the above by any nondecreasing sequence $\{\tau_n\}$ satisfying

$$\tau_n = o((\log \log n)^{\frac{1}{2}}), \tag{5.4}$$

since (5.3) is then still satisfied with c replaced by $2\tau_n^{\frac{1}{2}}$, and one thus obtains $T_n^*(n^{-\frac{1}{2}}\tau_n) = o(q_n)$ wp 1, as well as $t = o(K_n)$ for the crucial range of K_n in (5.2); thus, $T_n(n^{-\frac{1}{2}}t, p)$ can still be added to both the first and last expression of (5.2) without appreciably changing the expression λq_n on the left. Negative values of $K_n(p + n^{-\frac{1}{2}}t)$ or of t are treated similarly.

We turn now to the proof of Theorem 2. As before, by Lemma 1 we can and do assume the X_i uniform. For fixed value k_n of $K_n \equiv K_n(p)$, we shall for brevity treat in detail only the case $m = 2$, $t_1 = 0$, $t_2 > 0$, and write $|k_n|\, t_2 = s$ (in the notation just above (1.7)); it will be clear how larger values of m and arbitrary values of the t_i can be treated by a repetition of our steps. We shall also consider only the case $k_n > 0$, since the complementary case is handled in the same way.

Let u_0 and u_1 be fixed real numbers. By (1.12) and (5.2) with $-\lambda q_n$ replaced by $n^{-3/4}u_1$, we see that, given that $K_n(p) = k_n$, the event $U_n(s) < u_1$ is equivalent to

$$T_n(n^{-\frac{1}{2}}(s + k_n) - T_n(n^{-\frac{1}{2}}s, p) + n^{-3/4}u_1, p) - T_n(n^{-\frac{1}{2}}s, p) > -n^{-3/4}u_1. \tag{5.5}$$

The event $U_n(0) < u_0$ is of course

$$T_n(n^{-\frac{1}{2}}k_n + n^{-3/4}u_0, p) > -n^{-3/4}u_0. \tag{5.6}$$

Using our earlier techniques (as in the reduction of (5.2) using (5.3)), we see easily that, the probability limits being conditional on $K_n = k_n$,

$$\begin{aligned} &\operatorname{plim}_{n\to\infty} n^{3/4}[T_n(n^{-\frac{1}{2}}(s + k_n) - T_n(n^{-\frac{1}{2}}s, p) + n^{-3/4}u_1, p) \\ &\qquad - T_n(n^{-\frac{1}{2}}(s + k_n), p)] = 0, \\ &\operatorname{plim}_{n\to\infty} n^{3/4}[T_n(n^{-\frac{1}{2}}k_n + n^{-3/4}u_0, p) - T_n(n^{-\frac{1}{2}}k_n, p)] = 0, \end{aligned} \tag{5.7}$$

uniformly for $b^{-1} < k_n < b$, for each $b > 0$. Hence, the limiting conditional probability of the event $\{U_n(s) < u_1, U_n(0) < u_0\}$ is that of the inequalities

$$\begin{aligned} &-n^{3/4}[T_n(n^{-\frac{1}{2}}(s + k_n), p) - T_n(n^{-\frac{1}{2}}s, p)] < u_1, \\ &-n^{3/4}T_n(n^{-\frac{1}{2}}k_n, p) < u_0, \end{aligned} \tag{5.8}$$

providing these last limiting probabilities are continuous in u_0, u_1, which will turn out to be the case. But the two random variables on the left sides of the inequalities (5.8) are, by the Central Limit Theorem, asymptotically conditionally jointly normal (uniformly in $b^{-1} < k_n < b$) with means 0, variances k_n, and covariance max $(k_n - s, 0)$. Thus, $U_n(0)/k_n^{\frac{1}{2}}$ and $U_n(k_n t_2)/k_n^{\frac{1}{2}}$ have conditional limiting law equal to that described for $J(0)$ and $J(t_2)$ above (1.7). The general approach of finite-dimensional conditional laws of the U_n-process to the finite-dimensional laws of the J-process is obtained similarly.

Next, we show that, for each $c > 0$,

$$(5.9)\quad \lim_{\epsilon\to 0}\overline{\lim}_{n\to\infty}\sup_{|p'-p|\leqq \tau n^{-1/2}}\epsilon^{-1}P\{\sup_{|t|<\epsilon}n^{3/4}|R_n(p'+n^{-\frac{1}{2}}t) - R_n(p')| > c\} = 0.$$

From this it follows that, writing $B_\epsilon = \{j:|j| \leqq \tau/\epsilon, j = \text{integer}\}$, by choosing ϵ sufficiently small one can make

$$(5.10)\quad P\{\sup_{|t|\leqq\tau} \pm R_n(p + n^{-\frac{1}{2}}\tau) - \sup_{j\epsilon B_\epsilon} \pm R_n(p + n^{-\frac{1}{2}}j\epsilon) > n^{-3/4}c\}$$

as close to zero as desired for all large n. Since B_ϵ is finite, the proof that $\sup_{|t|\leqq\tau} U_n(t) \to \sup_{|t|\leqq\tau} J^*(t)$ in law follows at once from the convergence of finite-dimensional laws proved above, in the same manner that the analogous sample df deviation results follow from such convergence and the smallness of the analogue of (5.10) in [3] and [8]. Convergence in law of $f(U_n)$ to $f(J^*)$ for functionals f on the space of functions on $[-\tau, \tau]$ continuous except for finitely many finite jumps, and such that f is continuous in the uniform topology wp 1 according to the law of J^*, is then proved in the same way by approximating such functionals in the manner of Donsker [3], page 281, so that the result for the approximating functional can be obtained from the result for $\sup_{\tau\epsilon L} \pm U_n(t)$ for various intervals L.

By (1.1), relation (5.9) will be proved if we prove each of the two statements obtained from it by replacing $R_n(x)$ by (i) $S_n(x) - x$ and by (ii) $Y_{x,n} - x$. For (i), we thus consider

$$(5.11)\quad P\{T_n^*(\epsilon n^{-\frac{1}{2}}, p') > n^{-3/4}c\}.$$

(The corresponding expression with ϵ replaced by $-\epsilon$ is treated similarly.) As $n \to \infty$, the expression $[M/a(1 + w)]^{\frac{1}{2}}w(\delta - a)$ of (2.7), with $\delta = \frac{1}{2}$, approaches $c/2\epsilon^{\frac{1}{2}}$. Hence for all $\epsilon < c^2/20$ we can apply the first two lines of (2.10) and obtain, for all $n > N_0$ (where N_0 is independent of ϵ as long as $\epsilon < c^2/20$), that the expression (5.11) is no greater than

$$(5.12)\quad \begin{aligned} &\tfrac{4}{3}P\{|T_n(\epsilon n^{-\frac{1}{2}}, p')| > n^{-3/4}c\} \\ &\quad = \tfrac{4}{3}\{B^*(-n^{\frac{1}{2}}c-, n, \epsilon n^{-\frac{1}{2}}) + 1 - B^*(n^{\frac{1}{2}}c, n, \epsilon n^{-\frac{1}{2}})\} \\ &\quad \sim \tfrac{8}{3}\Phi(-c/\epsilon^{\frac{1}{2}}) \end{aligned}$$

uniformly for p' in any closed interval excluding 0 and 1, as $n \to \infty$, this last by the Central Limit Theorem with error term. Since $\lim_{\epsilon\to 0} \epsilon^{-1}\Phi(-c/\epsilon^{\frac{1}{2}}) = 0$, the result for (i) is complete.

As for (ii), consider first the expression

$$(5.13)\quad P\{\sup_{0<t<\epsilon}|Y_{p'+n^{-\frac{1}{2}}t,n} - Y_{p',n}| > n^{-3/4}c \mid Y_{p',n} = y, S_n(y) = p + n^{-1}\theta\},$$

where $0 \leqq \theta < 1$ accounts for excess of $S_n(y)$ over p due to the jump at y. A moment's reflection shows that under the conditioning of (5.13), the event

$$(5.14)\quad \{T_n^*(n^{-\frac{1}{2}}\epsilon + n^{-3/4}c, y) < n^{-3/4}c - n^{-1}\}$$

entails the complement of the main event in (5.13). The event (5.14) is very similar to the complement of the event of (5.11), except that we must now compute the conditional probability under the condition of (5.13), for y in a small neighborhood of p' of probability approaching 1 with n, e.g., for $|y - p'| < n^{-\frac{1}{4}}$. One obtains without difficulty, for the complement of (5.14), an analogue of (5.12). One must then consider (5.13) with $-t$ replacing t, and there is no essential difference. Thus, (5.9) is established.

Finally we show that the J^* sample paths are, wp 1, Lip α for all $\alpha < \frac{1}{2}$ by showing this for the paths of the J-process. From (1.7) we have $J(t + h) - J(t)$ distributed as $N(0, 2|h|)$, so that for m a positive integer we have

(5.15) $$E|X(t + h) - X(t)|^{2m} = (2m)!\,|h|^m/m!$$

From Loève (1960), page 519, we conclude that the J-paths are, wp 1, Lip $(m - 1)/2m$ for every $m > 0$. This completes the proof of Theorem 2. See also Remark 5 of Section 6.

6. Further results and remarks:

1. In view of the asymptotic normality of $U_n/|K_n|^{\frac{1}{2}}$ discussed below (1.5) (or that of the process $U(|K_n|\,t)/|K_n|^{\frac{1}{2}}$ discussed just above (1.7), or its more symmetric variant $U(|K_n(p + n^{-\frac{1}{2}}t)|\,t)/|K_n(p + n^{-\frac{1}{2}}t)|^{\frac{1}{2}})$, one may be led to inquire about the analogue of Theorem 1 or 3 for this process. The process resulting from this normalization of U by division by the chance variable $|K_n|^{\frac{1}{2}}$ seems much less natural and interesting to the author than does Bahadur's U. Much of the technique of Section 3 and 4 can still be applied to the altered problem; the values of K_n near 0, and their oscillation with n, now cause extra difficulties.

2. Of much greater interest than the result of Theorem 3 would be the analogue for $\sup_{0<p<1} \pm R_n(p)$, for simplicity when the X_i are uniform (or for $\sup_{-p<t<1-p} U_n(n^{\frac{1}{2}}t)$, to which one can apply an analogue of Lemma 1 if $\inf_{x\varepsilon J} F'(x) > 0$ on an interval J for which $P\{X_1 \varepsilon J\} = 1$). Here the methods we have used herein do not even yield a weak law, analogous to $\sup_{|t|\leq r} [\pm U_n(t)]$ approaching in law $\sup_{|t|\leq r} [\pm J^*(t)]$. (In particular, if $\sup_{0<p<1} c_n R_n(p) = O(1)$ in probability, then a separable process whose *finite*-dimensional distributions are the limiting ones for $c_n R_n(\cdot)$, is the process which has sample function identically 0, wp 1.) Some bounds on the law of $\sup_p \pm R_n(p)$ have, however, been obtained. For example it is exactly of order $(\log n)^{\frac{1}{2}}$ in probability. We shall return to this topic in another paper.

Much simpler is the consideration of $R_n(p)$ or the U_n processes corresponding to a fixed finite collection of values p, which are seen from Section 1 to be asymptotically independent. Moreover, Theorem 3 immediately yields its analogues for maxima of these quantities.

3. Bahadur has mentioned in [1] that the law of the iterated logarithm (L.I.L.) for sample quantiles follows at once from his estimate on R_n. In fact, much more is true: not only can one obtain in this way a classical form of the L.I.L. (Equation (4) of [1]), which can also be obtained directly but not as quickly by modifications of the standard proof for sums of Bernoulli random variables (as in [6]),

but also by using Bahadur's elegant device one can obtain at once the much more difficult strong form [5] from that which delimits the upper and lower classes for increasing sequences $\{c_n\}$ of positive values in the Bernoulli case as

$$(6.1)\quad \sum_n n^{-1}c_n e^{-\frac{1}{2}c_n^2} \begin{cases} = \infty \Leftrightarrow 1 = P\{\pm n^{\frac{1}{2}}(S_n(\xi_p) - p) > \sigma_p c_n \text{ i.o.}\}, \\ < \infty \Leftrightarrow 0 = P\{\pm n^{\frac{1}{2}}(S_n(\xi_p) - p) > \sigma_p c_n \text{ i.o.}\}. \end{cases}$$

For, writing $c_n' = \min(c_n, 10(\log\log n)^{\frac{1}{2}})$, one has $\{c_n'\}$ monotone and (by the ordinary L.I.L. for $S_n(\xi_p)$) in the same class as $\{c_n\}$. By Feller [5], $c_n' \pm 1/c_n'$ is in the same class as $\{c_n'\}$, as is therefore $\{c_n' \pm \frac{1}{10}(\log\log n)^{-\frac{1}{4}}\}$ and hence (again by the ordinary L.I.L. for $S_n(\xi_p)$) $\{c_n \pm \frac{1}{10}(\log\log n)^{-\frac{1}{4}}\}$. But (1.2) or (1.3) yields $|n^{\frac{1}{2}}F'(\xi_p)R_n(p)| < \frac{1}{10}(\log\log n)^{-\frac{1}{4}}$ for almost all n, wp 1. Hence, we obtain

THEOREM 4. *For $\{c_n\}$ positive and nondecreasing,*

$$(6.2)\quad \sum_n n^{-1}c_n e^{-\frac{1}{2}c_n^2} \begin{cases} = \infty \Leftrightarrow 1 = P\{\pm n^{\frac{1}{2}}F'(\xi_p)(Y_{p,n} - \xi_p) > \sigma_p c_n \text{ i.o.}\}, \\ < \infty \Leftrightarrow 0 = P\{\pm n^{\frac{1}{2}}F'(\xi_p)(Y_{p,n} - \xi_p) > \sigma_p c_n \text{ i.o.}\}. \end{cases}$$

4. At present a strong form of Theorem 1, analogous to (6.2), is unknown. The methods used herein give reasonably sharp probabilistic bounds of the type usually required, but the approach of [5] for cumulative sums, especially in the case of the lower class, will require some delicacy to be carried over to the present problem.

5. The conclusion of Theorem 2 regarding J^*-paths being Lip α for all $\alpha < \frac{1}{2}$ wp 1 suggests that one should be able to proceed in the elegant manner of Lamperti [9] to the convergence of Theorem 2 by working in the Lip α space, replacing S_n by the corresponding piecewise-linear continuous function S_n' (say) so that the resulting U_n is in Lip α. Unfortunately, this prescription cannot be carried out because Prokhorov's condition for the U_n to lie, with high probability in a common compact set in the Lip β/α space,

$$(6.3)\qquad E\,|U_n(p + n^{-\frac{1}{2}}t) - U_n(p)|^{\alpha} \leqq C\,|t|^{1+\beta}$$

with C, α, β *independent of* n, cannot be verified. This is true even if one attempts only the classical sample df results (such as Kolmogorov-Smirnov) which consider S_n rather than U_n in this manner.

Incidentally, replacing S_n by S_n' changes S_n by at most $1/n$ at each point and also, with high probability, changes $Y_{p,n}$ by little, since the largest sample spacing in the uniform case is well-known to be of order $n^{-1}\log n$ wp 1. Thus, Theorems 1, 2, and 3 still hold if S_n is replaced everywhere by S_n', and one sees also the validity of our remark in Section 1 concerning the irrelevancy to our results of the manner of definition of $Y_{p,n}$ when there is ambiguity due to np being an integer.

6. It would be interesting to investigate the efficacy of statistical procedures based on $R_n(p)$, $\sup_{0<p<1} R_n(p)$, and other variants of Bahadur's statistic; the "cancellation" of much of the information in $Y_{p,n}$ and $S_n(\xi_p)$, and the resulting smaller order of $R_n(p)$, make one wonder whether there are many meaningful

applications. As a trivial example, if $X_1, X_2, \cdots, X_n$ are independent and symmetrically distributed about 0, one could use $(\frac{1}{2} - S_n(0))/Y_{\frac{1}{2},n}$ to estimate the probability density function at 0, but this is not very efficient. See Section 3 of [1a] for Bahadur's use of $R_n(p)$ to reduce asymptotic consideration of certain nonlinear statistics to that of linear ones.

7. Let $\eta^{(1)}(n) = n(p - S_n(\xi_p))/\sigma_p$ and $\eta^{(2)}(n) = nF'(\xi_p)(Y_{p,n} - \xi_p)/\sigma_p$, and extend these to $\eta^{(i)}(t)$ for positive real t by linear interpolation between successive integer arguments. Write $\eta_n^{(i)}(t) = (2 \log \log n)^{-1/2}\eta^{(i)}(nt)$. The beautiful results of Strassen [11] include $\eta^{(1)}$ as a special case. Since our Theorem 1 or (1.2) implies that

$$\lim_{T\to\infty}(2T \log \log T)^{-1/2}\sup_{t\leq T} | \eta^{(1)}(t) - \eta^{(2)}(t) | = 0 \text{ wp } 1,$$

we obtain

THEOREM 5. *Strassen's Theorems* 2 *and* 3 *hold with* $\eta^{(1)}$ *(his* η*) replaced by* $\eta^{(2)}$. *Consequently, the results of Section* 3 *of* [11] *hold with* $\eta^{(1)}(n)$ *(his* S_n*) replaced by* $\eta^{(2)}(n)$.

The second part of Theorem 5 follows from the way in which Strassen's Theorems 2 and 3 and elementary integrability or moment considerations are used in the proofs of his Section 3. In particular, his paragraphs (i)–(vi), the Corollary to his Theorem 3, and his equations (3), (4), (11) are valid for $\eta^{(2)}(n)$, as are any other results corresponding to ones for $\eta^{(1)}(n)$ obtained from his Theorems 2 and 3 by such methods. Bahadur has informed the author that Bickel independently suggested that *direct* study of the $\eta^{(2)}(n)$ process might yield such results as the L.I.L. for $Y_{p,n}$. I have not considered here direct study of the $\eta^{(2)}$-process starting from scratch, since the indirect approach using the labors of Bahadur and Strassen yields our Theorem 5 so easily, and since results like those of [11] have not yielded the strong form (6.1)–(6.2) of the L.I.L. One can also obtain most of the analogues of Strassen's Section 3 by applying Bahadur's bound directly there without recourse to Theorems 2 and 3 of [11].

8. The author's attention has recently been called to an abstract of Eicker [4] concerning Bahadur's bound (1.2) and the fact that it cannot be improved using only the Borel-Cantelli Lemma as in [1] (to be contrasted with the present approach which yields (1.3) by using the n_r). Eicker states that he conditions on the value of $Y_{p,n}$ rather than on K_n or $S_n(\xi_p)$ as herein, but it seems unlikely that one would achieve any shortening of the present proof from the use of that conditioning.

The author is indebted to Roger Farrell for helpful discussions.

REFERENCES

[1] BAHADUR, R. R. (1966). "A note on quantiles in large samples". *Ann. Math. Stat.* **37** 577–580.

[1a] BAHADUR, R. R. (1967). "Rates of convergence of estimates and test statistics". *Ann. Math. Stat.* **38** 303–324.

[2] CHUNG, K. L. (1949). "An estimate concerning the Kolmogoroff limit distribution". *Trans. Amer. Math. Soc.* **67** 36–50.

[3] DONSKER, M. D. (1952). "Justification and extension of Doob's heuristic approach to the Kolmogorov-Smirnov Theorems". *Ann. Math. Stat.* **23** 277–281.

[4] EICKER, F. (1966). "On the asymptotic representation of sample quantiles" (abstract). *Ann. Math. Stat.* **37** 1425.

[5] FELLER, W. (1943). "The general form of the so-called law of the iterated logarithm". *Trans. Amer. Math. Soc.* **54** 373–402.

[6] FELLER, W. (1957). *An Introduction to Probability Theory and its Applications.* **1** (2nd edition). John Wiley & Sons, N. Y.

[7] KIEFER, J. (1961). "On large deviations of the empiric d.f. of vector chance variables and a law of the iterated logarithm". *Pac. Math.* **11** 649–660.

[8] KIEFER, J. and WOLFOWITZ, J. (1958). "On the deviations of the empiric distribution function of vector chance variables". *Trans. Amer. Math. Soc.* **87** 173–186.

[9] LAMPERTI, J. (1962). "On convergence of stochastic processes". *Trans. Amer. Math. Soc.* **104** 430–435.

[10] LOÈVE, M. (1960). *Probability Theory* (2nd edition). D. Van Nostrand Princeton.

[11] STRASSEN, V. (1964). "An invariance principle for the law of the iterated logarithm". *Z. Wahrscheinlichkeitstheorie und Verw. Gebiete* **3** 211–226.

STATISTICAL INFERENCE

CHAIRMAN: *G. A. Barnard*
UNIVERSITY OF ESSEX

PANEL: *J. C. Kiefer*
CORNELL UNIVERSITY

L. M. LeCam
UNIVERSITY OF CALIFORNIA, BERRELEY

L. J. Savage
YALE UNIVERSITY

J. C. Kiefer. Contrary to the expectations of Neyman-Pearson predictors or subjectivists in the audience, I'm not going to say anything about Bayesian inference. I think that what I would like to do this afternoon is to mention to you three problems of a fairly specific sort which I consider typical of the mathematical problems that are of theoretical interest in the future of statistics. I think it would be pretentious of me to tell you that these will definitely be important in the future or that they are, in any sense, representative of the other problems one could work on. I can just tell you that these are three specific areas which seem to me very challenging mathematically and which I hope also have some practical impact.

One area of statistical inference that has been of very great importance and has excited a tremendous amount of recent activity, especially by Erich Lehmann and some of his protégés, is in the area of certain nonparametric problems. To my way of thinking, there is to this time, from a theoretical point of view, very little in the way of a satisfactory theory of nonparametric inference. I'm not sure exactly what the right direction is in which to work. So far the attempt to find optimum procedures for particular nonparametric settings has been notoriously a failure, except in certain large sample problems. Recent work by Huber on robustness is an exception. People have worked a long time, for example, to try to find a test of the Kolmogorov-Smirnov variety, which in some minimax sense is very good for the two-sample problem, and it just seems to be a horrendous computation that none of us is able to do. So I think, without being specific, that there are a couple of directions in which one can still look regarding the foundations of nonparametric inference and why one uses procedures of certain types. One aspect of this, which I don't think is tremendously important in view of satisfactory empirical results but which is of some theoretical interest, is the basic rationale of using rank-order statistics, procedures associated with ranks. From the point of view of theoretical statistics, there is really very little justification for the general use of rank-order procedures, although we all know darned well that they behave very well in certain settings and will have increasing practical importance. And the reason that there's no theoretical justification, at least in small sample-size theories, is that the group of transformations that is present to reduce the data to the use of ranks is a very horrible group. From the point of view of the type of work Charles Stein has done in other settings, for this group of transformations, there's no reason to think that the invariant procedures, which in this case are the rank-order procedures, should turn out to be very good at all. Nevertheless, they do turn out clearly to have some good properties, and I think that it's an interesting little theoretical corner, to try to see why it is, that although the group is bad, the invariant procedures seem to be pretty good.

Something that I think is far more interesting in this first area I'm mentioning (nonparametric inference) has to do with Bayes's procedures. In the classical parametric case of regular estimation, there's a result I alluded to yesterday, the first example in this direction going back to von Mises in 1919; since then, Kolmogorov and LeCam and Wolfowitz have all done work on the subject. And the upshot is that, without dotting the *i*'s, if you have a smooth a priori distribution in a sufficiently regular setting, then the a posteriori distribution is asymptotically approximately normal around the maximum likelihood estimator, and in a sense, for large samples, this says that the maximum likelihood estimator or some other BAN estimator, is almost the only member of the complete class of procedures that you have to worry about. Now, in nonparametric inference, one would like to have a corresponding result, and the problem seems terrifically difficult. The person who knows most about this at the moment is David Freedman of Berkeley, and he has done some work on trying to see what happens when you put an a priori distribution on an infinite dimensional space of distributions in a nonparametric setting. And some really horrible things can happen. Without any question so fine as that of asymptotic normality or anything like that, it turns out that for most reasonable, widespread a priori distributions, when the sample size is large, the a posteriori distribution does not become concentrated in a neighborhood of the true distribution, so that even the basic consistency properties of Bayesian estimation which are attractive in the finite dimensional case are absent here. I think that that's really a challenging area, and if one could figure out what is going on there, maybe one could set the beginning of a good nonparametric theory, at least for large sample sizes.

Well, a second area of what seem to me important problems to work on has to do with the fact that we do have, in many settings, quite a good large-sample theory, but we don't know how large the sample sizes have to be for that theory to take hold. Now, I'm sure most of you are familiar with the error estimate one can give for the classical central-limit theorem, which goes by the name of the Berry-Esseen estimate, and which tells you that under certain assumptions one can actually give an explicit bound on the departure from the normal distribution of the sample mean for a given sample size, the error term being of order $1/\sqrt{n}$. For most other statistical problems, in fact for almost anything other than the use of the sample mean, we have nothing. The most obvious example of this (and this is not original with me; many people have been concerned with this), is the maximum likelihood estimator in the case of regular estimation. We all know what the asymptotic distribution is. Can you give explicitly some useful bound on the departure from the asymptotic normal distribution as a function of the sample size n? It seems to be a terrifically difficult problem. I might mention work on some problems of this variety, namely the case of a finite number of states of nature and a finite number of decisions, where for certain sequential settings one knows what asymptotically optimum procedures look like. (The first work on this asymptotic theory was due to Chernoff.) There has been some work by Lorden, who is at Northwestern, on the subject of getting explicit bounds in that case, that tell you for expected sample size fifty, one hundred, or something of the sort, if you use these asymptotically optimum stopping rules, how far from optimality you are for this finite expected sample size.

Finally, a third area that I think represents a good source of problems to work on goes under what I guess one could call by the name of "curve fitting." Now, to make matters simple, let me think of the case where we have polynomial regression in one variable, even though we all know that there may not be many settings where that's realistic, and suppose that one knows he wants to fit a polynomial of, at most, degree fifty. How do you choose what polynomial to fit? Well, there are various *ad hoc* procedures and, in fact, various computer programs that people have used to try to select a procedure for fitting the polynomial for a problem like this, but without what is a satisfactory theoretical background, in the way of an estimate of the error that would be satisfactory from, well, you Bayesians should pardon the expression, the Neyman-Pearson point of view. So it would be very nice to give a justified procedure, in that case, which might tell one, "Here is the data. We estimate this polynomial to be $3x^4 + 12x^{18} + x^{43}$." Now, it seems to me that a reasonable formulation of this problem from the theoretical point of view, and not unreasonable for giving practical insight, is the following: The error that one incurs should consist of two parts. One is an error due to inaccuracy in the form of your estimate. If we estimate the polynomial to be that, and if it's something else, so that there may be bias terms and what-not, then something like the integral or maximum of the mean-squared error might be a reasonable measure of our inaccuracy. But there's an additional term which is often not included, and that is payment for using a complicated function. As a first naive example of what one might do to this integrated or maximum mean-squared error of the estimator, one could add the number of nonzero coefficients you use in your estimating polynomial. So, of course, you can estimate more accurately if you use ten coefficients instead of three, but if the additional accuracy that you get is, so to speak, not worthwhile, then maybe you should use something with only three nonzero coefficients instead. The only theoretical work in this direction I know of is the work that Ted Anderson did a few years ago, which is really devoted to the question of what the highest degree coefficient is that you should keep. So in a problem like this, he would go down starting with the x^{50} coefficient, then the x^{49}. If he kept the x^{48} coefficient, then he would keep everything of lower degree, so that isn't really the problem at hand, and the problem of interest is really to select the *subset* of nonzero coefficients which is appropriate. What would be even nicer, I think, is that someone should eventually design on theoretical grounds, and maybe design a good computational scheme for doing this, a procedure that looks at combinations of a whole bunch of elementary functions, allowing simultaneously polynomials, exponentials, sines, and cosines, and see how few of these one can select and still do some sort of business. (Someday one would even aim at nonlinear regression.) Quite a few years ago, Neyman was interested in doing some work in this direction, and in the direction of smooth curve-fitting. But theoretically, it seems to be a very complicated problem, how to choose exactly the functions that go into your estimator. Well, I think that I will stop with this, and hope that I make my plane as a consequence.

L. M. LeCam. We came here to discuss problems which have to do with statistical inference, and also the future of statistics. Unfortunately, I gave a lot of thought to the problem, and I have strictly no idea what is in store for the future of statistics. The more I think about it, the less I know what inference means. All I will be able to say is something about some of the stuff I'm going to try to do in the next year or the following years, and even that may not turn out to be correct, because beyond the next year things will change.

About the problems of inference, since I presume that many of you expect that we would disagree on Bayesian statistics, I would like to state that I disagree with Bayesian statistics, and that I will devote some time in the future, if I can, trying to develop something which is more suitable. The point on which I disagree is mostly this: Bayesian statistics is intended for personal views, it does not provide any way of scientific discourse. For instance, all the pieces of equipment that statisticians talk about, experiments, models, and so forth, are somewhat irrelevant to the Bayesian. There is nothing in the Bayesian philosophy which allows you to say that a model is wrong, because it is something you think—you think about the probabilities, so you know them. I intend to work on that subject and work on trying to formalize some notion of concept of information, on how it is transmitted, some concept of experiment. I'm sure I will not be successful in the near future. One of the main reasons is that it is my feeling that, for instance, the logic which underlies statements about nature is not a Booleian logic. It is not even a modular logic, so that probability cannot cope with it. But this is all I have to say about inference. I might be totally mistaken about it, but that's my feeling.

Another area where I'm going to devote quite a bit of time, and with more hope of success, is an area that comes to me directly from application. In almost all the problems I had to deal with, which were either brought to me directly from applications or transmitted by my colleagues, the main difficulty was always to specify what the stochastic structure of the system was. If one tried to be realistic about the system, this structure was usually extremely complicated. I don't remember any particular case where I had to deal with a normal distribution. There was no such thing available. It was always much more complicated. So facing a problem which presumably has been stated in some form by thinking about it, trying to figure out what kind of stochastic structure would describe what happens there, you have some family, let's say, to make it simple, some family of densities which depend on some parameters. If you want to make it nonparametric, I'm willing to let the parameter to be in any kind of infinite dimensional space or whatever—it doesn't matter; but let's take some simple cases. You formulate it in terms of two, three, or four parameters. The densities may happen to be extremely complicated as a function of the parameters. As a function of the observations, it does not usually matter much what they are, except for the specific computing practices, which mean that if you have to use a Turing machine to compute the value of a particular real number, this might take you a few centuries, that's a different problem. But the way the density depends on the parameters seems to be very important. I have tried in the past few years and will keep trying to find ways of expressing approximations to those families of densities. So that the problem will not be of trying to find out what is the approximate distribution of some maximum likelihood estimates, for instance. But on the contrary, to try to find some family which is tractable—let's say an exponential family of densities (that's not always tractable, but it looks more accessible to the theoretician)—and try to see whether that family can be used as an approximation to the other one, the one that's presumed to be true. Approximation would be in the sense that if you derive from the approximation family the optimal procedure, or something approximately optimal there, that procedure will also be almost optimal for the true family, and conversely. As far as making derivations of optimality or finding out what the behavior of certain estimates is, the approximate family is a good replacement for the other. One can make that more formal; I had tried at one time to introduce a distance between experiments. An experiment consists, mostly, of a set, which is the set of possible results: a family of sets which are the ones for which you can determine whether the events they represent happen or not, and a family of probability measures on those sets. You can define a distance between two such experiments by just modifying slightly the definition of more informative experiments that was introduced by Blackwell and other people at the Rand Corporation. One can show that, for instance, if you have an experiment in which the parameter can take only a finite number of values at a time, in order that two experiments be close to each other in the sense of that distance, it's necessary and sufficient that the distribution of the likelihood ratios be close to each other in a suitable sense. There, again, there is a question of finding bounds, but it looks almost feasible for small dimensions. However, most of the models that one wants to look at are infinite-dimensional in the sense that the parameter can take an infinite number of values. Whether it is satisfactory to try to approximate that by a finite-dimensional model, finite number of values for the parameter, I do not know. I do not even know how to formulate that. This depends on the formulation of what is an adequate representation of the universe, and I don't know what that means, and I don't know that anybody else knows. Nevertheless, for the cases where you have a parameter that is real or two-dimensional or something similar to that, one can look at the likelihood functions as a stochastic process, and one can try to express, in various ways, that that stochastic process, for the true family of distributions, is close to the stochastic process we have for the approximate family of distributions. This leads to complicated things in stochastic processes which I don't know how to solve, but I presume that other people will try to help. I had tried, in particular, to find out what are the conditions from such families of distributions, to be approximable by an exponential family. I found that under some restrictive conditions, one can give such conditions. However, I found to my dismay this year that, if I remove one of the restrictive conditions, I just don't know what I'm talking about anymore. I don't know what is approximation by an exponential family anymore. I had some correspondence with various people on that subject, and this is something that may be of interest to those of you who think that mathematics may not be too important. I don't understand the subject, so I don't know whether I will be able to do anything with it, but Professor Linnik of Leningrad wrote to me that he believes that one of his students has a way of attacking the problem through applications of some theorems in the theory of categories. If you have heard of categories and functors that's what that is, and if a minor mathematician would tell me that, I would just be very skeptical. If Linnik tells me that, I will try to look at it.

These problems of approximation lead also to other problems, which, in my opinion, are very relevant for many applications. One of them is a problem of stability. You might call it robust estimates and so forth, but I don't mean that exactly. I mean stability of the method by which the estimate is obtained. Everybody knows that maximum likelihood is a very good method of estimation, but it is totally unstable. If you modify the distribution extremely slightly, it jumps around; it does not know what it's doing anymore. The nonparametric methods do not have this feature, by and large, and there is a question of why they might still be good and why they are somewhat stable. There are problems which are related to problems of stability, the problems of invariance. There is a large literature on finding invariant estimates, invariant tests. But what is meant there, is that you have a specific problem, you have a group of transformations that leaves the problem invariant, and you use only tests which have that particular invariance property. I'm thinking more of the invariance of the method by which the test is obtained. The usual justification that you should use invariant tests because the result you get should not depend on whether you are in continental Europe and use centimeters or whether you are in the United States and use inches is just plain delusion. There is no such thing involved in the invariance principle. That statement applies to the invariance of the methods by which the optimal procedure is obtained, not to the invariance of the test or the particular estimate you have. So this is what I will try to work on in the next future. I don't know that many other people will find it important, but I find it challenging and hope that some other people will help.

L. J. Savage. As you can see, we made no early preparation for this panel, but a day or so ago agreed that each would give a short talk on the general theme of inference, with more or less special reference to the future of inference. As each of us is saying, in his own way, we realize that our assignment doesn't make us prophets. We don't know the future of statistics; all we can do is talk about what the idea of the future puts into our minds. Even perception of the present is gigantic. Whoever perceives the present very well will be seeing about as far into the future as there is any hope of seeing. I feel license here to speak speculatively and informally; for it is not irresponsible to speculate in a panel like this. The object is to help start a conversation which will go on in this room and presumably somewhat later, not to confine ourselves to indubitable facts.

One theme that occurred to me in thinking about inference is that inference is suggestive of forming opinions or reaching decisions in the light of data—reaching conclusions, I meant to say. But there's a moral in my slip of the tongue, because data has many effects on us that aren't necessarily perfectly portrayed by the idea of coming to a conclusion or justifying a conclusion. Data may give us insight or understanding or bright ideas, and little attention has yet been given to these other psycholocgial or mental impacts of data, important though they are. It does not seem promising to set out directly to look for a formal theory of the ingenious, or a theory of how to feed ingenuity. How should data be set forth and analyzed so as to promote the ingenious reaction? Though, and indeed because, the subject seems to defy theory, it is worth much thought. Most theories of inference tend to stifle thinking about ingenuity and may indeed tend to stifle ingenuity itself. Recognition of this is one expression of the attitude conveyed by some of our brethren who are more empirical than thou and are always saying, " Look at the data." That is, their message seems to be, in part, " Break away from stereotyped theories that tend to stifle ingenious insights and do something else." Perhaps semi-formal progress in that direction is already coming.

Consider the invention of experiments. We statisticians are so used to teaching and studying the design of experiments that we forget what we may have first expected when we opened a book with such a title, namely, that it would be full of bright ideas for doing new kinds of experiments, and getting new bright ideas about experiments. We tend to teach, rather, a fairly well-developed theory for basically one kind of experiment, or some kind with a few accessories, whereas of course great progress in empirical science comes from somebody thinking up an experiment that goes in a different direction from those thought of before. Naturally that involves subject-matter knowledge, but I'm not sure that it's fruitful to draw a sharp line between statistical thinking and subject-matter thinking. To mention a related example, hypothesis-testing does not represent a stance conducive to the invention of hypotheses.

The themes of feeding ingenuity and evoking insight seem to lead to two sometimes discredited statistical activities. Thus, I'm rather surprised to see myself acquiring a great respect for descriptive statistics. Everybody knows that descriptive statistics is strictly for Psych 100, and yet—though maybe you haven't used just that word for it—many of the most tantalizing things in statistical work today could be called descriptive statistics. These are efforts to arrange and condense complicated bodies of data in ways that promise you a fighting chance to see what's essential. Factor analysis is one of the oldest theories of that sort. It has never been very popular with statisticians, in part because the inference problems associated with it are so repulsive. But the factor analysts have something to stand on when they say, "The data are often abundant. This way of organizing it may give us an insight, and we ought not to wait for somebody to tell us the small probabilities that this insight is due to a fortuitous configuration of the data." There are today, as you know, many exciting developments in factor analysis, cluster analysis, and other fancy kinds of scaling and fitting. These really, it seems to me, can be thought of as descriptive statistics, though of a sophisticated sort, yet not so sophisticated as to make very easy that kind of reliability analysis we have recently thought so necessary in statistical devices.

Not only does the search for insight encourage descriptive statistics, it also encourages the once cardinal sin of fooling around with the data. I don't imagine that anyone in this room will admit ever having taught that the way to do an experiment is first carefully to record the significance level then do the experiment, see if the significance level is attained, and if so, publish, and otherwise, perish. Yet, at one time we must have taught that; at any rate it has been extremely well learned in some quarters. And there is many a course outside of statistics departments today where the modern statistics of twenty or thirty years ago is taught in that rigid way. People think that's what they're suppose to do and are horribly embarrassed if they do something else, such as do the experiment, see what significance level would have been attained, and let other people know it. They do the better thing out of their good instincts, but think they're sinning and don't yet know that statistical theory has become much more tolerant.

But much further along the road is really fooling around with the data to see whether, looked at this way, that way, or the other way, it seems to spell "Merry Christmas." Of course, that really is dangerous. One can certainly bemuse himself, and one ought to be careful, whatever being careful means. At the same time there is increasing interest in, and respect for, fooling around and a recognition that we have often been too cautious in the recent past.

The areas for these activities of catching on, seeing inventive things, and so on, tend to be in places where there are very many parameters. If all that you are doing is weighing male and female robins selected at random, it will be rare that the data will present anything of much interest. The most you could have is two histograms, and there isn't ordinarily much richness of structure there. But we have always encountered, and perhaps now pay more attention to, problems in which there are ever so many parameters. Nonparametric inference is such a problem. Of course, it's called nonparametric inference because an early pioneer in the field wrote a paper called "Problems with too Goddamn Many Parameters." The press, being timid in those days, rendered this as "nonparametric inference." Recently a paper claiming to be about nonparametric inference was severely criticized by a referee, who said, "Of course, a problem in nonparametric inference is a problem with infinitely many parameters, and I don't see that many here."

Time series is another area where there are ever so many parameters. There is no such thing as imagining all the conceivable types of time series, or stochastic processes in time, that there could be. It's the sort of place where you want ingenuity all right, because sometimes a remarkable sort of time series could be present, and you would hope to know how to recognize that if you came upon it. Our tools here are still very, very primitive.

Another problem with too many parameters that fascinates me is the multiple questionnaire. If you've asked people dozens or hundreds of questions, how can the answers be organized? A naive way would be to report the frequency with which each combination comes forth, but if you just ask thirty dichotomous questions, then there are a billion possible combinations, so you might need a large sample indeed to get reliable frequencies in all the cells. Would you know what to do with them if you could get them?

A highly empirical kind of meteorology persists and seems likely to persist—no matter what advances theoretical meteorology makes. Like questionnaires, it involves very large amounts of data about each event. Similarly, the medical diagnostic problem tantalizes many people. Is there any hope of discovering interesting patterns of physical signs for prognosis or diagnosis, if the key is fairly complicated? Stepwise regression, which is in this same area, has already been mentioned.

In all these highly multiparametric problems, such progress as we have so far depends in part on what might be called preconception. That is, a general floating notion that some particular sort of great simplification, though implausible as an exact hypothesis, may have use as a rough hypothesis. We invoke preconception when we assume parametric models such as normality, or the Bernoulli or Poisson processes. We do it when we tentatively regard a time series as stationary. Or when we think it fruitful to think of a time series in the wide sense, which amounts, it seems to me, to entertaining one great big multivariate normal hypothesis that we know in our hearts couldn't be very true but may be true enough to justify somehow the bit of theory that we are trying to apply. (That this theory as mathematics needs no normality is beside the point.)

The intelligent man in the street employs preconceptions, whether he should or not. On hearing that 45 percent of the students in this university are girls and 30 percent of them come from Wisconsin he thinks that .45

times .3 has something to do with the putative fraction of girls from Wisconsin in the student body. He is well aware that the true fraction could be very different. (There could even conceivably be a law precluding girls from coming to this university from outside the state, to name an outlandish possibility.) But he nonetheless has independence somewhere in the center of the stage. The reason I'm stumbling over my tongue is that George Barnard has already told me that this is crazy and there's no way to say that it isn't crazy, so how can I speak with confidence. Yet people do think in terms of rough tentative independence, at the same time recognizing that there could well be greater (or smaller) tendency for boys to wander than girls.

But suppose we have data for several attributes: in-state, boy-girl, science, high grades in the freshman year, and so on. If we have frequencies for all pairs, what shall we conjecture about triples or quadruples? You may be saying, "Why conjecture about triples or quadruples? Investigate them. Why conjecture about anything?" But there is pressure to conjecture, because sooner or later if you carry on this line of thinking, you come back to the billion-cell table, and if you do have occasion to predict the frequencies in arbitrary cells of that table—as practical problems of prediction and diagnosis may require—you will have to bring some extrapolatory conjecture to bear. Thinking out loud about extrapolatory conjectures is becoming more popular and more pressing; it is one of the things that I would expect and hope to see in the future of statistical inference.

L. M. LeCam. I have very few comments to make. I'd like to make one comment on the problem that was raised by Professor Kiefer about fitting curves. I don't know how one should phrase the problem exactly, but sometime in my life I fitted a large number of curves, and we found, for instance, that if you put too many parameters in it, the result is likely to be disastrous. The curve is too flexible, too much of the randomness comes in, and when you try to make a prediction it's not very good. How to formalize that, I don't know, but it's somewhat in the same line.

G. A. Barnard. David Freedman's results about the inconsistency of the Bayes distribution, since they don't converge to positive density on the true parameter point, applies to cases where the sample space is not finite. It has always seemed to me that there are very profound reasons why we should always regard the infinite sample spaces that we deal with really as merely mathematical approximations to what always are, in fact, finite sample spaces. So to this extent one could argue that the Freedman result doesn't really apply in the actual problems we deal with. I wonder if he'd care to comment.

J. C. Kiefer. I think that one has to look here at the way in which infinite spaces or spaces with an infinite number of points crop into science in general. In physics we all know there are many situations where one is faced with a large finite number of molecules, and it is a mathematical convenience, quite often, to think of a continuum of matter and to deal with it mathematically. Now I think that in the present circumstance perhaps the difficulty is this—that although the sample space and the parameter space may be finite, they're still very large. I quite agree with you that, given that you were going to take some extremely large sample, David Freedman's results might well be irrelevant in many practical problems. Now there's a question of how relevant they are for small sample sizes, and I think it would be interesting to look into that. But my suspicion is that, for complicated statistical problems of the type that Professor Savage and Professor LeCam both alluded to also, where there are very many parameters present and where maybe an experiment looks just massive, a large sample theory based on how bad things can be, due to having infinite spaces, is perhaps not too far from what really happens, unless the actual sample sizes you have are quite tremendous in this finite (but large) dimensional situation.

C. Mallows. I don't know if Professor Kiefer is going to have time to say anything about this, but I wanted to ask whether he or the other members of the panel had considered the relevance of the growth of computing facilities to the future of statistical inference. It seems to me that as soon as you have more computing power available, you start thinking of different things to do, and this often leads you to entirely different ways of thinking about what the answer might be. This is relevant to what you've been talking about in connection with subset regression. A few years ago I started looking at this problem, and started to look at what happened if you fitted all possible subsets, which is quite feasible with a fast computer. I soon got to realize that I didn't want to pick a particular subset and fit that by least squares, but rather I wanted to consider several subsets and then use my judgment and whatever other things I knew about the situation to make some sort of conclusion which might not be just a subset-least-squares fit. And so, having the computing power available makes one think of doing different things. Do you have any feeling for how this is going to develop in the future? The other side of this question is, what should we now be teaching in inference?

J. C. Kiefer. Perhaps I will give my proxy to Professor LeCam to answer "What should we be teaching in inference?" As for the first question on computing, I think it's clearly true that with modern computing facilities one has tremendous capabilities one didn't have before. Just to mention one example, for years many of us hunted for exotic designs that had a tremendous amount of balance, so that the information matrix of the design would turn out to be diagonal or have very nice symmetry properties. Well, if one has high-speed computers it's highly questionable whether it's worthwhile to go to such pains just for the sake of having simple information matrices of that sort.

I would like to repeat one thing I mentioned yesterday about the danger that I really think is present in the use of computers. I am all for using them in statistical problems. But I think that the temptation to go to the machine and try this, that, and the other thing without having the patience to sit down and do a little bit of theoretical work first is very great. We all know the naive examples that were presented in the early days of computing machinery, of an engineer who puts a second-order linear differential equation with constant coefficients on a computing machine, or something like that, instead of staring at the problem for five seconds. I think on a more sophisticated level, it's possible to do exactly the same thing now in designing exotic, complicated machine techniques for very interesting and meaningful statistics problems. And I would make a plea that, if anything, it's even more important now to be careful and do some theoretical groundwork on just where these computational techniques are going to lead.

L. M. LeCam. I don't know. I'm a member of a department of statistics which consists of approximately twenty-nine people on the faculty, and we don't know what ought to be taught. What we feel is that the subject of statistics by itself has not been evolving too rapidly in the past few years. The subject of probability has been evolving extremely rapidly, and mathematics in general, has been evolving very, very rapidly. We do teach the various standard techniques. We try to get our students to get used to the computer. We have arrangements with out Computer Center. If the class wants to put something in the computer, they can do it free. Some of my colleagues try very hard to get the students to tackle some problems which require high-powered machinery. I must tell you that it might be that our computer outfit is not very good, but the results are often disastrous. It is often much quicker to do it by hand. So, in that respect, I have no great feeling for the future of statistics being helped by computers. I had a few problems myself in which I thought, well, since there is a machine down there, let's use it. After three weeks, I gave up, and I did it by hand. This has been the usual situation. The usual situation is that the computers tell me that they compute a log; it's not a log. They compute $e^{-x^2/2}$; it's not that—it's something else. When they multiply by some other number they get approximations where the decimals don't fit any more. It's all true; I don't trust a word of what they say.

G. A. Barnard. Sounds to me as if any computer workers who are interested in moving have somewhere to think of going to. I now throw the matter open for discussion from the floor.

DISCUSSION

G. H. Ball. One of the things Professor Savage said expresses a need I have. Maybe this is a special plea more than a question. In the graphics work, we're necessarily going to have to select some subset of the variables if we have more than five or six variables. How one selects these, using numerical techniques, is to me a very important question needing guiding procedures in addition to the guidance you may get from the particular problem.

I might just comment on your comment, Professor Kiefer. It seems to me that, like the adding machine made the person who could work the abacus very rapidly less useful, since it sort of spread that capability around by making it easier to learn, so maybe the computer spreads the theoretical capability around also.

J. C. Kiefer. Well, I hope you are right. I'm a little bit skeptical and I guess my feeling on the subject is to wish well for you and hope that you can emulate a very great philosopher by the name of Lefty Gomez who once said, "I'd rather be lucky than good." I think you have to be very lucky to go in blindly with the capabilities of modern computers and just automatically come out with something that is even of the right order of magnitude in efficiency of what you should be doing.

G. S. Mudholkar. As a statistician I always considered myself responsible for listening to, and/or attempting to solve, any problems involving data or uncertainty. Recently I met a physicist who works in mathematical theory of quantum mechanics. He had a space of states of nature. That it was a Hilbert space may not be relevant. The states assign probability distributions to observables. Well, his observables, as I discovered later, were simply random variables. I listened to his problems of translating observables into random variables. Then came a time when he claimed that his random variables would not necessarily have joint distributions. This somewhat shocked me in the beginning since I had not seen such a situation before. Later on I was convinced that it was true, because the space he was working with was not the usual probability space. Anyway we talked for a while. His problems seemed to fit very well into the usual decision theoretic formulation. I thought if I have not, some other statisticians must have attempted such problems. But the physicist had a very respectable background and had access to fairly wide information. It did not seem any statistician, except one, had cared for the problem. V. S. Varadrajan had looked into the problem of simultaneous observability.

Now here we have had a lot of discussion concerning status of statistics. No statistician attempted the problem. Physicists did not care whether statisticians had ability to do this. They did not consult statisticians because of a feeling that we would not understand what they were talking about; perhaps because we did not(?) possess mathematical tools. I wonder if this distinguished panel would consider statisticians responsible for dealing with such problems. They seem to be problems involving uncertainty. They involve genuine data coming out of experiments. I may be mistaken. Perhaps statisticians have attempted such problems. I wonder if this is a legitimate field of investigation for statisticians, say by way of model formation of formulating questions and attempting solutions. I wish to say that physicists have not asked proper questions. From a statistician's point of view they don't answer questions properly.

J. C. Kiefer. I will yield to Professor Barnard, who has done work with physicists.

Note added in proof: S. Kochen and E. Specter did the path-breaking work on lack of simultaneous observability in quantum mechanics; commenting on Professor Mudholkar's remark, this was not a problem the world had ignored, but rather was a difficult question on which many good scientists had tried and failed.

G. A. Barnard. Thank you Professor Kiefer. I'm sure we're very grateful to Professor Kiefer for the contribution this afternoon and for his early contributions to the discussion. On the point about the statistical analysis of physicists, it's true, the physicists are, I think, giving a great deal of attention to this. And I have made the comment elsewhere, that in my opinion statisticians have a great deal to learn from reading, in particular, a paper by Frank Solmnitz called "Statistical Analysis of Experiments in Particle Physics," which was published in the *Annual Review of Nuclear Science* in 1964. But I'll now pass on the rest of the panel. I'll ask Professor Savage now to come in, if you would, with comments on what's gone before, if you'd care to.

L. J. Savage. There is an isolated remark which would have been a little more appropriate while Kiefer was still here. He happened to mention the difficulty of a thorough-going predictive theory of how the likelihood function will behave. Asymptotically, it will probably present a nearly normal profile around the truth, and in any given experiment, if one is sufficiently interested and has the computational force, one can at least see what shape the likelihood function has got now that the data are in. While it is difficult to say how surely the likelihood will be nearly normal for a sample of a given size, if you've done the experiment, you can say how nearly normal the likelihood function you have in hand is, and how narrow it is. And if it is both narrow and nearly normal, inference is easier and more intelligible.

G. E. P. Box. One of the tendencies, not only in statistical inference, but in many other things, that I find perhaps holds back progress is that as soon as somebody talks about A being good, it's automatically assumed that this implies that B is bad, and it's automatically assumed that this means that one has to take an attitude with respect to B. And all that this does, I think, is hold back progress. Now in this business about computers and mathematics, there is no reason under the sun that I can see why somebody shouldn't in fact analyze the problem and use his mathematics on the problem and still use the computer as well, and also to play around with the data. He might use the data to give an idea of the sort of thing he might want to do, to perhaps do some calculations to see how a certain type of tentative model which he thought might be useful to pursue and to develop mathematically, what kind of results this might lead to, and so on. So again, I think that rather than be half a person, we ought to try and be a whole person. I think that to be a whole person, one could use all of these things and *should* use these things together. Of course, it's very easy to say that when the computer is used wrongly, you get wrong results. What we should do is try and use it right.

E. Parzen. I would like to present a brief problem and ask the members of the panel and audience how they would approach it. This problem was brought to my attention by a friend of mine who treats kidney stones. If you come to him, he'll ask you your medical history and he'll see how many kidney stones you've had and so on, so that for each patient he'll have a timed history of his kidney stones. Then he gives the patients massive doses of phosphorus. The people for whom this is true have been under treatment for about a couple years at most, and they have a history of either zero or at most one recurrence of kidney stones in this period. My friend has twenty-five patients for whom he has this timed history on their kidney stones up to the time he started treating them and their history since then. Now he wants to know first whether the data at this time is sufficient to publish the treatment as successful in arresting kidney stones, and if it's not yet ready to publish, keep track and let him know when to publish that he has a way to completely take care of kidney stones.

L. M. LeCam. He has a problem. In December 1965 and January 1966, we had a symposium at Berkeley—part of the Fifth Berkeley Symposium—in which there were at least three of four, maybe five or six, papers on that subject. If some of the people who know about the subject are in the audience, I would hope that they would speak up. However, the problem involved there is not a problem that we can solve in five minutes. If you look at the situation a little bit more carefully, it is an extremely complicated one, because if the fellow waits to announce his results and does not tell the medical society in time that he is actually doing something for the kidney stones, there will be lots of people suffering across the country from kidney stones that could be relieved otherwise. If, on the contrary, he claims to have a drug, phosphorus, for instance, which relieves kidney stones, but has some undesirable side effects. (The doctors, that is the doctors I know, all claim that there are no such undesirable side effects—it's always perfect, but that's not quite so.) Then if he claims that he has such a cure, he will do two things: he will poison a few people in the process; he will delude doctors into believing that there is a cure, so they might delay working on the problem further, and that might mean that the real cure might come in only a few years later. How do you take all the facts into account to decide when to stop the experiment, especially if the experiment may be dangerous to the patient? I don't really know, and I will not attempt to solve the problem in a few minutes.

G. A. Barnard. I'm tempted to come in and comment. If we are prepared to assume that the data are reasonably gospel, which one would very much doubt here, of course, but if one assumed that the occurrence of kidney stones would, which I think would not be too unreasonable, represent something like a Poisson process, then the intervals between the appearance of the kidney stones before and after treatment would be expected to differ by some kind of scale factor. And so, if one takes that, the likelihood function for that could be capable of being published at any stage, because this would be, of course, the appropriate information which could be combined with information from other physicians, which would eventually lead to a general consensus in the medical profession. In any case, I think it would never happen that one doctor's testimony would be accepted as the conclusive evidence for or against a given treatment.

O. Kempthorne. A very good answer, George. I think he should get a B.

G. A. Barnard. I know where to go for my kidney stones, now.

O. Kempthorne. I'm not sure that I really have anything intelligent to say or ever had, but let me try. I find myself in a rather peculiar position in that I really have to agree with Jimmy about some things. I'm highly in agreement with him that it is necessary for statistics to obtain some theory, some real understanding, of what we call descriptive statistics. Now, just consider a simple sort of experimental study. Let's say one is experimenting on humans, and one has five stimuli—I don't know what these are (use your imagination)—and one has a hundred subjects on each stimulus. Then one, shall we say, gives them tests, psychological tests—you know, the M.M.P.I. or something; I don't know; all these psychologists are full of tests—so you give them, say, twenty tests. Now, let's just think about this. Suppose you give them the tests before and after. Now you have five times a hundred, that's five hundred, times twenty tests, twenty times five hundred is 10,000, times two, 20,000. So you have in front of you twenty-thousand numbers. Now, how in the deuce do you look at twenty-thousand numbers? I think descriptive statistics has to be related to this question of how, in fact, you do look at large masses of numbers. It would seem that someone should be able to develop some sort of information theory with regard to this, because this sheet of twenty-thousand numbers will mean *nothing* to me. I will have to perform all sorts of calculations and do analyses of variance, not because I believe any distribution theory, and not because these numbers are random variables. These twenty-thousand numbers are *not* random variables, they are twenty-thousand numbers, definite numbers. And I want to form some understanding of these twenty-thousand numbers, so I will do an analysis of variance on them, for instance. In doing this, I have no idea of normal law theory with independent (μ, σ^2) errors and all that rubbish. I just want to look at the numbers, get some feel for them. Now, later, I may want to do some tests of significance or something like that, and then I will use some distributional assumptions. But many of our processes of statistics have to be looked at from the point of view of condensation of data. We have some sort of theory, I believe, on condensation of data, due of course to Fisher: the idea of sufficient statistics, and so on, for parametric situations. But this is a somehow rather inadequate affair, because it's only pathological distributions that have nice sufficient statistics. But can we develop some sort of nonparametric theory of condensation of data, for instance? I think perhaps this is relevant. I think one has to think in these ways.

I'd like to make a remark about *subjective* and *objective*. I think these words are sort of slopped around like mashed potatoes in a cafeteria. It seems to me rather obvious that the individual experimenter has to form subjective opinions. Well! All opinions are, indeed, subjective, and the individual research worker has to form opinions about models and so on. And how he does this, God knows. How he arrives at some prior distribution is completely irrelevant to science. I've felt this for a long time and I think I've read a statement by Popper quite recently where he expresses the same view, that it might be expressed in the following way: that science consists of objective validation by society of subjective opinions of individuals. It is critical to allow the individual scientist his subjective opinions and all the rest of it, and it is critical for society to say "I don't give a damn why you think this; I want to check up on it. And I am not interested in your 'prior.' I merely want to know what experimental studies, or whatever, I can do, which will confirm the theories that you are talking about."

L. J. Savage. I would like to say a few words, maybe a few hundred. This is, of course, the taboo subject, the one which we panelists did not allow ourselves. But these last remarks about society deciding, and so on, do invite comments.

In the first place, Oscar would not expect us to reason too closely about a metaphor. When society says, "I don't give a damn," who is she anyway? Society refers to a myriad of possible theoretical constructs—all of them imperfect.

One might entertain the idea that there are individual opinions in contrast with some objective way of concluding; that is a philosophical problem to be resolved. If there is an objective way to reach conclusions, let us delve it out and employ it. But none has been found and there seems to be a better picture and one more in tune with common sense. According to it, there is nothing in the world of knowledge but the opinions of individuals, though sometimes diverse opinions are coalesced by common evidence. I say this because I don't see where so-called objective knowledge is to come from. But, of course, it's a misunderstanding to imagine that I or other Bayesian statisticians wish to publish our own prior opinions or our own posterior opinions.

Naturally you, or you, or you, don't want to think something merely because I think it. The intended role of the concept of opinion as measured by a subjective distribution in this theory is that of a reasonably good model of each of us. A person publishing should so publish that each reader will have grist for his own subjective mill. Misunderstanding about this took an acute form in a review by van Dantzig. I had the honor of being the villain in the first of two reviews by him called "Statistical Priesthood," R. A. Fisher running a close second. The villainy, the priesthoodery, that van Dantzig thought I was up to was advocating a statistics of this sort: let a statistician (possibly without medical knowledge) decide what he thinks about kidney stones, before seeing the data, combine that with the data, publish the results, and expect the medical profession to take it all seriously. But the real tendency of Bayesian statistics is well portrayed in an excellent answer that Barnard gave, saying in effect, "If you publish the likelihood, then he who has eyes to see can study it and make his own inference from it." You don't need to be a full-fledged Bayesian to adopt that maxim, but the idea that you should publish what other intelligent minds would want to know about your work is the import of the Bayesian respect for personal opinion.

L. M. LeCam. I think I have to add a word to this business about likelihood since it came up twice already. Jimmy said one should compute the likelihood function in one particular case, and Prof. Barnard said that, in the case of the kidney stone, you should publish the likelihood function. I beg to disagree. You should publish the experiment, the whole thing as far as possible. It is not the privilege of the statistician to tell the doctor that formation of kidney stones is a Poisson process. He should publish the experiment, possibly not with the names of the patients since they might object, but with descriptions of age, ancestry, whatever you have—all the data, including the mistakes that were made in diagnosis. This cannot be replaced by any likelihood function.

Since we are now coming to the subject which we were trying to avoid, I have absolutely no disagreements with the Bayesian approach to statistics, as long as it applies to personal opinions of a person who has infinite memory and infinite mathematical capability. I feel, however, that it is a mistake to take that because there exists such a theory of behavior, because there is no available substitute for the time being, the case is closed. It is respectable in a way, but I think that it would be a disaster to leave it at that. The point that was mentioned about getting conclusions of an objective type is too complicated for me. I don't know how to do that. However, I would be satisfied, at least partly satisfied, if there was some way of formalizing, in some way or another, the transmission of information. The reason why I thought about the likelihood function a minute ago is that it seems to me that the likelihood function is not a sufficient vehicle to transmit information. It is a sufficient statistic. It is not a sufficient vehicle to transmit what the experiment was.

H. O. Hartley. I would like to ask two questions of Professor Savage. We know, of course, that there are many situations in which the old frequency concept of probability is irrelevant or, shall we say, not of interest. There are situations where it is rather silly to imagine that an experiment is repeated an infinite number or even a large number of times under similar conditions and that inferences can be formulated in terms of this infinite number of repetitions. There are, however, situations such as in quality control and inspection sampling, where the concept of drawing repeated samples from the same large, although not infinite, population is sensible. I would like to ask him, first of all, whether the Bayesians really claim that the concept of repeatedly sampling from a very large population is *never* sensible. The second question I'd like to ask is this: Some concepts of classical statistics can be defined in terms of frequencies where the infinite frequency does not refer to repeatedly conducting the same experiment, but where it refers to a great variety of experiments, say the experiments that are analyzed by a particular statistician during his professional life. Confidence intervals can be stated in terms of such repeated statements made by a particular statistician. That is to say, if he makes a particular calculation, well specified, and carries out that calculation for a great variety of different types of data, it can be stated that he'll be right in a specified proportion of analysis. It is, I believe, held by Bayesians that this frequency concept is, in practically all cases, an irrelevant concept. It is of no help, they say, because a biologist who is told about this says, "I don't care about what you do with other biologists, I want to know what you do with my experiment. I don't care about the others." I personally would like to remark that I haven't found a scientist to be so entirely unreasonable, I think they are more socially behaved! I think they have an interest in the other fellow's experiments. It is of some relevance to them to know that here is a statistician who is using tools which have that property. I have, therefore, indicated some disagreement, but I will very gladly listen to the opposite view.

L. J. Savage. Those are two big questions and we are hurried. As if that were not enough, I'm going to smuggle in a word, most illegally, about something else. I want to say—and dare say it for Barnard too—that we don't disagree with LeCam to the extent of thinking that one should not publish the data, but only the likelihood. Barnard was careful to specify the hypothesis, the almost make-believe hypothesis, that the data are what they seem to be. Language is a little bit too difficult, and I don't know just what Lucien and I do agree and disagree about; but I certainly don't think it would be wrong, in principle, with infinite channels of communication and with humans of infinite memory and all that, to publish, as it were, a wonderful 3-D picture of the whole experiment in all detail.

Do subjectivists (or do I) believe in frequencies? What I believe, what de Finetti believes, what a good number of modern personalistic Bayesians believe is that the phenomena to which we allude when we talk about frequency can be excellently explained as special cases of subjective probability. That is a technical matter on which there is much writing; it begins with a theorem proved and discussed by de Finetti. The essence is this: Rather than a dual theory of probability, one in which two different kinds are somehow stitched together, we think it more feasible and generally better to work with the one kind of probability that we understand clearly and that explains—gives us a foundation for—frequencies and also explains the old idea of symmetry probability, according to which the probability of drawing an ace "must be" one-fourth.

To the second question about some kind of justification of confidence intervals because they really are confidence intervals, I shall reply mainly by a fanciful example. Here is an experiment designed to measure the velocity of neon light in beer. The experiment is simple and above board. It involves measurements on a random number of steins of beer according to the Poisson distribution with mean 8. The situation is clear-cut and provides a 99% confidence interval of 2 meters per second. Now the experiment is over and we have read 3×10^{10} cm/sec $\pm$ $1/n$ cm/sec, in the sense that this experiment is so designed that intervals generated as this one was will capture the truth 99% of the time. But then, imagine a footnote saying, "In this particular experiment, the barmaid, whose stochastic behavior is entirely understood by us, brought only one stein of beer. Therefore, the experiment has been a bust, as we knew there was one chance in a thousand for it to be." Now what does it avail to know that the plan was one which had a good chance of capturing the truth, if the experiment says on the face of it, "This was a failure"? I am reminded of the statistician who refused to buy a certain kind of calculating machine, saying "I don't like it because it makes mistakes." The salesman said, "Hardly ever, and when it does, it rings a bell." Well, at least if my experiment makes a mistake and rings a bell, I want to hear the bell. The Bayesian analysis, proposed in lieu of confidence intervals, is one that has an ear cocked for that bell.

G. A. Barnard. Well, I am afraid we shall have to close on that. I must say I would like to take the privilege of commenting on the nature, not only of this discussion but also of the earlier discussions, that I've had the feeling that this perhaps attempted dispute concerning the foundations of statistics has not really generated the amount of heat that such disputes might once have been expected to. And it is a theme, I think, that has run through this conference that, in spite of the very much wider spectrum of persons attending, and speakers at, the conference, this conference has expressed the unity of the subject of statistics in a way that augurs very well indeed for its future in general. And I'm sure that you'll all join me in saying that we hope it augurs well for its future at Wisconsin.

Reprinted from
The Future of Statistics (Proceedings
of a Conference on the Future of Statistics)
Madison, WI, June 1967
Academic Press, New York, pp. 139-160 (1968)

Reprinted from
The Mathematical Sciences (1969), 60–71

The problem of drawing useful inferences from observations is something like probability theory inside out. In probability problems, you know how nature works and want to find out what is likely to happen. With statistical inference, on the other hand, you observe what has happened and then try to deduce the mechanism that made it happen. Statistical inference is inevitably fallible. One of the statistician's main tasks is to lessen the chance of guessing wrong. This requires him to recognize subtleties that go beyond purely mathematical considerations but must somehow be incorporated into mathematical procedures. In order to decide which of several procedures will best serve the user, the statistician must abandon time-honored rules of thumb and rely on precise analysis of the problem. That, the author explains, is what the modern mathematical theory of statistical inference is all about.

Statistical Inference

J. Kiefer[1]

The word "statistics" conjures up many meanings. It suggests the state of the economy, Detroit's quality-control methods, medical evidence linking lung cancer with smoking, and the physicist's configurations in statistical mechanics. In this paper, our interest is limited to *statistical inference*. This is the study of various possible procedures for analyzing data in order to guess the nature of the physical or biological mechanism that produced these data. It is an applied area of mathematics with particular conceptual difficulties not found in other applied areas. The mathematical theory tries to give a rationale for selecting a procedure for analyzing the data rather than relying on intuition or rule of thumb. We cannot hope to develop any deep theoretical results of the subject here. We hope to treat only a few examples that may give you some feeling for subtleties in even the simplest statistical problems where practitioners tread — usually fearlessly, frequently by intuition, and quite often disastrously.

How does statistical inference differ from probability theory? In probability theory one specifies a model (chance mechanism) and studies its consequences. For example, suppose a coin has probability p of coming up "heads" on a single toss, and $1 - p$ of coming up "tails." Here p is a

[1] This article was prepared with support from ONR Contract No. NONR 401(50). Reproduction in whole or in part is permitted for any purposes of the U.S. Government.

number between 0 and 1 that reflects the physical makeup of the coin; roughly speaking, it is the approximate proportion of heads you would expect in a long series of tosses, and it would be $\frac{1}{2}$ for a "fair coin." Our model is to flip the coin five times, with independent flips. "Independence" means that successive flips have no probabilistic effect on each other, so that the probability of obtaining a particular sequence of outcomes on the five tosses is computed by multiplying the probabilities of the outcomes for the individual tosses. Thus, the probability of obtaining "heads, tails, heads, tails, tails," which we hereafter abbreviate HTHTT, is

$$p \times (1 - p) \times p \times (1 - p) \times (1 - p) = p^2(1 - p)^3. \quad (1)$$

Clearly, we obtain the same probability $p^2(1 - p)^3$ for any other sequence of five tosses containing exactly two H's, for example TTHHT. There are ten different arrangements of two H's and three T's, and the event "two heads in five tosses" occurs if the five flips come up in any of these ten arrangements, each of probability $p^2(1 - p)^3$. Thus, the event "two heads in five tosses" has probability $10p^2(1 - p)^3$. Similarly, the probability of obtaining exactly k heads in five tosses, hereafter denoted $b_p(k)$, is

$$b_p(k) = \frac{5!}{k!(5 - k)!} p^k(1 - p)^{5-k} \quad (2)$$

if $k = 0, 1, 2, 3, 4$, or 5. (We have used the convention $0! = 1$.) This is the well-known binomial probability law, and of course $b_p(0) + b_p(1) + \cdots + b_p(5) = 1$ since this sum includes the probabilities of all possible sequences of five flips. The development leading to Equation 2, although very elementary, illustrates the computation of a consequence of a simple probability model, which typifies probability theory.

In contrast, in a statistical-inference model the chance mechanism is not completely known. Rather, the problem is to guess some of its features. A simple example is that of our coin when the value of p is unknown. The mint, it may be supposed, produces coins which, because of inhomogeneous makeup and irregular shape, range from those that almost always produce heads when flipped (p near 1) through those that produce about half heads (p near 1/2) to those that almost always produce tails (p near 0). It is not known which of these types of coin is the one we actually have, that is, which value of p between 0 and 1 characterizes the coin.

The emphasis of statistical inference is on choosing a procedure for using the observed outcome of the chance mechanism to make a guess regarding which of the *possible* mechanisms is the *actual* one at hand. In our example, this means using the observed sequence of five flips to make a guess as to the unknown value of p for our particular coin. In some reasonable sense, this guess should be as accurate as possible. However no procedure can guarantee to guess correctly. Any procedure we decide to use to make a guess about p will make the same guess in every experiment of five flips with outcome HTHTT (though perhaps some other guess whenever the outcome is TTHHH). Whatever the actual p is, the outcome HTHTT, and thus the guess obtained from it, really can arise. But it is easily seen that no meaningful statement made as this guess can be correct for all possible p. For example, if from HTHTT with two H's in five flips, we stated "I guess that p is 2/5," this would be correct if p were actually 2/5 and would be reasonably accurate if p were close to 2/5, but it would be an erroneous guess, for most practical purposes, if the coin were actually one for which $p = .9$. Similarly, any other procedure we could use can sometimes make bad quesses. As we shall see, the accuracy of different guessing procedures is a probabilistic computation.

What's the Question?

In precisely formulating a statistical problem about our coin, we begin by listing the possible relevant statements we could make about the coin. We shall decide on one of these statements after observing the outcome of the five tosses. This list can be regarded as the list of possible answers to a question about the nature of the coin.

For example, one might ask (a) "What value of p characterizes the coin at hand?" The possible answers are all real numbers between 0 and 1, and a statistical procedure for this question is a rule which associates, with each of the 32 possible sequences of five tosses, a corresponding guess as to the value of p.

A question requiring a less precise statement about p is (b) "Is the coin fair?" A statistical procedure for this question does not assert a numerical guess as to the value of p but, rather, makes one of the statements: "The coin is fair ($p = 1/2$)" or "The coin is unfair ($p \neq 1/2$)." If the mint wanted to test each coin it produced, remelting unfair ones to preclude their falling into the hands of scheming gamblers, this second question would be of interest; a more precise guess about the value of p for a given coin would be irrelevant. On the other hand, if the H and T of the coin are replaced by the life or death of a dying patient to whom a new drug is administered, then a numerical guess as to the probability p that the drug can save a life — an answer to question (a) — would often be called for.

You can think of other possible relevant questions. For example, a gambler would want to ask (c) "Is the coin fair, biased in favor of H, or biased in favor of T?" The answer tells him what use he can profitably make of a given coin: He can bet on the favored outcome in gambling games with unsuspecting victims if the coin is biased, or spend it if it is fair and, thus, has no such nefarious use. Of course, there are many similar examples of greater practical importance.

It is remarkable that, as natural as it may be to ask questions like (c), the classical development of statistics neglected such questions almost completely until Abraham Wald began his work on "statistical decision theory" in 1939 and emphasized the completely general nature of the questions one could ask. Prior to that, statisticians tended to try to push every problem into the formulation (a), called *point estimation*, or (b), called *hypothesis testing*, whether or not one of these formulations fitted the need of the customer. (A variant of (a), called *interval estimation*, was also sometimes used; here the guess took the form "I guess p to be in the interval from .37 to .45" instead of "I guess p to be .4.")

Here is a simple example of what happened when the wrong question was asked. An experimenter might be testing the productivity of several varieties of grain, wanting to decide what progress he had made in developing a new strain. The actual yield from any plant differs by a chance value, because of such effects as soil variation, from the "expected yield" for that variety, which is approximately the average yield per plant one would observe over a great many plants. The experimenter who felt he had to ask a question of type (a) or (b) would usually choose the latter and ask, "Are any of these varieties better than the standard variety most farmers now use?" If the data made him answer "No," he would return to the laboratory to try to develop other strains. But, if the answer were "Yes," he would realize he had asked the wrong question, for he would now want to know *which* of the varieties were superior ones in order to experiment further with these. He would therefore ask other questions of type (b), based on the same data: "Is variety number 1 a superior one? Are varieties 2 and 3 superior?" And so on. He would then try to combine the answers (which could even turn out to be logically incompatible) to reach a conclusion as to which varieties merited further study. By using such an *ad hoc* combination of statistical procedures which were designed to answer questions of type (b) rather than the question of real interest to him, he could end up making a rather inefficient use of the data. Moreover, he almost never knew the actual efficacy of this conglomerate procedure, that is to say, the probability that it would actually select just those varieties with expected yields appreciably greater than that of the standard variety.

Wald's approach departed from the tradition of artificially restricting attention to questions of types (a) and (b) and instead tried to ask the question of real concern. In our example that might be "Which varieties have expected yield at least 10 percent greater than that of the standard variety?" A procedure would then be designed to answer this question, with efficient use of the data and with a computation of the efficacy of the procedure in giving a correct answer.

Another interesting feature in the history of statistics is that, even in settings where a question of type (a) or (b) is appropriate, it has been less than forty years since statisticians worried much about making precise a reasonable measure of accuracy for statistical procedures and found how to construct procedures that use the data efficiently in terms of that measure. At the beginning of this century Karl Pearson and his school were the leading producers of statistical procedures, constructed largely on an intuitive basis, whose dangers we shall illustrate later in an estimation example. Beginning in 1912, R. A. Fisher showed how inefficient some of these intuitively appealing procedures could be. Over the next half century, Fisher contributed greatly to many mathematical developments in statistics and perhaps had more influence, especially on applied statisticians, than any other person. But some features of his work were unsatisfactory to those who felt that statistical inference should be based on a precise mathematical model with these two features: It makes explicit the penalties that can be incurred from reaching incorrect conclusions from the data, and it leads to the construction and use of a procedure that probably incurs small penalties by using the data efficiently. Jerzy Neyman, together with Karl Pearson's son Egon, was at the forefront of the resulting beginning of the modern mathematical theory of statistical inference.

The Cost of Being Wrong

Ideally, the precise formulation of a statistical-inference model must include not only the list of possible answers to the question asked but also a statement of the relative harm of making various incorrect answers. For example, for the question (a) of estimating the value p characterizing our coin, the loss incurred from misestimating p will presumably be larger, the further the guess is from the actual p. For our grain example of type (c), you may want to try to formulate, at least qualitatively, the way the penalty for an incorrect guess might reasonably depend on the actual average yields and the guess made. We shall simplify our subsequent calculations in this article by neglecting the precise penalty values. In particular, all incorrect guesses are regarded as equally serious in the hypothesis-testing example treated later. However, these values do, in fact, play an important part in determining which statistical procedures are good ones.

From a practical point of view, these penalty values, to be expressed in units of money or utility, are very hard even to approximate in most problems. A manufacturer may be able to assess the cost he would incur from misclassifying a defective light bulb as "good" in a quality control test. But what is the cost to an astronomer of misestimating by 20 percent the distance to a quasar he is studying? To go a step backward, in exploratory research one cannot always know in advance of the experiment the precise form of the question to be asked. Often a phenomenon shows up which has not occurred as a possibility to the experimenter. (The inviting practice of letting the data determine the question after the experiment and of answering it from these same data can lead to dangerous delusions about the accuracy of one's conclusions.) To go still another step backward in our pattern of formulation, it is often impossible to delimit precisely the class of *possible* probability mechanisms, one of which actually governs the experiment at hand.

What, then, is the worth of the theoretical developments of Neyman and Wald to the practical statistician? The answer is the same as in other areas of applied mathematics and science, where the careful study of a model that is not exactly correct can still lead to more useful decision-making or predictive procedures than will a formula or rule based on intuition or traditional rule of thumb. Moreover, the attempt to write down a precise model is often, in itself, remarkably helpful in clarifying the experimenter's thoughts about his problem. Incidentally, in view of the possible inaccuracy of the model, an important topic is the study of properties of a statistical procedure designed for use with one class of probability mechanisms when, because of incorrect formulation of the model, the actual mechanism is outside that class. This is beyond the realm of the present article.

Two simple examples illustrate some of the ideas arising earlier. In the first of these, we shall see that, even in the simplest models for a question of type (b), there are many procedures which use the data efficiently, and the choice among them is usually difficult.

Testing Between Simple Hypotheses

We shall continue to use the language of independent tosses of a coin, although you can, of course, imagine H and T to stand for the outcomes of any dichotomous experiment. Suppose our coin is known in advance to be characterized by either the value $p = 1/3$ or by the value $p = 2/3$; no other values are possible. On the basis of three independent flips, we are to guess which value of p actually characterizes the coin. Each of the two possible guesses specifies a single probability mechanism, unlike the example "The coin is unfair ($p \neq 1/2$)" discussed earlier, where the guess did not attempt to describe the exact value of p. When a "hypothesis" consists of just one possible mechanism, it is said to be "simple." Thus, statisticians refer to our present problem as one of testing between two simple hypotheses. The characterization of procedures that use the data efficiently will turn out to be particularly simple for such problems and much simpler than that for other hypothesis-testing problems.

Here are five possible prescriptions for making a guess from the data; you can easily write down others.

Procedure 1: guess "$p = 2/3$" if there are at least two H's in the three tosses; guess "$p = 1/3$" otherwise.

Procedure 2: guess "$p = 2/3$" if the first H precedes the first T or if no T occurs; guess "$p = 1/3$" otherwise.

Procedure 3: guess "$p = 2/3$" if there is at least one H in the three tosses; guess "$p = 1/3$" otherwise.

Procedure 4: guess "$p = 2/3$" if there is at most one H in the three tosses; guess "$p = 1/3$" otherwise.

Procedure 5: ignore the data and always guess "$p = 2/3$."

Which of these procedures would you use? Perhaps many people would feel intuitively that there is more tendency to obtain H's when $p = 2/3$ than when $p = 1/3$, so that Procedures 1, 2, and 3 do not seem too unreasonable and Procedures 4 and 5 seem unreasonable (the former because it works in the wrong direction; the latter, because it makes no use of the data). For a precise analysis, we first use a calculation like that of Equation 1, near the beginning of this essay, to list the probabilities of the eight possible sequences of three tosses under each of the two possible probability mechanisms (see Table 1).

Table 1

Outcome	Probability of Outcome when $p = 1/3$	Probability of Outcome when $p = 2/3$
HHH	1/27	8/27
HHT	2/27	4/27
HTH	2/27	4/27
THH	2/27	4/27
HTT	4/27	2/27
THT	4/27	2/27
TTH	4/27	2/27
TTT	8/27	1/27

Then, for each procedure, we use Table 1 to compute the probability that the chance outcome will be such as to lead to a correct guess when $p = 1/3$. Next, we perform the same computation with $p = 2/3$. For example, when $p = 1/3$, Procedure 2 makes the correct guess "$p = 1/3$" if there is a T on the first toss, that is, if the outcome is THH or THT or TTH or TTT, with a total probability of 18/27, computed from the middle column. Similarly, summing the probabilities in the last column for the four other possible outcomes which lead to the guess "$p = 2/3$" yields 18/27 for the probability of a correct guess when $p = 2/3$. A similar calculation for the other procedures gives Table 2.

What does this table show us about the appeal of various procedures? If a procedure were perfect, it would have probability one (certainty) of making a correct guess, whether the actual p is 1/3 or 2/3. It is not hard to see that no such procedure exists. Procedure 5 is always correct if $p = 2/3$, but never right if $p = 1/3$. The procedure which ignores the data and

Table 2

Procedure	Probability of Correct Guess when $p = 1/3$	Probability of Correct Guess when $p = 2/3$
1	20/27	20/27
2	18/27	18/27
3	8/27	26/27
4	7/27	7/27
5	0	1

always guesses "$p = 1/3$" would have the opposite behavior. No other procedure has probability one of making a correct guess, whatever p may be.

Moreover, we see that no procedure maximizes the probability of making a correct guess when $p = 1/3$ and also when $p = 2/3$: Procedure 5, alone among all procedures, maximizes the probability of making a correct guess if $p = 2/3$, but it minimizes this probability when $p = 1/3$. Since a perfect maximizing procedure does not exist, how are we to select the procedure to be used?

To start with, Procedures 2 and 4 can be eliminated from contention since, whatever the actual value of p, each of these has a smaller probability of making a correct guess than does Procedure 1. Thus, while we have not yet decided on which procedure to use, we can tell anyone who proposes to use Procedure 2 that he should not do so, since we can give him another procedure which has a higher probability of yielding a correct guess, whether the actual p is 1/3 or 2/3. A subject of considerable study among theoretical statisticians is the characterization in different problems of those procedures, called "admissible," which remain after the inferior procedures, such as Procedures 2 and 4, have been eliminated. The choice of the procedure to be used is thereafter restricted to the admissible procedures.

Procedures 1, 3, and 5 can be shown to be among the admissible procedures in our example; you can see for yourself that none of these three eliminates either of the others by having a higher probability in both columns.[2] The choice among Procedures 1, 3, 5, and the other admissible procedures must now involve some additional criterion which we have not yet mentioned. There are philosophical differences among statisticians as to the appropriateness of various criteria which have been suggested and used over the years. I shall next mention a few of these criteria.

[2] Each admissible procedure is characterized, by the Neyman–Pearson Lemma, to guess "p = 2/3" for those outcomes in Table 1 for which the ratio of the last to the middle column is greater than some constant c. You will see that c can be taken to be 1 for Procedure 1, 1/4 for Procedure 3, and 1/10 for Procedure 5. For technical reasons, theoretical statisticians also include procedures which, when the outcome yields a ratio c, perform an auxiliary experiment to decide which guess to make. These need not concern us here.

Basic Ways of Choosing

The approach of Fisher in hypothesis testing, usually in more complex examples than ours, was to use a procedure that attained a previously specified probability of making a correct guess under one particular probability mechanism. This mechanism was often an older theory whose validity was to be tested by the experiment or "no difference between new varieties and old" in the experiment we discussed in leading up to question (c). In our present example, let us suppose that $p = 1/3$ is the chosen mechanism. We must then specify the value a procedure is to yield in the middle column of Table 2. The Neyman–Pearson theory added to this formulation an aspect never treated by Fisher: Among all procedures with the specified value in the middle column, one should select that procedure with largest entry in the last column. Thus, among all tests with 8/27 in the second column, of which there are many in addition to Procedure 3 (for example, guess "$p = 1/3$" for outcomes THT and TTH, and "$p = 2/3$" otherwise), the Neyman–Pearson Lemma characterizes Procedure 3 as the one with maximum probability of making a correct guess when $p = 2/3$. If one accepts the figure 8/27, there is no doubt about using Procedure 3. The practical shortcoming is: Why 8/27 rather than some other figure (for example 20/27, which would have dictated using Procedure 1)? The value has often been chosen by tradition alone, in various ways in different fields of applications.

A second criterion is given by the "minimax" principle. For each procedure, we compute the minimum of the two probabilities listed in Table 2: 20/27 for Procedure 1, 8/27 for Procedure 3, 0 for Procedure 5. This figure gives a measure of the worst possible performance of the procedure. The pessimist, worried about the possibility of encountering a coin whose p yields this worst performance, may want to choose a procedure for which this minimum probability is as large as possible. In the present example, that is Procedure 1. There are also some objections to this approach, but space does not permit us to discuss them here.

A third possibility is to compute for each procedure some weighted average of the two probabilities listed in Table 2 and to choose a procedure that maximizes this average. For example, if we compute $\frac{1}{2}$ the middle column plus $\frac{1}{2}$ the last column for each procedure, we find that this average is largest for Procedure 1. On the other hand, $\frac{1}{4}$ of the middle column plus $\frac{3}{4}$ of the last column is largest for Procedure 3. Thus, different weighted averages favor the choice of different procedures. The practical difficulty here is: How do you choose the weights? In a very few problems, we know so-called *prior* probabilities that the actual physical mechanism will be of one form or another. For example, in our coin problem, we might know that, over a long period, $\frac{3}{4}$ of the coins have come out of the mint with $p = 2/3$, and $\frac{1}{4}$ with $p = 1/3$. In such a case, it is appropriate to use the prior probabilities $\frac{3}{4}$ and $\frac{1}{4}$ as weights, since the resulting average our procedure will maximize is then the total probability of a correct guess. This yields the choice of Procedure 3 in the example just given. Because the general scheme for computing such procedures uses the simple probabilistic formula known as Bayes' Theorem, the resulting procedure is called a Bayes procedure for the given prior probabilities. But there are few practical examples where such prior probabilities are known. In the absence of such knowledge, it is tempting to use equal prior probabilities to represent one's ignorance (which would lead to the use of Procedure 1 in our example) or to use the recently much-publicized subjectivist approach, which employs, in place of unknown physical prior probabilities, corresponding "subjective probabilities," which are supposed to reflect a quantitative measure of the customer's feelings about the problem. Both these possibilities have also received considerable criticism, which we cannot discuss here.

A fourth possibility is to take note of the symmetry of the problem, as reflected in the fact that relabeling H as T and vice versa means interchanging the values $p = 1/3$ and $p = 2/3$, since a coin with probability 1/3 of yielding H has probability 2/3 of yielding T (which becomes H under the relabeling). The symmetry of the problem suggests the criterion of using a symmetric procedure, one that guesses "$p = 1/3$" for a given sequence (for example, TTH) if and only if it guesses "$p = 2/3$" for the relabeled sequence (for example, HHT). In our example, only Procedure 1, among all admissible procedures, has this symmetry; in fact, it can be shown to be better than all other symmetric procedures. Thus, under the criterion of symmetry (usually called "invariance"), we would use Procedure 1. The difficulty with using this approach is that there are more complex problems in which no invariant procedure is admissible, so that the invariance criterion cannot lead to the choice of a satisfactory procedure; and there are other problems lacking symmetry, so that all procedures are equally "symmetric," and this criterion does not choose among them. An example of the latter is our coin problem with the two possible values $p = 1/3$ and $p = 2/3$ replaced by $p = 1/5$ and $p = 2/3$. Under relabeling, these become $p = 4/5$ and $p = 1/3$, which are not merely the original pair of values in reversed order. There is no symmetry to the problem, so all procedures are equally symmetric.

You may well wonder at this point which procedure to use in our example. I cannot tell you without further knowledge of the background of the problem, the real meaning of the two events we labeled H and T (they might mean "life" and "death," etc.), the use you will make of your guess, and a re-examination of the relative losses (which we have tacitly assumed equal) for the two possible types of incorrect guess. Even with this knowledge, I might find it difficult to tell you that a particular procedure is the one and only obvious one to use, although there are many circumstances where I would stop this hedging and use Procedure 1.

I have included such a long description of some of the criteria people use to select a procedure and have indicated that all of them have shortcomings, in order to emphasize the view of many theoretical statisticians — that there is no simple recipe which will tell you how to choose a statistical procedure in all possible settings. This is a frustrating aspect of the subject and perhaps one without parallel in other mathematical areas. It presents a large target for future theoretical work. At the same time it points up the danger of using some simple, superficially appealing recipe to select a statistical procedure, and the need for expertise in that selection.

An even more fundamental danger occurs in the use of an intuitively appealing rule of thumb to construct a procedure, without any consideration of admissibility. The statistical literature is full of suggestions of procedures like Procedure 2 and is almost devoid of mention of Procedure 5 (which should be used if, for example, one knows that the prior probability that $p = 1/3$ is quite small, say 1/10). In other settings where the mechanism is nothing as simple as coin-flipping, it is even easier to be led astray by intuition. We shall illustrate this now in an estimation problem that will exemplify the Pearson–Fisher controversy mentioned earlier.

An Estimation Example

A particle is moved along some fixed line by a force field around it. In each of three independent experiments, the particle, at rest, is placed at the same point P, and its position is measured one second later. Taking P as the origin, if the field were unaltered by outside influences and if the measurements were without error, the measured position of the particle after one second would be the same value, say θ millimeters, in each experiment. However, the experiments are not perfect, and we suppose that in each experiment the probability that the recorded measurement will be θ, $\theta - 1$, or $\theta + 1$ is 1/3 for each. (This law of errors is being chosen for arithmetic simplicity rather than physical reasonableness, but it will illustrate a phenomenon that could also exist for more reasonable but more complicated models.) Here the force field, and thus θ, is unknown in advance of experiments, the object of which is to "guess the value of θ." Let us call the three measurements X_1, X_2, X_3. A statistical procedure is a real-valued function of these three quantities, that is, a rule for computing from them a real number that will be stated as our guess as to the actual value of θ for the particular field at hand.

A common intuitive line of reasoning for selecting a procedure begins by remarking that the value θ is the mean, or first moment, of the mass distribution corresponding to our law of errors, which assigns mass $\frac{1}{3}$ to each of the points $\theta - 1$, θ and $\theta + 1$. This reasoning then proceeds to note that we can summarize the three measurements in the form of an "empiric mass distribution" that assigns to each real value the proportion of observations taking on that value. (For example, if $X_1 = 16.2$, $X_2 = 17.2$, $X_3 = 16.2$, this empiric mass distribution would assign mass $\frac{2}{3}$ to the value 16.2 and $\frac{1}{3}$ to the value 17.2.) Since this chance empiric mass distribution has some tendency to resemble the underlying mass distribution corresponding to the probability law, the reasoning concludes that a guess as to the mean θ of the latter should be obtained from the mean of the former, which is easily seen to be the "sample mean" $(X_1 + X_2 + X_3)/3$. This is a simple example of Pearson's "method of moments."

We can easily improve upon this procedure by altering the guessing rule whenever two of the three X_i take on a value 2 mm away from the third, guessing the midpoint between these two values rather than the sample mean in such cases. The sample mean will be in error by $\frac{1}{3}$ mm for such outcomes, while the new guess will be errorless. For example, if $X_1 = 14$, $X_2 = 16$, $X_3 = 14$, you can see that θ must be 15 because of the particular form of our law of errors, and this is what our new procedure guesses it to be. But the sample mean yields a guess of $\frac{44}{3}$, underestimating θ by $\frac{1}{3}$. Since the two procedures yield the same guess (and, hence, the same error) for all other types of outcomes, the new rule is certainly preferable to the sample mean, and it is not hard to see that the probability is $\frac{6}{27}$ that the outcomes will be of the type where this reduction of error occurs.

You may well think it an obvious improvement on the sample mean to replace it by the improved guess in the foregoing situations where the exact value of θ is obvious from the outcomes. But it is easy to alter the law of errors slightly, at the expense of arithmetic simplicity, to obtain a model where the sample mean can be improved upon without such an obvious motivation. The point is that the intuitively appealing guess, "sample mean," used blindly in so many experimental settings, may be appropriate for some assumed laws of errors and terribly inefficient for others. Only a precise probabilistic analysis, and not any intuitive rule of thumb, can determine what procedures are reasonable ones for a given statistical model.

That is what the modern mathematical theory of statistical inference is all about.

BIBLIOGRAPHY

Herman Chernoff and Lincoln E. Moses, *Elementary Decision Theory*, Wiley, New York, 1959.

J. L. Hodges, Jr., and E. L. Lehmann, *Basic Concepts of Probability and Statistics*, Holden-Day, San Francisco, 1964.

BIOGRAPHICAL NOTE

Jack Kiefer has spent most of his professional career at Cornell University, where he is now a professor of mathematics. As an undergraduate at Massachusetts Institute of Technology, he concentrated on engineering and economics until he became interested in mathematical statistics. After taking his S.B. and S.M. degrees there, he did graduate work and received his Ph.D. at Columbia University in 1952. His research has been primarily in statistics and probability theory. He has received a Guggenheim Fellowship.

Z. Wahrscheinlichkeitstheorie verw. Geb. 13, 321 – 332 (1969)

On the Deviations in the Skorokhod-Strassen Approximation Scheme

J. KIEFER*

Summary. In deriving his strong invariance principles, Strassen used a construction of Skorokhod: if the univariate d.f. F has first, second, and fourth moments 0, 1, and $\beta<\infty$, respectively, then there is a probability space on which are defined a standard Brownian motion $\{\xi(t), t\geqq 0\}$ and a sequence of nonnegative i.i.d. Skorokhod random variables $\{T_i, i>0\}$ such that

$$\left\{\xi\left(\sum_1^{n+1} T_i\right)-\xi\left(\sum_1^n T_i\right), n\geqq 0\right\}$$

are i.i.d. with d.f. F. Let

$$Z=\limsup_{n\to\infty} \pm\left[\xi\left(\sum_1^n T_i\right)-\xi(n)\right]\Big/[n(\log n)^2 \log\log n]^{\frac{1}{4}}.$$

Strassen showed $Z=O(1)$ wp 1. We prove $Z=(2\beta)^{\frac{1}{4}}$ wp 1. Consequently $Z=0$ wp 1 implies F is Gaussian, answering a special case of a question of Strassen. Analogous results hold for cases where $\xi\left(\sum_1^n T_i\right)$ is not a sum of independent random variables.

1. Introduction

Let F be a univariate d.f. satisfying

$$\begin{aligned} &\text{(a)} \quad \int_{-\infty}^{\infty} x\, dF(x)=0,\\ &\text{(b)} \quad \int_{-\infty}^{\infty} x^2\, dF(x)=1, \qquad\qquad (1)\\ &\text{(c)} \quad \int_{-\infty}^{\infty} x^4\, dF(x)<\infty. \end{aligned}$$

Skorokhod [18] showed how to construct a probability space on which are defined (i) a sequence $\{X_i, 1\leqq i<\infty\}$ of i.i.d. r.v.'s[1] with common d.f. F, (ii) a Brownian motion $\{\xi(t), t\geqq 0\}$ of standard normalization $(E\xi(t)\equiv 0, E\xi^2(t)=t)$, and (iii) a sequence of non-negative random variables $\{T_i, 1\leqq i<\infty\}$, such that, writing

$$\begin{aligned} &\beta=\operatorname{var}(T_1),\\ &U_0=0, \quad U_n=\sum_1^n T_i \qquad (n>0),\\ &S_n=\xi(U_n) \qquad (n\geqq 0),\\ &Y_n=S_n-S_{n-1} \qquad (n>0), \qquad\qquad (2)\\ &\Delta_n=\sigma\text{-field generated by } X_1,\dots,X_n \text{ and}\\ &\{\xi(t), t\leqq U_n\} \qquad (n>0), \end{aligned}$$

* Research under ONR Contract 401(50). Reproduction for any purpose of the U.S. Government is permitted.

1. Abbreviations used in this paper are listed in the last paragraph of this section.

21*

the following properties hold:

(a) the ξ-process is independent of $\{X_i\}$; T_n is Δ_n-measurable; $\{\xi(U_n+t)-\xi(U_n), t>0\}$ is independent of Δ_n;
(b) the T_i are i.i.d. with $E\{T_i\}=1$, (3)
(c) $\beta<\infty$,
(d) the sequence $\{Y_i, i>0\}$ has the same law as $\{X_i, i>0\}$.

We will refer to the setup described in the previous paragraph as the *i.i.d. case*, even though the general Skorokhod construction omits (1c) and (3c). Strassen [21], p. 333, has mentioned extensions, and has kindly informed the author that a more complete description is that Dambis and Dubins-Schwarz obtained analogues of Skorokhod's result for continuous parameter martingales, whereas Frank Jonas has carried out in detail the construction suggested by an observation (of Strassen and, independently, of David Freeman) that Skorokhod's construction extends to the case where

$$\left\{\sum_1^n X_i, 1\leqq n<\infty\right\}$$

is a martingale, with appropriate modification in the conclusion (3); subsequently Dubins [5] and Hall [9] obtained other constructions in the martingale case. We need not be more precise at this point, since the only properties we use of this *martingale case* (with restrictions whose consequences parallel those of (1c)) will be listed in (12) and (28) below; in particular, our results apply not just to Skorokhod's construction, but rather to general stopping variables, including those of Dubins and of Hall.

Strassen [19, 21] used the Skorokhod representation and its martingale extension in developing his beautiful strong invariance principles. Theorem 1.5 of [21] states that, in the i.i.d. case, as $n\to\infty$, with S_n obtained as in (2)–(3),

$$\xi(n)-S_n=O((n\log\log n)^{\frac{1}{4}}(\log n)^{\frac{1}{2}}) \qquad \text{wp 1.} \tag{4}$$

Strassen asks whether the existence of a probability space on which are defined a Brownian notion ξ and i.i.d. sequence $\{S_{n+1}-S_n, n\geqq 0\}$ with d.f. F, for which (4) holds with O replaced by o, implies that F is Gaussian. The answer in this generality, when S_n is not required to be obtained from stopping variables as in (2), is unknown. If S_n is assumed to be obtained as $\xi(U_n)$ from stopping times U_n as in (2)–(3), the question is answered affirmatively by the following result of the present paper:

Theorem 3. *In the i.i.d. case, under the assumptions of* (1), *if* $\beta>0$, *then for either choice of sign*

$$\limsup_{n\to\infty} \pm[\xi(n)-\xi(U_n)]/[2\beta n(\log n)^2\log\log n]^{\frac{1}{4}}=1 \qquad \text{wp 1.} \tag{5}$$

(Of course, when $\beta=0$ we have $T_n=1$ w.p. 1 and hence F is standard Gaussian.) Theorem 3 is obtained as a corollary of the more general Upper Class Theorem 1 and Lower Class Theorem 2, which do not require the Y_i to be i.i.d. (nor even for $\{\xi(U_n), n>0\}$ to be a martingale), but only the validity of certain properties (12) and (28) which are analogous to those of (1) or which are associated with usual proofs of the LIL.

In this paper it is only in the i.i.d. case (Theorem 3) that we verify in detail that such a simple condition as (1c) can be used to invoke Theorems 2 and 3. It is not hard to find sufficient conditions to invoke Theorems 2 and 3 for various cases where the Y_i are not i.i.d., but the known conditions are not sharp in the following sense that (1c) is: if (3c) fails in the i.i.d. case, the U_n do not obey the LIL (see Strassen [20]), and (4) will not generally hold.

We remark that Breiman [2] has considered the more general case of i.i.d. X_i not necessarily satisfying (1c), obtaining in this case bounds which are of a larger order than that of (4), and also necessary moment conditions for such behavior. While we do not consider such results in the present paper, we believe the techniques used herein may be adaptable to such cases where (1c) fails, through the use of analogues of standard LIL developments. To those who may be interested in pursuing such results, we will try to point out, where they occur, our few technical departures from usual patterns of proof of LIL-type theorems or from ideas of [12] and [13] which we have used.

Also, Breiman [3], Section 13.6, obtained a representation of the sample d.f. for uniformly distributed r.v.'s, in terms of the Brownian bridge, by using Skorokhod's construction for exponential r.v.'s $\{X_i\}$. It is easy to see that $n^{\frac{1}{2}}$ times the maximum error in this representation satisfies the right side of (4). Later Brillinger [4], evidently unaware of Breiman's book, obtained this result independently. See [15] for further comments.

The developments of this paper yield certain other results, as the interested reader will find it easy to verify. We now list some of these.

Theorem 4. *If* $\{\xi(t), t \geqq 0\}$ *is standard Brownian motion and* $0<\beta<\infty$, *then, for either choice of sign,*

$$\limsup_{t\to\infty} \Big\{ \sup_{|\tau-t|<[2\beta t \log\log t]^{\frac{1}{2}}} \pm[\xi(t)-\xi(\tau)]/[2\beta t(\log t)^2 \log\log t]^{\frac{1}{2}} \Big\} = 1 \quad \text{wp } 1$$

and

$$\limsup_{t\downarrow 0} \Big\{ \sup_{|\tau-t|<[2\beta t^3 \log|\log t|]^{\frac{1}{2}}} \pm[\xi(t)-\xi(\tau)]/[2\beta t^3(\log t)^2 \log|\log t|]^{\frac{1}{2}} \Big\} = 1 \quad \text{wp } 1.$$

The first of these results is obtained easily from the developments of the succeeding sections, and the second follows from the first on using time inversion ($\{t\,\xi(t^{-1}), t>0\}$ is a standard Brownian motion) and elementary estimates. Theorem 4 is related to the domain of Lévy's Hölder condition, the upper bound half of which in fact follows easily from (7) below. (See Ito-McKean [11], pp. 36–38).

The present paper also yields results concerning the *sample quantile process* $\eta^{(2)}$ introduced in Section 6 of [13], where we proved results for sample quantiles corresponding to the main results of [19] for sums of independent r.v.'s (Theorem 5 of [13]) and the *strong form of the LIL for sample quantiles* (Theorem 4 of [13]) corresponding to that of Feller [6] for sums of independent r.v.'s (strengthening the standard form of the LIL for sample quantiles obtained in [1]). Without taking the space here to define $\eta^{(2)}$, we remark that the result of [1] or Theorem 1 of [13] shows immediately that the order (with the right constant) of the deviation of ξ from $\eta^{(2)}$ is exactly the same as that of ξ from the random walk process $\eta^{(1)}$ (defined in Section 6.7 of [13]), and the latter is given at once by Theorem 3 of the present paper. See [15] regarding variation in p of p-tiles.

Finally, a development like that of Lemma 1 yields probability bounds for the difference between the maximum and minimum of partial sums and more general processes, analogous to Lévy-Kolmogorov-Hajek-Renyi bounds for the maximum or maximum norm. We will discuss this elsewhere [14].

It would be interesting to obtain "strong forms" corresponding to Theorems 3 and 4.

The following standard notation is used in this paper: The complement of an event A is denoted $\bar{A}$. The natural numbers and nonnegative reals are denoted by N and R^+, respectively. For real x, the greatest integer $\leqq x$ is denoted by int$\{x\}$. All "orders" (O or o) or asymptotic relations ($\sim$) refer to behavior as the exhibited dummy variable n or $r\to\infty$. We abbreviate "random variable" by "r.v.", "independent and identically distributed" by "i.i.d.", "distribution function" by "d.f.", "infinitely often" by "i.o.", "almost all n" (i.e., all n in N except for a finite number) by "a.a. n.", "law of the iterated logarithm" by "LIL", "with probability one" by "wp 1". Whenever we write such summation operations as $\sum_r$ it will be understood that the summation is over all r in N which are large enough that expressions like $\log\log n_r$ which appear in the summand are real.

The author is grateful to Volker Strassen for helpful comments.

2. Upper Class Result

In this section we suppose there is given a probability space on which are defined a standard Brownian motion $\{\xi(t), t\geqq 0\}$ and a sequence of nonnegative r.v.'s $\{T_i, i>0\}$. We define U_n as in (2). The only assumption we require is an upper class LIL-type estimate (12) on the sequence $\{U_n\}$; questions of dependence among the U_n and ξ do not otherwise enter.

As is often the case in LIL-type results, the upper class result is much easier to prove than the lower class result. Thus, Strassen rightfully calls (4) "easy" in the i.i.d. case, and our details are required only to obtain the right constant in (13). At the same time, a little delicacy is needed to obtain that constant. (It amounts to using (21) and (7) rather than an estimate like line 4, p. 22 of [2], and to replacing the n_r of [2] by numbers which yield our (13) rather than the larger order of Theorem 2 of [2].)

We remark also that, although Theorem 1 can be proved using approximately geometric times $n'_r\sim c^r$ as in the lower class proof of Theorem 2, instead of the n_r used below, it is not very enlightening to use such n'_r. This is because the desired result is concerned with the magnitude of oscillations of ξ over relatively short periods of time, which do not persist over long periods. In the lower class result we use such widely spaced n'_r and study what happens between succesive n'_r, because of such developments as the use of (50) to prove (43). (Other reasons for this choice will be found in the proof of Theorem 2 and the remarks which follow it.) But to use such n'_r in the upper class proof and then to subdivide the period between successive n'_r to look like the periods between the n_r below, is an unnecessary complication. Moreover, the direct use of the n_r makes the source of (4) and (5) transparent and points the way toward the right normalization in the cases where (1c) fails (as mentioned in the discussion of [2] in Section 1), although we have not investigated whether proof technicalities dictate the use of analogues of the n_r or the n'_r in such cases.

Before stating Theorem 1, we prove a simple lemma; it is related to the developments on pp. 36–38 of Ito-McKean [11], and it can also be obtained from the results on pp. 329–330 of Feller [7] (or, for large c, which is all that matters, from (3.6) of [8]), or from an application of the results of p. 651 of [12], although it will be shorter to prove it directly by a simple method which has further applications [14].

Lemma 1. *If ξ is standard Brownian motion and T, L, δ, c are positive values with $T<L$, then*

$$P\{\sup_{0\leqq t_1<t_2\leqq T}|\xi(t_1)-\xi(t_2)|\geqq c\}\leqq\frac{8\,T^{\frac{1}{2}}}{c(2\pi)^{\frac{1}{2}}}\,e^{-c^2/2T} \tag{6}$$

and

$$P\{\sup_{0\leqq t_1<t_2\leqq L,\,|t_1-t_2|\leqq T}|\xi(t_1)-\xi(t_2)|\geqq c\}\leqq\frac{8(L-T+\delta)(T+2\delta)^{\frac{1}{2}}}{\delta\,c(2\pi)^{\frac{1}{2}}}\,e^{-c^2/2(T+2\delta)} \tag{7}$$

Proof. We first prove (6). Let Γ be the event that for some v and w in $[0,T]$ with $v<w$ we have $|\xi(w)-\xi(v)|\geqq c$. The r.v. W is defined on Γ to be the least such w, and V is the least v corresponding to that w. Let $\Gamma^+=\Gamma\cap\{\xi(W)>\xi(V)\}$ and $\Gamma^-=\Gamma\cap\{\xi(W)<\xi(V)\}$. The event $W=w_1$ depends only on $\{\xi(t), 0\leqq t\leqq w_1\}$; hence, we have (as in Lévy's treatment of $\sup_{0\leqq t\leqq T}\xi(t)$)

$$P\{\xi(T)\geqq\xi(W)|\Gamma^+;\,W=w_1\}=\tfrac{1}{2} \tag{8}$$

and thus, since $\xi(W)-\xi(V)=c$ wp 1 on Γ^+,

$$\begin{aligned}P\{\Gamma^+\}&\leqq 2P\{\Gamma^+;\,\xi(T)\geqq\xi(W)\}\\&=2P\{\Gamma^+;\,\xi(T)-\xi(V)\geqq c\}\\&\leqq 2P\{\xi(T)-\min_{0\leqq t\leqq T}\xi(t)\geqq c\}\\&=2P\{\max_{0\leqq t\leqq T}[\xi(T)-\xi(T-t)]\geqq c\}.\end{aligned} \tag{9}$$

But the process $\{\xi(T)-\xi(T-t), 0\leqq t\leqq T\}$ is again a standard Brownian motion, so that Lévy's argument yields

$$P\{\Gamma^+\}\leqq 4P\{\xi(T)\geqq c\}. \tag{10}$$

The analogous result for Γ^- and the standard inequality for Gaussian tail probabilities yields (6).

We now turn to (7). Let $J=1+\mathrm{int}\{(L-T)/\delta\}$, and for $1\leqq j\leqq J$ define the event

$$F_j=\{\sup_{(j-1)\delta\leqq t_1,\,t_2\leqq T+j\delta}|\xi(t_1)-\xi(t_2)|\geqq c\}. \tag{11}$$

For every interval $[t_1,t_2]$ of length $\leqq T$ contained in $[0,L]$ there is at least one $j(1\leqq j\leqq J)$ such that $[t_1,t_2]\subset[(j-1)\delta, T+j\delta]$. Hence, the left side of (7) is no greater than $\sum_1^J P\{F_j\}$. Each $P\{F_j\}$ is bounded by (6) on replacing T there by $T+2\delta$. Since $J\leqq(L-T+\delta)/\delta$, we obtain (7).

We now state our upper class theorem.

Theorem 1. *Suppose there is a finite value $\beta>0$ such that*

$$\limsup_{n\to\infty} |U_n-n|/[2\beta n \log\log n]^{\frac{1}{2}} \leqq 1 \qquad \text{wp } 1. \tag{12}$$

Then

$$\limsup_{n\to\infty} |\xi(n)-\xi(U_n)|/[2\beta n(\log n)^2 \log\log n]^{\frac{1}{4}} \leqq 1 \qquad \text{wp } 1. \tag{13}$$

Proof. Let $\varepsilon>0$ be given, and write

$$q_n=[2\beta(1+\varepsilon)n\log\log n]^{\frac{1}{2}} \tag{14}$$

and

$$d_n=[2\beta(1+\varepsilon)^3 n(\log n)^2 \log\log n]^{\frac{1}{4}}. \tag{15}$$

Let $\{n_r\}$ be any increasing sequence of natural numbers satisfying

$$n_r=\{2^{-1}\beta(1+\varepsilon)r^2\log\log r\}[1+o(r^{-1})]. \tag{16}$$

It is easy to compute that

$$n_{r+1}-n_r \sim \beta(1+\varepsilon)r\log\log r \sim q_{n_r}. \tag{17}$$

Let

$$M_n=\{t\colon t\in R^+;\ |t-n|<q_n\} \tag{18}$$

and

$$M_r^*=\{(t,n)\colon t\in R^+;\ n\in N;\ |t-n|<q_{n_{r+1}};\ t,n\in[n_r-q_{n_{r+1}},\ n_{r+1}+q_{n_{r+1}}]\}. \tag{19}$$

We define the events

$$A_n=\{\sup_{t\in M_n}|\xi(t)-\xi(n)|>d_n\} \tag{20}$$

and

$$A_r^*=\{\sup_{(t,n)\in M_r^*}|\xi(t)-\xi(n)|>d_{n_r}\}. \tag{21}$$

Our theorem will be proved if we show that

$$P\{A_n \text{ occurs i.o.}\}=0. \tag{22}$$

If A_n occurs for some n satisfying $n_r \leqq n \leqq n_{r+1}$, then clearly A_r^* occurs. Hence, by the Borel-Cantelli lemma, (22) will follow from

$$\sum_r P\{A_r^*\}<\infty. \tag{23}$$

Let A_r^{**} be the event defined by (21) when N is replaced by R^+ in (19). Clearly A_r^* implies A_r^{**}. We bound $P\{A_r^{**}\}$ by applying (7) with

$$L=n_{r+1}-n_r+2q_{n_{r+1}} \sim 3\beta(1+\varepsilon)r\log\log r,$$

$$T=q_{n_{r+1}} \sim \beta(1+\varepsilon)r\log\log r,$$

$$c=d_{n_r} \sim [2\beta(1+\varepsilon)^2 r\log r\log\log r]^{\frac{1}{2}},$$

$$\delta=r,$$

and obtain

$$P\{A_r^*\} \leqq P\{A_r^{**}\} \leqq \frac{16\, q_{n_{r+1}}^{\frac{3}{2}} [1+o(1)]}{r\, d_{n_r} (2\pi)^{\frac{1}{2}}} \exp\{-d_{n_r}^2/2(q_{n_{r+1}}+2r)\}$$
$$\sim \frac{8\beta(1+\varepsilon)^{\frac{1}{2}} \log\log r}{(\pi \log r)^{\frac{1}{2}}} \exp\{-(1+\varepsilon)(1+o(1)) \log r\}. \tag{24}$$

This yields (23) and completes the proof of Theorem 1.

3. Lower Class Result

We again suppose there is given a probability space on which are defined a standard Brownian motion $\{\xi(t), t \geqq 0\}$ and a sequence of non-negative r.v.'s $\{T_i, i>0\}$, and define U_n as in (2). Our assumptions will be stated in (28) below in terms of certain positive values β, ε, and $\gamma>1$. For any choice of such values, let $\{n_r\}$ be a specific non-decreasing sequence satisfying

$$n_r \sim \gamma^r \tag{25}$$

(for example $n_r = \operatorname{int}\{\gamma^r\}$), define q_n by (14) and

$$m_r = n_r - n_{r-1}, \tag{26}$$

and define the events

$$D_r = \{|U_{n_r} - n_r| < q_{n_r}\},$$
$$B_r = \{(1-\varepsilon)\, q_{m_r} < (U_{n_r} - n_r) - (U_{n_{r-1}} - n_{r-1}) < q_{m_r}\}, \tag{27}$$
$$F_{r,\delta} = \{\max_{1 \leqq i \leqq \delta n_r} |U_{n_r+i} - U_{n_r} - i| < \varepsilon\, q_{m_r}\} \qquad \text{for } \delta>0.$$

Theorem 2. *Suppose there is a finite value $\beta>0$ such that for every sufficiently small $\varepsilon>0$ and sufficiently large $\gamma>0$ we have*

(a) $P\{D_r \text{ for a.a. } r\}=1$,
(b) $P\{B_r \text{ i.o.}\}=1$, (28)
(c) $P\{F_{r,\delta} \text{ for a.a. } r\}=1$ *for δ sufficiently small and positive.*

Then, for either choice of sign,

$$\limsup_{n\to\infty} \pm[\xi(n)-\xi(U_n)]/[2\beta\, n (\log n)^2 \log\log n]^{\frac{1}{2}} \geqq 1 \qquad \text{wp } 1. \tag{29}$$

Proof. Let ε be given, such that $0<\varepsilon<\frac{1}{400}$, and ε is small enough that (28) holds for all large γ. In particular,

$$(1-4\varepsilon)^2 (1+\varepsilon)^{-\frac{1}{2}} > \tfrac{16}{17}. \tag{30}$$

Choose γ so large that

$$\gamma - 1 > \varepsilon^{-2} \tag{31}$$

and that (28) holds. Choose δ so small that the equation of (28(c)) holds.

If the event

$$B_r \cap F_{r,\delta} \cap D_{r-1} \tag{32}$$

occurs, and if

$$n_r \leqq n \leqq n_r(1+\delta), \tag{33}$$

we have

$$(1-2\varepsilon)\, q_{m_r} - q_{n_{r-1}} < U_n - n < (1+\varepsilon)\, q_{m_r} + q_{n_{r-1}}. \tag{34}$$

Since $q_{n_{r-1}}/q_{m_r} \sim (\gamma-1)^{-\frac{1}{2}}$ and (31) holds, if r is larger than some constant (34) implies

$$(1-3\varepsilon)\, q_{m_r} < U_n - n < (1+2\varepsilon)\, q_{m_r}. \tag{35}$$

We now define

$$d'_n = [2\beta(1-4\varepsilon)^4\, n(\log n)^2 \log\log n]^{\frac{1}{2}}. \tag{36}$$

Let

$$J_r = \operatorname{int}\{\delta n_r/2 q_{m_r}\}, \tag{37}$$

and for $0 \leqq i < J_r$ define the numbers

$$\begin{aligned} n'_{r,i} &= n_r + \operatorname{int}\{i(1+5\varepsilon)\, q_{m_r}\}, \\ n''_{r,i} &= n_r + \operatorname{int}\{[i(1+5\varepsilon) + (1-4\varepsilon)]\, q_{m_r}\}, \end{aligned} \tag{38}$$

and the events

$$\begin{aligned} C'_{r,i} &= \{\xi(n''_{r,i}) - \xi(n'_{r,i}) > d'_{n_r}\}, \\ C''_{r,i} &= \{\sup_{0 \leqq x \leqq 8\varepsilon} |\xi(n''_{r,i} + x\, q_{m_r}) - \xi(n''_{r,i})| < 3\varepsilon^{\frac{1}{2}} d'_{n_r}\}, \\ Q'_r &= \bigcup_{0 \leqq i < J_r} C'_{r,i}, \\ Q''_r &= \bigcap_{0 \leqq i < J_r} C''_{r,i}. \end{aligned} \tag{39}$$

Suppose (35) holds for $n = n'_{r,i}$, where $0 \leqq i < J_r$ (so that (33) holds for large r). Then

$$n''_{r,i} < n'_{r,i} + (1-3\varepsilon)\, q_{m_r} < U_{n'_{r,i}} < n'_{r,i} + (1+2\varepsilon)\, q_{m_r} < n''_{r,i} + 8\varepsilon\, q_{m_r}, \tag{40}$$

the extreme inequalities requiring only $r >$ some constant. Obviously, (40) together with $C'_{r,i} \cap C''_{r,i}$ entails

$$\xi(U_{n'_{r,i}}) - \xi(n'_{r,i}) > (1 - 3\varepsilon^{\frac{1}{2}})\, d'_{n_r}. \tag{41}$$

Since ε is arbitrarily small, we conclude from (41) that Theorem 2 will follow from

$$P\{B_r \cap F_{r,\delta} \cap D_{r-1} \cap Q'_r \cap Q''_r \text{ i.o.}\} = 1. \tag{42}$$

We shall show below that

$$P\{Q'_r \text{ for a.a. } r\} = 1 \tag{43}$$

and

$$P\{Q''_r \text{ for a.a. } r\} = 1. \tag{44}$$

In view of (43), (44), (28a), and (28c), the condition (28b) thus will entail (42) and, thus, our theorem.

It remains to prove (44) and (43), which we do in that order. For the complement of $C''_{r,i}$ the familiar Lévy estimate and (30) yield, for all sufficiently large r, and $0 \leqq i < J_r$,

$$\begin{aligned}\log P\{\bar{C}''_{r,i}\} &\leqq -(3\varepsilon^{\frac{1}{2}} d'_{n_r})^2/2(8\varepsilon q_{m_r}) < -(3 d'_{n_r})^2/16 q_{n_r} \\ &= -(\tfrac{9}{16})(1-4\varepsilon)^2(1+\varepsilon)^{-\frac{1}{2}} \log n_r < -(\tfrac{9}{17}) \log n_r.\end{aligned} \tag{45}$$

Hence, for the complement of Q''_r we have from (37), (45), and (14), for all large r,

$$\begin{aligned}P\{\bar{Q}''_r\} &\leqq \sum_{i=0}^{J_r-1} P\{\bar{C}''_{r,i}\} < (\delta n_r/2 q_{m_r}) n_r^{-\frac{9}{17}} \\ &\sim (1-\gamma^{-1})^{-\frac{1}{2}} \delta n_r^{-\frac{1}{34}} [8\beta(1+\varepsilon) \log\log n_r]^{-\frac{1}{2}}.\end{aligned} \tag{46}$$

The Borel-Cantelli lemma and (46) yield $P\{\bar{Q}''_r \text{ i.o.}\} = 0$, which is equivalent to (44).

Next, since $1-5\varepsilon < [n''_{r,i} - n'_{r,i}]/q_{m_r} < 1-3\varepsilon$ for all sufficiently large r and for $0 \leqq i < J_r$, the standard Gaussian tail estimate then gives

$$P\{C'_{r,i}\} > \frac{[(1-6\varepsilon) q_{m_r}/2\pi]^{\frac{1}{2}}}{d'_{n_r}} \exp\{-(d'_{n_r})^2/2(1-3\varepsilon) q_{m_r}\}. \tag{47}$$

Hence, for all large r, and for $0 \leqq i < J_r$,

$$\begin{aligned}\log P\{C'_{r,i}\} &> -(d'_{n_r})^2/2 q_{m_r}(1-4\varepsilon) \\ &= -\tfrac{1}{2}(1-4\varepsilon)(1+\varepsilon)^{-\frac{1}{2}} \left\{\frac{n_r \log\log n_r}{m_r \log\log m_r}\right\}^{\frac{1}{2}} \log n_r.\end{aligned} \tag{48}$$

Since $n_r/m_r \sim \gamma/(\gamma-1) < [1+\varepsilon^2]$ by (31), and since $\varepsilon < \frac{1}{400}$, we obtain from (48) for all large r and for $0 \leqq i < J_r$,

$$\begin{aligned}\log P\{C'_{r,i}\} &> -\tfrac{1}{2}(1-4\varepsilon)[(1+\varepsilon^2)/(1+\varepsilon)]^{\frac{1}{2}} \log n_r \\ &> -\tfrac{1}{2}(1-\varepsilon) \log n_r.\end{aligned} \tag{49}$$

Since for each sufficiently large r the $C'_{r,i}$ $(0 \leqq i < J_r)$ are independent, we then obtain for the complement of Q'_r,

$$\begin{aligned}\log P\{\bar{Q}'_r\} &= \sum_{i=0}^{J_r-1} \log P\{\bar{C}'_{r,i}\} < (\delta n_r/2 q_{m_r}) \log[1 - n_r^{-(1-\varepsilon)/2}] \\ &< -(\delta n_r/2 q_{m_r}) n_r^{-(1-\varepsilon)/2} \\ &\sim -\delta[8(1+\varepsilon)\beta(\gamma-1)\gamma^{-1} \log\log n_r]^{-\frac{1}{2}} n_r^{\varepsilon/2}.\end{aligned} \tag{50}$$

The Borel-Cantelli lemma yields $P\{\bar{Q}'_r \text{ i.o.}\} = 0$, which is equivalent to (43). This completes the proof of Theorem 2.

Remarks on Theorem 2.

1. The conclusion of Theorem 2 obviously remains valid under such changes as replacing U_n by $-U_n$ in the definition of B_r, and replacing (28c) by $\lim_{\delta \downarrow 0} P = 1$ (which is pretty artificial in view of $0-1$ laws for $\{U_n\}$ which will usually hold in applications).

2. At first sight it may appear strange that no analogue of the last part of (3a) is assumed in Theorem 2. This is misleading, since the strong assumption (28c) allows us to avoid computing conditional probabilities of certain events, given B_r, which might require an analogue of (3a). Moreover, the verification of (43) by use of (50) makes strong use of the independent increments of ξ, and the use of (43)–(44) in our proof requires (28b) to hold; the verification of the latter in the application of Theorem 2 to Theorem 3 (Section 4) makes use of the structure of the $\{U_n\}$ process, and this can be expected in other applications. Thus, a weakening of (28), especially of (28c), can be given if one assumes conditions including analogues of the last part of (3a), but so far we have not obtained natural and useful conditions of this form.

3. A crucial aspect of the proof is that although Q'_r is concerned only with a small fraction $\delta/(\gamma-1)$ of the interval $[n_r, n_{r+1}]$, the number J_r is still large enough to yield (43). In generalizations and extensions one may want to replace (43)–(44) by an analogue of the weaker statement that $\bigcup_i (C'_{r,i} \cap C''_{r,i})$ occurs for a.a. r wp 1.

4. Proof of Theorem 3

We must verify the three parts of (28) in the i.i.d. case under the assumption (1) with $\beta > 0$.

Condition (28a) is a consequence of the Hartman-Wintner LIL [10].

To prove (28c), suppose $\delta < \gamma - 1$. Write $n_r^* = \text{int}\{n_r(1+\delta)\}$. There is a value ρ such that $n_r + 1 < n_r^* < n_{r+1}$ for $r \geqq \rho$. The r.v.'s T_i, $n_r < i \leqq n_r^*$, $r \geqq \rho$, are then distinct, and we relabel the sequence

$$T_{n_\rho+1}, T_{n_\rho+2}, \ldots, T_{n_\rho^*}, T_{n_{\rho+1}+1}, \ldots, T_{n_{\rho+1}^*}, T_{n_{\rho+2}+1}, \ldots$$

as

$$1+V_1, 1+V_2, 1+V_3, \ldots.$$

The Hartman-Wintner LIL for the V_i says that

$$\left|\sum_1^m V_i\right| < q_m \tag{51}$$

for a.a. m, wp 1. If (51) holds for two values $m = M$ and $m = k > M$, we have

$$\left|\sum_{M+1}^{k} V_i\right| \leqq \left|\sum_1^k V_i\right| + \left|\sum_1^M V_i\right| < q_k + q_M. \tag{52}$$

If f is any function from N into N with $f(n) > n$ for all n, we conclude that, for a.a. M in N, wp 1,

$$\begin{aligned} \max_{M<k\leqq f(M)} \left|\sum_{M+1}^{k} V_i\right| &< q_M + q_{f(M)} \\ &< 2q_{f(M)}. \end{aligned} \tag{53}$$

In particular, selecting only the values $M=n_r$, $r \geqq \rho$, with f chosen so that $f(n_r)=n_r^*$, and writing

$$k_r = \sum_{j=\rho}^{r} (n_r^* - n_r) \tag{54}$$

for $r \geqq \rho$, we obtain

$$\max_{1 \leqq i \leqq \delta n_r} |U_{n_r+i} - U_{n_r} - i| < 2q_{k_r} \qquad \text{for a.a. } r \geqq \rho, \text{ wp 1.} \tag{55}$$

Since $k_r \sim n_r \, \delta\gamma/(\gamma-1) \sim m_r \, \delta\gamma^2/(\gamma-1)^2$, we have

$$2q_{k_r} \sim [2\delta^{\frac{1}{2}} \gamma/\varepsilon(\gamma-1)] \, \varepsilon q_{m_r}. \tag{56}$$

Hence, if $\delta < \varepsilon^2(\gamma-1)^2/4\gamma^2$, we have $2q_{k_r} < \varepsilon q_{m_r}$ for all large r, and then (28c) follows from (55) and the definition (27).

We turn finally to (28b). While this condition can be verified in several ways, an expeditious proof relies on the Hartman-Wintner truncation scheme [10] and the validity of the analogue of (28b) in Kolmogorov's lower class proof [16], as we shall now show.

Firstly, given $0<\varepsilon<1$, let $\varepsilon'>0$ be such that

$$1-2\varepsilon' > (1-\varepsilon)(1+\varepsilon)^{\frac{1}{2}} \quad \text{and} \quad 1+3\varepsilon' < (1+\varepsilon)^{\frac{1}{2}}. \tag{57}$$

If T_i is truncated and centered (by constants depending only on i), say to Z_i, exactly as in [10], so that $|Z_n| = o(n/q_n)$, then it is shown in [10] that

$$\sum_1^n (T_i - Z_i) = o(q_n) \qquad \text{wp 1.} \tag{58}$$

Moreover, the developments of Kolmogorov's lower class proof [16] (also to be found in Loève [17], pp. 260–262) show, after minor arithmetic to verify the negligible difference (in the appropriate sense) between the first two moments of $\sum_1^n Z_i$ and U_n, that if

$$H_r = \sum_{i=n_{r-1}+1}^{n_r} Z_i - m_r, \tag{59}$$

then for each small positive ε' there is a small positive ε'' such that, for all large r,

$$P\{H_r > (1-\varepsilon')[2\beta m_r \log\log m_r]^{\frac{1}{2}}\} > r^{\varepsilon''-1}. \tag{60}$$

On the other hand, the analogue of the first inequality of (53) for the Z_i (and their LIL) rather than the V_i yields

$$|H_r| < (1+\varepsilon')(2\beta)^{\frac{1}{2}} \{(n_r \log\log n_r)^{\frac{1}{2}} + (n_{r-1} \log\log n_{r-1})^{\frac{1}{2}}\} \qquad \text{for a.a. } r, \quad \text{wp 1.} \tag{61}$$

If γ is so large that $(1+\varepsilon')(\gamma^{\frac{1}{2}}+1)/(\gamma-1)^{\frac{1}{2}} < (1+2\varepsilon')$, the Borel-Cantelli lemma for the independent H_r of (61), together with (60), yields

$$\sum_r P\{1-\varepsilon' < H_r[2\beta m_r \log\log m_r]^{-\frac{1}{2}} < 1+2\varepsilon'\} = +\infty. \tag{62}$$

By (58) and the analogue of (53) for the $T_i - Z_i$ instead of the V_i,

$$\left|\sum_{i=n_{r-1}+1}^{n_r} (T_i - Z_i)\right| < \varepsilon' [2\beta m_r \log\log m_r]^{\frac{1}{2}} \qquad \text{for a.a. } r, \text{ wp } 1. \tag{63}$$

The sequence (in r) of sums on the left side of (63) being independent, the Borel-Cantelli lemma and (63), with (62), yield

$$\sum_r P\{1 - 2\varepsilon' < [2\beta m_r \log\log m_r]^{-\frac{1}{2}} (U_{n_r} - U_{n_{r-1}} - m_r) < 1 + 3\varepsilon'\} = +\infty. \tag{64}$$

From (57), the definition of (27), and the Borel-Cantelli lemma applied to the independent events of (64), we obtain (28 b). This completes the proof of Theorem 3.

References

1. Bahadur, R.: A note on quantiles in large samples. Ann. math. Statistics **37**, 577–580 (1966).
2. Breiman, L.: On the tail behavior of sums of independent random variables. Z. Wahrscheinlichkeitstheorie verw. Geb. **9**, 20–25 (1967).
3. – Probability. Reading, Mass.: Addison-Wesley 1968.
4. Brillinger, D. R.: An asymptotic representation of the sample d.f. Bull. Amer. math. Soc. **75**, 545–547 (1969).
5. Dubins, L.: On a theorem of Skorokhod. Ann. math. Statistics **39**, 2094–2097 (1968).
6. Feller, W.: The general form of the so-called law of the iterated logarithm. Trans. Amer. math. Soc. **54**, 373–402 (1953).
7. – An introduction to probability theory and its applications, vol. II. New York: John Wiley 1966.
8. – The asymptotic distribution of the range of sums of independent random variables. Ann. math. Statistics **22**, 427–432 (1951).
9. Hall, W. J.: On the Skorokhod embedding theorem. (To be published.)
10. Hartman, P., Wintner, A.: On the law of the iterated logarithm. Amer. J. Math. **63**, 169–176 (1941).
11. Ito, K., McKean, H. P., Jr.: Diffusion processes and their sample paths. New York: Academic Press 1965.
12. Kiefer, J.: On large deviations of the empiric d.f. of vector chance variables and a law of the iterated logarithm. Pacific J. Math. **11**, 649–660 (1961).
13. – On Bahadur's representation of sample quantiles. Ann. math. Statistics **38**, 1323–1342 (1967).
14. – On the range of a general random walk. (To be published.)
15. – Old and new methods for studying order statistics and sample quantiles. Proc. 1st Intl. Conference on Nonparametric Inference.
16. Kolmogorov, A.: Das Gesetz des iterierten Logarithmus. Math. Ann. **101**, 126–135 (1929).
17. Loève, M.: Probability theory, 2nd edn. New York: Van Nostrand 1960.
18. Skorokhod, A.: Studies in the theory of Random processes. Reading, Mass.: Addison-Wesley 1965.
19. Strassen, V.: An invariance principle for the law of the iterated logarithm. Z. Wahrscheinlichkeitstheorie verw. Geb. **3**, 211–226 (1964).
20. – A converse to the law of the iterated logarithm. Z. Wahrscheinlichkeitstheorie verw. Geb. **4**, 265–268 (1966).
21. – Almost sure behavior of sums of independent random variables and martingales. Proc. Fifth Berkeley Sympos. math. Statist. Probab. Vol. II (Part I), 315–343 (1967).

Professor Dr. J. Kiefer
Department of Mathematics
Cornell University
Ithaca, N.Y. 14850, USA

(Received August 23, 1968)

Reprinted from *Nonparametric Techniques in Statistical Inference*, edited by Madan Lal Puri, C.U.P. 1970.

OLD AND NEW METHODS FOR STUDYING ORDER STATISTICS AND SAMPLE QUANTILES

J. KIEFER†

This paper, prepared for presentation with those of Weiss [21] and of Eicker [6] in this volume, contains some comments on those two papers, as well as related remarks on other results and unsolved problems in the area described by the present paper's title.

1 MULTINOMIAL EVENTS EQUIVALENT TO THOSE CONCERNING SAMPLE QUANTILES: DISTRIBUTION THEORY

It is remarkable that the years have seen such great neglect or ignorance of the simple approach used here by Weiss [21], of studying limit probabilities of events concerning sample quantiles by rewriting them as events concerning multinomial rv's. Presented with Weiss's paper, I decided to check once more on the approach used in standard texts in the simplest problem of this type—that of finding the limiting df of a single sample p-tile $Y_{p,n}$ $(0 < p < 1)$ of n iid rv's $X_1, \ldots, X_n$ when the df F of X_1 has positive continuous density f at the corresponding population p-tile, ξ_p. To my horror, each of the two dozen texts I found on my shelf which treated the subject at all, used a less wieldy proof in terms of the exact sample quantile density rather than one in terms of the limiting suitably normalized binomial df. Perhaps this should not have been too surprising in view of Weiss's comments and the need many instructors must have shared with me for years, of distributing supplementary notes on this topic when it arose in an elementary course; but one would have thought that the approach in Cramér or Wilks might have been abandoned in at least one standard text after all this time, in favor of the simpler approach.

Of course, this simpler approach can be found in earlier research papers of Weiss, as well as in David Moore's recent work on efficient multivariate location-parameter estimation; as Weiss indicates, the

† This paper was prepared as discussion of the papers of Weiss and Eicker at the First International Conference on Nonparametric Inference, under support from ONR Contract No. NONR 401 (50). Reproduction in whole or in part is permitted for any purpose of the United States Government.

[349]

computational advantage of using the multinomial technique becomes even greater in these more complex settings. What is the ancient history of the technique, which, while not deep, is so simple and useful? In a 'Classroom Note' to appear in the M.A.A. *Monthly*, in which Moore also gives an exposition of this technique, he mentions the passing reference to the method in Feller [10] (vol. 2, p. 23) and the use in Loève's paper [14]. There are undoubtedly other appearances of the method in research papers. (This morning I noticed a paper in the 1968 *JRSS(B)*, p. 570, by A. M. Walker, in which the multinomial treatment initiates a proof which for some reason then unnecessarily uses the characteristic function.) I do not know who first used the approach, but know that it was taught at Columbia (and, undoubtedly, elsewhere) twenty years ago; the earliest use I know of in print is in Smirnov's 1949 paper [18].

As Weiss remarks in his last section, this technique can also be used when $f(\xi_p) = 0$: indeed, (1) below is still completely trivial, and the arithmetic which follows is solely to illustrate the explicit computation of the approximate inverse df in particular cases. Suppose the df of X_1 is

$$\begin{cases} p - G_1(\xi_p^{(1)} - x) & \text{for } \; x \leqslant \xi_p^{(1)}, \\ p & \text{for } \; \xi_p^{(1)} \leqslant x \leqslant \xi_p^{(2)}, \\ p + G_2(x - \xi_p^{(2)}) & \text{for } \; \xi_p^{(2)} \leqslant x, \end{cases}$$

where $G_i(x) = 0$ only at $x = 0$ (and of course $\xi_p^{(1)} = \xi_p^{(2)}$ in the case of a unique p-tile). If the G_i are suitably regular, the domain of the asymptotic density of $Y_{p,n}$ can be conveniently broken up into two parts, as in Weiss's treatment (in addition to the part between the $\xi_p^{(i)}$ where this density is zero). We need only treat one part, corresponding to positive values of $Y_{p,n} - \xi_p^{(2)}$. Suppose G_2 is continuous and strictly increasing in an interval $[0, \epsilon]$ for some $\epsilon > 0$, and let G_2^{-1} be its inverse in that interval. Write $\sigma_p = [p(1-p)]^{\frac{1}{2}}$. For fixed $t > 0$ one then has, by the binomial equivalence,

$$\lim_{n \to \infty} P\{Y_{p,n} - \xi_p^{(2)} < G_2^{-1}(t\sigma_p n^{-\frac{1}{2}})\} = \Phi(t). \tag{1}$$

If G_2 is sufficiently regular, this can be made more explicit because G_2^{-1} will have a simple approximate form. Suppose $G_2(t)$ is regularly varying of exponent $r > 0$, as $t \downarrow 0$; that is, as $t \downarrow 0$,

$$G_2(t) \sim Ct^r L(t) \tag{2}$$

for some finite positive value C, where $L(t)$ is 'slowly varying', i.e. $\lim_{t \downarrow 0} L(t)/L(ct) = 1$ for each $c > 0$. We suppose also that, for $\alpha > 0$, $c > 0$, and β real,

$$\lim_{t \downarrow 0} L(ct^{\alpha}[L(t)]^{\beta})/L(t) = \gamma_{\alpha}, \tag{3}$$

where γ_α is a finite positive value not depending on c or β.† Then it is easily verified that, as $t \downarrow 0$,

$$G_2^{-1}(t) \sim [t/C\gamma_r L(t)]^{1/r}. \tag{4}$$

Define the positive value z by setting $zG_2^{-1}(n^{-\frac{1}{2}}\sigma_p \gamma_r \gamma_{1/r}^{-1})$ asymptotically equal to $G_2^{-1}(t\sigma_p n^{-\frac{1}{2}})$ in (1); it will be seen that z can be taken to be constant as $n \to \infty$. Abbreviating $n^{-\frac{1}{2}}\sigma_p \gamma_r \gamma_{1/r}^{-1}$ by u as $u \downarrow 0$, we have

$$\begin{aligned} G_2(zG_2^{-1}(u)) &\sim Cz^r u[C\gamma_r L(u)]^{-1} L(zu^{1/r}[C\gamma_r L(u)]^{-1/r}) \\ &\sim z^r u\gamma_{1/r}/\gamma_r = z^r n^{-\frac{1}{2}}\sigma_p, \end{aligned} \tag{5}$$

and from (5) as well as (4) with $t = u$ we obtain, for $z > 0$,

$$\lim_{n\to\infty} P\{[n^{\frac{1}{2}}C\gamma_{1/r}\gamma_{\frac{1}{2}} L(n^{-1})\,\sigma_p^{-1}]^{1/\alpha}(Y_{p,n} - \xi_p^{(2)}) < z\} = \Phi(z^r). \tag{6}$$

Another way of saying this, if we write a superscript 2 on C, L, r, and γ, and suppose a result corresponding to (6) to hold for $Y_{p,n} - \xi_p^{(1)} < 0$ (with superscripts 1 on C, L, γ, r), is that $\psi_n(Y_{p,n})$ is asymptotically standard normal, where

$$\psi_n(y) = \begin{cases} n^{\frac{1}{2}}C^{(2)}\gamma_{1/r^{(2)}}^{(2)}\gamma_{\frac{1}{2}}^{(2)} L^{(2)}(n^{-1})\,\sigma_p^{-1}(y - \xi_p^{(2)})^{r^{(2)}} & \text{if } y > \xi_p^{(2)}, \\ 0 \quad \text{if } \xi_p^{(1)} \leqslant y \leqslant \xi_p^{(2)}, \\ -n^{\frac{1}{2}}C^{(1)}\gamma_{1/r^{(1)}}^{(1)}\gamma_{\frac{1}{2}}^{(1)} L^{(1)}(n^{-1})\,\sigma_p^{-1}(\xi_p^{(1)} - y)^{r^{(1)}} & \text{if } y < \xi_p^{(1)}. \end{cases} \tag{7}$$

An important area for further research is that of establishing bounds on the difference between the actual distribution and the limiting df of a rv involving sample quantiles or order statistics, analogous to estimates of Berry–Esseen type for sums. Jumping ahead to the topic of the next section, we note that an invariance principle with error term, used in studies of certain functionals of the sample df S_n by such authors as Skorokhod [17] and Rosenkrantz [15], [16], can be a powerful tool.

† The condition (3) is not automatically satisfied by a slowly varying function, as can be seen from its failure for the function $L(t) = \exp\{(\log t^{-1})/\log\log t^{-1}\}$. It holds, for example, for $L(t) = \prod_{i=1}^{k} (\log_i t^{-1})^{a_i}$ where as usual $\log_1 = \log$ and $\log_{i+1} = \log \log_i$. For this function L, one has $\gamma_\alpha = \alpha^{a_1}$. Of course, the assumption (3) is only used for special values of α, β, c in the sequel. One can always take $C = 1$ in (2), but might not, for convenience.

The reason for developing this example in detail is to illustrate a simplication which often arises in statistics where inverses are concerned; the case $r = 0$ can be more difficult from this point of view, since in place of the simple expression (4) one must express the inverse of L in that case. The arithmetic of (4) is relevant in such other settings as nonregular ML examples of arbitrary normalization; for example, if the X_i have unknown location parameter θ and $P_\theta\{X_i - \theta \leqslant x\}$ is $G_2(x)$ for $x \geqslant 0$ and 0 for $x < 0$, where $G_2^{(x)}$ is concave for $x \geqslant 0$ and satisfies (2) and (3), then the ML estimator $\hat{T}_n = \min(X_i, \ldots, X_n)$ satisfies $P\{\hat{T}_n - \theta > G_2^{-1}(t/n)\} \sim e^{-t}$ for $t > 0$ as $n \to \infty$, as is well known, and (4) can be used to simplify this.

2 INVARIANCE PRINCIPLES; STRONG LAWS

Classical uses of the probabilistic invariance principle in limiting distribution theory are by now too familiar to be detailed here; it will suffice to recall a few of the early names such as Erdös, Kac, Donsker, Doob, Prohorov. Recent developments of interest here (in addition to those referred to in the previous paragraph) are (i) the introduction by Strassen [19], [20] of *strong invariance principles* which allow such results as the law-of-the-iterated-logarithm (LIL) to be obtained for sums $T_n = \sum_1^n U_i$ of fairly general sequences of rv's $\{U_i\}$ ($\{T_n\}$ can be a martingale satisfying appropriate conditions) from corresponding results for Brownian motion $\{B(t), t \geqslant 0\}$, as a consequence of a Skorokhod–Strassen representation of the form

$$T_n - ET_n - B(\operatorname{var} T_n) = O(g(\operatorname{var} T_n)) \quad wp\,1 \quad \text{as} \quad n \to \infty, \tag{8}$$

where $g(m)$ is of suitably small order as $m \to +\infty$; and (ii) the recognition by Bahadur of the close relationship between the *linear* process $\{nS_n(\xi_p)\}$ (a sequence of *sums* of iid rv's) and the *nonlinear* process $\{Y_{p,n}\}$ (no obvious transform of which is a martingale), for fixed p. Analogous to the use just mentioned of the development (i), the development (ii) allows strong laws about $\{nS_n(\xi_p)\}$ to be used to prove corresponding results about $\{Y_{p,n}\}$; this can also be thought of as a strong analogue to the corresponding distribution theory results discussed in §1, wherein the limiting df of normalized $Y_{p,n}$ is obtained from that of $S_n(\xi_p)$. Since strong convergence of $Y_{p,n}$ (in the case of unique ξ_p) is well known, we shall mention here LIL results.

Let us first backtrack to the direct application of the device of §1 for each n (as distinct from Bahadur's application of a representation for the whole sequence $\{Y_{p,n}\}$ at once). It is not too hard to see (and is probably fairly well known although not well publicized) that a LIL for $Y_{p,n}$ can be obtained in cases where f is positive and suitably regular† at ξ_p, by making appropriate alterations in the standard proof for Bernoulli rv's (as it appears, for example, in vol. 1 of Feller [10]), using the binomial equivalence of events concerning $Y_{p,n}$. (The main departure is that the study of $S_n(\xi_{p_n})$, with p_n varying slightly, is no longer quite that of events concerning an iid Bernoulli sequence.) Even easier is the ana-

† While some LIL-type results can be obtained without difficulty when $f(\xi_p) = 0$ but f is positive (except at ξ_p) and regular in a neighborhood of ξ_p, we omit description of these in this brief discussion.

logous result for an order statistic of fixed order, say $Z_{k,n} = k$th smallest of $X_1, X_2, \ldots, X_n$; for example, the bottom half is that, if the X_i are independent and uniformly distributed on [0, 1] and $nq_n \downarrow 0$, then

$$P\{Z_{k,n} < q_n \text{ for infinitely many } n\} = 1 \Leftrightarrow \sum_{n=1}^{\infty} n^{k-1} q_n^k = +\infty. \tag{9}$$

The domain of results between (9) and the LIL for $Y_{p,n}$ (where $|k-np| \leqslant 1$), namely,

$$\text{where} \quad k \to \infty \quad \text{as } n \to \infty \quad \text{but } k/n \to 0,$$

is not fully explored; we will return to this in § 3. For now, let us mention the much simpler way of proving the LIL for $Y_{p,n}$ which was perceived by Bahadur [1], who obtained it at once from the LIL for the binomial rv's $nS_n(\xi_p)$ upon invoking his result referred to in the previous paragraph, that

$$R_n(p) = Y_{p,n} - \xi_p + [S_n(\xi_p) - p]/f(\xi_p) \tag{10}$$

is of much smaller order than $n^{-\frac{1}{2}}$, wp 1. In fact, Bahadur's device enables one to obtain easily even the finer characterization of functions of upper and lower class for $\{Y_{p,n}\}$ [11], a result which would be very difficult to obtain (if, in fact, it is so obtainable) by modifying an existing proof [9] in the Bernoulli case in the manner mentioned above for the simple LIL result. The proof is like that at the end of § 3.

It would be very interesting to obtain a representation of the sample quantile process $\{Y_{p,n}\}$ for fixed p in terms of something like Brownian motion, in the manner of Skorokhod and Strassen for sums of rv's as in (8). One can do this *by using Bahadur's result* (see [11]), but that is the wrong direction, since one would like to *prove* Bahadur's result from a start with such a quantile process defined in terms of Brownian motion. A major difficulty, of course, is that $\{Y_{p,n}\}$ is not exactly Markovian, a martingale, or anything else that seems useful. I believe Miss Helen Finkelstein, a student of Strassen at Berkeley, worked on such a representation, but do not know the details. (We will return to this process in (15) below, but there p will also vary.)

One tempting attempt in this direction is to use Breiman's [3] Skorokhod–Strassen type construction of the Brownian bridge representation of $\{n^{\frac{1}{2}}(S_n(t) - t),\ 0 \leqslant t \leqslant 1\}$ in the case where the X_i are independent and uniform on [0, 1]; this uses (8) in the case where the U_i are iid and

$$P\{U_i > u\} = e^{-u} \quad \text{for} \quad u > 0.$$

It is well known in that case that $\{T_i/T_{n+1}, 1 \leqslant i \leqslant n\}$ has the same distribution as the order statistics from $\{X_1, X_2, \ldots, X_n\}$. Let S_n' be the 'sample df' constructed from $\{T_i/T_{n+1}, 1 \leqslant i \leqslant n\}$. Breiman uses this development to show that

$$\rho_n(t) = S'_n(t) - t - n^{-\frac{1}{2}}[B(nt) - tB(n)]$$

is suitably small.† Why, then, should we not use the sequence

$$\{T_{[np]}/T_{n+1}\}$$

(where $[x]$ is the greatest integer $\leqslant x$) and consequently

$$\{n^{-\frac{1}{2}}[B(np) - pB(n)]\} \quad \text{to approximate} \quad \{Y_{p,n}\}?$$

The difficulty is that, although the approximation is satisfactory for each fixed n, the joint distribution of the sequence is not appropriate; one can see this at once upon noting how T_i/T_{n+1} for each fixed i behaves as n varies.

This shortcoming also applies to the possibility of using Breiman's representation to obtain strong laws (such as Chung's LIL [5]) for $\{S_n\}$ from corresponding results for B. The construction of a Skorokhod–Strassen type representation for $\{S_n\}$, which is useful for obtaining strong laws, remains an open problem.

3 RECENT WORK ON BAHADUR'S REPRESENTATION

Eicker is to be commended for the fortitude he has shown in carrying out his delicate calculations. As he states, this delicacy appears in the precise nature of his combinatorial probabilities and in his using approximations only at a later stage of the proof than I did. However, since his proof (of the upper class result, which is all that is provided here) is somewhat longer, we should ask what its benefits are. There are several possibilities. First, perhaps the finer complete characterization of upper- and lower-class functions (that is, the description of those sequences $\{\phi_n\}$ for which $R_n(p) > \phi_n$ for finitely- or infinitely-many n, wp 1), which is as yet unknown,‡ requires such finer calculations. This is not really clear; on the one hand, both Feller and Chung have made it

† Subsequently Brillinger [4], evidently without knowing Breiman's result, used the same development. In addition, he computed in detail an upper bound on the order (wp 1) of $\sup_{0\leqslant t\leqslant 1} |\rho_n(t)|$. It is shown elsewhere [13] how this last result can be obtained in a few lines from Bahadur's representation and the order of $\sup_p |R_n(p)|$ discussed in the next section, and how $\rho_n(t)$ can thereby be more clearly described, and its exact order computed.

‡ Not to be confused with the corresponding complete characterization for $Y_{p,n}$, which was mentioned in §2.

very clear in analogous complete characterizations in LIL-type theorems they have proved, that asymptotic estimates of tail probabilities, with fairly rough error terms, suffice; on the other hand, the present setting contains notable differences from the earlier LIL problems, in that one is not considering sums of basic rv's but rather a smaller order difference between such a sum and a rv almost equal to that sum.

Secondly, perhaps such exact calculations may help in giving better demonstrations than now exist, of the limiting behavior of

$$R_n^* = \sup_{0 \leqslant p \leqslant 1} |R_n(p)| f(\xi_p).$$

As mentioned in [11], some results are known, but there are difficulties in obtaining such results from convergence results for the process $\{R_n(t), 0 \leqslant t \leqslant 1\}$, which cannot be normalized to approach weakly a nontrivial separable limiting process. The known results, proofs of which appear in [12], are that, if f is bounded away from 0 and f' is bounded on a finite interval to which f assigns probability 1, then

$$\left.\begin{aligned} &\lim_{n\to\infty} P\{R_n^* n^{\frac{3}{4}}(\log n)^{-\frac{1}{2}} > r\} = 2\sum_{m=1}^{\infty}(-1)^{m+1} e^{-2m^2r^4} \quad \text{for} \quad r > 0; \\ &\limsup_{n\to\infty} R_n^* n^{\frac{3}{4}}(\log n)^{-\frac{1}{2}}(\log\log n)^{-\frac{1}{4}} = 2^{-\frac{1}{4}} \quad \text{wp } 1. \end{aligned}\right\} \tag{11}$$

In fact, the first line of (11) is a trivial consequence of the fact that

$$R_n^* n^{\frac{1}{2}}(\log n)^{-\frac{1}{2}} / \sup_x |S_n(x) - F(x)|^{\frac{1}{2}} \to 1 \quad \text{in probability,}$$

which more fundamental result is proved in [12]. The second line of (11) follows from a corresponding strong law and the LIL [5] for the sample df deviations.

In connection with Eicker's Appendix B and his paper [7] quoted here, one should mentioned the related work on Takács and others on ballot problems.

Finally, let us describe the domain of usefulness of Bahadur's representation in studying the LIL for $\{Y_{p_n,n}\}$ when $p_n \downarrow 0$ as mentioned below (9). Of course, a bound on $R_n(p)$ for *fixed* p is no longer relevant as it was for determining the upper and lower classes for $\{Y_{p,n}\}$. Supposing again that the iid X_i are uniformly distributed on [0, 1], Eicker [8] recently showed that, if $p_n > (\log\log n)/o(n)$, then

$$\limsup_{n\to\infty} [n/2p_n \log\log(np_n)]^{\frac{1}{2}} [S_n(p_n) - p_n] = 1 \quad \text{wp } 1. \tag{12}$$

On the other hand, one of Baxter's results [2] is that, for fixed $c > 0$,

$$\limsup_{n\to\infty} [n(\log\log\log n)/\log\log n]\, S_n(cn^{-1}) = 1 \quad \text{wp } 1. \tag{13}$$

12-2

From the order (wp 1) of R_n^* given in (11), it follows at once from (12) that, if also $n^{\frac{1}{2}} p_n (\log n)^{-1} (\log\log n)^{\frac{1}{2}} \to \infty$,

$$\limsup_{n\to\infty} [n/2p_n \log\log(np_n)]^{\frac{1}{2}} [p_n - Y_{p_n, n}] = 1 \quad \text{wp } 1. \tag{14}$$

This is similar to the use of $R_n(p)$ to obtain the LIL for $Y_{p,n}$ as mentioned in §2. On the other hand, no corresponding conclusion about $Y_{c/n,n}$ is derivable from (13) from the estimate (11) on R_n^*; this suggests the problem of finding the order of $R_n(cn^{-1})$ or more generally of $R_n(p_n)$ with $p_n \downarrow 0$. Results on $S_n(p_n)$ and Y_{p_n}, n will appear elsewhere.

If one studies not $\{Y_{p_n,n}\}$ for a single sequence $\{p_n\}$, but rather the whole 'quantile process' $\{Y_{p,n}, 0 \leqslant p \leqslant 1\}$, then the result (11) again yields a LIL from the corresponding analogous result [5] for $\{S_n(t), a \leqslant t \leqslant b\}$ where $[a, b]$ is the interval where $f > 0$. The result is that, for $\lambda_n \uparrow \infty$,

$$P\{\sup_{0\leqslant p\leqslant 1} f(\xi_p)|Y_{p,n} - \xi_p| > \lambda_n n^{-\frac{1}{2}} \text{ infinitely often}\}$$
$$= \begin{Bmatrix} 1 \Leftrightarrow \infty = \\ 0 \Leftrightarrow \infty > \end{Bmatrix} \sum_n n^{-1} \lambda_n^2 e^{2\lambda_n^2}. \tag{15}$$

(An examination of the proof in [5] shows that the same criterion holds for the maximum deviation of one sign rather than maximum absolute deviation, and thus the same holds true in (15).) The proof of (15) develops along familiar lines from the fact [5] that the same criterion on $\{\lambda_n\}$ holds for $\sup_t |S_n(t) - t|$. One first notes, from [5] and (15), that

$$\lambda_n^* = [2^{-1}(1 * \epsilon) \log n]^{\frac{1}{2}}, \quad \text{for} \quad * = + \text{ or } -,$$

gives a sequence in each of the upper and lower classes for

$$\sup_p f(\xi_p)|Y_{p,n} - \xi_p|;$$

and that any $\{\lambda_n\}$ may be taken to lie between those two sequences by replacing λ_n by $(\lambda_n \wedge \lambda_n^+) \vee \lambda_n^-$ (which changes neither the summability class of (15) for $\{\lambda_n\}$, nor the monotonicity of $\{\lambda_n\}$, nor whether it is upper or lower class). Next, one notes from (11) that $R_n^* < \lambda_n^{-1}$ for all large n wp 1, for these restricted $\{\lambda_n\}$. Finally, it is trivial that $\{\lambda_n \pm \lambda_n^{-1}\}$ is monotone for large n and lies in the same summability class of (15) as $\{\lambda_n\}$.

REFERENCES

[1] R. R. Bahadur. A note on quantiles in large samples. *Ann. Math. Statist.* **37**, 577–80 (1966).

[2] G. Baxter. An analogue of the LIL. *Proc. Am. Math. Soc.* **6**, 177–81 (1955).

[3] L. Breiman. *Probability*. Addison–Wesley (1968).

[4] D. R. Brillinger. An asymptotic representation of the sample df. *Bull. Am. Math. Soc.* **75**, 545–7 (1969).

[5] K. L. Chung. An estimate concerning the Kolmogoroff limit distribution. *Trans. Am. Math. Soc.* **67**, 36–50 (1949).

[6] F. Eicker. A new proof of the Bahadur–Kiefer representation of sample quantiles. This volume, pp. 321–42.

[7] F. Eicker. On the probability that a sample df lies below a polygon. (To appear.)

[8] F. Eicker. A log log law for double sequences of random variables. (To appear.)

[9] W. Feller. The general form of the so-called law of the iterated logarithm. *Trans. Am. Math. Soc.* **54**, 373–402 (1943).

[10] W. Feller. *An Introduction to Probability Theory and its Applications*, vol. 1, second edn. (1957); vol. 2, first edn. (1966). New York: John Wiley and Sons.

[11] J. Kiefer. On Bahadur's representation of sample quantiles. *Ann. Math. Statist.* **38**, 1323–42 (1967).

[12] J. Kiefer. Deviations between the sample quantile process and the sample df. This volume, pp. 299–319.

[13] J. Kiefer. On a Brownian bridge approximation to the sample df. (To appear.)

[14] M. Loève. Ranking limit problems. *Proc. third Berkeley Symp.* vol. 2, 177–98 (1956).

[15] W. A. Rosenkrantz. On rates of convergence for the invariance principle. *Trans. Am. Math. Soc.* **129**, 542–52 (1967).

[16] W. A. Rosenkrantz. A rate of convergence for the Von Mises statistic. (To appear.)

[17] A. V. Skorokhod. *Studies in the Theory of Random Processes*. Addison–Wesley (1965).

[18] N. Smirnov. Limit distributions for the terms of a variational series. *Am. Math. Soc. Transl.* **67**, 1–64, Providence (1952); originally in *Trudy mat. Inst. V. A. Steklova* **25**, 60 pp. (1949).

[19] V. Strassen. An invariance principle for the LIL. *Z. Wahrscheinlichkeitstheorie und Verw. Gebiete* **3**, 211–26 (1964).

[20] V. Strassen. Almost sure behavior of sums of independent random variables and martingales. *Proc. fifth Berkeley Symp.* vol. 2, part 1, 315–43 (1966).

[21] L. Weiss. Asymptotic distribution of quantiles in some nonstandard cases. This volume, pp. 343–8.

Reprinted from *Nonparametric Techniques in Statistical Inference*, edited by Madan Lal Puri, C.U.P. 1970.

DEVIATIONS BETWEEN THE SAMPLE QUANTILE PROCESS AND THE SAMPLE DF

J. KIEFER†

1 INTRODUCTION AND SUMMARY

Let $X_1, X_2, \ldots$ be iid with common twice differentiable univariate df F on the unit interval I. We assume $\inf_{x\in I} F'(x) > 0$ and $\sup_{x\in I} F''(x) < \infty$, and write $\xi_p = F^{-1}(p)$ for the p-tile of F. Let S_n be the sample df and let $Y_{p,n}$ be the sample p-tile, both based on $(X_1, \ldots, X_n)$; i.e.

$$nS_n(x) = [\text{number of } X_i \leqslant x, 1 \leqslant i \leqslant n],$$

and
$$Y_{p,n} = \inf\{x\colon S_n(x) = p\}.$$

(The choice we have made in this last definition when np is an integer, is immaterial.) Write $\sigma_p = [p(1-p)]^{\frac{1}{2}}$. Let

$$R_n(p) = Y_{p,n} - \xi_p + [S_n(\xi_p) - p]/F'(\xi_p). \tag{1}$$

Bahadur (1966) initiated the study of $R_n(p)$, and it was shown in [8], among other things, that, for $u > 0$,

$$\lim_{n\to\infty} P\{n^{\frac{3}{4}}F'(\xi_p)\, R_n(p) \leqslant u\} = 2\int_0^\infty \Phi(k^{-\frac{1}{2}}u)\, d_k\, \Phi(k/\sigma_p)$$

(where Φ is the standard normal df), and that, for either choice of sign, $\# = +$ or $-$,

$$\limsup_{n\to\infty} \# \, F'(\xi_p)\, R_n(p)/[2^5 3^{-3}\sigma_p^2 n^{-3}(\log\log n)^3]^{\frac{1}{4}} = 1 \text{ wp } 1.$$

The behavior of $\sup_{|t|<T} R_n(p + n^{-\frac{1}{2}}t)$ was also studied, but results on the behavior of

$$\left.\begin{aligned} R_n^{\#} &= \sup_{p\in I} \# \, F'(\xi_p)\, R_n(p), \quad \text{where } \# = + \text{ or } -,\\ R_n^{*} &= \max(R_n^{+}, R_n^{-}), \end{aligned}\right\} \tag{2}$$

and the problem of studying these rv's, were only mentioned, without inclusion of proofs (§ 6.2 of [8]). In the present paper details will be given of the proofs of the following two results:

† Prepared under Contract No. NONR 401(50) for presentation at First International Conference on Nonparametric Inference, which the author was then unfortunately unable to attend. Reproduction is permitted for any purpose of the U.S. Government.

[299]

Theorem 1. *For* $\# = +, -$ or $*$, *and for* $t > 0$,

$$\lim_{n\to\infty} P\{n^{\frac{3}{4}}(\log n)^{-\frac{1}{2}}R_n^{\#} > t\} = 2\sum_{m=1}^{\infty}(-1)^{m+1}e^{-2m^2t^4}. \tag{3}$$

Theorem 2. *For* $\# = +, -$, *or* $*$,

$$\limsup_{n\to\infty} n^{\frac{3}{4}}(\log n)^{-\frac{1}{2}}(\log\log n)^{-\frac{1}{4}}R_n^{\#} = 2^{-\frac{1}{4}} \text{ wp } 1. \tag{4}$$

Readers familiar with the Kolmogorov–Smirnov statistic

$$D_n = n^{\frac{1}{2}}\sup_x |S_n(x) - F(x)|$$

will have noticed that (3) states that $n^{\frac{3}{4}}(\log n)^{-\frac{1}{2}}R_n^{\#}$ has the same limiting df as $D_n^{\frac{1}{2}}$. Indeed, we shall prove the more fundamental

Theorem 1 *A*. *For* $\# = +, -$, *or* $*$, *as* $n \to \infty$,

$$n^{\frac{3}{4}}R_n^{\#}/(D_n\log_n)^{\frac{1}{2}} \to 1 \quad \textit{in probability}, \tag{5}$$

from which Theorem 1 follows at once.

There is also a strong law corresponding to Theorem 1 A, and Theorem 2 follows at once from it and the law of the iterated logarithm (LIL) for D_n [5]. However, my proof of this strong law is at present so long and tedious that I will omit it here and prove Theorem 2 directly. The reader familiar with LIL proofs will see that (4) is easy to *guess* from (3), in that the right side of (3) for $t = [2^{-1}(1+\epsilon)\log\log n_r]^{\frac{1}{4}}$, with $n_r \sim \gamma^r$ and $\gamma > 1$, is summable in r if and only if $\epsilon > 0$; thus, roughly, one hopes as usual to prove (4) by first treating such a result with $\{n\}$ replaced by $\{n_r\}$.

Equation (5) explains why, perhaps unexpectedly, $n^{\frac{3}{4}}(\log n)^{-\frac{1}{2}}R_n^{\#}$ has the same limiting law in Theorem 1 as each of the two rv's of which it is the maximum.

Some of the consequences of Theorem 2 are treated in [9] and [10]. They include

Theorem 3. (*A consequence of Theorem* 2 *and* [5].) *If* $\lambda_n \uparrow \infty$, *then*

$$P\{\sup_{0\leqslant p\leqslant 1} F'(\xi_p)|Y_{p,n} - \xi_p| > \lambda_n n^{-\frac{1}{2}} \text{ i.o.}\}$$

$$= \begin{Bmatrix} 1 \Leftrightarrow \infty = \\ 0 \Leftrightarrow \infty > \end{Bmatrix} \sum_n n^{-1}\lambda_n^2 e^{-2\lambda_n^2}.$$

(The analogue for a single p is Theorem 4 of [8].)

Theorem 4. (*A consequence of Theorem* 2 *and* [6].) *If* $p_n \downarrow 0$ *and*

$$p_n > (\log\log n)/o(n),$$

then

$$P\{\limsup_{n\to\infty}[n^{-1}2p_n\log\log(np_n)]^{-\frac{1}{2}}F'(\xi_{p_n})[p_n - Y_{p_n,n}] = 1 \text{ wp } 1\}.$$

Theorem 5. *In the Breiman–Brillinger Brownian bridge representation of the sample df* [3], [4], [10] *with error* $\epsilon_n(p)$ *(say), there is a finite positive constant* C^* *such that*

$$\limsup_n \sup_p |\epsilon_n(p)|/[n(\log n)^2 (\log\log n)]^{\frac{1}{4}} = C^* \text{ wp } 1.$$

(Brillinger [4] showed that the order of $\sup_p |\epsilon_n(p)|$ was *no greater* than that just stated. The proof of this in [10] is much shorter and more elementary, and Theorem 2 makes it possible. Shortcomings of this representation are discussed in [9] and [10].)

The next paragraph justifies the fact that THE STATEMENT OF (4) AND THE STATEMENT OF (5) (AND HENCE, OF (3)) DO NOT DEPEND ON F OR THE 'SIGN' #.

We now mention a space-saving reduction. As in Lemma 1 of [8], we use Taylor's Theorem with remainder to obtain for the function R_n', defined to be the R_n of (1) when the X_i are replaced by the uniformly distributed rv's $F(X_i)$,

$$R_n'(p) - F'(\xi_p)\, R_n(p) = F''(\xi_p')\,(Y_{p,n} - \xi_p)^2/2F'(\xi_p), \tag{6}$$

where ξ_p' is a chance value between $Y_{p,n}$ and ξ_p.

The LIL for the sample df [5] easily shows that the right side of (6), uniformly in p, is $O(n^{-1}\log\log n)$ wp 1 as $n \to \infty$. Hence, IN THE PROOFS OF THEOREMS 1*A* AND 2 WE CAN AND DO HEREAFTER ASSUME F TO BE THE UNIFORM DF ON $[0,1]$. MOREOVER, WE ONLY PROVE THE STATED THEOREMS FOR THE CASE # = +. The last is justified by noting that the statement of the theorems for R_n^+ computed from $\{X_i\}$ is equivalent to their statement for R_n^- computed from $\{1 - X_i\}$, and the statements for R_n^+ and R_n^- imply that for R_n^*.

We recall some definitions and preliminary results from [8]. We define

$$\left.\begin{aligned} K_n(p) &= n^{\frac{1}{2}}[p - S_n(p)], \\ T_n(x,p) &= \begin{cases} S_n(x+p) - S_n(p) - x = S_n(x+p) - (x+p) + n^{-\frac{1}{2}}K_n(p) \\ \qquad\qquad\qquad\qquad\qquad\qquad\qquad \text{if } \quad (x+p) \in I, \\ 0 \quad \text{if } (x+p) \notin I. \end{cases} \end{aligned}\right\} \tag{7}$$

(The second line in the definition of T_n permits brevity in writing results like (10).) Then for F uniform it is clear, as in (1.12) of [8], that for $d > 0$, wp 1,

$$\left.\begin{aligned} R_n(p) > d &\Leftrightarrow T_n(n^{-\frac{1}{2}}K_n(p) + d, p) < -d, \\ R_n(p) < -d &\Leftrightarrow T_n(n^{-\frac{1}{2}}K_n(p) - d, p) \geqslant d. \end{aligned}\right\} \tag{8}$$

If Z_n is binomially distributed with parameters N and π_N, the central limit theorem asserts, as in Lemma 2 of [8] (obtained from pp. 168–79 of [7]), as $N \to \infty$,

$$\{N\pi_N \to \infty;\ \limsup_N \pi_N < 1;\ \Delta_N > 0;\ \Delta_N^2/N\pi_N \to \infty;\ \Delta_N^3/N^2\pi_N^2 \to 0\}$$
$$\Rightarrow \log P\{Z_n - N\pi_N > \Delta_N\} \sim -\Delta_N^2/2N\pi_N(1-\pi_N). \quad (9)$$

Finally, Lemma 5 of [8] contains the result

$$\{0 < \sigma < 1;\ a > 0;\ w > 0;\ [n/a(1+w)]^{\frac{1}{2}}\, w(\sigma - a) \geqslant 2\}$$
$$\Rightarrow P\{\sup_{0 \leqslant v \leqslant a} |T_n(v,p)| > w\} \leqslant \tfrac{4}{3} P\{|T_n(a^{(p)},p)| > w(1-\sigma)\}, \quad (10)$$

where $a^{(p)} = \min(a, 1-p)$. The last probabilistic expression is in fact a binomial probability. Using also the corresponding result for the rv's $1 - X_i$, we can obtain from (10) and (9), for δ fixed ($0 < \delta < 1$) and $n \to \infty$,

$$\{0 \leqslant p \leqslant 1;\ 0 < a_n \to 0;\ 0 < w_n \to 0;\ na_n \to \infty;\ w_n^2 n/a_n \to \infty;\ w_n^3 n/a_n^2 \to 0\}$$
$$\Rightarrow \log P\{\sup_{0 \leqslant |v| \leqslant a_n} |T_n(v,p)| > w_n\} < -(1-\delta)^3 w_n^2 n/2a_n \quad (11)$$

for all $n > N_0$ (say), where N_0 depends on $\{a_n\}$ and $\{w_n\}$ and δ, but not on p. In obtaining this result, we use the fifth condition of (11) to verify the fourth condition of both (9) and (10); and we use the fact that, if $a_n + p > 1$, then the conditions of (11) imply that, for $n >$ an N_0 of the above form, the last expression of (10) is no greater than twice that obtained by replacing $a_n^{(p)}$ by $-a_n$.

We abbreviate 'A_n infinitely often' (= 'infinitely many of the events $A_1, A_2, \ldots$ occur') by 'A_n i.o. (n)', or simply 'A_n i.o.' if no ambiguity is possible, and 'A_n for almost all n' (= 'all but a finite number of the events $A_1, A_2, \ldots$ occur') by 'A_n for a.a.n.' The complement of an event A is denoted $\bar{A}$. By $\operatorname{int}\{L\}$ we mean the greatest integer $\leqslant L$. Orders such as $O(n)$ or $o(r)$ hold as $n \to +\infty$ or $r \to +\infty$ (the latter when, in §§ 4 and 5, we deal with a subsequence $\{n_r\}$ of the natural numbers).

In §§ 2 and 3 we shall prove, respectively, the statement obtained by considering only the right and left inequality of

$$\lim_{n\to\infty} P\{(1-\bar{\epsilon})\, D_n^{\frac{1}{2}} < n^{\frac{3}{4}} (\log n)^{-\frac{1}{2}} R_n^+ < (1+\bar{\epsilon})\, D_n^{\frac{1}{2}}\} = 1 \quad (12)$$

for $\bar{\epsilon} > 0$. This and the paragraph containing (6) yield Theorem 1 A. Similarly, we obtain Theorem 2 by proving in §§ 4 and 5, respectively, the right and left halves of

$$P\{(1-\bar{\epsilon})\, 2^{-\frac{1}{4}} < \limsup_{n\to\infty} n^{\frac{3}{4}} (\log n)^{-\frac{1}{2}} (\log\log n)^{-\frac{1}{4}} R_n^+ < (1+\bar{\epsilon})\, 2^{-\frac{1}{4}}\} = 1. \quad (13)$$

Further comments (including remarks on the methods of proof) are contained in § 6.

2 WEAK UPPER BOUND

Fix $A > 0$. For $0 \leqslant p \leqslant 1$ and $d > 0$ and $L > 0$, we define the events

$$\left.\begin{aligned} B_n(p,d) &= \{\sup_{|t|\leqslant A} R_n(p+n^{-\frac{1}{2}}t) > n^{-\frac{3}{4}}d\}, \\ C_n(L) &= \{\sup_{p\in I} |K_n(p)| \leqslant L\} = \{D_n \leqslant L\}, \end{aligned}\right\} \tag{14}$$

and the set

$$M_n = \{p : p = 1 \text{ or } p = 2jn^{-\frac{1}{2}}A, j \text{ integral}, 0 \leqslant p < 1\}.$$

For $\# = +$, the statement obtained by considering only the right hand inequality in (12) will be shown in the next paragraph to follow from the fact (which we shall verify in this section) that, for each $L > 0$ and $c > L^{\frac{1}{2}}$,

$$\lim_{A\to 0}\lim_{n\to\infty} P\{\bigcup_{p\in M_n} [B_n(p, c[\log n]^{\frac{1}{2}}) \text{ and } |K_n(p)| \leqslant L]\} = 0. \tag{15}$$

We define

$$\psi(L) = \lim_{n\to\infty} P\{\overline{C_n(L)}\} = 2\sum_{m=1}^{\infty} (-1)^{m+1} e^{-2m^2L^2}.$$

This is of course the tail probability of Kolmogorov's limiting df. To see that (15) implies the right half of (12), let $\epsilon' > 0$ be given and let $h > 0$ be such that

$$\psi(h) - \psi(h^{-1}) > 1 - \epsilon'. \tag{16}$$

The subset $$Q = \{(x_1, x_2) : h \leqslant x_1 \leqslant h^{-1}, x_2 \geqslant (1+\bar{\epsilon})x_1\} \tag{17}$$

of the plane can clearly be covered by a finite number of sets Q_i of the form

$$Q_i = \{(x_1, x_2) : x_1 \leqslant c_i, x_2 > (1+\bar{\epsilon}/2)c_i\} \tag{18}$$

with $c_i > 0$. Since $\{D_n \leqslant L\} \subset \{|K_n(p)| \leqslant L\}$ for each p, we see that (15) with $L^{\frac{1}{2}} = c_i$ and $c = (1+\bar{\epsilon}/2)L^{\frac{1}{2}}$ implies that

$$\lim_{n\to\infty} P\{(D_n^{\frac{1}{2}}, n^{\frac{3}{4}}(\log n)^{-\frac{1}{2}} R_n^+) \in Q_i\} = 0. \tag{19}$$

Since finitely many Q_i cover Q, (19) holds with Q_i replaced by Q. Since ϵ' is arbitrarily small in (16), this establishes the right half of (12).

It remains to prove (15). Given $c > L^{\frac{1}{2}}$, choose $A > 0$ and δ $(0 < \delta < 1)$ so that

$$c^2(1-\delta)^4\delta^2 > A \quad\text{and}\quad c^2(1-\delta)^6 > A + L. \tag{20}$$

Using first the first line of (8), and then the relations

$$T_n(x, p+t) = T_n(x+t, p) - T_n(t, p)$$

and $$K_n(p+t) = K_n(p) - n^{\frac{1}{2}} T_n(t, p),$$

and lastly a trivial inclusion, we have

$$\begin{aligned} B_n(p,d) &= \{\exists t, |t| \leqslant A, \text{with } T_n(n^{-\frac{1}{2}}K_n(p+n^{-\frac{1}{2}}t) + n^{-\frac{3}{4}}d, p+n^{-\frac{1}{2}}t) < -n^{-\frac{3}{4}}d\} \\ &= \{\exists t, |t| \leqslant A, \text{ with } T_n(n^{-\frac{1}{2}}t + n^{-\frac{1}{2}}K_n(p) - T_n(n^{-\frac{1}{2}}t, p) \\ &\qquad\qquad + n^{-\frac{3}{4}}d, p) - T_n(n^{-\frac{1}{2}}t, p) < -n^{-\frac{3}{4}}d\} \\ &\subset \{\sup_{|t| \leqslant A} |T_n(n^{-\frac{1}{2}}t, p)| \geqslant n^{-\frac{3}{4}}\,\delta d\} \cup \\ &\quad \{\sup_{|t| \leqslant A + |K_n(p)| + n^{-\frac{1}{4}}(1+\delta)d} |T_n(n^{-\frac{1}{2}}t, p)| > n^{-\frac{3}{4}}(1-\delta)\,d\} \\ &= G_n(p,d) \cup H_n(p,d) \quad \text{(say).} \end{aligned} \tag{21}$$

Applying (11) with $a_n = An^{-\frac{1}{2}}$ and $w_n = n^{-\frac{3}{4}}\,\delta c(\log n)^{\frac{1}{2}}$, we have

$$\log G_n(p, c[\log n]^{\frac{1}{2}}) < -[(1-\delta)^4\,\delta^2 c^2 \log n]/2A \tag{22}$$

for all sufficiently large n, uniformly in p. On the other hand, applying (11) with

$$a_n = n^{-\frac{1}{2}}[A + L + n^{-\frac{1}{4}}(1+\delta)\,c(\log n)^{\frac{1}{2}}] \quad \text{and} \quad w_n = n^{-\frac{3}{4}}(1-\delta)\,c(\log n)^{\frac{1}{2}},$$

we have

$$\log P\{H_n(p, c[\log n]^{\frac{1}{2}}) \text{ and } |K_n(p)| \leqslant L\} < -[(1-\delta)^6 c^2 \log n]/2[A+L] \tag{23}$$

for all sufficiently large n, uniformly in p.

Since there are at most $(n^{\frac{1}{2}}/2A) + 2$ elements in M_n, equation (15) is a consequence of (22), (23), (20), and the inclusion between the first and last lines of (21).

3 WEAK LOWER BOUND

We now consider the expression consisting of only the lower bound inequality on R_n^+ in (12).

Let $\bar{\epsilon}$ be specified in (12) $(0 < \bar{\epsilon} < 1)$, and let g, H, b, ϵ be positive values, and let α be a positive integer, satisfying

$$H > 2; \quad b \geqslant g^2/(1-2H^{-1}) \geqslant g^2(1-\bar{\epsilon}/2)^{-2}; \quad \epsilon < 1/(\alpha+1)\,bH. \tag{24}$$

We shall devote this section to proving that

$$\lim_{n\to\infty} P\{D_n^{\frac{1}{2}} \geqslant b^{\frac{1}{2}};\, n^{\frac{3}{4}}(\log n)^{-\frac{1}{2}} R_n^+ \leqslant g\} = 0. \tag{25}$$

Since $g \leqslant (1-\bar{\varepsilon}/2)b^{\frac{1}{2}}$ by (24), the lower half of (12) follows from (25) in exactly the manner that the upper half followed from (19).

Let

$$\left.\begin{aligned} M_n^* &= \{j\colon 0 \leqslant j \leqslant \epsilon n^{\frac{1}{2}}, j \text{ integral}\},\\ M_n^{*\prime} &= \{j\colon 0 \leqslant j \leqslant \epsilon n^{\frac{1}{2}}-1, j \text{ integral}\},\\ p_{n,j}^i &= i/(\alpha+1)+n^{-\frac{1}{2}}jbH \quad \text{for} \quad j \in M_n^* \quad \text{and} \quad 1 \leqslant i \leqslant \alpha,\\ p_{n,j}^{*i} &= p_{n,j}^i + n^{-\frac{1}{2}}K_n(p_{n,j}^i)+n^{-\frac{3}{4}}g(\log n)^{\frac{1}{2}}. \end{aligned}\right\} \tag{26}$$

We write k_n^i for a vector with real components $k_{n,j}^i$, $j \in M_n^*$, abbreviate $K_n(p_{n,j}^i)$ as $K_{n,j}^i$, and write K_n^i for the random vector with components $K_{n,j}^i$, $j \in M_n^*$. We define the set

$$\Lambda_n^i = \{k_n^i\colon 0 < b \leqslant k_{n,j}^i < bH/2 \quad \text{for} \quad j \in M_n^*\}, \tag{27}$$

and the events

$$\left.\begin{aligned} &\Lambda_n^{i*} = \{K_n^i \in \Lambda_n^i\}, \quad \Lambda_n^{i**} = \{K_n^i \text{ or } -K_n^i \in \Lambda_n^i\},\\ &\Lambda_n^{**} = \bigcup_{1 \leqslant i \leqslant \alpha} \Lambda_n^{i**},\\ &A_{n,j}^i = \{R_n(p_{n,j}^i) > n^{-\frac{3}{4}}g(\log n)^{\frac{1}{2}}\} \quad \text{for} \quad j \in M_n^* \quad \text{and} \quad 1 \leqslant i \leqslant \alpha;\\ &A_n^i = \bigcup_{j \in M_n^{*\prime}} A_{n,j}^i. \end{aligned}\right\} \tag{28}$$

If Λ_n^{i*} occurs, then by (24), (26), and (27), for all sufficiently large n we have $p_{n,j}^i < p_{n,j}^{*i} < p_{n,j+1}^i$ whenever $j \in M_n^{*\prime}$. Hence, for such j, from (8) and (7),

$$\begin{aligned} P\{A_{n,j}^i \mid \Lambda_n^{i*};\, K_n^i = k_n^i\} &= P\{S_n(p_{n,j}^{*i}) - p_{n,j}^{*i} + n^{-\frac{1}{2}}k_{n,j}^i \\ &\quad < -n^{-\frac{3}{4}}g(\log n)^{\frac{1}{2}} \mid K_{n,j}^i = k_{n,j}^i;\, K_{n,j+1}^i = k_{n,j+1}^i\}. \end{aligned} \tag{29}$$

This last is a binomial probability; in the notation of (9), under the conditioning of (29), the number of X_t $(1 \leqslant t \leqslant n)$ falling between $p_{n,j}^i$ and $p_{n,j+1}^i$ is

$$N = n^{\frac{1}{2}}(bH + u_{n,j}^i), \tag{30}$$

where we have written (and will write)

$$u_{n,j}^i = k_{n,j}^i - k_{n,j+1}^i; \quad U_{n,j}^i = K_{n,j}^i - K_{n,j+1}^i; \tag{31}$$

the (conditional) probability that any of these N conditionally independent rv's X_t falls between $p_{n,j}^{*i}$ and $p_{n,j+1}^i$ is

$$\pi_N = (p_{n,j+1}^i - p_{n,j}^{*i})/(p_{n,j+1}^i - p_{n,j}^i) = 1-(bH)^{-1}[k_{n,j}^i + n^{-\frac{1}{4}}g(\log n)^{\frac{1}{2}}]; \tag{32}$$

and the associated binomial rv is

$$Z_N = n[S_n(p^i_{n,\,j+1}) - S_n(p^{*i}_{n,\,j})].$$

If we define

$$\Delta_N = (bH)^{-1}[n^{\frac{1}{2}}k^i_{n,\,j}u^i_{n,\,j} + (bH + u^i_{n,\,j})n^{\frac{1}{4}}g(\log n)^{\frac{1}{2}}], \qquad (33)$$

we can rewrite (29) in the notation of (9) as

$$P\{S_n(p^i_{n,\,j+1}) - S_n(p^{*i}_{n,\,j})$$
$$> n^{-\frac{1}{2}}(bH - k^i_{n,\,j+1})\,|\,K^i_{n,\,j} = k^i_{n,\,j};\ K^i_{n,\,j+1} = k^i_{n,\,j+1}\} = P\{Z_N - N\pi_N > \Delta_N\}. \qquad (34)$$

Define the subset of M^*_n,

$$J^i_n(k^i_n) = \{j: j \in M^{*\prime}_n;\ u^i_{n,\,j} \leqslant 0\},$$

and let $|J^i_n(k^i_n)|$ denote the cardinality of this set. Also define the set and the events

$$\left.\begin{array}{l} C^i_n = \{k^i_n: |J^i_n(k^i_n)| \geqslant \epsilon n^{\frac{1}{2}}/3\}, \\[2mm] C^{i*}_n = \{K^i_n \in C^i_n\}, \quad C^*_n = \bigcap_{1 \leqslant i \leqslant \alpha} C^{i*}_n. \end{array}\right\} \qquad (35)$$

At the end of this section we shall prove that, if $n^{-\frac{1}{2}}bH < \frac{1}{4}$, then

$$P\{\overline{C^{i*}_n}\} \leqslant c_0[\epsilon^{-1}n^{-\frac{1}{2}} + (bH)^{-\frac{1}{2}}n^{-\frac{1}{4}}], \qquad (36)$$

where c_0 is a universal constant.

In the next two sentences we consider N, π_N, Δ_N as functions of k and u without regard to the relation (31) between the latter. For fixed $k^i_{n,\,j}$, let $c^i_{n,\,j}$ be the value of $u^{i\prime}_{n,\,j}$ which makes the expression (33) equal zero. It is easy to verify that, for $k^i_n \in \Lambda^i_n$, the expression $\Delta^2_N/N\pi_N(1-\pi_N)$ obtained from (30)–(33) is strictly increasing in $u^i_{n,\,j}$ for $u^i_{n,\,j} \geqslant c^i_{n,\,j}$. Consequently, for $k^i_n \in \Lambda^{i\prime}_n$ and $j \in J^i_n(k^i_n)$, either $u^i_{n,\,j} < c^i_{n,\,j}$, in which case the probability of (34) is $< \frac{3}{4}$ for all sufficiently large n, or else $c^i_{n,\,j} \leqslant u^i_{n,\,j} \leqslant 0$, in which case (substituting this upper bound on $u^i_{n,\,j}$ into $\Delta^2_N/N\pi_N(1-\pi_N)$),

$$\Delta^2_N/N\pi_N(1-\pi_N) < g^2(\log n)\,[1+o(1)]/k^i_{n,\,j}(1-k^i_{n,\,j}/bH)$$
$$< g^2(\log n)\,[1+o(1)]/b(1-H^{-1}), \qquad (37)$$

the last inequality following from the fact that the concave function $k(1-k/bH)$ on the interval $b \leqslant k \leqslant bH/2$ attains its minimum at an end point, and that $0 < b(1-H^{-1}) < bH/4$ by (24). Moreover, it is easily seen that in the bounds of (37) the $o(1)$ term as $n \to \infty$ is uniform for $k^i_n \in \Lambda^i_n$ and j satisfying $c^i_{n,\,j} \leqslant u^i_{n,\,j} \leqslant 0$. This uniformity persists in the application of (9). Since the events $\{A^i_{n,\,j}, j \in M^{*\prime}_n\}$ for fixed n are conditionally independent, conditioned on the event $\{K_n^i = k_n^i\}$, we obtain from (35) and (37) (and the fact that half the last expression there exceeds $-\log(\frac{3}{4})$ for large n) for n sufficiently large, uniformly for $k_n^i \in \Lambda_n^i \cap C_n^i$,

$$\begin{aligned} P\{\overline{A_n^i} \mid \Lambda_n^{i*}; C_n^{i*}; K_n^i = k_n^i\} &= P\{\bigcap_{j \in M_n^{*\prime}} \overline{A_{n,j}^i} \mid \Lambda_n^{i*}; C_n^{i*}; K_n^i = k_n^i\} \\ &\leqslant \prod_{j \in J_n^i(k_n^i)} P\{\overline{A_{n,j}^i} \mid \Lambda_n^{i*}; C_n^{i*}; K_n^i = k_n^i\} \\ &\leqslant [1 - n^{-g^2[1+o(1)]/2b(1-H^{-1})}]^{\epsilon n^{\frac{1}{2}}/3}. \end{aligned} \tag{38}$$

The last expression approaches 0 as $n \to \infty$, provided that

$$g^2/2b(1-H^{-1}) < \tfrac{1}{2}, \tag{39}$$

and (39) follows at once from (24).

We note, using the second line of (28) and the elementary relation $(\bigcup_i \Lambda_i) \cap (\bigcup_i A_i) \supset \bigcup_i \Lambda_i - \bigcup_i (\Lambda_i \cap \overline{A_i})$, the simple probabilistic relations

$$\begin{aligned} P\{\Lambda_n^{**} \cap \bigcup_{1 \leqslant i \leqslant \alpha} A_n^i\} &\geqslant P\{\Lambda_n^{**}\} - P\{\bigcup_{1 \leqslant i \leqslant \alpha} [\overline{\Lambda_n^{i**}} \cap A_n^i]\} \\ &\geqslant P\{\Lambda_n^{**}\} - \left[P\{\overline{C_n^*}\} + \sum_{i=1}^{\alpha} P\{\overline{A_n^i} \mid \Lambda_n^{i**}; C_n^{i*}\}\right]. \end{aligned} \tag{40}$$

All of the above development was for fixed $g, H, b, \epsilon, \alpha$ satisfying (24), and of course the events appearing in (38) and (40) depend on those values. Still fixing those values, let $n \to \infty$ in (40). The previous paragraph, its analogue when Λ_n^{i*} is replaced by $\Lambda_n^{i**} - \Lambda_n^{i*}$, and (36), imply that the expression in square brackets, in the last line of (40), approaches zero. Thus, (40) yields

$$\lim_n P\{\Lambda_n^{**} \cap \overline{\bigcup_{1 \leqslant i \leqslant \alpha} A_n^i}\} = 0, \tag{41}$$

and hence, by inclusion,

$$\lim_n P\{\Lambda_n^{**} \cap \{R_n^+ \leqslant n^{-\frac{3}{4}} g(\log n)^{\frac{1}{2}}\}\} = 0. \tag{42}$$

It is well known that

$$\lim_{\epsilon \downarrow 0} \lim_n [P\{\bigcup_{1 \leqslant i \leqslant \alpha} \{b \leqslant |K_n(i/(\alpha+1))| < bH/2\}\} - P\{\Lambda_n^{**}\}] = 0. \tag{43}$$

Since Λ_n^{**} is a subset of the union in (43), we obtain from (43) on letting $\epsilon \downarrow 0$ in (42),

$$\lim_n P\{\{R_n^+ \leqslant n^{-\frac{3}{4}} g(\log n)^{\frac{1}{2}}\} \cap \bigcup_{1 \leqslant i \leqslant \alpha} \{b \leqslant |K_n(i/(\alpha+1))| < bH/2\}\} = 0. \tag{44}$$

Letting $H \to \infty$ in (44), and using the fact that the event expressed as a union there has a probability whose limit in H is attained uniformly in n, we obtain

$$\lim_n P\{R_n^+ \leqslant n^{-\frac{3}{4}}g(\log n)^{\frac{1}{2}};\ \max_{1\leqslant i\leqslant \alpha} |K_n(i/(\alpha+1))| \geqslant b\} = 0. \tag{45}$$

The event concerning K_n in the second half of (45) is a subset of the event $\{D_n \geqslant b\}$ and (invariance principle) their difference has a limiting probability, as $n \to \infty$, which can be made arbitrarily close to 0 by taking α sufficiently large. This fact and (45) yield (25).

This completes the proof of the lower inequality of (12), except for the proof of (36), to which we now turn.

Throughout this paragraph, symbols θ_i denote values satisfying $|\theta_i| \leqslant 1$ and symbols c_i denote universal constants. We write

$$\pi = n^{-\frac{1}{2}}bH = p^i_{n,j+1} - p^i_{n,j}$$

and assume $$\pi < \tfrac{1}{4},\quad c_1(n\pi)^{-\frac{1}{2}} < \tfrac{1}{12},\quad n^{\frac{1}{2}}\epsilon > 10, \tag{46}$$

where c_1 is the absolute constant of (48) below; the first restriction of (46) is that assumed just above (36), and the other two may be assumed because (perhaps with an enlarged c_0) (36) is trivial if either is violated. $U^i_{n,j}$ is of the form $n^{-\frac{1}{2}}(V - EV)$, where V is binomial, the sum of n iid Bernoulli rv's, each with expectation π. For $\pi < \frac{1}{4}$ (as specified in (46)), such a Bernoulli rv W_j satisfies

$$E|W_j - \pi|^3/[\mathrm{var}(W_j)]^{\frac{3}{2}} = [\pi^3(1-\pi) + \pi(1-\pi)^3]/[\pi(1-\pi)]^{\frac{3}{2}} \leqslant \pi^{-\frac{1}{2}}, \tag{47}$$

since $[\pi^2 + (1-\pi)^2](1-\pi)^{-\frac{1}{2}} \leqslant 1$ for $0 \leqslant \pi \leqslant \frac{1}{4}$. The Berry–Esseen bound and (47) yield, for $\pi < \frac{1}{4}$,

$$P\{U^i_{n,j} \leqslant 0\} = \tfrac{1}{2} + c_1\theta_1(n\pi)^{-\frac{1}{2}}. \tag{48}$$

For j_1, j_2 distinct integers in $M_n^{*\prime}$, if W_j is the indicator of the event $p^i_{n,j} < X_1 < p^i_{n,j+1}$, we have $E\{W_{j_1}W_{j_2}\} = 0$ and hence $\mathrm{cov}(W_{j_1}, W_{j_2}) = -\pi^2$. The inverse of the covariance matrix of W_{j_1}, W_{j_2} is hence

$$\begin{pmatrix} \pi(1-\pi) & -\pi^2 \\ -\pi^2 & \pi(1-\pi) \end{pmatrix}^{-1} = (1-2\pi)^{-1}\begin{pmatrix} \pi^{-1}-1 & 1 \\ 1 & \pi^{-1}-1 \end{pmatrix}, \tag{49}$$

which has maximum element $\Delta = (1-2\pi)^{-1}(\pi^{-1}-1)$. The bivariate normal law with means zero, variances $\pi(1-\pi)$, and covariance $-\pi^2$, assigns probability P_π(say) $< \frac{1}{4}$ to the first quadrant. (Elementary trigonometry in fact yields $P_\pi = \frac{1}{4} - c_5|\theta_5|\pi$, but this is neither helpful nor needed.) Bergstrom's [2] multivariate generalization of the Berry–Esseen bound asserts that

$$P\{U^i_{n,j_1} \leqslant 0,\ U^i_{n,j_2} \leqslant 0\} = P_\pi + c_2\theta_2\Delta^{\frac{3}{2}} \sum_{i=1}^{2} E\,|W_{j_i} - \pi|^3 n^{-\frac{1}{2}}$$
$$< \tfrac{1}{4} + c_3\theta_3(n\pi)^{-\frac{1}{2}}, \quad (50)$$

since $\Delta^{\frac{3}{2}} < 4\pi^{-\frac{3}{2}}$ for $0 < \pi < \frac{1}{4}$, and $E|W_{j_i} - \pi|^3$ is as in (47). If $Z^i_{n,j}$ is the indicator of the event $\{U^i_{n,j} \leqslant 0\}$, then (see (35)) the rv $|J^i_n(K^i_n)|$, hereafter denoted $|J|$, is simply $\sum_{j \in M^{*\prime}_n} Z^i_{n,j}$. For fixed n and i, the $Z^i_{n,j}$ are not independent; however, by (48) and (50), we can assert, for $j, j_1, j_2 \in M^{*\prime}_n$ and $j_1 \neq j_2$,

$$\left.\begin{aligned} &E\{Z^i_{n,j}\} = \tfrac{1}{2} + c_1\theta_1(n\pi)^{-\frac{1}{2}},\\ &\operatorname{var}(Z^i_{n,j}) = \tfrac{1}{4} - c_1^2\theta_1^2(n\pi)^{-1} \leqslant \tfrac{1}{4},\\ &\operatorname{cov}(Z^i_{n,j_1}, Z^i_{n,j_2}) < \tfrac{1}{4} + c_3\theta_3(n\pi)^{-\frac{1}{2}} - [\tfrac{1}{2} + c_1\theta_1(n\pi)^{-\frac{1}{2}}]^2\\ &\quad = c_4\theta_4(n\pi)^{-\frac{1}{2}} - c_1^2\theta_1^2(n\pi)^{-1} \leqslant c_4(n\pi)^{-\frac{1}{2}}. \end{aligned}\right\} \quad (51)$$

Since the number of elements m in $M^{*\prime}_n$ satisfies $n^{\frac{1}{2}}\epsilon - 1 \leqslant m \leqslant n^{\frac{1}{2}}\epsilon$, we have (using the last two inequalities of (46) in the next line)

$$\left.\begin{aligned} &E\{|J|\} = mE\{Z^i_{n,0}\} \geqslant (n^{\frac{1}{2}}\epsilon - 1)\,(\tfrac{1}{2} - c_1(n\pi)^{-\frac{1}{2}}) \geqslant 3n^{\frac{1}{2}}\epsilon/8;\\ &\operatorname{var}(|J|) < n^{\frac{1}{2}}\epsilon/4 + (n^{\frac{1}{2}}\epsilon)^2 c_4(n\pi)^{-\frac{1}{2}}. \end{aligned}\right\} \quad (52)$$

Hence,

$$\operatorname{var}(|J|)/[E\{|J| - n^{\frac{1}{2}}\epsilon/3\}]^2 < 576[n^{-\frac{1}{2}}\epsilon^{-1}/4 + c_4 n^{-\frac{1}{4}}(bH)^{-\frac{1}{2}}], \quad (53)$$

which, with Chebyshev's inequality, establishes (36).

4 STRONG UPPER BOUND

We now verify the statement obtained by considering only the right hand inequality of (13). Given a value $\bar{\epsilon} > 0$ there, fix $A > 0$ and choose $\delta\,(0 < \delta < 1)$ and $\gamma > 1$ to satisfy

$$(1-\delta)^8(1+\bar{\epsilon})^2/(1+2\delta) > 1 \quad \text{and} \quad \delta^2(1-\delta)^7(1+\bar{\epsilon})^2/4(\gamma-1)(1+2\delta) > 1 \quad (54)$$

We let $\{n_r, r = 1, 2, \ldots\}$ be any increasing sequence of positive integers for which, as $r \to \infty$,

$$n_r \sim \gamma^r. \quad (55)$$

We define $S_{n',n''}$ for $n' < n''$ as the sample df based on

$$(X_{n'+1}, X_{n'+2}, \ldots, X_{n''}),$$

and $T_{n',n''}$ is then obtained by replacing S_n by $S_{n',n''}$ in the first form on the right in the second line of (7). We define M_n as in § 2, and

$$\left.\begin{aligned} I_r &= \{n: n_r < n \leqslant n_{r+1}, n \text{ integral}\}, \\ d_n &= [2^{-1}(\log n)^2 \log\log n]^{\frac{1}{4}}(1+\bar{\epsilon}), \\ \Gamma_n(p) &= n_{r}^{-\frac{1}{2}}A + n^{-\frac{1}{2}}|K_n(p)| + n^{-\frac{3}{4}}(1+\delta)\,d_n \quad \text{for } n \in I_r, \\ \Gamma_r &= (1+2\delta)\,[(2n_r)^{-1}\log\log n_r]^{\frac{1}{2}}. \end{aligned}\right\} \tag{56}$$

Recalling (8), (14), and the first equation of (21), we see that our desired upper class result will follow from

$$\bigcup_{n\in I_r}\bigcup_{p\in M_{n_r}} \{\exists t, |t| \leqslant A, \text{ with } T_n(n^{-\frac{1}{2}}K_n(p+n_r^{-\frac{1}{2}}t)+n^{-\frac{3}{4}}d_n, p+n_r^{-\frac{1}{2}}t) < -n^{-\frac{3}{4}}d_n\} \text{ occurs only finitely often } (r), \text{ wp } 1. \tag{57}$$

(Note in (57) and in the sequel that the same sets

$$\{t: |n_r^{-\frac{1}{2}}t - p| \leqslant A\}, \quad p \in M_{n_r},$$

are used in studying what occurs for all $n \in I_r$.) From the remaining relations of (21), modified to take into account the departure of (57) from the second expression of (21), we see that (57) is a consequence of

$$P\{\bigcup_{n\in I_r}\bigcup_{p\in M_{n_r}}\{\sup_{|t|\leqslant A}|T_n(n_r^{-\frac{1}{2}}t, p)| \geqslant n^{-\frac{3}{4}}\delta d_n\} \text{ i.o. } (r)\} = 0 \tag{58}$$

and of

$$P\{\bigcup_{n\in I_r}\bigcup_{p\in M_{n_r}}\{\sup_{|t|\leqslant \Gamma_n(p)}|T_n(t,p)| > n^{-\frac{3}{4}}(1-\delta)\,d_n\} \text{ i.o. } (r)\} = 0. \tag{59}$$

Moreover, since $\sup_{p\in I}|K_n(p)| < (1+\delta)(2^{-1}\log\log n)^{\frac{1}{2}}$ for a.a. n wp 1 by the LIL for the sample df [5], elementary estimation yields

$$\sup_{n\in I_r,\, p\in M_{n_r}} \Gamma_n(p) < \Gamma_r \quad \text{for a.a. } r, \text{ wp } 1. \tag{60}$$

Since $n^{-\frac{3}{4}}d_n > n^{-1}n_r^{\frac{1}{4}}d_{n_r}$ for $n \in I_r$ and large r, we see that (59) will follow from

$$P\{\bigcup_{p\in M_{n_r}}\{\sup_{|t|\leqslant\Gamma_r}|T_{n_r}(t,p)| > n_r^{-\frac{3}{4}}(1-\delta)^2 d_{n_r}\} \text{ i.o. } (r)\} = 0, \tag{61}$$

$$P\{\bigcup_{p\in M_{n_r}}\{\max_{n\in I_r}\sup_{|t|\leqslant\Gamma_r}(n-n_r)|T_{n_r,n}(t,p)| \geqslant \delta(1-\delta)\,n_r^{\frac{1}{4}}d_{n_r}\} \text{ i.o. } (r)\} = 0, \tag{62}$$

and the obvious relation

$$(n-n_r)\,T_{n_r,n} + n_r T_{n_r} = nT_n \quad \text{for } n \in I_r. \tag{63}$$

It remains to prove (58), (61), and (62). In each of these (11) will be applied for a single sequence of pairs (a_n, w_n), so there is no difficulty from the increasing size of M_{n_r}. For fixed p and n, the probability of the event in the inside braces of (58) is estimated, as in (22) but now using $a_n = An_r^{-\frac{1}{2}}$ and $w_n = n^{-\frac{1}{2}}\delta d_n$ in (11), to be no greater than

$$\exp\{-(1-\delta)^4\delta^2(1+\bar{\varepsilon})^2[2^{-1}n^{-1}\log\log n]^{\frac{1}{2}}(\log n)/2An_r^{-\frac{1}{2}}\} \tag{64}$$

for all sufficiently large n (where, of course, $n \in I_r$ in (64)). The product of (64) with the upper bound $(n_r^{\frac{1}{2}}/2A)+2$ on the number of elements in M_{n_r} is clearly summable in n. This and the Borel–Cantelli lemma yield (58).

As for (61), the expression in braces there for fixed p and r is similarly seen to have probability no greater, for large r, than

$$\exp\{-(1-\delta)^8 n_r^{-\frac{1}{2}} d_{n_r}^2/2\Gamma_r\} = \exp\{-(1-\delta)^8(1+\bar{\varepsilon})^2(\log n_r)/2(1+2\delta)\}. \tag{65}$$

The product of (65) with $(n_r^{\frac{1}{2}}/2A)+2$ is summable in r by the first condition of (54), yielding (61).

Finally, we turn to (62). We shall carry through the details for one quarter of the proof, restricting attention to positive t and T in (62) and proving that

$$\bigcup_{p\in M_{n_r}}\{\max_{n\in I_r}\sup_{0\leqslant t\leqslant \Gamma_r}(n-n_r)T_{n_r,n}(t,p) \geqslant \delta(1-\delta)n_r^{\frac{1}{2}}d_{n_r}\} \tag{66}$$

occurs only finitely often (r) wp 1; the other three choices of sign are handled similarly, and (62) can be violated only if, for one of the four cases (of which (66) is one), the counterpart of (66) occurs i.o. (r) wp > 0.

Fix r and $p \in M_{n_r}$, and suppose the event in braces in (66) occurs. Call this event $D_r(p)$. Let ν be the smallest n in I_r for which the supremum in (66) satisfies the inequality there, and then let the rv Z be the smallest positive value of z $(0 < z \leqslant 1)$ for which

$$(\nu-n_r)T_{n_r,\nu}(z\Gamma_r,p) \geqslant \delta(1-\delta)n_r^{\frac{1}{2}}d_{n_r}. \tag{67}$$

The conditional probability

$$P\{(n_{r+1}-n)T_{n,n_{r+1}}(z\Gamma_r,p) < -\delta(1-\delta)n_r^{\frac{1}{2}}d_{n_r}/2 \mid D_r(p);\ \nu = n;\ Z = z\} \tag{68}$$

is a binomial probability $P\{Z_N < N\pi_N + \Delta_N\}$, where the binomial variate Z_N is the sum of N iid Bernoulli variates, each of expectation π_N, and

$$N = n_{r+1}-n, \quad \pi_N = z\Gamma_r, \quad \Delta_N = -\delta(1-\delta)n_r^{\frac{1}{2}}d_{n_r}/2,$$

so that $1-\pi_N \to 1$ uniformly in $z \in (0,1]$ as $r \to \infty$, and

$$\Delta_N^2/N\pi_N = \delta^2(1-\delta)^2(1+\bar{\varepsilon})^2(\log n_r)n_r/4(n_{r+1}-n)z(1+2\delta)$$

approaches infinity uniformly in $n \in I_r$ and $z \in (0,1]$ (and also in p) as $r \to \infty$. Hence, by Chebyshev's inequality, the probability (68) is $\leqslant \frac{1}{2}$

for all large r, uniformly in $n \in I_r$ and z and p. (The probability is zero if $n = n_{r+1}$.) From (67) and the event complementary to that of (68) we obtain, for large r,

$$P\{(n_{r+1}-n_r) \sup_{0\leqslant t\leqslant \Gamma_r} T_{n_r,n_{r+1}}(t,p) > \delta(1-\delta)\, n_r^{\frac{1}{2}}\, d_{n_r}/2 \,|\, D_r(p)\} \geqslant \tfrac{3}{4}. \quad (69)$$

Multiplying both sides of (69) by $P\{D_r(p)\}$, we obtain by inclusion,

$$P\{D_r(p)\} \leqslant (\tfrac{4}{3})\, P\{(n_{r+1}-n_r) \sup_{0\leqslant t\leqslant \Gamma_r} T_{n_r,n_{r+1}}(t,p) > \delta(1-\delta)\, n_r^{\frac{1}{2}}\, d_{n_r}/2\}. \quad (70)$$

The probability on the right side of (70) is, by (11), no greater than

$$\exp\{-\delta^2(1-\delta)^6 n_r^{\frac{1}{2}} d_{n_r}^2/8(n_{r+1}-n_r)\,\Gamma_r\}$$
$$< \exp\{-\delta^2(1-\delta)^7(1+\bar{\epsilon})^2(\log n_r)/8(\gamma-1)(1+2\delta)\} \quad (71)$$

for large r, uniformly in p. Multiplying this by $(n_r^{\frac{3}{4}}/2A)+2$, the result is summable in r by the second condition of (54). Thus, (70) and (71) imply that (66) occurs i.o. (r) wp 0, and the proof of this section is complete.

5 STRONG LOWER BOUND

For typographical simplification we write 4ϵ for the $\bar{\epsilon}$ of (13) and suppose ϵ is specified,

$$0 < \epsilon = \bar{\epsilon}/4 < \tfrac{1}{72}.$$

Let δ and γ be positive values satisfying the following conditions (written somewhat redundantly for the sake of easy reference):

$$\left.\begin{aligned} &\delta < 1;\ \gamma > \epsilon^{-1};\ \gamma > 2^{\frac{5}{2}}\epsilon^{-2}(1-\delta)^{-7};\ (1-\gamma^{-1})^{\frac{3}{2}} > 2\epsilon;\\ &\gamma > 36\epsilon^{-6};\ \gamma^{-1} < \epsilon^3/4;\ |\epsilon^4(1-\gamma^{-1})-\gamma^{-1}| < \epsilon^3/4. \end{aligned}\right\} \quad (72)$$

Let $\{n_r, r = 1, 2, \ldots\}$, as in §4, be any increasing sequence of positive integers satisfying (as $r \to \infty$)

$$n_r \sim \gamma_r. \quad (73)$$

Abbreviating the notation introduced below (55) (since n', n'' will always be n_r, n_{r+1} here), we write $m_r = n_{r+1}-n_r$, and let S_r' be the sample df based on the $n_{r+1}-n_r$ observations $X_{n_r+1}, X_{n_r+2}, \ldots, X_{n_{r+1}}$. Similarly, let $K_r'(p)$ be defined by (7) with S_n replaced by S_r' and $n^{\frac{1}{2}}$ replaced by $m_r^{\frac{1}{2}}$, and let T_r' be obtained by altering S_n, $n^{-\frac{1}{2}}$, and K_n in this way in (7). We shall repeatedly use asymptotic $(r \to \infty)$ relations such as

$$m_r \sim (1-\gamma^{-1})\, n_{r+1}, \quad \log n_r \sim \log n_{r+1}, \quad \log\log m_r \sim \log r, \quad \text{etc.,}$$

without further comment.

Let
$$d_n' = 2^{-\frac{1}{4}} n^{-\frac{3}{4}} (\log n)^{\frac{1}{2}} (\log\log n)^{\frac{1}{4}}. \tag{74}$$
We shall show
$$P\{\inf_{p\in I} T_{n_{r+1}}(n_{r+1}^{-\frac{1}{2}} K_{n_{r+1}}(p) + (1-4\epsilon)\, d'_{n_{r+1}}, p) < -(1-4\epsilon)\, d'_{n_{r+1}} \text{ i.o.}\} = 1, \tag{75}$$
which by (8) implies the left inequality of (13).

To obtain (75), we shall show
$$P\{\inf_{p\in I} T_r'(n_{r+1}^{-\frac{1}{2}} K_{n_{r+1}}(p) + (1-4\epsilon)\, d'_{n_{r+1}}, p) < -(1-2\epsilon)\, d'_{n_{r+1}} \text{ i.o.}\} = 1 \tag{76}$$
and
$$P\{\sup_{p\in I} T_{n_r}(n_{r+1}^{-\frac{1}{2}} K_{n_{r+1}}(p) + (1-4\epsilon)\, d'_{n_{r+1}}, p) \leqslant \epsilon\gamma d'_{n_{r+1}} \text{ for a.a. } r\} = 1; \tag{77}$$
(75) is an obvious consequence of (76), (77), and the facts that (as in (63)
$$n_r T_{n_r}(x,p) + m_r T_r'(x,p) = n_{r+1} T_{n_{r+1}}(x,p) \tag{78}$$
and (a consequence of the second condition of (72)), for large r,
$$n_r \epsilon\gamma + m_r(-1+2\epsilon) < -n_{r+1}(1-4\epsilon). \tag{79}$$

We first prove (77) (by a calculation similar to that used to obtain (61)). Consider the event in the first line of (80) below. Writing
$$L = \gamma^{-\frac{1}{2}} n_r^{-\frac{1}{2}} (\log\log n_r)^{\frac{1}{2}} + (1-4\epsilon)\, d'_{n_{r+1}} \quad \text{and} \quad \rho = \gamma\epsilon d'_{n_{r+1}},$$
we see that this event is contained in the event
$$\{\sup_{p\in I} \sup_{|\theta|\leqslant 1} |T_{n_r}(\theta L, p)| > \rho\},$$
which in turn entails the occurrence, for at least one p in the set
$$M'' = \{p\colon p = 1 \text{ or } p = jL,\ 0 \leqslant p < 1, j \text{ integral}\},$$
of the event $\{\sup_{|\theta|\leqslant 1} |T_{n_r}(\theta L, p)| > \rho/2\}$. Computing an upper bound on this last event by applying (11) with (n, a_n, w_n) of (11) $= (n_r, L, \rho/2)$ here, and noting that M'' contains fewer than $n_r^{\frac{1}{2}}$ elements when r is large and that $d'_{n_{r+1}} = o(L)$; we obtain, for r sufficiently large,
$$\begin{aligned} &P\{\sup_{p\in I} \sup_{|\theta|\leqslant 1} |T_{n_r}(\theta\gamma^{-\frac{1}{2}} n_r^{-\frac{1}{2}} (\log\log n_r)^{\frac{1}{2}} + (1-4\epsilon)\, d'_{n_{r+1}}, p)| > \gamma\epsilon d'_{n_{r+1}}\} \\ &\quad\leqslant n_r^{\frac{1}{2}} \exp\{-(1-\delta)^6 n_r (\gamma\epsilon d'_{n_{r+1}}/2)^2 / 2\gamma^{-\frac{1}{2}} (\log\log n_r)^{\frac{1}{2}} n_r^{-\frac{1}{2}}\} \\ &\quad\leqslant n_r^{\frac{1}{2}} \exp\{-\gamma\epsilon^2 (\log n_r)\, (1-\delta)^7 / 2^{\frac{7}{2}}\}. \end{aligned} \tag{80}$$
By the third condition of (72), the last expression is summable in r, which by the Borel–Cantelli lemma proves that the event of the first line

of (80) occurs i.o. (r) wp 0. But for each large r the event of the first line of (80) contains the *complement* of the event of (77), because

$$\sup_p |n_{r+1}^{-\frac{1}{2}} K_{n_{r+1}}(p)| < [(\log\log n_r)(1+\delta)/2\gamma n_r]^{\frac{1}{2}}$$

for a.a. r wp 1 (by the LIL for the sample df) and $[(1+\delta)/2]^{\frac{1}{2}} < 1$ (by (72)). Hence, the complement of the event of (77) occurs i.o. (r) wp 0. Thus, (77) is verified.

It remains to prove (76). To this end we modify the definitions of (26)–(28) as follows (in order that we may consider values of K'_r of order $(\log\log m_r)^{\frac{1}{2}}$):

$$\left.\begin{aligned} M'_r &= \{j\colon 0 \leqslant j < \epsilon^8 m_r^{\frac{1}{2}}/(\log\log m_r)^{\frac{1}{2}}, j \text{ integral}\},\\ p'_{r,j} &= 2^{-1} + m_r^{-\frac{1}{2}} j b \epsilon^{-1} (\log\log m_r)^{\frac{1}{2}} \quad \text{for } j \in M'_r,\\ p^{*\prime}_{r,j} &= p'_{r,j} + b m_r^{-\frac{1}{2}} (\log\log m_r)^{\frac{1}{2}},\\ \Lambda^{*\prime}_r &= \{b < K'_r(\tfrac{1}{2})(\log\log m_r)^{-\frac{1}{2}} < b(1+\epsilon^4)\},\\ A'_{r,j} &= \{T'_r(b[m_r^{-1}\log\log m_r]^{\frac{1}{2}}, p'_{r,j}) < -(1-\epsilon)\, d'_{n_{r+1}}\}. \end{aligned}\right\} \tag{81}$$

Here the value ϵ is as specified above (72), and b is chosen so that

$$2^{-\frac{1}{2}} > b > 2^{-\frac{1}{2}}(1-\gamma^{-1})^{\frac{1}{2}}, \quad b > \tfrac{1}{2}, \tag{82}$$

where γ was chosen to satisfy (72). We shall prove the following three results, from which (76) follows at once:

$$\{P\Lambda^{*\prime}_r \cap \bigcup_{j\in M'_r} A'_{r,j} \text{ occurs i.o. } (r)\} = 1; \tag{83}$$

$$P\{\sup_{|\theta|\leqslant 1,\, j\in M'_r} [T'_r(b[1+\theta\epsilon^3][m_r^{-1}\log\log m_r]^{\frac{1}{2}}, p'_{r,j}) - T'_r(b[m_r^{-1}\log\log m_r]^{\frac{1}{2}}, p'_{r,j})] \leqslant \epsilon d'_{n_{r+1}} \quad \text{for a.a. } r\} = 1; \tag{84}$$

$$P\{\Lambda^{*\prime}_r \cap \{\max_{j\in M'_r} |[n_{r+1}^{-\frac{1}{2}} K_{n_{r+1}}(p'_{r,j}) + (1-4\epsilon)\, d'_{n_{r+1}}]/b[m_r^{-1}\log\log m_r]^{\frac{1}{2}} - 1| > \epsilon^3\} \text{ i.o. } (r)\} = 0. \tag{85}$$

To prove (83), we follow a development like (but simpler than) that leading from (29) to (45). Given that $\Lambda^{*\prime}_r$ occurs and that $K'_r(p'_{r,j}) = k'_{r,j}$ for $j \in M'_r$, we have $p'_{r,j} < p^{*\prime}_{r,j} < p'_{r,j+1}$, and $A'_{r,j}$ is conditionally a binomial event with

$$\left.\begin{aligned} N &= m_r^{\frac{1}{2}}(u'_{r,j} + b\epsilon^{-1}(\log\log m_r)^{\frac{1}{2}}),\\ u'_{r,j} &= k'_{r,j} - k'_{r,j+1}, \quad U'_{r,j} = K'_{r,j} - K'_{r,j+1},\\ \pi_N &= (p'_{r,j+1} - p^{*\prime}_{r,j})/(p'_{r,j+1} - p'_{r,j}) = 1-\epsilon,\\ Z_N &= m_r[S'_r(p'_{r,j+1}) - S'_r(p^{*\prime}_{r,j})],\\ A'_{r,j} &\Leftrightarrow m_r^{-1} Z_N > (1-\epsilon)\, d'_{n_{r+1}} + m_r^{-\frac{1}{2}} u'_{r,j} + b(\epsilon^{-1}-1)[m_r^{-1}\log\log m_r]^{\frac{1}{2}}\\ &\Leftrightarrow Z_N - N\pi_N > (1-\epsilon)\, m_r d'_{n_{r+1}} + \epsilon m_r^{\frac{1}{2}} u'_{r,j} \equiv \Delta_N. \end{aligned}\right\} \tag{86}$$

Define

$$\left.\begin{aligned} \pi'_r &= m_r^{-\frac{1}{2}} b \epsilon^{-1} (\log\log m_r)^{\frac{1}{2}}, \\ \mu_r &= \epsilon^8 m_r^{\frac{1}{2}} (\log\log m_r)^{-\frac{1}{2}}, \\ |J_r| &= \text{number of nonpositive } U'_{r,j} \text{ for } j \text{ satisfying } j, j+1 \in M'_r, \\ C_r^{*\prime} &= \{|J_r| \geqslant \mu_r/3\}. \end{aligned}\right\} \tag{87}$$

The argument of the last paragraph of § 3 applies, essentially intact, with $p^i_{n,j}$, $U^i_{n,j}$, π, n, $n^{\frac{1}{2}}\epsilon$ of that section replaced by $p'_{r,j}$, $U'_{r,j}$, π'_r, m_r, μ_r, here. (The analogues of (46) cause no difficulty, since $\pi'_r \to 0$, $(m_r \pi'_r)^{-\frac{1}{2}} \to 0$, and $\mu_r \to \infty$, as $r \to 0$.) One obtains, in analogy with (52), that for all sufficiently large r,

$$\left.\begin{aligned} E|J_r| &\geqslant 3\mu_r/8, \\ \operatorname{var}(|J_r|) &\leqslant \mu_r/4 + \mu_r^2 c_4 (m_r \pi'_r)^{-\frac{1}{2}}. \end{aligned}\right\} \tag{88}$$

Hence, the analogue of (53)–(36) is that

$$P\{\overline{C_r^{*\prime}}\} \leqslant c'_0\{\mu_r^{-1} + (m_r \pi'_r)^{-\frac{1}{2}}\} \tag{89}$$

for some constant c'_0, and therefore

$$\Sigma_r P\{\overline{C_r^{*\prime}}\} < \infty. \tag{90}$$

Just as in § 3, it is again easily established that $\Delta_N^2/N\pi_N(1-\pi_N)$ is a strictly increasing function of $u'_{r,j}$, for $u'_{r,j} \geqq$ the value $c'_{r,j}$ of $u'_{r,j}$ at which this function vanishes. If $u'_{r,j} \leqslant 0$, then by using (82) we have for large r and for $c'_{r,j} \leqslant u'_{r,j}$, upon substituting into $\Delta_N^2/N\pi_N(1-\pi_N)$ the upper bound $0 \geqslant u'_{r,j}$ obtained above (in parallel with (37)),

$$\begin{aligned} \Delta_N^2/N\pi_N(1-\pi_N) &\leqq 2^{-\frac{1}{2}}(1-\epsilon)(1-\gamma^{-1})^{\frac{3}{2}} b^{-1}[1+o(1)] \log n_{r+1} \\ &\leqq (1-\epsilon)\log n_{r+1}. \end{aligned} \tag{91}$$

From the relation between the first expression of (40) and the last line of (40) we thus have for r large, by using a calculation like that of (38),

$$\begin{aligned} P\{\Lambda_r^{*\prime} \cap \bigcap_{j \in M'_r} A'_{r,j}\} &\geqq P\{\Lambda_r^{*\prime}\} - P\{\bigcap_{j \in M'_r} \overline{A'_{r,j}} | \Lambda_r^{*\prime}; C_r^{*\prime}\} - P\{\overline{C_r^{*\prime}}\} \\ &\geqq P\{\Lambda_r^{*\prime}\} - [1 - e^{-2^{-1}(1-\epsilon/2)\log n_{r+1}}]^{\epsilon^8 m_r^{\frac{1}{2}}/3(\log\log m_r)^{\frac{1}{2}}} \\ &\qquad - P\{\overline{C_r^{*\prime}}\}. \end{aligned} \tag{92}$$

By the first inequality of (82) and (9) (with

$$N = m_r, \pi_N = \tfrac{1}{2}, \quad Z_N = m_r[1 - S'_r(\tfrac{1}{2})],$$

$$\Delta_N[m_r \log\log m_r]^{-\frac{1}{2}} = b \quad \text{or} \quad b(1+\epsilon^4)),$$

we see that $\sum_r P\{\Lambda_r^{*\prime}\} = +\infty$; on the other hand (recall (90)), the remaining terms of the last line of (92) are summable. Since the events of (83) for different values of r depend on different X_i's, they are independent, and hence (83) follows from (92) and the Borel–Cantelli lemma.

As for (84), the event considered there for fixed r and j (the $\sup_j$ operation being temporarily deleted) can be written, by using the T_n relation just above (21), as

$$\sup_{|\theta|\leqslant 1} T_r'(\theta b\epsilon^3[m_r^{-1}\log\log m_r]^{\frac{1}{2}}, p'_{r,j} + b[m_r^{-1}\log\log m_r]^{\frac{1}{2}}) \leqslant \epsilon d'_{n_{r+1}}. \quad (93)$$

By (11) (with n replaced by m_r and $(1-\delta)^3$ of (11) equal to $\frac{1}{2}$) the probability of the *complement* of the event (93), for all sufficiently large r and for each $j \in M'_r$, is at most

$$\exp\{-m_r(\epsilon d'_{n_{r+1}})^2/4b\epsilon^3[m_r^{-1}\log\log m_r]^{\frac{1}{2}}\}$$
$$= \exp\{-b^{-1}\epsilon^{-1}(32)^{-\frac{1}{2}}(1-\gamma^{-1})^{\frac{3}{2}}[1+o(1)]\log n_{r+1}\}. \quad (94)$$

There are fewer than $n_{r+1}^{\frac{1}{2}}$ elements in M'_r for large r, and we see by the fourth condition of (72) and the first condition of (82) that $n_{r+1}^{\frac{1}{2}}$ times the last expression of (94) is summable. Hence, by the Borel–Cantelli lemma, the *complement* of the event considered in (84) for fixed r (but including now the $\sup_j$ operation) occurs for only finitely many r, wp 1. This proves (84).

It remains to prove (85). This will clearly follow from adding the inequalities in the following three results which we shall prove next:

$$\sup_{0\leq t\leq \epsilon^7 b} n_{r+1}^{-\frac{1}{2}}|K_{n_{r+1}}(\tfrac{1}{2}+t) - K_{n_{r+1}}(\tfrac{1}{2})|$$
$$\leqq 3^{-1}b[m_r^{-1}\log\log m_r]^{\frac{1}{2}}\epsilon^3 \quad \text{for a.a. } r, \text{wp } 1; \quad (95)$$

$$|n_{r+1}^{-\frac{1}{2}}K_{n_{r+1}}(\tfrac{1}{2}) - m_r^{\frac{1}{2}}n_{r+1}^{-1}K'_r(\tfrac{1}{2})|$$
$$< 3^{-1}b[m_r^{-1}\log\log m_r]^{\frac{1}{2}}\epsilon^3 \quad \text{for a.a. } r, \text{wp } 1; \quad (96)$$

$$\Lambda_r^{*\prime} \Rightarrow |m_r^{\frac{1}{2}}n_{r+1}^{-1}K'_r(\tfrac{1}{2}) + (1-4\epsilon)d'_{n_{r+1}} - b[m_r^{-1}\log\log m_r]^{\frac{1}{2}}|$$
$$< 3^{-1}b[m_r^{-1}\log\log m_r]^{\frac{1}{2}}\epsilon^3 \quad \text{for a.a. } r. \quad (97)$$

For fixed r, the *complement* of the event of (95) can be written

$$\sup_{0\leqslant t\leqslant \epsilon^7 b}|T_{n_{r+1}}(t,\tfrac{1}{2})| > 3^{-1}\epsilon^3 b[m_r^{-1}\log\log m_r]^{\frac{1}{2}}, \quad (98)$$

and by (9) and (10) (with the δ of (10), not that of (72), now chosen to satisfy $(1-\epsilon^7 b)^{-1}(1-\delta)^2 > \frac{1}{2}$) the probability of the event (98) for large r is at most

$$\exp\{-[n_{r+1}/4\epsilon^7 b]\,3^{-2}\epsilon^6 b^2[m_r^{-1}\log\log m_r]\}, \quad (99)$$

which is summable since $\epsilon < \frac{1}{72}$ and $b > \frac{1}{2}$. This and the Borel–Cantelli lemma yield (95).

As for (96), since $n_{r+1}^{\frac{1}{2}} K_{n_{r+1}} = n_r^{\frac{1}{2}} K_{n_r} + m_r^{\frac{1}{2}} K_r'$, the rv considered on the left side of (96) is

$$|n_{r+1}^{-\frac{1}{2}} K_{n_{r+1}}(\tfrac{1}{2}) - m_r^{\frac{1}{2}} n_{r+1}^{-1} K_r'(\tfrac{1}{2})| = |n_r^{\frac{1}{2}} n_{r+1}^{-1} K_{n_r}(\tfrac{1}{2})|. \qquad (100)$$

By the LIL, $|K_{n_r}(\frac{1}{2})| < [\log\log n_r]^{\frac{1}{2}}$ for a.a. r, wp 1. This fact, together with (100), the fifth condition of (72), and the last condition of (82), yields (96).

Finally, the event $\Lambda_r^{*\prime}$ can be written

$$m_r n_{r+1}^{-1} - 1 < \frac{m_r^{\frac{1}{2}} n_{r+1}^{-1} K_r'(\frac{1}{2}) - b[m_r^{-1} \log\log m_r]^{\frac{1}{2}}}{b[m_r^{-1} \log\log m_r]^{\frac{1}{2}}} < (1+\epsilon^4)\, m_r n_{r+1}^{-1} - 1. \qquad (101)$$

Thus, the last two conditions of (72), and the fact that $d'_{n_{r+1}} = o(m_r^{-\frac{1}{2}})$, yield (97), completing the proof of this section, and hence of Theorem 2.

6 COMMENTS

A. On the proofs

1. While the weak law of this paper is more difficult to obtain than the straightforward limit law (displayed below (1)) for a single p, the strong law here requires less delicate technique than that of [8] (displayed above (2)). The reason for the latter is that the critical values of K_n here can be obtained crudely from the LIL, while in [8] they are only a fraction as large, and the value of the fraction is crucial.

2. The crude device of (16)–(19), used to prove (12) from (15) and (25), is used to avoid the more difficult computation of relevant conditional probabilities given D_n. The latter computation is not helped by passing to the limiting Brownian bridge, since the limiting R_n deviations are not then present; this reflects the failure, mentioned in paragraph 6.2 of [8], of usual invariance arguments in the case of the R_n process. We remark that the complications attending the use of the device of (16)–(19) when the number of sets Q_i one must consider is unbounded in n, accounts for part of the difficulty, mentioned in § 1, of finding a simple proof of a strong counterpart of (5).

3. We note that, as in the case of the LIL for D_n, large deviations of $R_n^{\#}$ in the strong law (4) are accounted for essentially by behavior near $p = \frac{1}{2}$; this is reflected in the definitions of (81).

4. The right *order of magnitude* in the strong upper class conclusion of (4) can be obtained in a few lines, in contrast with the development of

§ 4; one considers (21), (22), (23) as before and notes that, with d_n multiplied by $3^{\frac{1}{2}}$, the expression (65) with n for n_r, when multiplied by $n^{\frac{1}{2}}$, is summable in n (as is, even more obviously, (64)). This is of interest for establishing a brief treatment [10] of the subject of [3], [4] mentioned in Theorem 5.

5. Since the crude bound of (36)–(89) suffices for the proofs of §§ 3 and 5, we have not taken the space to give sharper (exponential) bounds, but it is not hard to do so.

6. The reduction which uses (6) breaks down without the assumptions of the first paragraph, but it is not clear to what extent the *conclusions* of [1], [8], and the present paper fail.

B. Open problems

1. The previous paragraph A 6 suggests some obvious problems, of studying $R_n^{\#}$ for F not satisfying the assumptions used herein. Other variants to be studied include the results of [8] quoted in the first paragraph of Section 1 but with fixed p replaced by $\{p_n\}$. (For example, with F uniform, $R_n(\frac{1}{2}n)$ is of order n^{-1} rather than $n^{-\frac{3}{4}}$, in probability.) The consideration of (2) with I replaced by $\{0 \leqslant p \leqslant p_n\}$ is relevant to the extension of Theorem 4 to cases where $p_n \downarrow 0$ at a faster rate.

2. Similarly, one can study (2) without multiplying by F' there.

3. Computations like that of paragraph B 2 just above, and of other variants of $R_n^{\#}$ analogous to test statistics like ω^2 or other functionals of sample spacings, are of interest for purposes of inference. It is striking that, at least in the sense of (5), $R_n^{\#}$ does not exhibit the drastic loss of information evident for $R_n(p)$ for a single p, as discussed in § 6.6 of [8].

4. The finer characterization of functions of upper and lower class, for $R_n(p)$ as well as $R_n^{\#}$, still seems beyond the present methods.

5. A theoretical development which allows functionals of the R_n process to be treated in terms of a limiting process (see paragraph A 2 of this section, and paragraph 6.2 of [8]) would be very interesting.

C. Corrections of [8]

It seems in place to record these here:

In (1.4), lower '$\frac{1}{2}$' to yield subscript $p+n^{-\frac{1}{2}}t$ on ξ.

In proof of Lemma 1, write U_1, U_2 for H_1, H_2.

Just above (1.20), write $\lambda > 0$ for $\lambda > 1$.

In (1.22), write $\cap$ for $\cup$.

In (4.2), definition of D_r, the $+$ sign should be a superscript.

Seven lines below (2.2), replace '(2.1)' by '(2.2)'.

REFERENCES

[1] R. R. Bahadur. A note on quantiles in large samples. *Ann. Math. Statist.* **37**, 577–80 (1966).

[2] H. Bergstrom. On the central limit theorem in the space R^k, $k > 1$. *Sk. and Akt.* **28**, 106–27 (1945).

[3] L. Breiman. *Probability*. Addison-Wesley (1968).

[4] D. R. Brillinger. An asymptotic representation of the sample df. *Bull. AMS* **75**, 545–7 (1969).

[5] K. L. Chung. An estimate concerning the Kolmogoroff limit distribution. *Trans. Amer. Math. Soc.* **67**, 36–50 (1949).

[6] F. Eicker. A log log law for double sequences of random variables. (To appear.)

[7] W. Feller. *An introduction to probability theory and its applications*, vol. 1, second edn. (1957); vol. 2, first edn. (1966). New York: John Wiley and Sons.

[8] J. Kiefer. On Bahadur's representation of sample quantiles. *Ann. Math. Statist.* **38**, 1323–42 (1967).

[9] J. Kiefer. Old and new methods for studying order statistics and sample quantiles. This volume, pp. 349–357. (Discussion of Weiss and Eicker papers.)

[10] J. Kiefer. On a Brownian bridge approximation to the sample df. (To appear.)

ITERATED LOGARITHM ANALOGUES FOR SAMPLE QUANTILES WHEN $p_n \downarrow 0$

J. KIEFER
CORNELL UNIVERSITY

1. Introduction

This paper is concerned with behavior of the Law of Iterated Logarithm (LIL) type for sample p_n-tiles, $p_n > 0$, when $p_n \downarrow 0$. The results are all stated for uniformly distributed random variables, from which they may easily be translated into results for general laws.

Let $X_1, X_2, \cdots$ be independent identically distributed random variables, uniformly distributed on $[0, 1]$. Let $T_n(x) = \{\text{number of } X_i \leq x, 1 \leq i \leq n\}$, so that $n^{-1}T_n$ is the right continuous *sample distribution function* based on $X_1, X_2, \cdots, X_n$. Define the *sample p_n-tile* $Z_n(p_n)$ as $\min\{z: T_n(z) \geq np_n\}$. This makes $Z_n(p_n) = np_n$-th order statistic when np_n is a fixed integer. (When $np_n \to \infty$ our results do not depend on the choice of definition of $Z_n(p_n)$ in cases of ambiguity.)

The earliest nontrivial result in this area, due to Baxter [2], is that, for any positive constant c,

$$\text{(1.1)} \qquad \limsup_n T_n(c/n) \log\log\log n (\log\log n)^{-1} = 1, \qquad \text{wp } 1.$$

On the other hand, it is trivial (and a consequence of Theorem 2 herein, with $k = 1$) that

$$\text{(1.2)} \qquad \liminf_n T_n(c/n) = 0, \qquad \text{wp } 1.$$

We thus no longer have the symmetry in asymptotic behavior of positive and negative deviations of $T_n(\pi_n) - ET_n(\pi_n)$ that prevails when π_n is constant; indeed, why should we, when $nT_n(c/n)$ is asymptotically Poisson rather than normal?

This difference in behavior means we will have to state results for the two directions of oscillations separately, and (since the analogue of (1.2) will not always be so simple to state) dictates a choice of nomenclature which we had best introduce at the outset: to eliminate possible confusion with reference to the two *directions* of oscillation, we drop the usual "upper or lower class" LIL terms completely, replacing these by "outer or inner class" for sequences $\{f_n\}$ beyond which $T_n(\pi_n)$ moves (in a direction away from $ET_n(\pi_n)$) finitely or infinitely often

Research carried out under ONR contract Nonr 401 (50) and NSF Grant GP 9297.

with probability one. Then, "top or bottom" bounds will refer to the most or least positive oscillations of $T_n(\pi_n)$. Thus, for example, $\{f_n\}$ is a bound of top inner class if $f_n > n\pi_n$ and $T_n(\pi_n) > f_n$ i.o. with probability one. If $\{(1 \pm \varepsilon)f_n\}$ gives top bounds of the two classes, we shall for brevity simply call $\{f_n\}$ a top bound. (If, as when π_n is constant, this yields too gross a result for T_n, we would instead specify a top bound on $T_n(\pi_n) - n\pi_n$.) Thus, $\log\log n/\log\log\log n$ is a top bound for $T_n(c/n)$ in Baxter's case mentioned above. Bounds for $Z_n(p_n)$ are described similarly. Of course, top bounds for Z_n are related to bottom bounds for T_n and vice versa, by the well known relation

$$T_n(\pi_n) \geqq k_n \Leftrightarrow Z_n(k_n/n) \leqq \pi_n, \tag{1.3}$$

which follows from the definitions.

The proofs of the present paper employ standard techniques and estimates of binomial probabilities, and we have sometimes introduced inessential assumptions to maintain simplicity and brevity. The results are mainly about first order deviations of T_n or Z_n from their expectations; while some "strong form" results are known, with few exceptions (for example, Theorem 1) they entail much longer proofs and I do not presently know the conclusions for the full spectrum of sequences $\{\pi_n\}$ considered herein. Theorems 1 and 2 cover the behavior of $Z_n(k/n)$ with k fixed; Theorem 5 covers the domain of normal limiting behavior and resulting classical LIL form for $T_n(\pi_n)$, and Theorems 3 and 4 cover the behavior in between these two extremes, including Baxter's case; Theorem 6 translates some of the conclusions for T_n into conclusions for Z_n.

We now mention work related to that of the present paper, other than that of Baxter described above. Bahadur [2] used his relation between the Z_n and T_n processes to obtain the LIL for $Z_n(p)$ with p fixed from the classical LIL for binomial $T_n(p)$. The same method can be used to obtain the strong form of inner and outer classes [6]. While the classical techniques of Theorems 5 and 6 herein also yield the LIL for $Z_n(p)$ (the main departure from the usual LIL proof for $T_n(p)$ being that one is now led by (1.3) to the LIL for $T_n(p_n)$ for varying p_n), Bahadur's technique provides a great saving of effort when it comes to the strong form.

Eicker [4] obtained top outer bounds (analogous to (1.1)) for $T_n(\pi_n)$ when $\pi_n \downarrow 0$ and $n\pi_n/\log\log n \to \infty$, but not bottom outer bounds, and also obtained both inner bounds. This is the domain treated in Theorem 5 of the present paper, where the usual binomial-normal LIL form holds. Eicker's inner class proof follows essentially the classical lines which are therefore sketched only in brief outline herein; this proof applies also to certain cases where π_n is bounded away from zero and one. His top outer bound proof uses fine estimates of the probability that the T_n process exceeds certain polygonal bounds; our proof, while much more routine, is considerably shorter, and treats also the bottom outer bound.

Robbins and Siegmund [10] have just announced the use of an interpolating process and the derivation of probability estimates for ever exceeding certain bounds (in the spirit of their earlier work with sums of random variables), in obtaining strong top bounds for the first order statistics $Z_n(1/n)$. Our Theorem 2 with $k = 1$ states only the first order term of this strong form.

In [1], [6], [7], and [8] relations between the $T_n(p)$ and $Z_n(p)$ processes were studied (with respect to varying p as well as n). For example, in the spirit of [1], one thereby obtains in [8] (where some of the results of the present paper were also described) the strong form for $\sup_{0<p<1}|Z_n(p) - p|$ oscillations from the corresponding sample of distribution function results of Chung [3]. Large deviations of either these processes or of their difference, over the domain $0 < p < p_n \downarrow 0$, are related to the present results and will be treated elsewhere; some such considerations have appeared in the work of Chibisov, LeCam, and others. Also related are the numerous papers on weak laws for order statistics, about which we mention only the appearance therein, not surprisingly, of the "fundamental equation" (2.24) which arises below. (See, for example, [9].)

Section 2 contains definitions and relevant binomial tail probability estimates. Statements of the main results are contained in Section 3, along with proofs of the simple first two theorems. The remainder of the paper contains the other proofs.

2. Preliminaries

We shall use i.o., f.o., and a.a.n. in their customary meaning of infinitely often, finitely often, and for almost all (all but finitely many) n. We treat events indexed by the natural numbers $\{n\}$ or a subsequence $\{n_j\}$, and the usual expressions of limiting behavior ($\to$, $\sim$) or of order (such as $O(z(n))$ or $o(g(j))$) refer to behavior as $n \to +\infty$ or $j \to +\infty$. The symbols $\uparrow$, $\downarrow$ are used for monotone, not strict. approach.

We let $\log_1$ denote the natural logarithm and $\log_{j+1} = \log\log_j$. Also, $\log_j^i x = (\log_j x)^i$. In summations and other appearances of such an expression as $\log_j n$, the domain of n is understood to begin where the expression is meaningful.

We shall try to reserve π_n for the argument of T_n and p_n for that of Z_n, with $k_n \geqq 0$ being used for bounds on T_n. We define $T_{n_1,n_2} = T_{n_2} - T_{n_1}$, the observation counter based on $X_{n_1+1}, \cdots, X_{n_2}$. Limiting behavior will often be conveniently described in terms of

$$h_n = n\pi_n/\log_2 n, \qquad H_n = np_n/\log_2 n. \tag{2.1}$$

Either of the Borel-Cantelli lemmas is denoted by BC.

We use int $\{x\}$ to denote the largest integer $\leqq x$, and int$^+$ $\{x\}$ for the smallest integer $\geqq x$.

When we consider a subsequence $\{n_j\}$ of the natural numbers, we write

$$I_j = \{n\colon n_j < n \leqq n_{j+1}\} \quad\text{and}\quad I_j^- = \{n\colon n_j < n < n_{j+1}\}. \tag{2.2}$$

We write $m_j = n_j - n_{j-1}$. The two subsequences $\{n_j\}$ we shall consider, with typical estimates they imply, are

$$(2.3)\qquad \begin{aligned} &n_j \sim \lambda^j \qquad \text{where} \quad \lambda > 1,\ m_j/n_j \sim (\lambda - 1)/\lambda, \\ &\log_2 n_j \sim \log j, \end{aligned}$$

and, for $\alpha > 0$,

$$(2.4)\qquad \begin{aligned} &n_j \sim e^{\alpha j \log j} \qquad\qquad m_j/n_j = 1 - O(j^{-\alpha}), \\ &\log_2 n_j \sim \log j. \end{aligned}$$

Much of our treatment can be carried out in terms of very simple events, essentially as employed in [2] for top bounds on T_n. Suppose $\pi_n \downarrow 0$ and $k_n \uparrow$, and let

$$(2.5)\qquad A_n = \{T_n(\pi_n) \geqq k_n\}.$$

Let $\{n_j\}$ be *any* increasing sequence of natural numbers, and define

$$(2.6)\qquad B_j^* = \{T_{n_j, n_{j+1}}(\pi_{n_{j+1}}) \geqq k_{n_{j+1}}\}$$

and

$$(2.7)\qquad C_j^* = \{T_{n_{j+1}}(\pi_{n_j}) < k_{n_j}\}.$$

Then, since $T_n(\pi)$ is nondecreasing in n and π, and $\pi_n \downarrow$, $k_n \uparrow$,

$$(2.8)\qquad C_j^* \subset \{T_n(\pi_n) < k_n,\ n \in I_j\},$$

and hence

$$(2.9)\qquad \{C_j^*, \text{a.a. } j\} \Rightarrow \{A_n \text{ f.o.}\}, \qquad \text{wp } 1.$$

In the other direction, obviously

$$(2.10)\qquad \{B_j^* \text{ i.o.}\} \Rightarrow \{A_n \text{ i.o.}\}, \qquad \text{wp } 1,$$

the events of (2.6) of course being useful because they are independent.

We shall see that, in the top outer class proofs of Theorems 1 and 3, we can even avoid the use of subsequences as employed in [2] and in "normal case" proofs such as that of Theorem 5 below, and, as an alternative, work with the events

$$(2.11)\qquad A_n' = \{X_n \leqq \pi_n,\ T_{n-1}(\pi_n) \geqq k_n - 1\}.$$

As long as $1 \leqq k_n \uparrow$ and $\pi_n \downarrow 0$, it is evident that

$$(2.12)\qquad \{A_n \text{ i.o.}\} \Rightarrow \{A_n' \text{ i.o.}\}, \qquad \text{wp } 1.$$

Similarly, if $\pi_n \downarrow$ and $k_n \uparrow$, for bottom outer bounds on T_n we use

$$(2.13)\qquad \{T_{n_j}(\pi_{n_{j+1}}) < k_{n_{j+1}} \text{ f.o.}\} \Rightarrow \{T_n(\pi_n) < k_n \text{ f.o.}\}, \qquad \text{wp } 1.$$

The bottom inner bound treatment is only slightly less simple. If

$$(2.14)\qquad \begin{aligned} Q_j^* &= \{T_{n_j, n_{j+1}}(\pi_{n_{j+1}}) \leqq k_{n_{j+1}} - \gamma_j\}, \\ R_j^* &= \{T_{n_j}(\pi_{n_{j+1}}) > \gamma_j\}, \end{aligned}$$

where the γ_j are arbitrary nonnegative values, then

$$\{Q_j^* \text{ i.o.}\} \cap \{Q_j R_j^* \text{ f.o.}\} \Rightarrow \{T_{n_{j+1}}(\pi_{n_{j+1}}) \leqq k_{n_{j+1}} \text{ i.o.}\}, \qquad \text{wp } 1. \tag{2.15}$$

In fact, in the simple first order proofs of Theorems 2 and 4 it suffices to take $\gamma_j = 0$, and to show $P\{R_j^* \text{ f.o.}\} = 1$ for the second half of the left side of (2.15).

In using the simple devices of (2.5) through (2.15) we require (in addition to $\pi_n \downarrow 0$) $k_n \uparrow$. This is not always too convenient: as described below the statement of Theorem 3, for given $\{\pi_n\}$ the natural formula for a bound k_n in terms of π_n in Theorems 3 and 4 may not yield a monotone k_n. However, in such cases we will be able to replace the nonmonotone $\{k_n\}$ by a sequence $\{k_n^*\}$ such that

$$k_n^* \uparrow, \; k_n^* = [1 + O(1)]k_n, \tag{2.16}$$

and then use the appropriate device of (2.5) through (2.15) on k_n^*; by virtue of our considerations being first order (so that we prove $(1 \pm \varepsilon)k_n^*$ lies in the appropriate class), $\{k_n\}$ is then by definition in the same class as $\{k_n^*\}$.

The study of sequences $\{\pi_n\}$ for which π_n is not monotone, or for which the technique of (2.16) fails, is more complex, requiring in place of (2.5) through (2.15) calculations which are somewhat similar to those stemming from (4.18) and (4.19) in the proof of Theorem 5. We omit such cases. It is obvious that bounds for some $\{\pi_n\}$ for which the departure from monotonicity is slight enough, may be obtained from those of majorizing and minorizing monotone sequences when corresponding bounds for the latter sequences coincide.

To give a little relief from the burden of memory or page turning, we shall reserve further definitions until they are encountered in the proof of Theorem 5.

We now list our binomial estimates. In Theorems 1 and 2 we consider $Z_n(k/n)$ with k fixed, and require only the following simple and familiar estimates ([5], p. 140) for nonnegative integral $\bar{k}$:

$$\begin{aligned} &\limsup_n n\pi_n < \bar{k} \Rightarrow \\ &\log P\{T_n(\pi_n) \geqq \bar{k}\} - O(1) = \log P\{T_n(\pi_n) = \bar{k}\} \\ &\qquad = \bar{k} \log (n\pi_n) - n\pi_n - \log (\bar{k}!) + o(1). \end{aligned} \tag{2.17}$$

and

$$\begin{aligned} &\{\pi_n = o(n^{-1/2}), \liminf_n n\pi_n > \bar{k}\} \Rightarrow \\ &\log P\{T_n(\pi_n) \leqq \bar{k}\} - O(1) = \log P\{T_n(\pi_n) = \bar{k}\} \\ &\qquad = \bar{k} \log (n\pi_n) - n\pi_n - \log (\bar{k}!) + o(1). \end{aligned} \tag{2.18}$$

In the domain of Theorems 3 and 4, we must take account of the fact that $k_n \to \infty$. Writing

$$\begin{aligned} b(z, N, P) &= \binom{N}{z} P^z (1 - P)^{N-z}, \qquad z \text{ integral}, \\ B(z, N, P) &= \sum_{y \leqq z} b(y, N, P), \\ B^+(z, N, P) &= \sum_{y \geqq z} b(y, N, P), \end{aligned} \tag{2.19}$$

we state the required binomial estimate as

LEMMA 1. *Suppose* $N \to \infty$, $P_N \to 0$, $z_N \to +\infty$. *If*

$$P_N = o(N^{-1/2}), \qquad z_N = o(N^{1/2}),$$
$$\limsup_N NP_N/z_N < 1, \tag{2.20}$$

then

$$\log B^+(z_N, N, P_N) = z_N[\log (NP_N/z_N) - (NP_N/z_N) + 1 + o(1)]. \tag{2.21}$$

Moreover, (2.21) *holds with* B *replacing* B^+, *provided the last condition of* (2.20) *is replaced by* $\liminf_N NP_N/z_N > 1$; *and the right side of* (2.21) *equals* $\log b(\text{int}^+ z_N, N, P_N)$ *or* $\log b(\text{int}\, z_N, N, P)$, *even without any third condition of* (2.20).

PROOF. The last condition of (2.20) implies that B^+/b is bounded ([5], p. 140). Putting $z' = \text{int}^+ z_N$ and writing out $\log b(z', N, P_N)$ and using Stirling's approximation, we see that the first condition of (2.20) allows us to neglect $NP_N + \log(1 - P_N)^N$, the second allows us to neglect $\log[N^{z'}/N(N-1)\cdots(N-z'+1)]$, and the two together imply $z_N P_N = o(1)$ and thus allow us to neglect $z' \log(1 - P_N)$. Since $z_N \sim z'$ and also we can absorb $\log z'$ into the $o(1)z_N$ term, we obtain (2.21). The result for B is obtained in the same way.

We now specialize the parameter values in Lemma 1, as used in the proofs of Theorems 3 and 4. Firstly, as we shall explain after Theorems 1 and 2 where k_n is bounded, we subsequently insure that k_n is unbounded by assuming

$$\liminf_n [\log(n\pi_n)/\log_2 n] \geqq 0 \tag{2.22}$$

in the top bound considerations for $T_n(\pi_n)$ of Theorem 3, and

$$\liminf_n [n\pi_n/\log_2 n] > 1 \tag{2.23}$$

in the bottom bound considerations of Theorem 4.

Secondly, in the domain where (2.22) or (2.23) is satisfied and where also $\pi_n = O(n^{-1}\log_2 n)$, the bounds on $T_n(\pi_n)$ are described in terms of the solutions of a certain transcendental equation.

The fundamental equation. We consider the solutions, for $c > 0$, of the equation

$$\beta(\log \beta - 1) = (1 - c)/c. \tag{2.24}$$

The left side of (2.24) is convex in $\beta > 0$ and attains its minimum value -1 at $\beta = 1$. Hence, (2.24) has a solution $\beta_c' > 1$ if $c > 0$ and a second positive solution $\beta_c'' < 1$ if $c > 1$. For future reference we note that β_c' is decreasing in c while β_c'' is increasing, and that

$$c \to 0 \Leftrightarrow \beta_c' \to \infty \Rightarrow \beta_c' \sim c^{-1}/\log c^{-1} \to +\infty,\ c\beta_c' \sim 1/\log c^{-1} \to 0; \tag{2.25}$$
$$c \to +\infty \Leftrightarrow \beta_c' \to 1 \Rightarrow c\beta_c' \to +\infty;$$

$$c\downarrow 1 \Leftrightarrow \beta_c'' \to 0 \Rightarrow \beta_c'' \sim (c-1)/\log (c-1)^{-1}, \qquad c\beta_c'' \to 0;$$

$$c \to +\infty \Leftrightarrow \beta_c'' \to 1, \quad c\beta_c'' \to +\infty.$$

Finally, on either part $\beta < 1$ or $\beta > 1$ of (2.24), $d(c\beta)/dc = (\beta - 1)/\log \beta > 0$.

We now state the specialization of Lemma 1 used in proving Theorems 3 and 4. Recall the definition (2.1) of h_n.

LEMMA 2. *Suppose $n \to \infty$, $\pi_n \to 0$, $h_n = O(1)$, and that (2.22) is satisfied. Let $\{\rho_n\}$ and $\{d_n\}$ be sequences of positive values which are bounded away from 0 and ∞ and for which $n\rho_n$ is integral. Assume*

$$\limsup_n \rho_n/d_n\beta_{h_n}' < 1, \tag{2.26}$$

where β_c' is defined below (2.24). Then

$$\begin{aligned}(2.27)\quad \log P\{T_{n\rho_n}(\pi_n) \geqq d_n h_n \beta_{h_n}' \log_2 n\}\\ = \{-d_n + h_n[d_n - \rho_n - d_n\beta_{h_n}' \log (d_n/\rho_n)] + o(1)\} \log_2 n;\end{aligned}$$

moreover, $\log P\{T_{n\rho_n}(\pi_n) = \text{int}^+ [d_n h_n \beta_{h_n}' \log_2 n]\}$ *satisfies the same relation.*

If (2.22) and (2.26) are replaced in the above by (2.23) and

$$\liminf_n \rho_n/d_n\beta_{h_n}'' > 1, \tag{2.28}$$

then

$$\begin{aligned}(2.29)\quad \log P\{T_{n\rho_n}(\pi_n) \leqq d_n h_n \beta_{h_n}'' \log_2 n\}\\ = \{-d_n + h_n[d_n - \rho_n - d_n\beta_{h_n}'' \log (d_n/\rho_n)] + o(1)\} \log_2 n.\end{aligned}$$

PROOF. We shall demonstrate (2.27); the proof of (2.29) is almost identical. We put $N = n\rho_n$, $P_N = \pi_n$, $z_N = d_n h_n \beta_{h_n}' \log_2 n$ in Lemma 1. Then $z_N \to +\infty$ unless there is a sequence $\{n_j\}$ for which $h_{n_j} \to 0$ and $h_{n_j}\beta_{h_n}' = O(1/\log_2 n_j)$, which by the first line of (2.25) would entail $1/\log h_{n_j}^{-1} = O(1/\log_2 n_j)$; this last is contradicted by the fact that (2.22) (with $h_{n_j} \to 0$) implies $\log h_{n_j}^{-1} = o(\log_2 n_j)$; we conclude that $z_N \to +\infty$. Next, $h_n = O(1)$ implies the first condition of (2.20) as well as $z_N = O(\log_2 n) = O(\log_2 N)$, and this last yields the second condition of (2.20). Also, $NP_N/z_N = \rho_n/d_n\beta_{h_n}'$, so that (2.26) implies the last condition of (2.20). Thus, Lemma 1 applies, and substitution into (2.21) gives

$$\begin{aligned}(2.30)\quad d_n h_n \beta_{h_n}'\{-\log \beta_{h_n}' + 1 - 1/\beta_{h_n}' + \log (\rho_n/d_n)\\ + [1 - \rho_n/d_n]/\beta_{h_n}' + o(1)\} \log_2 n.\end{aligned}$$

Since the first three terms in braces in (2.30) sum to $-1/h_n\beta_{h_n}'$ by (2.24), and since $d_n h_n \beta_{h_n}' o(1) = o(1)$, we obtain (2.27).

When $h_n \to \infty$, the approximations of Lemma 2 are insufficient, and we need the normal approximation instead. This tool is also well known; a careful reading of Feller ([5], pp. 168–173, 178–181) shows that the development there actually applies with only minor and obvious modifications when $\pi_N \to 0$ sufficiently slowly, and we state this result as

LEMMA 3. *If* $x_N = (k_N - N\pi_N)[N\pi_N(1 - \pi_N)]^{-1/2}$, *then*

$$\{N \to \infty, x_N \to +\infty, x_N[N\pi_N(1 - \pi_N)]^{-1/2} \to 0\} \Rightarrow B^+(k_N, N, \pi_N) = (2\pi)^{-1/2} x_N^{-1} \exp\{-x_N^2[1 + o(1)]/2\}. \tag{2.31}$$

(Feller's expression [5], p. 181, (6.11) for the $\frac{1}{2}o(1)x_N^2$ term in (2.31) is also correct under the present conditions; it is conveniently expressed as

$$\tfrac{1}{2}o(1)x_N^2 = \tfrac{1}{6}(2\pi_N - 1)x_N^3[N\pi_N(1 - \pi_N)]^{-1/2} + O(x_N^4[N\pi_N(1 - \pi_N)]^{-1}) \tag{2.32}$$

for use in other limit laws, but this will not be required herein.)

Finally, we shall also use the elementary fact that

$$\inf_{N,\pi} B^+(N\pi - 1, N, \pi) > 0; \tag{2.33}$$

the "bad case" where the $N\pi - 1$ is required rather than $N\pi$ is of course $N \to \infty$, $N\pi \downarrow 0$. Similarly,

$$\inf_{N,\pi} B(N\pi + 1, N, \pi) > 0. \tag{2.34}$$

In each of these expressions we include zero in the domain of N; both probabilities are then one, corresponding to the interpretation $b(0, 0, \pi) = 1$ which is appropriate in the application.

It seems essential that the proofs, as carried out in the present paper, be divided into the several cases as treated. For, the estimates of Lemma 2 are useless in the "normal" case of Theorem 5, just as those of Lemma 3 are useless for proving Theorems 3 and 4. Again, the geometric $\{n_j\}$ of (2.3) is inadequate in the inner class bottom bound proofs of Theorems 2 and 4, where (2.4) is used, but the latter cannot be used in the corresponding outer class proofs; in Theorem 5 we again use geometric $\{n_j\}$ where, also, it is impossible to avoid using subsequences by using (2.11) and (2.12) as in parts of Theorems 1 and 3. (For certain strong form results, other sequences $\{n_j\}$ must of course be considered.)

3. Main results

The short proofs of Theorems 1 and 2 will be given in this section, but proofs of the other theorems stated in this section will be deferred in favor of discussion here. For *bounded* top bounds on $T_n(\pi_n)$ or corresponding bottom bounds on the kth order statistic $Z_n(k/n)$ (k fixed), the situation is completely known [8], and is elementary to verify. We forego the artificial generality of bounded but varying k_n, or oscillatory $n\pi_n$, and state the result simply as

THEOREM 1. *If k is a positive integer and $n\pi_n \downarrow 0$, then*

$$P\{T_n(\pi_n) \geqq k \text{ i.o.}\} = P\{Z_n(k/n) \leqq \pi_n \text{ i.o.}\} = \begin{Bmatrix} 0 \Leftrightarrow \infty > \\ 1 \Leftrightarrow \infty = \end{Bmatrix} \sum_n n^{k-1}\pi_n^k \tag{3.1}$$

(*so that* $\log [nZ_n(k/n)]$ *has*

(3.2) $$-k^{-1}\{\log_2 n + \log_3 n + \cdots + (1+\varepsilon)\log_j n\}$$

as bottom outer or inner bound, depending on whether $\varepsilon > 0$ *or* $\varepsilon \leqq 0$).

PROOF. *Outer class.* By (1.3) we are concerned with the events A_n of (2.5), with $k_n = k$. The geometric sequence $\{n_j\}$ of (2.3) can be used with (2.7) and (2.9) to give a proof, in standard manner. However, we emphasize the simplicity of the present case by giving a proof without a subsequence, using the events $\{A'_n\}$ of (2.11) with $k_n = k$ and the relation (2.12). If the series of (3.1) converges, then BC and the estimate (2.17) with $\bar{k} = k - 1$ imply $P\{A'_n \text{ i.o.}\} = 1$.

Inner class. Let $n_j = 2^j$. Since $n\pi_n \downarrow 0$, divergence of the series of (3.1) implies divergence of $\Sigma'_j(n_j\pi_{n_j})^k$. The estimate (2.17) for $P\{B^*_j\}$, BC, and (2.10) complete the proof of Theorem 1.

In view of Theorem 1, our further concern with top bounds k_n on $T_n(\pi_n)$ is with the case $k_n \to +\infty$. Thus, π_n should be such that the series of (3.1) diverges for each fixed k. This is obviously the case if (2.22) holds, but not if $\limsup_n [\log(n\pi_n)/\log_2 n] < 0$. Hence, ignoring oscillatory behavior where neither of these holds, we hereafter assume π_n satisfies (2.22) in discussing these top bounds. This condition is used in order to apply Lemma 2 in the proof of Theorem 3.

Even the first order lower bounds on $Z_n(k/n)$ in Theorem 1 depend on k, and thereby exhibit quite a different behavior from the upper bounds, to which we now turn.

THEOREM 2. *If k is a positive integer, then*

(3.3) $$P\{nZ_n(k/n) > (1+\varepsilon)\log_2 n \text{ i.o.}\} = \begin{cases} 0 & \text{if } \varepsilon > 0, \\ 1 & \text{if } \varepsilon \leqq 0. \end{cases}$$

PROOF. *Outer class.* By BC, (1.3), and (2.13) with $k_n = k$, it suffices to show that, for $d > 1$, there is a $\lambda > 1$ for which $P\{T_{n_j}(dn^{-1}_{j+1}\log_2 n_{j+1}) \leqq k-1\}$ is summable, where $n_j = \text{int } \lambda^j$. But, by (2.3) and (2.18) this probability is

(3.4) $$\exp\{(k-1)\log[d\lambda^{-1}\log j] - d\lambda^{-1}\log j + O(1)\},$$

which is summable if $1 < \lambda < d$.

Inner class. We now put $n_j = \text{int}\{e^{\alpha j \log j}\}$ with $\alpha > 0$. Then, by (2.4) and (2.18),

(3.5) $$\begin{aligned} P\{T_{n_j, n_{j+1}}(n^{-1}_{j+1}\log_2 n_{j+1}) \leqq k-1\} \\ = \exp\{(k-1)\log_3 n_{j+1} - [1 - O(j^{-\alpha})]\log_2 n_{j+1} + O(1)\}, \end{aligned}$$

whose sum diverges. By (2.14) and (2.15) with $\gamma_j = 0$ and $k_n = k - 1$, the proof is completed upon computing, again from (2.4) and (2.18) (now with $\bar{k} = 0$),

(3.6) $$\begin{aligned} P\{T_{n_j}(n^{-1}_{j+1}\log_2 n_{j+1}) \geqq 1\} \\ = 1 - \exp\{-O(1)j^{-\alpha}\log_2 n_{j+1} + O(1) = O(j^{-\alpha}\log j). \end{aligned}$$

REMARKS. In (2.15), $\{Q^*_j \text{ i.o. wp } 1\}$ is essentially automatic in this case, and it is the event $\{Q^*_j R^*_j \text{ f.o. wp } 1\}$ for which geometric $n_j \sim \lambda^j$ is inadequate. With

n_j as chosen in the proof and $\alpha > 1$, not merely $\{Q_j^* R_j^* \text{ f.o. wp } 1\}$, but even $\{R_j^* \text{ f.o. wp } 1\}$ is satisfied; however, in strong form analogues one cannot always be so cavalier.

In view of Theorem 2, our further investigation of bottom bounds k_n on $T_n(\pi_n)$, with $k_n \to +\infty$, will be made under the assumption (analogous to (2.22)) that (2.23) holds. This will be discussed further, just after the statement of Theorem 4.

We now turn to the domain of behavior between that of the kth order statistic (fixed k) and that of "normal" LIL. This includes Baxter's case. We recall the definition (2.1) of h_n.

THEOREM 3. *Suppose* $\pi_n \downarrow 0$, (2.22) *is satisfied*, $h_n = O(1)$, *and that there is a* k_n^* *satisfying* (2.16) *for* $k_n = h_n \beta'_{h_n} \log_2 n$. *Then*

$$\limsup_n T_n(\pi_n)/h_n \beta'_{h_n} \log_2 n = 1 \quad \text{wp } 1. \tag{3.7}$$

In particular, if $\pi_n \downarrow 0$, *a top bound* k_n^* *on* $T_n(\pi_n)$ *is given in various ranges of* π_n *by*:

$$h_n \to c > 0 \Rightarrow k_n^* = c\beta'_c \log_2 n; \tag{3.8}$$

$$h_n \to 0 \quad \text{and} \quad \log_2 n/\log h_n^{-1} \sim g_n \uparrow +\infty \Rightarrow k_n^* = g_n; \tag{3.9}$$

in particular,

$$\begin{aligned} k_n^* &= \log_2 n/\log_3 n \quad \text{if} \quad \log (n\pi_n) = o(\log_3 n), \\ k_n^* &= \log_2 n/(B+1)\log_3 n \quad \text{if} \quad \pi_n \sim An^{-1}\log_2^{-B} n, \quad B > -1, \\ k_n^* &= \log_2 n/\log (n\pi_n)^{-1} \quad \text{if} \quad \begin{cases} \log_2 n/\log (n\pi_n)^{-1} \uparrow +\infty, \\ \log_3 n/\log (n\pi_n)^{-1} \to 0. \end{cases} \end{aligned} \tag{3.10}$$

The use of (2.16) here and in Theorem 4 is not as unnatural as it may first appear. For example, with the positive constant c near zero in Theorem 3 or near one in Theorem 4, and $h_n = c + (-1)^n/3n \log n \log_2 n$, we have $\pi_n \downarrow$, $h_n \to c$ (and, in the case of Theorem 4, even $n\pi_n \uparrow$); but the "natural" k_n given in the denominator of (3.7) or (3.11) is not monotone, which of course $k_n^* = c\beta_c \log_2 n$ is, and the conclusion of each theorem is still valid. The particular cases of (3.8), (3.9), (3.10) have been stated in terms of a simple increasing k_n^* rather than $h_n \beta'_{h_n} \log_2 n \sim \log_2 n/\log h_n^{-1}$.

We also note that, in both Theorem 3 and Theorem 4, h_n of exactly order one does not imply that $\lim h_n$ exists. One can obtain $1 < \liminf_n h_n < \limsup_n h_n < +\infty$ while $\pi_n \downarrow$, $n\pi_n \uparrow$, and even the "natural" $k_n \uparrow$ (discussed in the previous paragraph), by letting $h_{n+1} - h_n$ take successive blocks of positive and negative steps of size $\varepsilon/n \log n \log_2 n$ with ε sufficiently small.

THEOREM 4. *Suppose* $\pi_n \downarrow 0$, (2.23) *is satisfied*, $h_n = O(1)$, *and that there is a* k_n^* *satisfying* (2.16) *for* $k_n = h_n \beta''_{h_n} \log_2 n$. *Then*

$$\liminf_n T_n(\pi_n)/h_n \beta''_{h_n} \log_2 \text{n} = 1 \quad \text{wp } 1. \tag{3.11}$$

In particular, if $\pi_n \downarrow 0$,

(3.12) $$h_n \to c > 1 \Rightarrow k_n^* = c\beta_c'' \log_2 n$$

is a bottom bound on $T_n(\pi_n)$.

By putting $h_n = 1 + \delta$ with $\delta > 0$ in (3.12), and then letting $\delta \to 0$ and using the third line of (2.25), we obtain that (in a case where (2.23) is not satisfied)

(3.13) $$\limsup_n h_n \leqq 1 \Rightarrow \liminf_n T_n(\pi_n)/\log_2 n = 0, \qquad \text{wp } 1.$$

In fact, when $h_n \equiv 1$ we can use (3.4) or (1.3) and (3.3) with $k = 1$ to obtain

(3.14) $$\pi_n = n^{-1} \log_2 n \Rightarrow \liminf_n T_n(\pi_n) = 0, \qquad \text{wp } 1.$$

Since we shall see that the behavior of $c\beta_c'$ as $c \to 0$ in (2.25) yields the top bounds (3.8), (3.9), (3.10) as $h_n \to 0$, it is tempting to try to use the third line of (2.25) to obtain seemingly analogous *bottom* bounds in terms of $h_n - 1 \downarrow 0$. However, the latter are less accurately viewed as first order results in $h_n - 1$ than as second order results in h_n, which cannot be obtained from the behavior of $c\beta_c''$ as $c \downarrow 1$ without more effort, because of the failure of (2.23). We do not have complete results in this domain, and shall not discuss it further except to mention here, as an example of what is involved in the subdomain of smallest values of $h_n - 1$ of interest, that the determination of which sequences $h_n - 1$ of the particular form $L \log_3 n/\log_2 n$ (L constant) continue to imply the conclusion of (3.14) when $h_n \downarrow 1$, requires a more delicate argument than that used in proving Theorems 2 and 4. (Of course, the strong top bounds on $Z_n(1/n)$ [10] imply this second order consequence, but yield nothing about sequences $h_n - 1$ which vanish more slowly.)

We now turn to sequences $\{\pi_n\}$ which vanish slowly enough that "normal" LIL behavior prevails. Of course, we may write $\pi_n(1 - \pi_n) \sim \pi_n$.

THEOREM 5. *If* $\pi_n \downarrow 0$, $h_n \to +\infty$ *and* $n\pi_n \uparrow$, *then, for either choice of sign,*

(3.15) $$\limsup_n \pm [T_n(\pi_n) - n\pi_n][2n\pi_n \log_2 n]^{-1/2} = 1, \qquad \text{wp } 1.$$

REMARKS. The assumption that $n\pi_n$ is nondecreasing, although natural enough, is not essential, but is used to simplify the outer class proofs. For example, if $n^L\pi_n$ is increasing for any positive value L, it is only necessary to put $n_j \sim (1 + \varepsilon/3)^{2j/(L+1)}$ to use the same proof. However, if $n\pi_n$ oscillates too much, the right side of (4.14) need not be bounded away from zero, and a longer proof is needed. It will be evident that only minor changes in the proof are required to cover various other cases, for example, $\pi_n \downarrow \pi_0 > 0$, in which case the coefficient of $\log_2 n$ in (3.15) must of course be altered to $2n\pi_0(1 - \pi_0)$. It will be seen that the inner class proof does not use monotonicity of π_n or of $n\pi_n$. Eicker points out that his inner class results apply to more general sequences of sets than $[0, \pi_n]$; here, if J_n is a subset of $[0, 1]$ of Lebesgue measure π_n, we consider $T_n(J_n) = \{\text{number of } X_i \text{ in } J_n, 1 \leqq i \leqq n\}$. One must note, however, that such inner bounds on $T_n(J_n)$ may not be sharp if J_n moves too rapidly, and may even give the wrong order. One need only cite the familiar example of J_n chosen so that the random variables $T_n(J_n)$ are independent, in which case $\pm [T_n(J_n) - n\pi_n]$ has $[2n\pi_n \log n]^{1/2}$ as top bound under our assumptions on π_n.

In (3.1) equivalent T_n and Z_n bounds were treated, and (3.14) gives a consequence of Theorem 2 for T_n. It remains to translate the results of Theorems 3 to 5 into conclusions for the sample quantiles $Z_n(p_n)$ beyond the domain of Theorems 1 and 2. For each positive value v, we define c'_v to be the positive value satisfying the pair of equations

$$\beta'_{c'_v}(\log \beta'_{c'_v} - 1) = (1 - c'_v)/c'_v, \qquad c'_v\beta'_{c'_v} = v, \tag{3.16}$$

with $\beta' > 1$. We define c''_v by the analogue of (3.16) obtained by replacing β' by $\beta'' < 1$. The existence and uniqueness of c'_v and c''_v follows from (2.25) and the sentence following it. We recall the definition (2.1) of H_n.

THEOREM 6. *Suppose $p_n \downarrow 0$. If $H_n \uparrow +\infty$, then, for either choice of sign,*

$$\limsup_n \pm [Z_n(p_n) - p_n][2p_n n^{-1} \log_2 n]^{-1/2} = 1, \quad \text{wp } 1. \tag{3.17}$$

If $0 < v < \infty$ and $H_n \to v$, then

$$\begin{aligned} c''_v &= \limsup_n nZ_n(p_n)/\log_2 n, \quad &\text{wp } 1 \\ c'_v &= \liminf_n nZ_n(p_n)/\log_2 n, \quad &\text{wp } 1 \end{aligned} \tag{3.18}$$

If $H_n \to 0$ (and $np_n \geqq 1$ to avoid trivialities), then

$$\limsup_n nZ_n(p_n)/\log_2 n = 1, \quad \text{wp } 1; \tag{3.19}$$

while if $h_n \downarrow 0$ and $np_n \uparrow +\infty$, then

$$\liminf_n H_n \log [nZ_n(p_n)/\log_2 n] = -1, \quad \text{wp } 1. \tag{3.20}$$

REMARKS. As in the case of Theorem 5, the assumption $H_n \uparrow$ is stronger than needed, but it is made to simplify the proof. Also, if $p_n \downarrow p_0 > 0$, essentially the same proof yields (3.17) with $2p_n$ replaced by $2p_0(1 - p_0)$. (Compare the remarks below Theorem 5.) In particular, when $p_n = p_0$ we obtain the LIL for sample p_0-tiles, but not the strong form [6], which would require considerably more effort using the present route. Part of (3.17) was stated in [8] under unnecessary restrictions. More satisfactory forms than (3.19) and (3.20) are obviously related to strong forms of Theorems 3 and 4. As they stand, (3.19) and (3.20) are also correct for the case $p_n = k/n$ of Theorems 2 and 1, and the grossness of (3.20) as a description for the latter is evident.

Proofs of Theorems 3 through 6

PROOF OF THEOREM 3. In each particular case of (3.8), (3.9), (3.10), the assumptions preceding (3.7) as well as the correctness of the stated k_n^* follow easily from the first line of (2.25) and from the validity of (3.7), and we turn to the proof of the latter.

Inner class. Let $\varepsilon > 0$ be specified, $\varepsilon < 1$. Let $n_j \sim \varepsilon^{-j}$. Then $m_j \sim (1-\varepsilon)n_j$ by (2.3). Hence, we can compute

$$\log P\{T_{n_{j-1},n_j}(\pi_{n_j}) \geqq d_{n_j} h_{n_j} \beta'_{h_{n-j}} \log_2 n_j\} \tag{4.1}$$

by using (2.27) with $\rho_n \sim d_n \sim 1-\varepsilon$ and with n replaced by n_j in (2.27); here d_n is chosen so that (2.16) is satisfied, where $k_n^* = (1-\varepsilon)^{-1} d_n h_n \beta'_{h_{n_j}} \log_2 n$. Note that (2.26) is satisfied because (2.24) and (2.25) and $h_n = O(1)$ imply

$$\liminf_n \beta'_{h_n} > 1. \tag{4.2}$$

Also, $h_n \beta'_{h_n}$ is bounded, by (2.25). Hence, from (2.27), the expression (4.1) equals $-(1-\varepsilon)[1+o(1)]\log_2 n_j$, so (2.10) and BC yield the desired result. (The last clause of Lemma 1 indicates that we do not need to verify (2.26) for the above half of the proof, but (4.1) is needed below, anyway.)

Outer class. As in the case of Theorem 1, there is a proof using (2.7) to (2.9) with the n_j of (2.3), which we omit in order to demonstrate the simplicity of the situation by working directly with (2.12) and the A'_n of (2.11) with k_n replaced there by

$$k'_n = d_n h_n \beta'_{h_n} \log_2 n, \tag{4.3}$$

and $d_n - 1 \to \varepsilon$, small and positive, with d_n chosen so that $k'_n \uparrow$ and k'_n is integral. By (4.2), for small enough ε and with $\rho_n = 1$, (2.26) is satisfied. Because of (2.26) (as used in (2.20)) we have for the event of (2.11) with k'_n for k_n,

$$\begin{aligned} \log P\{A'_n\} &= \log[\pi_n B^+(k'_n - 1, n-1, \pi_n)] \\ &\sim \log[\pi_n b(k'_n - 1, n-1, \pi_n)] \\ &= \log[k'_n n^{-1} b(k'_n, n, \pi_n)]. \end{aligned} \tag{4.4}$$

From the previously obtained boundedness of $h_n \beta'_{h_n}$, we have $\log k'_n = o(\log_2 n)$. Consequently, from (4.4) and (2.27),

$$\begin{aligned} \log P\{A'_n\} = -\log n + \{&-(1+\varepsilon) \\ &+ h_n[\varepsilon - (1+\varepsilon)\beta'_{h_n} \log(1+\varepsilon)] + o(1)\}\log_2 n. \end{aligned} \tag{4.5}$$

By (4.2), the quantity in square brackets in (4.5) is negative for ε sufficiently small (and positive) and for all large n. This, (2.12), and BC complete the proof.

PROOF OF THEOREM 4. Equation (3.12) follows from (3.11).

Outer class. Given ε, small and positive, let d_n be chosen so that $d_n h_n \beta'_{h_n} \log_2 n \uparrow$ and $d_n \to 1-\varepsilon$. Put $n_j \sim \lambda^j$. Because of (2.13), we compute, using (2.29) with n replaced by n_{j+1} and $\rho_{n_{j+1}} = n_j/n_{j+1} \sim \lambda^{-1}$,

$$\begin{aligned} &\log P\{T_{n_j}(\pi_{n_{j+1}}) < d_{n_{j+1}} h_{n_{j+1}} h_{n_{j+1}} \beta''_{h_{n_{j+1}}} \log_2 n_{j+1}\} \\ &\quad = \{-(1-\varepsilon) + h_{n_{j+1}}[1 - \varepsilon - \lambda^{-1} - (1-\varepsilon)\beta''_{h_n} \\ &\qquad \log(\lambda(1-\varepsilon))] + o(1)\}\log_2 n_{j+1}, \end{aligned} \tag{4.6}$$

provided (2.28) is satisfied. Write $\liminf_n h_n = \bar{h}$. Since $\bar{h} > 1$ by (2.23), the structure (2.24)—(2.25) of the functional equation implies that $1 < 1 - \bar{h}\beta''_{\bar{h}} \log \beta''_{\bar{h}} = \bar{h}(1 - \beta''_{\bar{h}})$. Moreover, by the comment following (2.25) and the fact that $(\beta - 1)/\log \beta < 1$ for $0 < \beta < 1$, we conclude that $h - h\beta''_h$ increases in h. Hence, $h_n(1 - \beta''_{h_n}) > 1 + 2\delta$ for some $\delta > 0$ and all large n. Now let $\lambda > 1$ be chosen so close to one that $\lambda d\beta_h < 1$ (which yields (2.28)) and such that $1 - \lambda^{-1} < \varepsilon\delta/(1 + 2\delta)$. Regarding the expression in braces on the right side of (4.6), other than the $o(1)$ term, as a function of $\varepsilon > 0$, upon expanding it in powers of ε and noting that $\log \lambda = 1 - \lambda^{-1} + O(\varepsilon^2)$ we obtain

$$\begin{aligned} &-(1-\varepsilon) + h_{h_{j+1}}[1 - \varepsilon - \lambda^{-1} \\ &\qquad - (1-\varepsilon)\beta''_{h_{n_{j+1}}}(-\varepsilon + 1 + \lambda^{-1} + O(\varepsilon^2))] \\ &= -(1-\varepsilon) + h_{n_{j+1}}(1 - \beta''_{h_{n_{j+1}}})(1 - \lambda^{-1} - \varepsilon) + O(\varepsilon^2) \\ &< -(1-\varepsilon) + (1+2\delta)(-\varepsilon + \varepsilon\delta/(1+2\delta)) + O(\varepsilon^2) \\ &< -1 - \delta\varepsilon + O(\varepsilon^2). \end{aligned} \tag{4.7}$$

Thus, for ε sufficiently small and positive, the probability of (4.6) is summable.

Inner class. As in the proof of Theorem 2, we now use the $n_j = \text{int}\,\{e^{\alpha j/\log j}\}$ with $\alpha > 0$, of (2.4). Hence, choosing $d_n - 1 \to \varepsilon$ small and positive and such that $d_n h_n \beta''_{h_n} \log_2 n \uparrow$, we obtain from (2.29) with n_{j+1} for n and $\rho_{n_j} = m_j/n_j = 1 - O(j^{-\alpha})$,

$$\begin{aligned} \log P\{T_{n_j, n_{j+1}}(\pi_{n_{j+1}}) &\leqq d_{n_{j+1}} h_{n_{j+1}} \beta''_{h_{n_{j+1}}} \log_2 n_{j+1}\} \\ = \{-(1+\varepsilon) &+ h_{n_{j+1}}[\varepsilon - (1+\varepsilon)\beta''_{h_{n_{j+1}}} \log(1+\varepsilon)] \\ &\qquad + o(1)\} \log_2 n_{j+1}. \end{aligned} \tag{4.8}$$

(While the last comment in the statement of Lemma 1 implies that this half of the proof does not require (2.28), the latter is in fact satisfied provided $(1 + \varepsilon)^{-1} > \limsup_n \beta''_{h_n}$, the last quantity being less than one by (2.24) and (2.25) since $h_n = O(1)$.) As in the outer class proof, we again have $h_n(1 - \beta''_{h_n}) > 1 + 2\delta$ for some $\delta > 0$ and all large n, and consequently the expression in braces on the right side of (4.8) is greater than $-1 + \delta\varepsilon$ for ε sufficiently small and positive and for all large n. Hence, the probabilities of (4.8) have divergent sum. In view of (2.14) and (2.15), it remains to compute

$$\begin{aligned} P\{T_{n_j}(\pi_{n_{j+1}}) \geqq 1\} &= 1 - (1 - \pi_{n_{j+1}})^{n_j} \\ &\sim (n_j/n_{j+1}) h_{n_{j+1}} \log_2 n_{j+1} \\ &= O(j^{-\alpha} \log j), \end{aligned} \tag{4.9}$$

which is summable if $\alpha > 1$. This completes the proof of Theorem 4.

PROOF OF THEOREM 5. *Outer class*. Given $\varepsilon > 0$, put $n_j \sim (1 + \varepsilon/3)^j$ and modify the previous notation by writing

$$\begin{aligned} k_n(\varepsilon) &= n\pi_n + (1+\varepsilon)[2n\pi_n \log_2 n]^{1/2}, \\ D_n(\varepsilon) &= \{T_n(\pi_n) \geqq k_n(\varepsilon)\}, \quad D^*_j(\varepsilon) = \bigcup_{n \in I_j} D_n(\varepsilon). \end{aligned} \tag{4.10}$$

The desired outer top bound result is

$$P\{D_n(\varepsilon) \text{ i.o.}\} = 0. \tag{4.11}$$

The first step is common in such proofs. If we show that

$$\liminf_j P\{D_{n_{j+1}}(\varepsilon/3) | D_j^*(\varepsilon)\} > 0, \tag{4.12}$$

then $P\{D_n(\varepsilon) \text{ i.o.}\} = P\{D_j^*(\varepsilon) \text{ i.o.}\} = 0$, by BC if $P\{D_{n_{j+1}}(\varepsilon/3)\}$ is summable, which it is by Lemma 3.

For brevity, we hereafter write k_n for $k_n(\varepsilon)$ (never for $k_n(\varepsilon/3)$). For $\nu \in I_j$, we define

$$G_\nu = \{T_\nu(\pi_{n_{j+1}}) \geqq k_\nu \pi_{n_{j+1}}/\pi_\nu - 1\}. \tag{4.13}$$

If the event $D_j^*(\varepsilon)$ of (4.10) occurs, define the random variable N_j^* by $N_j^* = \min\{n : D_n(\varepsilon) \text{ occurs}, n \in I_j\}$. Clearly,

$$\begin{aligned} P\{D_{n_{j+1}}(\varepsilon/3) \mid D_j^*(\varepsilon)\} &= EP\{D_{n_{j+1}}(\varepsilon/3) \mid D_j^*(\varepsilon); N_j^*; T_{N_j^*}(\pi_{N_j^*})\} \\ &\geqq \inf_{z \geqq k, \nu \in I_j^-} P\{D_{n_{j+1}}(\varepsilon/3) \mid T_\nu(\pi_\nu) = z\}. \end{aligned} \tag{4.14}$$

Since

$$\begin{aligned} P\{D_{n_{j+1}}(\varepsilon/3) | T_\nu(\pi_\nu) = z\} &\geqq P\{G_\nu D_{n_{j+1}}(\varepsilon/3) | T_\nu(\pi_\nu) = z\} \\ &= P\{D_{n_{j+1}}(\varepsilon/3) \mid G_\nu; T_\nu(\pi_\nu) = z\} \\ &\qquad \cdot P\{G_\nu \mid T_\nu(\pi_\nu) = z\}, \end{aligned} \tag{4.15}$$

we obtain (4.12) from (4.14) if there is a $\delta > 0$ such that, for all large j and $\nu \in I_j^-$,

$$\inf_{z \geqq k_\nu} P\{D_{n_{j+1}}(\varepsilon/3) | G_\nu; T_\nu(\pi_\nu) = z\} > \delta \tag{4.16}$$

and

$$\inf_{z \geqq k_\nu} P\{G_\nu \mid T_\nu(\pi_\nu) = z\} > \delta. \tag{4.17}$$

The conditional probability of (4.17) is clearly

$$B^+(k_\nu \pi_{n_{j+1}}/\pi_\nu - 1, z, \pi_{n_{j+1}}/\pi_\nu), \tag{4.18}$$

which is a minimum for $z = \text{int}^+ \{k_\nu\}$. This and (2.33) yield (4.17).

Since $\pi_{n_{j+1}} \leqq \pi_\nu$, the probability of $D_{n_{j+1}}(\varepsilon/3)$ conditioned on values of $T_\nu(\pi_\nu)$ and $T_\nu(\pi_{n_{j+1}})$ is the same as that conditioned only on the last. Hence abbreviating $\text{int}^+ \{k_\nu \pi_{n_{j+1}}/\pi_\nu - 1\}$ by μ, we see that the left side of (4.16) is at least

$$\begin{aligned} &\inf_{z \geqq k_\nu, z \geqq y \geqq \mu} P\{D_{n_{j+1}}(\varepsilon/3) \mid T_\nu(\pi_{n_{j+1}}) = y; T_\nu(\pi_\nu) = z\} \\ &= \inf_{y \geqq \mu} P\{D_{n_{j+1}}(\varepsilon/3) | T_\nu(\pi_{n_{j+1}}) = y\} \\ &= \inf_{y \geqq \mu} P\{T_{\nu, n_{j+1}}(\pi_{n_{j+1}}) \geqq k_{n_{j+1}}(\varepsilon/3) - y\} \\ &= B^+(k_{n_{j+1}}(\varepsilon/3) - \text{int}^+ \{k_\nu \pi_{n_{j+1}}/\pi_\nu - 1\}, n_{j+1} - \nu, \pi_{n_{j+1}}). \end{aligned} \tag{4.19}$$

By (2.33) and the fact that $\text{int}^+ \{x\} \geqq x$, we will thus establish (4.16) if we show that

$$k_{n_{j+1}}(\varepsilon/3) - [k_\nu(\varepsilon)\pi_{n_{j+1}}/\pi_\nu - 1] \leqq (n_{j+1} - \nu)\pi_{n_{j+1}} - 1 \tag{4.20}$$

for all large j and $\nu \in I_j^-$. Dividing both sides of (4.20) by $\pi_{n_{j+1}}$ and using (4.10), we obtain that (4.20) is equivalent to

$$\begin{aligned}(1 + \varepsilon/3)n_{j+1}[2(\log_2 n_{j+1})/n_{j+1}\pi_{n_{j+1}}]^{1/2} + 2\pi_{n_{j+1}}^{-1} \\ \leqq (1 + \varepsilon)\nu[2(\log_2 \nu)/\nu\pi_\nu]^{1/2}.\end{aligned} \tag{4.21}$$

The ratio of $2\pi_{n_{j+1}}^{-1}$ to the term preceding it approaches zero. Also, $n_{j+1}\pi_{n_{j+1}} > \nu\pi_\nu$ and $n_{j+1}(\log_2^{1/2} n_{j+1})/\nu \log_2^{1/2} \nu < n_{j+1}(\log_2^{1/2} n_{j+1})/n_j \log_2^{1/2} n_j \sim 1 + \varepsilon/3$. We conclude that (4.21) is satisfied for all large j and $\nu \in I_j^-$, completing the proof of (4.11).

The proof of the outer bottom result is very similar, so we shall merely list the changes. In (4.10) we replace $(1 + \varepsilon)$ by $-(1 + \varepsilon)$ in the definition of $k_n(\varepsilon)$, and $\geqq$ by $\leqq$ in the definition of $D_n(\varepsilon)$, as well as in the domain of z in (4.14), (4.16), and (4.17). The event G_ν of (4.13) is replaced by

$$\{T_\nu(\pi_{n_{j+1}}) \leqq k_\nu\pi_{n_{j+1}}/\pi_\nu + 1\}; \tag{4.22}$$

also, int^+ is replaced everywhere by int. The probability (4.18) is replaced by

$$B(k_\nu\pi_{n_{j+1}}/\pi_\nu + 1, z, \pi_{n_{j+1}}/\pi_\nu), \tag{4.23}$$

whose minimum subject to $z \leqq k_\nu$ is at $z = \text{int}\{k_\nu\}$; this minimum is bounded away from zero, by (2.34). Finally, in (4.19) μ becomes $\text{int}\{k_\nu\pi_{n_{j+1}}/\pi_\nu + 1\}$, and the domain of the first infimum is $\{y \leqq z \leqq k_\nu, y \leqq \mu\}$; when we replace the resulting domain $\{y \leqq \min(k_\nu, \mu)\}$ by $\{y \leqq \mu\}$, we cannot increase the infimum, and the analogues of last two expressions of (4.19) give, for a lower bound on the analogue of (4.16),

$$\begin{aligned}\inf_{y \leqq \mu} P\{T_{\nu, n_{j+1}}(\pi_{n_{j+1}}) \leqq k_{n_{j+1}}(\varepsilon/3) - y\} \\ = B(k_{n_{j+1}}(\varepsilon/3) - \text{int}\{k_\nu\pi_{n_{j+1}}/\pi_\nu + 1\}, n_{j+1} - \nu, \pi_{n_{j+1}}).\end{aligned} \tag{4.24}$$

By (2.34) and the fact that $\text{int}\{x\} \leqq x$, we obtain as the analogue of (4.20),

$$k_{n_{j+1}}(\varepsilon/3) - [k_\nu\pi_{n_{j+1}}/\pi_\nu + 1] \geqq (n_{j+1} - \nu)\pi_{n_{j+1}} + 1. \tag{4.25}$$

Recalling that $(1 + \varepsilon)$ has been replaced by $-(1 + \varepsilon)$ in the definition (4.10) of $k_n(\varepsilon)$, we see that (4.25) is again equivalent to (4.21).

Inner class. The proof follows usual LIL lines and has been given by Eicker [4] in essentially this form, so we only sketch it. For the top inner bound, one shows, with $n_j \sim \lambda^j$, that

$$P\{T_{n_j}(\pi_{n_j}) > n_j\pi_{n_j} + (1 - \varepsilon)[2h_{n_j}]^{1/2} \log_2 n_j \text{ i.o.}\} = 1, \tag{4.26}$$

by showing that

$$(4.27)\qquad P\{T_{n_{j-1},n_j}(\pi_{n_j}) > [n_j - (1 - [4\lambda/h_{n_j}]^{1/2})n_{j-1}]\pi_{n_j} + (1-\varepsilon)[2h_{n_j}]^{1/2}\log_2 n_j \text{ i.o.}\} = 1,$$

and

$$(4.28)\qquad P\{T_{n_{j-1}}(\pi_{n_j}) > (1 - [4\lambda/h_{n_j}]^{1/2})n_{j-1}\pi_{n_j}, \text{ a.a. } j\} = 1.$$

Of these, (4.27) is a consequence of (2.31) and BC provided λ is chosen so large that $[\lambda/(\lambda - 1)]^{1/2}[(4/\lambda)^{1/2} + (1 - \varepsilon)2^{1/2}] < 2^{1/2}$. The probability complementary to (4.28) is proved summable by using the standard Markov-Cramér inequality with abbreviations $T = T_{n_{j-1}}(\pi_{n_j})$, $\delta = [4\lambda/h_{n_j}]^{1/2}$, to obtain

$$(4.29)\qquad \begin{aligned} P\{T \leqq (1-\delta)ET\} &= P\{e^{-\delta T} \geqq e^{-(1-\delta)\delta ET}\} \\ &\leqq e^{(1-\delta)\delta ET}\, Ee^{-\delta T} \\ &= \exp\{n_{j-1}[\pi_{n_j}(1-\delta)\delta \\ &\qquad + \log(1 - \pi_{n_j} + \pi_{n_j}e^{-\delta})]\} \\ &< \exp\{-\delta^2 n_{j-1}\pi_{n_j}/3\}, \end{aligned}$$

the last inequality for δ and π_{n_j} sufficiently small. For the bottom inner bound, replace $>$ by $<$, $(1 - \varepsilon)$ by $-(1 - \varepsilon)$, and $-[4\lambda/h_{n_j}]^{1/2}$ by $[4\lambda/h_{n_j}]^{1/2}$ in (4.26), (4.27), (4.28); and replace (4.29) by

$$(4.30)\qquad P\{T \geqq (1+\delta)ET\} \leqq e^{-(1+\delta)\delta ET}Ee^{\delta T},$$

with the same final estimate as in (4.29).

REMARK. The coefficient δ of T and ET in the second expression of (4.29) and (4.30) is not the usual minimizing value for exponential binomial bounds, but is less cumbersome and is close enough to yield the desired conclusions.

PROOF OF THEOREM 6. If $H_n \uparrow +\infty$ and $h_n = H_n \pm \lambda(2H_n)^{1/2}$ where $\lambda = 1 + \varepsilon$ or $1 - \varepsilon$, then $h_n \uparrow$ for large n. Applying Theorem 5 for these four possible choices of h_n, and noting that $h_n \mp \lambda(2h_n)^{1/2} = H_n + O(1)$, yields (3.17). Similarly, (3.18) follows from Theorems 3 and 4. If $H_n \to 0$, we obtain (3.19) from Theorem 2 and from (3.18) for v small and positive; by (2.24) and (2.25), $\lim_{v\downarrow 0} c_v'' = 1$. Finally, (3.20) follows from (1.3) with $k_n = np_n$ and $h_n = \exp\{-(1 \pm \varepsilon)H_n^{-1}\}$ upon invoking Theorem 3, the condition $np_n \to +\infty$ of the latter then being equivalent to (2.22).

REFERENCES

[1] R. R. BAHADUR. "A note on quantiles in large samples," *Ann. Math. Statist.*, Vol. 37 (1966), pp. 577–580.

[2] G. BAXTER. "An analogue of the law of the iterated logarithm," *Proc. Amer. Math. Soc.*, Vol. 6 (1955), pp. 177–181.

[3] K. L. CHUNG. "An estimate concerning the Kolmogoroff limit distribution," *Trans. Amer. Math. Soc.*, Vol. 67 (1949), pp. 36–50.

[4] F. EICKER, "A log log law for double sequences," to appear.

[5] W. FELLER, *An Introduction to Probability Theory and its Applications*, Vol. 1, New York, Wiley, 1957 (2nd ed.).

[6] J. Kiefer, "On Bahadur's representation of sample quantiles," *Ann. Math. Statist.*, Vol. 38 (1967), pp. 1323–1342.

[7] ———, "Deviations between the sample quantile process and the sample df," *Proceedings of the First International Conference on Nonparametric Inference*, Cambridge, Cambridge University Press, 1969, pp. 299–319.

[8] ———, "Old and new methods for studying order statistics and sample quantiles," *Proceedings of the First International Conference on Nonparametric Inference*, Cambridge, Cambridge University Press, 1969, pp. 349–357.

[9] V. F. Kolchin, "On the limiting behavior of extreme order statistics in a polynomial scheme," *Theor. Probability Appl.*, Vol. 14 (1969), pp. 458–469.

[10] H. Robbins and D. Siegmund, "An iterated logarithm law for maxima and minima" (abstract), *Ann. Math. Statist.*, Vol. 41 (1970), p. 757.

Reprinted from
Proc. Sixth Berkeley Sympos. on Math. Statist. and Probability
1, 227–244 (1970)

Z. Wahrscheinlichkeitstheorie verw. Geb. 24, 1–35 (1972)

Skorohod Embedding of Multivariate RV's, and the Sample DF

J. Kiefer*

The main purpose of this paper is to study certain representations of sums of iid k-vector rv's as embeddings in k-dimensional Brownian motion by vectors of stopping times, in extension of Skorohod's scheme [20], and consequent error estimates for weak and strong invariance principles. In particular, letting $k \to \infty$ we embed the sample df in the Gaussian process with 2-dimensional time to which it has long been known to converge weakly. We discuss previous sample df embeddings, which have yielded related results; while some of our estimates are slight improvements, the emphasis here will be on the naturality of the embedding per se (although it will be indicated why it is probably far from the final word on the subject.

1. Introduction

Skorohod's embedding scheme has been used and extended in a number of directions. In particular, the original use [20] to obtain error estimates for weak convergence of certain functionals of a sequence of summands of iidrv's has been broadened by Rosenkrantz [15] and Sawyer [17, 18, 19]. On the other hand, Strassen [21, 22] used such embeddings to obtain his strong invariance principles for martingales.

Turning to questions involving the sample df, a number of authors have worked on analogous schemes. The first of these was Breiman [3], who used Skorohod embedding for iidrv's Y_i with $P\{Y_i > y\} = e^{-(y+1)^+}$, and the familiar fact that $\left\{\sum_1^i (Y_j+1) \Big/ \sum_1^{n+1} (Y_j+1),\ 1 \leqq i \leqq n\right\}$ have the same joint df as the order statistics $\{X_{n,i}, 1 \leqq i \leqq n\}$ from n iidrv's with uniform density on $[0, 1]$, to approximate the sample df deviations by a Brownian bridge. Then Brillinger [4] independently used the same scheme and also gave an upper bound wp 1 on the error. Rosenkrantz [16, Section III] gave essentially the same bound, although his emphasis was on certain weak convergence problems, e.g., for the v. Mises statistic.

The disadvantage of this representation was evidently pointed out by Pyke (according to [4]) and has been discussed further in [12]: while it yields a satisfactory approximation for a single large n, it does not yield the right joint distribution for several large n's at once. Related weak convergence results have been obtained by Bickel, Billingsley, Pyke and Root, among others.

Subsequently Müller [14] gave a proof of the convergence in law of the sample df process to the Gaussian process with two-dimensional time and the proper covariances, that is, with independent "Brownian bridge" increments, and in a striking analysis he gave the first estimate of the error for certain functionals of the *sequence* of sample df's, analogous to the results of [20, 15, 17, 18, 19]. This estimate is based on an embedding of the sample df which uses a well-known

* Research supported by NSF Grant GP 9297.

representation of the $X_{n,i}$ in terms of exponential rv's, different from Breiman's: $\{(n-i)\log(X_{n,n-i+1}/X_{n,n-i}), 0\leqq i\leqq n-1\}$ $(X_{n,n+1}=1)$ may be taken to be standard iid exponential rv's like Y_i+1 above. This representation, like Breiman's, depends on n in such a way that it also cannot be used for joint (in n) distributions; however, Müller cleverly adds roughly $tN^{1/3-\varepsilon}$ of these, each approximating the sample df for a different set of $n=N^{2/3+\varepsilon}$ observations, to obtain (using further estimates) an approximation of the joint law of the sample df for each integral number tN of observations.

We shall discuss the embeddings of [3, 4, 14] in more detail in Section 5, where some results concerning them will be stated and proved.

The main development of the present paper is the representation of the sample df by a Skorohod-type embedding in the appropriate two-dimensional Gaussian process. Denote the unit interval by I, the reals by R, the nonnegative reals by R^+, the positive integers by Z^+. Let $\xi^*(\cdot,\cdot)$ be a Gaussian process on $I\times R^+$ with continuous sample functions, zero expectation, and

$$E\,\xi^*(s_1,t_1)\,\xi^*(s_2,t_2)=\min(t_1,t_2)\,[\min(s_1,s_2)-s_1 s_2] \tag{1.1}$$

(so that there are independent increments in t and a Brownian bridge in s for fixed t). For future reference we define the closely related

$$\xi(z,t)=(z+1)\,\xi^*(z/(z+1),t),$$

a continuous Gaussian process on $R^+\times R^+$ with zero expectations and independent increments in both directions:

$$E\,\xi(z_1,t_1)\,\xi(z_2,t_2)=\min(t_1,t_2)\min(z_1,z_2). \tag{1.2}$$

(We shall always use ξ^* and ξ as in (1.1)–(1.2); univariate Brownian motions, will be denoted ξ_i, ξ_i', etc.) Let S_n be the sample df based on the first n of the iidrv's $\{X_i, i\geqq 1\}$, uniformly distributed on I, with tS_t defined as usual by linear interpolation from nS_n and $(n+1)S_{n+1}$ if $n<t<n+1$, and $S_0=0$. We take the left-continuous version of S_n to conform with the embeddings of Section 2. Let

$$\hat{\xi}(s,t)=t[S_t(s)-s], \qquad s\in I,\ t\in R^+. \tag{1.3}$$

A main consequence of our embedding is

Theorem 1. *ξ^* can be defined on a probability space on which there is defined a random function T: $I\times R^+\to R^+$ such that*

$$\xi^*(s,T(s,t)) \text{ has the same joint law as } \hat{\xi} \text{ of (1.3)}; \tag{1.4}$$

and, as $t\to\infty$,

$$t^{-1/2}\sup_{0\leqq s\leqq 1}|\xi^*(s,T(s,t))-\xi^*(s,t)|=O(t^{-1/6}(\log t)^{2/3}) \text{ wp}1. \tag{1.5}$$

(It will become clear from the use made of (3.23) that all our results such as (1.5) and (1.6) can be stated in terms of either continuous t or discrete n.)

A corresponding weak law, essentially Müller's Theorem 3 with obvious changes in some of his assumptions (described just above our Lemma 6) and replacement of n^ε by $(\log n)^\lambda$, is

Corollary 1 (to Theorem 2). *For G_i continuous on $I \times R^*$ where R^* is a subinterval of* $(0, +\infty)$ *of positive (possibly infinite) length and γ is given by* (3.34),

$$P\left\{G_1\left(s,\frac{k}{n}\right)<n^{-1/2}k[S_k(s)-s]<G_2\left(s,\frac{k}{n}\right) \text{ for } s\in I,\ k\in Z^+,\ k/n\in R^*\right\}$$
$$-P\{G_1(s,t)<\xi^*(s,t)<G_2(s,t) \text{ for } s\in I,\ t\in R^*\} \qquad (1.6)$$
$$=O(n^{-1/6}(\log n)^{\gamma}).$$

The result also holds if only one G_i is present, of course. (Müller does not state an analogue of (1.5), but one can be obtained from his work.) As indicated above, one reason for giving (1.6) is that the embedding is somewhat more natural, the computations simpler, and the source of error perhaps more transparent, than in Müller's ingenious development. Also, the proof of Theorem 2 helps one to understand the more intricate Theorem 1. In Section 6 we will discuss the possibility of improving the estimate from either approach.

The developments of Lemmas 1′ and 4′ allow explicit constants to be obtained for the bounds of (1.5), (1.6), etc. In an already long paper we have not taken the extra space to do this.

The result (1.6) is one of a spectrum of possible results in terms of the degree of fineness (in s) at which the sample df is considered. At the other extreme we have

Corollary 2 (to Lemma 4′). *Under assumption* (3.34) *with $\alpha>1/4$ (but ignoring conditions in $|s-s'|$ there), for any finite subset I' of I and some λ',*

$$[\text{left side of (1.6) with } I' \text{ for } I]=O(n^{-1/4}(\log n)^{\lambda'}) \qquad (1.7)$$

and (1.4) *holds with*

$$[\text{left side of (1.5) with } I' \text{ for } I]=O(t^{-1/4}(\log t)^{\lambda'}) \ \text{wp}\,1. \qquad (1.8)$$

When one considers the error term given in multivariate Berry-Esseen results, or in the considerations of Section 7.3 of [20], or of [15, 17, 18] when I' is a single point and G_i = constant and $R^*=(0,1]$, it is impossible to believe that the bound in (1.7) cannot be replaced by $n^{-1/2}(\log n)^{\lambda'}$. In Section 6 the inability of the Skorohod technique, as used here, to achieve such a better rate, will be analyzed.

In an attempt to extend the results of Section 7.4 of [20] or of [19], one considers $f(t,x,s)$ continuously differentiable in t, x with partial derivatives slowly varying in x (of order $O(|x|^A)$ for some A). Denote the cardinality of I' by $|I'|$. It is then a routine application (which we shall omit) of Lemma 4′ to prove

Corollary 3. *If the second probability of* (1.9) *has a bounded derivative in λ, then there is a value β such that, uniformly in λ,*

$$P\left\{\frac{1}{n|I'|}\sum_{k=1}^{n}\sum_{s\in I'}f\left(\frac{k}{n},n^{-1/2}k[S_k(s)-s],s\right)\leq\lambda\right\}$$
$$-P\left\{\frac{1}{n|I'|}\sum_{s\in I'}\int_0^1 f(t,\xi^*(s,t),s)\,dt\leq\lambda\right\}=O(n^{-1/4}(\log n)^{\beta}). \qquad (1.9)$$

When I' consists of a single element, a very special case of [19] yields (1.9) with the exponent 1/4 replaced by 1/2. (The form of the corresponding statement

1*

in [20] is slightly different, but the appropriate value is again 1/2.) Thus, again, the Skorohod technique fails to yield what we believe to be the correct result in (1.9), for reasons described in Section 6. Moreover, it is surely tempting to conjecture that, extending the differentiability assumption to s, one can obtain an error estimate $O(n^{-1/2}(\log n)^{\beta})$ with $\sum_s$ replaced by $\int ds$ in (1.9). Using the embedding of the present paper and second derivatives slowly varying, the author can at present only prove

Theorem 3. *Under the above assumptions, if also the second probability below has bounded derivative in λ, then there is a β such that, uniformly in λ,*

$$\begin{aligned} &P\left\{\int_0^1 n^{-1}\sum_{k=1}^{n} f\left(\frac{k}{n}, n^{-1/2}k[S_k(s)-s], s\right) ds \leqq \lambda\right\} \\ &\quad - P\left\{\int_0^1\int_0^1 f(t, \xi^*(s,t), s)\, dt\, ds \leqq \lambda\right\} = O(n^{-1/5}(\log n)^{\beta}). \end{aligned} \tag{1.10}$$

This is slightly better than what one would get directly from Theorem 1 or [14]; the additional strength comes from the possibility of using $B_n = O(n^{1/5}(\log n)^{\beta'})$ in place of the $O([n/\log n]^{1/3})$ which will be used in proving Theorem 2, and this in turn is made possible by using Lemma 1′ and two derivatives in s to estimate the difference of the rv's obtained by integrating over s for fixed k (or t) and corresponding sums over B_n points. The harder part of the proof is essentially in Sawyer's work [19], and we will not include the details in this already long paper, especially since the chief novelty of our development is present in the embedding and proof of Theorem 2, and the conclusion (1.10) seems far from definitive.

The above can be viewed as an extension of Rosenkrantz's result for fixed $k=n$ (or t); while stated in [16] for the von Mises statistic, the result is more general:

Theorem 4 (Rosenkrantz). *If $f \in C^2(R \times I)$ with partial derivatives of slow growth in the unbounded variable, and if the second probability below has bounded derivative in λ, then there is a value $\beta > 0$ such that*

$$\begin{aligned} &P\left\{\int_0^1 f(n^{1/2}[S_n(s)-s], s)\, ds \leqq \lambda\right\} - P\left\{\int_0^1 f(\xi^*(s,1), s)\, ds \leqq \lambda\right\} \\ &\quad = O(n^{-1/4}(\log n)^{\beta}). \end{aligned} \tag{1.11}$$

This can be proved using either Breiman's embedding as in [16], or Müller's. Either method also yields the order of (1.11) for the modification of (1.6) obtained by restricting k to the single value n; this conclusion is contained in [14]. A defect of the embedding of the present paper is that, although it yielded a better result than Theorem 1 or direct application of [14] in Theorem 3, it yields the same order as in (1.10) if used in the context of Theorem 4 where the other embeddings do better. As Rosenkrantz points out, the result of Chan-Li-Tsian [7] for approximating the Kolmogorov distribution suggests that $n^{-1/2}(\log n)^{\beta}$ is again the desired result, and none of the methods yields that at present. This will be discussed further in Section 6.

Remarks on Theorems 3 and 4. While we have not obtained definitive results in this domain, the subject seems important enough to deserve certain comments.

(1) Values of β can be discerned from [19] and the proofs described above. The condition that $f(t, x, s)$ has partials in t and s that are $O(|x|^4)$ can be much weakened without affecting the conclusion. Moreover, the assumption on $\partial f/\partial x$ in (1.9) and (1.10) can be weakened greatly at the expense of obtaining the slightly weaker conclusion $O(n^{-1/5+o(1)})$.

(2) Theorems 3 and 4 are easily modified to allow a finite measure, $\mu(ds)$, to replace ds (and similarly for t). We shall discuss elsewhere the analogue of (1.11) when S_n is replaced by the sample quantile process (essentially S_n^{-1}); (1.11) and a result of [11] immediately yield $O(n^{-1/4}(\log n)^\beta)$ as an estimate of error in this case. In particular, Chernoff-Savage (linear rank) statistics and corresponding location parameter estimators linear in the sample quantiles, for 1− and 2−sample problems, can be treated in this way. One would hope for better bounds, as illustrated in the easily obtained $O(n^{-1/2})$ for linear combinations of a fixed number of sample quantiles.

(3) We forego the statement and proof of almost sure analogues of (1.9)–(1.11), which are obvious via Borel-Cantelli. As with (1.5), these are statements about imperfect methods, more than anything of intrinsic meaning. See also [19], Corollary 2, for the more definitive result obtained in the case studied there, and [10] regarding limitations of the Skorohod technique discussed elsewhere herein.

(4) One can often obtain the required boundedness of the limiting density function in (1.9)–(1.11) by well known Fourier-analytic techniques we shall not discuss. Lemma 6 treats the corresponding problem for the results of (1.6)–(1.7).

(5) Just as (1.9) with I' a single point is a very special case of Section 7.4 of [20] or of [19], so (1.10) and (1.11) have extensions to cases where $\{S_1(s), s\in I\}$ is replaced by another continuous time process, and $n S_n$ by the sum of n such iid processes. As mentioned in the next section, the crucial thing is that the martingale $\{(z+1)\,S_1(z/(z+1))-z, z\in R^+\}$ be replaced by another martingale, the bounds obtainable depending on continuity properties of the latter. We treat this further in [13]. If, instead, one considers a martingale whose time domain is a discrete set of h points (possibly $h=\infty$), the embedding of Section 2.2 yields approximation theorems for partial sums of the corresponding iid h-vectors with finite fourth moments. Unfortunately, the error term obtained for the normal approximation by this method is again limited to $O(n^{-1/4}(\log n)^\beta)$, compared with the $n^{-1/2}$ of the multivariate Berry-Esseen bound. Section 2.2 also treats k-dimensional time.

The proof of (1.6), once T has been defined, is not too difficult, involving only a slightly delicate balance of several error terms, which are estimated by adopting techniques used in [8] and [9]. The main difficulty is in defining T properly. This involves a Skorohod-type embedding in which we consider simultaneously several stopping times in order to get a vector rv (Section 2.2); in the sample df case, we are interested in the infinite-dimensional rv $\{S_1(s)-s, s\in I\}$. This technique is not in itself so surprising, and we shall elsewhere treat other such multivariate embeddings [13]. The difficulty here is not so much in defining stopping times which yield a representation of S_n, as in identifying where $\xi^*(s, t)$ sits in the resulting picture. We shall discuss the embedding in Section 2.3; an alternate one can be obtained corresponding to the T_p^* of Section 2.1. Theorem 2 will be proved in Section 3; Theorem 1, whose development requires slightly more technical complication, is treated in Section 4. We have tried to spare the reader as much pain as possible by omitting straightforward but long arithmetic in proofs of Lemmas 1′ and 4′, Theorem 1, etc. whenever the ideas are present in earlier proofs.

We turn now to a simple result about approximating S_n, which has nothing to do with embedding but is used in proving the theorems of this paper. (Indeed, we have previously used such estimates elsewhere.) Let B be a positive integer and define $S_{B,n}$ and ξ_B^* by

$$S_{B,n}(x)=\begin{cases} S_n(x) & \text{if } x=i/B, \quad 0\leqq i\leqq B;\\ \text{linear for} & i/B\leqq x\leqq (i+1)/B, \quad 0\leqq i<B.\end{cases} \tag{1.12}$$

$$\xi_B^*(x,t)=\text{same with } \xi^*(x,t) \quad \text{for } S_n(x).$$

Lemma 1. *Suppose* $0<\varepsilon_n<1, B_n\in Z^+, c>1/2$. *There are positive constants* C_1, C_2 *(independent of n) such that, if* $B_n^{-1}<C_1(c-\frac{1}{2})$ *and* $n^{-1}B_n\log(B_n/\varepsilon_n)<C_1(c-\frac{1}{2})^2/c$, then

$$P\{\sup_{x\in I} n^{1/2}|S_n(x)-S_{B_n,n}(x)|\geqq [c\,B_n^{-1}\log(B_n/\varepsilon_n)]^{1/2}\}\leqq C_2\,\varepsilon_n. \tag{1.13}$$

In particular, (1.13) *holds if* $B_n\to\infty$ *and* $n^{-1}B_n\log(B_n/\varepsilon_n)\to 0$. *If* $n^{1/2}[S_n(x)-x]$ *is replaced by* $\xi^*(x,1)$ *in* (1.12) *and* (1.13), *then* (1.13) *remains valid.*

Proof. Given that $nS_n(B^{-1})=m>0$, the rv

$$\bar{D}_n=\sup_{0\leqq x\leqq B^{-1}} n\,m^{-1}|S_n(x)-S_{B,n}(x)| \tag{1.14}$$

is clearly distributed as $\sup_{x\in I}|S_m(x)-x|$. Hence [6], for all positive m and d, and some constant c',

$$P\{\bar{D}_n>m^{-1/2}\,d\,|\,n\,S_n(B^{-1})=m\}\leqq c'\,e^{-2d^2}. \tag{1.15}$$

On the other hand, the standard Markov exponential bound for the binomial case yields, for $\delta>0$,

$$\begin{aligned} P\{S_n(B^{-1})\geqq(1+\delta)\,B^{-1}\}&\leqq e^{-nB^{-1}\delta(1+\delta)}\,E\,e^{\delta n S_n(B^{-1})}\\ &\leqq e^{-nB^{-1}\delta^2[1+O(\delta+B^{-1})]/2}\end{aligned} \tag{1.16}$$

as $\delta, B^{-1}\to 0$. Suppose $0<\varepsilon<1\leqq B$. Set $d=[c\,n\,m^{-1}B^{-1}\log(B/\varepsilon)]^{1/2}$ in (1.15) and $\delta=2[c\,n^{-1}B\log(B/\varepsilon)]^{1/2}$ in (1.16). We then obtain

$$\begin{aligned} &P\{n^{1/2}\sup_{0\leqq x\leqq B^{-1}}|S_n(x)-S_{B,n}(x)|\geqq[c\,B^{-1}\log(B\,\varepsilon)]^{1/2}\}\\ &\quad\leqq c'\,e^{-[2c/(1+\delta)]\log(B/\varepsilon)}+e^{-2c[1+O(\delta+B^{-1})]\log(B/\varepsilon)}. \end{aligned} \tag{1.17}$$

Substituting B_n, ε_n for B, ε, we note that our hypothesis implies that $\delta<2C_1^{1/2}(c-\frac{1}{2})$. Also, the term $O(\delta+B^{-1})$ in (1.16) and (1.17) depends only on δ and B^{-1}, not on n. Thus, for $2c>1$ and C_1 fixed at a suitably small value, the right side of (1.17) is bounded by $(1+c')(\varepsilon_n/B_n)$. Exactly the same development holds if $\{0\leqq x\leqq B_n^{-1}\}$ is replaced in (1.17) by $\{B_n^{-1}i\leqq x\leqq B_n^{-1}(i+1)\}$, $1\leqq i<B_n^{-1}$. Since there are B_n such regions making up I, (1.13) is proved. The remainder of the Lemma is now obvious.

Remarks on Lemma 1. (1) If B_n is constant, it is clear how to modify (1.16) to obtain (1.13) once more, with a change in c there. Also, the uniform spacing of the B_n intervals can be modified. (2) For the purpose of obtaining numerical results for approximating S_n by $S_{B,n}$, the constants C_i can be made explicit with slight additional effort. (3) Finer estimates can be obtained upon replacing c by $2^{-1}+a_n$, with slowly varying $a_n \downarrow 0$; this and corresponding lower and upper class characterizations for the almost sure (in n) analogue of (1.13) are of no concern to us here, although the gross first order result for the latter is essentially present in the proof of Theorem 1. (4) Finally, it is simple to see that the methods of [8] can be used to obtain bounds of the same type for $\sup_{k\leqq n,\, x\in I} n^{-1/2} k|S_k(x)-S_{B_n,k}(x)|$ (or, for probability $\varepsilon_n = n^{-r}, r>0$, as we shall require, the result can even be obtained by summing probability bounds obtained for fixed k, $1\leqq k\leqq n$). Moreover, if we replace the domain $k\leqq n$ here by $k\leqq \bar{c}\, n(\log n)^{\beta}$ for fixed positive $\bar{c}$ and β, it is easy to see that $[B_n^{-1} n \log n]^{1/2}$ need only be multiplied by $\bar{c}^{1/2}(\log n)^{\beta/2}$ to yield the probability bound n^{-1}. Thus, in the form we shall require in Section 3, we state

Lemma 1′. *There are positive values ε', C_1', and c such that, for $\beta\geqq 0$,*

$$C_1'\leqq B_n \leqq \varepsilon' n/\log n$$
$$\Rightarrow P\{\sup_{0\leqq t\leqq \bar{c}n(\log n)^{\beta},\, x\in I} t|S_t(x)-S_{B_n,t}(x)|\geqq c'[\bar{c}B_n^{-1}n(\log n)^{\beta+1}]^{1/2}\}\leqq n^{-1}. \tag{1.18}$$

Again, the same results hold if $t[S_t(x)-x]$ is replaced by $\xi^*(x,t)$ in (1.12) *and* (1.18).

We have used here our repeated notational convention of piecewise linear interpolation in t to obtain such functions as $tS_{B_n,t}$ (from $nS_{B_n,n}$). As we have remarked, (3.23) implies the equivalence of results we use, for continuous t and discrete n.

2. Embeddings

2.1. Some Possibilities

We will depart from the sound principle of not wasting the reader's time with the author's false tries and negative results, for a simple reason: as mentioned in Section 1, the methods finally used herein do not, in their present form, yield the desired (conjectured) results, and it may save time for future workers to see what some of the alternative possibilities are and are not.

The process $\{S_1(p)-p,\ p\in I\}$ is not a martingale, but it is easily checked that

$$L_1(p)=(1-p)^{-1}[S_1(p)-p], \quad 0\leqq p<1, \tag{2.1}$$

is (by our choice above (1.3)) a left-continuous martingale of a very simple form, and hence is well known to have a Skorohod embedding in a standard Brownian motion $\bar{\xi}$. In fact, there are several ways of generating S_1 from $\bar{\xi}$, as described in the next two paragraphs. Since trouble with joint distributions (when marginals behave satisfactorily) will be seen to be a difficulty throughout this work, as in fact has already been mentioned in Section 1 in connection with [3–4], and since the construction of the next paragraph is not the standard Skorohod martingale embedding, let us illustrate by two brief examples the pitfalls one must avoid if one does not mechanically follow one of the standard constructions. For simplicity, let p_1 and p_2 be fixed values, $0<p_1<p_2<1$, and consider the problem of generating rv's with the same distribution as $(S_1(p_1), S_1(p_2))$, from Brownian motion. If we stop a standard Brownian motion $\bar{\xi}$ at the first time T_1' that $\bar{\xi}(T_1')=1-p_1$ or $-p_1$, and then at the first subsequent time T_2' that $\bar{\xi}(T_1'+T_2')-\bar{\xi}(T_1')=1-p_2$ or $-p_2$, we see that $(\bar{\xi}(T_1')+p_1, \bar{\xi}(T_1'+T_2')-\bar{\xi}(T_1')+p_2)$ has the right marginal distributions but the wrong joint distribution. Similarly, if T_2'' is the first time that $\bar{\xi}(T_2'')=p_2^{-1}$ or $-(1-p_2)^{-1}$, then $\bar{\xi}(T_1')+p_1$ and $p_2(1-p_2)\,\bar{\xi}(T_2'')+p_2$ have the right marginals but the wrong joint law since, with positive probability, the first is 1 and the second is 0.

It is not hard to guess natural schemes that work. Motivated by the above, let $T_p^*=\inf\{t\colon \bar{\xi}(t)=1-p \text{ or } -p\}$. Then $\{\bar{\xi}(T_p^*)+p,\ p=p_1 \text{ or } p_2\}$ has the correct joint law of $(S_1(p_1), S_1(p_2))$ precisely because $\bar{\xi}(T_{p_1}^*)=1-p_1$ implies $\bar{\xi}(T_{p_2}^*)=1-p_2$ if $p_1<p_2$. Similarly, we see that $\{\bar{\xi}(T_p^*)+p, p\in I\}$ has exactly the law of $\{S_1(p), p\in I\}$ (or $\{(1-p)^{-1}\bar{\xi}(T_p^*), 0\leqq p\leqq 1\}$ has the same law as (2.1)); note that $T_0^*=T_1^*=0$ wp 1.

The details of the remainder of this section and the next two sections can be carried out using $\{T_p^*\}$. However, it is technically somewhat simpler to work in terms of a different infinite set $\{T_z, z\in R_1^+\}$ of stopping times, because of the fact that $T_{p_1}^*$ can be either greater or less than $T_{p_2}^*$ (which is why $\{T_p^*, p\in I\}$ are not the stopping times usually encountered in Skorohod embeddings), while we shall have T_z nondecreasing in z wp 1. We define

$$T_z=\inf\{t\colon \bar{\xi}(t)=1 \text{ or } -z\}. \tag{2.2}$$

Then it is easily checked that $\{(1-p)\,\bar{\xi}(T_{p/(1-p)})+p, p\in I\}$ $(=1$ if $p=1)$ has the same law as $\{S_1(p), p\in I\}$. In fact, this is just the simplest Skorohod embedding of the martingale L_1. Working with T_z rather than T_p^*, we will find it convenient to replace $I-\{1\}=\{p\}$ by $R_1^+=\{z\}$, and we shall repeatedly use z, p for variables related, as above, by $p=z/(z+1)$ and $z=p/(1-p)$. Thus, $\{\bar{\xi}(T_z), z\in R_1^+\}$ has the same law as the martingale $L_1(z/(z+1))$, which we hereafter denote by

$$Q_1(z)=(z+1)\,S_1(z/(z+1))-z, \qquad z\in R_1^+. \tag{2.3}$$

While the almost sure limit, 1, of $Q(z)$ or $\bar{\xi}(T_z)$ as $z\to+\infty$ cannot be adjoined while maintaining a martingale, this causes absolutely no trouble at $p=1$, since what matters there is that $\lim_{z\to\infty}(z+1)^{-1}Q_1(z)=0=S_1(1)-1$ wp. 1. We shall hereafter usually work in terms of Q_1 rather than S_1 or L_1, and correspondingly in terms of the ξ of (1.2) rather than the ξ^* of (1.1). We shall write Q_n for (2.3) with S_1 replaced by S_n, and $t\,Q_t$ is obtained from $n\,Q_n$ by piecewise linear interpolation.

A natural way to continue the development is now this: Let $\xi_1', \xi_2', \ldots$ be independent standard Brownian motions. Let $T_{i,z}'$ be the first time that $\xi_i'(T_{i,z}')=1$ or $-z$, so that $\{\xi_i'(T_{i,z}'), z\in R_1^+\}$ are iid (in i) processes, each distributed as Q_1. Then $\left\{\sum_1^n \xi_i'(T_{i,z}'), z\in R_1^+\right\}$ has exactly the distribution of $\{nQ_n(z), z\in R_1^+\}$, from which we obtain at once an exact representation of $n\,S_n$.

But where is $\xi(z,t)$? Nothing so obvious as $\xi(z,n)=\sum_1^n \xi_i(z)$ will be useful, since by the central limit theorem that could not even yield $o(1)$ in (1.5)–(1.6). A natural attempt (at least to me) is to let $U'_{m,z}=\sum_1^m T'_{i,z}$ (with $U'_{0,z}=0$) and define

$$\xi'(z,t)=\sum_1^m \xi'_i(T'_{i,z})+\xi'_{m+1}(t-U'_{m,z}) \quad \text{whenever } U'_{m,z}\leqq t\leqq U'_{m+1,z}. \tag{2.4}$$

For then, for fixed $z>0$, by the strong Markov property of the $T'_{i,z}$, we have that $\{\xi'(z,t), t\geqq 0\}$ is a standard Brownian motion and thus $\{\xi'(z,zt), t\geqq 0\}$ has the desired law of $\{\xi(z,t), t\geqq 0\}$ for each z. Unfortunately, the strong Markov property does not yield the right joint distribution: for $0<z_1<z_2$ and $\varepsilon>0$, choose t_1 so small that $P\{T'_{1,z_1}>t_1\}>1-\varepsilon$, and note that $T'_{1,z_1}>t_1$ implies $\xi'(z_1,t_1)=\xi'(z_2,t_1)$. In fact, we even lose joint normality. (A reason for mentioning this "failure" ξ' is that, as discussed in Section 6, it offers some promise for eventual success.)

However, realizing that the joint behavior of the processes $\xi(z_i,\cdot)$ is reflected in the independence, for fixed $z_1<z_2\leqq z'_1<z'_2$, of the processes $\eta(z_1,z_2;\cdot)$ and $\eta(z'_1,z'_2;\cdot)$ defined by

$$\eta(z_1,z_2;t)=\xi(z_2,t)-\xi(z_1,t), \tag{2.5}$$

we are motivated to define rv's distributed like the $\xi'_i(T'_{i,z_j})$ above in terms of the $\eta(z_{j-1},z_j;\cdot)$ rather than in terms of the ξ'_i. This can be done in a more general context, to which we now temporarily turn.

2.2. *Skorohod Vector Embeddings*

A number of workers have been concerned with the possibility of a Skorohod representation of a vector rv. One obvious piece of wishful thinking must be dispelled at the outset: we cannot hope to succeed with a single stopping time. This is apparent if one tries to embed, in 2-dimensional Brownian motion, a 2-vector taking on the four equiprobable values $(\pm 1, \pm 1)$. On the other hand, if one looks at each of the two coordinates of the Brownian motion at its obvious stopping time to yield the values ± 1, we have a representation.

Let $A_m=(A_{m1}, A_{m2}, \ldots, A_{mh})$ be iid (in m) h-vectors such that $A_{11}, A_{12}, \ldots, A_{1h}$ is a 0-expectation martingale, $E\{A_{1j}|F_{j-1}\}=A_{1,j-1}$ wp 1 for $1\leqq j\leqq h$ where $A_{10}=0$ and F_j is the σ-field generated by $\{A_{11},\ldots,A_{1j}\}$. Without further mention we can and will use conditional probability measures

$$P\{A_{1,j+1}\leqq u|A_{11}=a_1,\ldots,A_{1j}=a_j\}.$$

Assume also, for the moment, that A_1 is bounded wp 1. Let $\{\eta(i-1,i;t), t\geqq 0\}$, $1\leqq i\leqq h$, be h independent standard Brownian motions. (This notation is motivated by the sample df results and (2.5); think of $\eta(i-1,i;t)=\xi(i,t)-\xi(i-1,t)$.)

Now represent $A_{11},\ldots,A_{1h}$ by Skorohod embedding, *except* that instead of using a single Brownian motion for all $A_{1,j+1}-A_{1j}$ we use $\eta(0,1;\cdot)$ for A_{11}, $\eta(1,2;\cdot)$ for $A_{12}-A_{11}$, and so on up to $\eta(h-1,h;\cdot)$ for $A_{1h}-A_{1,h-1}$. Formally, $T^{(1)}_{i-1,i}$ is a stopping time defined inductively on i for $1\leqq i\leqq h$ (and possibly using additional randomization as in Skorohod's original method) such that, if $B^{(1)}_i=\{\eta(j-1,j;T^{(1)}_{j-1,j})=a_j-a_{j-1}, 1\leqq j<i\}$ $(a_0=0)$, then

$$P\{\eta(i-1,i;T^{(1)}_{i-1,i})\leqq u|B^{(1)}_i\}=P\{A_{1i}-A_{1,i-1}\leqq u|A_{1j}=a_j, j<i\}. \tag{2.6}$$

This yields $\{\eta(i-1,i;T^{(1)}_{i-1,i}), 1\leqq i\leqq h\}$ distributed as $\{A_{1i}-A_{1,i-1}, 1\leqq i\leqq h\}$.

Then, given $T^{(1)}_{i-1,i}=t_i$ (say), we obtain an h-vector independent of that of the previous sentence, but with the same distribution, by writing $\eta(j-1,j;T^{(2)}_{j-1,j}+t_j)-\eta(j-1,j;t_j)$ in place of $\eta(j-1,j;T^{(1)}_{j-1,j})$ on the left side of (2.6) (including $B^{(1)}_i$). That is, we observe the $\eta(j-1,j;\cdot)$ process from time $T^{(1)}_{j-1,j}$ to $T^{(1)}_{j-1,j}+T^{(2)}_{j-1,j}$ to get a rv which represents $A_{2j}-A_{2,j-1}$, successively for $j=1,2,\ldots,h$.

We continue in this fashion. Thus, define $U^{(0)}_{i,i+1}=T^{(0)}_{i,i+1}=0$ and then $U^{(m)}_{i,i+1}=\sum_{q=1}^{m} T^{(q)}_{i,i+1}$ inductively for $m\geqq 0$ and $0\leqq i<h$ by letting the vectors $(T^{(q)}_{0,1},\ldots,T^{(q)}_{h-1,h})$ be iid in q and such that, with

$$B^{(q)}_i=\{\eta(j-1,j;U^{(q)}_{j-1,j})-\eta(j-1,j;U^{(q-1)}_{j-1,j})=a_j-a_{j-1}, 1\leqq j<i\} \quad (a_0=0), \tag{2.7}$$

$T^{(q)}_{i,i+1}$ is any stopping time satisfying

$$\begin{aligned} &P\{\eta(i-1,i;U^{(q)}_{i-1,i})-\eta(i-1,i;U^{(q-1)}_{i-1,i})\leqq u|B^{(q)}_i\}\\ &\quad=P\{A_{q,i}-A_{q,i-1}\leqq u|A_{qj}=a_j, j<i\}. \end{aligned} \tag{2.8}$$

Again, we can assume our probability structure allows external randomization if needed.

We thus obtain

Lemma 2. *Skorohod vector embedding:*

$$\left\{\left\{\sum_{j=1}^{i}\eta(j-1,j;U^{(n)}_{j-1,j}), 1\leqq i\leqq h\right\}, n\in Z^+\right\} \tag{2.9}$$

has the same joint law as $\left\{\sum_{m=1}^{n} A_m, n\in Z^+\right\}$.

Remarks on the Embedding (2.9). 1. It is now obvious how to study, for each j, the difference between $\eta(j-1,j;U^{(n)}_{j-1,j})$ and $\eta(j-1,j;EU^{(n)}_{j-1,j})$, and thus obtain precise almost sure results on this difference as in [10], and also the k-dimensional analogues of Strassen's results [21, 22]. (See Section 6A for further comments.) Also, the analogues of the results of (1.7) and (1.9) (described there for the special case $A_{1i}=L_1(z_i)$) follow at once from using our developments with the techniques of [20] and [19] for more general summands. (See, however, Section 6C regarding inapplicability of some of these techniques.) Moreover, the representation (2.9) does not require boundedness of A_1, and the error term one obtains in these approximations will depend on (conditional) moments in a manner made clear in [2, 10, 19].

Thus, (1.7), (1.8), (1.9) *carry over, with appropriate modification of their right sides as described in* [2, 10, 19] *under various moment conditions, to general* $\{A_m\}$ *with* $\{A_{1j}\}$ *a martingale, upon replacing*

$$\{k(S_k(s)-s) \text{ and } \xi^*(s,t);\ 1\leqq k\leqq n, t\in I, s\in I'\} \tag{2.10}$$

by

$$\left\{\sum_{m=1}^{k}\sum_{j=1}^{i} A_{mj} \text{ and } \sum_{j=1}^{i}\eta(j-1,j;tE(A_{1j}-A_{1,j-1})^2),\ 1\leqq k\leqq n, t\in I, 1\leqq i\leqq h\right\}. \tag{2.11}$$

(As remarked earlier and as is explained in Section 6, the exponent 1/4 in (1.7) and (1.9) cannot be replaced by 1/2 with the present scheme, even if $h=2$.)

2. The iid structure in m of $\{A_m\}$ is inessential. In fact, there is no difficulty in embedding processes $\left\{\sum_{m=1}^{n} A_{mj}; j, m \in Z^+\right\}$ which are martingales in n as well as in j. Moreover, there are embeddings for which, as with the T_p^* of the previous section, the stopping times for $A_{1j}-A_{1,j-1}$ need not be ordered in j.

3. Of great interest is the possibility of replacing j (and, possibly, n) by a continuous time parameter in (2.9). Some circumstances where this is possible will be treated in [13]. In Section 2.3, below, we will see what goes wrong with this attempt in the sample df example.

4. Another direction of extension is to the case where the index set of j in A_{mj} is not 1-dimensional, e.g., to the sample df of chance k-vectors. The arithmetic of [8] shows the kind of modifications that are needed in Lemmas 1′ and 4′. One obtains somewhat worse orders than in (1.5), (1.6), (1.10), (1.11), in terms of a process $\xi^{*(k)}(s,t)$ with k-dimensional s. Results can be obtained for general $\{A_{1j}\}$ with j in a lattice of R^k, a setting considered more extensively by J. Zinn, a student of J. Kuelbs.

2.3. *Back to the Sample* df

Let I' be any subset of I, as in Corollary 2, (1.7). In view of the nature of $S_n(1)$, we can limit our treatment to the case $1 \notin I'$. Let $R' = \{z\colon z = p/(1-p),\ p \in I'\}$ be the corresponding set of non-negative values under the correspondence described above (2.3). Adjoining $z_0 = 0$ (unless $0 \in R'$ already), we write $R' = \{z_0, z_1, \ldots, z_k\}$ with $z_i < z_{i+1}$. Now, with ξ the process of (1.2), replace the $\eta(i-1, i; t)$ of Section 2.2 by the $\eta(z_{i-1}, z_i; (z_i - z_{i-1})^{-1} t)$ of (2.5), which have the same distribution. (The convenience of this normalization of the η of (2.5) will be seen below to be that the analogue here of $T_{i-1,i}^{(1)}$ of (2.6) will have expectation $z_i - z_{i-1}$.) The rv's A_{1i} are the $Q_1(z_i)$ of (2.3). Thus, for the embedding of Section 2.2, (2.7)–(2.8) with $\eta(j-1, j; U_{j-1,j}^{(q)})$ there replaced by $\eta(z_{j-1}, z_j; (z_j - z_{j-1})^{-1} U_{z_{j-1}, z_j}^{(q)})$ here, can be realized by treating $\eta(z_{j-1}, z_j; (z_j - z_{j-1})^{-1} T_{z_{j-1}, z_j}^{(1)})$ exactly as $\bar{\xi}(T_{z_j}) - \bar{\xi}(T_{z_{j-1}})$ of (2.2) or $\xi_1'(T_{1,z_j}') - \xi_1'(T_{1,z_{j-1}}')$ below (2.3):

$$T_{z_{j-1},z_j}^{(1)} = \inf\Big\{t\colon t \geqq 0, \sum_{i=1}^{j-1} \eta(z_{i-1}, z_i; (z_i - z_{i-1})^{-1} T_{z_{i-1},z_i}^{(1)}) + \eta(z_{j-1}, z_j; (z_j - z_{j-1})^{-1} t) = 1 \text{ or } -z_j\Big\} \tag{2.12}$$

and, similarly,

$$\begin{aligned} T_{z_{j-1},z_j}^{(q)} = \inf\{t\colon t \geqq 0, &\sum_{i=1}^{j-1} [\eta(z_{i-1}, z_i; (z_i - z_{i-1})^{-1} U_{z_{i-1},z_i}^{(q)}) \\ &- \eta(z_{i-1}, z_i; (z_i - z_{i-1})^{-1} U_{z_{i-1},z_i}^{(q-1)})] \\ &+ [\eta(z_{j-1}, z_j; (z_j - z_{j-1})^{-1} (U_{z_{j-1},z_j}^{(q)} + t)) \\ &- \eta(z_{j-1}, z_j; (z_j - z_{j-1})^{-1} U_{z_{j-1},z_j}^{(q-1)})] = 1 \text{ or } -z_j\}. \end{aligned} \tag{2.13}$$

Since $\{\eta(z_{j-1}, z_j; (z_j - z_{j-1})^{-1} t), t \geqq 0\}$ has variance 1 per unit of time t, we see that $T_{z_{i-1},z_i}^{(q)}$ has the same expectation as the $T_{z_j} - T_{z_{j-1}}$ of (2.2) or $T_{1,z_j}' - T_{1,z_{j-1}}'$

below (2.3), namely

$$E\,T^{(q)}_{z_{j-1},z_j}=z_j-z_{j-1}; \tag{2.14}$$

thus, our normalization assures that the *expectation* of the third argument of $\eta(z_{i-1},z_i;(z_i-z_{i-1})^{-1}U^{(q)}_{z_{i-1},z_i})$ is q.

We have eliminated the difficulty of not finding the ξ of (1.2) in the ξ' of (2.4) obtained from the $T'_{i,z}$, by now embedding $\{Q_n(z), z\in R', n\in Z^+\}$ in ξ. As we shall indicate in the next paragraph, all of $\{Q_n(z), z\leqq z_k\}$ has in fact been embedded in ξ, but $Q_n(z)$ for all z is not as evident as one might like in the above embedding using R'. (The R' above can be modified to include, for example, an infinite sequence approaching 0 or ∞, but this will not help achieve what we want here.) For simplicity, consider $\{Q_1(z), z\leqq 1\}$. We would like to replace the above embedding of $\{Q_1(z), z\in R'\}$ by $\{Q_1(z), z\leqq 1\}$, by piecing together (roughly) stopped parts of a continuum of differential processes $\xi(z, z+dz; t)$. One might try to get at this by considering a sequence of nested finite sets whose union is dense, e.g., $R'_L=\{z: z=i/2^L, 0\leqq i\leqq 2^L\}$, hoping somehow to take a limit with L of the previous embeddings to obtain $\{Q_1(z), z\leqq 1\}$. We do not see how to make this work. Roughly, although $2^L T^{(1)}_{z_j,z_{j+1}}$ has expectation 1 for all L and $z_{j+1}-z_j=2^{-L}$, it follows from (2.19) that it has variance 2^L and must itself be very large for some (chance) j. The latter is certainly the case for an interval (z_j, z_{j+1}) containing the z for which Q_1 jumps, and it will also be true for (chance) values $\bar z$ for which, in terms of (2.2), ξ has a local minimum at $t=T_{\bar z}$ so that $\inf_{z>\bar z}(T_z-T_{\bar z})>0$.

This failure reflects our feeling that the present method of embedding, while it has proved a useful tool, is not an ultimate one. Actually, all of $\{Q_n(z), 0\leqq z\leqq z_k\}$ (and, with a slight modification, all of $\{Q_n(z), z\in R_1^+\}$) *does* sit in ξ in the above embedding with R', but not in a form which yields the desired Skorohod-type estimates without further calculations. For example, consider $R'\{0,1,2\}$. It is convenient to think of ξ as being defined on the domain of planar Borel sets, with, as usual, $\xi(A)$ having variance A and being independent over disjoint A's; thus, we simply write $\xi([0,z]\times[0,t])$ for $\xi(z,t)$. In Fig. 1 is shown a possible realization of the set

$$A=\{[0,1]\times[0,U^{(2)}_{0,1}]\}\cup\{[1,2]\times[0,U^{(2)}_{1,2}]\}$$

for which $\xi(A)=2Q_2(2)$. The shaded subset B of A, for which $\xi(B)=2Q_2(1.6)$, is obtained by letting

$$\begin{aligned}\tau^{(i)}=\inf\Big\{t\colon t\geqq 0, &\,[\eta(0,1;U^{(i)}_{0,1})-\eta(0,1;U^{(i-1)}_{0,1})]\\ &+\Big[\eta\Big(1,2;\sum_{j=1}^{i-1}\tau^{(j)}+t\Big)-\eta\Big(1,2;\sum_{j=1}^{i-1}\tau^{(j)}\Big)\Big]=1 \text{ or } -1.6\Big\},\end{aligned} \tag{2.15}$$

and $2Q_2(z)$ is obtained similarly for other values z.

In similar fashion, by using an unbounded sequence $\{z_i\}$ one obtains an explicit embedding of $\{Q_n(z), z\in R^+\}$; however, the large z_j's play no role in the proofs and will simply be omitted. Thus, in Section 3 one can define z_j's for $j\geqq B_n$, to exhibit all of Q_n, but it is unnecessary to consider them in proving Theorem 2.

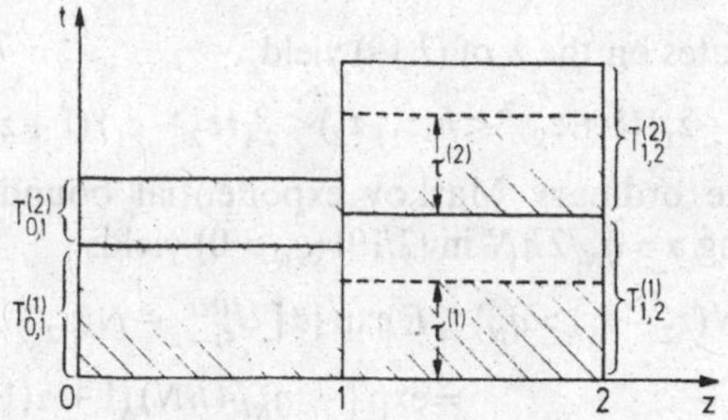

Fig. 1. Examples of $2Q_2(2)$ and $2Q_2(1.6)$

A similar remark will apply in Section 4 (Theorem 1), where for technical simplicity the values $z > B_n$ are treated slightly differently.

In terms of the discussion, above, of the difficulties accompanying a process of subdivision, the reader might find it instructive to see what happens to this already complicated picture when one subdivides the simple R' of Fig. 1!

What we shall do, then, is to use Lemma 1′ to approximate S_n by $S_{B_n,n}$ for an appropriate B_n, and then use the embedding of Sections 2.2–2.3 for $S_{B_n,n}$ in ξ, to prove Theorem 2.

We conclude this section by computing some elementary properties of the $T^{(i)}$ and $U^{(i)}$. It is well known that, for $\alpha < \pi^2/2(1+z)^2$ and $z > 0$,

$$E\exp\{\alpha T^{(1)}_{0,z}\} = \frac{\cos[(\alpha/2)^{1/2}(1-z)]}{\cos[(\alpha/2)^{1/2}(1+z)]}. \tag{2.16}$$

Hence, as $\alpha \to 0$,

$$E\exp\{\alpha T^{(1)}_{0,z}\} = 1 + \alpha z + \alpha^2(z + 3z^2 + z^3)/6 + O(\alpha^3(z+1)^4 z). \tag{2.17}$$

From the above or Wald's equation, as already noted in (2.14), $E T^{(1)}_{z_1,z_2} = z_2 - z_1$. It will simplify notation and yield the same result if we work in terms of the $\bar{\xi}$ of (2.2) rather than the η of (2.12)–(2.13). We note now that $P\{T^{(1)}_{z_1,z_2} = 0 \mid \bar{\xi}(T^{(1)}_{0,z_1}) = 1\} = 1$; and from looking at stopping boundaries we see that, given that $\bar{\xi}(T^{(1)}_{0,z_1}) = -z_1$, the conditional distribution of

$$\{(1+z_1)^{-1}[\bar{\xi}([1+z_1]^2 t + T^{(1)}_{0,z_1}) + z_1],\ 0 \leq [1+z_1]^2 t \leq T^{(1)}_{z_1,z_2}\}$$

is that of the unconditional distribution of $\{\bar{\xi}(t), 0 \leq t \leq T^{(1)}_{0,(1+z_1)^{-1}(z_2-z_1)}\}$. Since $P\{T^{(1)}_{0,z_1} = 1\} = z_1/(1+z_1)$, we obtain from (2.17),

$$\begin{aligned} E\exp\{\alpha T^{(1)}_{z_1,z_2}\} &= (1+z_1)^{-1}[z_1 + E\exp\{\alpha(1+z_1)^2\, T^{(1)}_{0,(1+z_1)^{-1}(z_2-z_1)}\}] \\ &= 1 + \alpha(z_2 - z_1) + \alpha^2\{(z_2 - z_1)(1+z_1)^2 + 3(z_2 - z_1)^2(1+z_1) \\ &\quad + (z_2 - z_1)^3\}/6 + O(\alpha^3(1+z_2)^4(z_2 - z_1)). \end{aligned} \tag{2.18}$$

Consequently, routine calculation shows that

$$\begin{aligned} &\log E\exp\{\alpha[T^{(1)}_{z_1,z_2} - (z_2 - z_1)]\} \\ &\quad \sim \alpha^2\{(z_2 - z_1)(1+z_1)^2 + 3(z_2 - z_1)^2 z_1 + (z_2 - z_1)^3\}/6 \\ &\quad = \alpha^2 h(z_1, z_2) \text{ (say)}, \end{aligned} \tag{2.19}$$

provided that this last expression is $o(1)$ and that the ratio of $\alpha^3(1+z_2)^4(z_2 - z_1)$ to this expression is $o(1)$.

Elementary estimates on the h of (2.19) yield

$$\tfrac{1}{12}(z_2-z_1)(1+z_2)^2<h(z_1,z_2)<\tfrac{5}{24}(z_2-z_1)(1+z_2)^2. \tag{2.20}$$

We now compute ordinary Markov exponential bounds (as in (1.16) for Bernoulli rv's). Putting $\alpha=q_N/2hN$ in (2.19) ($q_N>0$) yields

$$\begin{aligned} P\{U^{(N)}_{z_1,z_2}-N(z_2-z_1)>q_N\} &\leqq E\exp\{\alpha[U^{(N)}_{z_1,z_2}-N(z_2-z_1)-q_N]\} \\ &=\exp\{-(q_N^2/4hN)(1+o(1))\}. \end{aligned} \tag{2.21}$$

The negative deviations of $U^{(N)}$ are treated similarly. Moreover, considering the conditions just below (2.19), from (2.20) we have $\alpha^2 h$ of exactly order

$$N^{-2}q_N^2(z_2-z_1)^{-1}(1+z_2)^{-2}$$

and the ratio of $\alpha^3(1+z_2)^4(z_2-z_1)$ to $\alpha^2 h$ of exactly order $N^{-1}q_N(z_2-z_1)^{-1}$. If the latter is $o(1)$, so is the former. Thus, we have

$$\begin{aligned} &q_N/N(z_2-z_1)=o(1) \\ &\quad\Rightarrow P\{|U^{(N)}_{z_1,z_2}-N(z_2-z_1)|>q_N\}\leqq\exp\{-(q_N^2/4hN)(1+o(1))\}. \end{aligned} \tag{2.22}$$

Finally, at slight expense in sharpness we will put the estimate (2.22) in a form useful in the proof of Theorem 2 of the next section. We will use (2.22) with z_1, z_2 replaced by z_{j-1}, z_j where $z_j>z_{j-1}$ and $1\leqq j<B_n$ ($z_0=0$). From (2.22) and the right side of (2.20) we obtain

Lemma 3. *There is a positive constant C_3 such that*

$$\begin{aligned} q_N/N\leqq C_3(z_j-z_{j-1})&\Rightarrow P\{|U^{(N)}_{z_{j-1},z_j}-N(z_j-z_{j-1})|>q_n\} \\ &\leqq\exp\{-q_N^2/N(z_j-z_{j-1})(1+z_j)^2\}. \end{aligned} \tag{2.23}$$

In fact, in applying this in Section 3 we shall use the particular z_j of (3.2) ($1\leqq j<B_N$) with

$$q_N=2(z_j-z_{j-1})[NB_n\log(nB_n)]^{1/2} \tag{2.24}$$

for various large integers N. From (3.2), $B_n(z_j-z_{j-1})/(1+z_j)^2=(B_n-j)/(B_n-j+1)$ $\geqq 1/2$. Hence, (2.23) becomes

$$\begin{aligned} &(3.2)\text{ with } N^{-1}B_n\log(nB_n)\leqq C_3^2/4 \\ &\quad\Rightarrow P\{|U^{(N)}_{z_{j-1},z_j}-N(z_j-z_{j-1})|>q_N\text{ of (2.24)}\}\leqq(nB_n)^{-2}. \end{aligned} \tag{2.25}$$

3. Statement and Proof of Theorem 2

It is convenient to divide the considerations into two parts – bounding in probability the difference between ξ and the embedded process, and then bounding the limiting density of the functional.

To emphasize the essentials of the proof, we postpone the statement of the required Lemma 4′ and first prove Lemma 4, which is of no interest in itself in view of the existence of embeddings for fixed n mentioned in Section 1, but which contains the main ideas of Lemma 4′ with simpler arithmetic. We use the notation of Lemma 1 and of Section 2.3, and write $\varepsilon=\min(C_3^2, 1/2)$ where C_3 is as described in (2.25).

Lemma 4. *With the embedding of nQ_n in ξ described in Section 2.3, with the p_j equally spaced, and with ε as defined just above, there is a positive constant C_4 such that, for n sufficiently large,*

$$\begin{aligned} 2 \leqq B_n \leqq \varepsilon n/8 \log n \Rightarrow P\{ &\max_{1 \leqq j < B_n} (z_j+1)^{-1} |nQ_n(z_j) - \xi(z_j, n)| \\ &\leqq C_4 [(nB_n)^{1/4} (\log n)^{3/4} + B_n \log n]\} \geqq 1 - n^{-1}. \end{aligned} \tag{3.1}$$

Proof. Corresponding to equally spaced $p_j = B_n^{-1} j$ $(1 \leqq j < B_n)$, we let $z_j = p_j/(1-p_j) = j/(B_n - j)$ and thus

$$\begin{aligned} z_j - z_{j-1} &= B_n/(B_n - j)(B_n - j + 1), \\ z_j + 1 &= B_n/(B_n - j). \end{aligned} \tag{3.2}$$

Fixing n and B_n and dropping the subscript on the latter, for positive integral N we define

$$f(N) = 2[NB \log(nB)]^{1/2}. \tag{3.3}$$

Let ε be as defined above. We define inductively

$$\begin{aligned} m_0 &= n_0 = n, \\ m_{i+1} &= f(m_i), \\ n_{i+1} &= m_i - m_{i+1} \quad \left(\text{and hence } \sum_1^i n_r = n - m_i\right). \end{aligned} \tag{3.4}$$

This yields

$$\begin{aligned} m_i &= n[4n^{-1} B \log(nB)]^{1-2^{-i}} \quad &\text{for } i \geqq 0, \\ m_i/m_{i-1} &= [4Bn^{-1} \log(nB)]^{2^{-i}} \quad &\text{for } i \geqq 1. \end{aligned} \tag{3.5}$$

Let

$$K = \begin{cases} 0 & \text{if } \ m_1/m_0 > \varepsilon, \\ \max\{i\colon m_i/m_{i-1} \leqq \varepsilon\} & \text{otherwise.} \end{cases} \tag{3.6}$$

Then

$$m_i < m_{i-1} \quad \text{and} \quad n_i \geqq (1-\varepsilon)\, m_{i-1} \qquad \text{for } 1 \leqq i \leqq K \tag{3.7}$$

and, from (3.5) with $B > 1$, if $n \geqq 3$,

$$K \leqq [\log^+(\log n/\log \varepsilon^{-1})]/\log 2. \tag{3.8}$$

In the remainder of this proof we will simplify notation by writing m_i and n_i, rather than integers close to them, as numbers of observations. It can be seen that proper reinterpretation of these symbols yields (3.1) without difficulty when n is large. Also for brevity, we shall write

$$\bar{\eta}_j(t) = \eta(z_{j-1}, z_j;\, t),$$

$$V_j^{(i)} = \begin{cases} 0 & \text{if } \ i = 0, \\ (z_j - z_{j-1})^{-1}\, U_{z_{j-1}, z_j}^{(n_1 + \cdots + n_i)} & \text{if } \ 1 \leqq i \leqq K, \\ (z_j - z_{j-1})^{-1}\, U_{z_{j-1}, z_j}^{(n)} & \text{if } \ i = K+1. \end{cases} \tag{3.9}$$

(In particular, if $K=0$ we have $V_j^{(1)}=(z_j-z_{j-1})^{-1}U^{(n)}$; we drop obvious subscripts on $U^{(n)}$.)

To estimate $\bar\eta_j((z_j-z_{j-1})^{-1}U^{(n)})-\bar\eta((z_j-z_{j-1})^{-1}EU^{(n)})=\bar\eta_j(V_j^{(K+1)})-\bar\eta_j(n)$ (recall (2.14): $(z_j-z_{j-1})^{-1}EU^{(n)}=n$), we write the telescoping sum

$$\begin{aligned}\bar\eta_j(V_j^{(K+1)})-\bar\eta_j(n)&=\bar\eta_j(V_j^{(K)}+[V_j^{(K+1)}-V_j^{(K)}])-\bar\eta_j(V_j^{(K)}+m_K)\\&\quad+\sum_{i=1}^{K}\{\bar\eta_j(V_j^{(i)}+m_i)-\bar\eta_j(V_j^{(i)}+m_i-[V_j^{(i)}-V_j^{(i-1)}-n_i])\}.\end{aligned}\tag{3.10}$$

We first consider the first two terms on the right side of (3.10); in the next paragraph it will be seen why this difference requires a separate treatment. We see that, whether or not $K>0$,

$$4\varepsilon^{-1}B\log(nB)\leqq m_K<4\varepsilon^{-2}B\log(nB);\tag{3.11}$$

the right hand inequality follows from substituting $m_{K+1}/m_K>\varepsilon$ and the second line of (3.5) into the first line; the left hand inequality follows similarly from $m_K/m_{K-1}\leqq\varepsilon$ if $K>0$, and from the condition on B_n in (3.1) if $K=0$. Thus, conditional on $V_j^{(K)}$, the $V_j^{(K+1)}-V_j^{(K)}=(z_j-z_{j-1})^{-1}[U^{(n)}-U^{(n-m_K)}]$ of (3.10) has the same distribution as $(z_j-z_{j-1})^{-1}U^{(N)}$ of (2.25) with $N=m_K$ satisfying the condition of (2.25). We now recall (2.24) and let j vary, and consider the sum of probabilities (2.25) over $1\leqq j<B_n$. We obtain, wp 1,

$$P\{|(V_j^{(K+1)}-V_j^{(K)})-m_K|\leqq 2[m_K B\log(nB)]^{1/2},1\leqq j<B|\{V_j^{(K)}\}\}\geqq 1-n^{-2}B^{-1}.\tag{3.12}$$

Let A be the event of (3.12). Under the same conditioning as in (3.12), $\eta'_j(\tau)=\bar\eta_j((z_j-z_{j-1})^{-1}\tau+V_j^{(K)})$ is a standard Brownian motion for $\tau\geqq 0$. Write

$$\begin{aligned}\Gamma_j&=\{\tau\colon|\tau|\leqq 2(z_j-z_{j-1})[m_K B\log(nB)]^{1/2},\ \tau+(z_j-z_{j-1})m_K\geqq 0\},\\q'_j&=3(z_j-z_{j-1})^{1/2}(m_K B)^{1/4}[\log(nB)]^{3/4}.\end{aligned}\tag{3.13}$$

By the standard inequality for the tail probabilities of the maximum of a Brownian motion, we have for $n\geqq 10$,

$$\begin{aligned}&P\{|\bar\eta_j(V_j^{(K)}+[V_j^{(K+1)}-V_j^{(K)}])-\bar\eta_j(V_j^{(K)}+m_K)|>q'_j|A,\{V_j^{(K)}\}\}\\&\leqq P\{\sup_{\tau\in\Gamma_j}|\eta'_j(\tau+(z_j-z_{j-1})m_K)-\eta'_j((z_j-z_{j-1})m_K)|>q'_j|\{V_j^{(K)}\}\}<(nB)^{-2}.\end{aligned}\tag{3.14}$$

From (3.13), (3.11), and (3.2) we have

$$\begin{aligned}\sum_{r=1}^{j}q'_r&<5B^{1/2}\varepsilon^{1/2}\log(nB)\sum_{1}^{j}(z_r-z_{r-1})^{1/2}\\&<5B\varepsilon^{-1/2}\log(nB)\log(2[z_j+1]).\end{aligned}\tag{3.15}$$

Thus, from (3.12), (3.13), (3.14), and (3.15), for n sufficiently large,

$$\begin{aligned}P\Big\{&\sum_{r=1}^{j}|\bar\eta(V_r^{(K+1)})-\bar\eta(V_r^{(K)}+m_K)|\\&\leqq 5B\varepsilon^{-1/2}\log(nB)\log(2[z_j+1]),1\leqq j<B\Big\}\geqq 1-n^{-2}.\end{aligned}\tag{3.16}$$

We next consider the i-th term of the sum in (3.10). First note that, by (3.6)–(3.7), $n_i \geqq (1-\varepsilon) m_{i-1} \geqq \varepsilon^{-1}(1-\varepsilon) m_i \geqq m_K$; thus, since m_K satisfied the condition for N in (2.25), so does n_i. Hence, from (2.25) with $N = n_i$, we obtain, as in (3.12),

$$P\{|V_j^{(i)} - V_j^{(i-1)} - n_i| \leqq 2[n_i B \log(nB)]^{1/2}, 1 \leqq j < B\} \geqq 1 - n^{-2} B^{-1}. \quad (3.17)$$

Since $n_i < m_{i-1}$, the product of the two equations of (3.5) yields

$$2[n_i B \log n B]^{1/2} < m_{i-1}^{1/2} [4B \log n B]^{1/2} = m_{i-1}^{1/2} \left[m_i \frac{m_i}{m_{i-1}} \right]^{1/2} = m_i. \quad (3.18)$$

This means that, if the event A' (say) of (3.17) occurs, then, conditional on $\{V_j^{(i)}\}$ and $\{V_j^{(i-1)}\}$, the arguments of the two $\bar{\eta}_j$ terms in the i-th difference of the sum of (3.10) *are both* $> V_j^{(i)}$. That difference is thus conditionally of the form $\bar{\eta}_j(t_1) - \bar{\eta}_j(t_2)$ where t_1 and t_2 are determined by $\{\bar{\eta}_j(t), t \leqq \min(t_1, t_2)\}$. (The absence of this property is what entailed separate treatment of the difference studied in the previous paragraph; that treatment used here would yield a bound inferior to (3.1).) The event of (3.17) consequently implies a corresponding change in the argument of the η'_j process (defined below (3.12)), of $(z_j - z_{j-1})[V_j^{(i)} - V_j^{(i-1)} - n_i]$ time units. Since the η'_j are standard, we obtain in terms of a standard η'_0,

$$P\left\{\left|\sum_{r=1}^{j} \{\bar{\eta}_r(V_r^{(i)} + m_i) - \bar{\eta}_r(V_r^{(i-1)} + m_i - n_i)\}\right| > \bar{q}_j \,|\, A'_j, \{V_j^{(i)}\}, (V_j^{(i-1)})\right\}$$
$$= P\{|\eta'_0(\tau_j)| > \bar{q}_j\} \leqq 2\tau_j^{1/2} \bar{q}_j^{-1} e^{-\bar{q}_j^2/2\tau_j}, \quad (3.19)$$
$$\tau_j = \sum_{r=1}^{j} (z_r - z_{r-1}) |V_r^{(i)} - V_r^{(i-1)} - n_i| \leqq z_j m_i.$$

Thus, putting $\bar{q}_j = 2[z_j m_i \log(nB)]^{1/2}$, (3.17) and (3.19) yield for n sufficiently large,

$$P\left\{\left|\sum_{r=1}^{j} \{\bar{\eta}_r(V_r^{(i)} + m_i) - \bar{\eta}(V_r^{(i-1)} + m_i - n_i)\}\right| \leqq 2[z_j m_i \log(nB)]^{1/2}, 1 \leqq j < B\right\} \geqq 1 - n^{-2}. \quad (3.20)$$

By (3.6), $\sum_1^K m_i^{1/2} < m_i^{1/2}/[1-\varepsilon^{1/2}] < 8[nB \log(nB)]^{1/4}$. Thus, finally, from (3.8), (3.10), (3.16), (3.20), and the condition on B in (3.1), we have for n sufficiently large and some positive constant C_4,

$$P\left\{\left|\sum_{r=1}^{j} \{\bar{\eta}_r(V_r^{(K+1)}) - \bar{\eta}_r(n)\}\right| \leqq C_4 \{z_j^{1/2} (nB)^{1/4} (\log n)^{3/4} + B(\log n) \log[2[z_j + 1]]\}, 1 \leqq j < B\right\} > 1 - n^{-1}. \quad (3.21)$$

The fact that $(z+1)^{-1} \max[z^{1/2}, \log 2(z+1)] < 1$, with (3.21), yields (3.1).

Remarks on Lemma 4. (1) This is our crucial estimate concerning the embedding of Section 2.3. The martingale structure of the $U_z^{(n)} - nz$, in both n and z, can be used to give an alternate proof, but that approach is also long due to the complication of $n Q_n(z) - \xi(z, n)$ being determined by stopping times dependent on ξ. This last effect is minimized in the present proof by the device which enables the sum in (3.10) to be treated in terms of Brownian motion deviations for fixed epochs which can be summed on j as independent normal rv's, leaving only the first difference on the right side of (3.10) for grosser path-dependent estimation. (2) We have been somewhat cavalier in the choice of constants in the limits entering into the proof, but a more careful choice would only alter C_4, not the order of the deviation in (3.1). However, the behavior of the embeddings as $p \to 0$ or 1 can obviously be sharpened; as one might expect, the behavior for $|p-1/2|<\varepsilon'$ accounts for most of the deviation. (3) We have given the proof for equally spaced p_j, for the sake of Corollary 1. The changes in estimates for other choices of the p_j should be clear. In particular, if B_n is bounded in n and the p_j are fixed, the bounds stated in (3.1) are valid; this is used in Corollary 2. (4) It is easily seen that, with a slight change in the values of ε and C_4 and an accompanying increase in the coefficient 2 in the definition of q_N in (2.24), one obtains (3.1) with $nQ_n(z_j)-\xi(z_j,n)$, replaced by $mQ_m(z_j)-\xi(z_j,m)$, for each $m\leqq n$, and with the probability bound $1-n^{-1}$ replaced by $1-n^{-3}$. As in the statement of Lemma 1', we require such a result for $m\leqq \bar{c}\,n(\log n)^\beta$, and it is easily verified that this essentially involves substituting this larger value for n, while leaving B_n unchanged, in the upper bound on $|t\,Q_t-\xi|$ of (3.21). With a slight additional argument this last comment yields

Lemma 4'. *There is an $\varepsilon'>0$ and $C_4'>0$ such that, for $\beta\geqq 0$, $\bar{c}>0$, and n sufficiently large,*

$$2\leqq B_n\leqq \varepsilon' n/\log n$$
$$\Rightarrow P\{(z_j+1)^{-1}|t\,Q_t(z_j)-\xi(z_j,t)|\leqq 2C_4'\left[(\bar{c}\,n[\log n]^{\beta+3}B_n)^{1/4}+B_n\log n\right] \quad \text{for } 0\leqq t\leqq \bar{c}\,n(\log n)^\beta \text{ and } 1\leqq j<B_n\}\geqq 1-n^{-1}. \tag{3.22}$$

The required additional argument is in fact given by

$$\begin{aligned} 0\leqq \sup_{m\leqq t\leqq m+1,\,0\leqq z<\infty} (z+1)^{-1}|t\,Q_t(z)-m\,Q_m(z)|\leqq 1 \text{ wp } 1,\\ P\{\sup_{m\leqq t\leqq m+1,\,0\leqq z<\infty}(z+1)^{-1}|\xi(z,t)-\xi(z,m)|\geqq 2\log n\}\leqq C_5 n^{-3}\end{aligned} \tag{3.23}$$

for some constant C_5; the latter is easily proved by the methods of [8].

We can think of $\{t(z_j+1)^{-1}Q_t(z_j/(z_j+1))\}$ of (3.22) as an embedding of $\{t\,S_{B_n,t}(j/B_n)-t\,j/B_n,\ 1\leqq j<B_n,\ 0\leqq t\leqq \bar{c}\,n(\log n)^\beta\}$ of (1.18). It is easily seen that the maximum of the rates in (1.18) and (3.22) has its minimum when the two rates have the common value $n^{1/3}(\log n)^{(\beta+2)/3}$, at $B_n=n^{1/3}(\log n)^{(\beta-1)/3}$. Recalling how, in connection with Fig. 1, we could consider all of $k[S_k(x)-x]$, $x\in I$, to be embedded in ξ^*, we conclude easily from Lemma 1' and Lemma 4' (with, in fact, any power of n in place of n^{-1} on the right side of (3.24)),

Theorem 2. *For $\beta\geqq 0$ and $\bar{c}>0$ there is an embedding of $\{k[S_k(x)-x],\ k\leqq \bar{c}\,n(\log n)^\beta,\ x\in I\}$ in ξ^* such that, for some constant c^*,*

$$\begin{aligned} P\{\sup_{t\leqq \bar{c}n(\log n)^\beta,\,x\in I} n^{-1/2}|t[S_t(x)-x]-\xi^*(x,t)|\\ \geqq c^* n^{-1/6}(\log n)^{(\beta+2)/3}\}=O(n^{-1}).\end{aligned} \tag{3.24}$$

We have stated (3.24) with continuous t, for comparison with Theorem 1. As in Corollary 1, the main interest is of course in integral t for S_t and continuous t for ξ^*, although Corollary 1 remains essentially unchanged for any of the four possible combinations. (See Remark 3 on Lemma 6.) In Lemma 5 we consider only the combination stated in Corollary 1.

To obtain Corollary 1 from Theorem 2, we need only (3.23) and the familiar device used in such weak convergence results, e.g. by Skorohod, Müller, Sawyer, of using boundedness near 0 of the density of an appropriate functional of ξ^*. Let R^* be a subinterval of R^+ and write

$$\begin{aligned} R_n^* &= \{t\colon t\in R^* \text{ and } nt\in Z^+\},\\ H_2 &= \sup_{s\in I,\, t\in R^*} [\xi^*(s,t)-G_2(s,t)], \\ H_1 &= -\inf_{s\in I,\, t\in R^*} [\xi^*(s,t)-G_1(s,t)]. \end{aligned} \tag{3.25}$$

The random function $\bar{\xi}(s,t)$ in (3.26) below can be any function defined on the same probability space as ξ^*, although in proving Corollary 1 we take $\bar{\xi}(s,t)$ to be our embedding of $n^{1/2}\,t[S_{nt}(s)-s]$.

Lemma 5. *Suppose* $P\{0\leqq H_i<\delta+\delta'\}<C(\delta+\delta')$ *for* $i=1$ *or* 2, *that* R_n^* *is non-empty, and that* $|G_i(s,t_1)-G_i(s,t_2)|<C'|t_1-t_2|^\alpha$ *whenever* $C'|t_1-t_2|^\alpha<\delta'$, *for* $s\in I$, $t_j\in R^*$. *Then*

$$\begin{aligned} 2n^{-1/2}\log n+C'n^{-\alpha}<\delta' \Rightarrow &|P\{G_1(s,t)<\xi^*(s,t)<G_2(s,t),(s,t)\in I\times R^*\}\\ &-P\{G_1(s,t)<\bar{\xi}(s,t)<G_2(s,t),(s,t)\in I\times R_n^*\}| \\ &<2C_5|R_n^*|\,n^{-3}+2C(\delta+\delta')+P\{\sup_{I\times R_n^*}|\xi^*(s,t)-\bar{\xi}(s,t)|\geqq\delta\}. \end{aligned}\tag{3.26}$$

Proof. A quarter of the demonstration is that, if $|\xi^*(s,t)-\bar{\xi}(s,t)|<\delta$ on $I\times R_n^*$ and $\bar{\xi}(s,t)<G_2(s,t)$ on that set, then $\xi^*(s,t)<G_2(s,t)+\delta$ thereon and hence (with $\xi^*(s,t)$ here $=n^{-1/2}(1-s)\,\xi(s(1-s)^{-1},nt)$ of (3.23)), except on a set of probability $2C_5|R_n^*|\,n^{-3}$, we have $\xi^*(s,t)<G_2(s,t)+\delta+2n^{-1/2}\log n+C'n^{-\alpha}$ on $I\times R^*$; consequently, $\xi^*(s,t)$ can be $\geqq G_2(s,t)$ somewhere on $I\times R^*$ with probability at most $C(\delta+\delta')$. For $\xi^*(s,t)<G_2(s,t)\leqq\bar{\xi}(s,t)$ on R_n^* the only contribution is from the last term of (3.26).

What remains to be verified in order that Theorem 2 and Lemma 5 can be applied in particular cases is of course the boundedness near 0 of the densities of the H_i. This has been proved in particular cases by Skorohod (for I replaced by a point and $R^*=(0,A]$) and by Müller (for $R^*=[1,\infty)$, essentially with $b_\infty\geqq 1/2$ in (3.34) below while assuming no Lipschitz condition but certain other restrictions described in Remark (1) on Lemma 6). A simple condition and proof (more elementary than Müller's) are given in

Lemma 6. *Suppose the* G_i *are continuous with* $|G_i(s,t)|<G(t)$ *and* $|G_i(s,t)-G_i(s',t)|<\bar{G}(t)\,|s-s'|$ *for* $s,s'\in I$ *and* $t\in R^*$. *Also (if* $0\in\bar{R^*}$*) suppose* $\inf\{t\colon \xi^*(s,t)=G_i(s,t)$ *for some* i *and* $s\}>0$ wp 1. *Then, for* $\bar{\delta}>0$,

$$P\{0\leqq H_i<\bar{\delta}\}\leqq 6\bar{\delta}\inf_{t\in R^*}\{t^{-1/2}+2t^{-1}[G(t)+\bar{G}(t)]\}. \tag{3.27}$$

2*

Proof. We shall give the proof for H_2. Let

$$\begin{aligned}\gamma_1 &= \inf\{s: \sup_{t\in R^*}[\xi^*(s,t)-G_2(s,t)]=0\},\\ 1-\gamma_2 &= \inf\{s: \sup_{t\in R^*}[\xi^*(1-s,t)-G_2(1-s,t)]=0\}.\end{aligned}\tag{3.28}$$

The γ_i are defined on the same set Γ (say) of sample paths, on which $\gamma_1 \leqq \gamma_2$. On Γ, define

$$\tau_i = \inf_{t\in R^*}\{t: [\xi^*(\gamma_i,\tau_i)-G_2(\gamma_i,\tau_i)]=0\}. \tag{3.29}$$

Then, given Γ and $\gamma_1 = s_1 \leqq 1/2$, $\tau_1 = t_1 > 0$, the process $\{t_1^{-1/2}\xi^*(s,t_1),\ s_1 \leqq s \leqq 1\}$ has the same conditional law as $\{\xi^*(s,1),\ s_1 \leqq s < 1\}$ given that $\xi^*(s_1,1) = t_1^{-1/2} G_2(s_1,t_1)$. The obvious analogue holds for time variable $1-s$ given $\gamma_2 = s_2 \geqq 1/2$, $\tau_2 = t_2 > 0$. Since always $\gamma_1 \leqq 1/2$ or $\gamma_2 \geqq 1/2$ on Γ,

$$\begin{aligned}P\{0\leqq H_2<\bar\delta\} &\leqq P\{0\leqq H_2<\bar\delta\,|\,\Gamma\}\\ &\leqq P\{0\leqq H_2<\bar\delta,\gamma_1\leqq 1/2\,|\,\Gamma\}+P\{0\leqq H_2<\bar\delta,\gamma_2\geqq 1/2\,|\,\Gamma\}.\end{aligned}\tag{3.30}$$

We treat in detail only the first term on the right side of (3.30), which is no greater than the infimum over $s_1 \leqq 1/2$ and t_1 of

$$P\{0\leqq \sup_{s_1\leqq s\leqq 2/3}[\xi^*(s,t_1)-G_2(s,t_1)]<\bar\delta\,|\,\Gamma,\gamma_1=s_1,\tau_1=t_1\}. \tag{3.31}$$

Writing, as earlier, $(z+1)\,\xi^*(s,t)=\xi(z,t)$ and $z_1 = s_1(1-s_1)^{-1}$, and using

$$|G_2(s,t)-G_2(s_1,t)| < \bar G(t)\,|z-z_1|/(1+z)(1+z_1),$$

we have

$$\begin{aligned}\xi^*(s,t_1)-G_2(s,t_1) &= (z+1)^{-1}\{\xi(z,t_1)-(z-z_1)\,G_2(s,t_1)\\ &\qquad -(z_1+1)[G_2(s,t_1)-G_2(s_1,t_1)]-\xi(z_1,t_1)\}\\ &\geqq (z+1)^{-1}\{\xi(z,t_1)-\xi(z_1,t_1)-(z-z_1)[G(t)+\bar G(t)]\}.\end{aligned}\tag{3.32}$$

Substituting $z_1 \leqq 1$, $z \leqq 2$, we see that the probability of (3.31) is no greater than

$$\begin{aligned}P\{&\sup_{0\leqq z-z_1\leqq 1} t_1^{-1/2}\{\xi(z,t_1)-\xi(z_1,t_1)-(z-z_1)[G(t)+\bar G(t)]\}<3\bar\delta t_1^{-1/2}\}\\ &\leqq 3\bar\delta t_1^{-1/2}\{1+2t_1^{-1/2}[G(t)+\bar G(t)]\},\end{aligned}\tag{3.33}$$

the last by a Brownian motion estimate (e.g., [20], p. 173).

Remarks on Lemma 6. (1) The dependence on s can be weakened, both in allowing a weaker Lipschitz condition and also in allowing G_i and its differences to vanish, but sufficiently slowly, near $s=0$ or 1 (in particular, Müller's assumption that $\liminf_{s(1-s)\downarrow 0} G_i > 0$ is unnecessary); we forego the altered statement of Lemma 6 and Corollary 1. (2) Somewhat different conditions can be given using the "first t" in place of (3.28), but the present conditions and proof seem, on the whole, more expeditious. (3) If also $\inf_{I\times R^*} G_2 > \bar\delta > 0$, the analogous result is obtained for $P\{-\bar\delta < H_2 \leqq 0\}$, from the above with G_2 replaced by $G_2 - \bar\delta$; this would be needed only for the somewhat artificial combination (in the paragraph following (3.24)) of continuous t for S_t and discrete t for ξ^*.

Assumptions and Proof of Corollary 1. We adopt the following notation to describe the behavior of the G_i:

For positive constants $C_0, L, a_\infty, b_\infty, b_0$, and real a_0,

$$|G_i(s,t)|+|s-s'|^{-1}|G_i(s,t)-G_i(s',t)|<\begin{cases} C_0\, t^{-a_0}, & t\leqq 1,\\ C_0\, t^{a_\infty}, & t\geqq 1;\end{cases}$$

$$|G_i(s,t_1)-G_i(s,t_2)|<C_0(t_1^L+t_1^{-L})\,|t_2-t_1|^\alpha \tag{3.34}$$

for $0<t_2-t_1<1$ and some $\alpha>1/6$;

$$(-1)^i\, G_i(s,t)>\begin{cases} C_0^{-1}\, t^{1/2-b_0}, & t\to 0,\\ C_0^{-1}\, t^{1/2+b_\infty}, & t\to\infty;\end{cases}$$

$$\gamma=2/3+1/6\,b_\infty+\max\left[(1+a_0)/2b_0,\ 1/4b_\infty,\ (a_\infty-1)/2b_\infty\right],$$

where the terms with subscript 0 (resp., ∞) are omitted if $0\notin R^*$ (resp., if R^* is bounded).

The last assumption on the G_i implies (from standard estimates for $\sup_{s\in I,\,0\leqq t\leqq T}[\xi^*(s,t)$ or $n^{1/2}\,t|S_t(s)-s|]$, as in [6, 8]), that for some $c'>0$

$$P\{G_1(s,t)<\xi^*(s,t),\ n^{1/2}\,t[S_{nt}(s)-s]<G_2(s,t) \text{ for } s\in I \text{ and } t<c'(\log n)^{-1/2b_0} \text{ or } t>c'(\log n)^{1/2b_\infty}\}>1-n^{-1}. \tag{3.35}$$

It suffices, therefore, to prove (1.6) for $R^*\cap[c'(\log n)^{-1/2b_0},\ c'(\log n)^{1/2b_\infty}]$. Consequently, the coefficient of $\bar\delta$ in (3.27) (and hence the C of Lemma 5) is of order $(\log n)^\gamma$ where $\gamma=\max[1/4b_0,\ (1+a_0)/2b_0,\ 1/4b_\infty,\ (a_\infty-1)/2b_\infty]$. Note that $1/2-b_0\geqq -a_0$, so that $(1+a_0)/2b_0\geqq 1/4b_0$. The second assumption of (3.34) allows C' to be taken to be a power of $\log n$ in Lemma 5, and thus, since $\alpha>1/6$, δ' can be taken as $n^{-1/6}$. From Theorem 2 with $\beta=1/2b_\infty$, we thus obtain that the right side of (3.26) (with $\delta=c^*n^{-1/6}(\log n)^{(\beta+2)/3}$) is of order

$$\delta C=n^{-1/6}(\log n)^{\gamma+(\beta+2)/3},$$

which yields (1.6).

Remark. The modifications for dependence in s (near 0 and 1) and bounds of orders other than monomials in (3.34), are straightforward. The third condition, where we have used positive powers b_0 and b_∞ rather than suitable slowly varying functions, is but the simplest condition to yield a result like (3.35) and avoid the trivial case where the second probability of (1.6) is 1; in the latter degenerate case one can also obtain estimates like (1.6). Of course, $a_\infty\geqq 1/2+b_\infty$.

Proof of Corollary 2. (Of course, the $|s-s'|$ term in (3.34) can be omitted.) B_n is now fixed in (3.22), and the proof of Lemma 4′ obviously holds even though the p_j need not be uniformly spaced in I'. The analogue of Lemma 5 is valid, and Lemma 6 can be replaced by an analogous result following the lines of Skorohod's corresponding result for I' a point. The remainder of the proof of (1.7) is as for

Corollary 1, and (1.8) follows from the fact that probabilities corresponding to the n^{-1} of (3.35) can be made n^{-2} by merely multiplying the deviations considered in Lemma 1′, 4′, etc. by suitable constants.

4. Proof of Theorem 1

We shall give details of the features not present in the proof of Theorem 2, but will merely sketch the differences when parts of the argument are similar to those of the proofs of Lemma 1′ and 4′, etc.

We must first alter the embedding by letting B_n and the z_j vary with the n-th observation. (The use of Section 3, with n varying, would only yield the existence of $\{T_n(s,t), n\in Z^+\}$ so that (1.5) held with T_n for T there whenever $t\leqq n$.) We shall make a single choice of the parameters of the embedding from the outset, to yield the best order obtainable in (1.5) from the analogues of Lemmas 1′ and 4′. As in Section 3, we shall save space by not distinguishing notationally between large real values and their integral parts.

We shall keep even less track of constants than in Section 3, and will denote by $\Omega(1)$ any positive function (whose domain will depend on the context) which is bounded away from 0 and ∞ and whose definition may change from one usage to the next. $O(1)$ is used similarly.

Define

$$\begin{aligned} M_r &= r8^r \quad \text{for } r\geqq 0,\\ B_n &= 2^r \quad \text{for } M_{r-1}<n\leqq M_r, \end{aligned} \tag{4.1}$$

so that

$$B_n=\Omega(1)\,(n/\log n)^{1/3}. \tag{4.2}$$

Whenever we are using 2^r subdivision of I, they are of equal length, and the corresponding z_j, here termed $z_j^{(r)}$, are given by (3.2) with $B_n=2^r$.

For $0<n\leqq M_1$ we use the embedding of Section 2 at the single value $z_1^{(1)}=1\,(p_1=1/2)$, and denote the embedding by $\eta(0,z_1^{(1)};\,U_{[1,1]}^{(n)})$; the subscript $[r;j]$ abbreviates $(z_{j-1}^{(r)},z_j^{(r)})$. Since the "last $z_j^{(r)}$", which by (4.1) is $B_n-1=2^r-1$, is going to vary with n, it is convenient (although it is not strictly necessary), here and below, to adjoin $\{\eta(z',z'';n),\ z',z''\geqq B_n-1\}$ to the $\{\eta(z_{j-1}^{(r)},z_j^{(r)};\,W_{[r,j]}^{(n)}),\ j<B_n\}$ (with $W=U$ above and as defined generally below), so as to have the expected value of the last argument of η be n regardless of the first two arguments. We call $A=\{(p,n)\colon p>1-B_n^{-1}\}$ the "adjoined region". We then define, for $n\leqq M_1$,

$$n\tilde{Q}_n(z)=\begin{cases}\eta(0,z;\,U_{[1,1]}^{(n)}) & \text{for } 0<z\leqq z_1^{(1)},\\ \eta(0,z_1^{(1)};\,U_{[1,1]}^{(n)})+\eta(z_1^{(1)},z;\,n) & \text{for } z\geqq z_1^{(1)}.\end{cases} \tag{4.3}$$

(An alternative to adjoining for $z\geqq B_n-1$ as above is in fact to define $Q_n(z)=Q_n(B_n-1)$ for $z\geqq B_n-1$. This means deleting the rectangle $[1,3]\times[0,M_1]$ in Fig. 2 and lowering what is above it. A subsequent modification five paragraphs below must then be altered accordingly.)

We write $j'=b$ if $j=2b$ or $2b-1$. This means the two intervals $[z_{2j'-2}^{(r)},z_{2j'-1}^{(r)}]$ and $[z_{2j'-1}^{(r)},z_{2j'}^{(r)}]$ make up $[z_{j'-1}^{(r-1)},z_{j'}^{(r-1)}]$.

We continue as follows: Suppose $\tilde{Q}_n$ is defined for $n\leqq M_{r-1}$. Then, for $M_{r-1}<n\leqq M_r$, we define $n\tilde{Q}_n-M_{r-1}\tilde{Q}_{M_{r-1}}$ at the 2^r-1 arguments $z_j^{(r)}$ by using

stopping times $W_{[r,j]}^{(n)} - W_{[r-1,j']}^{(M_{r-1})}$ on $\{\eta(z_{j-1}^{(r)}, z_j^{(r)}; t + W_{[r-1,j']}^{(M_{r-1})}), t \geq 0\}$ in the same way $(z_j^{(r)} - z_{j-1}^{(r)})\, U_{[r,j]}^{(n-M_{r-1})}$ was used to obtain $(n-M_{r-1})\, Q_{n-M_{r-1}}$ from the $\eta(z_{j-1}, z_j; t)$ for $B = 2^r - 1$ in Section 3. Our "adjoining" described above (4.3) makes $W_{[r-1,(2^r-1)']}^{(M_{r-1})} \equiv M_{r-1}$ for all r. Finally, we write, for $M_{r-1} < n \leq M_r$ and $z_{j-1} \leq z \leq z_j$,

$$n\tilde{Q}_n(z) - M_{r-1}\tilde{Q}_{M_{r-1}}(z) = \sum_{\alpha=1}^{j-1} [\eta(z_{\alpha-1}^{(r)}, z_\alpha^{(r)}; W_{[r;\alpha]}^{(n)}) - \eta(z_{\alpha-1}^{(r)}, z_\alpha^{(r)}; W_{[r-1,\alpha']}^{(M_{r-1})})] + [\eta(z_{j-1}^{(r)}, z; W_{[r,j]}^{(n)}) - \eta(z_{j-1}^{(r)}, z_j^{(r)}; W_{[r-1,\alpha']}^{(M_{r-1})})], \tag{4.4}$$

with the last term in square brackets replaced by

$$[\eta(z_{2^r-1}^{(r)}, z; n) - \eta(z_{2^r-1}^{(r)}, z; M_{r-1})] \quad \text{if} \quad z > z_{2^r-1} = 2^r - 1.$$

Here is a diagram of $\tilde{Q}_{M_1+1}(2)$, in Fig. 2: it is ξ (shaded area).

Note especially: as r increases, bases of rectangles are piled on top of broader, lower ones; the region $[1,3] \times [0, M_1]$ comes from the "adjoined" definition which makes the *expected* height of the shaded part equal $M_1 + 1$ for each z; the value 2 is not a $z_j^{(2)}$, and so we must split $[1,3] \times [M_1, W_{[2,3]}^{(M_1+1)}]$ vertically although this rectangle was determined by stopping the process with argument $[1,3] \times [M_1, M_1 + t]$ by moving up its horizontal upper boundary. This last is a technical convenience in defining $\tilde{Q}$, but will necessitate proving in (4.8) that $\tilde{Q}$ is suitably close to a corresponding interpolated process analogous to that defined in (1.12). Thus, the full $\{n[(z+1)\, S_n(z/(z+1)) - z], z \leq 1, n \in Z^+\}$ (which can be modified in an obvious way from $\{z \leq 1\}$ to $\{z \leq z^{R_0} - 1\}$ for any fixed R_0) still sits in ξ in the above picture, but it is given in terms of the horizontal cuts of Fig. 1 rather than the vertical cuts used in $\tilde{Q}_n$, of "whole rectangles" like $[0,1] \times [0, W_{[1,1]}^{(1)}]$.

Theorem 1 requires this embedding for $\{z < \infty\}$, not just $\{z \leq \text{large constant}\}$. The additions we must make to achieve this turn out to be trivial, since it will turn out that we do not need to estimate *differences* between ξ and the embedded $n S_n(p) - M_{r-1} S_{M_{r-1}}(p)$ for $p > 1 - B_n^{-1}$ (when $M_{r-1} < n \leq M_r$) in obtaining our estimates; this will be seen explicitly in connection with the discussion of (4.16). Consequently, our embedding can be completed in $A = \{(p, n): p > 1 - B_n^{-1}\}$ in any convenient way that exhibits the presence of $S_n(p)$. One possibility is that men-

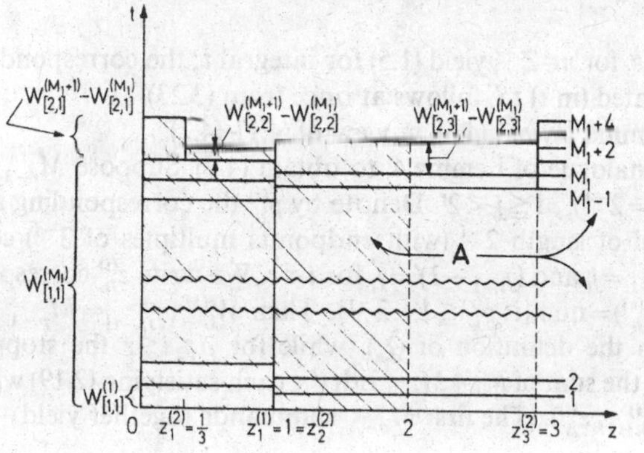

Fig. 2. Example of $Q_{M_1+1}(2) = \xi$ (shaded area)

tioned in connection with Fig. 1, of adjoining any sequence of unbounded $z_j^{(r)} \geqq B_n$. A simpler alternative is to use the standard process $\{\eta(2^r-1, 2^r-1+t; n) - \eta(2^r-1, 2^r-1+t; n-1), t \geqq 0\}$ already exhibited in the "adjoined" region A of our definition of $\tilde{Q}_n$, in the manner the ξ_i' were used in Section 2.1, stopping this process the first time the resulting sample df of the single n-th observation reaches 1. The deficiencies noted for this scheme in Section 2 are now irrelevant for the reason stated earlier in this paragraph.

We define $\tilde{\xi}(s, n) = (1-s)\, n\, \tilde{Q}_n(s/(1-s))$ (with $\tilde{\xi}(1, n) = 0$), including region A in the definition. Recall that $\tilde{\xi}(s, n) - \tilde{\xi}(s, M_{i-1})$ gives an exact embedding of $[n\, S_n(s) - M_{i-1}\, S_{M_{i-1}}(s) - (n - M_{i-1})\, s]$ *for* $s =$ *integral multiples of* 2^{-i}, if $M_{i-1} < n \leqq M_i$. For $M_{i-1} < n \leqq M_i$ define

$$\xi^*(s, n) - \xi^*(s, M_{i-1}) = \begin{cases} \tilde{\xi}(s, n) - \tilde{\xi}(s, M_{i-1}) & \text{if } \; s = 2^{-i} j, j \text{ integral}, \\ \text{linearly interpolated from the above if} & \\ \qquad 2^{-i} j < s < 2^{-i}(j+1). & \end{cases} \tag{4.5}$$

Thus, if $M_{r-1} < n \leqq M_r$, the rv $\xi^*(2^{-r} j, n)$ is a sum of r independent rv's, one with exactly the distribution of $[n S_n(2^{-r} j) - M_{r-1} S_{M_{r-1}}(2^{-r} j) - (n - M_{r-1})\, 2^{-r} j]$, and the i-th of the others with the distribution of the rv linearly interpolated from $[M_i S_{M_i}(p) - M_{i-1} S_{M_{i-1}}(p) - (M_i - M_{i-1})\, p]$ at the values of $p = 2^{-i} \bar{j}$ closest to $2^{-r} j$. Finally, in conformity with (1.3), we define $\{\hat{\xi}(s, n)\}$ to be the embedded process distributed *exactly* as $\{n S_n(s) - n s, s \in I, n \in Z^+\}$; this is obtained from the horizontal cuts described at the end of the second preceding paragraph for the i-th observation when $s/(1-s) \leqq B_i - 1$, and from any exact representation used in the manner of the paragraph just above, for $s/(1-s) > B_i - 1$ (region A).

Our proof is divided into four parts:

$$P\{\sup_{s \in I} |\xi^*(s, n) - \xi^*_{B_n}(s, n)| = O(n^{1/3} (\log n)^{2/3})\} > 1 - n^{-2}, \tag{4.6}$$

$$P\{\max_{0 < j < B_n} |\xi^*(B_n^{-1} j, n) - \tilde{\xi}(B_n^{-1} j, n)| = O(n^{1/3} (\log n)^{2/3})\} > 1 - n^{-2}, \tag{4.7}$$

$$P\{\max_{0 < j < B_n} |\hat{\xi}(B_n^{-1} j, n) - \xi^*(B_n^{-1} j, n)| = O(n^{1/3} (\log n)^{2/3})\} > 1 - n^{-2}, \tag{4.8}$$

$$P\{\sup_{s \in I} |\xi^*(s, n) - \hat{\xi}(s, n)| = O(n^{1/3} (\log n)^{2/3})\} > 1 - n^{-2}. \tag{4.9}$$

These four equations, for $n \in Z^+$, yield (1.5) for integral t; the corresponding result for linearly interpolated (in t) $t S_t$ follows at once from (3.23).

Remark 4 on Lemma 1 yield (4.6), in view of (4.1)–(4.2).

We require an analogue of Lemma 4, to obtain (4.7). Suppose $M_{r-1} < n \leqq M_r$. Temporarily fix $p_{j_r}^{(r)} = 2^{-r} j_r$, $1 \leqq j_r < 2^r$. Denote by $p_{j_i}^{(i)}$ the corresponding right endpoint of the interval of length 2^{-i} (with endpoints multiples of 2^{-i}) containing $[p_{j_r-1}^{(r)}, p_{j_r}^{(r)}]$; thus, $j'_{i+1} = j_i$ and $(j_{i+1} - 1)' = j_i$ for $i < r$. We write $z_{j_i}^{(i)}$ correspondingly. Also, define $\rho \equiv \rho(p_{j_r}^{(r)}) = \min\{i : p_{j_r}^{(r)} \leqq 1 - 2^{-i}\}$. Then $W_{[\rho-1, j_\rho - 1]}^{(M_{\rho-1})} = M_{\rho-1}$ (the "adjoined" region A in the definition of $\tilde{Q}_n$), while for $\rho \leqq i \leqq r$ the stopping time $W_{[i, j_i]}^{(M_i)} - W_{[i-1, j_i - 1]}^{(M_{i-1})}$ is the sum of $M_i - M_{i-1}$ iidrv's, each satisfying (2.19) with (z_1, z_2) there replaced by $(z_{j_i-1}^{(i)}, z_{j_i}^{(i)})$. (The first $M_{\rho-1}$ summands together yield

$$\log E \exp\{\alpha(W^{(M_{\rho-1})} - M_{\rho-1})\} = 0$$

in place of (2.19).) All of these intervals $[z^{(i)}_{j_i-1}, z^{(i)}_{j_i}]$ are contained in $[z^{(\rho)}_{j_\rho-1}, z^{(\rho)}_{j_\rho}]$, from which it follows that, in the expression h of (2.19)–(2.20), all factors $(1+z^{(i)}_{j_i})$ (replacing $(1+z_2)$ there) are $\Omega(1)(1+z^{(r)}_{j_r})$ for $\rho \leqq i \leqq r$. (The $\Omega(1)$ here is independent of n, r, j_r.) Also, $z_{j_i}-z_{j_i-1}=\Omega(1)(1+z_{j_i})^2 2^{-i}$. Hence, from (2.20), we have as analogue of (2.19) for $M_{i-1}<v\leqq M_i$, in terms of stopping times with unit expectation,

$$\begin{aligned}\log E \exp\{\alpha[W^{(v)}_{[i,j_i]}-W^{(v-1)}_{[i,j_i]}-1]\} &\sim \alpha^2 (z^{(i)}_{j_i}-z^{(i)}_{j_i-1})^{-2} h(z^{(i)}_{j_i-1}, z^{(i)}_{j_i}) \\ &= \alpha^2 \Omega(1) 2^i \leqq \alpha^2 \Omega(1) B_n . \end{aligned} \tag{4.10}$$

A sum $\overline{W}^{(N)}$ (say) of N such terms (some replaced by 0 if $i<\rho$) yields, in place of (2.21), with $\alpha = q/2\Omega(1)NB_n$,

$$\begin{aligned} P\{\overline{W}^{(N)}>q\} &\leqq E\{\exp[\alpha \overline{W}^{(N)}-q]\} \\ &\leqq \exp\{N\alpha^2 \Omega(1)B_n-\alpha q\} \\ &= \exp\{-q^2/4\Omega(1)NB_n\}. \end{aligned} \tag{4.11}$$

This last is $<n^{-3}$ if $q=\Omega(1)(NB_n \log n)^{1/2}$; this and the corresponding result for $\{W^{(N)}<q\}$ make up the analogue of (2.24)–(2.25) (with q here for $(z_j-z_{j-1})^{-1} q_N$ there), and the condition of the first line of (2.25), inherited from those following (2.19), is again seen to be that $N^{-1} B_n \log n <$ some small positive value.

We use the above to prove (4.7) by going through the proof of Lemma 4 with $W^{(n_1+\cdots+n_i)}_{[r,j_r]}$ in place of the $V^{(i)}_j$ of (3.9). We use (3.3)–(3.8) as before (and in view of (4.1) we have $K \sim (\log\log n)/\log 2$). The analogue of (3.12) is valid, since the proof of (3.11) depends only on (3.5). We replace (z_j-z_{j-1}) in (3.13) by $(z^{(r)}_{j_r}-z^{(r)}_{j_r-1})$ and now sum on j_r to obtain the obvious counterparts of (3.14), (3.15), (3.16). These and the analogues of (3.17)–(3.20), with the coefficients 2 and 2^2 altered slightly, yield the analogue of (3.21) with the probability changed to $1-n^{-2}$. Thus, (4.7) follows from the fact that

$$(nB_n)^{1/4}(\log n)^{3/4}+B_n \log n = 2n^{1/3}(\log n)^{2/3}. \tag{4.12}$$

We turn to (4.8). We define $\rho(p)$ just as above (4.10). Still supposing $M_{r-1}<n\leqq M_r$, fix $p^{(r)}_{j_r}$ and abbreviate $(p^{(i)}_{j_i-1}, p^{(r)}_{j_r}, p^{(i)}_{j_i})$ by $(p', \bar{p}, p'')$. We first look at that portion Ξ (say) of $\tilde{\xi}(\bar{p}, n)-\xi^*(\bar{p}, n)$ that comes only from values $t \geqq M_{\rho(p)}$ for each p, $0<p\leqq \bar{p}$. This is the portion from those (s, t) of $I \times R^+$ outside (above and to the left of) the "adjoined" region A. Since, moreover, the (noninterpolated) contributions from values $p \leqq p'$ cancel in the difference Ξ, we are left with

$$\begin{aligned} \Xi = &\{[\tilde{\xi}(\bar{p}, n)-\tilde{\xi}(\bar{p}, M_{\bar{\rho}-1})]-[\tilde{\xi}(p', n)-\tilde{\xi}(p', M_{\bar{\rho}-1})]\} \\ &-\{[\xi^*(\bar{p}, n)-\xi^*(\bar{p}, M_{\bar{\rho}-1})]-[\xi^*(p', n)-\xi^*(p', M_{\bar{\rho}-1})]\}, \end{aligned} \tag{4.13}$$

where $\bar{\rho}=\rho(\bar{p})$. This difference Ξ may be thought of as a sum of $M_{r-1}-M_{\bar{\rho}-1}$ rv's, corresponding to the $(M_{\bar{\rho}-1}+1)$-th to M_{r-1}-th observation; there is no interpolation, linear or by vertical cuts, from the $(M_{r-1}+1)$-th to n-th observation. For each i, $\bar{\rho} \leqq i \leqq r-1$, the M_i-M_{i-1} rv's corresponding to the observations in that i-th group are iid, and each has a distribution which is the same as that of $\xi^{***}(\pi_i, W^*)$ of (4.14) (below), which we now describe in terms of the notation of Section 2, translating time to begin at $t=0$ for notational ease: For a process ξ^{**} distributed like ξ^*, stop $\xi^{**}(p'',t)-\xi^{**}(p',t)$ at time $t=(z_{j_i}-z_{j_i-1})^{-1}\,T^{(1)}_{[i,j_i]}=W^*$ (say). The contribution of ξ^{**} from the time rectangle $[p',p'']\times[0,W^*]$ to the $\tilde{\xi}$ part of (4.13) is $\xi^{**}((1-\pi_i)\,p'+\pi_i\,p'',W^*)-\xi^{**}(p',W^*)$, where $\pi_i=(\bar{p}-p')/(p''-p')$ is the interpolating proportion. Similarly, the contribution to the $\xi^{\#}$ part of (4.13) is $(1-\pi_i)\,\xi^{**}(p',W^*)+\pi_i\,\xi^{**}(p'',W^*)-\xi^{**}(p',W^*)$. The difference between these two contributions can be written as $\xi^{***}(\pi_i,W^*)$, where

$$\begin{aligned}\xi^{***}(\pi,W^*)=\xi^{**}&((1-\pi)\,p'+\pi_i\,p'',W^*)\\&-(1-\pi)\,\xi^{**}(p',W^*)-\pi\,\xi^{**}(p'',W^*).\end{aligned}\tag{4.14}$$

Recall that W^* was defined in terms of the values of $\xi^{**}(p'',t)-\xi^{**}(p',t)$. A simple computation in terms of the Gaussian distribution (more easily done in terms of the corresponding independent increment z-process for which, however, one does not have the simple linear interpolation obtained by putting z', z'' for p', p'' in (4.14)) now shows that the conditional law of $\{\xi^{***}(\pi,w), 0\leqq\pi\leqq1\}$, given $W^*=w_0$ and even the whole path $\{\xi^*(p'',t)-\xi^{**}(p',t), t\leqq W^*\}$, is the same as the unconditional law of $\{\xi^{***}(\pi,w_0), 0\leqq\pi\leqq1\}$, namely, it is that of $(p''-p')^{1/2}\,w_0^{1/2}$ times a standard Brownian bridge. (This simple dependence on w_0 of course differs strikingly from the conditional distribution of $\tilde{\xi}$, obtained from the same horizontal cuts that defined W^*.) Consequently, given $W^{(M_i)}_{[i,j_i]}-W^{(M_i-1)}_{[i,j_i]}=w_i$, the total contribution to $\tilde{\xi}(\bar{p},n)-\xi^{\#}_{B_n}(\bar{p},n)$ from the $(M_{i-1}+1)$-th to M_i-th observation is that of $w_i^{1/2}\,2^{-i/2}\,\xi_i^{**}(\pi_i)$ where ξ_i^{**} is a standard Brownian bridge. If the ξ_i^{**}'s are taken to be independent in i, we thus have, finally, the representation (given the w_i as above for $i<r$)

$$\Xi=\sum_{i=\bar{p}}^{r-1} w_i^{1/2}\,2^{-i/2}\,\xi_i^{**}(\pi_i).\tag{4.15}$$

It is now simple to use exponential bounds in the manner of the proof of Lemma 1: $(W^{(M_i)}_{[i,j_i]}-W^{(M_i-1)}_{[i,j_i]})/(M_i-M_{i-1})$ is close to 1 for all large $i<r$, with high probability; with that probability, the rv's taking on the values w_i are such that

$$\sum_{i=\bar{p}}^{r-1} w_i\,2^{-i}=O(n^{2/3}(\log n)^{1/3});$$

also, $E\exp\alpha\,\xi_i^{**}(\pi_i)\leqq e^{\Omega(1)\alpha^2}$. We obtain that the r.v. Ξ of (4.13) and (4.15) is $O(n^{1/3}(\log n)^{2/3})$ for all $\bar{p}=p^{(r)}_{j_r}$ $(1\leqq j_r<B_n)$, with probability $>1-(2n)^{-2}$.

To complete the proof of (4.8) we shall show

$$\begin{aligned}P\{\sum_{p=2^{-r}j<\bar{p}}[\tilde{\xi}(p,M_{\rho(p)-1})-\tilde{\xi}(p-2^{-r},M_{\rho(p)-1})]=O(n^{1/3}(\log n)^{2/3})\\ \text{for } \bar{p}=2^{-r}\bar{j},\ 0<\bar{j}<2^r\}>1-(4n)^{-2}\end{aligned}\tag{4.16}$$

and the corresponding result for $\xi^{\#}$ replacing $\tilde{\xi}$ in (4.16). (This separate treatment of the $\tilde{\xi}$ and $\xi^{\#}$ parts confirms our earlier assertion that the method of embedding in the "adjoined" region A was irrelevant.) Recalling that $M^{-1/2}\,\tilde{\xi}(p,M)$ is a standard Brownian bridge in the adjoined region A, and that (p,M_i) and (p,M_{i-1}) are in that region if and only if $p>1-2^{-i}$, we see that the sum in (4.16) has variance no greater than

$$\sum_{i=1}^{r}(M_i-M_{i-1})2^{-i}(1-2^{-i})<\Omega(1)\,n^{2/3}(\log n)^{1/3}. \tag{4.17}$$

Moreover, if we denote the sum in (4.16) by $\eta_n(1-\bar{p})$, it is easy to see, as in [8] Lemma 2, that $P\{\sup_{0\leqq s\leqq 1/2}|\eta_n(s)|>\lambda\}\leqq O(1)\,P\{|\eta_n(1/2)|>\lambda/2\}$. Consequently, (4.17) and the usual normal exponential bound yield (4.16). The corresponding result for $\xi^{\#}$ is very similar; the crucial analogue of the abovementioned parallel of [8] now uses not just sample d.f.'s, but sums of random functions *obtained from sample df's linearly interpolated in s at various different spacings.* In the next paragraph we compute bounds of the required type for such sums which arise in treating the more difficult region complementary to A, and the result for $\xi^{\#}$ in (4.16) follows similarly.

We thus turn finally to (4.9). The discussion of the previous paragraph enables us to limit discussion to the complement of A; that is, corresponding to (4.15), to sums of random functions

$$H(p)=\sum_{i=\rho(p)}^{r-1}(M_i-M_{i-1})[S_{(i)}(p)-S_{(i)}^{\#}(p)],\qquad p_{j_r-1}^{(r)}\leqq p\leqq p_{j_r}^{(r)}, \tag{4.18}$$

where the $S_{(i)}$ are independent sample d.f.'s, $S_{(i)}$ based on M_i-M_{i-1} observations, and $S_i^{\#}$ is the piecewise linear function interpolated from values $S_{(i)}(2^{-i}j)$, j integral. As in (4.13), it is only the contribution from $p_{j_i}^{(i)}$ to $\bar{p}$ that doesn't cancel out in the i-th summand of (4.18). Suppose we set $\varepsilon^{-1}=2^{r+4}$ and can show that, for $\lambda\geqq 8$,

$$\begin{aligned}P\{&\sup_{0\leqq\pi\leqq 1}|H((1-\pi)\,p_{j_r-1}^{(r)}+\pi\,p_{j_r}^{(r)})|\geqq\lambda\}\\&=O(1)\,P\{\max_{0\leqq j\leqq\varepsilon^{-1}}|H(1-j\,\varepsilon)\,p_{j_r-1}^{(r)}+j\,\varepsilon\,p_{j_r}^{(r)}|\geqq\lambda/2\},\end{aligned} \tag{4.19}$$

with $O(1)$ independent of r. Then, as in [8] and [12a], it is easily seen that the right side of (4.19) can be estimated by exponential bounds of the type we have used repeatedly, to yield (4.9). We suppose the first p in $[p_{j_r-1}^{(r)},p_{j_r}^{(r)}]$ where $|H(p)|\geqq\lambda$, say $p=\bar{p}_0$, is in the left half of this interval; time reversal handles the other half. If $\bar{p}_0+\bar{\varepsilon}_0$ is the least of the ε^{-1} possible arguments of H on the right side of (4.19) for which $\bar{p}_0+\bar{\varepsilon}_0\geqq\bar{p}_0$, we will show

$$P\{|H(p_0+\varepsilon_0)|\geqq\lambda/2\,|\,\bar{p}_0=p_0,\,\bar{\varepsilon}_0=\varepsilon_0,\,|H(p_0)|=\lambda_0>\lambda\}=\Omega(1), \tag{4.20}$$

which yields (4.19).

Let u_i be the number of observations in $[p_{j_i-1}^{(i)},p_{j_i}^{(i)}]$ among the $(M_{i-1}+1)$-th to M_i-th observation, and let v_i be the number of these in $[p_{j_r-1}^{(r)},p_{j_r}^{(r)}]$. If $p_0=p_{j_r-1}^{(r)}$, (4.20) is trivial; so we assume $H(p_{j_r-1}^{(r)})=h_0$ (say) with $|h_0|<\lambda$. We also treat explicitly only the case $H(p_0)=\lambda_0$ of positive deviations. We obviously have $\sum_i(M_i-M_{i-1})\,[S_i^{\#}(p_0+\varepsilon_0)-S_i^{\#}(p_0)]=\sum_i\varepsilon_0\,2^{-i}\mu_i$. Since also $\lambda_0\geqq\lambda$, we obtain the desired $H(p_0+\varepsilon_0)\geqq\lambda/2$ provided that

$$\sum_i(M_i-M_{i-1})\,[S_i(p_0+\varepsilon_0)-S_i(p_0)]\geqq\sum_i\varepsilon_0\,2^{-i}\mu_i\quad\lambda/2. \tag{4.21}$$

The total number of observations in $[p_0, p_{j_r}^{(r)}]$ from the $(M_\rho+1)$-th to M_{r-1}-th observation (where $\rho=\rho(p_{j_r}^{(r)})$) is $N_0=\sum_i v_i 2^{-r}(p_{j_r}^{(r)}-p_0)-(\lambda-h_0)$, and the probability that any specified one of these falls in $[p_0, p_0+\varepsilon_0]$ is $P_0=\varepsilon_0/(p_{j_r}^{(r)}-p_0)$. Thus, conditional on the v_i, $\bar{p}_0=p_0$, $\bar{\varepsilon}_0=\varepsilon_0$, and $H(p_0)=\lambda_0$, the left side of (4.21) is binomial with mean $N_0 P_0$, and $N_0 P_0$ exceeds the right side of (4.21) by

$$N_0 P_0-\sum_i \varepsilon_0 2^{-i}\mu_i+\lambda_0/2=\varepsilon_0\sum_i(2^{-r}v_i-2^{-i}\mu_i)+\left\{\frac{\lambda}{2}-\frac{(\lambda-h_0)\,\varepsilon_0}{p_{j_r}^{(r)}-p_0}\right\}. \tag{4.22}$$

Since $\lambda-h_0<2\lambda$, $\varepsilon_0\leqq 2^{-r-4}$, and $p_{j_r}^{(r)}-p_0\geqq 2^{-r-1}$, the term in braces in (4.22) is $>\lambda/4$. The unconditional probability that the sum in (4.22) (which has expectation 0) is $\geqq -1$ is easily seen to be bounded away from 0, uniformly in r and $p_{j_r}^{(r)}$, by using elementary estimates similar to those used in [12a]. (The latter shows that a binomial rv exceeds its mean by at least -1 with probability bounded away from 0.) Hence, (4.21) is established, as is thus (4.9).

Finally, $T(1,t)$ is defined arbitrarily in (1.4)–(1.5), e.g., as 0.

This completes the proof of Theorem 1.

Müller's method can also apparently be used to give a strong analogue of his weak convergence result.

The technique used in (4.18)–(4.22) to treat sums of non-iid processes has broader usage, whose statement we forego.

In view of our comment in Section 1 that (1.5) is a statement about an imperfect method rather than anything intrinsic about $\hat{\xi}$ and ξ^*, we forego any lower class considerations.

5. Breiman-Brillinger Brownian Bridge

In the second paragraph of Section 1 we mentioned Breiman's representation in terms of iidrv's Y_i with df $1-e^{(y+1)^+}$. In view of the shortcomings we have described for this representation, it will not be worth while to do more than sketch the results alluded to in [12], although some elements of the proofs may warrant mention. Also, as discussed in the next section, certain results about Breiman's scheme, used with Müller's approach, could possibly improve on some of our results; unfortunately, those results will not be found here.

Let $Z_0=0$, $Z_n=\sum_1^n Y_i$, and $Z'_n=Z_n+n$. From our remarks about $\{Z'_i/Z'_{n+1}, 1\leqq i\leqq n\}$ in Section 1, the random function

$$G_n(t)=\begin{cases} i/(n+1) & \text{if } \ t=Z'_i/Z'_{n+1}, \quad 0\leqq i\leqq n+1, \\ \text{linear for } & t\in[Z'_i/Z'_{n+1}, Z'_{i+1}/Z'_{n+1}], \quad 0\leqq i\leqq n, \end{cases} \tag{5.1}$$

is distributed as the continuous strictly increasing (wp 1) piecewise linear version of the sample df for uniform rv's; it will be obvious that the conclusions contained herein are true for the common discontinuous versions, but the use of the invertible G_n of (5.1) simplifies notation and arithmetic.

We hereafter write $m=n+1$.

Let $\{\xi_1(t), t \geqq 0\}$ be a standard Brownian motion defined on a probability space on which are also defined nonnegative iidrv's $\{T_i\}$ which yield a Skorohod embedding of $\{Z_n\}$ in ξ_1; that is, if $U_n = \sum_1^n T_i (U_0 = 0)$, then $\{\xi_1(U_n), n \geqq 0\}$ is distributed as $\{Z_n\}$. Here $ET_1 = EZ_1^2 = 1$. A number of different embeddings are available, as mentioned in Section 6, with various finite values of $\beta = \operatorname{var}(T_1)$ (see Section 6D). We assume such an embedding and corresponding β are chosen, and hereafter identify Z_n with $\xi_1(U_n)$. We also define Z_t to be linear in $t \in [n, n+1]$, and write $Z'_t = Z_t + t$. Also define $r(t) = Z_t - \xi_1(t)$. As usual, the development can be stated in terms of continuous t or discrete n, and we shall not distinguish (e.g., in the subscript of $U_{n\alpha}$) between a real t and its integral part. From (5.1) and the LIL for Z_m, we have, uniformly for $0 \leqq t \leqq 1$ as $m \to \infty$, wp 1,

$$\begin{aligned} m^{1/2}[G_n^{-1}(t) - t] &= m^{1/2}[Z'_{mt}/Z'_m - t] = m^{1/2}\left\{\frac{Z_{mt} - tZ_m}{Z_m + m}\right\} \\ &= m^{-1/2}(Z_{mt} - tZ_m) + O(n^{-1/2}(\log\log n)^{1/2}) \\ &= m^{-1/2}[\xi_1(mt) - t\,\xi_1(m)] + m^{-1/2}[r(mt) - t\,r(m)] \\ &\quad + O(n^{-1/2}(\log\log n)^{1/2}). \end{aligned} \tag{5.2}$$

It is known from [10] that

$$\limsup_{t \to \infty} \pm [2\beta\, t(\log t)^2 \log\log t]^{-1/4}\, r(t) = 1 \text{ wp } 1. \tag{5.3}$$

This would yield an estimate of the term $m^{-1/2}[r(mt) - t\,r(m)]$ in (5.2) for fixed t, but not the right constant; nor would it yield an estimate for the supremum over t of this term. However, slight modifications of the proof of (5.3) give the desired result, upon which we shall comment after a sketch of the proof. We apologize in advance for the sketch which follows, which can be made intelligent only by reading it with [10]; this results from our decision to document the assertions in as little additional space as possible[1].

Theorem 5. *For fixed* α, $0 < \alpha < 1$,

$$\begin{aligned} \limsup_{n \to \infty} \pm [r(\alpha n) - \alpha\, r(n)]/[2\beta\, n(\log n)^2 \log\log n]^{1/4} \\ = [\alpha^4 + 2\alpha^3 + \alpha]^{1/4} \text{ wp } 1. \end{aligned} \tag{5.4}$$

Moreover,

$$\limsup_{n \to \infty} \pm \sup_{0 \leqq \alpha \leqq 1} [r(\alpha n) - \alpha\, r(n)]/[2\beta\, n(\log n)^2 \log\log n]^{1/4} = 2 \text{ wp } 1. \tag{5.5}$$

Proof of (5.4). Mainly, one studies the variations in ξ_1 produced by both deviations $U_{\alpha n} - \alpha n$ and $U_n - n$, rather than by just the latter as in (5.3). (Similar considerations have just appeared in [6b]; see also Section 6A.) For the *upper class proof*, we use a LIL for linear combination of partical sums (which has an obvious extension to more than the two summands for which we state and use it):

[1] We simplify this discussion by using the same nongeometric n_r for upper class results, as in Theorem 1 of [10]. (In the lower class proofs here and in Theorem 2 of [10], geometric n_r are used.) However, it should be noted that geometric n_r can be used, both in [10] and the present paper, with corresponding changes in the values chosen on the bottom of p. 326 of [10] to apply Lemma 1 there. We also take this opportunity to apologize for the misprints in the Summary of [10], where β was erroneously defined as EX_1^4 instead of as $\operatorname{var}(T_1)$.

Lemma 7. *For real constants a_1, a_α and α, with $0<\alpha<1$,*

$$\limsup_{n\to\infty} [a_1(U_n - n) + a_\alpha(U_{n\alpha} - n\alpha)]/[2\beta n \log\log n]^{1/2} = [\alpha(a_1+a_\alpha)^2 + (1-\alpha)a_1^2]^{1/2} \text{ wp } 1. \tag{5.6}$$

This is proved in standard fashion, either by direct treatment of the rv in square brackets, or by approximating the event in question by a union of events which are quadrants in the space of $\{U_n - U_{n\alpha}, U_{n\alpha}\}$; a finite number of such events suffices by the marginal LIL's. For use in the upper class proof of Theorem 5, it is critical also to establish the analogues of the finer conditions used in (2.8) of [10].

We continue with the proof of Theorem 5. Using the $\{n_r\}$ of Theorem 1 of [10], if $U_{hn_r} - h n_r = u_h [2\beta n_r \log\log n_r]^{1/2}$ for $h=1, \alpha$, one computes the conditional probability of the event $\{|\xi_1(T_{hn}) - \xi_1(hn)| > (1+8\varepsilon) c_h [2\beta n_r (\log n_r)^2 \log\log n_r]^{1/4}$ for hn suitably close to hn_r (made precise in (18)–(21) of [10]), $h=1,\alpha\}$ to be $<\exp\{-(1+2\varepsilon)(\log r)(c_1^2/|u_1| + c_\alpha^2/|u_\alpha|)\}$. A finite number of quadrants of the form $\{(x_1, x_\alpha)\colon \pm x_1 > c_1,\ \pm x_\alpha > c_\alpha\}$ with $c_1^2/|u_1| + c_\alpha^2/|u_\alpha| = 1$ covers the region $\pm(x_\alpha - \alpha x_1) > (|u_\alpha| + \alpha^2 |u_1|)^{1/2}(1+\varepsilon)$ in the (x_1, x_α)-plane. Using (5.6) with $a_1 = \pm\alpha^2$ and $a_\alpha = \pm 1$, one obtains the upper class result. The *lower class proof* entails similar modifications of the proof of Theorem 2 of [10]; it is important to understand that one works with u_h which the U_{hn_r} oscillations exceed infinitely often, and then shows that oscillations in ξ_1 produced by these u_h occur for almost all r. One obtains $-(1-\varepsilon)[c_1^2/u_1 + c_\alpha^2/u_\alpha](\log n_r)/2$ for the joint probability replacing (49); the detailed changes in definitions (38)–(39) require space which is not warranted here; one crucial change will be alluded to just below, since it reflects a difference between (5.4) and (5.5).

Proof of (5.5). This turns out to be somewhat easier. The upper class result follows at once upon noting that the function $[2\beta n(\log n)^2 \log\log n]^{1/4}(1+\varepsilon)$ is in the upper class for each of $|\xi_1(U_n) - \xi_1(n)|$ and $\sup_{0<\alpha<1} |\xi_1(U_{\alpha n}) - \xi_1(\alpha n)|$. The crucial feature in the lower class proof, which can only be understood by reading [10] in detail, is that when we vary α near 1 we no longer need (as one does in the lower class proof of (5.4)) the same i for the $C'_{r,i,h}$ associated with $h=1$ and α in the analogue of Q'_r of (39). The event $\cap_h \cup_i C'_{r,i,h}$ used instead yields, for $\log P\{C'_{r,i,h}\}$, twice the value obtained just above for (49) as $\alpha\to 1$, and this yields the desired result.

Remarks on Theorem 5. (1) Many applications (e.g., statistics of Kolmogorov-Smirnov type) can be phrased in terms of the deviations of G_n^{-1} rather than of G. For this purpose, Theorem 5 gives the asymptotic maximum deviation (wp 1) of $m^{1/2}[G_n^{-1}(t) - t] - m^{1/2}[\xi_1(mt) - t\,\xi_1(m)]$, for a fixed t or for the maximum over t. (2) The weak law corresponding to the above is not difficult; that corresponding to (5.8) below is harder. The comments made in connection with Theorem 4 indicate why it does not seem worthwhile to spend more space on this. (3) For the deviations of $m^{1/2}[G_n(t) - t] - m^{-1/2}[\xi_1(mt) - t\,\xi_1(m)]$, there remains to be studied the difference

$$R_n(t) = m^{1/2}[G_n(t) - t] - m^{1/2}[t - G_n^{-1}(t)], \tag{5.7}$$

which is the deviation between the sample df and sample quantile process first studied by Bahadur [1], and the exact behavior of whose oscillations was determined for fixed t in [9] and for $\sup_t \pm R_n(t)$ in [11]. The oscillations of the latter are of the same order as those of $n^{-1/2} \sup_\alpha [r(\alpha n) - \alpha r(n)]$, but the constant on the right side of (5.5) gets replaced by $2^{-1/4}$. From this, one obtains the result

$$\begin{aligned} &\limsup_{n\to\infty} \sup_t |m^{1/2}[G_n(t) - t] \\ &\qquad - m^{-1/2}[t\,\xi_1(n) - \xi_1(nt)]|/[n^{-1}(\log n)^2 \log\log n]^{1/4} = \Omega(1) \text{ wp } 1. \end{aligned} \tag{5.8}$$

The upper bound part of (5.8) was given by Brillinger. His proof is certainly succinct; however, we have given the present analysis, with a different breakup from that of [4] of the components of the difference of (5.8), because it may offer better insight as to the source of deviations. We now sketch very briefly how the somewhat complicated correct constant on the right side of (5.8) can be computed. Roughly, for the more difficult lower class proof, one notes that, with $n_r \sim \gamma^r$ and γ large as in [11], a deviation $G_{n_r}(p_0) - p_0 \sim \pm c_{p_0}[2n_r^{-1} p_0(1-p_0) \log\log n_r]^{1/2}$ for infinitely many r produces a deviation of $\sup_{|p-p_0|<\varepsilon} \pm R_{n_r}(p) > (1-\varepsilon) c_{p_0}^{1/2} [n_r (\log n_r)^2 \log\log n_r]^{1/4}$ for almost all of those r, wp 1. This and the proof of (5.5), which shows that values α near 1 are crucial for $\sup_\alpha [r(\alpha n) - \alpha r(n)]$, shows that the event

$$\Big\{ \sup_{|p-p_0|<\varepsilon} \pm R_{n_r}(p) > c_{p_0}^{1/2} [n_r (\log n_r)^2 \log\log n_r]^{1/4},$$
$$\sup_\alpha \pm [r(\alpha n_r) - \alpha r(n_r)] > 2c_1 [n_r (\log n_r)^2 \log\log n_r] \Big\}$$

occurs infinitely often wp 1 if $\exp\{-(1+\varepsilon)[c_1^2/|u_1| + c_{p_0}^2/|u_{p_0}|] \log r\}$ is not summable, where the u_h's are exceeded infinitely often as normalized deviations of the U_{hn_r}, as in the proof of Theorem 5. Maximizing $c_{p_0}^{1/2} + 2c_1$ subject to $c_1^2/|u_1| + c_{p_0}^2/|u_{p_0}| = 1 - 2\varepsilon$ involves solving a cubic, and then p_0 must finally be chosen to give the overall maximum C^* (say) of $c_{p_0}^{1/2} + 2c_1$, using Lemma 7 or quadrants for the U_{hn_r}. The geometry of the quadrants for ξ_1 deviations, used as in the proof of (5.4), makes the upper class proof follow for this C^*. For fixed p the conclusion is simpler, since [9] $R_n(p)$ has smaller order than $\sup_\alpha [r(\alpha n) - \alpha r(n)]$. Thus, the upshot of this paragraph is, finally,

Theorem 6. *There is a positive constant* C^* such that

$$\begin{aligned} &\limsup_{n\to\infty} \sup_t |m^{1/2}[G_n(t) - t] \\ &\qquad - m^{-1/2}[t\,\xi_1(n) - \xi_1(nt)]|/[n^{-1}(\log n)^2 \log\log n]^{1/4} = C^* \text{ wp } 1; \end{aligned} \tag{5.9}$$

moreover,

$$\begin{aligned} &\limsup_{n\to\infty} \pm \{n^{1/2}[G_n(t) - t] - n^{-1/2}[t\,\xi_1(n) - \xi_1(nt)]\}/[2\beta n (\log n)^2 \log\log n]^{1/4} \\ &\qquad = [t^4 + 2t^3 + t]^{1/4} \text{ wp } 1. \end{aligned} \tag{5.10}$$

Remark (4). For application to distribution-free functionals, the results as stated in Theorems 5 and 6 suffice. For other functionals (e.g., linear combinations of sample quantiles, as mentioned in Remark 2 to Theorem 4), corresponding results can be stated by transforming from uniform rv's s as in [9] and [11].

In the next section we shall return to Breiman's representation in conjunction with Müller's proof.

6. Other Results and Possible Directions

A. Theorem 1, or its analogue for the embedding of Lemma 2, implies that various results obtained by Strassen [21, 22] from his strong invariance principle for sums of iid random variables from corresponding Brownian motion results, have analogues for the sample df process (1.3) in terms of ξ^*. Thus, if there were a simple Motoo-type proof of the upper-lower class results for ξ^*, we would have an immediate proof of Chung's corresponding result for S_n. Strassen's elegant characterization of the closure of the functions $\{n^{-1/2}\,\xi_1(n\,t), n\in Z^+, t\in I\}$ where ξ_1 is standard Brownian motion in any number of dimensions, has no startlingly different counterpart for $\{n^{-1/2}\,\xi^*(s, n\,t)\}$: as in [21], after dividing by $(2\log\log n)^{1/2}$, one again obtains that the closure consists of integrals (now in s, t) of functions of L_2 norm $\leqq 1$, and this yields results for $\{n^{1/2}\,\hat{\xi}(s, n\,t)\}$ as mentioned at the beginning of this section. Such calculations can be viewed as extensions of those which appear in the proof of Theorem 5 and in Section 1.8 of [6b]. For fixed t, the corresponding closure and iterated logarithm results for $\{n^{1/2}\,\hat{\xi}(s, n)\}$ have recently been published by Finkelstein [6a]. Recently Wichura [22a] has obtained very general results which include Strassen-type conclusions for the sample df for vector rv's of any dimension r. Wichura's approach uses classical Kolmogorov and Hartman-Wintner approximation techniques rather than Skorohod embedding. While it does not yield explicit error estimates like ours, it appears much more expeditious for its purpose than the embedding techniques of the present paper, which can apparently be extended to higher dimensional cases (and to the independent increment cases of [22a]), but which then require even lengthier calculations than those herein. Bickel has recently applied an estimate with exponent 1/4 (on n) replaced by $1/2(r+1)$ in Theorem 5, in connection with density estimation in dimension r.

My belief is that techniques used in recent work of Nagaev will yield sharper estimates than those obtainable by embedding, especially in higher dimensional analogues of (1.6)–(1.11).

Asymptotic properties of processes with multidimensional time are surveyed in depth in two papers of Pyke [14a, 14b], which also contain new results. The idea of using embeddings like the one arising from (2.2) surely occurred to Pyke and Root and to Brillinger, and probably to others.

B. We have mentioned in Section 1 the limitations of Skorohod embeddings for the problems we consider, and we shall return to this in C below. Nevertheless, it may be worthwhile knowing how far such methods can be pushed. There is certainly no obvious invitation for improvement of the exponent 1/6 in our proof of Theorem 1 and 2. On the other hand, it is conceivable that the exponent could be improved to 1/4 by either of two attacks. Firstly, it is clear from Müller's proof ([14], p. 207) that, if only one had suitable "exponential bounds" for adding his sample df embedding error ([14], p. 199), the exponent could be increased from 1/6 to 1/4. We do not now see how to obtain such bounds; and in fact equation (4) of [14] uses a break-up in s the need for which might not be unrelated to ours.

Perhaps such bounds are more easily obtained for Breiman's representation, which, incidentally, can be used in place of Müller's in the latter's proof outlined in Section 1. Secondly, the "near miss" of (2.4) offers some hope, in that the joint distribution for large m will be very close to normal; this embedding did not evidence ξ explicitly, but it does not have the disadvantageous delicacy of balancing errors from small B_n in Lemma 1′ against those from large B_n in Lemma 4′. Incidentally, Professor Müller has given convincing heuristics against the possibility of improvement beyond $n^{-1/4}$, and in view of the comment below (1.11) this strengthens my belief, mentioned earlier, that Nagaev's approach will yield more than embedding does for weak laws.

C. We have described in Section 1 why the order of error in the results we obtain does not seem sharp. We now describe why we think this is inherent in vector Skorohod embeddings. The idea is explained in a remarkable trick of arithmetic extracted and simplified from Skorohod's proof [20], pp. 177–178. Suppose ξ_1 is a standard Brownian motion and we want to estimate $P_n = P\{\xi_1(t) < g(t),\ 0 < t < 1 + \varepsilon_n\}$ in terms of $P\{\xi_1(t) < g(t),\ 0 < t < 1\}$ as $\varepsilon_n \to 0$. One possibility is to begin by estimating the difference between $\xi_1(t)$ and $\xi_1(t(1+\varepsilon_n)) = \xi_1(t')$ (say), and then estimating P_n in terms of $\xi_1(t')$, $0 < t' < 1$. Since $\xi_1(t) - \xi_1(t(1+\varepsilon_n))$ is $\Omega_p(\varepsilon_n^{1/2})$, this last is a lower limit on the result. Skorohod avoids using this. Instead, he writes

$$P_n = P\{(1+\varepsilon_n)^{-1/2}\, \xi_1(t(1+\varepsilon_n)) < (1+\varepsilon_n)^{-1/2}\, g(t(1+\varepsilon_n)),\ 0 < t < 1\},$$

and then estimates the right side of the inequality as $g(t) + O(\varepsilon_n)$, and this $O(\varepsilon_n)$ then yields (with a logarithmic term) the final error. In Skorohod's context the ε_n corresponds to our $n^{-1}(z_2 - z_1)^{-1}\, U^{(n)}_{z_1, z_2} - 1$, which is $O_p(n^{-1/2})$. But in our vector embeddings of Sections 2.2 and 2.3 we do not have a single counterpart of $(1+\varepsilon_n)$ to use in Skorohod's manner. Rather, we have a different such value for each (z_{i-1}, z_i), and must thus use the first, inferior, arithmetical scheme, based on the differences $\eta(z_{i-1}, z_i;\ (z_{i-1} - z_i)^{-1}\, U^{(n)}_{z_{i-1}, z_i}) - \eta(z_{i-1}, z_i;\ n)$, the counterpart of which is what Skorohod avoids. In the absence of the ingenuity to circumvent this difficulty, we cannot improve the exponent 1/4 of (1.7) by the Skorohod technique.

D. If one is using a Skorohod embedding, say for iidrv's, should one use that of Skorohod, Dubins, Breiman, Hall, Root, or some other? If ξ_1 is standard and T_1 is the Skorohod rv such that $\xi_1(T_1)$ has the desired distribution of a given rv X_1 with $EX_1 = 0$, $EX_1^2 = \sigma^2$, $EX_1^4 < \infty$, then all of the above methods have $ET_1 = \sigma^2$ and $\operatorname{var}(T_1) < \infty$, and in view of the results of [10] quoted in Section 5, and analogous weak laws such as [19], it seems desirable to use the method with smallest $\operatorname{var}(T_1)$. Intuitively, Root's nonrandomized stopping boundary (unique according to results of Loynes) would be guessed best, but we do not know how to prove this. However, an easy comparison of two of the methods is sometimes possible. Skorohod's method differs from Breiman's [3] except when X_1 takes on only two values. Since the former chooses at random a pair of functionally related constant stopping bounds for ξ_1, while the latter chooses them with additional randomness so that they are not functionally related, we guessed the latter was inferior. To the contrary, a simple computation shows

Theorem 7. *If* X_1 *has a symmetric law and* $E|X_1|^r = v_r$, *then*

$$ET^2_{1,\,\mathrm{BREIMAN}} = \left[v_4 + \frac{4v_2 v_3}{v_1}\right]\Big/3 \leqq 5v_4/3 = ET^2_{1,\,\mathrm{SKOROHOD}}, \tag{6.1}$$

with equality if and only if $|X_1|$ *takes on only one nonzero value.*

The superiority of Breiman's method is actually somewhat more general than for symmetric X_1. The supremum of the ratio of the two sides of (6.1) is of course 5. (Sawyer [19] has given bounds on $ET_{1,SK}$ in general.) Should (6.1) shake one's intuition that Root's least randomized T_1 is best?

References

1. Bahadur, R. R.: A note on quantiles in large samples. Ann. math. Statistics **37**, 577–580 (1966).
2. Breiman, L.: On the tail behavior of sums of independent random variables. Z. Wahrscheinlichkeitstheorie verw. Geb. **9**, 20–25 (1967).
3. Breiman, L.: Probability. Reading, Mass. Addison-Wesley 1968.
4. Brillinger, D. R.: An asymptotic representation of the sample df. Bull. Amer. math. Soc. **75**, 545–547 (1969).
5. Chung, K. L.: An estimate concerning the Kolmogoroff limit distribution. Trans. Amer. math. Soc. **67**, 36–50 (1949).
6. Dvoretzky, A., Kiefer, J., and Wolfowitz, J.: Asymptotic minimax character of the sample df and classical multinomial estimator. Ann. Math. Statist. **27**, 642–669 (1956).

6a. Finkelstein, H.: The law of the iterated logarithm for empirical distributions. Ann. math. Statistics **42**, 607–615 (1971).

6b. Freedman, D.: Brownian Motion and Diffusion. San Francisco: Holden-Day 1971.

7. Gnedenko, B., Korolyuk, V. S., and Skorohod, A. V.: Asymptotic expansions in probability theory, Proc. 4th Berkeley Sympos. math. Stat. and Probab. (1960), Vol. II, pp. 153–170. Berkeley: Univ. of Calif. Press 1961.
8. Kiefer, J.: On large deviations of the empiric df of vector chance variables and a law of the iterated logarithm. Pacific J. Math. **11**, 649–660 (1961).
9. Kiefer, J.: On Bahadur's representation of sample quantiles. Ann. math. Statist. **38**, 1323–1342 (1967).
10. Kiefer, J.: On the deviations in the Skorohod-Strassen approximation scheme. Z. Wahrscheinlichkeitstheorie verw. Geb. **13**, 321–332 (1969).
11. Kiefer, J.: Deviations between the sample quantile process and the sample df. Proc. First internat. Conf. Nonpar. Inf. (1969), 299–319, Cambridge Univ. Press 1970.
12. Kiefer, J.: Old and new methods for studying order statistics and sample quantiles. Proc. First theorie verw. Geb. **3**, 211–226 (1964).

12a. Kiefer, J.: Iterated logarithm analogues for sample quantiles when $p_n \downarrow 0$, Proc. 6th Berkeley Sympos. math. Statist. Probab. (1970), Vol. 1. Berkeley: Univ. of Calif. Press 1971.

13. Kiefer, J.: Skorohod embedding of multivariate processes: sums of martingales. (To appear.)
14. Müller, D. W.: On Glivenko-Cantelli convergence. Z. Wahrscheinlichkeitstheorie verw. Geb. **16**, 195–210 (1970).

14a. Pyke, R.: Empirical processes. (To appear.)

14b. Pyke, R.: Partial sums of matrix arrays and Brownian sheets. (To appear.)

15. Rosenkrantz, W. A.: On rates of convergence for the invariance principle. Trans Amer. math. Soc. **129**, 542–552 (1967).
16. Rosenkrantz, W. A.: A rate of convergence for the von Mises statistic. Trans. Amer. math. Soc. **139**, 329–337 (1970).
17. Sawyer, S.: A uniform rate of convergence for the maximum absolute value of partial sums in probability. Commun. pure appl. Math. **20**, 647–658 (1967).
18. Sawyer, S.: Uniform limit theorems for the maximum cumulative sum in probability. Trans. Amer. math. Soc. **132**, 363–367 (1968).

19. Sawyer, S.: Rates of convergence for some functionals in probability. (To appear.)
20. Skorohod. A. V.: Studies in the theory of random processes. Reading, Mass.: Addison-Wesley 1965.
21. Strassen, V.: An invariance principle for the law of the iterated logarithm. Z. Wahrscheinlichkeitstheorie verw. Geb. **3**, 211–226 (1964).
22. Strassen, V.: Almost sure behavior of sums of independent random variables and martingales. Proc. 5th Berkeley Sympos. Math. Statist. Probab. (1965), Vol. II (part 1), 315-343. Berkeley: Univ. of Calif. Press 1967.
22a. Wichura, M.: Some Strassen-type laws of the iterated logarithm for multiparameter stochastic processes with independent increments. (To appear.)

J. Kiefer
White Hall
Department of Mathematics
Cornell University
Ithaca, N.Y. 14850
USA

(Received June 1, 1971)

Reprinted from MYCOLOGIA, Vol. LXVII, No. 1, pp. 203–205, Jan.–Feb., 1975

Bayesian Analysis of Generic Relations in Agaricales, by R. E. Machol and R. Singer. Nova Hedwigia 21: 753–787. 1971. Price, not given.

This short paper deserves review for several reasons. There is the claim of a new numercial technique of taxonomy, described as eliminating shortcomings of various previous methods. There are striking numerical values of the "odds ratios" that measure the authors' degree of belief in the validity of the classifications they obtain by this methodology. And of course there is Singer's stature. These could well combine to make the work a point of departure for much future taxonomic research by mycologists.

I believe there are serious deficiencies in the statistical foundations of the paper that will not be evident to most of its readers, and that mechanical use of the technique could produce volumes of meaningless taxonomic results at the hands of mycologists less skilled and experienced than Singer. In all fairness, I must state that the authors, in a friendly and cooperative correspondence with me, show awareness of many of the problems that arise in using their technique, and of the possibility of its misuse. But they are less concerned than I am that their presentation invites such use among statistically uninitiated biologists; and they do not share my feeling of the degree to which the assumption of "independence" (discussed below) used in their calculations makes the odds ratios meaningless. It is impossible in this brief review to list all of my statistical objections, or to do more than outline my concern about independence. I hope to discuss these matters in more detail elsewhere.

The technique itself classifies a lower taxon between two (or more) higher taxa. Three examples are carried out, representing "cases where genera of Agaricales have been inserted in their respective families with some degree of doubt." The most striking of the conclusions is that *Ripartites,* which had been inserted in the Crepidotaceae by Singer (1951) and in the Tricholomataceae by Kühner and Romagnesi (1953) and Singer (1962), is more appropriately classified in Paxillaceae, conforming to the disregarded assignment by Quélet (1886). I will not discuss this or the other taxonomic conclusions, which the authors describe intelligently. The results (as distinguished from the detailed method and the asserted odds ratios) seem very reasonable qualitatively. But I am not really professionally competent to judge the biology, and have indicated that the purpose of this review lies elsewhere.

The authors' technique combines the character frequencies associated with numerical taxonomy into a precise probability model that yields an appealing form of conclusion: the odds that genus A belongs to family B rather than to family C. The stated odds ratio is only as meaningful as the model. An aspect of this model is the assumption of independence of characters. This assumption states, for example, that if a species randomly chosen from genus A has probability $\frac{1}{2}$ of having spore shape ellipsoid to ovoid, $\frac{2}{3}$ of having brown spores, and $\frac{1}{3}$ of having hygrophanous pileus, then such a randomly chosen species has probability $\frac{1}{2} \times \frac{2}{3} \times \frac{1}{3}$ of having spore shape, spore color, and pileus appearance of the stated description. If there were 90 species in the genus, 10 are then asserted to have this combination of character states. Now, it is easy to specify any number of species between 0 and 30 as having this combination, and still to satisfy the hypothesized $\frac{1}{2}$, $\frac{2}{3}$, $\frac{1}{3}$ frequencies for *single* character specifications; thus, in an extreme case, all 30 species with hygrophanous pileus could have ellipsoidal brown spores. It may well be that the actual degree of dependence is not as great as in this extreme case. Perhaps there is only a slight tendency for these characters to be correlated; maybe only 14 or 15 species have the stated combination of three characters. The points to be emphasized are, firstly, that even such less extreme cases of dependence can make the odds ratios, computed assuming independence, grossly incorrect; and, secondly, that not even a biologist of Singer's experience can hope to choose characters that satisfy this independence assumption adequately for the validity of the authors' computations, especially where not three but dozens of factors are involved in the product formula. While the problem of independence of characters is alluded to briefly on page 764, the illustration there treats a fairly extreme case of dependence, and it occurs several pages after the product formula is introduced without any mention of the independence assumption. Thus, this presentation is unlikely to make the typical reader aware that "independence of characters" is not a qualitative notion but a precise and complicated mathematical assumption. (A small sampling of readers confirms my fears.)

Given sufficiently detailed data (counts of character combinations, rather than simply frequencies for individual characters), one could detect very *large* departures from independence; however, the number of species under observation is much too small to verify independence to the extent needed to validate the authors' computation of odds ratios, even to within several orders of magnitude. That same small number of observations eliminates any possibility of using a more complex but realistic probability model that does not assume independence. What, then, would most statisticians do with these data?

Those I have asked, people experienced in such classification problems, would use some method intuitively appealing to them to reduce from several dozen characters to perhaps five most informative combinations, upon which to base the classification. None of them would assume independence. Above all, they would refrain from stating conclusions in terms of such numerical odds with these data, aware that such precisely stated figures, even when presented with warnings like the authors', invite the reader to accept them as meaningful. (If they were not to be read in this way, why present them?) I share these views. This amount of data simply can not support a useful precise analysis in terms of a sufficiently realistic probability model.

Among the topics I will discuss in a longer treatment are the Bayesian polemics in the present paper, the previous use elsewhere of this and similar methods, some questionable details of the method, some errors in the comparison of the technique with previous methods of numerical taxonomy. I can only expect fruitful disagreement from the authors, at least on some of these matters. For now, let me summarize that this early attempt among mycological taxonomists to give examples of quantitative classification assessments contains a useful discussion of the problems the authors encountered, as well as very interesting character counts and taxonomic conclusions. As for the detailed model and the meaningfulness of the odds ratios, you now have one statistician's warning.—J. KIEFER, Department of Mathematics, Cornell University.

Z. Wahrscheinlichkeitstheorie verw. Gebiete
34, 73–85 (1976)

Zeitschrift für
Wahrscheinlichkeitstheorie
und verwandte Gebiete

Asymptotically Minimax Estimation of Concave and Convex Distribution Functions

J. Kiefer* and J. Wolfowitz**

Department of Mathematics, Cornell University, Ithaca, N.Y. 14853, USA and
Department of Mathematics, University of Illinois, Urbana, Ill. 61801, USA

1. Introduction

Roughly speaking, this paper deals with the problem of efficiently estimating a distribution function when essentially nothing is known about it except that it is concave (or convex). A precise formulation of the problem is given in Section 3. The entire paper is written for concave distributions, but, mutatis mutandis, all results hold for convex distributions. The problem is one of a class with a large literature.

To the best of our knowledge our problem was first treated by Grenander [5]. He proved that the maximum likelihood estimator of a concave distribution F is the least concave majorant C_n of the empiric distribution function F_n of the n independent observations. It follows immediately from Marshall's lemma (Lemma 3 below, which asserts that $\sup_x |C_n(x)-F(x)| \leq \sup_x |F_n(x)-F(x)|$) that C_n is a uniformly consistent estimator of F. (Throughout we use "estimator" as an abbreviation for "sequence of estimators". Also we omit the phrase "on its interval of support" when describing a d.f. as "concave" or "convex".)

As a consequence of studies of reliability theory there arose an interest in related problems, e.g., estimating increasing (decreasing) failure rate distributions, unimodal distributions, and others. There is now a very large literature, much of which is cited in [6]. The latter book is a good introduction to the subject and a guide to further reading. It would be invidious for us to cite a limited number of papers and ignore others equally worthy, and any attempt at completeness in our references would be out of the question for practical reasons. The authors take this opportunity to express their gratitude to Professor Frank Proschan, for guidance to the literature to which he himself is a distinguished contributor.

In many of the papers of this large literature, some of them very ingenious, the authors obtain estimators by application of the maximum likelihood method

* Research under NSF Grant MPS72-04998 A02.

** Research supported by the U.S. Air Force under Grant AFOSR-70-1947, monitored by the Office of Scientific Research.

or some other method (often quite difficult to do) and prove that their estimators are consistent. Obviously, efficiency is the important property, and proving consistency is only a first step in shrinking the class of possible estimators among which the asymptotically efficient estimators are to be found. The only reason for using maximum likelihood estimators is the hope that the success of this method in the classical case may carry over to these difficult non-parametric problems. No successful attempt has ever been made to show that the maximum likelihood method or any other method will produce asymptotically efficient estimators in these non-parametric problems.

The essential reason (and intuitive basis) for the asymptotic efficiency of the maximum likelihood estimator in the classical case is now well understood ([7], pp. 3.11–12). It is that the maximum likelihood estimator is asymptotically equivalent to the Bayes estimator with respect to an a priori distribution which is uniform on a small interval (of length $O(n^{-1/2})$) centered at the true value of the parameter. One verifies easily that the proof of this fact does not carry over to the present problem. The fact that, even when estimating a single parameter, any serious departure from the classical conditions has as a consequence that the maximum likelihood estimator is no longer efficient, does not bode well for the efficiency of the maximum likelihood estimator for our non-parametric problem.

In [1] and [2] the present authors (in [1] in collaboration with A. Dvoretzky), proved that F_n is asymptotically minimax, in several reasonable senses there precisely defined, as an estimator of F when F is known only to belong to the class of all distribution functions (d.f.'s) or to the class of all continuous d.f.'s. In the present paper we prove (in Section 4) that this is also true when F is known only to belong to the class of all concave (convex) d.f.'s. Now F_n need not be concave (convex), and may therefore be considered by the statistician as an unsuitable estimator of F. However, it follows immediately from Marshall's lemma cited earlier that C_n, which is concave (convex) by definition (in the convex case C_n is defined to be the greatest convex minorant of F_n), and hence suitable to be used as an estimator, is also asymptotically minimax for estimating F in the senses defined in [1, 2], and below in the present paper.

In Section 3 we prove, under certain additional restrictions, that

$$\sup_x |C_n(x) - F_n(x)| = o_P(n^{-1/2}).$$

Consequently, in this case, C_n is essentially no better than F_n for estimating F, in spite of the Marshall lemma, except of course for the fact that C_n is concave (convex).

2. Preliminaries

In this section we give some definitions and probability estimates that may be useful in a variety of applications.

The rv's $X_1, X_2, \ldots$ are i.i.d. according to some d.f. F on the reals. We define B_F to be the smallest closed interval to which F assigns probability one, and define

$$\alpha_0(F)=\sup\{x: F(x)=0\},$$
$$\alpha_1(F)=\inf\{x: F(x)=1\}, \tag{2.1}$$

with the convention that $\alpha_0=-\infty$ or $\alpha_1=+\infty$ if there is no x satisfying the appropriate condition in braces. By f we denote a derivative of the absolutely continuous part of F. For any real function g on B_F, we define

$$\|g\|=\sup_{x\in B_F}|g(x)|. \tag{2.2}$$

The words "convex" and "concave" are always used with the phrase "and continuous in the interior of B_F" being understood. The empiric d.f. based on n observations is denoted by F_n. The functions F and F_n are defined so as to be right continuous.

Although the normal approximation can be verified to hold for the binomial tail probabilities that arise in the sequel, we shall use the more elementary estimates obtained from Markov's inequality, since the best power of n obtainable in Theorem 1 by the present methods is the same for the two estimates. (The logarithms of these probabilities are asymptotically the same, in the domain we encounter.) Also, as in [8], Lemma 1, it suffices to use below an algebraically simple approximation to the best choice of the coefficient t, rather than the more complex best choice.

Lemma 1. *For positive $p_n\to 0$ and positive $\delta_n\to 0$, if F is the uniform* d.f. *on* $[0,1]$,

$$P\{|F_n(p_n)-p_n|\geqq\delta_n p_n\}\leqq 2e^{-np_n\delta_n^2[1+o(1)]/2}, \tag{2.3}$$

where the $o(1)$ term depends only on δ_n and p_n.

Proof. For $t>0$, we have

$$\begin{aligned}P\{F_n(p)\geqq(1+\delta)p\}&=P\{e^{tnF_n(p)}\geqq e^{nt(1+\delta)p}\}\\&\leqq e^{-nt(1+\delta)p}Ee^{tnF_n(p)}\\&=[e^{-t(1+\delta)p}(pe^t+1-p)]^n.\end{aligned} \tag{2.4}$$

The substitution $t=\delta$ and expansion of each exponential in the last expression to terms in t^2 plus remainder, yields the result for positive deviations. The result for negative deviations is obtained, similarly, by substituting $t=-\delta$ into

$$P\{F_n(p)\leqq(1-\delta)p\}\leqq[e^{-t(1-\delta)p}(pe^t+1-p)]^n, \tag{2.5}$$

valid for $t<0$.

We now define and study an interpolating process for any fixed, continuous F on R^1 and, correspondingly, for F_n. For each positive integer k, let $a_j^{(k)}$ be any values satisfying

$$\begin{aligned}F(a_j^{(k)})&=j/k \quad \text{if } 0<j<k,\\a_0^{(k)}&=\alpha_0(F),\\a_k^{(k)}&=\alpha_1(F).\end{aligned} \tag{2.6}$$

Each interval $[a_j^{(k)}, a_{j+1}^{(k)}]$ is assigned the same probability $1/k$ in (2.6); we do not require the obvious generalization to unequal probabilities in our applications.

Let $L^{(k)}$ be any nondecreasing function on R^1 satisfying

$$L^{(k)}(a_j^{(k)}) = F(a_j^{(k)}), \quad 0 \leqq j \leqq k. \tag{2.7}$$

The choice of $L^{(k)}$ just above Lemma 4 in Section 3 is piecewise linear, but this is not assumed in Lemma 2, and other choices may prove convenient in other applications. We define $L_n^{(k)}$ by

$$L_n^{(k)}(x) = F_n(a_j^{(k)}) + k[F_n(a_{j+1}^{(k)}) - F_n(a_j^{(k)})][L^{(k)}(x) - F(a_j^{(k)})]$$
$$\text{for } a_j^{(k)} \leqq x \leqq a_{j+1}^{(k)}, \ 0 \leqq j \leqq k. \tag{2.8}$$

Thus, $L_n^{(k)}$ is also nondecreasing, and

$$L_n^{(k)}(a_j^{(k)}) = F_n(a_j^{(k)}), \quad 0 < j < k. \tag{2.9}$$

In the application where $L^{(k)}$ is piecewise linear, so is $L_n^{(k)}$.

In the sequel, $\{k_n, n \geqq 1\}$ will always denote a sequence of positive integers satisfying, for large n,

$$n^{b_1} < k_n < n^{b_2} \quad \text{for some } b_i > 0 \text{ with } b_2 < 1. \tag{2.10}$$

We shall write $p_n = 1/k_n$. The condition (2.10) can be weakened in Lemma 2, but that would not improve the power of n on the left side of (3.18). This lemma sharpens some estimates of [9].

Lemma 2. *Under* (2.10) *there is a positive value* C_0 *such that, for sufficiently large* n, *for all continuous* F,

$$P_F\{n^{1/2} \sup_x |F_n(x) - F(x) - L_n^{(k_n)}(x) + L^{(k_n)}(x)|$$
$$> [C_0 k_n^{-1} \log k_n]^{1/2}\} < n^{-2}. \tag{2.11}$$

Proof. For typographic simplicity, we drop the superscript k_n throughout. On the interval $[a_j, a_{j+1}]$, we obtain from (2.7)–(2.8)

$$\begin{aligned} &F_n(x) - F(x) - L_n(x) + L(x) \\ &= \{F_n(x) - F_n(a_j) - p_n^{-1}[F(x) - F(a_j)][F_n(a_{j+1}) - F_n(a_j)]\} \\ &\quad + \{[F(x) - L(x)](p_n^{-1}[F_n(a_{j+1}) - F_n(a_j)] - 1)\} \\ &= G_n(x) + H_n(x) \quad \text{(say)}. \end{aligned} \tag{2.12}$$

We shall prove that, for a suitable C_0, and for n sufficiently large, we have, uniformly in j $(0 \leqq j \leqq k_n)$,

$$P_F\{n^{1/2} \sup_{a_j \leqq x \leqq a_{j+1}} |G_n(x)| > \tfrac{1}{2}[C_0 k_n^{-1} \log k_n]^{1/2}\} < 4^{-1} k_n^{-1} n^{-2}, \tag{2.13}$$

and also (2.13) with H_n replacing G_n. Summing over these two sources of deviation and the $k_n + 1$ intervals (a_j, a_{j+1}) then yields (2.11).

Given the event $n[F_n(a_{j+1}) - F_n(a_j)] = n_0 > 0$, the process $\{n n_0^{-1} G_n(x), a_j \leqq x \leqq a_{j+1}\}$ has the law of "empiric minus true continuous d.f.", on an interval to which the latter assigns probability one. Consequently ([1]), for all positive n_0, all $d_n > 0$, and some constant C,

$$P\{n^{1/2} \sup_{a_j \leqq x \leqq a_{j+1}} |G_n(x)| > d_n | n[F_n(a_{j+1}) - F_n(a_j)] = n_0\} \leqq C e^{-2d_n^2 n/n_0}. \tag{2.14}$$

Now fix $\delta_n < 1$ at a small enough positive value so that the $o(1)$ term in (2.3) is $> -1/2$ for k_n sufficiently large. The estimate (2.3), in terms of the present setting, yields

$$P\{n[F_n(a_{j+1}) - F_n(a_j)] \geqq 2np_n\} \leqq 2e^{-C' n p_n} \tag{2.15}$$

for some positive C'. Consequently, from (2.15) and (2.14),

$$P\{n^{1/2} \sup_{a_j \leqq x \leqq a_{j+1}} |G_n(x)| \geqq d_n\} \leqq C e^{-d_n^2/p_n} + 2e^{-C' n p_n}. \tag{2.16}$$

From the fact that $\sup_{a_j \leqq x \leqq a_{j+1}} |F(x) - L(x)| \leqq p_n$, and from Lemma 1, we have

$$P\{n^{1/2} \sup_{a_j \leqq x \leqq a_{j+1}} |H_n(x)| \geqq d_n\} \leqq P\{n^{1/2} |F_n(a_{j+1}) - F_n(a_j) - p_n| \geqq d_n\}$$
$$\leqq 2e^{-d_n^2[1+o(1)]/2p_n}, \tag{2.17}$$

provided that $\delta_n = d_n/n^{1/2} p_n \to 0$. Because of (2.10), the $o(1)$ term depends on δ_n and n, since p_n is a function of n.

Now substitude $d_n^2 = 4^{-1} C_0^2 p_n \log p_n^{-1}$ into (2.16) and (2.17). From (2.10), for C_0 sufficiently large, each of the right sides of (2.16) and (2.17) is seen to be $< 1/4n^2 k_n$, and the condition below (2.17) is satisfied. This completes the proof of (2.13), and thus of Lemma 2.

3. Estimating Concave F: Closeness of Two Estimators

Throughout this section we assume $X_1, X_2, \ldots$ are i.i.d. rv's with common d.f. F, concave on the interval $[\alpha_0(F), \infty)$; note that $\alpha_0(F)$ is finite, and F may have a jump at $\alpha_0(F)$. Denote the class of all such F by $\mathscr{F}$. Many results of this subject, in the literature or the present paper, are valid only for a subset of $\mathscr{F}$. Thus, F might also be assumed continuous, $\alpha_0(F)$ might be assumed known, and/or $\alpha_1(F)$ might be assumed finite, or an upper bound on $\alpha_1(F)$ might be assumed known. We shall state such restrictions where they are used. When $\alpha_0(F)$ is known, we shall take it to be zero, without loss of generality.

Assuming F continuous and $\alpha_0(F) = 0$, the maximum likelihood estimator of F, based on n observations, was shown by Grenander [4] to be C_n, the smallest concave majorant of F_n satisfying $C_n(0) = 0$. In the present section we show that, under certain assumptions, F_n and C_n are quite close for large n. A theoretical consequence is that the two estimators enjoy similar optimum properties (treated in the next section). A practical consequence is that, instead of computing C_n exactly, the statistician may often find it satisfactory to use any concave function sufficiently close to F_n; if he does not care whether his estimator is concave, he can even use F_n itself.

The ML estimator remains the same if $\alpha_1(F)$ is assumed finite, as is done in (3.2).

For simplicity of presentation, we shall prove the next four lemmas and Theorem 1 under a single set of assumptions, and thereafter discuss the assumptions and remark on which of the conclusions can be modified to hold under other assumptions. We write, for twice differentiable F with $\alpha_0(F)=0$,

$$\begin{aligned}\gamma(F)&=\sup_{0<x<\alpha_1(F)} |-f'(x)|/\inf_{0<x<\alpha_1(F)} f^2(x),\\ \beta(F)&=\inf_{0<x<\alpha_1(F)} |-f'(x)/f^2(x)|.\end{aligned} \tag{3.1}$$

Note that γ and β are invariant under changes of scale and location. We hereafter assume

$$\begin{aligned}&F \text{ continuous with } \alpha_0(F) \text{ known to be 0 and } \alpha_1(F) \text{ unknown but known to be } <\infty;\\ &F \text{ concave and twice continuously differentiable on } (0,\alpha_1(F));\ \beta(F)>0,\ \gamma(F)<\infty.\end{aligned} \tag{3.2}$$

We remind the reader that $\|g\|=\sup_{0\leqq x\leqq \alpha_1(F)} |g(x)|$.

We now give an outline of our method of estimating the magnitude of C_n-F_n:

(A) For every concave function h on $[0,a_1(F)]$, it is proved ([5]) that

$$\|C_n-h\|\leqq\|F_n-h\|. \tag{3.3}$$

(B) Under the assumption (3.2), Lemma 1 is used to show that, for suitable k_n and functions $L^{(k)}$ and $L_n^{(k)}$ of (2.7)–(2.8), the event

$$A_n=\{L_n^{(k_n)} \text{ is concave on } [0,\infty)\} \tag{3.4}$$

has probability near one.

(C) Under A_n, we can use (A) with $h=L_n^{(k_n)}$ to show that

$$\|C_n-F_n\|\leqq 2\|F_n-L_n^{(k_n)}+L^{(k_n)}-F\|+2\|L^{(k_n)}-F\|. \tag{3.5}$$

(D) The first term on the right side of (3.5) is estimated by using Lemma 2; the second term is estimated by a simple analytic argument.

Lemma 3 (Marshall [5]). *If h is concave,* (3.3) *holds.*

We now define $L^{(k)}$ to be linear on each of the intervals $[a_j^{(k)}, a_{j+1}^{(k)}]$, for $0\leqq j\leqq k-1$.

Lemma 4. *If* (3.2) *holds, for k_n sufficiently large (depending only on $\beta(F)$),*

$$1-P\{A_n\}\leqq 2k_n e^{-n\beta^2(F)/80k_n^2}. \tag{3.6}$$

Proof. For $0\leqq j\leqq k_n-1$, write

$$\begin{aligned}T_{n,j}&=F_n(a_{j+1}^{(k_n)})-F_n(a_j^{(k_n)})\\ \Delta_{n,j}&=a_{j+1}^{(k_n)}-a_j^{(k_n)}.\end{aligned} \tag{3.7}$$

Since $L^{(k_n)}$ is linear on each of the k_n successive intervals of length $\Delta_{n,j}$ $(0\leqq j\leqq k_n-1)$, it follows from (2.8) and the computation of the derivative of $L_n^{(k_n)}$ in each interval that

$$A_n = \bigcap_{j=0}^{k_n-2} \{T_{n,j}/\Delta_{n,j} \geqq T_{n,j+1}/\Delta_{n,j+1}\}$$
$$= \bigcap_{j=0}^{k_n-2} B_{n,j} \quad \text{(say)}. \tag{3.8}$$

For $0 \leqq j \leqq k_n - 2$, and $0 < \delta_n < 1/3$, the event $B_{n,j}$ is a consequence of

$$|T_{n,i} - k_n^{-1}| \leqq \delta_n/k_n \quad \text{for } i = j, j+1 \tag{3.9}$$

and (since $1 + 3\delta_n > (1+\delta_n)/(1-\delta_n)$)

$$\Delta_{n,j+1}/\Delta_{n,j} \geqq 1 + 3\delta_n. \tag{3.10}$$

We next verify that (3.10) holds for $0 \leqq j \leqq k_n - 2$, provided that $\delta_n \leqq \beta(F)/6k_n < 1/3$. Since $dF^{-1}(t)/dt = 1/f(F^{-1}(t))$ and $d^2F^{-1}(t)/dt^2 = (-f'/f^3)(F^{-1}(t))$, a second order Taylor expansion about $(j+1)/k_n$ yields

$$\Delta_{n,j+1} = F^{-1}((j+2)/k_n) - F^{-1}((j+1)/k_n)$$
$$= k_n^{-1}/f(a_{j+1}^{(k_n)}) + (2k_n^2)^{-1}(-f'(\xi)/f^3(\xi)) \tag{3.11}$$

for some ξ between $a_{j+1}^{(k_n)}$ and $a_{j+2}^{(k_n)}$. Since f is decreasing, we obtain

$$\Delta_{n,j} \leqq k_n^{-1}/f(a_{j+1}^{(k_n)})$$

and also

$$\Delta_{n,j+1}/\Delta_{n,j} \geqq 1 + (2k_n)^{-1} f(a_{j+1}^{(k_n)})(-f'(\xi)/f^3(\xi))$$
$$\geqq 1 + (2k_n)^{-1}(-f'(\xi)/f^2(\xi))$$
$$\geqq 1 + (2k_n)^{-1}\beta(F). \tag{3.12}$$

Hence (3.10) holds if $\delta_n = \beta(F)/6k_n$. With this choice of δ_n we consequently obtain (3.6) from (3.9), (3.10), and Lemma 1, since the conditions $\delta_n \to 0$ and $p_n \to 0$ of that lemma are automatically satisfied as $k_n \to \infty$.

Lemma 5. *If* A_n *occurs* (3.5) *holds.*

Proof. From Lemma 3 and (3.3) with $h = L_n^{(k_n)}$,

$$0 \leqq C_n - F_n \leqq \|C_n - L_n^{(k_n)}\| + \|L_n^{(k_n)} - F_n\|$$
$$\leqq 2\|L_n^{(k_n)} - F_n\|$$
$$\leqq 2\|L_n^{(k_n)} - F_n + F - L^{(k_n)}\| + 2\|L^{(k_n)} - F\|. \tag{3.13}$$

Lemma 6. *If* F *satisfies* (3.2),

$$\|F - L^{(k)}\| \leqq \gamma(F)/2k^2. \tag{3.14}$$

Proof. Fix j, $0 \leqq j \leqq k-1$, and define $g(x) = F(x + a_j^{(k)}) - L(x + a_j^{(k)})$ on the interval $[0, \Delta_j]$, where $\Delta_j = a_{j+1}^{(k)} - a_j^{(k)}$. Then $g \geqq 0$, and

$$g(0) = g(\Delta_j) = 0 \tag{3.15}$$

Hence, a Taylor expansion about 0 gives, on $[0, \Delta_j]$,

$$g(x) = g'(0+)x + g''(\xi_x)x^2/2, \tag{3.16}$$

where $0 < \xi_x < x$. Evaluating (3.16) at $x = \Delta_j$ and again using (3.15) yields $g'(0+) = -\Delta_j g''(\xi_{\Delta_j})/2$. Since $g''(\xi_x) \leqq 0$, (3.16) becomes

$$\begin{aligned} g(x) &= -g''(\xi_{\Delta_j})\Delta_j x/2 + g''(\xi_x)x^2/2 \\ &\leqq -g''(\xi_{\Delta_j})\Delta_j^2/2. \end{aligned} \tag{3.17}$$

Since $g''(x) = f'(x + a_j^{(k)})$, and since $\Delta_j = 1/kf(a_j^{(k)} + \xi')$ for some ξ' in $(0, \Delta_j)$ by Rolle's theorem, the result (3.14) follows from (3.17).

Theorem 1. *If F satisfies* (3.2), *for all sufficiently large n (depending only on $\beta(F)$ and $\gamma(F)$)*

$$P_F\{\|C_n - F_n\| > n^{-2/3}(\log n)^{5/6}\} < 2n^{-2}, \tag{3.18}$$

so that

$$P_F\{\lim_{n\to\infty} [n^{2/3}/\log n] \|C_n - F_n\| = 0\} = 1. \tag{3.19}$$

Proof. Let $k_n = [\beta^2(F)n/200 \log n]^{1/3}$. By Lemma 4, for n sufficiently large (depending on $\beta(F)$) $1 - P\{A_n\} \leqq n^{-2}$. If A_n occurs, by Lemma 5, (3.5) holds. Note that (2.10) is satisfied. By Lemma 2, with probability at least $1 - n^{-2}$, the first term on the right side of (3.5) is at most $2C_0^{1/2}(200/\beta^2(F))^{1/6} n^{-2/3}(\log n)^{2/3}$, for all sufficiently large n (independent of F). By Lemma 6, the second term on the right side of (3.5) is at most $\gamma(F)(200/\beta^2(F))^{2/3} n^{-2/3}(\log n)^{2/3}$. The combination of these estimates yields (3.18).

Under (3.18) the limiting laws of $n^{1/2}\|C_n - F\|$ and $n^{1/2}\|F_n - F\|$ are of course the same.

Remarks. 1. In order to discuss the extent to which our results hold under assumptions other than (3.2), we must first describe the way in which the definition of C_n is altered under the various possible restrictions mentioned in the first paragraph of the present section. If $\alpha_0(F) = 0$ but the assumption of continuity of F is dropped, C_n is defined to be the smallest concave majorant of F_n, subject to $C_n(0-) = 0$; thus, $C_n(0) = F_n(0)$. This is the ML estimator in an extended sense [3]. If also $\alpha_0(F)$ is unknown, C_n is the smallest concave majorant of F_n on the interval $[\min_{1 \leqq i \leqq n} x_i, \infty)$, subject to being 0 to the left of this interval. This last estimator, although discontinuous, may also be used when $\alpha_0(F)$ is unknown but F is assumed continuous, in which case an ML estimator does not exist; alternatively, it may be modified to be a "neighborhood ML estimator" [3].

The above definitions are unchanged if $\alpha_1(F)$ is assumed finite or if an upper bound on $\alpha_1(F)$ is assumed known. If $\alpha_1(F)$ is known exactly, wp 1 the ML estimator does not exist, but these estimators, or slight modifications seem appropriate.

2. It is not difficult to see that Lemmas 3 and 5 remain valid for any of the alterations of Remark 1.

3. The proofs of Lemmas 4 and 6, and thus of Theorem 1, can be seen to hold under any or all of the following modifications of (3.2): F can be permitted to

have a jump at $\alpha_0(F)$; $\alpha_0(F)$ can be unknown; $\alpha_1(F)$ can be known. In addition, if f' fails to exist on a finite set H (where f can even have jumps) and (3.2) holds when the set H is removed in the definition of β and γ, Theorem 1 remains valid; it is only necessary to adjoin H to the set of $a_j^{(k)}$, for each k, removing the original $a_j^{(k)}$ closest to each member of H in order to make sure the analogue of Lemma 4 still holds.

4. The proofs of Lemmas 1, 2, and 4 have been kept short by using simple rather than sharp estimates; however, such estimates, or the weakening of (3.18) to require only that the probability approach zero, would not improve the power of n attainable in (3.18)–(3.19) by the present approach.

Of more interest is an extension of Theorem 1 obtainable when $\beta(F)=0$, due to f' vanishing on a finite set. We indicate the construction for the simple case where that set consists of the single point $\alpha_1(F)$. Thus, we assume that $f'(\alpha_1(F)-)=0$ and that, for $0<\varepsilon<1$,

$$0<\beta(F,\varepsilon)\overset{\text{def}}{=}\inf_{0<x<F^{-1}(1-\varepsilon)}[-f'(x)/f^2(x)]. \tag{3.20}$$

In order that $\gamma(F)$ still be finite, we assume also that $0<f(\alpha_1(F))=\bar{C}(F)$ (say).

The idea is to use $\beta(F,\varepsilon)$ in place of β in the last line of (3.12), which entails having ξ bounded away from $\alpha_1(F)$ for each k; this last necessitates changing the definition of $L^{(k)}$ slightly. Writing $a_*^{(k)}=F^{-1}(1-1/2k)$, we alter the definition of $L^{(k)}$ just above Lemma 4 *only* on the interval $[a_{k-1}^{(k)},\alpha_1(F)]$, where we define

$$L^{(k)}(x)=\begin{cases}1-1/2k & \text{at } x=a_*^{(k)},\\ \text{linear} & \text{on } [a_{k-1}^{(k)},a_*^{(k)}],\\ F(x) & \text{on } [a_*^{(k)},\alpha_1(F)].\end{cases} \tag{3.21}$$

It is clear that $L^{(k)}$ is concave, and we examine the proof of Lemma 4. For $j\leqq k_n-3$, the development of (3.11) and (3.12) still holds, with $\xi<a_{k-1}^{(k)}$. The event B_{n,k_n-2} is now seen to occur if, in analogy to (3.9)–(3.10),

$$\begin{aligned}&|T_{n,k_n-2}-k_n^{-1}|\leqq\delta_n/k_n,\\ &|F_n(a_*^{(k_n)})-F_n(a_{k_n-1}^{(k_n)})-(2k_n)^{-1}|<\delta_n/2k_n,\\ &2\Delta_{k_n-1}/\Delta_{k_n-2}\geqq 1+3\delta_n.\end{aligned} \tag{3.22}$$

The last of these is seen to hold (upon replacing k_n by $2k_n$ in the last expression of (3.11), etc.) if $\delta_n=\beta(F,1/2k_n)/12k_n$, which choice of δ_n can of course also be used for $j<k_n-2$. We obtain (3.6) with $\beta^2(F)/80$ replaced by $\beta^2(F,1/2k_n)/300$.

The conclusion of Lemma 6 is unaltered, and the proof of Theorem 1 is changed by replacing $200/\beta^2(F)$ everywhere in the proof by approximately $800/\beta_n^2(F)$. This entails solving approximately, for each large n, the relationship

$$\frac{k_n^3}{\beta^2(F,1/2k_n)}=n/800\log n \tag{3.23}$$

for k_n; since the denominator of the left side is non-increasing in k_n, there is a unique positive solution, and k_n may be taken to be the closest integer to it. This determination makes $1-P\{A_n\}<n^{-2}$ for n large. Assuming (2.10) holds, we apply Lemma 2 as before. The first term on the right side of (3.5), of order $(\log n/n)^{7/3}\beta_n^{-1/6}$, is of smaller order than the second term, of order $(\log n/\beta_n^2 n)^{2/3}$.

Thus, (3.18) is replaced by

$$P_F\{\|C_n-F_n\|\geqq 2\gamma(F)k_n^{-2}\}<n^{-2}, \tag{3.24}$$

where k_n is determined by (3.23), with the obvious analogue for (3.19).

To indicate the domain of this extension, we consider an example. Suppose $-f'(x)\sim A(F)(\alpha_1(F)-x)^q$ as $x\uparrow\alpha_1(F)$, for some $q>0$. Then $\beta(F,\varepsilon)\sim -f'(\alpha_1(F)-\varepsilon/\bar{C}(F))/\bar{C}^2(F)\sim A(F)\bar{C}^{q-2}(F)\varepsilon^q$ as $\varepsilon\downarrow 0$. Thus, (3.23) yields k_n of order $(n/\log n)^{1/(3+2q)}$, so that (2.10) holds, and consequently the term k_n^{-2} in (3.24) is of order $(n/\log n)^{-2/(3+2q)}$. Since the conclusion of Theorem 1, which gives only an upper bound on $\|C_n-F_n\|$, is of interest only when that bound is $o(n^{-1/2})<O_p(\|F_n-F\|)$, we see that $q<1/2$ is the domain of interest.

5. Although the device just described works equally well for modifiying Lemma 4 when $\alpha_1(F)=+\infty$, neither our original proof nor our modification yields a useful extension of Theorem 1 in that case because of the failure of Lemma 6. In fact, whether or not $\alpha_1(F)<\infty$, if $f(x)\downarrow 0$ as $x\uparrow\alpha_1(F)$ it is not difficult to show that $-f'/f^2$ is unbounded and hence $\gamma(F)=+\infty$. Much more is true, as we now show in demonstrating why the modification of Remark 4 does not work for Lemma 6. Supposing $\alpha_1(F)=+\infty$, we shall show that it is impossible that

$$-f'(x)/f^2(x)<\delta/[1-F(x)] \tag{3.25}$$

for any $\delta<1$ and all large x; putting $x=F^{-1}(1-1/2k)$, this shows that Lemma 6 would yield the inadequate estimate $\sup_{0<x<F^{-1}(1-1/2k)}[F(x)-L(x)]=O(k^{-1})$. If (3.25) held, we would have $-f'/f<\delta f/[1-F]$ on an interval $[t_\delta,\infty)$. Integration yields $-\log[f(x)f(t_\delta)]<-\delta\log\{[1-F(x)]/[1-F(t_\delta)]\}$, or $f(x)/[1-F(x)]^\delta$ bounded below by a value $C_\delta>0$. A second integration yields $(1-\delta)^{-1}\{[1-F(t_\delta)]^{1-\delta}-[1-F(x)]^{1-\delta}\}\geqq C_\delta(x-x_\delta)$, presenting a contradiction as $x\to\infty$. In similar fashion, if $\alpha_1(F)<\infty$ one can show, in place of (3.25), that it is impossible for $-f'[1-F]^\delta/f^2$ to be bounded for any $\delta<1$, and this in turn implies the inadequacy of Lemma 6, in that it cannot provide an estimate of order k^{-r} for $F-L$, for any $r>1$. An analogous result holds if f is unbounded at 0 and truncation at $F^{-1}(1/2k)$ is attempted.

6. If there is an interval $[r_1,r_2]$ of positive length on which f is a positive constant, and if $\bar{L}_n$ is the chord between $(r_1,F_n(r_1))$ and $(r_2,F_n(r_2))$, the probability that $F_n(x)<\bar{L}_n(x)-n^{-1/2}$ for some x in $[r_1,r_2]$ is bounded away from 0 as $n\to\infty$. Thus, there is no useful extension of Theorem 1 if F is not *strictly* concave on $(0,\alpha_1(F))$. However, if F is concave and (3.2) is satisfied when $(0,\alpha_1(F))$ is replaced by the complement therein of such an interval $[r_1,r_2]$, the limiting law of $n^{1/2}\|C_n-F_n\|$ can be expressed as the law of the deviation of a Brownian bridge *below* a chord between points of its graph on a corresponding interval. The limiting law of $\|C_n-F\|$, and the case where there is more than one such interval, can be treated similarly.

4. Asymtotic Minimax Character of F_n and C_n for Estimating a Concave d.f.

Throughout this section we consider

$$\mathscr{F}=\{F\colon F \text{ is continuous and concave on } [0,\infty)\}. \tag{4.1}$$

The conclusions stated below, regarding asymptotic optimality of F_n for estimating F in $\mathscr{F}$, will easily be seen to hold for the more restricted classes of concave d.f.'s discussed in Sections 2 and 3. Moreover, for every concave d.f. F there is a sequence of strictly concave d.f.'s $\{H_j\}$ such that $P_F\{\lim_{j\to\infty} dH_j(x)/dF(x)=1\}=1$, from which it follows that for each sample size n, each estimator of F has the same supremum risk over $\mathscr{F}$ as over

$$\mathscr{F}^*=\{F: F \text{ is continuous and strictly concave on } [0,\infty)\}. \tag{4.2}$$

Consequently, our conclusions hold if (4.1) or any of its variants is reduced by requiring strict concavity.

The space D of decisions (possible estimates of F) is any collection of real functions on R^1 that includes the functions F_n. If D is restricted by demanding that the estimate of F be continuous, our optimality conclusions still hold with F_n replaced by any of several possible continuous modifications F_n^* that have been suggested and for which $\sup_x |F_n(x)-F_n^*(x)|\leqq c_n$ wp 1 under every F, where $c_n=o(n^{-1/2})$. If D is further restricted by demanding that the estimate be concave, no such F_n or F_n^* is usable. However, the simple fact (Lemma 3) that $\|C_n-F\|\leqq \|F_n-F\|$, together with the monotone form of W below and the optimality conclusion when use of F_n *is* permitted, imply that C_n is still asymptotically minimax for D restricted to concave estimates.

Let W be any nonnegative nondecreasing function on the nonnegative reals for which

$$\int_0^\infty W(r)re^{-2r^2}\,dr<\infty. \tag{4.3}$$

We assume W is not identically zero. A nonrandomized estimator g_n of F, based on n observations, takes on values in D and has risk function

$$r_n(F; g_n)=E_F W(n^{1/2}\|g_n-F\|). \tag{4.4}$$

We will not discuss the routine measure-theoretic background and consideration of randomized estimators which are treated in detail in Section 1 of [1]. In the statements that follow, g_n is permitted to be randomized. Our main result is

Theorem 2. *Under the above assumptions,*

$$\lim_{n\to\infty}\frac{\sup_{F\in\mathscr{F}} r_n(F; F_n)}{\inf_{g_n}\sup_{F\in\mathscr{F}} r_n(F; g_n)}=1. \tag{4.5}$$

We remark that it is much simpler to prove asymptotic "optimality" results by replacing r_n by its limit, restricting one's self to regular sequences $\{g_n\}$ for which such limits exist. This is a weaker type of result than ours, and does not imply uniformity in F as $n\to\infty$ of the type exhibited in (4.5). From a practical point of view, such uniformity is important.

Before outlining the proof of Theorem 2, we mention that many other loss and risk functions could be considered in place of those of (4.4). For example,

$$\int_0^\infty E_F W(n^{1/2}|g_n(x)-F(x)|)\,dF(x) \tag{4.6}$$

gives a risk function of integrated form rather than a function of the maximum deviation. More general forms are discussed in [1] and [2], and the optimality results for such loss functions in our setting of concave F are proved by using the developments of those earlier papers in the same way that we now use them to prove (4.5) under (4.4). The reader who is unacquainted with those papers may want to consult pp. 649–650 of [1] and 477–478 of [2] for some explanatory intuitive comments.

Proof of Theorem 2. The required mathematical developments are contained in [1] and [2], and it is only necessary to fit them into the present context. It will be simpler to use the second of these references for part of the results (even though a multivariate d.f. is the ultimate interest there), because of the usable form in which the required preliminary results of [2] are set forth, and because the corresponding results of [1], although more refined in their explicit presentation of an error estimate, are based on Bayes procedure calculations for a particular form of prior law which must be modified (as it is in [2]) to fit the present problem.

For each positive integer h, let $\mathscr{F}^{(h)}$ be the class of all absolutely continuous d.f.'s F of the following form: $F(0)=0$, $F(1)=1$, and the density F' is some constant p_i (say) on each interval $([i-1]/[h+1], i/[h+1])$ for $1 \leqq i \leqq h+1$; furthermore,

$$p_1 \geqq p_2 \geqq \cdots \geqq p_{h+1}. \tag{4.7}$$

This last implies that $\mathscr{F}^{(h)} \subset \mathscr{F}$.

Let U denote the uniform d.f. on [0, 1], and write $\|\Psi\|_h = \max_{1 \leqq i \leqq h+1} |\Psi(i/[h+1])|$. Working in the spirit of [1] and [2], we shall prove (4.5) by showing (a) that the (constant) risk function of F_n for the original problem of this section is close to the value $E_U W(n^{1/2}\|F_n - U\|_h)$ when n and h are large, and that (b) for fixed h and large n, the procedure F_n is approximately minimax for the problem obtained by replacing $\mathscr{F}$ by $\mathscr{F}^{(h)}$ and $\|\cdot\|$ in (4.4) by $\|\cdot\|_h$. More precisely, writing

$$r_n^{(h)}(F; g_n) = E_F W(n^{1/2}\|g_n - F\|_h) \tag{4.8}$$

for F in $\mathscr{F}^{(h)}$, and noting that $r_n^{(h)} \leqq r_n$, and that $U \in \mathscr{F}^{(h)}$, we see that (4.5) is implied by

$$\lim_{h\to\infty} \lim_{n\to\infty} r_n^{(h)}(U; F_n) = \lim_{n\to\infty} r_n(U; F_n) \tag{4.9}$$

and

$$\lim_{n\to\infty} \frac{\sup_{F\in\mathscr{F}^{(h)}} r_n^{(h)}(F; F_n)}{\inf_{g_n} \sup_{F\in\mathscr{F}^{(h)}} r_n^{(h)}(F; g_n)} = 1. \tag{4.10}$$

The result (4.9) is precisely the result (4.5) of [1], and we turn to (4.10).

For fixed h, the problem with $\mathscr{F}$ replaced by $\mathscr{F}^{(h)}$ and r_n replaced by $r_n^{(h)}$ is treated in Section 3 of [2] as a multinomial problem; the multinomial $(h+1)$-vector with i-th-component $nF_n(i/[h+1]) - nF_n([i-1]/[h+1])$ is sufficient for $\mathscr{F}^{(h)}$, and this allows (4.10) to be proved by analyzing the multinomial problem. The result (4.10) is then obtained from Lemma 10 of [2], once one notes two aspects of the developments there. First, the integrability assumption (3.6) of [2], used there because of the consideration of multivariate d.f.'s, can be replaced by our (4.3) in the present univariate setting. Second, the set B_h' of [2] is our (4.7), and thus satisfies the assumption, made there, requiring it to be the closure of its interior. This completes the proof of Theorem 2.

We have not attempted here to obtain a more precise estimate of the departure from minimaxity (departure from 1 of the ratio in (4.5)) as was done in [1]. Indeed, such an estimate seems more difficult to obtain here because the concavity restriction does not permit us to use a sequence of prior laws centered at the uniform law (all $p_i = 1/[h+1]$) for the multinomial problem as we did in [1]; the uniform law is now an extreme point of the set (4.7) on which the prior laws must be supported.

References

1. Dvoretzky, A., Kiefer, J., Wolfowitz, J.: Asymptotic minimax character of the sample distribution function and of the classical multinomial estimator. Ann. Math. Statist. **27**, 642–669 (1956)
2. Kiefer, J., Wolfowitz, J.: Asymptotic minimax character of the sample distribution function for vector chance variables. Ann. Math. Statist. **30**, 463–489 (1959)
3. Kiefer, J., Wolfowitz, J.: Consistency of the maximum likelihood estimator in the presence of infinitely many incidental parameters. Ann. Math. Statist. **27**, 887–906 (1956)
4. Grenander, U.: On the theory of mortality measurement. Part II. Skand. Akt. Tid. **39**, 125–153 (1956)
5. Marshall, A.W.: Discussion of Barlow and van Zwet's papers in M.L. Puri (Ed.). Nonparametric techniques in statistical inference. p. 175–176. Cambridge University Press, 1970
6. Barlow, R.E., Bartholomew, D.J., Bremner, J., M., Brunk. H.D.: Statistical inference under order restrictions. New York: Wiley 1972
7. Weiss, L., Wolfowitz, J.: Maximum probability estimators and related topics. Lecture Notes in Math. **424**. Berlin, Heidelberg, New York: Springer 1974
8. Kiefer, J.: Skorokhod embedding of multivariate r.v.'s, and the sample d.f. Z. Wahrscheinlichkeitstheorie verw. Gebiete **24**, 1–35 (1972)
9. Kiefer, J., Wolfowitz, J.: On the deviations of the empiric distribution function of vector chance variables. Trans. Amer. Math. Soc. **87**, 173–186 (1958)

Received October 13, 1975

17.

LARGE SAMPLE COMPARISON OF TESTS AND EMPIRICAL BAYES PROCEDURES

JACK C. KIEFER was born January 25, 1924 in Cincinnati, Ohio. His professional life has been spent in the Mathematics Department of Cornell University after receiving a Ph.D. degree from Columbia University in 1952. He is a member of the National Academy of Sciences and is also a Fellow of the Institute of Mathematical Statistics (President, 1969-1970).

DAVID S. MOORE received his A.B. degree from Princeton University in 1962 and his Ph.D. degree in Mathematics from Cornell University in 1967. Since that time he has been at Purdue University where he is now an Associate Professor of Statistics. He is Associate Editor of the Journal of the American Statistical Association and author of a number of papers on large sample theory of tests of fit and procedures using order statistics.

JACK C. KIEFER was born January 25, 1924, in Cincinnati, Ohio. His professional life has been spent in the Mathematics Department of Cornell University after receiving a Ph.D. degree from Columbia University in 1952. He is a member of the National Academy of Sciences and is also a Fellow of the Institute of Mathematical Statistics (President, 1969-1970).

DAVID S. MOORE received his A.B. degree from Princeton University in 1962 and his Ph.D. degree in Mathematics from Cornell University in 1967. Since that time he has been at Purdue University where he is now an Associate Professor in Statistics. He is Associate Editor of the *Journal of the American Statistical Association* and author of a number of papers on large sample theory of tests of fit and procedures using order statistics.

17.

LARGE SAMPLE COMPARISON OF TESTS AND EMPIRICAL BAYES PROCEDURES

Jack C. Kiefer

Mathematics Department
Cornell University
Ithaca, New York

David S. Moore

Department of Statistics
Purdue University
Lafayette, Indiana

The organizers of this conference have made a selection of recent influential statistical ideas, and have asked us to present an exposition of the two topics of the title. The emphasis of the first of these is to be on the use of limit theorems other than the central limit theorem in large sample comparison of tests, in contrast with the now more familiar "local" comparison treated by Pitman, Wilks, Wald, LeCam, Neyman, Weiss, and Wolfowitz, among others. The nonlocal comparison of tests was developed by Chernoff, Hodges and Lehmann, and Bahadur, producing a striking result in a paper of Hoeffding (1965). Empirical Bayes procedures were introduced by Robbins (1955).

LARGE SAMPLE COMPARISON OF TESTS

Introduction

We begin with a simple testing model: one observes independent and identically distributed random variables $X_1, \ldots, X_n$. The probability density function of X_i is unknown, but belongs to a known class

349

$\{f_\theta, \theta \in \Theta\}$ labeled in terms of an index set Θ. For example, Θ might be the upper half-plane and $f_{(\theta',\theta'')}$ the normal density with mean θ' and variance θ''. Or the class $\{f_\theta\}$ might consist of every symmetric density and $\theta = (\theta', g)$, where θ' is the median of f_θ and g the density of the "error" $X_1 - \theta'$, symmetric about 0.

It is desired to test the null hypothesis that $\theta \in \Theta_0$ for $\Theta_0 \subset \Theta$ against the alternative that $\theta \in \Theta - \Theta_0$. For simplicity, we shall assume throughout the first two sections of this paper that (1) the parameter space Θ is a subset of Euclidean k-dimensional space; (2) $\Theta_0 = \{\theta_0\}$, so that we are testing the simple null hypothesis H_0: $\theta = \theta_0$; (3) all critical regions considered are defined in terms of sums of iid random variables standardized to approach a normal distribution (under θ_0) by the central limit theorem. (The central limit theorem is not used in approaches that involve a computation like Eq. [2].) Given a sequence of critical regions $\{T_n\}$, we have two probabilities of error: the significance level or probability of erroneous rejection of H_0

$$\alpha_n = P_{\theta_0}[(X_1, \ldots, X_n) \in T_n]$$

and the probability of erroneous acceptance of H_0 when an alternative θ is true

$$\beta_n(\theta) = 1 - P_\theta[(X_1, \ldots, X_n) \in T_n] \qquad \theta \neq \theta_0$$

Suppose we have two competing families of critical regions for the same problem. (We say "family of critical regions" because the region actually used depends on the sample size n and the level α selected. Thus, one might compare--as Hoeffding did--the χ^2 and likelihood ratio families for a multinomial testing problem.) How shall we compare their performance? Given sequences $\{T_n\}$ and $\{T'_n\}$ of critical regions, if T_n has $\alpha_n = \alpha$ and $\beta_n(\theta) = \beta$ for a fixed alternative $\theta \neq \theta_0$, we may ask how many observations m are required for the critical region T'_m to attain $\alpha'_m = \alpha$ and $\beta'_m(\theta) = \beta$. The ratio n/m is the *efficiency* of $\{T'_n\}$ relative to $\{T_n\}$. Unfortunately, this relative efficiency usually depends on several of α, β, θ_0, θ, and n. It is therefore natural to seek some large sample simplification by investigating the behavior of the error probabilities as the sample size n increases. In addition to identifying good statistical procedures for large samples, such studies may suggest the form of good procedures for small n.

Any reasonable sequence of tests has the property that as n increases and the level α remains fixed, the probability $\beta_n(\theta)$ of erroneous acceptance approaches zero for any alternative θ. Thus, some quantity (α, β, or θ) in addition to n must change as n increases, and various approaches to large sample comparison of tests can be distinguished by the constraints placed on these quantities.

We wish to stress two themes in the development of this area: First, the use of tools other than the central limit theorem to compare tests. (This may be called the "mathematical front.") Second, establishment of the large sample optimality of procedures based on likelihood, in this case the likelihood ratio (LR) family of tests. (This is the "likelihood front," on which there have been significant advances in the theory of estimation as well as testing.) When the alternative as well as the hypothesis is simple, the Neyman-Pearson Fundamental Lemma, of course, states that any LR test is most powerful of its level for any sample size. Large sample optimality of LR tests (as of the analogous maximum likelihood estimators) has since been established with respect to a number of criteria. Hoeffding's contribution was to show that even families of tests that differ by little from LR tests and that are asymptotically equivalent to them in one sense (Pitman efficiency) may be inferior in another sense, in large samples.

Approaches to Large Sample Comparison

We will mention three approaches. The earliest of these was to study the relative efficiency of tests when α is fixed (or $\alpha_n \to \alpha$ and $0 < \alpha < 1$) and the alternative θ_n varies with n in such a way that $\beta(\theta_n) \to \beta$ for a fixed β, $0 < \beta < 1$. In the cases we are discussing, θ_n must approach θ_0 at rate $n^{-1/2}$ to obtain nontrivial α and β. This "local comparison" was systematized by Pitman in the one-dimensional case and bears his name. The central limit theorem is the essential mathematical tool in studying local alternatives. Early work in this setting is attributable to Wilks; Wald's definitive paper (1943) established several optimum properties of the LR and related families. Further work on local properties is contained in the work of LeCam, Neyman, and Weiss and Wolfowitz; this includes the more complex case of composite null hypotheses, various optimality criteria, and families of procedures other than LR tests. We omit details, as local comparisons are not our concern here.

[We remark that some of these last-mentioned developments are counterparts of the asymptotically efficient estimation results for Bayes and maximum likelihood estimators (LeCam, Wolfowitz), which are relevant to the discussion in the section on the standard Bayesian model (vide infra), and for Wolfowitz's maximum probability estimator.]

Other comparisons of tests leave the alternative θ fixed. Calculation of the probabilities of error can then no longer be handled by the central limit theorem, but requires results on probabilities of large deviations. (If t_n is a normalized sample mean, so that t_n converges in law to the standard normal distribution by the central limit theorem, $\{t_n \geq a_n\}$ is a *large deviation* of t_n if $n^{-1/2} a_n \to a$ for $0 < a < \infty$. In this case, the central limit theorem says only that $p_n = P[t_n \geq a_n]$ approaches 0, which is uninformative. We want to know the speed with which this probability approaches 0.) Cramér began the study of this probabilistic problem in 1938, and the literature now contains many sources for both order results (of the form $n^{-1} \log p_n \to c$) and asymptotic results (of the form $p_n/c_n \to 1$) for probabilities of large deviations. Only order results are required for the large sample comparisons of tests done to date.

Chernoff (1952) first used probabilities of large deviations to compare tests. We will mention two contrasting "fixed alternative" approaches. Hodges and Lehmann (1956) fixed θ and $\alpha (0 < \alpha < 1)$ and studied the rate of convergence to 0 of $\beta_n(\theta)$. In the usual cases

$$\beta_n(\theta) = e^{-nc(\theta)[1+o(1)]} \quad \text{for all } \alpha \qquad [1]$$

so that an asymptotic relative efficiency can be defined as the ratio of the indices $c(\theta)$ for competing families of tests.

Beginning in 1960, R. R. Bahadur produced an extensive theory of large sample properties of statistical procedures, which he recently summarized in a monograph, Bahadur (1971). His approach to tests can be stated as follows: fix θ and β and study the rate of convergence of α_n to 0. Again, one usually obtains

$$\alpha_n = e^{-nb(\theta)[1+o(1)]} \quad \text{for all } \beta \qquad [2]$$

so that an asymptotic relative efficiency can again be defined. Bahadur's approach has borne more fruit than has that of Hodges and Lehmann for two reasons. First, it is easier; Eq. [2] requires an order result for probabilities of large deviations under θ_0, whereas Eq. [1] requires a similar result under the alternative θ. The Bahadur index $b(\theta)$ has therefore been computed for many more families of tests than has the Hodges-Lehmann index $c(\theta)$. Second, we have done Bahadur an injustice to have described his work in this framework. His basic idea was to study the behavior of the actually attained level of the test as a random variable. This natural "stochastic comparison" of tests turns out to be equivalent to the nonstochastic comparison based on Eq. [2]. Bahadur has shown in some generality that LR tests have maximum $b(\theta)$ and are therefore asymptotically optimal by his criterion.

Hoeffdings' Contribution

Hoeffding (1965) considered several testing problems involving the multinomial distribution with k cells and unknown vector $\theta = (\theta_1, \ldots, \theta_k)$ of cell probabilities. We will discuss only the problem of testing the simple null hypothesis $\theta = \theta_0$ for a fixed probability vector θ_0. Hoeffding made an advance on both the mathematical front and the LR front. Mathematically, he built on work of Sanov to give an order result for probabilities of large deviations in this k-dimensional multinomial case. Previous comparisons of tests had used such results only for sums of univariate random variables.

On the LR front, Hoeffding succeeded in distinguishing the large sample performance of the LR family of tests for $\theta = \theta_0$ from that of the familiar Pearson χ^2 tests for this problem. If $\bar{X}_{in}$ for i = 1, ..., k is the proportion of n observations falling in the ith cell, the LR test is based on the information-distance statistic

$$L_n = \Sigma_i \bar{X}_{in} \log \frac{\bar{X}_{in}}{\theta_{0i}}$$

The χ^2 statistic is of course

$$Q_n^2 = \Sigma_i \frac{(\bar{X}_{in} - \theta_{0i})^2}{\theta_{0i}}$$

These tests had long been treated as being asymptotically equivalent because of their equivalence under local comparison. Q_n^2 is the dominant term in the Taylor's series expansion of L_n about θ_0; $2nL_n$ and nQ_n^2 have the same χ^2 limiting distribution under the null hypothesis; the two families of tests have the same large sample performance against local alternatives.

The spirit and nature of Hoeffding's comparison can be demonstrated with minimal mathematics in the two-cell (k = 2) case. This we do in the next section, which may be omitted without loss of continuity. Here, we content ourselves with observing that although Hoeffding's precise comparison was not one of those discussed above, he implicitly showed that the LR and χ^2 families have the same Hodges-Lehmann performance for all alternatives θ, but that *the LR family has strictly better Bahadur performance for "most" alternatives*. As Bahadur's theory has become a standard tool in the decade since Hoeffding's work, the latter work is now most easily understood in Bahadur's framework. That it could be so understood was shown in detail by J. C. Gupta (1972).

Fixed-alternative comparisons ask more of the χ^2 test than its creators probably intended. The coincidence of power results for local alternatives is closer to the motivation for Q_n^2, which involves the relevance of the expected value of Q_n^2 and hence of normal theory. Nevertheless, this common test has been discredited for large samples and fixed alternatives by Hoeffding's result.

Progress on the LR front has, of course, continued. Brown (1971) has shown in considerable generality that appropriate tests of LR type (actually LR tests of possibly larger hypotheses) are at least as good as any given sequence of tests in both the Hodges-Lehmann and Bahadur senses. The more difficult task of analyzing what makes an apparently equivalent test strictly inferior to a LR test for large samples in the generality of Brown's setting awaits another advance on the mathematical front--more general large deviation results for multivariate problems. Herr (1967) has done this in certain multivariate normal cases, but much work remains. It would also be valuable to investigate the sample size required for LR tests to be close to optimal (or, alternatively, to be superior to χ^2 tests in the multinomial case).

Hoeffding's Result for Two Cells

To illuminate Hoeffding's discovery that Q_n^2 is inferior to L_n, we will consider the special case k = 2. This amounts to observing n independent Bernoulli random variables $X_1, \ldots, X_n$ with $\theta = P[X_i = 1]$ unknown. For $0 < p < 1$, the test statistics for the hypothesis $\theta = p$ are $L_n = I(\bar{X},p)$ and $Q_n^2 = Q^2(\bar{X},p)$, where $\bar{X}$ is the sample mean of $X_1, \ldots, X_n$ and

$$I(\theta,p) = \theta \log \frac{\theta}{p} + (1 - \theta)\log \frac{1 - \theta}{1 - p}$$

$$Q^2(\theta,p) = \frac{(\theta - p)^2}{p(1 - p)}$$

The information distance $I(\theta,p)$ between two probability vectors [here between $(\theta,1-\theta)$ and $(p,1-p)$] plays a central role in all large deviation comparisons of tests.

Analysis of this special case requires only one mathematical tool, an order result for probabilities of large deviations of binomial random variables, obtainable from Cramér's inequality. Specifically, if B_n is a binomial random variable with mean np, $q = 1 - p$ and

$$p_n = P\left[\frac{B_n - np}{(npq)^{1/2}} \geq bn^{1/2}\right] \qquad b > 0$$

then, for $p + b(pq)^{1/2} < 1$,

$$\frac{1}{n}\log p_n \to -I[p + b(pq)^{1/2}, p] \qquad [3]$$

From Eq. [3], one may first calculate that the Hodges-Lehmann index $c(\theta)$ defined in Eq. [1] is $I(p,\theta)$ for both L_n and Q_n. Thus, the Hodges-Lehmann approach also fails to distinguish Q_n from L_n.

The Bahadur index $b(\theta)$ defined in Eq. [2] depends on whether $p > 1/2$ or $p < 1/2$. (When $p = 1/2$, the tests based on L_n and Q_n are identical.) For the remainder of this discussion, we assume that the hypothesized p exceeds 1/2. Another application of Eq. [3] then shows that for Q_n

$$\begin{aligned} b_Q(\theta) &= I(\theta,p) & 0 \leq \theta \leq p \qquad [4]\\ &= I[\theta - 2(\theta - p), p] & p \leq \theta \leq 1 \end{aligned}$$

whereas it is known that for the LR statistic L_n

$$b_L(\theta) = I(\theta,p) \qquad 0 \leq \theta \leq 1 \qquad [5]$$

The situation is illustrated in Fig. 1, where $Q^2(\theta,p)$ and $I(\theta,p)$ are drawn for $p = 3/4$. Note that $I(\theta,p)$ is not symmetric about p, but increases more slowly for $\theta < p$ when $p > 1/2$. Thus, Eqs. [4] and [5] say that $b_Q(\theta) < b_L(\theta)$ for $\theta > p$, so that Q_n is inferior to L_n against alternatives $\theta > p$.

Let us now look at this comparison as Hoeffding did. For sufficiently regular sets A, he showed (for general k) that

$$\frac{1}{n} \log P[\bar{X} \in A|\theta] \to - I(A,\theta) \qquad [6]$$

where

$$I(A,\theta) = \inf_{\omega \in A} I(\omega,\theta)$$

is the information distance of A from θ. Suppose, next, that for $\delta > 0$

$$A(\delta) = \{\theta: Q^2(\theta,p) \geq \delta\}$$

is a χ^2 critical region and

$$B(\delta) = \{\theta: I(\theta,p) \geq I(A(\delta),p)\}$$

is a corresponding LR critical region. These regions are illustrated in Fig. 1.

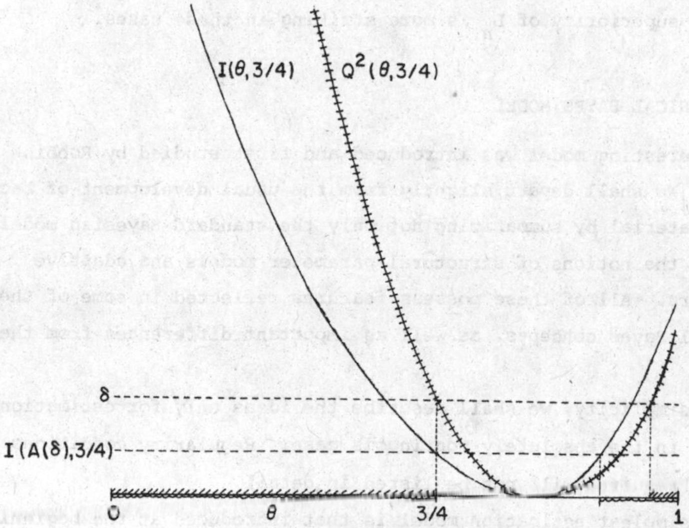

FIG. 1. $I(\theta,3/4)$ and $Q^2(\theta,3/4)$. The hatched region above the axis indicates the set $A(\delta)$ and the marked region below the axis indicates the set $B(\delta)$.

Applying Eq. [6] with $\theta = p$ and $A = A(\delta)$ or $B(\delta)$ shows that both critical regions have asymptotically the same log α_n, as both are the same information distance from p. The LR critical region includes the χ^2 region and is thus at least as powerful. More specifically, alternatives $\theta < p$ are the same information distance from both acceptance regions $A(\delta)^c$ and $B(\delta)^c$, and hence have asymptotically the same log β_n. [In this heuristic sketch, we consider only alternatives in $A(\delta)$; this includes any given $\theta \neq p$ for sufficiently small δ.] But alternatives $\theta > p$ are strictly closer to the χ^2 acceptance region and therefore Q_n has larger log β_n than does L_n for these alternatives.

The geometry of Fig. 1 is indicative of the general case. Hoeffding showed that for any k the analogs of $A(\delta)$ and $B(\delta)$ have only finitely many common boundary points, one on each line segment joining $p = (p_1, \ldots, p_k)$ to the unit vector in the direction of smallest components p_i. L_n is superior to Q_n for all θ not lying on these line segments, by arguments indicated above. When $k > 2$, the exceptional line segments form a small portion of the parameter space, so that the superiority of L_n is more striking in these cases.

THE EMPIRICAL BAYES MODEL

This interesting model was introduced and first studied by Robbins (1955). We shall depart slightly from the usual development of background material by summarizing not only the standard Bayesian model, but also the notions of structural parameter models and adaptive estimators. All of these possess features reflected in some of the empirical Bayes concepts, as well as important differences from the latter.

For simplicity, we shall describe the ideas only for estimation problems in the absolutely continuous case. Regularity conditions that are required will not be listed in detail.

The simplest estimation model is that introduced at the beginning of the first section, except that the object is now to *estimate* some function ϕ of the unknown θ governing the probability law of the X_i. A common example is $\phi(\theta) = \theta'$ in either of the examples of the first section. An estimator t_n is a rule for guessing $\phi(\theta)$ on the basis of $X_1, \ldots, X_n$. As in the case of testing, it is often difficult to compute an estimator which is "optimal" in some prescribed sense for a given sample size n. It is again natural to study sequences $\{t_n\}$ of estimators as the sample size n increases in the hope of establishing desirable large sample properties.

The Standard Bayesian Model

The Bayesian model adds two assumptions to the estimation problem described above: (1) that the parameter θ can be regarded as a random variable, and (2) that the prior distribution G of this random variable is known. Note that the value of θ, once it is chosen according to G, remains the same in the density f_θ of each X_i.

Bayes' theorem combines G with the observed data to produce the posterior distribution of θ. Comparison of estimators in this model is based on the posterior expected loss $R(t_n, G)$ of an estimator t_n. A *Bayes estimator* $t^*_{G,n}$ of $\phi(\theta)$ is an estimator that minimizes this expected loss. For example, if (as in the examples of the first section) a real parameter $\phi(\theta)$ is to be estimated and loss is measured by squared error, then $t^*_{G,n}$ is the posterior expectation of $\phi(\theta)$.

A feature of interest to us in this Bayesian formulation is that the desired performance of the Bayes procedure is relatively insensitive to slight errors in the specification of G. More precisely

$$\frac{R(t^*_{G',n}, G)}{R(t^*_{G,n}, G)} \quad [7]$$

is close to 1 when G is close to G' (under reasonable regularity conditions, as usual). Thus, using $t^*_{G',n}$ when the actual prior law is G (close to G') is almost as good as using the Bayes procedure $t^*_{G,n}$ relative to G. This is just the asymptotic optimality of Bayes estimators referred to in the section on large sample comparison of tests (vide supra).

If the Bayesian model as stated above is correct, there is no disagreement about using $t^*_{G,n}$. One source of controversy arises because doubt may be thrown on the simple-minded form assumed for $\{f_\theta\}$, or for the assumed loss function, or on the stated aim of the inference [estimation of $\phi(\theta)$]. Another source of controversy lies in the Bayesian assumptions (1) and (2). Bayesian statisticians feel that a description of rational thought legitimizes the use of a subjective guess for G in the absence of knowledge of an actual G; others disagree strongly, but we need not discuss this controversy in detail here.

The Empirical Bayes Model

We now turn to Robbins' model. We are faced with a sequence of independent estimation problems, each of which must be acted on as it arises. These problems are, however, related as follows: the observed X_i in the ith problem has density f_{θ_i} once θ_i is given, but the θ_i are themselves iid with unknown distribution G. So at the nth inference, we have available $X_1, \ldots, X_n$ and we can hope that if n is large some information about the unknown prior law G can be wrung from the past observations $X_1, \ldots, X_{n-1}$. If we knew G exactly, we would estimate θ_n by $t^*_{G,1}(X_n)$, and in the absence of such exact knowledge it seems reasonable (as discussed below in the section on adaptive estimators) to use this estimator with G replaced by an estimator of G; this is Robbins' proposal, which we now describe in further detail.

One can construct an empirical Bayes estimator of θ_n by (1) finding an estimator $\hat{G}_{n-1}(X_1, \ldots, X_{n-1})$ of G; and (2) using the Bayes estimator $t'_n \overset{\text{def}}{=} t^*_{\hat{G}_{n-1},1}(X_n)$. One thus acts as if $\hat{G}_{n-1}$ were the known prior law. If the estimator $\hat{G}_{n-1}$ of G is a good one, then on the basis of Eq. [7] we expect that for large n

$$\frac{R(t'_n, G)}{R(t^*_{G,1}, G)} \qquad [8]$$

is close to 1. Robbins found appropriate $\hat{G}_{n-1}$ for several problems and established Eq. [8] in these cases. Thus, we do as well asymptotically in estimating θ_n when G is unknown as we would if we knew the prior law G.

The proof of Eq. [8] in various settings and the study of the rate of convergence to 1 of the ratio in Eq. [8] has produced a body of literature by Hannan, Johns, van Ryzin, Samuel, Gilliland, and others. One can expect further research to yield reasonably efficient procedures for small n.

Of interest to many observers will be the extent to which Bayesians are able in practice to depart from the standard Bayesian model with a subjective guess of G, and can instead imbed the problem at hand as the nth one in the empirical Bayes model, to yield and use a more formally described guess $\hat{G}_{n-1}$. To non-Bayesians, Robbins' model will seem much more acceptable in many practical settings than the original Bayesian formulation. For example, X_i might be an observation of some biological characteristic of an organism at location i, governed by a parameter θ_i of which distribution G is characteristic of the species but is unknown. Or X_i might be the result of a diagnostic test on individual i made at a preventive medicine clinic run for workers in a large plant, and θ_i an index of the underlying condition having an unknown distribution characteristic of this population of workers.

Structural Parameter Models

Robbins' model may be compared with one which had already been the subject of a large body of literature by 1955, estimation of structural parameters. This is a non-Bayesian framework.

Here again the observations $X_1, \ldots, X_n$ are independent, but the density of X_i is indexed by (α, θ_i). The structural parameter α is to be estimated, while θ_i is an incidental parameter which varies from observation to observation. In the most common example, the X_i are points in the plane derived from an unknown line α by adding independent error vectors with zero means to points on α with abscissas θ_i. This simplest line-fitting situation is illustrated in Fig. 2. One approach to the structural parameter problem is to consider the θ_i to

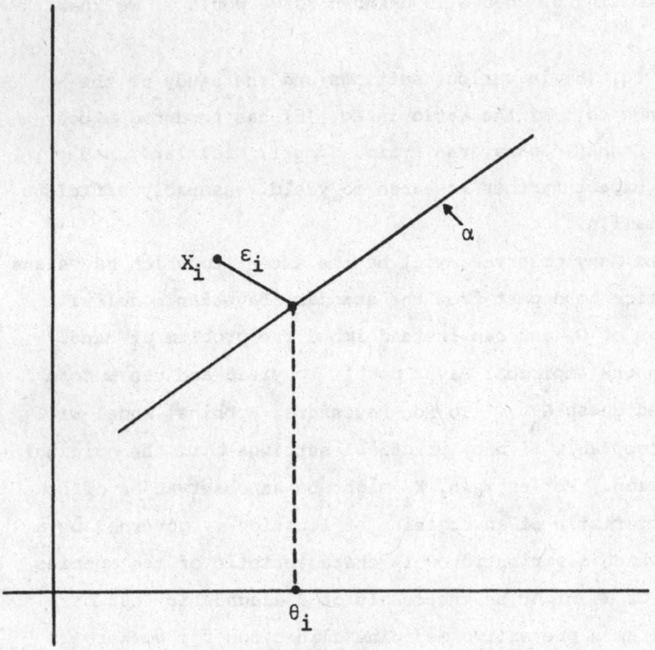

FIG. 2. The structural parameter model with α an unknown line, and observation X_i determined by adding a random error vector ε_i to the point on α with abscissa θ_i.

be independent random variables with the same law G. Indeed, the term "structural model" is sometimes reserved for this case. Such models have been studied by Geary, Reiersol, Wald, Neyman and Scott, Wolfowitz, and others. A survey of their work is given by Moran (1971).

The frameworks of the empirical Bayes model and the structural parameter model have been exhibited in Fig. 3 to point out their considerable similarities. The difference between the models lies primarily in the inference to be made.

EMPIRICAL BAYES MODEL

$X_1, \ldots, X_n$ independent observations

X_i has parameter θ_i

$\theta_i, \ldots, \theta_n$ iid with law G unknown

X_i has marginal density

$$f_G(X) = \int f_\theta(x)\, dG(\theta)$$

Use $X_1, \ldots, X_n$ to estimate θ_n

STRUCTURAL PARAMETER MODEL

$X_1, \ldots, X_n$ independent observations

X_i has parameter (α, θ_i)

$\theta_1, \ldots, \theta_n$ iid with law G unknown

X_i has marginal density

$$f_{\alpha,G}(x) = \int f(x|\alpha,\theta)\, dG(\theta)$$

Use $X_1, \ldots, X_n$ to estimate α

FIG. 3. Comparison of the empirical Bayes and structural parameter models for estimation.

Adaptive Estimators

The methodology of estimation used in the empirical Bayes setting is related to a methodology arising in the example $\theta = (\theta', g)$ of the first section and which can also be employed in the structural parameter model. If we knew g in the nonparametric problem given in the first section, we could use some known good estimator, e.g., Pitman's location parameter estimator $t_{g,n}$ (say), for estimating θ'. The form of this estimator of course depends on g. Because we do not know g, we find an appropriate estimator $\hat{g}_n$ of it, based on $X_1, X_2, \ldots, X_n$ and then use $t_{\hat{g}_n,n}$ to estimate θ'. This rough recipe requires care in its execution, but such an approach has been carried out in various settings by Weiss and Wolfowitz, LeCam, Hajek, and others. Under suitable assumptions on the unknown g, one can find an estimator of the form $t_{\hat{g}_n,n}$ or something similar, whose accuracy is asymptotically the same as that of the $t_{g,n}$ we would use if we *knew* g.

The spirit of this approach, of constructing procedures by *adapting* their form to what the data seems to say about the error law, is also used in a number of small sample-size studies of "robust" estimators.

Empirical Bayes estimators make use of adaptive estimation in the estimation of G by $\hat{G}_n$. It is also clear that a possible approach to the construction of "good" estimators of α in the structural model is to first estimate G by some $\hat{G}_n(X_1, \ldots, X_n)$, then substitute $\hat{G}_n$ into the estimator $t_{G,n}$ of α that we would use were G known. In some of the work in this setting, G is only estimated implicity; in other work, explicit estimates are given (e.g., Wolfowitz's minimum distance estimator).

Thus, the empirical Bayes model is not only connected with the Bayesian formulation of inference problems but is tied in spirit to structural models and adaptive estimators. One may even ask if there are some practical structural parameter problems in which the successive θ_i are of enough interest to be estimated along with α. Empirical Bayes methods can then be used.

ACKNOWLEDGMENT

The work of the first author was written under National Science Foundation Grant 35816 GPX. The work of the second author was sponsored, in part, by the Air Force Office of Scientific Research, Air Force Systems Command, USAF, under Grant No. AFOSR-72-2350. The U.S. government is authorized to reproduce and distribute reprints for governmental purposes notwithstanding any copyright notation hereon.

REFERENCES

Bahadur, R. R. (1971). *Some limit theorems in statistics*. *Reg. Conf. Ser. Appl. Math. 4*, Soc. Ind. Appl. Math., Philadelphia.

Brown, L. D. (1971). Non-local asymptotic optimality of appropriate likelihood ratio tests. *Ann. Math. Statist. 42*, 1206-1240.

Chernoff, H. (1952). A measure of asymptotic efficiency for tests of a hypothesis based on the sum of observations. *Ann. Math. Statist. 23*, 493-507.

Gupta, J. C. (1972). Probabilities of medium and large deviations with statistical applications. Ph.D. thesis, University of Chicago.

Herr, D. G. (1967). Asymptotically optimal tests for multivariate normal distributions. *Ann. Math. Statist. 38*, 1829-1844.

Hodges, Jr., J. L. and Lehmann, E. L. (1956). The efficiency of some nonparametric competitors of the t-test. *Ann. Math. Statist. 27*, 324-335.

Hoeffding, W. (1965). Asymptotically optimal tests for multinomial distributions. *Ann. Math. Statist. 36*, 369-408.

LeCam, L. (1956). On the asymptotic theory of estimation and testing hypotheses. *Proc. 3rd Berkeley Symp. Math. Statist. Prob. 1*, 129-156.

Moran, P. A. P. (1971). Estimating structural and functional relationships. *J. Multivariate Anal. 1*, 232-255.

Neyman, J. (1959). Optimal tests of composite statistical hypotheses. In *The Harold Cramér Volume*, pp. 213-234. Almquist & Wiksell, Stockholm.

Robbins, H (1955). An empirical Bayes approach to statistics. *Proc. 3rd Berkeley Symp. Math. Statist. Prob. 1*, 157-163

Wald, A. (1943). Tests of hypotheses concerning many parameters when the number of observations is large. *Trans. Amer. Math. Soc. 54*, 426-482.

Weiss, L., and Wolfowitz, J. (1969). Asymptotically minimax tests of composite hypotheses. *Z. Wahrschein. Verw. Geb. 14*, 161-168.

Reprinted from
On the History of Statistics and Probability.
Proc. of a Symp. on the American Mathematical Heritage,
(1976), 347–365.
M. Dekker, New York

The Annals of Statistics
1976, Vol. 4, No. 5, 836–865

ADMISSIBILITY OF CONDITIONAL CONFIDENCE PROCEDURES

BY J. KIEFER[1]
Cornell University

A formal structure for conditional confidence (cc) procedures is investigated. Underlying principles are a (conditional) frequentist interpretation of the cc coefficient Γ, and highly variable Γ. The latter allows the stated measure of conclusiveness to reflect how intuitively clear-cut the outcome of the experiment is. The methodology may thus answer some criticisms of the Neyman–Pearson–Wald approach, but is in the spirit of the latter and includes it. Example: X has one of k densities f_ω wrt ν. A nonempty set of decisions $D_\omega \subset D$ is "correct" for state ω. A nonrandomized cc procedure consists of a pair (δ, Z) where δ is a nonrandomized decision function and Z is a conditioning rv. The cc coefficient is $\Gamma_\omega = P_\omega\{\delta^{-1}(D_\omega) \mid Z\}$. If $X = x_0$, we make decision $\delta(x_0)$ with "cc $\Gamma_\omega(x_0)$ of being correct if ω is true"; it is unnecessary, but often a practical convenience (as for un-cc intervals), to have Γ_ω independent of ω. Possible notions of "goodness" are discussed; e.g., $(\bar{\bar{\delta}}, \bar{\bar{Z}})$ at least as good as $(\bar{\delta}, \bar{Z})$ if $P_\omega\{\bar{\bar{\delta}}(X) \in D_\omega$ and $\bar{\bar{\Gamma}}_\omega > t\} \geqq P_\omega\{\bar{\delta}(X) \in D_\omega$ and $\bar{\Gamma}_\omega > t\}$ $\forall\, t, \omega$, and $P_\omega\{\bar{\bar{\Gamma}}_\omega = 0\} \leqq P_\omega\{\bar{\Gamma}_\omega = 0\}$ $\forall\, \omega$. It is proved that cc procedure $(\bar{\delta}, Z)$ is admissible if the non-cc δ is admissible. For 2-hypothesis problems the converse is true; otherwise, "star-shaped" partitions of the likelihood ratio space are needed. Other loss structures are also treated.

0. Introduction. Although there is a large literature of conditioning in statistical inference, there has been no methodical presentation of a frequentist non-Bayesian framework that considers possible criteria of goodness of such procedures, and methods for constructing them, in general statistical settings. The present paper is devoted to decision-theoretic admissibility considerations, in such a framework.

Various illustrations of this approach are contained in [4] and [2]. Its relationship with other work on conditioning, and some discussion of foundations, will be found in [2]. Still, it may be appropriate here to indicate that the present approach was motivated by the observation that many critics of the Neyman–Pearson–Wald (NPW) approach to statistics seem disturbed at the prospect of making a decision that is not accompanied by some data-dependent measure of "conclusiveness" of the experimental outcome. For example, in deciding whether a normal rv X with unit variance has mean -1 or $+1$, where the standard symmetric NP test makes a decision accompanied only by the assessment of error probabilities $\Phi(-1)$, some statisticians may be disturbed by an

Received April 1975; revised February 1976.
[1] Research under NSF Grant MPS 72-04998 A02.
AMS 1970 *subject classifications.* 62C07, 62C15.
Key words and phrases. Conditional inference, conditional confidence coefficients, admissibility, complete classes, canonical procedure.

intuitive feeling that they are much surer of the conclusion when $X = 10$ than when $X = .5$. The authors of various axiomatic studies of statistical foundations (Bayesian, likelihood, fiducial, evidential, etc.) have criticized other aspects of the NP approach as well; but this notion of wanting a measure of conclusiveness that depends on the experimental outcome has appealed also to many practitioners, as is evidenced by the old and continued usage of the methodology that states the level at which a singificance test would "just reject" a null hypothesis (and which is historically a starting point for an extensive theory constructed by Bahadur).

A principle adopted in our developing of a methodology that may satisfy the objection mentioned in the previous paragraph, is that our measure of conclusiveness should have a frequentist (law of large numbers) meaning similar to that emphasized by Neyman for classical NP tests and confidence intervals. As discussed in [2], this implies that, except in rare cases of symmetry, our approach and form of conclusion cannot agree with those of the critics mentioned above.

The methodology we propose is a procedure consisting of a decision rule and a conditioning random variable. The frequentist properties of such a procedure are studied in terms of the conditional probability of a correct decision given the conditioning random variable. This quantity is called the "conditional confidence," extending the usage of "confidence" from its traditional meaning in unconditional interval estimation to the present setting, where it has an analogous frequentist interpretation.

In conformity with our motivating comments, this conditional confidence coefficient should in practice be highly variable as a function of the conditioning random variable; otherwise, we may as well use an unconditional NPW procedure. There is a brief discussion of such variability near the end of Section 2. The basic notions are introduced in Section 1; the ideas are illustrated in examples in Section 1.3. In Section 2 some possible admissibility criteria are defined. The simplest of these regards a procedure as inadmissible if there is another procedure with smaller probability of yielding a zero conditional confidence coefficient, and with stochastically larger coefficient on the set where a correct decision is made.

In Section 3, sufficient conditions for admissibility are obtained. It turns out, perhaps surprisingly, that if a NPW unconditional procedure is admissible in a classical sense, then every conditioning used in conjunction with it yields a procedure admissible in the ordering considered here. Just as in classical decision theory, additional criteria are needed to choose among these; some such criteria are discussed in Section 2 and, more extensively, in [2].

Section 5 contains complete class theorems for the k-hypothesis problem. The result is especially simple when $k = 2$, where the minimal complete class coincides with that obtained in Section 3; for larger k, additional procedures are included. In leading up to these results, Section 4 studies special "canonical" procedures in terms of which closure results are proved. Section 6 discusses some other possible definitions of admissibility.

1. Notation, definitions, examples.

1.1. *Basic ideas and notation.* It is convenient to assume the underlying measure space of possible outcomes of the experiment, $(\mathscr{X}, \mathscr{B}, \nu)$, to be of countable type with compactly generated σ-finite ν with respect to which the possible states of nature $\Omega = \{\omega\}$ have densities f_ω. This implies existence of regular conditional probabilities in the sequel. (See, e.g., [6], pages 193–194.) We write X for the rv representing the experiment with outcome in $\mathscr{X}$. By I_A we denote the characteristic function of the subset A of the domain of this function; in particular, it may be the indicator of the event A.

In the decision space D, we assume that for each ω there is specified a nonempty subset D_ω of decisions that are "correct" when ω is true. We write $\Omega_d = \{\omega : d \in D_\omega\}$. The confidence "flavor" is best exhibited in terms of the simple loss function implied by consideration in these terms; other possible treatments will be described in Section 6.

A (nonrandomized) decision rule $\delta : \mathscr{X} \to D$ is required to have $\delta^{-1}(D_\omega) \in \mathscr{B}$ for each ω. (This may easily be translated in terms of a structure on D.) The apparent neglect of randomization is intended to aid in clarity of exposition. It is not a genuine neglect, since, as will be discussed presently, $\mathscr{X}$ will be regarded as the product of a more primitive sample space $\mathscr{X}'$ and a randomization space.

Any subfield $\mathscr{B}_0$ of $\mathscr{B}$ may be called a *conditioning subfield*. It is convenient to consider only those subfields generated by statistics on $(\mathscr{X}, \mathscr{B})$. If Z is such a statistic, with $Z(\mathscr{X}) = B$,

$$\mathscr{B}_0 = \{S : S \in \mathscr{B}, S = Z^{-1}(A) \text{ for some } A \subset B\}.$$

The relationship between $\mathscr{B}_0$ and the partition of $\mathscr{X}$ induced by Z is discussed in [1]. We denote the resulting (conditioning) partition of $\mathscr{X}$ by $\{C^b, b \in B\}$. We also write, for a given δ and $\mathscr{B}_0$,

$$\begin{aligned} C_d{}^b &= C^b \cap \delta^{-1}(d), \\ C_\omega &= \delta^{-1}(D_\omega) = \{x : \delta(x) \text{ is correct when } \omega \text{ is true}\}, \\ K_\omega{}^b &= C^b \cap C_\omega. \end{aligned} \tag{1.1}$$

The $K_\omega{}^b$ are not necessarily disjoint, but the $C_d{}^b$ $(b \in B, d \in D)$ are, and constitute the *partition* C of $\mathscr{X}$. Note that subscripts always index D or Ω; superscripts index B; symbols such as overbars or left superscripts distinguish different partitions and their operating characteristics.

A *conditional confidence procedure* is a pair $(\delta, \mathscr{B}_0)$ or (δ, Z) or, equivalently, the corresponding partition C. Its associated *conditional confidence function* is the set $\Gamma = \{\Gamma_\omega, \omega \in \Omega\}$ of conditional probabilities

$$\Gamma_\omega = P_\omega\{C_\omega \mid \mathscr{B}_0\}, \qquad \omega \in \Omega. \tag{1.2}$$

For each ω, this $\mathscr{B}_0$-measurable function on $\mathscr{X}$ can be regarded as a function of Z, and we write $\Gamma_\omega^{\ b}$ for its value on the set $Z = b$. A *conditional confidence statement* associated with $(\delta, \mathscr{B}_0)$ is the pair (δ, Γ). It is used as follows: if $X = x_0$, and $Z(x_0) = z_0$, we state that "for each ω, we have confidence $\Gamma_\omega(x_0) = \Gamma_\omega^{\ z_0}$ of being correct if ω is true." We make decision $\delta(x_0)$.

Such statements have a conditional frequentist interpretation analogous to that of the NPW setting $\mathscr{B}_0 = \{\mathscr{X}, \emptyset\}$, where Γ_ω is simply the probability of a correct decision when ω is true. In a sequence of n independent experiments, if ω_i is true and X_i is observed in the ith experiment, then the proportion of correct decisions made will be close to $n^{-1} \sum_1^n \Gamma_{\omega_i}(X_i)$ with probability near one when n is large. This is true even if different $(\delta, \mathscr{B}_0)$'s are used in the experiments, but the frequentist meaning is of most intuitive value in cases where the dependence of Γ_ω on ω can be removed, just as in the usual NP treatment of confidence intervals or composite hypotheses with specified minimum power. Thus, if there is a function ψ on $\mathscr{X}$ such that $\Gamma_\omega(x) \geqq \psi(x)$ for all ω and all x, we may make the more succinct statement that "we have conditional confidence at least $\psi(x_0)$ that $\delta(x_0)$ is a correct decision," if $X = x_0$. The frequentist interpretation in terms of the law of large numbers is thereby simplified, and its practical meaning is made clearer if we look only at those experiments in which $\Gamma(X_i) \geqq .9$ (for example): a correct decision is very likely made in about 90% or more of such experiments, if their number is large. Whenever there is a $\mathscr{B}_0$-measurable set A of positive ν measure on which $\Gamma_\omega^{\ Z(x)}$ can be chosen not to depend on ω, we write Γ^b in place of $\Gamma_\omega^{\ b}$ for b in $Z(A)$.

The class of all conditional confidence procedures C is denoted by $\mathscr{C}$. Additional notation will be introduced as required.

1.2. *More about conditioning partitions.* In Section 2 we shall discuss goodness criteria for partitions. One element of those considerations is the set

$$Q_\omega = \{x : \Gamma_\omega(x) = 0\} . \tag{1.3}$$

This is a (f_ω) maximal $\mathscr{B}_0$-measurable set on which we *never* make a confidence statement correct for ω. A procedure is called nondegenerate if $P_\omega\{Q_\omega\} = 0$ $\forall\, \omega$, and the class of all nondegenerate procedures is denoted by $\mathscr{C}^+$. The intuitive appeal of using a nondegenerate procedure can be seen in the NP setting of testing between two simple hypotheses whose densities are positive throughout $\mathscr{X}$: it amounts there to prohibiting use of the trivial critical regions, $\mathscr{X}$ and $\emptyset$.

We denote by $\mathscr{C}^{B_0}$ all partitions with $B = B_0$, and write $\mathscr{C}^{B+} = \mathscr{C}^B \cap \mathscr{C}^+$. For positive integral L we denote by B^L the finite label set $\{1, 2, \cdots, L\}$ and abbreviate $\mathscr{C}^{B^L}$ by $\mathscr{C}^L$. We have found it convenient not to demand essential minimality of B for a given partition, but instead to permit a procedure in $\mathscr{C}^L$ to have $\nu(C^b) = 0$ for some b; thus, $\mathscr{C}^{L-1}$ can be regarded as essentially a subset of $\mathscr{C}^L$. However, the *nondegenerate partitions* $\mathscr{C}^{L+}$ are disjoint. Between $\mathscr{C}^L$ and $\mathscr{C}^{L+}$ are the *proper partitions* of $\mathscr{C}^L$, those for which each C^b has positive probability for some ω.

Thus, $\mathscr{C}^1$ is the classical NPW case, the Wald framework of unconditional statistical decision theory. It is a useful reference mark against which to compare the various notions we use.

For finite (or denumerably infinite) B, we interpret (1.2) through the simple conditional probability formula: if $P_\omega\{C^b\} > 0$,

$$\Gamma_\omega^{\ b} = P_\omega\{K_\omega^{\ b}\}/P_\omega\{C^b\}, \tag{1.4}$$

and Q_ω is the union of all C^b's where $P_\omega\{K_\omega^{\ b}\} = 0$. We shall see that other B's are also of interest, for example, B = a real interval.

1.3. *Examples.* We now illustrate the notions introduced above in terms of a simple 2-*hypothesis setting*, where $\Omega = D = \{1, 2\}$ and $D_i = \{i\}$. To avoid questions of randomization, suppose ν is Lebesgue measure on a real interval $\mathscr{X}$ and that $\nu(\{x: f_2(x)/f_1(x) = c\}) = 0$ for every c. Then Corollary 5.4, together with the comments of Section 4 on elimination of randomization, assert that every admissible procedure (in a sense described in Section 2) is obtained essentially by subdividing the $\{C_1, C_2\}$ of a NP $\mathscr{C}^1$-partition into arbitrary $C_i^{\ b}$. For an example of computational simplicity, suppose f_i is normal with mean $2i$ and variance 1. The symmetric NP $\mathscr{C}^1$ partition $C_1 = (-\infty, 3)$, $C_2 = [3, \infty)$ has $\Gamma = \Phi(1) = .84$. Using the same C_i with $C^1 = \{x: |x - 3| < 1.5\} = \mathscr{X} - C^2$, we obtain a symmetric $\mathscr{C}^2$ partition with $\Gamma^1 = [\Phi(1) - \Phi(-.5)]/[\Phi(2.5) - \Phi(-.5)] = .78$ and with $\Gamma^2 = \Phi(-.5)/[\Phi(-.5) + \Phi(-2.5)] = .98$. Out of the probability .84 of making a correct decision, a portion .31 is associated with the "more conclusive" C^2 statement for which the conditional confidence coefficient is .98. Asymmetric examples of f_i's and examples in which it is deemed more important to make correct statements under f_1 than f_2, will be given in [2].

For a further 2-hypothesis illustration, suppose $\mathscr{X}$ = real line and that $f_2(x)/f_1(x)$ is nondecreasing, positive and finite. The finest conditioning (continuum B) that makes $\Gamma_1 = \Gamma_2$ w.p. 1 under both f_i can be shown to be determined by letting $\int_{-\infty}^{c_0} [f_1(x)f_2(x)]^{\frac{1}{2}}\, dx = \int_{c_0}^{\infty} [f_1(x)f_2(x)]^{\frac{1}{2}}\, dx$ and $C_1 = (-\infty, c_0)$; and by making $c(x) < c_0$ satisfy, for $x > c_0$, the relation $\int_{-\infty}^{c(x)} [f_1(t)f_2(t)]^{\frac{1}{2}}\, dt = \int_x^{\infty} [f_1(t)f_2(t)]^{\frac{1}{2}}\, dt$. Then $B = [c_0, \infty)$ and, except for the exceptional value $b = c_0$, we set $C^b = \{c(b), b\}$. This construction is illustrated and discussed in [2].

It will often be a practical convenience in examples like the above, where $\Gamma_\omega^{\ b}$ is independent of ω, to relabel B in such a way that $\Gamma^b = b$.

1.4. *Randomization.* We now discuss the representation of randomization and its role in the theory. A standard device in $\mathscr{C}^1$ developments is the representation of the randomization device as Lebesgue measure μ^1 on the Borel sets $\mathscr{L}$ of the unit interval I. (We denote Lebesgue measure on I^k by μ^k.) A randomized procedure for a problem with underlying measure space $(\mathscr{X}', \mathscr{B}', \nu')$ can then be regarded as a nonrandomized procedure on $(\mathscr{X}, \mathscr{B}, \nu) = (\mathscr{X}' \times I, \mathscr{B}' \times \mathscr{L}, \nu' \times \mu^1)$, which can be thought of as the background of our model. In particular, the underlying ν is then atomless, and certain sets of operating characteristics considered later are convex. As will be discussed in Section 4, if the original

ν' was atomless, results of Dvoretzky, Wald and Wolfowitz (1951) imply that such an introduction of randomization is unnecessary to achieve these operating characteristics, under assumptions that include $\mathscr{C}^L$ procedures for the k-hypothesis problem treated herein.

In practical terms, the use of randomized procedures is subject to all the criticisms present in $\mathscr{C}^1$ developments, but such criticism is perhaps even more pointed for conditional procedures, where one may find it unappealing to condition on the value of $Z(X)$ but not on the outcome of the randomization scheme. In many discrete case settings, the practitioner will do what he does in $\mathscr{C}^1$ considerations, and sacrifice exact fulfillment of some stated probabilistic objectives in favor of using a nonrandomized procedure. As in the case of $\mathscr{C}^1$, we consider the randomization enlargement from $\mathscr{X}'$ to $\mathscr{X}$ in order to achieve the convexity of some set of operating characteristics.

We also note that, although the set of operating characteristics of procedures in $\mathscr{C}$ or $\mathscr{C}^+$ turns out to be convex (Theorem 4.3), that of procedures in $\mathscr{C}^L$ (or $\mathscr{C}^{L+}$) does not. This can be seen from the fact that a 50–50 random choice between two procedures in $\mathscr{C}^L$ can be viewed in $\mathscr{X}' \times I$ as a $\mathscr{C}^L$ partition of $\mathscr{X}' \times [0, \frac{1}{2})$ together with another partition of $\mathscr{X}' \times [\frac{1}{2}, 1]$, and is thus in $\mathscr{C}^{2L}$ rather than in $\mathscr{C}^L$.

On the other hand, $\mathscr{C}$ and $\mathscr{C}^L$ are "closed" in a natural sense, while $\mathscr{C}^+$ and $\mathscr{C}^{L+}$ are not (Sections 4 and 5).

Further discussion of randomization, and an illustrative example of what is lost by not randomizing in a discrete case, is found in [2].

It has seemed best for the purpose of trying to make the ideas of this paper understandable, not to use the most general notation throughout. Thus, in the previous exposition and in development of the simple sufficient conditions for admissibility of Section 3, we can use nonrandomized (δ, Z), understood to be based on $\mathscr{X} = \mathscr{X}' \times I$ as described above, and thus to be equivalent to use of randomized procedures on $\mathscr{X}'$. Such randomized procedures η are defined in general terms in Section 4, where also the notion of a "canonical procedure" is introduced in order to prove convergence and compactness properties of $\mathscr{C}$. Section 5, where necessary conditions for admissibility are obtained, can then be read without detailed reading of Section 4. Section 5 and the end of Section 3 are written in a special form convenient for the k-hypothesis problem, in terms of the sufficient statistic $\{f_i/\sum_j f_j, 1 \leq i \leq k\}$ that takes on values in the $(k-1)$-simplex $\mathscr{S}_{k-1}$. (The role of sufficiency is the same as in $\mathscr{C}^1$ theory and will not be discussed further.) Accordingly, a procedure can be regarded either as a randomized procedure on $\mathscr{S}_{k-1}$ or as a nonrandomized partition of $\mathscr{T}_k = \mathscr{S}_{k-1} \times I$, and we try to use the two points of view in such a manner as to aid understanding: the former when proofs require it, the latter when it gives a simple geometric picture of a procedure, especially when the partition is "almost" a cartesian product of I with a partition of $\mathscr{S}_{k-1}$. In using $\mathscr{T}_k$ we denote partitions by $\hat{C}$.

The classes $\mathscr{C}$, $\mathscr{C}^+$, $\mathscr{C}^B$ are to be viewed as consisting of all randomized η on $\mathscr{X}'$, or equivalently of all partitions of $\mathscr{X}' \times I$ that have the appropriate properties, possibly with $\mathscr{S}_{k-1}$ for $\mathscr{X}'$.

2. Goodness criteria. A natural beginning for optimality considerations is the intuitive notion that, if ω is true, we prefer a procedure that states this truth in a form that reflects strong conclusiveness, to one that asserts it only as a weakly felt conclusion. This comparison is meaningful if the first procedure has at least as much probability of making its assertion as the second; otherwise, a direction of preference is not so clearly natural. We are led, then, to consider the tail probability law

$$G_\omega(t) = P_\omega\{\Gamma_\omega > t;\, \delta(X) \in D_\omega\} \tag{2.1}$$

for $0 \leqq t \leqq 1$. This function is right continuous. It is also defined at $t = 0-$, and we now verify its behavior at 0, for future reference.

Lemma 2.1. *We have $G_\omega(0-) = G_\omega(0)$, and consequently*

$$G_\omega(0) = P_\omega\{\delta(X) \in D_\omega\} = 1 - P_\omega\{Q_\omega\}\,. \tag{2.2}$$

Proof. We use the fact that Γ_ω is $\mathscr{B}_0$-measurable to compute that the "jump" in question is

$$\begin{aligned} P_\omega\{\delta(X) \in D_\omega, \Gamma_\omega = 0\} &= E_\omega\{E_\omega\{I_{\{\delta(X)\in D_\omega\}} I_{\{\Gamma_\omega = 0\}} \mid \mathscr{B}_0\}\} \\ &= E_\omega\{I_{\{\Gamma_\omega=0\}} E_\omega\{I_{\{\delta(X)\in D_\omega\}} \mid \mathscr{B}_0\}\} \\ &= E_\omega\{I_{\{\Gamma_\omega=0\}} \Gamma_\omega\} = 0\,. \end{aligned} \tag{2.3}$$ □

A possible notion of admissibility, that quantifies the intuitive notion that led to consideration of (2.1), is to regard $\bar{\bar{C}}$ as better than $\bar{C}$ if

$$\bar{\bar{G}}_\omega(t) \geqq \bar{G}_\omega(t) \quad \forall\, \omega, t\,, \tag{2.4}$$

with strict inequality somewhere.

In Section 6 we discuss some of the possible modifications of this criterion. While (2.4), with the adjunction of (2.6) or (2.8) as described hereafter, seems to the author the simplest criterion that conforms with the earlier motivating comments, it is not the aim of this paper to insist on the use of any one such criterion. Rather, these first considerations are intended to typify the formulation and study of reasonable conditional admissibility criteria. Many of the comments of Section 2 apply qualitatively to other criteria of similar character. Some criteria lead to further complications or simplifications. For example, the use of $E_\omega \Gamma_\omega = P_\omega\{C_\omega\}$ in place of (2.1) reduces the admissibility question to a known one of $\mathscr{C}_1$ decision theory, but (Section 6(B)) this criterion may not reflect the aims that motivated use of conditional confidence procedures.

For a procedure in $\mathscr{C}^1$, the function $G_\omega(t)$ has only a single jump, at $t = \Gamma_\omega = P_\omega\{\delta(X) \in D_\omega\}$, of magnitude Γ_ω. Hence, (2.4) is the usual notion of domination in terms of the risk function for zero-one loss function. However, for a procedure outside $\mathscr{C}^1$, (2.4) alone is not satisfactory. To see this, consider the k-hypothesis problem, where $\Omega = \{1, 2, \cdots, k\} = D$ and $D_i = \{i\}$, so that

$C_i^b = K_i^b$. Let us fix the C_i's and consider only partitions $\{C_i^b\}$ with those C_i's. Every such procedure has $G_i(0) = P_i\{C_i\}$, but essentially only the partition $\tilde{C}$ with $\tilde{C}_i^i = C_i$ has $G_i(1-) = P_i\{C_i\}$, and thus it is essentially the only partition with the given C_i that is admissible in the sense of (2.4).

This partition $\tilde{C}$ is highly unappealing intuitively. It always asserts $\Gamma_i = 1$ when decision d_i is reached, but does so in the form of a trivial conditioning, "when ω is true, the conditional probability of asserting it is true, given that it is asserted to be true, is one." It does not come close to achieving an aim that seems essential to the usefulness of conditional confidence procedures: that, for each b, the Γ_ω^b be of comparable magnitude (and perhaps even equal). This property is important for two reasons. Firstly, it is useful to be able to regard different values of Z as indexing different levels of conclusiveness; thus, in $\mathscr{C}^2$ we might like to regard decisions made when $X \in C^1$ as being "weakly conclusive" and those made when $X \in C^2$ as being "strongly conclusive." Secondly, from a practical point of view, it is useful (just as it is in the case of $\mathscr{C}^1$ confidence intervals) to be able to state a (conditional) confidence coefficient or useful lower bound, that is independent of ω—the $\Gamma^{Z(x)}$ or $\psi(x)$ of Section 1.

There are two fairly obvious possible modifications to using (2.4) alone. One of these is to impose on the class of procedures being considered a restriction that eliminates procedures like $\tilde{C}$. The restriction to procedures in $\mathscr{C}^+$ achieves this, since $\tilde{C}$ is degenerate. Because of the simplicity of this restriction, it is useful theoretically in admissibility developments. However, in using it one should keep in mind that the difficulty presented by comparison of procedures with $\tilde{C}$ is not completely eliminated by restriction to $\mathscr{C}^+$. This is because, in this k-hypothesis setting, it is easy to modify $\tilde{C}$ very slightly to obtain a procedure $\tilde{\tilde{C}}$ for which $P_i\{\tilde{\tilde{C}}_i^j\}$ is positive but small for $i \neq j$; this $\tilde{\tilde{C}}$ is in $\mathscr{C}^+$, and the Lévy distance between $\tilde{G}_\omega$ and $\tilde{\tilde{G}}_\omega$ is small. A procedure C in $\mathscr{C}^+$ with the same C_i's as $\tilde{\tilde{C}}$, and with almost equal Γ_ω^b's for each fixed b, will then not be *strictly* worse than $\tilde{\tilde{C}}$ in the sense (2.4), by Corollary 3.2; but C is *close* to being "very much worse" than $\tilde{\tilde{C}}$, in that its G_ω is much worse than $\tilde{G}_\omega$ which is close to $\tilde{\tilde{G}}_\omega$; thus, we have a "subadmissibility" phenomenon. Having found the admissible procedures in $\mathscr{C}^+$, then, further considerations in the form of the next paragraph, or in achieving reasonable closeness of the Γ_ω^b's for each fixed b (which rules out $\tilde{\tilde{C}}$), are necessary in choosing satisfactory procedures.

A second possible alteration of (2.4), even without *imposing* the restriction to $\mathscr{C}^+$, involves examination of incorrect assertions: if ω is true, and if we state that it is not true (by making a decision outside D_ω), it is better not to sound so sure about that untruth. This leads to consideration of laws associated with Γ and variants of it, under each possible true ω. Such possibilities will be discussed in Section 6(B), near the end of which the difficulty of using one of the most natural possibilities is discussed. For now, we mention only the simple

$$H_\omega(t) = P_\omega\{\delta(X) \notin D_\omega, 1 - \Gamma_\omega > t\}, \tag{2.5}$$

motivated by the fact that $1 - \Gamma_\omega$ is the conditional probability, under ω, of making an incorrect decision. We then define $\bar{\bar{C}}$ as *better than* $\bar{C}$ *in the sense* (2.4) $\wedge$ (2.6) if (2.4) is satisfied and also

$$(2.6) \qquad\qquad \bar{\bar{H}}_\omega(t) \leqq \bar{H}_\omega(t) \quad \forall\, t, \omega\,,$$

with strict inequality somewhere in either (2.4) or (2.6). The procedures $\tilde{C}$ and $\tilde{\tilde{C}}$ perform as poorly in terms of (2.6) as they did well in terms of (2.4), so the anomaly obtained from using (2.4) alone is thereby eliminated. It is somewhat simpler to consider only the value $t = 1-$ in (2.6), yielding

$$(2.7) \qquad P_\omega\{\bar{\bar{\delta}}(X) \notin D_\omega,\, \bar{\bar{\Gamma}}_\omega = 0\} \leqq P_\omega\{\bar{\delta}(X) \notin D_\omega,\, \bar{\Gamma}_\omega = 0\} \qquad \text{for all} \quad \omega\,,$$

which (by Lemma 2.1) is equivalent to

$$(2.8) \qquad\qquad P_\omega\{\bar{\bar{Q}}_\omega\} \leqq P_\omega\{\bar{Q}_\omega\} \qquad \text{for all} \quad \omega\,.$$

This leads to another notion of admissibility, obtained by defining $\bar{\bar{C}}$ to be better than $\bar{C}$ if (2.4) and (2.8) hold with strict inequality somewhere. The simplicity of this definition makes it, too, a useful theoretical notion. Although it eliminates the domination of $\tilde{C}$, it cannot eliminate the subadmissibility of $\tilde{C}$, since all procedures $\bar{\bar{C}}$ in $\mathscr{C}^+$ trivially satisfy (2.8) for all $\bar{C}$. Nevertheless, it has proved useful to delimit admissible procedures in the sense of (2.4) $\wedge$ (2.8) and then to consider functions such as those of (2.5) as parts of a supplementary operating characteristic to be studied further along with those of (2.1) in selecting a procedure. As far as *admissibility* is concerned, adjoining something like (2.6) to (2.4) and (2.8) cannot decrease the class of admissible procedures. Since a subset of that class will be seen in Section 3 already to be quite large *without* adjoining (2.6), and since examples studied thus far do not reveal any severe shortcoming of restricting selection of a procedure to those procedures proved in Section 3 to be admissible in the sense of (2.4) $\wedge$ (2.8) (or even to the smaller class of $\mathscr{C}^+$ procedures admissible in the sense of (2.4)), we shall not pursue such weaker admissibility notions further here. If other notions of "loss" are appropriate, this limitation to admissibility in the sense of (2.4) $\wedge$ (2.8) may prove unwise. In Section 5 we characterize (2.4) $\wedge$ (2.6)-admissible procedures partly because of the difficulty of using (2.8) directly as part of the admissibility criterion in a complete class proof that proceeds by maximizing certain linear functionals.

Choosing a conditional confidence procedure from the admissible ones is perhaps more difficult here than in $\mathscr{C}^1$ decision theory, because of the additional properties of interest. Some considerations additional to performance in terms of (1.3), (2.1), (2.5) as described above are: (1) In many problems the $\Gamma_\omega{}^b$ for each fixed b should be equal or close, for reasons mentioned earlier (although, as in the $\mathscr{C}^1$ NP 2-hypothesis setting, this is not always desirable). (2) The $\Gamma_\omega{}^b$ should have a large variation in b (possibly reflected through some specific measure of dispersion of G_ω); otherwise, a simpler procedure, with smaller B—perhaps even a $\mathscr{C}^1$ procedure—could as well be used. (3) Usually it is unattractive to a practitioner to use a procedure in which any Γ_ω^b is too small; this may, however, motivate enlargement of D to include decisions reflecting indifference for borderline data.

Other aspects enter when the simple loss structure does not accurately represent all the experimenter's potential rewards and penalties. For example, in interval estimation or problems with complex D (see (3) above), some measure of "size" associated with the decision may be relevant. (See Section 6 (C).) The computational complexity of large B might be another concern. Also, since there are so many admissible procedures in $\mathscr{C}^{L+}$, the imposition of some simplifying structure on the form of the procedure may help computationally in the construction and comparison of partitions; for example, in the 2-hypothesis problem one might, in $\mathscr{C}^{L+}$, restrict consideration to procedures with *monotone interval structure*, those in which, if S_i^b is the convex closure of the range of $f_2(x)/f_1(x)$ for x in C_i^b, any two S_i^b's have at most one point in common, and S_1^b is to the left of $S_1^{b'}$ if and only if S_2^b is to the right of $S_2^{b'}$. (According to Theorem 3.1, some of the intuitively less appealing procedures that violate this form are also admissible.) The operating characteristics of partitions of such a restricted form may not constitute so large a set as those of all $\mathscr{C}^{L+}$ procedures. On the other hand, at least in simple cases (Ω, D finite), there are typically infinitely many procedures with a given operating characteristic (functions of (1.3), (2.1), (2.5)), and it is desirable computationally to be able to achieve any such operating characteristic, at least approximately, with a procedure of simple form.

These aspects of computing and choosing among conditional confidence procedures (in particular, of justifying the restriction of the previous paragraph) are clearly quite involved and will not be treated further here. Illustrations and further discussion will be contained in [2]. Here we remark only that $\mathscr{C}^L$ procedures are more robust for small L.

In the succeeding sections, where various admissibility concepts are used, "admissibility in the sense of (A)" or "(A)-admissibility" will be used to refer to the criterion obtained from the "at least as good as" inequalities (A). We write $\mathbf{G} = \{G_\omega, \omega \in \Omega\}$ and $\mathbf{H} = \{H_\omega, \omega \in \Omega\}$.

3. Sufficient condition for admissibility. For any partition $C = \{C_d^b\}$ in $\mathscr{C}$, there is an associated $\mathscr{C}^1$ partition $|C| = \{|C_d|\}$ defined by $|C_d| = \bigcup_b C_d^b = \{x: \delta(x) = d\}$. We call $|C|$ the *decision partition* or *underlying* $\mathscr{C}^1$ *partition* of C. For procedures $\mathscr{C}^1$, admissibility in the sense of (2.4) $\wedge$ (2.8) (or (2.4) alone) is the standard decision-theoretic admissibility for risk function $P_\omega\{\delta(X) \notin D_\omega\}$. In this section we shall prove

THEOREM 3.1. *If* $|C|$ *is admissible among* $\mathscr{C}^1$ *procedures, then* C *is admissible in* $\mathscr{C}$ *in the sense of* (2.4) $\wedge$ (2.8) (*and hence in the sense of* (2.4) $\wedge$ (2.6))

Since all $\mathscr{C}^+$ procedures have $P_\omega\{Q_\omega\} = 0$, we have

COROLLARY 3.2. *If $C \in \mathscr{C}^+$ and $|C|$ is admissible in $\mathscr{C}^1$, then C is admissible in $\mathscr{C}^+$ in the sense of* (2.4).

By restricting competitors in the proof of Theorem 3.1, we obtain

THEOREM 3.3. *If $\mathscr{C}$, $\mathscr{C}^+$ are replaced by $\mathscr{C}^B$, $\mathscr{C}^{B+}$ for any fixed B, the statements of Theorem* 3.1 *and Corollary* 3.2 *remain true.*

Since (Section 1.4) $\mathscr{X}$ can be thought of as containing a randomization component I, the admissibility asserted above and in Corollary 3.4 allows also for competitors that randomize among conditionings on the original $\mathscr{X}'$.

Note that it has not been necessary in the above statements to make regularity assumptions on D and Ω (in particular, on their finiteness, so the conclusions apply to such settings as interval estimation), because we do not touch on the detailed characterization of $\mathscr{C}^1$-admissible procedures; under various assumptions there are well-known results in terms of Bayes procedures and finer structure (e.g., [7]). In contrast, in Section 5 we treat the k-hypothesis problem in order to obtain explicit *necessary* conditions for admissibility. The results of the present section would be even more obvious if the operating characteristic for admissibility were $\{P_i\{C_i^b\}\}$; it is not that, but the proof in terms of (2.1) is not much more complicated. The results here establish the unfortunately large collection of admissible procedures mentioned in Section 2. Section 5 indicates that there are in general even more admissible procedures than those characterized in the present section. However, the admittedly limited calculation of examples to date indicates no great practical advantage to be found in using procedures outside those characterized above.

PROOF OF THEOREM 3.1. We begin with a simple computation (3.2) that holds for arbitrary partitions C. Let B_ω be the σ-field induced on $\mathscr{X}$ by Γ_ω. Define the measures K_ω^* and G_ω^* on $(\mathscr{X}, \mathscr{B}_\omega)$ by

$$K_\omega^*(A) = P_\omega\{A\}, \qquad G_\omega^*(A) = P_\omega\{A; \delta(X) \in D_\omega\}, \tag{3.1}$$

and let $K_\omega^\sharp$ and $G_\omega^\sharp$ be corresponding Borel measures on $[0, 1]$ induced by Γ_ω; that is $G_\omega^\sharp(L) = G_\omega^*(\Gamma_\omega^{-1}(L))$, and similarly for $K_\omega^\sharp$. A straightforward conditional expectation calculation shows that $dG_\omega/dK_\omega^* = \Gamma_\omega$ a.e. (K_ω^*) and (since $-\int_L dG_\omega(t) = G_\omega^\sharp(L)$) consequently that $(dG_\omega^\sharp/dK_\omega^\sharp)(t) = t$ a.e. $(K_\omega^\sharp)$. Moreover, $K_\omega^\sharp \ll G_\omega^\sharp$ on $(0, 1]$, where $(dK_\omega^\sharp/dG_\omega^\sharp)(t) = t^{-1}$. Thus,

$$\begin{aligned} -\textstyle\int_{0+}^1 t^{-1}\, dG_\omega(t) &= \textstyle\int_{0+}^1 t^{-1} G_\omega^\sharp(dt) \\ &= \textstyle\int_{0+}^1 K_\omega^\sharp(dt) = 1 - P_\omega\{Q_\omega\}. \end{aligned} \tag{3.2}$$

We also recall that, if $\bar{F}$ and $\bar{\bar{F}}$ are two df's on $(0, \infty)$, then

$$\bar{F}(t) \geqq \bar{\bar{F}}(t) \quad \forall\, t \tag{3.3}$$

implies, for every strictly increasing function h integrable under both $\bar{F}$ and $\bar{\bar{F}}$,

$$\textstyle\int_{0+}^\infty h(t)\, d\bar{F}(t) \leqq \int_{0+}^\infty h(t)\, d\bar{\bar{F}}(t), \tag{3.4}$$

and strict inequality for some t in (3.3) implies strict inequality in (3.4).

Now suppose $\bar{C}$ is a partition whose underlying $\mathscr{C}^1$ partition is $\mathscr{C}^1$-admissible, but which is not itself admissible in the sense (2.4) $\wedge$ (2.8); let $\bar{\bar{C}}$ be better. Thus, (2.4) and (2.8) hold, with strict inequality somewhere. At $t = 0$, (2.4) becomes (by Lemma 2.1) $P_\omega\{\bar{\bar{\delta}}(X) \in D_\omega\} \geqq P_\omega\{\bar{\delta}(X) \in D_\omega\}\ \forall\, \omega$, and equality must hold because the underlying $\mathscr{C}^1$ partition of $\bar{C}$ is $\mathscr{C}^1$-admissible. Thus, $\bar{G}_\omega(0) = \bar{\bar{G}}_\omega(0)\ \forall\, \omega$. For each ω for which $\bar{G}_\omega(0) > 0$ (and it is easily seen that we need consider only such ω), we consequently have from (3.1) that

$$\bar{F}_\omega(t) \equiv 1 - \bar{G}_\omega(t)/\bar{G}_\omega(0)\,, \qquad \bar{\bar{F}}_\omega(t) \equiv 1 - \bar{\bar{G}}_\omega(t)/\bar{\bar{G}}_\omega(0) \tag{3.5}$$

are both df's on (0, 1]. By (2.4) we have (3.3) and thus (3.4) for these df's. For $h(t) = -t^{-1}$, we obtain from (3.2)

$$P_\omega\{\bar{Q}_\omega\} \leqq P_\omega\{\bar{\bar{Q}}_\omega\} \quad \forall\, \omega\,, \tag{3.6}$$

and since this h is strictly increasing we conclude that strict inequality in (2.4) for any ω (for some t) implies strict inequality in (3.6). Since we cannot have both (3.6) and (2.8) with a strict inequality somewhere, this contradicts existence of the assumed $\bar{\bar{C}}$. □

Before turning to preliminaries used in establishing necessary conditions for admissibility, we describe some of the above results in other terms that will be useful later. For that purpose, we need only treat the case where Ω and D are finite, say $\Omega = \{1, 2, \cdots, k\}$ and $D = \{1, 2, \cdots, r\}$. In terms of the discussion of Section 1, we think of procedures as partitions of $\mathscr{X} = \mathscr{X}' \times I$. A familiar description of $\mathscr{C}^1$-Bayes procedures is given in terms of the probability k-vectors, whose range is the $(k-1)$-simplex $\mathscr{S}_{k-1} = \{(s_1, \cdots, s_k) : \sum_1^k s_i = 1;\ s_i \geqq 0\ \forall i\}$. In order to work in terms of nonrandomized procedures (as discussed in Section 1), we replace $\mathscr{S}_{k-1}$ by $\mathscr{T}_k = \mathscr{S}_{k-1} \times I$. Write $f^* = (f_1^*, \cdots, f_k^*)$ for the mapping of $\mathscr{X}'$ into $\mathscr{S}_{k-1}$ defined (a.e. $\sum_1^k f_i\, d\nu$) by $f_\omega^*(x) = f_\omega(x)/\sum_1^k f_i(x)$, and let

$$\hat{\nu}(A) = \int_{f^{*-1}(A)} \textstyle\sum_1^k f_i(x)\nu(dx)\,. \tag{3.7}$$

The sufficient statistic $S = f^*(X)$ then has density $\hat{f}_i(s) = s_i$ on $\mathscr{S}_{k-1}$ with respect to $\hat{\nu}$, when i is true. It does not matter that this reduction from $\mathscr{X}'$ to $\mathscr{S}_{k-1}$ may introduce atoms, since we still have μ^1 on I (the last component of $\mathscr{T}_k$) to represent "randomization," as described in Section 1. In place of partitions of $\mathscr{X}$ we then consider partitions $\hat{C}$ of $\mathscr{T}_k$. Every (nonrandomized) $\mathscr{C}^1$-Bayes procedure on $\mathscr{T}_k$ in these terms is equivalent to a partition $|\hat{C}|$ of $\mathscr{T}_k$ which, for some probability vector $p = \{p_i\}$, satisfies

$$\hat{C}_j \subset \{(s, u) : \textstyle\sum_{i \in \Omega_j} p_i s_i = \max_h \sum_{i \in \Omega_h} p_i s_i\} \quad \forall j\,. \tag{3.8}$$

An important feature is the well-known convexity of the sets on the right side of (3.8). The $\mathscr{C}^1$-Bayes procedures may include some inadmissible procedures, but that cannot occur if all f_i have the same support; and Bayes procedures relative to strictly positive p are admissible in any case. Characterization of admissible $\mathscr{C}^1$ Bayes procedures in general can be found in [7]. To shorten further discussion, we paraphrase Corollary 3.2 and Theorem 3.3 in the simplest case, in terms of partitions of $\mathscr{T}_k$. We hereafter denote by cl A, int A, and cc A, the closure, interior, and convex closure of a subset A of $\mathscr{S}_{k-1}$ or $\mathscr{T}_k$ in the relative Euclidean topology of either of these. By $s^{(i)}$ we denote the ith vertex of $\mathscr{S}_{k-1}$, where $s_i^{(i)} = 1$. The *border* of $\mathscr{S}_{k-1}$ is the subset $\{s: \text{some } s_i = 0\}$.

COROLLARY 3.4. *For the k-hypothesis problem where all f_i have the same domain of positivity, a partition $\hat{C}$ of $\mathscr{T}_k$ in $\mathscr{C}^+$ is admissible in $\mathscr{C}^+$ in the sense of* (2.4), *if for some strictly positive p the decision partition $|\hat{C}|$ satisfies*

$$\text{(3.9)} \qquad \text{(i) for each } i, \quad \text{int } \hat{C}_i \text{ contains } s^{(i)} \times I;$$
$$\text{(ii)} \quad \text{cc } \hat{C}_i \cap \text{cc } \hat{C}_j \subset \{(s, u): p_i s_i = p_j s_j\}, \qquad \text{for} \quad i \neq j.$$

The conclusion holds if $\mathscr{C}^+$ is replaced by $\mathscr{C}^{B+}$ for fixed B.

We hereafter say a partition $\hat{C}$ of $\mathscr{T}_k$ has *structure W* if its $|\hat{C}|$ satisfies (3.9). Under the assumptions of Corollary 3.4, the only procedures admissible in $\mathscr{C}^+$ for the k-sample problem that are covered in Corollary 3.2 are those of structure W. The procedures $\hat{C}$ with structure W have $|\hat{C}|$ in $\mathscr{C}^{1+}$, but of course require further conditions on the subpartition of $\hat{C}_j$ into the $\hat{C}_j{}^b$, in order that $\hat{C}$ be in $\mathscr{C}^+$.

When the f_i do not have common domain of positivity, it is possible to extend Corollary 3.4 to include some procedures that do not satisfy (3.9)(i), if $\nu > 0$ on the border of $\mathscr{S}_{k-1}$, e.g., if $\nu(s^{(i)}) > 0$ for some i.

4. Canonical procedures, randomization and topology of $\mathscr{C}$. Throughout this section Ω and D are finite, although Lemma 4.1 will be seen at once to hold more generally and the other results can be proved under compactness assumptions on D and (except for the result taken from [3]) on Ω. We begin by treating a difficulty related to but beyond the phenomena that motivate the notion of regular convergence of $\mathscr{C}^1$ decision functions. In the latter, in such a simple setting as the two (simple)-hypothesis problem with $\nu = \mu^1$ on $\mathscr{X}' = [0, 1)$, a sequence of nonrandomized decision functions need not converge a.e. to a nonrandomized decision function (or anything else). For example, define the measure ξ_n on $\mathscr{X}' = [0, 1)$ by its density

$$\text{(4.1)} \qquad d\xi_n(x)/dx = 1 \quad \text{if} \quad 2i/n \leq x < (2i+1)/n \quad \text{for any integer} \quad i,$$
$$= 0 \quad \text{otherwise.}$$

Then, if ${}^n\bar{\eta}$ is the $\mathscr{C}^1$ decision function whose probability of making d_1 when $X = x$ is ${}^n\bar{\eta}(d_1 \mid x) = d\xi_n(x)/dx$, so that ${}^n\bar{\eta}$ is nonrandomized, the sequence of ${}^n\bar{\eta}$ converges almost nowhere pointwise, but converges "regularly" to $\bar{\eta}^*(d_1 \mid x) \equiv \frac{1}{2}$. (Formal definitions will be given below.) Now, with $B = \mathscr{X}'$ and $Z(x) = x$, consider the nonrandomized conditional confidence procedure ${}^n\eta$ for which the probability of making decision d_i and conditioning with label $b = Z(x)$ when

$X = x$ is given by

$$^n\eta(d_1, Z(x)\,|\,x) = 1 - {}^n\eta(d_2, Z(x)\,|\,x) = d\xi_n(x)/dx\,. \tag{4.2}$$

($^n\eta(d_i, Z(x)\,|\,x)$ is simply the characteristic function of $^nC_{d_i}^{Z(x)}$. The degeneracy of these ^nC's is inessential and only for the purpose of simplifying the illustration; $\mathscr{C}^+$ examples are easily given.) It might seem natural to consider these $^n\eta$ to converge "regularly," if at all, to $\eta^*(d_i, Z(x)\,|\,x) \equiv \frac{1}{2}\,\forall\, i$. But if $f_1(x) \equiv 1$, the use of a fair coin to choose between d_1 and d_2 for each x would make $P_i\{\text{make } d_i\,|\,X = x\} = \frac{1}{2}\,\forall\, x$ whereas $^n\Gamma_1(x) = P_1\{^n\delta(x) = d_1\,|\,X = x\}$ is only 0 or 1. We would thus not have the desired convergence of $^nG_1$. The difficulty of course stems from the fact that $\xi_n \to \xi^*$ weakly does not imply $d\xi_n/dx \to d\xi^*/dx$ anywhere, and this last, of no consequence in $\mathscr{C}^1$ theory, matters here because of our use of $^n\Gamma_i$ and ^nG as measures of conditional confidence performance. A way out of the difficulty (which is more obvious in the case of finite B) is directed by the comments in Section 1 alluding to conditioning on the outcome of a randomization. If we let $B^* = [0, 1) \times M$ where $M = \{1, 2\}$, and replace $\mathscr{X}'$ by $\mathscr{X}'^* = \mathscr{X}' \times M$ with ν replaced by $\nu^* = \nu \times \rho$ with $\rho(1) = \rho(2) = \frac{1}{2}$ (the randomization), we obtain $^*\Gamma_i = 0$ or 1 with the desired probabilities. Here we replaced B by two copies of it in a manner that might not be so obvious to imitate in general. We now introduce a structure that makes such a construction mechanical (and which yields a different but simpler solution to the difficulty in the above example).

For fixed B with σ-field $\mathscr{B}_0'$, let $J = D \times B$ with σ-field $\mathscr{J} = 2^D \times \mathscr{B}_0'$. It may help motivative the definition that follows to note that a nonrandomized (δ, Z) can be thought of as a mapping from $\mathscr{X}'$ to J; in this interpretation, $\Gamma_\omega = P_\omega\{(\delta, Z) \in (D_\omega \times B)\,|\,\mathscr{D}_0 \times \mathscr{B}_0'\}$ where $\mathscr{D}_0$ is the trivial σ-field on D. A conditional confidence procedure, expressed in terms that allow for randomization, is a function η on $\mathscr{J} \times \mathscr{X}'$ such that $\eta(A'\,|\,x)$ is a probability measure $(\mathscr{J})$ in A' for each x and is measurable $(\mathscr{B})$ for each A'. If η is used and $X = x$, and if the randomization according to $\eta(\cdot\,|\,x)$ produces outcome (d, b), we make decision d and regard the conditioning label as b. Define $P_{\omega,\eta}\{A'\} = E_\omega \eta(A'\,|\,X)$. The conditional probability $P_{\omega,\eta}\{D_\omega \times B\,|\,\mathscr{D}_0 \times \mathscr{B}_0'\}$ is well defined and essentially unique $(P_{\omega,\eta}, \forall\, \omega)$. We define

$$\gamma_\omega = P_{\omega,\eta}^{\prime}\{D_\omega \times B\,|\,\mathscr{D}_0 \times \mathscr{B}_0'\}\,, \tag{4.3}$$

so that γ_ω is a $\mathscr{D}_0 \times \mathscr{B}_0'$-measurable function on $D \times B$, which we can thus view as a $\mathscr{B}_0'$-measurable function on B. We denote its value at b by $\gamma_\omega{}^b$. If one begins with a nonrandomized δ with its Γ_ω, the γ_ω of the corresponding η is seen to satisfy $\Gamma_\omega(x) = \gamma_\omega{}^{Z(x)}$.

A *canonical conditional confidence procedure* is one that satisfies (with labels distinguished by *)

$$B^* = I^k = \{b^*\} = \{(b_1^*, b_2^*, \cdots, b_k^*)\}\,, \tag{4.4}$$
$$\gamma_\omega{}^{b^*} = b_\omega{}^* \quad \text{w.p. 1} \quad \text{under} \quad \omega\,, \quad \forall\, \omega\,,$$

with $\mathscr{B}_0^{*\prime}$ the Borel sets. (This formalizes the practically convenient relabeling mentioned after the examples of Section 1, in which we regard B as the set of attainable conditional confidence values; but now we relabel whether or not Γ_ω^b is independent of ω.)

Note that, for a canonical η^*, the conditional probability γ_ω is measurable relative to the cylinder sets over Borel sets in the ωth coordinate.

The set of operating characteristics G of all η^* is easily seen to be convex. We now prove

LEMMA 4.1. *For every η, there is a canonical η^* with the same operating characteristic* G.

PROOF. For an arbitrary η, define $g_\eta: B \to L^k$ by $g_\eta(b) = ({}^\eta\gamma_1^b, \cdots, {}^\eta\gamma_k^b)$. Let η^* be the (canonical) procedure given by

$$\eta^*(\Delta \times A \mid x) = \eta(\Delta \times g_\eta^{-1}(A) \mid x) \tag{4.5}$$

for $\Delta \subset D$ and $A \in \mathscr{B}_0^{*\prime}$ in I^k, and then extended from such cylinder sets $\Delta \times A$. To see that η^* is a canonical procedure, we note that, for $A \in \mathscr{B}_0^{*\prime}$, from (4.5),

$$\begin{aligned} &\textstyle\int_A b_\omega^* P_{\eta^*,\omega}\{D \times db^*\} \\ &= \textstyle\int_{g_\eta^{-1}(A)} P_{\eta,\omega}\{D_\omega \times B \mid \mathscr{D}_0 \times g_\eta^{-1}(\mathscr{B}_0^{*\prime})\} P_{\eta,\omega}\{D \times db\} \\ &= P_{\eta,\omega}\{D_\omega \times g_\eta^{-1}(A)\} = P_{\eta^*,\omega}\{D_\omega \times A\} \\ &= \textstyle\int_A P_{\eta^*,\omega}\{D_\omega \times B \mid \mathscr{D}_0 \times \mathscr{B}_0^{*\prime}\} P_{\eta^*,\omega}\{D \times db^*\}, \end{aligned} \tag{4.6}$$

so that the first integrand is indeed a version of the last. Note that the possible lumping of several b's with the same vector $\{{}^\eta\gamma_\omega^b, \omega \in \Omega\}$ into a single $g_\eta(b)$ is exhibited in the second integrand of (4.6), where $g_\eta^{-1}(\mathscr{B}_0^{*\prime}) \subset {}^\eta\mathscr{B}_0'$.

Since ${}^\eta\gamma_\omega^b = {}^{\eta^*}\gamma_\omega^{g(b)}$ and $P_{\eta,\omega}\{D_\omega \times g^{-1}(A)\} = P_{\eta^*,\omega}\{D_\omega \times A\}$ by construction, putting $A = \{b^* : b_\omega^* > t\}$ we obtain ${}^\eta G_\omega = {}^{\eta^*}G_\omega \; \forall\, \omega$. □

We call the η^* of Lemma 4.1 a canonical procedure coresponding to η.

In any convergence or compactness proofs concerning $\{G_\omega\}$ we may hereafter limit consideration to canonical procedures.

On the space of operating characteristics G we use the usual notion of *weak convergence* of a sequence ^{N}G to $\bar{\text{G}}$, as meaning $\int_0^1 c\, d\, {}^N G_\omega \to \int_0^1 c\, d\bar{G}_\omega$ for each ω and continuous c. This is equivalent to convergence in the sense of the Lévy metric of weak convergence; noting that G_ω is nonincreasing, we define $\|\bar{G}_\omega - \bar{\bar{G}}_\omega\|$ here to be the maximum distance along the lines of slope 1 between the graphs (including vertical segments at jumps) of $\bar{G}_\omega$ and $\bar{\bar{G}}_\omega$, and the Lévy metric is $\|\bar{\text{G}} - \bar{\bar{\text{G}}}\| = \sum_1^k \|\bar{G}_\omega - \bar{\bar{G}}_\omega\|$. It will suffice to consider sequences in studying compactness. On the space of canonical procedures η^*, we adopt Wald's $\mathscr{C}^1$ notion of *regular convergence of* $\{{}^N\eta^*\}$ *to* $\bar{\eta}^*$, defining it in the present context (by substituting J for the usual D) to mean

$$\textstyle\int_{\mathscr{X}} g(x)\{\int_J c(j)\, {}^N\eta^*(dj \mid x)\}\nu(dx) \to \int_{\mathscr{X}} g(x)\{\int_J c(j)\bar{\eta}^*(dj \mid x)\}\nu(dx) \tag{4.7}$$

for every continuous real function c on J and every g in $\mathscr{L}_1(\mathscr{X}, \mathscr{B}, \nu)$. It is well known ([7] or [5]) that the set of procedures $\{\eta\}$ is sequentially compact in this topology.

The purpose of reduction to canonical procedures is given in the following:

LEMMA 4.2. *If a sequence ${}^N\eta^*$ of canonical procedures converges regularly to a procedure $\bar{\eta}$, then $\bar{\eta}$ is canonical.*

PROOF. By definition, η^* is canonical if and only if $b_\omega = P_\omega\{D_\omega \times B \mid \mathscr{D}_0 \times \mathscr{B}_0'\}$; that is,

$$\int_{\mathscr{X}} \int_J I_A(b) b_\omega \eta^*(dj \mid x) f_\omega(x)\nu(dx) = \int_{\mathscr{X}} \eta^*(D_\omega \times A \mid x) f_\omega(x)\nu(dx) = \int_{\mathscr{X}} \int_J I_{D_\omega}(d) I_A(b) \eta^*(dj \mid x) f_\omega(x)\nu(dx) \tag{4.8}$$

for each ω and every A in $\mathscr{B}_0^{*\prime}$. Since $\mathscr{B}_0^{*\prime}$ is the Borel field on $B^* = I^k$, (4.8) is equivalent to the equation obtained from the two extreme members of (4.8) if I_A is replaced there by an arbitrary continuous function c on B^*. Since both $c(b)b_\omega$ and $I_{D_\omega}(d)c(b)$ represent continuous functions on J, the equation with these two functions as integrands is preserved under weak convergence of a sequence of η^*'s to a limiting $\bar{\eta}^*$, so the latter is canonical. □

From Lemma 4.2 and the previously stated compactness of the set of all η in the sense of regular convergence, we conclude that the set $\{\eta^*\}$ of canonical procedures is compact in that sense. We now prove the rest of

THEOREM 4.3. *The set of all canonical η^* is compact in the sense of regular convergence. The set of operating characteristics ${}^\eta G$ of the family of all procedures η, or of all canonical η^*, is convex and is compact in the sense of weak convergence. These results, except for convexity, remain valid if η and η^* are restricted to $\mathscr{C}^L$.*

PROOF. We observe that, for all continuous c on I,

$$\int_0^1 c(t)\, d\, {}^{\eta^*}G_\omega(t) = \int_{\mathscr{X}} \int_J I_{D_\omega}(d) c(b_\omega) \eta^*(dj \mid x) f_\omega(x)\nu(dx)\,. \tag{4.9}$$

Since $I_{D_\omega}(d)c(b_\omega)$ represents a continuous function on J, we infer the continuity of the map $\eta^* \to {}^{\eta^*}G$ from the set of canonical procedures in the topology of regular convergence to the set of operating characteristics in the topology of weak convergence. By compactness of the sub-df's on [0, 1], for any sequence $\{{}^N G\}$ corresponding to canonical $\{{}^N\eta^*\}$ there is a subsequence for which ${}^{N'}\eta \to \bar{\eta}^*$ regularly, with $\bar{\eta}^*$ canonical, and by the previous sentence, ${}^{N'}G \to {}^{\bar{\eta}^*}G$ weakly. Finally, regular convergence of a sequence of canonical procedures in $\mathscr{C}^L$ to a limit clearly implies that the limit is in $\mathscr{C}^L$. □

We now enlarge the operating characteristic. Consideration of a sequence of procedures ${}^N\eta$ like the $\tilde{\tilde{C}}$ of Section 2, converging regularly to $\tilde{C}$ there, shows that $P_\omega\{{}^N Q_\omega\}$ does not converge to $P_\omega\{{}^{\tilde{C}} Q_\omega\}$. However, everything we have developed for G_ω can be carried out for H_ω, now defined by

$$H_\omega(t) = E_\omega \eta([D - D_\omega] \times \{b : 1 - \gamma_\omega{}^b > t\} \mid X)\,, \tag{4.10}$$

with $\|H_\omega - \bar{H}_\omega\|$ being obtained as the maximum distance along a line of slope

-1. Lemma 4.2 requires no change, and the straightforward changes in the remainder of our development yield

THEOREM 4.4. *The statements of Lemma* 4.1 *and Theorem* 4.3 *remain valid with* **G** *replaced by* (**G**, **H**).

Theorem 5.1 implies such results as completeness of the admissible procedures (in the sense of (2.4) $\wedge$ (2.6)).

Another decision-theoretic consideration (not used in the sequel) is that of "elimination of randomization." The development of pages 5–8 of Dvoretzky, Wald and Wolfowitz (1951) shows that, for our compact metric $D \times B$ (= their D), since Ω is finite, if ν is atomless every randomized η can be replaced by a nonrandomized $^*\eta$ with the same **G**, **H**. It is only necessary to interpret their δ^l as an $^l\eta$ in our scheme, so that the displays on page 7 give $P_{^l\eta,\omega}\{d \times {}^l\beta\} = P_{\eta,\omega}\{d \times {}^l\beta\}$ for each set $^l\beta$ in an increasingly fine net of partitions of B. The weak convergence of $\{^l\eta\}$ to a nonrandomized $^*\eta$ and the remainder of their proof yield $P_{^*\eta,\omega}\{d \times A\} = P_{\eta,\omega}\{d \times A\}$ for all d and $A \in \mathscr{B}_0'$, as desired.

The remaining result we need in the next section is one of approximating $\mathscr{C}$ by $\mathscr{C}^L$ in the sense of **G** and **H**. Recall that any η may be regarded as a partition of $\mathscr{X}' \times I$.

LEMMA 4.5. *For every* $\varepsilon > 0$ *there is a finite* $L(\varepsilon)$ *such that, for every* η *in* $\mathscr{C}$, *there is an* $\bar{\eta}$ *in* $\mathscr{C}^{L(\varepsilon)}$ *with* $\|{}^\eta\mathbf{G} - {}^{\bar\eta}\mathbf{G}\| + \|{}^\eta\mathbf{H} - {}^{\bar\eta}\mathbf{H}\| < \varepsilon$.

PROOF. Partition I^k into a finite number of Borel sets $\{U^r; r = 1, 2, \cdots, L(\varepsilon)\}$, each with $\max\{|b_\omega{}^* - \bar{b}_\omega| : b^*, \bar{b}^* \in U^r, \omega \in \Omega\} < \varepsilon/6k$. Given η, let η^* be the correponding canonical procedure. Define $\bar\eta$ (in $\mathscr{C}^{L(\varepsilon)}$) by $\bar\eta(d \times r \mid x) = \eta^*(d \times U^r \mid x)$. Let b^{*r} be any point in U_r. On U_r we have $b_\omega{}^{*r} - \varepsilon/6k < {}^{\eta^*}\gamma_\omega < b_\omega{}^{*r} + \varepsilon/6k$ w.p. 1. Integrating over U_r with respect to $P_{\eta^*,\omega}\{D \times db^*\}$, we obtain

$$(4.11) \qquad (b_\omega{}^{*r} - \varepsilon/6k)P_{\eta^*,\omega}\{D \times U_r\} < P_{\eta^*,\omega}\{D_\omega \times U_r\} < (b_\omega{}^{*r} + \varepsilon/6k)P_{\eta^*,\omega}\{D \times U_r\}\,.$$

By construction of $\bar\eta$, (4.11) is the same as

$$(4.12) \qquad (b_\omega{}^{*r} - \varepsilon/6k)P_{\bar\eta,\omega}\{D \times r\} < P_{\bar\eta,\omega}\{D_\omega \times r\} < (b_\omega{}^{*r} + \varepsilon/6k)P_{\bar\eta,\omega}\{D \times r\}\,,$$

so that $|{}^{\bar\eta}\gamma_\omega{}^r - b_\omega{}^{*r}| < \varepsilon/6k$. The range of values of the rv ${}^{\eta^*}\gamma_\omega$ on the set U_r is thus entirely within $\varepsilon/3k$ of the value of ${}^{\bar\eta}\gamma_\omega$ at r, and these sets of values, coupled with any d, have the same probability $P_{\eta^*,\omega}\{d \times U^r\} = P_{\bar\eta,\omega}\{d \times r\}$. It follows that $\|{}^{\eta^*}G_\omega - {}^{\bar\eta}G_\omega\| < \varepsilon/2k$ and $\|{}^{\eta^*}H_\omega - {}^{\bar\eta}H_\omega\| < \varepsilon/2k$. $\square$

5. Necessary condition for admissibility in the k-hypothesis problem. Unfortunately, there seems to be many more admissible partitions than the simple ones obtained in Section 3. Equally unfortunately, the essentially complete class obtained in the present section appears to be far from minimal, except in the case $k = 2$ (Corollary 5.3). This will be discussed at the end of the section. Our aim is to characterize admissibility in terms of underlying $\mathscr{C}^1$-partitions.

Since the results thus far obtained seem far from definitive, we shall give them here only for the k-sample problem. A corresponding statement for general D can be obtained by the same methods, but is even less satisfactory; some particular multidecision problems are treated in [4].

For the sake of easy comparison with the results of Section 3, we state our results in language of Corollary 3.4. In the k-hypothesis problem, the decision partition $|\hat{C}|$ of a partition $\hat{C}$ of $\mathscr{T}_k$ will be said to be *star-shaped* if

(i) $s^{(i)} \times I \in \hat{C}_i$ for each i;

(5.1) (ii) for each i and line $L = \{s: \sum_j c_{ij}(s_j - s_j^{(i)}) = 0\}$ through $s^{(i)}$, there is a closed segment L' of L such that $(\text{int } L') \times I \subset (L \times I) \cap \hat{C}_i \subset L' \times I$,

where int L' refers to the topology of R^1. The partition will be called *polyhedral* if each int $\hat{C}_i$ is the interior of a finite union of closed nondegenerate (irregular) k-simplices. Thus, each star-shaped decision partition of $\mathscr{T}_k$ is obtained by taking the cartesian product with I of a star-shaped partition of $\mathscr{S}_{k-1}$ (defined by removing I everywhere in (5.1)), possibly modifying it by introducing randomization on common boundaries of the $\hat{C}_i$. A decision partition satisfying (5.1) and also (3.9)(i) will be termed *positive star-shaped*. This last property is possessed by structure W partitions, but the projections on $\mathscr{S}_{k-1}$ of the cl $\hat{C}_i$ of a positive star-shaped partition are not even necessarily convex. The possibility of replacing star-shapedness by a more special structure will be discussed at the end of this section.

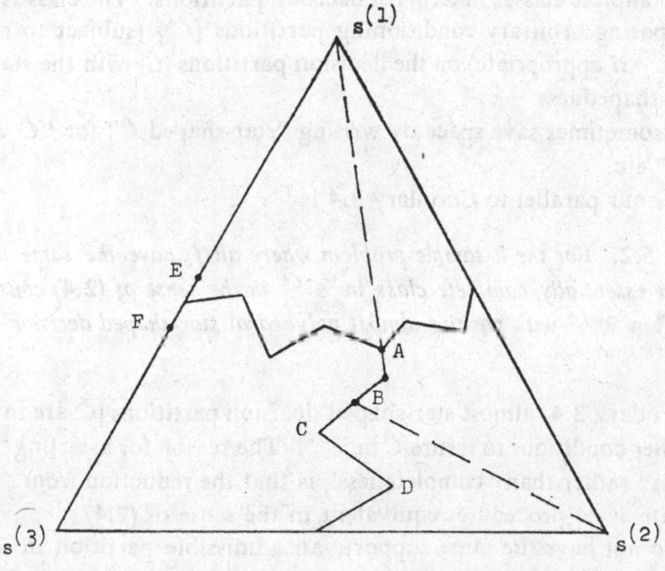

FIG. 1. A positive polyhedral almost star-shaped partition, shown on $\mathscr{S}_2$ when $k = 3$. Solid lines border stars; dashed lines are portions of lines L of the definition. C is the vertex of star-point BCD of the star around $s^{(2)}$. The exceptions to star-sharpedness, of Theorem 5.4, can occur on segments such as AB and EF.

In discussing phenomena that can occur for polyhedral star-shaped regions, we shall find it convenient to call the projection of cl $\hat{C}_i$ onto $\mathscr{S}_{k-1}$ the *star around* $s^{(i)}$. Of course, this may even be convex. However, in cases where this star contains long narrow protrusions into the remainder of $\mathscr{S}_{k-1}$ (a very qualitative notion), it will be useful to call these *star-points*. We will be more precise when $k = 3$: a triangular star-point of the star about $s^{(i)}$ is any triangle of that star two sides of which are on the boundary of the star. The angle between them is at the *vertex* ot the star point. Some of these definitions are illustrated in Figure 1.

We also require a slight extension of these notions. The motivation is that, if a sequence of procedures with polyhedral star-shaped (positive or not) decision partitions, with bounded number of sides on the polyhedra, converges regularly to a limiting procedure, that limit clearly has the following *property A*: it is equivalent ($\mathfrak{v} \times \mu^1$) to a procedure whose decision partition differs from (5.1) on at most a finite number of sets of the form $H \times I$ where each H is a $(k-2)$-dimensional hyperplane through some $s^{(i)}$. We call any decision partition of $\mathscr{T}_k$ (positive or not, polyhedral or not) *almost star-shaped* if it has property A. In the motivating polyhedral case, that part H' of such an H on which (5.1) is violated typically arises as the limit of a union of sharp star-points of a star around a given $s^{(i)}$, which union is in a narrowing dihedral angle (about $s^{(i)}$) approaching H. If $\mathfrak{v}$ is absolutely continuous with respect to μ^{k-1} on $\mathscr{S}_{k-1}$, every almost star-shaped region is equivalent ($\mathfrak{v} \times \mu^1$) to a star-shaped region.

It is to be noted that the theorems and corollaries of this section characterize essentially complete classes in terms of decision partitions. The class is obtained by superimposing arbitrary conditioning partitions $\{\hat{C}^b\}$ (subject to restriction to $\mathscr{C}^+$ or $\mathscr{C}^L$ if appropriate) on the decision partitions $|\hat{C}|$ with the stated property of star-shapedness.

We shall sometimes save space by writing "star-shaped $\hat{C}$" for "$\hat{C}$ with star-shaped $|\hat{C}|$," etc.

For $\mathscr{C}^{L+}$, our parallel to Corollary 3.4 is

THEOREM 5.2. *For the k-sample problem where all f_i have the same domain of positivity, an essentially complete class in $\mathscr{C}^{L+}$ in the sense of* (2.4) *consists of all procedures $\hat{C}$ in $\mathscr{C}^{L+}$ with positive almost polyhedral star-shaped decision partitions $|\hat{C}|$ of $\mathscr{T}_k$.*

As in Corollary 3.4, almost star-shaped decision partitions $|\hat{C}|$ are in $\mathscr{C}^{1+}$ but require further conditions to insure $\hat{C}$ in $\mathscr{C}^+$. The reason for asserting "essential completeness" rather than "completeness" is that the reduction from $\mathscr{X}$ to $\mathscr{T}^k$ may eliminate some procedures equivalent in the sense of (2.4).

If the f_i do not have the same support, an admissible partition in $\mathscr{C}^{L+}$ can have $\hat{C}_i = s^{(i)} \times I$ for some i, and thus not satisfy (3.9)(i); Theorem 5.2 can be restated for this case. For $\mathscr{C}^+$ (general B) we are unable, even when all f_i have the same support, to verify (3.9)(i) for admissible $\hat{C}$ with the present proof, and can only improve slightly on the result for $\mathscr{C}$. We now state the latter (without the assumption of Theorem 5.2):

THEOREM 5.1. *For the k-sample problem, an essentially complete class in* $\mathscr{C}$ *in the sense of* (2.4) $\wedge$ (2.6), *and hence in the sense of* (2.4) $\wedge$ (2.8), *consists of the procedures* $\hat{C}$ *for which* $|\hat{C}|$ *is in the regular closure of the polyhedral positive star-shaped partitions of* $\mathscr{T}_k$.

The shortcoming of this result over that of Theorem 5.2 will be discussed in Remark 1 at the end of this section: more extensive randomization may be required in some of the procedures of the Theorem 5.1 essentially complete class, than in star-shaped procedures.

In the 2-hypothesis problem, the star-shapedness of (5.1), as well as almost star-shapedness, reduces to division of $\mathscr{S}_1$ into two intervals with randomization on the boundary of the two intervals, but never any randomization at an $s^{(i)}$ if that is the boundary. These are exactly the $\mathscr{C}^1$-admissible partitions for $|\hat{C}|$, and Theorem 5.1 combined with the results of Section 3 thus yields the first sentence of the following:

COROLLARY 5.3. *For the 2-hypothesis problem, an essentially complete class of admissible procedures in* $\mathscr{C}$ *in the sense* (2.4) $\wedge$ (2.6) *or* (2.4) $\wedge$ (2.8) *consists of all partitions in* $\mathscr{C}$ *whose underlying* $\mathscr{C}^1$*-partitions have the interval structure* (5.1). *This conclusion remains true if* $\mathscr{C}$ *is replaced everywhere by* $\mathscr{C}^L$, *or by* $\mathscr{C}^+$ *or* $\mathscr{C}^{L+}$ *with* (2.4) $\wedge$ (2.8) *replaced by* (2.4).

The last sentence of this corollary is obvious for $\mathscr{C}^+$, and an examination of that part of the proof of Theorem 5.1 that pertains to $\mathscr{C}^L$ shows the validity of the remainder when $k = 2$.

When $k = 3$, we cannot argue for the analogue of Corollary 5.3, but we can somewhat sharpen Theorem 5.1:

COROLLARY 5.4. *For the 3-hypothesis problem, an essentially complete class in* $\mathscr{C}$ *in the sense of* (2.4) $\wedge$ (2.6), *and hence in the sense of* (2.4) $\wedge$ (2.8), *consists of all procedures* $\hat{C}$ *for which* $|\hat{C}|$ *is an almost star-shaped partition of* $\mathscr{T}_3$, *and for which the exceptions to star-shapedness lie on at most the three lines* $\{s_i = 0\}$ *and one additional line through a single* $s^{(i)}$, *part of which borders the stars of the other two* $s^{(j)}$.

This will be proved below.

PROOF OF THEOREM 5.1. We shall work entirely in terms of partitions of $\mathscr{T}_k$ (which are equivalent to η's on $\mathscr{S}_{k-1}$ as discussed earlier), and therefore simplify notation by dropping carats. Let $\bar{C}$ be a given partition of $\mathscr{T}_k$. Our steps in finding a $\bar{\bar{C}}$ that is at least as good as $\bar{C}$ in the sense of (2.4) $\wedge$ (2.6), and that satisfies (5.1), are these: (a) For each integer $n > 0$, there is by Lemma 4.5 a finite L_n and a procedure ${}^n\bar{C}$ in $\mathscr{C}^{L_n}$ such that $\|\bar{G}_\omega - {}^n\bar{G}_\omega\|$ and $\|\bar{H}_\omega - {}^n\bar{H}_\omega\|$ are $< n^{-1}$ for all ω, where the norm is that of Section 4. (b) For each fixed n, there is a partition ${}^n\bar{\bar{C}}$ in $\mathscr{C}^{L_n}$ that is at least as good as ${}^n\bar{C}$ in the sense of

(2.4) $\wedge$ (2.6), and that is a regular limit, as $m \to \infty$, of a sequence $\{^{(n,m)}C\}$ of procedures with positive polyhedral star-shaped decision partitions. (c) We may assume the $^{(n,m)}C$ are canonical, since that is merely a relabeling (with possible lumping) of B and yields the same operating characteristics. For each fixed n, a subsequence $^{(n,m')}C$ converges regularly and its operating characteristics G, H converge weakly by Theorem 4.4. A diagonalization of (n, m') produces a subsequence $\{^N\bar{\bar{C}}\}$ of the $^{(n,m')}C$'s that, by Theorem 4.4, converges regularly to a partition $\bar{\bar{C}}$ of $\mathscr{T}_k$, with $^N\bar{\bar{G}}$, $^N\bar{\bar{H}}$ converging weakly to $\bar{\bar{G}}$, $\bar{\bar{H}}$. The conclusion of Theorem 5.1 follows from this last together with (a) and (b), and it remains to prove the latter.

For fixed L, we shall show that the admissible procedures in $\mathscr{C}^L$ are regular limits of procedures with positive polyhedral star-shaped decision partitions. Since the operating characteristics of procedures in $\mathscr{C}^L$ are compact (Theorem 4.4), this yields (b). Suppose that, for two partitions $\check{C}$ and $\tilde{\tilde{C}}$ of $\mathscr{T}_k$ in $\mathscr{C}^L$ we have

$$(5.2) \qquad \begin{aligned} P_i\{\tilde{\tilde{C}}_i{}^b\} &\geqq P_i\{\check{C}_i{}^b\} \quad \forall\, b, i\,, \\ P_i\{\tilde{\tilde{C}}^b - \tilde{\tilde{C}}_i{}^b\} &\leqq P_i\{\check{C}^b - \check{C}_i{}^b\} \quad \forall\, b, i\,. \end{aligned}$$

Then, whenever $P_i\{\check{C}_i{}^b\}$ is positive, so is $P_i\{\tilde{\tilde{C}}_i{}^b\}$. Since $1 - (\Gamma_i{}^b)^{-1} = P_i\{C^b - C_i{}^b\}/P_i\{C_i{}^b\}$, (5.2) implies that $\tilde{\tilde{\Gamma}}_i{}^b \geqq \check{\Gamma}_i{}^b$. Hence, $\tilde{\tilde{C}}^b$ contributes at least as large an atom of probability as does $\check{C}^b$ to the sub-df $G_i(1 - t)$, and that atom corresponds to at least as large a value of Γ_i. Thus, (5.2) implies $\tilde{\tilde{G}}_i(t) \geqq \check{G}_i(t)\ \forall\, i, t$, and similarly $\tilde{\tilde{H}}_i(t) \leqq \check{H}_i(t)\ \forall\, i, t$. Moreover, strict inequality holds somewhere in one of these last two sets of inequalities if it holds somewhere in (5.2). We conclude that (2.4) $\wedge$ (2.6)-admissibility in $\mathscr{C}^L$ implies (5.2)-admissibility in $\mathscr{C}^L$, and it remains to show that (5.2)-admissible partitions in $\mathscr{C}^L$ have the structure stated at the outset of this paragraph.

For C a $\mathscr{C}^L$-partition of $\mathscr{T}_k$, let r_c be the point in R^{2kL} with coordinates $P_i\{C_i{}^b\}$, $-P_i\{C^b - C_i{}^b\}$ for $i \in \Omega$, $b \in B_L$. These points can be analyzed using standard decision theoretic results. The set of all such r_c's is convex and compact, and ([7], Theorem 3.19) the set of (5.2)-admissible elements is contained in the closure of the set of all "anti-Bayes" elements with respect to strictly positive prior vectors; such an anti-Bayes element with respect to a $2kL$-vector $\{a_i{}^b, \bar{a}_i{}^b\}$ of positive coordinates is a partition in $\mathscr{C}^L$ that *maximizes*

$$(5.3) \qquad \textstyle\sum_{i,b} [a_i{}^b P_i\{C_i{}^b\} - \bar{a}_i{}^b P_i\{C^b - C_i{}^b\}]\,.$$

The representation of the partition C of $\mathscr{T}_k$ as a randomized procedure η on $\mathscr{S}_{k-1}$ is

$$(5.4) \qquad \eta(i, b \mid s) = \textstyle\int_0^1 I_{C_i{}^b}(s, u)\, du$$

as the (randomization) probability, given $S = s$, that $(\delta, Z) = (i, b)$; of course $\sum_{i,b} \eta_c(i, b \mid s) = 1$. Write

$$(5.5) \qquad H_i{}^b(s) = a_i{}^b s_i - \textstyle\sum_{j \neq i} \bar{a}_j{}^b s_j\,.$$

We can then rewrite (5.3) as

$$(5.6)\qquad \sum_{i,b} \int_{\mathscr{S}_{k-1}} [a_i{}^b\eta(i, b \mid s) - \bar{a}_i{}^b \sum_{j\neq i} \eta(j, b \mid s)]\hat{f}_i(s)\hat{\nu}(ds) = \sum_{i,b} \int_{\mathscr{S}_{k-1}} H_i{}^b(s)\eta(i, b \mid s)\hat{\nu}(ds)\,.$$

From (5.5) we see that, for fixed (i_0, b_0) and $i \neq i_0$, the hyperplane $\{s\colon H_{i_0}^{b_0}(s) = H_i{}^b(s)\}$ divides $\mathscr{S}_{k-1}$ into two regions with $s^{(i_0)}$ and $s^{(i)}$ in their interiors. Hence, the set

$$(5.7)\qquad M_{i_0}^{b_0} = \{s\colon H_{i_0}^{b_0}(s) \geq H_i{}^b(s) \text{ for all } (i, b) \text{ with } i \neq i_0\}$$

is convex, polyhedral, and contains $s^{(i_0)}$ in its interior. We conclude that $M_{i_0} = \bigcup_b M_{i_0}^b$ is the i_0-star of a positive polyhedral star-shaped partition of $\mathscr{S}_{k-1}$, and upon replacing the weak inequality by strict inequality in (5.7) one obtains sets whose union over b_0 is int M_{i_0}. A standard argument on (5.6) shows that, a.e. $\hat{\nu}$, every maximizer of (5.3) satisfies, for each i_0,

$$(5.8)\qquad \eta(i_0, B_L \mid s) = 1 \quad \text{if } s \in \operatorname{int} M_{i_0}\,, \qquad = 0 \quad \text{if } s \notin M_{i_0}\,.$$

Thus, every such η is equivalent to a positive polyhedral star-shaped partition of $\mathscr{T}_k$. □

Proof of Theorem 5.2. Although the set $\{r_C\colon C \in \mathscr{C}^{L+}\}$ is not closed, a trivial geometric observation shows that, for each $\bar{C}$ in $\mathscr{C}^{L+}$, there is an anti-Bayes partition $\bar{\bar{C}}$ in $\mathscr{C}^{L+}$ (relative to $\{a_i{}^b, \bar{a}_i{}^b\}$ of nonnegative components, some but not all of which may now be zero) that is at least as good as $\bar{C}$ in the sense of (5.2) and hence (2.4). Moreover, the development of the previous paragraphs shows that these $\bar{\bar{C}}$ are all limits of sequences procedures with positive polyhedral star-shaped decision partitions having a bounded number of faces, and hence they have decision partitions that are almost polyhedral star-shaped. It remains to show that these anti-Bayes partitions $\bar{\bar{C}}$ in $\mathscr{C}^L$ have $|\bar{\bar{C}}|$ *positive* star-shaped; we must verify (3.9)(i) for them.

Fix i_0. There are three cases to consider. (i) If there is a b_0 for which $a_{i_0}^{b_0} > 0$, then clearly $s^{(i_0)} \in \operatorname{int} M_{i_0}^{b_0}$. (ii) If there is a b' for which $\bar{a}_{i_0}^{b_0'} > 0$, then for every b and $i \neq i_0$ the set $M_i{}^b$ excludes $\{s\colon H_i{}^b(s) < H_{i_0}^{b'}(s)\}$, and the latter clearly contains a neighourhood of $s^{(i_0)}$. (iii) If no $a_{i_0}^{b_0}$ or $\bar{a}_{i_0}^{b_0'}$ is positive, then (since some $a_i{}^b$ or $\bar{a}_i{}^b$ is positive) one computes easily that M_{i_0} contains no points of $\{s\colon s_i > 0\ \forall\, i\} = \mathscr{S}_{k-1}^*$ (say). But $\hat{\nu}(\mathscr{S}_{k-1} - \mathscr{S}_{k-1}^*) = 0$ if all f_i have the same support, so $\hat{\nu}(M_{i_0}) = 0$ in this last case. Thus, either there is an i_0 for which (iii) holds, in which case the resulting anti-Bayes partition is not in $\mathscr{C}^{L+}$, or else for each i_0 $(1 \leq i_0 \leq k)$ either (i) or (ii) holds, in which case the resulting anti-Bayes partition is in $\mathscr{C}^{L+}$. □

Proof of Corollary 5.4. Fix $\varepsilon > 0$. Let C have a $|C|$ that is a positive star-shaped partition of $\mathscr{T}_3$, and let V be any triangular star point of the star about $s^{(1)}$, with vertex P. Suppose cl V is surrounded by C_2, in a neighborhood of P. The angle θ of cl V facing $s^{(1)}$ at P, because of (5.1) with $i = 1$, has the property that it contains part of the segment $s^{(1)}P$ near P; but it is also such that its side closest to $s^{(3)}$, when extended beyond P away from $s^{(1)}$, intersects the segment $s^{(1)}s^{(3)}$, because of (5.1) with $i = 2$. If $P \in N_\varepsilon = \{s: \text{all } s_i \geqq \varepsilon\}$, elementary trigonometry shows that $\theta \geqq 2 \tan^{-1}(2\varepsilon/3^{\frac{1}{2}}) \geqq c_0\varepsilon$, where $c_0 > 0$. It follows (essentially from compactness of graphs satisfying the same Lipschitz condition) that, for each sequence ${}^{n}C$ of procedures with positive polyhedral star-shaped decision partitions of $\mathscr{S}_2$ and each $\varepsilon > 0$, there is a subsequence that converges regularly on N_ε to a procedure with star-shaped decision partition on N_ε, except for at most a single exceptional H in the definition of "almost star-shaped": for each n, it is easy to see that at most one ${}^{n}C_{i_0}$ can have a star-point V each of whose two sides is bordered near the vertex by a *different* one of the other two C_i. This single V alone can be in N_ε and have an angle $\theta < c_0\varepsilon$, and in the limit a subsequence of such V's can collapse toward an H. A subsequence $\varepsilon_m \to 0$ and diagonalization completes the proof, in view of Theorem 5.1. □

Remarks on the necessary conditions for admissibility.

1. For $k = 2$ or 3, elementary geometry has shown (Corollaries 5.3 and 5.4) that taking the regular closure in Theorem 5.1 does not introduce much randomization over that present in Theorem 5.2 for $\mathscr{C}^{L+}$. We have been unable to verify the corresponding fact for $k = 4$, although it is natural to conjecture that such randomization should not be necessary. The difficulty is that, viewing the boundary of a polyhedral star $\hat{C}_{i_0}$ as the graph of distance from $s^{(i_0)}$ (as a function of angular position), we might have a sequence of such graphs with increasing number of increasingly frequent oscillations (sharp star-points) whose magnitudes stay bounded away from zero. Such a sequence ${}^{n}\hat{C}$ could converge regularly to a procedure ${}^{\infty}\hat{C}$ for which $|{}^{\infty}\hat{C}|$ randomizes (i.e., for which $0 < \eta(i \times B \mid s) < 1$ on a set of positive μ^{k-1} measure in $\mathscr{S}_{k-1}$, similar to what occurs in $\mathscr{C}^1$ decision theory for sequences like (4.1)). When $k = 3$, the star-shapedness for $i \neq i_0$ limited the existence of such sharp points, for positive polyhedral procedures, and ${}^{\infty}\hat{C}$ could only differ from star-shapedness on one line segment H (in the definition of almost star-shapedness). Moreover, that H was associated with the limit of a sequence of points of ${}^{n}\hat{C}_{i_0}$ that had each of the other two ${}^{n}\hat{C}_i$'s on one of its sides. When $k = 4$ the failure of this last property seems to make possible the existence of a large number of sharp points of nonvanishing magnitude in ${}^{n}\hat{C}_{i_0}$. To see what can happen, we now give an example, when $k = 4$, of a ${}^{n}\hat{C}_1$ that has a single large sharp point a large portion of which is completely surrounded by ${}^{n}\hat{C}_2$. It is not clear at this writing that one cannot have such an increasing number of points that ${}^{\infty}\hat{C}$ has the undesirable extent of randomization noted above. If so, we still do not know whether such an eventuality must be part of a complete class for general B, or whether the difficulty lies in the reduction to using (5.2) as discussed in Remark 2 below, so that such ${}^{\infty}\hat{C}$ could be eliminated by a different admissibility proof.

For this example, suppose $L = 3$. The star about $s^{(1)}$ will be our concern. It will be convenient notationally to label the elements of B as $\{1, 3, 4\}$. To simplify the arithmetic, we make most of the 24 coefficients zero; adding a small positive value to all of them does not change the substance of the example. We now construct ${}^rC$ for r large. Let ε be small; for definiteness, take $r\varepsilon = \frac{1}{2}$. Let

$$a_1^{\ 1} = a_2^{\ 4} = a_2^{\ 3} = 1\,,$$
$$\bar{a}_1^{\ 1} = \bar{a}_2^{\ 4} = \bar{a}_2^{\ 3} = r^2\,,$$
$$\bar{a}_3^{\ 1} = \bar{a}_4^{\ 1} = r(\tfrac{1}{2} - \varepsilon)\,,$$
$$\bar{a}_3^{\ 4} = \bar{a}_4^{\ 3} = r\,,$$

and let all other $a_i^{\ b}$ and $\bar{a}_i^{\ b}$ be zero. A simple geometric picture may be obtained as follows: Define, in $\mathscr{S}_3$, the lines

$$L' = \{s\colon s_1 + s_2 = 1\}\,,$$
$$L'' = \{s\colon s_3 + s_4 = 1\}\,,$$
$$L = \{s\colon s_1 = s_2,\, s_3 = s_4\}\,,$$
$$L^* = \{s\colon s_3 = s_4,\, s_2 - s_1 = 2r\varepsilon s_3\}\,,$$

the planes

$$\Pi^1 = \{s\colon s_1 = s_2\}\,,$$
$$\Pi^4 = \{s\colon H_1^{\ 1}(s) = H_2^{\ 4}(s)\} = \{s\colon (s_2 - s_1) = r[(\tfrac{1}{2} + \varepsilon)s_3 - (\tfrac{1}{2} - \varepsilon)s_4]\}\,,$$
$$\Pi^3 = \{s\colon H_1^{\ 1}(s) = H_2^{\ 3}(s)\} = \{s\colon (s_1 - s_2) = r[(\tfrac{1}{2} - \varepsilon)s_3 - (\tfrac{1}{2} + \varepsilon)s_4]\}\,,$$

and the points

$$p' = (\tfrac{1}{2}, \tfrac{1}{2}, 0, 0)\,,$$
$$p'' = (0, 0, \tfrac{1}{2}, \tfrac{1}{2})\,,$$
$$p^* = (\tfrac{1}{4}, \tfrac{1}{4}, \tfrac{1}{4}, \tfrac{1}{4})\,,$$
$$p_3 = (0, 0, \tfrac{1}{2} - \varepsilon, \tfrac{1}{2} + \varepsilon)\,,$$
$$p_4 = (0, 0, \tfrac{1}{2} + \varepsilon, \tfrac{1}{2} - \varepsilon)\,.$$

Take L' to be vertical with $s^{(2)}$ at its top. L' is bisected by Π^1, which contains L and L'', with $L \cap L' = p'$ and $L \cap L'' = p''$. The points p_3 and p_3 are very close to p'', and hence the lines $\Pi^4 \cap \Pi^1$ and $\Pi^3 \cap \Pi^1$ are close to L. On the other hand, the line $L^* = \Pi^4 \cap \Pi^3$ is not everywhere close to L; e.g., the point on L^* above p^* is $((1 - r\varepsilon)/4, (1 + r\varepsilon)/4, \frac{1}{4}, \frac{1}{4})$. Hence, the region V above Π^1, under Π^4 and Π^3, and with $\frac{1}{4} \leqq s_3 + s_4 \leqq \frac{3}{4}$, is a very thin wedge of length $\frac{1}{2}$ and thickness $< 2\varepsilon$, whose height is $r\varepsilon/4 = \frac{1}{8}$ at p^*.

The coefficients $a_i^{\ b}$ and $\bar{a}_i^{\ b}$ have been chosen so that int $V \subset {}^rC_1$ but so that the region just on the other side of Π^4 and Π^3 is in ${}^rC_2$, with the part bordering Π^b in ${}^rC_2^{\ b}$. To see this, one computes that $H_1^{\ 1} - H_2^{\ b}$ has the right behavior at Π^b (for $b = 4, 3$), and that $H_2^{\ 4}(s) - H_2^{\ 3}(s) \geqq 0 \Leftrightarrow s_4 \geqq s_3$. All that remains is then to verify that $H_i - H_i^{\ b} > 0$ on a neighbourhood of V for all (i, b) other than (2, 3) and (2, 4). This is the case because the dominating term in $H_1^1 - H_1^b$, of order r^2, is $\bar{a}_1^1 s_1$ in the case $b = 1$ and $\bar{a}_1^b s_2$ otherwise.

This completes the example.

2. Even in the restricted context of Theorem 5.2, except when $k = 2$ it is generally false that the star-shaped partitions are all admissible, even if they are positive and polyhedral. The main cause of this is the use of (5.2), in terms of which it is more difficult for a given $\tilde{C}$ to be improved upon by a $\tilde{\tilde{C}}$ than in the original sense of (2.4) $\wedge$ (2.6), because the improvement must be *on each* $\tilde{C}^b$. In terms of G_ω or H_ω, we can have domination without it occuring for each b. For example, one can imagine $L = 2$ and $\tilde{G}_1$ with jumps of .3, .3 at .5, .7 while $\tilde{\tilde{G}}_1$ has jumps of .2, .4 at .6, .8; then $\tilde{G}_1 \leqq \tilde{\tilde{G}}_1$, but (5.2) cannot be satisfied for any labeling of C_i^b's. (It is easily seen that inadmissibility of $\tilde{C}$ in the sense of (5.2) does permit any relabeling of $\tilde{\tilde{C}}$ by permutation of B while fixing the labels for $\tilde{C}$.) The difficulty is also reflected in the fact that the operating characteristics (G, H) of $\mathscr{C}^L$ partitions are not a convex set, unlike the set of vectors r_C.

3. We now mention briefly some possible improvements on the necessary conditions of this section. There are special cases where small support of $\mathfrak{v}$ can yield superficially better descriptions of admissible $\hat{C}$ that cannot hold in general. More interesting improvements may be sought in the characterization of $|\hat{C}|$ or in the breakup into $\{\hat{C}^b\}$. For example, the development of (5.3)—(5.7) for $\mathscr{C}^L$ yields C_i^b with convex interiors provided that the a_i^b and $\bar{a}_i^b$ are so different that $H_{i_0}^{b'}(s) = H_{i_0}^{b''}(s)$ only on a hyperplane of $\mathscr{S}_{k-1}$. This is always the case if the star around $s^{(i_0)}$ has L star-points. If, instead, $a_{i_0}^{b'} = a_{i_0}^{b''}$ and $\bar{a}_j^{b'} = \bar{a}_j^{b''}$ for $j \neq i_0$, then any measurable breakup of $M_{i_0}^{b'} - \bigcup_{b \neq b', b''} M_{i_0}^b$ can be used for $C_{i_0}^{b'}$, $C_{i_0}^{b''}$; they need not be convex. Such a degeneracy in the $\bar{a}_j^b$ and $a_{i_0}^b$ is reflected in M_{i_0} being a union of fewer than L convex sets. When C_{i_0} has fewer than L star-points (or analogous points on the boundary of $\mathscr{S}_{k-1}$), our development yields little about the breakup into $\{C_{i_0}^b\}$ outside of a neighborhood of the boundary of C_{i_0}. In short, the greatest possible departures from convexity (most points on stars) of the $\hat{C}_i$ can occur only with a breakup into the simplest $\hat{C}_i^b$ (those with convex closure).

4. It is unclear to what extent the star-shapedness can be reduced in Theorem 5.2 (aiming toward less departure from convexity), but simple examples when $k = 3$ and $L = 2$ show that it cannot be eliminated for the sense (5.2). (For example, if $a_3^1 = a_3^2 = \bar{a}_1^1 = \bar{a}_2^2 = 1$ and all other coefficients are ε, small and positive, then C_3 differs little from the set $\{s: s_3 > \min(s_1, s_2)\}$.) A different proof, taking account of Remark 2, would apparently be needed for significant reduction. One possibility that the author has not been able to implement is the maximization of $\sum_i \int h_i(t)d[-G_i(t)]$ (together with a linear combination reflecting (2.6) or (2.8)) for increasing functions other than the $h_i(t) = -c_i t^{-1}$ used in the sufficient condition of Section 3. For another possibility, we refer the reader to an "exchange" argument in the case $k = 2$ that appears in [2], and in which points of a $C_1^{b'}$ can be exchanged with those of larger s_1 in a $C_2^{b''}$ to

improve both G_i; this argument fails to yield convexity for $k > 2$, and thus far has yielded only minor information bounding the slope of the boundary of the stars as a function of position in $\mathscr{S}_{k-1}$.

The use of (2.6) rather than (2.8) alone in the proof, even when (2.4) $\wedge$ (2.8) is the criterion of admissibility, reflects the difficulty of treating the $P_\omega\{Q_\omega\}$ as part of a vector like r_c.

5. One *can* introduce a more detailed operating characteristic, and accompanying notion of admissibility, relative to which the star-shaped anti-Bayes procedures of the present section become admissible except for modifications on the border of $\mathscr{S}_{k-1}$ analogous to those of [8] (so that, for example, one must have $\hat{C}_i \cap \{(s, u) : s_i = 0\} = \emptyset$). Such a more detailed operating characteristic for $\mathscr{C}^L$ is given by the equivalence class r_c^* of all vectors obtained from r_c by permutations (relabeling) of B^L; the partition $\bar{C}$ is at least as good as $\bar{\bar{C}}$ in this sense if there is a permutation σ for which

$$(5.9) \qquad P_i\{\bar{C}_i^{\,b}\} \geq P_i\{\bar{\bar{C}}_i^{\,\sigma b}\} \quad \text{and} \quad P_i\{\bar{C}^b - \bar{C}_i^{\,b}\} \leq P_i\{\bar{\bar{C}}^{\sigma b} - \bar{\bar{C}}_i^{\,\sigma b}\} \quad \forall\, b, i\,.$$

Equivalently, one can adjoin to (5.9) the condition $\bar{\Gamma}_i^{\,b} \geq \bar{\bar{\Gamma}}_i^{\,\sigma b}$ $\forall\, b, i$, since the latter follows from (5.9). Then the star-shaped partitions that are anti-Bayes relative to strictly positive vectors (as in our development (5.3)—(5.8)) are admissible; and so are the star-shaped partitions obtained as limits of these, except for the indicated restrictions on the border of $\mathscr{S}_{k-1}$. Remark 1 is however still relevant.

This criterion seems intuitively too "fine," and yields too many admissible procedures to be taken as justification for blindly using an arbitrary (positive) anti-Bayes procedure. The G_i obtained by "lumping" of $\{\Gamma_i^{\,b}, P_i\{C_i^{\,b}\}, b \in B\}$ reflect the notion of goodness in a preferred degree of detail, and represent the conditional confidence aim better than does r_c^*.

6. Since the results for $k \geq 4$ are more informative for $\mathscr{C}^L$ than for $\mathscr{C}$, the use of finite approximations of general partitions ((a) of proof of Theorem 5.1) was expeditious. However, anti-Bayes procedures for general B can also be computed directly, the role of r_c being induced by the probability measures on $B \times D$ determined by rectangles, i.e., by $P_i\{Z \in A, \delta = d\}$ for $A \in \mathscr{B}_0$ and $d \in D$. As in the case of $\mathscr{C}^L$, this yields a weaker notion of admissibility than (2.4) $\wedge$ (2.6). Thus far we have not been able to reduce the huge complete class for $\mathscr{C}$ through this approach.

7. The remarks of Section 4 on elimination of randomization can be used to simplify the results of Theorem 5.2 and Corollaries 5.3 and 5.4. However, when applied to Theorem 5.1 they yield nonrandomized partitions whose structure is quite unclear.

6. Other loss and risk structures. We mention briefly here only four of the many possible modifications to using (2.4) $\wedge$ (2.6) with the implied 0–1 loss structure from decisions in D_ω and its complement when ω is true.

(A) In Section 3, if we replace the conditional probability $\Gamma_i^{\,b}$ by a *conditional gain* $W_i^b = \sum_{j \in D_i} w_{ij} P_i\{K_j^b\}$ where the given numbers w_{ij} are positive for $j \in D_i$, we see that the proofs of that section are easily modified, *provided* w_{ij} *is independent of* j *in* D_i. In the k-hypothesis setting of Section 5, consideration of $w_{ii}\Gamma_i^b$ in place of Γ_i^b does not alter the conclusions. If performance is measured instead in terms of conditional expected *loss* $\sum_{j \notin D_i} m_{ij} P_i\{C_j^b\}$ where $m_{ij} > 0$ for $j \notin D_i$, the conclusion of Section 3 remains valid if the m_{ij} are equal for $j \notin D_i$; the development of Section 5 proceeds with replacement of r_c by the $2kL$-vector $\{\sum_{j \neq i} m_{ij} P_i\{C_j^b\}, -P_i\{C^b\}\}$, and the changes thereafter are obvious.

(B) For brevity, the next modification will be described only for the k-hypothesis problem and in terms of the original 0–1 losses. A more detailed picture of conditional probabilities of *incorrect* decisions than is present in the H_ω of (2.5) can be given for $j \neq i$, in terms of the functions

$$H_{ij}(t) = P_i\{\delta = d_j; \Gamma_{ij} > t\}, \tag{6.1}$$

where

$$\Gamma_{ij} = P_i\{C_j \mid \mathscr{B}_0\}. \tag{6.2}$$

The Γ_{ij} have an obvious frequentist interpretation, and in practice if $Z(X) = b_0$ and $\delta(x) = d_{j_0}$ the vector $\{\Gamma_{ij_0}^{b_0}, i \neq j_0\}$ of confidences associated with the states i for which d_{j_0} is incorrect, could be stated in addition to $\Gamma_{j_0}^b$. Also define, in analogy with (1.3),

$$Q_{ij} = \{x : \Gamma_{ij}(x) = 0\}. \tag{6.3}$$

Intuitively, we prefer the $P_i\{Q_{ij}\}$ to be large rather than small. Thus we are led to the possibility of defining $\bar{\bar{C}}$ to be at least as good as $\bar{C}$ if, for all unequal i and j,

$$\bar{\bar{H}}_{ij}(t) \leqq \bar{H}_{ij}(t) \quad \forall\, t \tag{6.4}$$

and

$$P_i\{\bar{\bar{Q}}_{ij}\} \geqq P_i\{\bar{Q}_{ij}\}. \tag{6.5}$$

(It is unnecessary to adjoin (2.4) $\wedge$ (2.8), but the conclusion below will be the same with that addition.)

In the development of Section 3, we now replace classical $\mathscr{C}^1$-admissibility of $|C|$ by admissibility in the sense of the classical vector risk $\{P_i\{C_j\}, i \neq j\}$, in terms of which $\bar{\bar{C}}$ is at least as good as $\bar{C}$ if $P_i\{\bar{\bar{C}}_j\} \leqq P_i\{\bar{C}_j\}$ for all unequal i, j. (Alternatively, this sense of admissibility may be replaced by one that implies it, that there is some set of losses $m_{ij} > 0$ for which $|C|$ is admissible for the risk function $r(i) = \sum_{j \neq i} m_{ij} P_i\{C_j\}$.) In analogy with (3.1)—(3.2), we obtain $P_i\{C_j; \Gamma_{ij} = 0\} = 0$ and $-\int_{0+}^{\infty} t^{-1} d[-G_{ij}(t)] = P_i\{Q_{ij}\} - 1$. Parallel to the development of the paragraph following (3.4), we have (6.4) at $t = 0$ yielding $P_i\{\bar{\bar{C}}_j\} \leqq P_i\{\bar{C}_j\}$ for all $i \neq j$; if $|\bar{C}|$ is classically vector-admissible, these must be equalities. The remainder of the proof parallels (3.5) and (3.6), leading to the conclusion that $|\bar{C}|$ vector-admissible implies $\bar{C}$ admissible in the sense induced by (6.4) $\wedge$ (6.5).

The sense (6.4) $\wedge$ (6.5) seems somewhat less ad hoc than that of the r_c^* of Section 5 (Remark 5), but it would be more natural still to work with the law of Γ_j under state $i \neq j$, in place of that of Γ_{ij}. A useful admissibility result has not been obtained for the resulting criterion because of the failure thus far to find a substitute for (3.2).

The development of Section 5 for the criterion (6.4) is straightforward.

(C) Another consideration is a measure of "size" of a decision, analogous to length of a confidence interval. For example, the making of a decision d for which Ω_d contains many elements should intuitively be penalized in terms of some measure of the size of Ω_d. Let $e_j > 0$ be the "size" penalty incurred when d_j is made. The question of whether conditional or unconditional expectation of e is more suitable to consider, with illustrations, will be discussed in [2], [4]. For the moment, we treat unconditional expectation, and define $\bar{\bar{C}}$ to be at least as good as $\bar{C}$ if (2.4), (2.8), and

$$\sum_j e_j P_\omega\{\bar{\bar{\delta}} = d_j\} \leq \sum_j e_j P_\omega\{\bar{\delta} = d_j\} \quad \forall\, \omega \tag{6.6}$$

hold. The corresponding $\mathscr{C}^1$ notion for $|C|$ is (6.6) together with $P_\omega\{\bar{\bar{\delta}} \in D_\omega\} \geq P_\omega\{\bar{\delta} \in D_\omega\}\ \forall\, \omega$. If $|\bar{C}|$ is $\mathscr{C}^1$-admissible but $\bar{\bar{C}}$ is better than $\bar{C}$ in the sense of (2.4) $\wedge$ (2.8) $\wedge$ (6.6), either (i) G_ω, $P_\omega\{Q_\omega\}$ are the same for both procedures $\forall\, \omega$, in which case the $\mathscr{C}^1$-admissibility of $|\bar{C}|$ is contradicted by the strict inequality holding somewhere in (6.6); or else (ii) equality holds in (6.6), in which case the proof of Section 3 again leads to a contradiction. Thus, Theorems 3.1 and 3.3 and Corollary 3.2 remain valid for this notion of admissibility.

In Theorem 3.4, a somewhat larger class is now obtained than would have been obtained in Section 3 for the general (not necessarily k-hypothesis) decision problem. The (2.4) $\wedge$ (2.8)-admissible procedures are still admissible, but we also obtain the "anti-Bayes" partitions which, for any set of positive values $\{a_\omega, \bar{a}_\omega\}$, maximize

$$\sum_\omega a_\omega P_\omega\{\delta \in D_\omega\} - \sum_{\omega,j} \bar{a}_\omega P_\omega\{d_j\} e_j\,. \tag{6.7}$$

The standard argument then yields $\delta(x)$ to be a value j maximizing (a.e. ν)

$$\sum_{\omega \in \Omega_j} a_\omega f_\omega(x) - e_j \sum_\omega \bar{a}_\omega f_\omega(x)\,, \tag{6.8}$$

which is easily translated into the language of Theorem 3.4. The interior of each region $\delta^{-1}(d_j)$ is convex and polyhedral, but is of a more complex form than that obtained for (2.4) $\wedge$ (2.8), where all $\bar{a}_\omega = 0$ in (6.8).

The developments analogous to those of Section 5 proceed in a similar manner for (2.4) $\wedge$ (2.6) $\wedge$ (6.6) to those for (2.4) $\wedge$ (2.6).

A similar analysis applies for admissibility in terms of such operating characteristics as (analogous to G_ω) the law of $\Gamma_\omega(x) - e_{\delta(x)}$ on $\delta^{-1}(D_\omega)$ and of $-e_{\delta(x)}$ on the complement of $\delta^{-1}(D_\omega)$, when ω is true.

Illustrations of these approaches for ranking and selection problems are contained in [4].

It would be interesting to treat losses that are general functions of the Γ_ω. As indicated in Remark 4 of Section 5, we do not know how to do this.

(D) It has been suggested by Larry Brown that our admissibility in terms of the measure G_ω of performance of Γ_ω on $\delta^{-1}(D_\omega)$ is no more natural than that in terms of the df Π_ω of Γ_ω on all of $\mathscr{X}$. The present paper's considerations are motivated by the possible attractiveness to the practitioner of knowing that Γ_ω has a tendency to be large on the set where a correct decision is made, in contrast with the questionable appeal of its being large on the complementary set where the decision is incorrect. This can be thought of as a rudimentary loss consideration that attempts to have the same "shape" as that of a loss function L on $\Omega \times D \times I^\Omega$ that reflects the combined effect of decision and conditional confidence on loss, to yield a traditional unconditional risk computed as $E_\omega L(\omega, \delta(X), \Gamma(X))$ (which, as remarked in (c) above, we are unable to handle). On the other hand, Brown cites possible practical goals of meeting specified confidence coefficient standards, that could lead to his formulation. In any event, the author does not have very strong preference for any particular admissibility scheme to the exclusion of others that seem reasonable and lead to the identification of sensible conditional confidence procedures.

Brown's admissibility considerations are related to ours by $G_\omega(t) = \int_{\tau>t} \tau \, d\Pi_\omega(\tau)$ and $H_\omega(t) = \int_{\tau<1-t} (1-\tau) \, d\Pi_\omega(\tau)$. These relations imply that admissibility as defined by replacing G by $-\Pi$ in (2.4) ($\bar{\bar{\Gamma}}_\omega$ stochastically larger than $\bar{\Gamma}_\omega$) implies (2.4)—(2.6)-admissibility, but there are additional admissible procedures in Brown's sense. His considerations relate to increasing convex functions c to be integrated with respect to Π_ω, the counterpart of our increasing functions integrated with respect to G_ω. He treats such problems as finding, for a given Z, the δ that maximizes a function of the quantities $E_\omega c(\min_\omega \Gamma_\omega)$, motivated by the idea that $\min_\omega \Gamma_\omega$ is all one can claim as guaranteed confidence for all ω. (This formalizes the spirit of the remarks in the paragraph below (1.2).) He obtains results on procedures with monotone structure, described near the end of Section 2. These results will appear in a paper by Brown.

Acknowledgment. The author is grateful to Larry Brown, both for the material of Section 6(D), and also for extensive comments that were used in rewriting this paper.

REFERENCES

[1] Bahadur, R. R. and Lehmann, E. L. (1955). Two comments on "Sufficiency and statistical decision function." *Ann. Math. Statist.* **26** 139-142.

[2] Brownie, C. and Kiefer, J. (1975). Conditional confidence statements. To appear.

[3] Dvoretzky, A., Wald, A. and Wolfowitz, J. (1951). Elimination of randomization in certain statistical decision procedures and zero-sum two-person games. *Ann. Math. Statist.* **22** 1-21.

[4] Kiefer, J. (1975). Conditional confidence approach in multidecision problems. *Proc. Fourth Dayton Multivariate Conference.* To appear.

[5] Le Cam, L. (1955). An extension of Wald's theory of statistical decision functions. *Ann. Math. Statist.* **26** 69-81.

[6] NEVEU, J. (1965). *Mathematical Foundations of the Calculus of Probabilities.* Holden-Day, San Francisco.

[7] WALD, A. (1950). *Statistical Decision Functions.* Wiley, New York.

[8] WALD, A. and WOLFOWITZ, J. (1951). Characterization of the minimal complete class of decision functions when the number of distributions and decisions is finite. *Proc. Second Berkeley Symp. Math. Statist. Prob.* 149–158, Univ. of California Press.

DEPARTMENT OF MATHEMATICS
WHITE HALL
CORNELL UNIVERSITY
ITHACA, NEW YORK 14853

Reprinted from:
STATISTICAL DECISION THEORY AND RELATED TOPICS, II
© 1977
ACADEMIC PRESS, INC
NEW YORK SAN FRANCISCO LONDON

ASYMPTOTICALLY MINIMAX ESTIMATION OF CONCAVE AND CONVEX DISTRIBUTION FUNCTIONS. II

By J. Kiefer* and J. Wolfowitz**

Cornell University and University of Illinois

0. *Introduction.* The present paper (II) is a sequel to our paper (I) of the same name, which appeared in the *Zeitschrift fuer Wahrscheinlichkeitstheorie und Verwandte Gebiete*, 34 (1976), 73-85. The notation and definitions of I will be adopted in toto, and familiarity with I is required for the understanding of II. To make II self-contained would require an intolerable amount of repetition from I. We therefore number the following section as the fifth (of the combined paper).

The purpose of the present paper is to prove Theorem 3 below. This result may be considered as essentially a generalization of Theorem 1 (this is not literally true). Conditions (3.2) are now replaced by (5.3) below, and the conclusions are very little different. Conditions (5.3) are still more restrictive than we would like, although a fairly formidable amount of argument is nevertheless required. To prove the theorem under essentially weaker conditions would require even more argument or a completely new idea or both.

Let $X_1,\ldots,X_n$ be independent chance variables with the common concave (respectively, convex) distribution function F, let F_n be their empiric distribution function, and let C_n be the least concave majorant (resp., greatest convex minorant) of F_n. The conclusions of both Theorems 1 and 3 may be stated approximately and simply as follows:

$$\sup_x |C_n(x) - F_n(x)| = o_p(n^{-\frac{1}{2}}).$$

*Research under NSF Grant MCS 75-22481.

**Research supported by the U. S. Air Force under Grant AFOSR-76-2877, monitored by the Office of Scientific Research.

The precise results are stated in Theorems 1 and 3, which explain the role of our result (Theorem 2) on estimation in the case described in the title.

In Section 5 we proceed as we did in I, and state our results and arguments for concave distribution functions. These apply, mutatis mutandis, to convex distribution functions.

To make things easier for the reader, we now restate Theorem 2 from I, with the additions described in the text of I. By concave (resp. convex) d.f. we mean a d.f. for which that property holds on the smallest interval $[\alpha_0(F), \alpha_1(F)]$ of F-probability one.

THEOREM 2. <u>Let</u> $\mathfrak{F}$ <u>be the set of all continuous concave</u> (resp. convex) d.f.'s <u>on</u> R^1. <u>Then</u>

$$\lim_{n\to\infty} \frac{\sup_{F\in\mathfrak{F}} r_n(F;F_n)}{\inf_{g_n} \sup_{F\in\mathfrak{F}} r_n(F;g_n)} = 1 \tag{4.5}$$

Because of Marshall's lemma (I, Lemma 3), F_n may be replaced by C_n in (4.5) above. This result also holds with any or all of the following modifications of $\mathfrak{F}$: (1) F can be permitted to have a jump at the left (resp., at the right) end of the interval of support; (2) finiteness of $\alpha_1(F)$ (resp., $\alpha_0(F)$), boundedness of α_0 and/or α_1 with known bounds, or exact knowledge of α_0 and/or α_1 can be assumed.

A usual treatment of the concave case is that in life testing, where $\alpha_0(F)$ is assumed to be zero. The obvious modifications in the definition of C_n, required under these variations, are discussed in Remark 1 of Section 3 of I.

5. *Estimating concave* F *under conditions* (5.3). In this section we weaken assumption (3.2) and conclusion (3.18). It is trivial that $||F_n-C_n|| \le n^{-\frac{1}{2}}(\log n)^c$, $c > 0$, with high probability, and we shall only treat here cases where an exponent $< -\frac{1}{2}$ is attainable.

The outline of the proof is the same as for Theorem 1. The

principal change in detail is a more complex definition of $L^{(k)}$ (and, hence, $L_n^{(k)}$) in order to improve the estimate of Lemma 6. This redefinition in turn necessitates changes in the proof of Lemma 4. However, the form of $L^{(k)}$ here will be such that we will be able to refer to the estimates of Section 3 for much of the calculation.

The change in the definition of $L^{(k)}$ is made in order to handle behavior in a neighborhood of any point where $-f'(x)/f^2(x)$ approaches 0 or $+_-\infty$, corresponding to $\beta = 0$ or $\gamma = +\infty$ in (3.2). Near such a point, we use "local" analogues of β and γ to obtain a better bound. (The global values used in the simpler proof of Theorem 1 do not suffice to treat most of the examples listed at the end of this section, including those with $\alpha_1 = +\infty$.) At the same time, the probability assigned to intervals $[a_j^{(k)}, a_{j+1}^{(k)}]$ near such a point will depend on detailed behavior near the point, and cannot be taken as k^{-1} as it could under (3.2).

We introduce the concepts used in the replacement (5.3) of (3.2).

We define x' to be an exceptional point for F in $[\alpha_0(F), \alpha_1(F)]$ if f or f' is not both bounded and bounded away from 0 as $x \to x'$. In this definition, if x' is α_0 or α_1, the approach $x \to x'$ is one-sided. If $\alpha_1 = +\infty$, the value $x' = \alpha_1$ is always exceptional.

The development is made simpler if f or f' does not oscillate too wildly or approach 0 or ∞ too rapidly at an exceptional point. For $0 < \rho + F(x') < 1$, let x'_ρ satisfy $F(x'_\rho) = F(x')+\rho$. Here ρ can be positive or negative unless x' is α_0 or α_1. Write $I(x',\rho)$ for the open interval with end points x'_ρ and $x'_{2\rho}$. We shall say that F is regular at the isolated exceptional point x' (or that x' is a regular exceptional point) if

$$\text{(5.1)}\quad \begin{aligned} &\lim\sup_{\rho\to 0}\sup_{\xi',\xi''\in I(x',\rho)} f'(\xi')/f'(\xi'') < \infty,\\ &\lim\sup_{\rho\to 0}\sup_{\xi',\xi''\in I(x',\rho)} f(\xi')/f(\xi'') < \infty,\\ &\lim\inf_{\rho\to 0}\inf_{\xi',\xi''\in I(x',\rho)} f'(\xi')/f'(\xi'') > 0,\\ &\lim\inf_{\rho\to 0}\inf_{\xi',\xi''\in I(x',\rho)} f(\xi')/f(\xi'') > 0. \end{aligned}$$

The last two conditions are trivially satisfied except at $x' = \alpha_0$ or α_1; there is some redundancy in (5.1) at those points, which we shall not take the space to discuss.

In order to phrase our final simplifying regularity condition, we define, for ε small and positive and x' an isolated exceptional point,

$$\text{(5.2)}\quad \begin{aligned} \beta^+(\varepsilon;x',F) &= -f'(x'_\varepsilon)/f^2(x'_\varepsilon),\\ \beta^-(\varepsilon;x',F) &= -f'(x'_{-\varepsilon})/f^2(x'_{-\varepsilon}). \end{aligned}$$

Half of the definition is omitted if $x' = \alpha_0$ or α_1.

In place of (3.2) we now assume

(5.3)
- $\alpha_0(F) = 0$;
- F is continuous;
- F has only finitely many exceptional points, all regular;
- F is concave on $(0,\alpha_1(F))$, and is twice continuously differentiable there except possibly at the exceptional points;
- for each exceptional point x', there are values $\mu^+(x',F) > -1$ and $\mu^-(x',F) > -1$ such that, for $* = +$ or $-$, $\beta^*(\varepsilon;x',F) = \varepsilon^{\mu^*(x',F)+q(\varepsilon)}$ as $\varepsilon \downarrow 0$, where $q(\varepsilon) = o(1)$ as $\varepsilon \downarrow 0$.

We define $\bar{\mu}(F)$ to be the maximum of zero and of the quantities $\mu^+(x',F)$ and $\mu^-(x',F)$ over all exceptional points. Note that $\bar{\mu}(F) = 0$ for F satisfying (3.2). The possible values of the quantities μ^*, and the need for the inequality in the last condition of (5.3), will be seen later. Our result, weakening (3.2) but obtaining a weaker conclusion in place of (3.18) if $\bar{\mu}(F) > 0$, is

THEOREM 3. <u>If</u> F <u>satisfies</u> (5.3), <u>there is a function</u> τ_F <u>with</u> $\lim_{n\to\infty} \tau_F(n) = 0$ <u>such that</u>

(5.4) $$P_F\{||C_n - F_n|| > n^{-(\bar{\mu}(F)+2)/(2\bar{\mu}(F)+3)+\tau_F(n)}\} < 2n^{-2},$$

<u>so that, for every function</u> h <u>with</u> $\lim_{n\to\infty} h(n) = 0$,

(5.5) $$P_F[h(n)n^{(\bar{\mu}(F)+2)/(2\bar{\mu}(F)+3)-\tau_F(n)}||C_n - F_n|| = 0] = 1.$$

We note that the case $\bar{\mu}(F) = 0$ essentially includes Theorem 1, if one bothers to relate the function τ_F to the o(1) term in the last condition of (5.3). Before turning to the proof, we remark on the way in which the μ^* enter into (5.4), to aid in understanding our approach.

First, if the μ^* are all < 0, a smaller order than $n^{-2/3+\tau}$ is indeed achievable for the maximum of $|C_n - F_n|$ in a neighborhood (shrinking as $n \to \infty$) of each x'; however, at nonexceptional points the proof of Theorem 1 shows that the best order obtainable by our approach is $n^{-2/3+\tau}$, and this accounts for the presence of $\bar{\mu}$ rather than max u^* in this case. The development in the neighborhood of x' in the case of negative μ^* (Part II of the proof) is thus aimed at achieving order $n^{-2/3+\tau}$ rather than the best order there.

Second, the proof is divided into four parts to help in simplifying the notation and understanding the detailed calculations by isolating the various modifications needed in the proof of Theorem 1. The first part contains the simplest modifications. It is also the longest since, having gone through those modifications fully in the case of a single exceptional x' and simplest μ^* as treated there, we can limit ourselves in the other three parts to describing the additional changes.

Third, the order of our development is to find a partition of $[\alpha_0, \alpha_1]$, with associated (varying) interval probabilities that are as small as possible, yet yield the analogue of Lemma 4, to see what this yields in Lemma 6 and Lemma 2, and to assemble the bounds as in the paragraph of proof below the statement of Theorem 1.

Throughout the proof we write c_i for positive constants and $\Lambda(n)$ or $\Lambda_i(n)$ for positive functions of the form $n^{o(1)}$. The symbol $o(1)$ always denotes a function which approaches 0 as $n \to \infty$; if it depends on other variables, the approach is uniform. The symbol $\Omega(1)$ denotes a positive function that is bounded away from 0 and ∞. The precise meaning of these symbols can vary even within the same equation; in contrast, Λ is sometimes used with a subscript when it is necessary to keep track of detailed relationships.

PART I. In this part of the proof we assume <u>there is a single exceptional point</u> $x' = \alpha_0(F) = 0$, with $0 \le \mu^+(x',F) = \bar{\mu}$. As will be discussed in Part III, the choice of x' is inessential and is made to simplify notation; the treatment of the behavior to the <u>left</u> of an x' involves slight additional modification, described in Part III. (Part II treats a single $x' = \alpha_0$ when $\mu^+ < 0$, and Part IV treats the case of several exceptional points.) We shall now prove Theorem 3 under the above restriction.

In our proof, it will be necessary to vary the probabilities of intervals $[a_j^{(k)}, a_{j+1}^{(k)}]$. We alter the definitions of Section 2 as follows: For each k we consider a vector $(p_0^{(k)},\ldots,p_{k-1}^{(k)})$ of positive values which sum to 1, and determine the $a_j^{(k)}$ by $a_0^{(k)} = \alpha_0, a_k^{(k)} = \alpha_1$, and, in place of the rest of (2.6),

$$F(a_{j+1}^{(k)}) - F(a_j^{(k)}) = p_j^{(k)}, \quad 0 \le j \le k-1. \tag{5.6}$$

We define

$$\sigma_j^{(k)} = \sum_{i=0}^{j} p_i^{(k)}, \quad a_*^{(k)} = F^{-1}(p_0^{(k)}/2).$$

The function $L^{(k)}$ again satisfies (2.7) (for the $a_j^{(k)}$ of (5.6)), and, in addition,

(5.7) $$L^{(k)}(a_*^{(k)}) = F(a_*^{(k)}) = p_0^{(k)}/2.$$

The remainder of the definition of $L^{(k)}$, replacing that just above Lemma 4, is that $L^{(k)}$ is linear on each of the intervals $[a_*^{(k)}, a_1^{(k)}]$ and $[a_j^{(k)}, a_{j+1}^{(k)}]$, $1 \le j \le k-1$, and that

(5.8) $$L^{(k)}(x) = F(x) \quad \text{for} \quad 0 \le x \le a_*^{(k)}.$$

The use of this last modification will become apparent in the discussion of Lemma 6.

The function $L_n^{(k)}$ is defined by the following analogue of (2.8), in terms of the $L^{(k)}$ just given:

(5.9) $$L_n^{(k)}(x) = F_n(a_j^{(k)})+(p_j^{(k)})^{-1}[F_n(a_{j+1}^{(k)})-F_n(a_j^{(k)})][L^{(k)}(x) - F(a_j^{(k)})]$$

$$\text{for } a_j^{(k)} \le x \le a_{j+1}^{(k)},\ 0 \le j \le k.$$

Note that for j = 0 this is $F_n(a_1^{(k)})L^{(k)}(x)/p_0^{(k)}$; although the definition of $L^{(k)}$ is broken into two parts at $a_*^{(k)}$, the random multiplier $F_n(a_1^{(k)})$ is the same for the whole interval. Once more, $L_n^{(k)}$ is nondecreasing, and (2.9) holds for $0 \le j \le k$.

We define $\gamma_j^{(k)}$ and $\beta_j^{(k)}$, for $1 \le j \le k-1$, to be the values obtained if the supremum and infimum in (3.1) are taken over $(a_j^{(k)}, a_{j+1}^{(k)})$ instead of over $(0,\alpha_1(F))$. Similarly, $\gamma_0^{(k)}$ and $\beta_0^{(k)}$ are defined by computing these extrema over $(a_*^{(k)}, a_1^{(k)})$.

We begin by looking at the modifications needed in Lemma 4. For $j \ge 0$, the definition of $T_{n,j}$ is as given in (3.7) for the present $a_j^{(k_n)}$; for $\Delta_{n,j}$ the definition in (3.7) will still be used for $j \ge 1$, but in considering the event $B_{n,0}$ we will now use

(5.10) $$\Delta_{n,*} = a_1^{(k_n)} - a_*^{(k_n)}.$$

By differentiating (5.9) to the right and left of each

$a_j^{(k_n)}$, $1 \le j \le k_n-1$, we see that the event A_n of (3.4) is now $(\bigcap_{j=1}^{k_n-2} B_{n,j}) \cap B_{n,*}$, where the $B_{n,j}$ are as defined in (3.8) and

(5.11) $$B_{n,*} = \{T_{n,0}/2\Delta_{n,*} \ge T_{n,1}/\Delta_{n,1}\}.$$

(The form of $L^{(k_n)}$ on $[0,a_1^{(k_n)}]$ insures that $L_n^{(k_n)}$ is concave there.)

For $1 \le j \le k_n-2$, we now obtain, in place of (3.9) - (3.10), that $B_{n,j}$ follows from

(5.12) $$|T_{n,i}-p_i^{(k_n)}| \le \delta_{n,j}p_i^{(k_n)} \text{ for } i = j,\ j+1$$

and

(5.13) $$p_j^{(k_n)}\Delta_{n,j+1}/p_{j+1}^{(k_n)}\Delta_{n,j} \ge 1+3\delta_{n,j},$$

where $0 < \delta_{n,j} < 1/3$. The relation (3.11) becomes

(5.14) $$\Delta_{n,j+1} = p_{j+1}^{(k_n)}/f(a_{j+1}^{(k_n)})-(p_{j+1}^{(k_n)})^2 f'(\xi)/2f^3(\xi);$$

and $\Delta_{n,j} \le p_{j+1}^{(k_n)}/f(a_{j+1}^{(k_n)})$ yields, in place of (3.12),

(5.15) $$\Delta_{n,j+1}p_j^{(k_n)}/\Delta_{n,j}p_{j+1}^{(k_n)} \ge 1 + \beta_{j+1}^{(k_n)} p_{j+1}^{(k_n)}/2.$$

Similarly, $B_{n,*}$ follows from (5.12) for $i = 0$ and, replacing (5.13),

(5.16) $$p_0^{(k_n)}\Delta_{n,1}/2p_1^{(k_n)}\Delta_{n,*} \ge 1+3\delta_{n,0},$$

the line after (5.14) becomes $2\Delta_{n,*} \le p_1^{(k_n)}/f(a_1^{(k_n)})$, and we thus also obtain, in place of (5.15),

(5.17) $$\Delta_{n,1}p_0^{(k_n)}/2\Delta_{n,*}p_1^{(k_n)} \ge 1+\beta_1^{(k_n)} p_1^{(k_n)}/2.$$

Thus, defining $\bar{\delta}_0 = \beta_1^{(k_n)} p_1^{(k_n)}/6$, $\bar{\delta}_{k_n-1} = \beta_{k_n-1}p_{k_n-1}/6$, and

$\bar{\delta}_i = \min_{j=i,i-1} \beta_{j+1}^{(k_n)} p_{j+1}^{(k_n)}/6$ for $1 \le i \le k_n-2$, we conclude that, if all $\bar{\delta}_i < 1/3$, the event A_n follows from

$$(5.18) \quad |T_{n,i} - p_i^{(k_n)}| \le \bar{\delta}_i p_i^{(k_n)}, \quad 0 \le i \le k_n-1.$$

If $\max_{0 \le i \le k_n-1} \bar{\delta}_i$ and $\max_{0 \le i \le k_n-1} p_i^{(k_n)}$ approach 0, we can thus use Lemma 1 and obtain, in place of (3.7),

$$(5.19), \quad 1-P\{A_n\} \le 2 \sum_{i=0}^{k_n-1} e^{-np_i^{(k_n)}\bar{\delta}_i^2/3}$$

for large n.

In order to make the right side of (5.19) $\le n^{-2}$, we shall choose the $p_i^{(k_n)}$ to be about as small as possible, so as to obtain the best advantage in the analogue of Lemma 6. In the calculations that follow, it will turn out that the first term (i = 0) of (5.19) already establishes the power of n attainable in Lemma 6, and the remaining terms are then merely chosen so as not to destroy that order.

We now establish that, for appropriate Λ_i, the choices $k_n = \Lambda_1(n)n^{1/3}$ and

$$(5.20) \quad p_j^{(k_n)} = \begin{cases} n^{-1/(3+2\bar{\mu})} j^{-2\mu^+/(3+2\mu^+)} \Lambda_2(n) & \text{for } j \le \Lambda_3(n)n^{1/3}, \\ n^{-1/3}\Lambda_4(n) & \text{otherwise} \end{cases}$$

make $1-P\{A_n\} \le n^{-2}$. This calculation will occupy us through the paragraph of (5.26).

First, for $j \le \Lambda_3(n)n^{1/3} = m_n$ (say) we obtain, since $3/(3+2\mu^+) > 0$,

$$(5.21) \quad \sigma_j^{(k_n)} = n^{-1/(3+2\mu^+)} j^{3/(3+2\mu^+)} \Omega(1)\Lambda_2(n).$$

Hence, supposing m_n an integer (a matter of definition of Λ_3),

$$p_{m_n}^{(k_n)} = n^{-1/3}[\Lambda_3(n)]^{-2\mu^+/(3+2\mu^+)}\Lambda_2(n),$$

(5.22)
$$\sigma_{m_n}^{(k_n)} = [\Lambda_3(n)]^{3/(3+2\mu^+)}\Omega(1)\Lambda_2(n).$$

Whatever the choice of Λ_2 (to be determined below, from (5.25)), we choose Λ_3 so as to make $\sigma_{m_n}^{(k_n)} = 1/\log n$. (Any other sequence $\sigma_{m_n}^{(k_n)} = \Lambda(n) \to 0$ would work as well; Λ_2 can be chosen for large n to depend on this sequence and on q, but not circularly on Λ_3.) This step is not even necessary for the present case of a single x' = 0, but we have introduced it here to illustrate the procedure used in Part IV, of "localizing" the behavior of F and choice of $p_j^{(k_n)}$ in each of a number of small non-overlapping intervals containing an exceptional point as endpoint.

The function Λ_4 (and hence Λ_1) is determined so that $n^{-1/3}\Lambda_4(n)$ is as close as possible to $p_{m_n}^{(k_n)}$; exact equality may be impossible because $p_j^{(k_n)}$ is constant for $j \geq m_n$ and $\sum_0^{k_n-1} p_j^{(k_n)} = 1$. However, this choice makes $p_{m_n+1}^{(k_n)}/p_{m_n}^{(k_n)} \to 1$, so that $p_j^{(k_n)}$ varies "smoothly" on j at $j = m_n$. Thus, $p_j^{(k_n)}/p_{j+1}^{(k_n)} = \Omega(1)$ uniformly for $0 \leq j \leq k_n-1$. It follows from (5.3) and the fact that $\sigma_{m_n}^{(k_n)} \to 0$ that, for $1 \leq j \leq m_n$,

(5.23)
$$\beta_j^{(k_n)} = (\sigma_j^{(k_n)})^{\mu^+ + q(\sigma_j^{(k_n)})},$$

and hence that the exponent $-np_i^{(k_n)}\bar{\delta}_i^2/3$ of (5.19) is, uniformly for $0 \leq i \leq m_n-1$,

$$(5.24)\quad -n(p_{i+1}^{(k_n)})^3(\beta_{i+1}^{(k_n)})^2\Omega(1) =$$

$$- \Omega(1)[\Lambda_2(n)]^{3+2\mu^+ +o(1)} n^{o(1)} (1+i)^{o(1)},$$

where the three expressions o(1), obtained from q and (5.21), are not the same. The function Λ_2 is chosen large enough so that (5.24) will be $\leq -(\log n)^2$ for $0 \leq i \leq m_n-1$.

Since $\beta_{m_n}^{(k_n)} = (\log n)^{-\mu^+ +o(1)}$ (possibly unbounded if $\mu^+ = 0$) and $-f'/f^2$ is bounded, we have, uniformly for $j \geq m_n$,

$$(5.25)\qquad \beta_j^{(k_n)}/\beta_{m_n}^{(k_n)} \leq (\log n)^{\mu^+ +o(1)}.$$

Consequently, from (5.25), uniformly for $i \geq m_n$,

$$(5.26)\qquad -np_i^{(k_n)}\frac{\bar{\delta}_i^2}{3} = -n(p_{m_n}^{(k_n)})^3(\beta_{i+1}^{(k_n)})^2\Omega(1)$$

$$\leq - (\log n)^2(\beta_{i+1}^{(k_n)}/\beta_{m_n}^{(k_n)})^2\Omega(1)$$

$$\leq - (\log n)^{2+2\mu^+ +o(1)}$$

From this and the end of the previous paragraph we conclude that each of the k_n terms in the sum (5.19) is $\leq n^{-3}$ for n sufficiently large, uniformly in i, provided that $\max_i \bar{\delta}_i \to 0$ and $\max_i p_i^{(k_n)} \to 0$, and these are easily verified, completing the proof of the assertion containing (5.20).

Next, we turn to the analogue of Lemma 6. The proof of that lemma demonstrates that, for $1 \leq j \leq k-1$,

$$(5.27)\quad \sup_{a_j^{(k)} \leq x \leq a_{j+1}^{(k)}} |F(x)-L^{(k)}(x)| \leq \gamma_j^{(k)}(p_j^{(k)})^2/2.$$

Similarly,

$$(5.28)\qquad \sup_{a_0^{(k)} \le x \le a_1^{(k)}} |F(x) - L^{(k)}(x)| = \sup_{a_*^{(k)} \le x \le a_1^{(k)}} |F(x)-L^{(k)}(x)| \le \gamma_*^{(k)}(p_1^{(k)})^2/8.$$

From (5.1) and (5.2), we have $\gamma_*^{(k)} = \beta_*^{(k)}\Omega(1)$ (which is where (5.8) is used), and $\gamma_j^{(k)} = \beta_j^{(k)}\Omega(1)$ uniformly in j. Consequently,

$$(5.29)\qquad ||F-L^{(k)}|| \le \Omega(1)\max_{j=*,1,\ldots,k_n-1}\beta_j^{(k)}(p_j^{(k)})^2.$$

While using k_n in (5.29) we recall the remark just above (5.23) (which, with the choice of $p_j^{(k_n)}$, allows us to limit the range of j to values $\le m_n$ in what follows) and the calculations of p_{j+1}/p_j and β_{j+1}/β_j which in like manner apply for p_1/p_* and β_1/β_*. From (5.20), (5.21), and (5.23) we obtain, for a suitable Λ_5,

$$||F-L^{(k_n)}|| \le \Omega(1)\max_{1\le j\le m_n}\beta_j^{(k_n)}(p_j^{(k_n)})^2$$

$$(5.30)\qquad = \max_{1\le j\le m_n} n^{-(2+\mu^+)/(3+2\mu^+)}j^{-\mu^+/(3+2\mu^+)}\Lambda_5(n)$$

$$= n^{-(2+\mu^+)/(3+2\mu^+)}.$$

(We explain briefly why it is sufficient to maximize over $1 \le j \le m_n$ in (5.30). Since 0 is the only exceptional point so far, f'/f^2 and thus $\beta_j^{(k)}$ is bounded and bounded away from 0 when j is large. Specifically, by (5.22) $\sigma_{m_n}^{(k_n)} = n^{o(1)}$ and hence $\beta_j^{(k_n)} = n^{o(1)}$ for $j \ge m_n$. Hence, for $j \ge m_n$,

$$\beta_j^{(k_n)}(p_j^{(k_n)})^2 = n^{o(1)} x [n^{-1/3}\Lambda_4(n)]^2 = n^{-2/3+o(1)}$$

$$\le n^{o(1)} x[\text{bound on } \beta_{m_n}^{(k_n)}(p_{m_n}^{(k_n)})^2 \text{ used in (5.30)}]).$$

Finally, for the choice (5.20) of $p_j^{(k_n)}$ it is easily checked that, for an appropriate Λ_6, the proof of Lemma 2 is modified (treating intervals of varying probabilities and the modification (5.8)) to yield, with $\bar{p}^{(k_n)} = \max_j p_j^{(k_n)}$,

(5.31) $$P_F\{||F_n - F - L_n^{(k_n)} + L^{(k_n)}|| > \Lambda_6(n)[n^{-1}\bar{p}^{(k_n)}]^{\frac{1}{2}}\} < n^{-2}.$$

In the present case $\bar{p}^{(k_n)} = p_1^{(k_n)}$ and hence

$$\Lambda_6(n)[n^{-1}\bar{p}^{(k_n)}]^{1/2} = n^{-(2+\mu^+)/(3+2\mu^+)}\Lambda_7(n).$$

The final conclusions of the last three paragraphs, when substituted for the roles of Lemma 4, Lemma 6, and Lemma 2 in the paragraph of proof just below (3.19), yield the proof of Theorem 3 in our special case of Part I.

PART II. We again suppose <u>there is a single exceptional point</u> $x' = \alpha_0(F) = 0$, but now $-1 \le \mu^+(x',F) < 0$. We shall now indicate as succinctly as possible the way in which the proof of Part I must be altered to prove Theorem 3 under the present restriction.

This is the case in which, as mentioned earlier, one can achieve a "local" improvement in the order of $|C_n - F_n|$ near x', but it is pointless to do so because the behavior at nonexceptional points makes -2/3 the most negative possible power of n (for our method of proof) in (5.4). This fact means that, rather than using the analogue of Lemma 4 to find the smallest order of $p_j^{(k_n)}$ that will make (5.19) small enough, and having that determine the best order obtainable in Lemmas 6 and 2, we could instead now let the order $n^{-2/3+\tau}$ in Lemmas 6 and 2 determine the $p_j^{(k_n)}$ and then check Lemma 4. Since then less is "wasted" in making the deviation obtained from Lemmas 6 and 2 smaller than needed near x', it might be thought that the argument could apply for a larger range of μ^+ values than does the determination based on Lemma 4. Formally this is true except for one crucial point. In order for the argument of the analogue of Lemma 4 to make sense, $\beta_j^{(k_n)} p_j^{(k_n)}$ must not be unbounded, as we see from (5.15) and the fact that $\delta_{n,i}$ must be < 1/3 (or, with alterations, < 1). But it turns out that the minimum μ^+ value for which this is so is -1, whichever way of determining the $p_j^{(k_n)}$ is used. So the extra applicability of the second method is illusory, and for the sake of uniformity we use the method of Part I.

Thus, the $p_j^{(k_n)}$ are determined so as to yield (5.26). It turns out that such a determination is then again given by (5.20), with possible changes of the Λ_i. The proof of Part I is now used, with minor exceptions which we shall list.

The development through (5.19) remains intact.

Since the exponent of j in (5.21) is still > 0, that formula is still correct.

The only other place in which the sign of μ^+ enters is in the development leading to (5.26). The exponent of the lower bound $(\log n)^{\mu^+ + o(1)}$ is now negative, and at the end of the previous paragraph Λ_2 must consequently be chosen to make (5.25) $\leq -(\log n)^{2-2\mu^+}$. The result (5.26) (in particular, the inequality) is then valid.

Once more $\max_i p_i^{(k_n)} \to 0$ is clear, and for the essential domain $1 \leq j \leq m_n$ we compute, from (5.20), (5.21), and (5.23),

$$\beta_j^{(k_n)} p_j^{(k_n)}$$

(5.32)
$$= \Omega(1) n^{-[\mu^+ + 1 + o(1)]/(3+2\mu^+)} j^{(\mu^+ + 3[o(1)])/(3+2\mu^+)} \times [\Lambda_2(n)]^{1+\mu^+ o(1)},$$

where o(1) has the same meaning in all three appearances; namely, it is $q(\sigma_j)$. Thus, $\max_i \bar{\delta}_i \to 0$, provided $\mu^+ > -1$. (We have written out (5.32) in order to indicate the difficulty when $\mu^+ \leq -1$. A more careful analysis might take care of the value $\mu^+ = -1$ for some q, but a completely different device is needed when $\mu^+ < -1$.) This completes the proof in our special case of Part II.

PART III. We now suppose that <u>there is a single exceptional</u> $x' = \alpha_1(F)$ with $-1 < \mu^-(x',F) < \infty$. In order to use the previous developments with a minimum of rewriting, it is convenient, <u>for this Part III only</u>, to denote the interval endpoints by $\alpha_0(F) = b_k^{(k)} < b_{k-1}^{(k)} < \ldots < b_1^{(k)} < b_*^{(k)} < b_0^{(k)} = \alpha_1(F)$, with

(5.33)
$$F(b_j^{(k)}) - F(b_{j+1}^{(k)}) = \pi_j^{(k)}, \quad 0 \leq j \leq k-1,$$
$$F(b_0^{(k)}) - F(b_*^{(k)}) = \pi_0^{(k)}/2,$$

where $(\pi_0^{(k)}, \ldots, \pi_{k-1}^{(k)})$ is again a vector of positive values summing to 1. Essentially all of Parts I and II proceeds for the present case with minor notational changes, upon replacing $[a_j^{(k)}, a_{j+1}^{(k)}]$ and $p_j^{(k)}$ there by $[b_{j+1}^{(k)}, b_j^{(k)}]$ and $\pi_j^{(k)}$ (with $b_*^{(k)}$ for $a_*^{(k)}$, etc.) in the present setting. We let $\Delta_{m,j}$, $\Omega_j^{(k)}$, and $\gamma_j^{(k)}$ refer to the interval $(b_{j+1}^{(k)}, b_j^{(k)})$ here. The one major change is caused by the fact that, because of the monotonicity of f that led to the line just below (5.14), the right side of the relation (5.15) refers to the interval <u>to the right</u> of $a_j^{(k_n)}$. Thus,

$\beta_*^{(k_n)} p_*^{(k_n)}$ did not occur in the analogue of Lemma 4 in Part I, but in the present case the quantity $\beta_*^{(k_n)} \pi_o^{(k_n)}$ occurs. The analogue of the development (5.14), (5.16), and (5.17) is now, with $b_1^{(k_n)} < \xi < b_*^{(k_n)}$,

$$\Delta_{n,*} = (\pi_o^{(k_n)}/2)f(b_1^{(k_n)}) - (\pi_o^{(k_n)}/2)^2 f'(\xi)/2f^3(\xi),$$

$$2\Delta_{n,*}\pi_1^{(k_n)}/\Delta_{n,1}\pi_o^{(k_n)} \geq 1 + 3\delta_{n,o},$$

(5.34) $$\Delta_{n,1} \leq \pi_1^{(k_n)}/f(b_1^{(k_n)}),$$

$$2\Delta_{n,*}\pi_1^{(k_n)}/\Delta_{n,1}\pi_o^{(k_n)} \geq 1+\beta_*^{(k_n)}\pi_o^{(k_n)}/4.$$

This and the previously argued regularity of p_j/p_{j+1} and β_j/β_{j+1} make the previous arguments apply with obvious changes.

We note here the second role of the form (5.8); without that adjustment of the piecewise linear $L^{(k)}$ of Section 3, we would have $\beta_o^{(k_n)}$ in place of $\beta_*^{(k_n)}$ in (5.34), and this would be infinite if $\mu^- < 0$. The other role of (5.8), as in Part I, is to replace $\gamma_o^{(k_n)}$ by $\gamma_*^{(k_n)}$ in the analogue of Lemma 6.

PART IV. We <u>now treat the general case assumed in Theorem 3</u>. For any exceptional point x' interior to $[\alpha_o,\alpha_1)$, we use the construction of Parts I and II with $[0,F^{-1}(1/\log n)]$ replaced by the interval $[x',F^{-1}(F(x') + 1/\log n)]$, and that of Part III with $[F^{-1}(1 - 1/\log n), \alpha_1(F)]$ replaced by $[F^{-1}(F(x') - 1/\log n), x']$, to obtain on each of these a set of $n^{1/3}\Lambda(n)$ subintervals and corresponding probabilities that yield n^{-3} for the probability of (5.4) when $\sup|F_n - C_n|$ is computed only on either of these two intervals. The construction is such that, near each of the interval endpoints other than x', the subinterval has probability $\Lambda(n)n^{-1/3}$. Thus, if x" is the next exceptional point to the right of x', for large n the interval $J = [F^{-1}(F(x')+1/\log n), F^{-1}(F(x'')-1/\log n)]$ can be subdivided into subintervals of varying probability uniformly of the form $n^{-1/3}\Lambda(n)$ in such a way as to maintain the smoothness of p's and β's for neighboring intervals throughout $[0,\alpha_1(F))$. As in the paragraph containing (5.26), we obtain $n^{-2/3+o(1)}$ for the supremum on J of $|F_n - C_n|$ that is exceeded with probability $< n^{-1/3}$. Adding the contributions to (5.4) from the various J's and sequences of shrinking intervals with exceptional endpoints, we obtain Theorem 3.

REMARKS. The remarks of the previous section apply to the present setting (5.3) as well as to (3.2). We now give examples of F's that exhibit the various values of μ^*.

(a) The point $x' = 0$. We go through only one calculation in detail, and merely state the results in other cases. Suppose that, for x sufficiently small,

$$F(x) = x - x^{r+1}/(r+1) \tag{5.35}$$

for some $r > 0$. Then $f(x) = 1-x^r$ and $f'(x) = -rx^{r-1}$, so that $-f'(x)/f^2(x) \sim rx^{r-1}$ for x small. Since $F^{-1}(\varepsilon) \sim \varepsilon$ as $\varepsilon \downarrow 0$, we obtain $\mu^+ = r-1$. In similar fashion it can be checked that (again with $r > 0$ and the behavior studied being near $x = 0$)

$$\begin{aligned} F(x) &= (\log x^{-1})^{-r} \implies \mu^+ = -r^{-1}-1, \\ F(x) &= x(\log x^{-1})^{r} \implies \mu^+ = -1, \qquad (5.36) \\ F(x) &= x^{r'}(\log x^{-1})^{r''} \implies \mu^+ = -1, \end{aligned}$$

the last with r" arbitrary and $0 < r' < 1$. To summarize, all real values are possible for μ^+.

(b) The cases $0 < x' < \alpha_1$ and $x' = \alpha_1 < \infty$ with $f(\alpha_1) > 0$. These two cases exhibit the same behavior, and we consider the latter. If $\alpha_1 = 1$, an analogue of (5.35) is (with $r > 0$, $c > 0$, and stated behavior near $x = 1$)

$$F(x) = (x+c)/(1+c) - (1-x)^{r+1}/(r+1) \implies \mu^- = r-1. \tag{5.37}$$

The values $\mu^- < -1$ cannot occur in this case if $f(\alpha_1) > 0$ (or, similarly, for $0 < x' < \alpha_1$), as a simple argument shows. Suppose $\alpha_1 = 1$. Then, writing $H = -(\mu^- + 1)/2 > 0$, we have, for u sufficiently near 1, from (5.3),

$$(5.38) \qquad -f'(u)/f(u) > f(u)[1-F(u)]^{-(H+1)}.$$

Integration, with $u_1 < u_2 < 1$, yields

$$(5.39) \qquad \log \frac{f(u_1)}{f(u_2)} > H^{-1}\{[1-F(u_2)]^{-H} - [1-F(u_1)]^{-H}\}.$$

Letting $u_2 \to 1$, the left side remains bounded and the right side is unbounded, yielding a contradiction. If $\mu^- = -1$ and $q \leq 0$, a similar proof of impossibility can be given.

(c) <u>The case</u> $\alpha_1 < \infty$, $f(\alpha_1) = 0$. Here μ^- is completely determined. Suppose $\alpha_1 = 1$, and, if $\mu^- < -1$, rewrite (5.39) as $f(u_2) < ce^{-H^{-1}[1-F(u_2)]^{-H}}$. Integrating from $F^{-1}(1-\varepsilon)$ to 1, we obtain

$$(5.40) \quad \varepsilon < \int_{F^{-1}(1-\varepsilon)}^{1} ce^{-H^{-1}[1-F(u_2)]^{-H}} du_2 < \int_{F^{-1}(1-\varepsilon)}^{1} ce^{-H^{-1}\varepsilon^{-H}} du_2 < ce^{-H^{-1}\varepsilon^{-H}},$$

giving a contradiction as $\varepsilon \downarrow 0$. Thus, we cannot have $\mu^- < -1$. Similarly, if $\mu^- > -1$, write $H = (\mu^- + 1)/2 > 0$, and (5.3) yields, for u near 1,

$$(5.41) \qquad -f'(u)/f(u) < f(u)[1-F(u)]^{-1+H},$$

which, after integration from u_1 to u_2, gives

$$(5.42) \qquad \log[f(u_1)/f(u_2)] < H^{-1}\{[1-F(u_1)]^{H} - [1-F(u_2)]^{H}\},$$

again producing a contradiction as $u_2 \uparrow 1$. We conclude that $\mu^- = -1$ in this case.

(d) <u>The case</u> $\alpha_1 = +\infty$. Here $\mu^- \leq -1$ again. In fact, one cannot have $-f'(x)/f^2(x) < c[1-F(x)]^{-1}$ for all $x > x_0$ (say) and constant $c < 1$. To see this, integrate $-f'/f$ as before to obtain, after slight manipulation, $f(x)/[1-F(x)]^c > c_1$ for $x > x_0$, where $c_1 > 0$. Integration from x_0 to $+\infty$ yields a contradiction. In the other direction, if $1-F(x) = x^{-r}$ for $r > 1$ as $x \to \infty$ we obtain $\mu^- = -1$ and $\varepsilon^{q(\varepsilon)}$ constant. Unlike the situation in (c), we can now have $\mu < -1$: if $F(x) = 1-(\log x)^{-r}$ for $r > 0$ and large x, we obtain $\mu^- = -1-r^{-1}$.

Unfortunately, the interesting case $\alpha_1 = +\infty$, $\mu^- = -1$ requires further research.

Conditional Confidence Statements and Confidence Estimators

J. KIEFER*

Procedures are given for assessing the conclusiveness of a decision in terms of a (chance) conditional confidence coefficient Γ that has frequentist interpretability analogous to that of a traditional confidence coefficient. Properties of such procedures are compared in terms of the distribution of Γ. This leads to recommendations on the choice of conditioning. Also, a methodology for estimating the confidence when it depends on unknown parameter values is given. The notions of confidence are not limited to interval estimation; examples are also given in hypothesis testing and selection problems and in nonparametric and sequential settings.

KEY WORDS: Conditional confidence; Estimated confidence; Conditioning partition; Ancillarity; Estimated risk; Hypothesis testing; Interval estimation; Selection and ranking; Nonparametric procedures; Sequential procedures.

1. INTRODUCTION

For some time now, especially in the last 20 years, a number of attacks have been raised against the Neyman-Pearson-Wald (NPW) approach to statistical problems. For example, in the realm of hypothesis testing, there are advocates of the likelihood principle (e.g., Barnard 1949), of Bayesian inference and modifications of it (Savage 1954, Dempster 1968, Shafer 1976), of fiducial or structural inference (Fisher 1956, Fraser 1968), of notions of evidence (especially Birnbaum, e.g., 1961a, 1977). There is a vast literature on this subject in addition to the preceding list, and we shall limit our further references to works with sufficient relationship to the present article. Common to these developments is dissatisfaction with the form of conclusion, including assessment of its conclusiveness, in NPW methodology. For example, in the elementary setting of testing between simple hypotheses H_0 and H_1, the specification only of the type I and II probabilities of errors (α, β) of the test being used, and of the decision reached, is often viewed as an unsatisfactory form of the conclusion to the problem. Thus de Finetti (1974, p. 128), a leader of the Bayesian school, says that "accept or reject is the unhappy formulation which I consider as the principal cause of the fogginess widespread all over the field of statistical inference and general reasoning."

The framers of these objections appeal to various axioms of rational behavior and to arguments about what they view as reasonable properties of a statement of statistical conclusions in various settings, including attempts to differentiate between "inference" and "decision." But one motivation for some of these developments is surely the practical man's feeling that classical NPW procedures often assign the same decision and numerical measure of conclusiveness for two different sample values, one of which actually seems intuitively much more conclusive than the other. Following are two illustrations:

Example 1(a): Suppose $n \geq 2$ and we observe independent and identically distributed (iid) random variables (rv's) $X_1, X_2, \ldots, X_n$, with uniform density from $\omega - \frac{1}{2}$ to $\omega + \frac{1}{2}$, where ω is real and unknown. Write U_n and V_n for the smallest and largest observations, $T_n = (V_n + U_n)/2$, and $W_n = V_n - U_n$. A classical two-sided confidence interval procedure of confidence γ is the interval $[\max(U_n + d_n,\ V_n) - \frac{1}{2},\ \min(U_n,\ V_n - d_n) + \frac{1}{2}]$, where $d_n = [(1-\gamma)/2]^{1/n}$ if $\gamma > 1 - 2^{1-n}$, and otherwise d_n satisfies $2d_n{}^n - (2d_n - 1)^n = 1 - \gamma$ and $0 < d_n < 1$. This procedure has various optimum properties, as described by Pratt (1961a), who also describes certain intuitively unappealing features of the procedure: If $W_n > d_n$, the interval *must* contain ω, and similarly if W_n/d_n is only slightly less than one, we feel fairly strongly that the interval covers the true ω; the opposite holds if W_n/d_n is small. Yet, regardless of the value of W_n, we state the same confidence coefficient γ.

Example 2(a): Suppose we observe a normally distributed rv X with mean θ and unit variance, and must decide between the two simple hypotheses H_0: $\theta = -1$ and H_1: $\theta = 1$. The symmetric NP test rejects H_0 if $X \geq 0$ and has $(\alpha, \beta) = (.16, .16)$. Thus we make the same decision d_1 in favor of H_1 whether $X = 0.5$ or $X = 5$, but the statement of error probabilities and decision reached, "$(\alpha, \beta, d) = (.16, .16, d_1)$," that we make for either of these sample values, does not exhibit any detailed data-dependent measure of conclusiveness that conveys our stronger feeling in favor of H_1 when $X = 5$, than when $X = 0.5$.

We shall return to both of these examples later. (Examples are labeled consecutively by arabic numerals

* J. Kiefer is Horace White Professor of Mathematics, Cornell University, Ithaca, NY 14853. Research was supported by NSF grant MCS 75-22481. The author is grateful, for helpful comments and criticisms on earlier versions of this article, to A. Birnbaum, L. Brown, R. Buehler, D. Moore, J. Pratt, J. Wolfowitz, and, above all, to Cavell Brownie. Several Editors, N.L. Johnson, E.L. Crow, and M.H. DeGroot, also deserve thanks for trying to make the article more readable.

Reprinted from: © Journal of the American Statistical Association
December 1977, Volume 72, Number 360
Invited Paper
Theory and Methods Section
Pages 789–827

followed by (a), (b), etc. for ease of further discussion in different sections.) For now, we mention as an illustration of the attempt to give a data-dependent measure of the sample's conclusiveness, both before and after the advent of the Neyman-Pearson (NP) theory and its critics, the practice of stating the level at which a null hypothesis would "just be rejected." (Indeed, this practice led to Bahadur's theoretical treatment of large-sample optimality considerations in a long sequence of papers of which Bahadur and Raghavachari (1970) is a recent member.) In the simplest setting of testing between simple H_0 and H_1, adherents of the various non-NPW schools take advantage of "lucky observations" to make more conclusive sounding statements than they would for "unlucky outcomes." We propose to achieve this end in a frequentist extension of the NPW framework.

In the present article, we give an exposition and discussion of a systematic approach to stating statistical conclusions which, by incorporating a measure of conclusiveness that depends on the sample, may assuage the uneasiness, just described, that some practitioners have with the NP (α, β, d) statement of error probabilities and decision. (The relationship of this article to the large body of work on conditioning will be discussed presently. Some of the more mathematical considerations are introduced and developed further in Kiefer (1975, 1976) and Brownie and Kiefer (1977).) We shall not enter into any discussion of the axiomatics and distinction between inference and decision, in terms of which the criticisms mentioned previously are written. Rather, we describe methodologies of conditional and estimated confidence whose justification lies solely in the classical frequency interpretation of probability (law of large numbers), just as does that of the NP approach (and which has been emphasized by Neyman since his earliest work, e.g., 1937) in which repeated use of level .05 tests will very probably result in mistatements in approximately five percent of a large number of independent applications where H_0 is true. The other alternatives to the NP approach, previously mentioned, do not have this property, except in special cases of symmetry, described later.

Our approach, then, is in the NPW spirit (and the (α, β, d) form of conclusion will be seen to be an extreme case of it). We shall now explain our first departure from the classical NPW development in the simplest context. That setting is not that of confidence intervals in which "confidence" is such a common notion but rather that of testing between two simple hypotheses, where the term "confidence" is rarely used, but the concept of confidence is present. The complexity of the interval estimation setting, where there are infinitely many possible decisions, makes it a less attractive one in which to introduce the ideas of our conditional confidence structure than the simple hypotheses setting in which we merely have to get used to thinking of $1 - \alpha$ and $1 - \beta$ as confidences in correct decisions, as described in the following.

Thus we now consider the setting of testing H_0: X has density f_0 against H_1: X has density f_1. Also in this introduction, we consider only the simplest form of our methodology, which is analogous to that where $\alpha = \beta$. Our basic approach is (1) that we partition the sample space $\mathfrak{X}$ into a family of subsets $C = \{C_i^b, b \in B, i = 0, 1\}$, where B is a set of labels; (2) when the sample X falls in C_i^b, we state a conclusion (Γ^b, d_i) that "H_i is true, with (conditional) confidence Γ^b"; and (3) Γ^b (like the level α or an ordinary confidence coefficient) has a frequency interpretation, but it is now a conditional one: writing $C^b = C_0^b \cup C_1^b$, we shall have, for both $i = 0$ and $i = 1$,

$$P_i\{\text{making the correct decision } d_i | C^b\} = \Gamma^b . \quad (1.1)$$

Here P_i denotes probability computed when H_i is true. We now illustrate these ideas in the context of an earlier illustration.

Example 2(b): The unconditional NP procedure described in Example 2(a) has a single element in B, say b, and $C_0^b = (-\infty, 0)$, while $C_1^b = [1, \infty)$. The set C^b is the entire real line, and thus the probabilities of (1.1) are unconditional. We see that the (unconditional) confidence coefficient Γ^b is simply $.84 = 1 - \alpha = 1 - \beta$, which is the complement of the common error probability value. As an example of a partition of the type described in (1) above, we might let B consist of two elements, $B = \{w, s\}$, this labeling being used to suggest weak and strong feelings of conclusiveness about the decision. A simple partition C is given by choosing a value $c > 0$ and setting

$$\begin{aligned} C_0^s &= (-\infty, -c] , \quad C_1^s = [c, \infty) , \\ C_0^w &= (-c, 0) , \quad C_1^w = [0, c) . \end{aligned} \quad (1.2)$$

Thus the set C^s, where a strongly conclusive decision is made, is $\{x: |x| \geq c\}$, while in $C^w = \{x: |x| < c\}$ we make a weakly conclusive decision. For example, if we choose $c = 1$, the conditional confidence coefficient of (1.1) is easily computed in terms of the standard normal df Φ to be

$$\begin{aligned} \Gamma^w &= P_1\{C_1^w | C^w\} = [\Phi(0) - \Phi(-1)]/ \\ &\qquad [\Phi(0) - \Phi(-2)] = .71 , \quad (1.3) \\ \Gamma^s &= P_1\{C_1^s | C^s\} = \Phi(0)/[\Phi(0) + \Phi(-2)] = .96 . \end{aligned}$$

(The same values of Γ^w and Γ^s are obtained if the computation is made under P_0.) Consequently, if we use this partition with $c = 1$ and we obtain the observed value $X = 0.5$ given in the illustration, we state "H_1 is true with conditional confidence .71"; while if $X = 5$, we state "H_1 is true with conditional confidence .96." We have thus achieved our goal of stating a larger measure of conclusiveness in the latter case, than in the former.

The frequency interpretation of such a procedure is clear: Suppose for simplicity that, as in Example 2(b), B is discrete and $P_i\{C^b\} > 0$ for all b and i. Then, in repeated use of this approach in a long sequence of independent experiments, if there are many experiments in which the sample falls in C^b, the correct inference will very probably be made in approximately a proportion Γ^b of those experiments; the precise meaning is that of the law of large numbers in the Bernoulli case. In Example 2(b), approximately 96 percent of the experiments for which $|X| \geq 1$ will lead to a correct conclusion, while only 71 percent of those for which $|X| < 1$ will do so. An analogous interpretation holds even when the experiments are not identical (in which case the partition, of course, depends on the experiment).

Intuitively, a crucial idea of our approach is that one "attaches" Γ^b to C^b. In fact, a convenient labeling (not adopted in general because of cases we shall encounter in which the probability in (1.1) depends on the true hypothesis H_i) is that in which B is taken to be the set of attainable values Γ^b, and the C^b's are labeled so that $\Gamma^b = b$. In Example 2(b), we would replace $\{w, s\}$ by $\{.71, .96\}$; then, e.g., if the sample X falls in $C^{.96}$ we are conditionally 96 percent confident that our decision is correct. Thus when $X = 5 \in C_1{}^{.96}$, we would be 96 percent confident (conditionally) that H_1 is true. Regardless of the labeling, we define the *conditional confidence coefficient* $\Gamma(X)$ by

$$\Gamma(X) = \Gamma^b \quad \text{if} \quad X \in C^b . \tag{1.4}$$

This, then, has a conditional frequency interpretation analogous to that of an ordinary confidence coefficient or probability of correct decision.

In Example 2(b) it was pointed out how the NP theory of testing, hereafter called the NP *setting*, fits into this framework. Similarly, in other examples of testing between two simple hypotheses, if B consists of a single element b, then in the NP setting $\mathfrak{X}$ is partitioned into $\mathfrak{X} = C^b = C_0{}^b \cup C_1{}^b$ where $C_1{}^b$ is the critical region, chosen to make $\alpha = \beta = 1 - \Gamma^b$. Then (1.1) expresses this structure. Notice that, in this simple case, one does not condition on a single $C_j{}^b$, which would lead to the useless

$$P_i\{\text{correct decision} \mid C_j{}^b\} = \begin{cases} 1, & \text{if } i = j \\ 0, & \text{if } i \neq j . \end{cases} \tag{1.5}$$

This same conditioning would yield (1.5) for general B, and it indicates why we condition as we do in (1.1) instead. This discussion should help to explain that the present conditional approach is in some sense an extension of that of NPW.

Conditional inference is of course an old idea, applied in many settings beginning (at least) with Fisher, who introduced the notion of an ancillary statistic partly as a basis for conditioning. There is a large body of literature on that topic; more recently, there is work by Buehler, Wallace, Brown, Pierce, and others (discussed in greater detail in Section 6) that is best known for studying the possibility of conditioning in such settings as common normal testing and interval estimation examples, so as to obtain an analog of (1.1) with two different values Γ^b corresponding to two conditioning sets. Anderson (1973) gives a comprehensive treatment of conditional inference from the viewpoint of unconditional asymptotic properties. Pratt (1961a) uses Example 1(a), due originally to Welch (1939), to illustrate some features he considers unappealing in an unconditional NP confidence interval construction, but it typifies the literature of this subject that, although Welch uses conditioning, he makes the Γ^b's identical in b. Similarly, the paper of Bartholomew (1967) discusses conditioning concentrating on what it achieves for the unconditional operating characteristic; and Cox's interesting discussion (1958) on the advisability of conditioning on the outcome of a fair coin used to choose which of two experiments is performed, uses tests of the same conditional level. Also, among both Fisher's followers and others, many statisticians have undoubtedly conditioned as we would in particular invariant problems such as interval estimation of location parameters, but usually this is done with a conditional confidence coefficient independent of the conditioning. (An exception to this pattern is Lehmann's treatment of conditional tests (1959), whose relationship to the considerations of Section 3 is described in Brownie and Kiefer (1977).) Our emphasis, however, is on constructing procedures with *conditional confidence coefficient that is highly variable*, and on giving a coherent framework for construction and theoretical assessment of a wide variety of conditional confidence procedures that includes many statistical settings. Despite the extensive work on conditioning, we have not encountered such a treatment in the literature, even in the simple hypothesis setting used in the preceding illustration (and in terms of which many of our ideas are introduced in Section 3).

We now describe briefly the relationship of our methodology to the works of some of the NP critics mentioned earlier. We begin with a development contained in a number of papers of Birnbaum (e.g., 1961a, 1962a). It is to be emphasized that Birnbaum's views in this area changed over the years, as he discussed in a number of papers of which (1977) is the most recent. The mention of the likelihood function approach of Birnbaum (1961a) in what follows is thus only for the purpose of clarifying its relationship to the conditional confidence approach which it resembles somewhat in appearance. In this discussion it is also illustrated that B, the set of conditioning labels, can be an interval of values.

The axiomatic development of "evidence" by Birnbaum (1961a) is mainly pointed toward justifying the "likelihood principle," that evidence (in our hypothesis-testing example) should be given in the form of the likelihood ratio $\lambda(X) = f_1(X)/f_0(X)$ or some one-to-one function of λ. A central concept in the development is the intrinsic significance level $\alpha(X) = 1/[1 + \lambda(X)]$ if $\lambda(X) > 1$ (or $\lambda(X)/[1 + \lambda(X)]$ if $\lambda(X) < 1$). In symmetric binary experiments (those in which $\lambda(X)$ has the same law under H_0 that $1/\lambda(X)$ has under H_1) the structure can be thought of as our setup with $C^b = \{x: 1/[1 + \lambda(x)] = b \text{ or } 1 - b\}$ and $\Gamma^b = b$ for $b > \frac{1}{2}$, assuming we decide that H_1 is true if $\lambda(X) > 1$ and that H_0 is true if $\lambda(X) < 1$. With this interpretation, we obtain $\alpha(x) = 1 - \Gamma(x)$ for $\alpha(x) > \frac{1}{2}$, so that $\alpha(X)$ has a frequency interpretation in such symmetric settings. (We are not concerned here with the precise definition of the decision rule when $\lambda(X) = 1$, nominally the set $C^{\frac{1}{2}}$. In many continuous case examples, e.g., Example 2(c), this is a single point of probability zero, where, therefore, the definition cannot affect the operating characteristic of a conditional procedure described in Section 2.)

Example 2(c): For the normal densities of this example, $\lambda(x) = e^{2x}$. The labeling that makes $\Gamma^b = b$ as just described makes C^b consist of the two points $\pm\frac{1}{2}\log(b^{-1} - 1)$ for $b > \frac{1}{2}$. (The single point $X = 0$ is in $C^{\frac{1}{2}}$.) Thus $\Gamma(x) = 1/[1 + e^{-2|x|}]$ for $x \neq 0$. For this continuum B, the two values $X = 0.5$ and $X = 5$ considered earlier yield $\Gamma(0.5) = .73$ and $\Gamma(5) = .99995$.

In general (asymmetric) experiments such as the Bernoulli Example 3 and exponential Example 4 of Section 3, the "intrinsic significance level" $\alpha(X)$ does *not* have the desired frequency interpretation. This was illustrated by Birnbaum (1961a, p. 434; see also the next paragraph). He gave a representation of such an asymmetric experiment as part of a mixture of symmetric experiments (that are not actually performed), but this does not yield a satisfactory frequency interpretation for $\alpha(x)$. As we shall see below, and as is well-known in the NP setting, this is manifested in the fact that a choice of $\Gamma(X)$ that yields (1.1) (or (1.1) modified by letting the probability depend on i in a specified way) is not a universal function of $\lambda(X)$ or $\alpha(X)$ alone, but depends also on f_0 and f_1.

In fact, no exact (conditional) frequency interpretation (i.e., no evaluation of the left side of (1.1)), except for the trivial (1.5), can be made on the basis of the observed value of X alone. Any such interpretation must be made relative to the procedure C being used, just as is the case for the (α, β) of a test. When $X = x_0$, we must know in which C^b the point x_0 lies, and not just x_0 itself, to make a useful frequency interpretation. The statement of the value taken on by $\lambda(X)$ or $\alpha(X)$ alone cannot yield such an interpretation. Thus although it is tempting to think in terms of the conditioning obtained by lumping together the values in $\mathfrak{X}$ yielding a given value $\lambda_0 > 1$ of either λ or $1/\lambda$, corresponding to likelihood preference for H_1 or H_0, the frequency interpretation of such a partition unfortunately depends in detail on the form of f_0 and f_1. For example, in the discrete case, if $C' = \{x', x''\}$ where $\lambda_0 = f_1(x')/f_0(x') = f_0(x'')/f_1(x'')$, then $P_1\{\text{correct decision}\,|\,C'\} = f_1(x')/[f_1(x') + f_1(x'')]$, *which, regardless of* λ_0, *can take on any value between zero and one* for suitable f_i's on three or more points. (Birnbaum gives inequalities between $\alpha(x)$ and $\lambda(x)$ in some of his examples.)

In short, if one regards (as we do) the presence of a frequency interpretation as a primary characteristic that a measure of conclusiveness should possess, one must forego the idea that such a measure depends on λ alone. Without considering the axiomatics that lead to the likelihood principle, we simply remark that we find a frequency interpretation intuitively more important than consistency with such axioms, from a practical point of view, and suspect that many practitioners agree. The lack of a frequency interpretation (except in special cases) of an approach like that of Birnbaum (1961a) makes it unacceptable to me.

It should be apparent from the previous discussion that our development will not generally conform to Bayesian statistics. The quantity $\Gamma(X)$ will not represent a posterior probability, and posterior probabilities do not have frequency interpretations except when the assumed prior law actually governs the physical experiment (which is familiar from discussions of the NP case). Nor does fiducial inference (including Fraser's structural setup) or Dempster's upper and lower posterior inference achieve our frequentist interpretation. An exception is the symmetric experiment with symmetric prior law and the C^b the same as in our discussion of Birnbaum's $\alpha(X)$ preceding Example 2(c), where we showed that $1 - \alpha(X) = \Gamma(X)$. More generally than in the hypothesis-testing setting, there are statistical problems invariant under a group transitive on the parameter space, in which Birnbaum's approach of (1961a), Bayesian analysis with an invariant prior law (if the group is compact), fiducial inference, and our approach, can lead to consistent interpretations. (In fact, the maximal invariant function of a minimal sufficient statistic is well-known to be ancillary in such settings and can be used for conditioning.) In other examples, such as those of Section 4, invariant statistics that are not ancillary are used for conditioning. The construction of unconditional invariant confidence sets is discussed from a decision-theoretic viewpoint in Kiefer (1966), and Birnbaum (1962a, b) treats a special case of a finite parameter space and finitely many decisions, obtaining a confidence set with fixed confidence coefficient. Our emphasis will be on the possibility of obtaining a conditional confidence coefficient that varies with X, in both such symmetric settings and also asymmetric settings (like Example 4) where the various other approaches already mentioned do not yield the frequentist interpretation of conditional confidence.

We remark also that, even in symmetric testing situations with Γ ranging over a continuum of values, $1 - \Gamma$ is not simply the usual significance level $s(X)$ (say) at which H_0 would just be accepted (or rejected). For example, if the f_i are the normal laws of Example 2(c) and $x > 0$, we saw that $1 - \Gamma(x) = \alpha(x) = [1 + \lambda(x)]^{-1} = [1 + e^{2x}]^{-1}$ whereas

$$s(x) = \int_x^\infty f_0(u)du = \Phi(-x-1) \ .$$

The method of nested confidence sets, one for each confidence level, has been proposed by Cox (1958) and Birnbaum (1961b). This does have the frequentist interpretation of confidence, but without conditioning it suffers from the same defect as the procedures of Examples 1(a) and 2(a). Moreover, whatever their appeal, the nested confidence sets cannot be regarded as making a decision with associated single measure of conclusiveness $\Gamma_w(X)$.

As a final introductory comment, we emphasize that we do not believe it possible to give a simple prescription for choosing among the possible partitions C but leave that choice up to the experimenter, to be based on the context of his experiment. If he must take one of two immediate actions, he may well choose B to have just one element as in the NP theory. In other settings, such as an exploratory first stage of a scientific investigation, he might choose the finest possible partition, illustrated in Example 2(c) and defined in general in Section 3.5 (and for which, in a symmetric experiment, $\Gamma(X) = 1 - \alpha(X)$). The resulting maximum detail in the set of possible values of Γ^b, and absence of need to choose among various possible finite sizes for the set B, may make this the intuitively most appealing conditioning. Still, other experimenters may find occasion to use an in-between partitioning, as in the weak-strong partition of Example 2(b); or they may simply prefer using instead the corresponding unconditional four-decision procedure without any confidence coefficient, as discussed at the end of Section 3.2. Thus we have skirted consideration of "axiomatics" by letting the experimenter choose his own; or, more likely, by letting him choose the set B that is intuitively appealing to him and yields satisfactory probabilistic characteristics, whether or not it is easily expressable in terms of such simple axioms. We do not dictate his level of conditioning or whether he should condition at all.

We believe that any attempt to prescribe a general partitioning rule, applicable to all experimental settings, must prove unsatisfactory. Similarly, one should not blindly stick to the restriction we previously made for the sake of introductory simplicity, to partitions for which the conditional probability of (1.1) is the same for both i. The possible consequences of always using such a restriction are familiar in the geometry of decision theory, where in the NP setting it is easy to construct subminimax or subadmissible examples in which a change of procedure from that dictated by a particular criterion (Bayes, minimax, etc.) can decrease the risk function greatly under some states of nature at the expense of only a slight increase under the others. Statistics is not so simple a subject that it can be codified in terms of a simple recipe that yields satisfactory methods in all problems. As a minimum exercise of caution, the practitioner who approaches every problem with the same rule for constructing a procedure should look at the operating characteristics of nearby competitors to the procedure his rule produces.

Notation and the definitions of elements of our structure for comparing conditional confidence procedures is given in Section 2. Section 3 presents this approach in the simple hypothesis testing framework where we believe the ideas are most transparent; more extensive discussion and examples in this setting appear in Brownie and Kiefer (1977). More complicated but practical examples of multidecision (ranking and selection) and composite hypothesis settings are treated in Section 4. A phenomenon that occurs for composite but not simple hypotheses motivates consideration in Section 5 of a methodology of estimated confidence, sometimes less satisfactory than conditional confidence, but which may provide a variable data-dependent measure of conclusiveness in cases where no suitable procedure with variable Γ^b exists. Interval estimation is discussed in Section 6. Section 7 mentions briefly a number of extensions and variations, including sequential and nonparametric models and the use of conditional loss in place of conditional probability.

2. GENERAL NOTATION AND DEFINITIONS

While the simple specialization of our notation in the two-hypothesis setting will be described in Section 3, it is convenient to introduce our general framework at this point.

All measure-theoretic discussion will be omitted from the present article. We assume that, for each of the possible states of nature ω in Ω, the rv X has either an absolutely continuous distribution with density f_ω on $\mathfrak{X}$ or a discrete distribution with frequency function f_ω. (The interested reader may consult Kiefer (1976), Bahadur and Lehmann (1955), and Neveu (1965, pp. 193–194) for a rigorous treatment of conditional probabilities, in terms of subfields.) Possibly X is a sufficient statistic for a more basic model. In the decision space D, we assume that for each ω there is specified a nonempty subset D_ω of decisions that are viewed as "correct" when ω is true. (This is generalized in Section 7.4, but we believe the present setup captures the main confidence flavor.) Write $\Omega(d) = \{\omega : d \in D_\omega\}$.

We consider nonrandomized *decision rules* $\delta : \mathfrak{X} \in D$, and define $C_\omega = \{x : \delta(x) \in D_\omega\}$. The role of randomization, and the reason we usually neglect it, are discussed in Section 3.4.

Any function Z on $\mathfrak{X}$ may be termed a *conditioning statistic*. The range of Z is then the index set B of Section 1, and $C^b = \{x : Z(x) = b\}$. We write $C_d{}^b = C^b \cap \{x : \delta(x) = d\}$ and $K_\omega{}^b = C^b \cap C_\omega$. The $K_\omega{}^b$ are not necessarily disjoint, but the $C_d{}^b (b \in B, d \in D)$ are, and they constitute the partition C of $\mathfrak{X}$. It is useful to think of C as the intersection of two partitions, one induced by δ and the other by Z.

A conditional confidence procedure is a pair (δ, Z). Its associated conditional confidence function (of b and ω) is the conditional probability

$$\Gamma_\omega{}^b = P_\omega\{C_\omega | Z = b\} \ . \tag{2.1}$$

The rv Γ_ω is defined by its value $\Gamma_\omega(x) = \Gamma_\omega{}^b$ for x in C^b, for each b. A conditional confidence statement associated with the procedure (δ, Z) is the pair (δ, Γ_ω). It is used as follows: if $X = x_0$, with $Z(x_0) = b_0$, we state that for each ω we have conditional confidence $\Gamma_\omega{}^{b_0}$ of being correct, if ω is true; we make decision $\delta(x_0)$. In the unconditional NP setting of confidence intervals or tests, this prescription reduces to what we state and do; thus for testing between simple hypotheses as described in Section 1, we have confidence $\Gamma_0{}^b = 1 - \alpha$ of being correct if H_0

is true and confidence $\Gamma_1{}^b = 1 - \beta$ of being correct if H_1 is true.

The procedure has the (conditional) frequentist interpretation described in the introduction.

The dependence of the conditional confidence coefficient on ω is no different from the possible inequality between α and β in the simple hypothesis setting, or from the more general dependence of the power on the true ω in testing between composite hypotheses. However, if there is a function γ on $\mathfrak{X}$ such that $\Gamma_\omega(x) \geq$ (respectively, $=$) $\gamma(x)$ for all ω and x, we may simply state more succinctly that "we have conditional confidence at least (respectively, exactly) $\gamma(x_0)$ of making a correct decision," when $X = x_0$. Such a function γ is illustrated in Section 4. An extension of the author's admissibility theory for conditional procedures (Kiefer 1976) has been developed by Brown (1977) in terms of $\min_\omega \Gamma_\omega$, the smallest "guaranteed confidence." The use of the lower bound γ in a confidence statement has obvious practical appeal, but so does the idea of estimating the unknown ω and thus Γ_ω (to be discussed in Section 5), especially in examples in which $\min_\omega \Gamma_\omega$ can theoretically never be reasonably large (first paragraph of Section 5).

We shall write C for the associated pair (δ, Z) as well as for the equivalent partition $\{C_d{}^b\}$. Note that a subscript (usually d, i, or ω on C, Γ, etc.) always refers to a decision or state of nature, and a superscript (usually b, y, etc.) refers to the partition $\{C^b\}$ induced by Z. Different partitions C will be denoted by symbols such as overbars; for example, $\bar{C}$ will have $\bar{\delta}$, $\bar{\Gamma}$, and $\bar{G}$ (defined below) associated with it. The class of all partitions will be denoted by $\mathfrak{C}$.

Often a particular value of the rv Γ, or a preassigned value for it, is denoted by γ. Whenever $\Gamma_\omega{}^b$ is independent of ω, we may write it as Γ^b.

The probabilistic behavior of Γ_ω is naturally most meaningful from the frequentist interpretation of confidence statements, when ω is the true state of nature. This behavior is summarized (as discussed in Section 3.1.) in the tail probability law

$$G_\omega(t) = P_\omega\{\Gamma_\omega(X) > t \; ; \; \delta(X) \in D_\omega\} \; . \tag{2.2}$$

We could also examine the law of Γ_ω on $\{\delta \in D_\omega\}$ when the true state is $\omega' \neq \omega$, or even joint laws of Γ_ω, $\Gamma_{\omega'}$. This aspect of incorrectly asserted confidence statements will be discussed further in our admissibility considerations of Section 3.1. It suffices for now to mention that, in the simplest problems such as the k-hypothesis problem, one does not require such detailed properties in considering admissibility; rather, we shall only need $P_\omega\{Q_\omega\}$, where

$$Q_\omega = \{\Gamma_\omega = 0\} \; . \tag{2.3}$$

This is the union of all sets C^b on which we *never* make a confidence statement correct for ω and with positive $\Gamma_\omega{}^b$.

The inclusion of the event $\{\delta(x) \in D_\omega\}$ in (2.2) is based on the intuitive idea that it is not comforting to know that Γ_ω is large for a value of X for which we do not make a decision correct for the true ω. Still, arguments can be given for comparing procedures on the basis of $P_\omega\{\Gamma_\omega > t\}$ in place of (2.2); see Brown (1977). Analogous comments hold for (3.4).

Example 2(d): In the symmetric hypothesis-testing setting of Example 2, the procedures were also symmetric, so $G_0 = G_1$. For the unconditional NP procedure of Example 2(a), Γ_ω can only take on the value .84, so $G_\omega(t)$ represents the assignment to the value $t = .84$ of probability $P_\omega\{\delta(X) \in D_\omega\} = .84$; i.e.,

$$\begin{aligned} G_\omega(t) &= .84 \; , \quad \text{if} \quad t < .84 \; , \\ &= 0 \; , \quad \text{if} \quad t \geq .84 \; . \end{aligned} \tag{2.4}$$

Similarly, for the set B of Example 2(b) with two elements, we have $P_\omega\{\Gamma_\omega(X) = .71; \delta(X) = d_\omega\} = P_1\{C_1{}^w\} = .34$ and $P_\omega\{\Gamma_\omega(X) = .96; \; \delta(X) = d_\omega\} = P_1\{C_1{}^s\} = .50$. Thus for this procedure,

$$\begin{aligned} G_\omega(t) &= .84 \; , \quad \text{if} \quad t < .71 \; , \\ &= .50 \; , \quad \text{if} \quad .71 \leq t < .96 \; , \\ &= 0 \; , \quad \text{if} \quad t \geq .96 \; . \end{aligned} \tag{2.5}$$

In the choice between the NP partition of (2.4) and the partition of (2.5), the experimenter must decide whether it is desirable to split up the .84 probability of a correct decision with unconditional confidence .84, into probability .50 of a correct decision with the higher conditional confidence .96 and probability .34 of a correct decision with the lower conditional confidence .71. Analogously, for the continuum B of Example 2(c), one can use the fact that $\Gamma_\omega(x) > t > \frac{1}{2}$ when $|x| > -\frac{1}{2}\log(t^{-1} - 1)$ to show that

$$\begin{aligned} G_\omega(t) &= .84 \; , && \text{if} \quad t \leq .5 \; , \\ &= \Phi(1 + \tfrac{1}{2}\log(t^{-1} - 1)) \; , && \text{if} \quad .5 \leq t < 1 \; , \\ &= 0 \; , && \text{if} \quad t \geq 1 \; . \end{aligned} \tag{2.6}$$

We note that these partitions, being based on the same C_0 and C_1, have the same value of $G_\omega(0) = P_\omega\{\delta(X)$ is correct$\}$. For the first two of these partitions, both Q_i are empty; while, for the continuum B partition, $Q_0 \cup Q_1$ contains the single point $X = 0$.

This nature of the sets Q_i for these partitions supports the intuitive feeling that it is undesirable to use degenerate conditioning sets C^b on which only one of the two decisions can be made. In fact, when B has two elements, if the two conditioning sets are to have the "weak" and "strong" meaning of our development of Example 2(b), $\Gamma_0{}^b$ and $\Gamma_1{}^b$ should at least be of roughly the same magnitude for each b, and in particular it seems undesirable that one should be positive and the other zero. Thus we are led, for general Ω and D, to the next definition.

A procedure will be called *nondegenerate* if $P_\omega\{Q_\omega\} = 0$ for all ω, and the class of all nondegenerate procedures will be denoted by $\mathfrak{C}^+$. This notion will be discussed in Section 3.1.

We note that, from the frequentist viewpoint of

making a correct decision, the relevant part of the definition of Γ_ω is on C_ω, where we make a decision in D_ω. This is reflected in (2.2). Thus $1 - P_\omega\{Q_\omega\}$ is not simply $G_\omega(0)$, and $1 - G_\omega$ is not a proper df.

The *operating characteristic* of a procedure in simple cases can be taken to consist of the values of the $\Gamma_\omega{}^b$ and the probabilities under ω of attaining them, as reflected in G_ω. In more complex settings such as those of interval estimation, any operating characteristic that is sensible for the treatment of optimality or admissibility requires considerations additional to (2.2), e.g., reflecting either the law of $\Gamma_{\omega'}$ when ω is true (which may also be relevant in simple settings, as in (3.4)), or the length of the interval (see. Section 7). Further notation will be introduced as needed.

For each positive integer L, we denote by $\mathcal{C}^L$ the partitions with label set consisting of L elements. It is convenient to regard $\mathcal{C}^L$ as essentially a subset of $\mathcal{C}^{L+1}$ by allowing the latter to include sets C^L of probability 0 for all ω. (This question does not arise when B is a continuum.) We then designate by $\mathcal{C}^{L*}$ the *proper* partitions in $\mathcal{C}^L$, namely, those for which each C^b has positive probability under some ω. We also write $\mathcal{C}^{L+} = \mathcal{C}^L \cap \mathcal{C}^+$. Clearly, $\mathcal{C}^{L+} \subset \mathcal{C}^{L*}$. These notions can easily be extended to the case where L is the set of all positive integers.

The precedures treated in (2.4) and (2.5) are thus in $\mathcal{C}^{1+}$ and $\mathcal{C}^{2+}$, respectively.

A case of particular interest in which B is infinite, termed *continuum* B, is that where B is an interval as in Example 2(c) or (in Example 6) a higher dimensional Euclidean analog, and the rv Z has continuous df under each ω with almost every point of B having positive density for some ω. (In the simplest two-hypothesis settings of Section 3 or Example 2(c), almost every $C_i{}^b$ is a single point.) It will also be of practical value to consider examples of mixed B, where the above definition is altered by allowing some points of B to have positive probability. Partitions in $\mathcal{C}^1$ are the unconditional ones of classical decision theory (in particular, in Section 3, of the NP setting), and will often be referred to as such, in our discussion of properties of partitions.

In many of our developments, we shall emphasize ideas and avoid measure theory by considering only partitions in $\mathcal{C}^L$ rather than $\mathcal{C}$. We then interpret the rv Γ_ω of (2.1), as in the introduction, to take on the value $\Gamma_\omega{}^b = P_\omega\{K_\omega{}^b\}/P_\omega\{C^b\}$, on the set $C_\omega{}^b$ (only relevant if $P_\omega\{C^b\} > 0$); and Q_ω of (2.3) can be thought of as the union of all C^b's where $P_\omega\{C_\omega{}^b\} = 0$. Although a continuum B often seems of greatest value to us, our illustrations will often be in $\mathcal{C}^2$, to minimize arithmetic.

In examples involving the normal distribution, the law with mean μ and variance σ^2 will be denoted by $\mathfrak{N}(\mu, \sigma^2)$, the df of the standard $\mathfrak{N}(0, 1)$ will be denoted by Φ, and its density by ϕ.

3. CHOOSING BETWEEN TWO SIMPLE HYPOTHESES

Although this is not the most important setting for practical applications, many of the phenomena characteristic of our approach occur in it in the clearest and simplest form, and we shall, therefore, introduce them here. In later sections, we shall then state only briefly, or even omit as obvious, the close parallels found in other settings, reserving longer discussion for new phenomena.

The setup of Section 2 now specializes to $\Omega = \{0, 1\}$. Throughout this section, except in Section 3.4, we shall simplify discussion by assuming the f_i are densities of absolutely continuous distributions and are positive on the same set. We denote by P_i the probability measure on $\mathfrak{X}$ when the true law is f_i. As in Example 2, the decision space has two elements, $D = \{d_0, d_1\}$ and $\Omega(d_i) = i$. We abbreviate C_{d_i} by C_i, etc. Thus $C_i{}^b \equiv C_{d_i}{}^b = K_i{}^b$ here.

Some of our examples will treat symmetric settings (discussed briefly in Section 1), those in which $\lambda(X)$ under P_0 has the same law as $1/\lambda(X)$ under P_1. We often transform such examples to the form $f_1(x) = f_0(-x)$ on $\mathfrak{X} = R^1$. Then $|X|$ is an ancillary statistic (as discussed more extensively in the setting of Section 6); i.e., its law is the same under both f_i. In fact, if $P_i\{\lambda = 0\} = 0$, $|X|$ is maximal ancillary and maximal invariant since $X = |X| \operatorname{sgn} X$. Any Z which is a function of $|X|$ yields, with the symmetric decision rule δ, a partition with $\Gamma_0{}^b = \Gamma_1{}^b$. (The exceptional point 0 has probability zero under both ω.)

The reader may find it helpful to the understanding of many of the concepts that arise herein to check their meaning in the case of the unconditional NP setting $\mathcal{C}^1$.

3.1 Admissibility Considerations

We now give motivation for our admissibility criterion. It seems natural at first glance to regard a partition $\bar{C}$ as superior to $\hat{C}$ from the conditional confidence viewpoint if, in the notation of (2.2), the tail probabilities G_ω satisfy

$$\bar{G}_\omega(t) \geq \hat{G}_\omega(t) \quad \text{for all } \omega \quad \text{and} \quad t \geq 0 \,, \tag{3.1}$$

with strict inequality somewhere. The relation (3.1) means that $\bar{\Gamma}_\omega$ on $\bar{C}_\omega$ is stochastically at least as large as $\hat{\Gamma}_\omega$ on $\hat{C}_\omega$ for both ω, and larger under at least one, presumably a situation where most experimenters would prefer to use $\bar{C}$. If we restrict consideration to the Neyman-Pearson setting $\mathcal{C}^1$, we see (as illustrated in (2.4)) that G_ω has a single jump for $t > 0$, of height $1 - \alpha_\omega$ at $1 - \alpha_\omega$, where α_ω is the probability of error when ω is true; thus (3.1) yields the usual notion of admissibility in that case.

However, (3.1) alone is not sensible when B can have more than one element, as we shall now illustrate.

Example 2(e): In Example 2(d), suppose we again use $C_0 = (-\infty, 0)$ and $C_1 = [0, +\infty)$ but define the partition $\bar{C}$ of $\mathcal{C}^2$, with $B = \{0, 1\}$, by

$$\begin{aligned} \bar{C}_0{}^0 &= C_0 \,, \quad \bar{C}_1{}^0 = \phi \,, \\ \bar{C}_0{}^1 &= \phi \,, \quad \bar{C}_1{}^1 = C_1 \,. \end{aligned} \tag{3.2}$$

Then $P_i\{C_i{}^i | C^i\} = 1$, so that $\bar{\Gamma}^i = 1$ with probability .84 when f_i is true. (In effect, this is the same conditioning that led to (1.5).) Hence,

$$G_i(t) = .84 \ , \quad \text{if} \quad t < 1 \ , \\ \quad = 0 \ , \quad \text{if} \quad t \geq 1 \ . \tag{3.3}$$

Thus in the sense of (3.1) alone, $\bar{C}$ is superior to all three of the partitions of Example 2(d), even the NP $\mathcal{C}^1$ partition.

This example shows the need to consider a criterion additional to (3.1). Intuitively, when ω is true, if $\omega' \neq \omega$, it should not be very probable that we assert that ω' is true with strong assurance (high $\Gamma_{\omega'}$). This might suggest amending (3.1) to state that $\bar{C}$ is better than $\hat{C}$ if (3.1) holds and also, with $\omega' = 1 - \omega$ (the other state),

$$P_\omega\{\bar{\Gamma}_{\omega'} > t; \ \bar{\delta}(X) \in D_{\omega'}\} \\ \leq P_\omega\{\hat{\Gamma}_{\omega'} > t; \ \hat{\delta}(X) \in D_{\omega'}\} \quad \text{for all } \omega, \omega', t \ , \tag{3.4}$$

with strict inequality somewhere in either (3.1) or (3.4). However, we shall simplify the ensuing development by using instead

$$P_\omega\{\bar{Q}_\omega\} \leq P_\omega\{\hat{Q}_\omega\} \quad \text{for all } \omega \ , \tag{3.5}$$

which can be shown to be a consequence of (3.4).

Thus we define $\bar{C}$ to be *better than* $\hat{C}$ if (3.1) and (3.5) hold, with strict inequality somewhere, and the notions of admissibility and complete class are derived from this. A partition is admissible if no better one exists. Note that, in Example 2(e), $\bar{C}$ has $P_\omega\{\bar{Q}_\omega\} = .16$ while $P_\omega\{Q_\omega\} = 0$ for each of the three partitions of Example 2(d). Thus $\bar{C}$ is *not* better than these three partitions (as it was in the sense of (3.1) alone).

We call C a NP *partition* if $\{C_0, C_1\}$ constitutes an admissible $\mathcal{C}^1$ NP procedure (test). In that case, $\{C_0, C_1\}$ will be called the *underlying* NP *partition* of C.

Theorem 3.1: A partition C is an admissible partition in $\mathcal{C}$ if and only if it is a Neyman-Pearson partition.

Thus although very few partitions are admissible in the sense of (3.1) alone, there is a very large collection of admissible procedures when (3.5) is added. For example, all three partitions of Example 2(d), as well as $\bar{C}$ of (3.2), are admissible. More generally, for any NP partition $\{C_0, C_1\}$ in $\mathcal{C}^1$, an admissible partition in $\mathcal{C}^2$ can be obtained by pairing any measurable subset of C_0 with any measurable subset of C_1 to make up C^1, and letting their complements in the C_i then make up C^2. (The more detailed consequences of such pairing are discussed in Section 3.3 and Brownie and Kiefer (1977).)

An alternate rationale for the adoption of a criterion additional to (3.1) is contained in the discussion that led to the definition of nondegenerate partitions in Section 2. All such partitions have $P_\omega\{Q_\omega\} = 0$ for all ω; we see that, under this condition, partitions such as (3.2) are removed from contention, and (3.1) alone suffices for admissibility.

Theorem 3.2: A nondegenerate partition C in $\mathcal{C}^{L+}$ (or $\mathcal{C}^+$) is admissible among partitions in $\mathcal{C}^{L+}$ (or $\mathcal{C}^+$) in the sense of (3.1) alone, if and only if it is a NP partition.

It should be noted that Theorem 3.2 does not completely eliminate examples similar to (3.2). For, if we modify the $\bar{C}$ of (3.2) by transferring a set of small positive measure from $\bar{C}_0^0$ to $\bar{C}_0^1$ and also from $\bar{C}_1^1$ to $\bar{C}_1^0$, we obtain a partition $\ddot{C}$ whose $\ddot{G}_i$'s are close to the $\bar{G}_i$'s but which is in $\mathcal{C}^{2+}$. Thus $\ddot{C}$ is not strictly better than the partitions of Example 2(d) (as was $\bar{C}$) in the sense of (3.1), but it almost is: There is a subadmissibility phenomenon, in that the $\ddot{G}_i$'s, when altered by very small probability shifts, can be appreciably better than the G_i's of (2.4), (2.5), and (2.6).

Such a phenomenon is not unusual in decision theory, where it often points to the need of considering operating characteristics in more detail, or of imposing an additional criterion reflecting one's aims, to select among admissible partitions.

For the first of these, if (3.5) is replaced by the finer (3.4) it is easily seen that $\bar{C}$ of (3.2) assigns appreciable probability to the unfortunate event $\Gamma_{\omega'} = 1$ when the true state is $\omega \neq \omega'$; hence $\ddot{C}$ assigns such probability to $\Gamma_{\omega'}$ near 1, which the partitions of Example 2(c), better in this respect, do not. There is a trade-off between these latter partitions and $\ddot{C}$ (or $\bar{C}$) in the behaviors of the laws of Γ_ω and $\Gamma_{\omega'}$ when ω is true, and if both of these laws are considered there is no subadmissibility phenomenon, at least of the trivial kind just described.

As to the other possibility, we might impose such an additional criterion as that of fulfilling our intuitive aim of achieving roughly equal values $\Gamma_0{}^b$ and $\Gamma_1{}^b$. The partitions of Example 2(c) achieve this goal, unlike $\bar{C}$ and $\ddot{C}$. Further discussion is in Section 3.3.

The second possibility will perhaps appeal more to those statisticians who, in problems of interval estimation, are more concerned with the correct coverage property of estimators than with the property of not including incorrect parameter values. In such settings (Section 6), consideration of $P_\omega\{Q_\omega\}$ will often be replaced by something like expected volume of the conditional confidence set (Section 7). In the traditional unconditional setting, such a notion of volume has received more recent attention than the (sometimes equivalent; see Pratt (1963)) notion of covering incorrect parameter values.

3.2 Relationship to Another Decision-Theoretic Framework

We have mentioned in Section 1 differences between our development and those of the likelihood principle, the Bayesian approach, etc. There is a superficial similarity between our framework and another decision-theoretic one, which we now discuss, for simplicity, in terms of the $\mathcal{C}^{2+}$ partition of Example 2(c) and (2.5).

That partition, with $B = \{w, s\}$, resembles the $\mathcal{C}^1$ partition for a four-decision problem of the type suggested by our motivating comments, with possible decisions d_{0s}, d_{0w}, d_{1s}, d_{1w}, where d_{is} means "strong belief that i is true" and d_{iw} means "weak belief that state i is true." (See also Quesenberry and Gessaman (1968).) There are several aspects that do not correspond to our treatment. A difference in emphasis is the form of our confidence statement as consisting not only of a d_i but also of a value Γ_i between zero and one, with its frequency interpretation. This is not remedied merely by including with a procedure δ for the four-decision setup its operating characteristic $P_i\{\delta(X) = d_{jq}\}$ from which conditional probabilities Γ_i^w and Γ_i^s could be obtained, because the probabilities associated with a given partition in $\mathcal{C}^{2+}$, and not the labeling by index b of the sets, determine whether a strong or weak confidence statement is truly being made. Thus labeling the sets of our partition as C_i^s and C_i^w is really meaningful to us only if both the Γ_i^s are somewhat larger than both the Γ_i^w. As is illustrated by the partitions $\tilde{C}$ and $\ddot{C}$ discussed in Example 2(e) and just after the statement of Theorem 3.2, not all admissible partitions even allow such a labeling.

In addition, "$\tilde{C}$ better than $\hat{C}$" means something quite different here from what it means in the four-decision problem, so that the familiar complete class theorems in terms of operating characteristics for four-decision problems (e.g., Weiss 1954) do not yield results for our approach.

The preceding discussion points out the difference between the conditional confidence framework and that of $\mathcal{C}^1$ decision theory with an enlarged decision space. It does not choose between them, and there are settings where each may be preferred. If one of four definite actions must be taken, with known loss structure, a $\mathcal{C}^1$ procedure would be chosen. A framework that yields, instead, a data-dependent measure of conclusiveness, is perhaps justified by the observed tendency of practitioners not to be able to list all possible decisions and losses in advance of an experiment (as desirable as it may be to do so). Rather, they often seem to find that the next action or stage of experimentation crystallizes in their minds, even as a *possible* decision, only after the data have been recorded. A smaller D (of two elements rather than four or more in the present example), together with Γ, may then be of greater practical value. It has the additional merit of offering subsequent readers of the results of a scientific investigation a numerical frequentist measure of conclusiveness that can help them decide their own future research experiments. A continuum B (Example 2(c)) may offer the most natural and appealing conditioning for these purposes.

3.3 Choice of a Partition

The admissibility considerations of Section 3.1 yield a class of admissible procedures which is too large to be very helpful in the choice of a partition. We now consider some additional properties that an experimenter might regard as desirable in a procedure.

In deciding between the two possible densities of X, an experimenter will often regard f_0 and f_1 as qualitatively the same (thus not assigning f_0 the special role it sometimes has in traditional hypothesis testing, as discussed in Brownie and Kiefer (1977)); this includes regarding as roughly comparable the rewards from the two types of correct decisions and the penalties from the two incorrect ones. Under such circumstances, the experimenter who chooses to use a classical $\mathcal{C}^1$ procedure will often find that a partition that minimaxes the error probability (a NP test with $\Gamma_1 = \Gamma_0$), or that comes close to it, is satisfactory, unless a subminimax phenomenon exists for the given f_i's. Correspondingly, if he chooses to use a procedure in $\mathcal{C}^L$, he will find it desirable to use a partition with similar properties under the two f_i's. Combining this with the previously described requirement that each C^b is characterized by a different degree of conclusiveness, we find that a procedure should satisfy, at least approximately, the constraints

$$\Gamma_0^b = \Gamma_1^b\,, \quad \text{for all } b \text{ in } B\,, \tag{3.6}$$

$$G_0(t) = G_1(t)\,, \quad 0 \le t \le 1\,. \tag{3.7}$$

In addition, for each i the values Γ_i^b should be appreciably different as b ranges over B, and the probability masses assigned to these values by G_i should be spread out. (If the masses are not spread out, one might as well use a $\mathcal{C}^1$ partition!)

Imposing the constraint (3.6) is the $\mathcal{C}^L$ analog of the conventional $\mathcal{C}^1$ restriction to confidence intervals with constant confidence coefficient. As mentioned in Section 1, a practical advantage of doing this in $\mathcal{C}^L$ as well as $\mathcal{C}^1$ is that it is simpler to state a single confidence level, than one which depends on the unknown true state of nature.

In symmetric testing settings, partitions satisfying both (3.6) and (3.7) exactly are easily constructed in an obvious way, as was illustrated in Example 2(c–d). However, when symmetry is absent it may be difficult, if not impossible, to obtain a partition C in $\mathcal{C}^{L+}$, for $L > 1$, which satisfies (3.6) and (3.7) exactly. In this case, in view of the preceding considerations, it often seems reasonable to choose (3.6) as the primary requirement to be satisfied.

The $\mathcal{C}^2$ partition considered in Example 2(b) has the property that all the C_i^b are nonempty and that C_1^b is to the left of $C_1^{b'}$ if and only if $C_0^{b'}$ is to the left of C_0^b. A similar property holds for the continuum B partition of Example 2(c). We call such partitions *monotone*. For other hypothesis-testing examples in which λ is not monotone as it is here, such partitions can be constructed in terms of intervals of values of the likelihood ratio $\lambda(X)$ rather than X itself. Aside from their intuitive appeal, these partitions can be shown in many settings to produce a wider spread in the laws G_i of the Γ_i^b than do nonmonotone partitions. The desirability of such spread was described earlier.

Detailed comparison of the G_i of various monotone and nonmonotone partitions is given by Brownie and Kiefer (1977).

3.4 Discreteness and Randomization

When the f_i are not densities of absolutely continuous distributions as assumed in Sections 3.1–3.3, the role of randomization in $\mathcal{C}^1$ decision theory is to break up lumps of probability so as to obtain operating characteristics not obtainable for nonrandomized procedures. The same is true in our conditional confidence framework, but the practical objections to the use of randomization are perhaps even greater since the conditional confidence as well as the decision would depend on the randomization outcome. Fortunately, for typical examples with a moderate number of observations, we can usually satisfy (3.6) and the other aims described in Section 3.3, to a sufficient approximation, without randomizing, as we now illustrate.

Example 3: Let f_i be the frequency function of a binomial rv X based on 20 Bernoulli observations, with parameter value $p_0 = .35$ or $p_1 = .5$. Because of the discreteness and asymmetry, we cannot hope to satisfy $\Gamma_0^b = \Gamma_1^b$ exactly for all b, with a nonrandomized partition. A reasonable monotone $\mathcal{C}^{2+}$ partition C, defined by $C_0^2 = \{0, 1, \ldots, 5\}$, $C_0^1 = \{6, 7, 8\}$, $C_1^1 = \{9, 10, 11\}$, $C_1^2 = \{12, 13, \ldots, 20\}$, has operating characteristic

$$\begin{aligned} \Gamma_0^1 &= .703, & P_0\{C_0^1\} &= .517; \\ \Gamma_1^1 &= .683, & P_1\{C_1^1\} &= .497; \\ \Gamma_0^2 &= .926, & P_0\{C_0^2\} &= .245; \\ \Gamma_1^2 &= .924, & P_1\{C_1^2\} &= .252. \end{aligned} \tag{3.8}$$

The differences $|\Gamma_1^b - \Gamma_0^b|$ are smaller than the differences $\Gamma_i^2 - \Gamma_i^1$, reflecting approximate satisfaction of (3.6). Also, (3.7) is approximately satisfied.

Another partition for this setting, a discrete analog of continuum B, takes most of the sets C^b to consist of two points: $\{8, 9\}, \{7, 10\}, \ldots, \{2, 15\}, \{1, 16\}, \{0, 17, 18, 19, 20\}$, again with $C_0 = \{0, 1, \ldots, 8\}$. The corresponding values of (Γ_0^b, Γ_1^b) are (.582, .572), (.729, .705), ..., (.975, .988), (.978, .996), (.968, .999). The anomaly of nonmonotone Γ_0^b, caused by asymmetry and discreteness, can be remedied if desired by lumping the last two C^b into $\{0, 1, 16, 17, 18, 19, 20\}$, with $(\Gamma_0^b, \Gamma_1^b) = (.977, .997)$.

3.5 Construction of Partitions

When λ does not exhibit the symmetry described in Section 2, the construction of partitions satisfying (3.6) is not as simple as it was for the normal distributions in Example 2. Moreover, even in symmetric settings some thought is required about the specifications one makes in constructing a procedure. For example, in the setting of Example 2, if we sought a symmetric monotone procedure in $\mathcal{C}^{2+}$, we could not specify both the values of Γ^s and Γ^w we desire, since one determines the other. Analogously, in $\mathcal{C}^{3+}$ we can specify two of the three Γ^b values, but these are subject to certain constraints that depend on the f_i: The Γ^b values cannot be too close or far apart. Also, one may find, in analogy with the possible consequences of too rigid a requirement of the properties desired for a procedure in $\mathcal{C}^1$ decision theory, that satisfying such a specification as that of the exact value of Γ^s or $P_i\{C_i^s\}$ produces an unsatisfactory operating characteristic in other respects. In the final choice of procedure, the full operating characteristic should be examined, and several partitions should be compared.

We illustrate with a construction in $\mathcal{C}^{2+}$ satisfying (3.6), with Γ^s equal to a specified value γ. If $\lambda(x) = f_1(x)/f_0(x)$ is strictly increasing in the continuous case on a real interval $\mathfrak{X}$, it is easily seen that this entails setting $C_0^s = (-\infty, a_0]$ and $C_1^s = [a_1, +\infty)$, where a_0 and a_1 are the unique values satisfying

$$\int_{-\infty}^{a_0} f_1(y)dy \Big/ \int_{a_1}^{+\infty} f_1(y)dy = \gamma^{-1} - 1 = \int_{a_1}^{+\infty} f_0(y)dy \Big/ \int_{-\infty}^{a_0} f_0(y)dy \ . \tag{3.9}$$

Then (in order that $\Gamma_0^w = \Gamma_1^w$) we must have $C_0^w = (a_0, a^*)$ and $C_1^w = [a^*, a_1)$, where a^*, between a_0 and a_1, is the unique value satisfying

$$\int_{a_0}^{a^*} f_1(y)dy \Big/ \int_{a^*}^{a_1} f_1(y)dy = \int_{a^*}^{a_1} f_0(y)dy \Big/ \int_{a_0}^{a^*} f_0(y)dy \ . \tag{3.10}$$

The resulting common value of the two sides of (3.10) is then $(\Gamma^w)^{-1} - 1$.

In symmetric settings, the computations are simplified since (as in Example 2(b)) we have $a_0 = a_1$ and $a^* = 0$. The resulting simplified (3.9) is then easy to solve. In asymmetric settings trial-and-error may be required to solve the two equations (3.9). We now illustrate the results of such a computation.

Example 4(a): Suppose f_0 and f_1 are exponential densities,

$$\begin{aligned} f_0(x) &= 5e^{-5x}, & \text{for } x > 0, \\ f_1(x) &= e^{-x}, & \text{for } x > 0, \end{aligned} \tag{3.11}$$

with both $f_i(x) = 0$ for $x \leq 0$. Suppose we want the strong conditional confidence coefficient to be $\gamma = .9$. The solution to the two equations of (3.9) is $a_0 = .0554$, $a_1 = .723$. Then (3.10) yields $a^* = .248$. For this partition, of course $\Gamma^s = .9$, and $\Gamma^w = .64$. The operating characteristic is given by $P_0\{C_0^s\} = .24$, $P_0\{C_0^w\} = .47$, $P_1\{C_1^s\} = .49$, $P_1\{C_1^w\} = .30$. The lack of symmetry has thus produced rather different G_i; the large difference between $P_0\{C_0^s\}$ and $P_1\{C_1^s\}$ (or between $P_0\{C_0^w\}$ and $P_1\{C_1^w\}$) may make it more desirable to some practitioners to sacrifice exact equality in (3.6) in order to achieve G_i that come closer to satisfying (3.7). Finally, the above $\mathcal{C}^2$ partition should be compared in the manner of Example 2(d) (just below (2.5)) with $\mathcal{C}^1$ partitions. For example, the $\mathcal{C}^1$ partition satisfying (3.6) has $C_0 = (-\infty, .281)$, $C_1 = [.281, +\infty)$ and $\Gamma = .755$. Note that, in contrast to the symmetric setting, we no longer have the same C_0 and C_1 here as in the $\mathcal{C}^2$ partition previously obtained.

We now describe, for continuum B, the analog of the partition of Example 2(c) for general continuous f_i with

λ again strictly increasing on a real interval $\mathfrak{X}$. We shall first give the result in terms of a labeling of B that is simply related to $\mathfrak{X}$, rather than in the form for which $\Gamma^b = b$. Define a^{**} to satisfy

$$\int_{-\infty}^{a^{**}} [f_0(x)f_1(x)]^{\frac{1}{2}}dx = \int_{a^{**}}^{+\infty} [f_0(x)f_1(x)]^{\frac{1}{2}}dx \,, \quad (3.12)$$

and for $x \in \mathfrak{X}$ and $x > a^{**}$, define $H(x)$ (nonincreasing and with derivative $H'(x)$) by

$$\int_{-\infty}^{H(x)} [f_0(t)f_1(t)]^{\frac{1}{2}}dt = \int_{x}^{+\infty} [f_0(t)f_1(t)]^{\frac{1}{2}}dt \,. \quad (3.13)$$

Then with $B = \{x\colon x \geq a^{**},\ x \in \mathfrak{X}\}$ and the pairing $C^x = \{H(x), x\}$ for $x > a^{**}$ (and $C^{a^{**}} = \{a^{**}\}$), it can be shown that, for $x > a^{**}$,

$$\Gamma_0^x = \Gamma_1^x = [1 + |H'(x)|\, f_1(H(x))/f_1(x)]^{-1}$$
$$= [1 + f_0(x)/f_0(H(x))\,|H'(x)|]^{-1} \,. \quad (3.14)$$

If it is desired to relabel B as $\bar{B} = \{\bar{b}\}$ (say) for which $\Gamma^{\bar{b}} = \bar{b}$, one simply lets $C^{\bar{b}}$ consist of the $\{x, H(x)\}$ for which the expression of (3.14) equals $\bar{b}$.

Example 4(b): In the setting of (3.11), we obtain $a^{**} = (\log 2)/3 = .23$ from (3.12), and

$$H(x) = -[\log (1 - e^{-3x})]/3$$

from (3.13). Then from (3.14) we have $\Gamma^x = \Gamma(x) = \Gamma(H(x)) = [1 + (e^{3x} - 1)^{-\frac{1}{2}}]^{-1}$ for $x > a^{**}$. For numerical illustrations, if $X = .250$, the decision and confidence (or conditional confidence) are $(d_0, .755)$ for the $\mathfrak{C}^1$ procedure of Example 4(a); $(d_1, .64)$ for the $\mathfrak{C}^2$ procedure; and $(d_1, .52)$ for the continuum B procedure. If $X = .5$, the corresponding results are $(d_1, .755)$, $(d_1, .64)$, $(d_1, .70)$; and if $X = 1$, they are $(d_1, .755)$, $(d_1, .9)$, $(d_1, .98)$.

In many respects the conditioning with continuum B illustrated here is the most appealing one. It is characterized by being the finest monotone partition for which (3.6) is satisfied. Thus it is the most data-dependent conditioning, and in this respect most closely resembles the nonfrequentist measures of conclusiveness mentioned in the introduction. It may be asked, if one is willing to depart from the unconditional $\mathfrak{C}^1$ NP procedures, why stop at $\mathfrak{C}^2$, rather than to use the partition of continuum B.

A feature of the partition of continuum B that may be unattractive to some is the nearness of $\Gamma(x)$ to the uninformative value .5 when x is near a^{**}. While we feel this is not a bad feature since (especially in symmetric settings) such a sample value is intrinsically uninformative, we remark that one can achieve some of the features of the different partitions above by introducing mixed partitions, which include the analogs of the C_i^w of $\mathfrak{C}^2$ partitions for uninformative x and of the continuum B C^{b}'s for more informative values. This is illustrated in Example 5(c) of Section 4, and is discussed further by Brownie and Kiefer (1977), as are such other modifications as partitions constructed to satisfy particular probability requirements that do not treat the two hypotheses symmetrically.

It is not hard to give examples of asymmetric settings where, at the a^* and a^{**} of the above constructions, λ is far from one. This will disturb adherents of foundational persuasions other than the frequentist NPW position, and we remark only that this is not merely a phenomenon of conditioning since it can also occur near the critical value for unconditional $\mathfrak{C}^1$ NP tests. In asymmetric settings, one cannot maintain a frequentist approach without sometimes encountering what will be anomalies to other viewpoints.

4. COMPOSITE HYPOTHESIS AND MULTIDECISION EXAMPLE: SELECTION AND RANKING

The general multidecision framework is treated elsewhere (Kiefer 1975a, 1975b). Problems of selection and ranking offer a simple and natural setting for our approach, and we will illustrate this briefly with a few of the simplest cases. In this section, we treat only nonsequential examples involving the selection from two or three normal populations of the one with the largest mean. When two populations are involved, the framework of Examples 5(a), (b), and (c) is that of composite hypotheses and illustrates features additional to those of Section 3. For fixed b, the minimum of Γ_ω^b, discussed shortly after (2.1) and which is of interest just as in the NP $\mathfrak{C}^1$ theory for composite hypotheses, must now be computed over more than two elements. Our examples are symmetric ones; in other settings it might be of interest to compute $\min_{\omega \in \Omega_i} \Gamma_\omega^b$ for each of the two hypotheses Ω_i, for procedures for which these two quantities differ; these quantities are analogous to the differing Γ_0^b, Γ_1^b we encountered in such situations in Section 3 and to the possibly different values of the maximum risk over each Ω_i in NP $\mathfrak{C}^1$ theory.

Example 5: Two normal populations. We first consider the conditional confidence parallel of the Bechhofer (1954) formulation, in the following parts (a), (b), and (c). For simplicity, assume the variances and sample sizes n for the two populations are equal. After a scale transformation and reduction by sufficiency, the setup becomes $X' = (X_1', X_2')$ where the X_i' are independent and normal with means θ_i and variances n^{-1}. A value $\Delta^* > 0$ is specified, the difference in means that is worth detecting. The decision d_i' (asserting θ_i is larger) is correct if $\theta_i - \theta_{3-i} > -\Delta^*$; thus we are indifferent to differences of no more than Δ^* in $|\theta_1 - \theta_2|$. The main interest is of course in (θ_1, θ_2) for which only one d_i' is correct, outside the indifference region. A further reduction by invariance transforms the problem to $X \sim \mathfrak{N}(\theta, 1)$ and decisions d_i, where $\Omega(d_1) = \{\theta\colon \theta > -\bar{\Delta}\}$ and $\Omega(d_0) = \{\theta\colon \theta < \bar{\Delta}\}$; here X, θ, $\bar{\Delta}$, are $(n/2)^{\frac{1}{2}}$ times $X_1 - X_2$, $\theta_1 - \theta_2$, Δ^*, respectively. The operating characteristic is of interest when $|\theta| \geq \bar{\Delta}$.

The problem is thus reduced to one of choosing between two composite hypotheses.

The classical symmetric $\mathcal{C}^1$ rule is $\delta(X) = d_1$ when $X \geq 0$; i.e., $C_1 = \{x: x \geq 0\}$, and $P_\theta\{\delta(X) \in D_\theta\} = \Phi(|\theta|)$ for $|\theta| \geq \bar{\Delta}$. Often a value P^* (close to one) is also specified, and the problem is phrased of selecting n (in terms of P^*, Δ^*) to yield $P_\theta\{\delta(X) \in D_\theta\} \geq P^*$ if $|\theta| \geq \bar{\Delta}$, which is satisfied if $P_{\bar{\Delta}}\{X > 0\} \geq P^*$.

We now illustrate some of the conditional confidence variations that are possible here.

Example 5(a): A procedure in $\mathcal{C}^2$. With the same (C_0, C_1) as above, let $C^1 = \{x: |x| < r\}$ for some positive value r. Thus $b = 1$ and 2 fulfill the spirit of weaker and stronger conclusions as to which θ_i is larger. An easy computation using the monotone likelihood structure $(f_\theta(x)/f_\theta(-x) \uparrow$ in $\theta > 0$ for each $x > 0)$ yields

$$|\theta| \geq \bar{\Delta} \Rightarrow \begin{cases} \Gamma_\theta^1 = [\Phi(r - |\theta|) - \Phi(-|\theta|)]/ \\ \qquad [\Phi(r - |\theta|) - \Phi(-r - |\theta|)] \geq \Gamma_{\bar{\Delta}}^1, \\ \Gamma_\theta^2 = [1 + \Phi(-|\theta| - r)/ \\ \qquad \Phi(|\theta| - r)]^{-1} \geq \Gamma_{\bar{\Delta}}^2 \end{cases} \quad (4.1)$$

Hence the consumer can speak of conditional confidence "at least $\Gamma_{\bar{\Delta}}^b$ of selecting the larger mean when $|\theta| \geq \bar{\Delta}$," if $X \in C^b$. The values of n and r can be selected to yield satisfactory values of $\Gamma_{\bar{\Delta}}^1$ and $\Gamma_{\bar{\Delta}}^2$ and of the probabilities of correctly asserting these values.

For comparison, we note that the corresponding result for the simple hypothesis setting of Section 3, with Ω now the two values $\theta = \pm\bar{\Delta}$, would consist only of the values $\Gamma_{\bar{\Delta}}^b$ in (4.1).

Example 5(b): A procedure with continuum B. In terms of the reduced setup of Example 5(a), the approach of Section 3.5 now yields, for $Z(x) = |x|$,

$$\begin{aligned} |\theta| \geq \bar{\Delta} \Rightarrow \Gamma_\theta^{|x|} &= [1 + \phi(-|x| - |\theta|)/\phi(|x| - |\theta|)]^{-1} \\ &= [1 + e^{-2|x\theta|}]^{-1} \geq [1 + e^{-2\bar{\Delta}|x|}]^{-1} = \Gamma_{\bar{\Delta}}^{|x|}. \quad (4.2) \end{aligned}$$

Thus we have conditional confidence at least $[1 + e^{-2\bar{\Delta}|x|}]^{-1}$ of selecting the larger mean whenever $|\theta| \geq \bar{\Delta}$, if $X = x$.

Example 5(c): Mixed procedure. The procedure of Example 5(b) may appeal to some practitioners who want their conditional confidence statement to reflect the fact that, for x near 0, they do not feel very conclusive about their selection. As mentioned in Section 3, others may want to achieve the benefits of Example 5(b) for large $|x|$ without completely sacrificing the value of $\Gamma_{\bar{\Delta}}^1$ of 5(a) when $|x|$ is small. One possible procedure, with the same (C_0, C_1), is given by

$$\begin{aligned} Z(x) &= 0, \quad \text{if} \quad |x| < r, \\ &= |x|, \quad \text{if} \quad |x| \geq r, \end{aligned} \quad (4.3)$$

for which Γ_θ^0 is the Γ_θ^1 of (4.1) and $\Gamma_\theta^{|x|}$ is given by (4.2) for $|x| \geq r$.

Example 5(d): Subset selection approach. In this approach of Gupta (1956), the $\mathcal{C}^1$ format allowed for three possible decisions (the nonempty subsets of $\{0, 1\}$); the label of the larger population mean is asserted to be in the chosen subset. After reduction to (X, θ), this amounts to deciding whether θ is positive (d_1), negative (d_0), or real (d_2, say). The classical procedure modifies that of Example 5(a) by making d_2 if $|X| < q$, for some positive q. If $Z(x) = |x|$, we obtain for the analog of (4.2),

$$\begin{aligned} \Gamma_\theta^{|x|} &= 1, \quad \text{if} \quad |x| < q, \\ &= [1 + \phi(-|x| - |\theta|)/\phi(|x| - |\theta|)]^{-1}, \\ &\qquad \text{if} \quad |x| \geq q. \end{aligned} \quad (4.4)$$

The possible shortcoming of (4.2) as $|x| \to 0$ (which was the motivation for (4.3)) is of course absent in (4.4). However, some practitioners may feel uncomfortable with the first line of (4.4). I do not feel that the value $\Gamma = 1$ here is defective in the way it was in Example 2(e), where the fact that $\bar{Q}_i \neq \emptyset$ was to blame for making a "$\Gamma = 1$" statement about a *decision which might be wrong;* in the case of the first line of (4.4), the *decision must be correct* when $|X| < q$, and I believe many practitioners may feel more comfortable about saying $\Gamma = 1$ in this event, than in using the $\mathcal{C}^1$ procedure that yields a value $\Gamma < 1$ for $|X| < q$. In fact, the latter possibility is one of the subjects of criticism of $\mathcal{C}^1$ confidence statements by authors such as Pratt (1961a). Thus (4.4) does not seem unreasonable to me.

Nevertheless, as an indication of the variety of conditional confidence procedures that is available, we mention that a procedure with most of the features of (4.4) but with $\Gamma < 1$ is obtained by using the Z of (4.3) with $r > q$, with the δ of the previous paragraph.

Example 6(a): More than 2 populations. We illustrate some of the features that differ from those for two populations, by considering independent samples from three normal populations with equal variances and sample sizes. After transformation, this amounts to the model $X = (X_1, X_2, X_3)$ with $X_i \sim \mathfrak{N}(\theta_i, 1)$ and d_i the decision that θ_i is largest. There are now a number of different possibilities for parameter regions where only d_1, e.g., is correct, and we choose the simplest for the sake of brevity: $\theta_1 - \bar{\Delta} \geq \theta_2, \theta_3$. We regard other cases as points of complete indifference, which of course reflects the oversimplification when $\theta_1 = \theta_2$ and θ_3 is much smaller. See Kiefer (1975) for other models.

The Bechhofer $\mathcal{C}^1$ procedure selects the d_i for which X_i is largest. We shall use that δ in what follows. For brevity, we discuss only analogs of the continuum B of Example 5(b). Write $X_{[1]} \leq X_{[2]} \leq X_{[3]}$ for the ordered X_i's, and similarly for the θ_i's. For $i = 1$ or 2, write $b_i = x_{[3]} - x_{[i]}$ and $\Delta_i = \theta_{[3]} - \theta_{[i]}$.

The finest invariant B is $\{(b_1, b_2): 0 \leq b_2 \leq b_1\}$. If $\Delta_2 \geq \bar{\Delta}$, this is easily seen to yield, in an obvious notation,

$$\begin{aligned} \Gamma_{(\Delta_1, \Delta_2)}^{(b_1, b_2)} &= [1 + (\exp\{b_1\Delta_2\} + \exp\{b_1\Delta_1\} \\ &\quad + \exp\{b_2\Delta_1\} + \exp\{b_2\Delta_2\})/(\exp\{b_1\Delta_1 + b_2\Delta_2\} \\ &\quad + \exp\{b_1\Delta_2 + b_2\Delta_1\})]^{-1}; \end{aligned} \quad (4.5)$$

and $\Delta_2 \geq \bar{\Delta}$ can be shown to imply

$$\Gamma^{(b_1, b_2)}_{(\Delta_1, \Delta_2)} \geq \Gamma^{(b_1, b_2)}_{(\bar{\Delta}, \bar{\Delta})} = [1 + (\exp\{b_1\bar{\Delta}\} + \exp\{b_2\bar{\Delta}\})/(\exp\{b_1\bar{\Delta} + b_2\bar{\Delta}\})]^{-1}. \quad (4.6)$$

The simpler $B = \{b_2 : b_2 \geq 0\}$ is easier to use, for three or more populations. From the fact that the density $\bar{g}$ (say) of $X_3 - X_1$ and $X_3 - X_2$ is jointly normal with variances two and correlation .5, we find (e.g.) the contribution from the event $X_3 > X_2 > X_1$ to the density of $Z(X)$ at b_2 by integrating $\bar{g}(t, b_2)$ over $b_2 \leq t < \infty$. Such calculations yield, in analogy to (4.6),

$$\Gamma_{\bar{\Delta}, \bar{\Delta}}^{b_2} = \left\{1 + \frac{\Phi(-6^{-\frac{1}{2}}b_2)\phi(2^{-\frac{1}{2}}(b_2 + \bar{\Delta})) + \Phi(-6^{-\frac{1}{2}}(b_2 + \bar{\Delta}))\phi(2^{-\frac{1}{2}}b_2)}{\Phi(6^{-\frac{1}{2}}(\bar{\Delta} - b_2))\phi(2^{-\frac{1}{2}}(\bar{\Delta} - b_2))}\right\}^{-1}. \quad (4.7)$$

Modifications of either of the above B's can be made in the spirit of Example 5(c). We consider elsewhere (Kiefer 1975a, 1975b) such modifications, as well as other ranking and selection goals for an arbitrary number of populations.

5. ESTIMATED CONFIDENCE

In the composite hypothesis setting of Section 4 there can occur an unpleasant feature that was absent for simple hypotheses. Suppose, as in Example 5, that the rv X is distributed as $\mathfrak{N}(\theta, 1)$, but that we now want to test the hypothesis $\theta > 0$ against the alternative $\theta < 0$. Formally, this means setting $\Delta^* = \bar{\Delta} = 0$, and it is certainly a practical possibility that the customer does not find it easy to specify the value Δ^* that is just worth detecting. It is a familiar fact that, as $\theta \to 0$, the probability $\Phi(|\theta|)$ of a correct decision for the symmetric NP $\mathcal{C}^1$ procedure approaches the useless value $\frac{1}{2}$, and this value is also assumed as each of lower bounds $\Gamma_{\bar{\Delta}}{}^1$, $\Gamma_{\bar{\Delta}}{}^2$, $\Gamma_{\bar{\Delta}}{}^{|x|}$ of (4.1) and (4.2). And yet, intuitively the larger $|X|$ is, the more certain we feel about our decision. For example, why not estimate θ by X and thus estimate $\Phi(|\theta|)$ by $\Phi(|X|)$, with corresponding estimators for the Γ_θ of (4.1) and (4.2)?

This last possibility is in the spirit of estimating the power or risk function, a subject which has been treated earlier for unconditional procedures (e.g., in Lehmann 1959, Sandved 1968); $\Phi(|X|)$ is simply the maximum likelihood (ML) estimator of the $\Gamma_\theta = \Phi(|\theta|)$ of the $\mathcal{C}^1$ procedure. This estimator is not unbiased, and no unbiased estimator exists. However, estimators $\hat{\Gamma}(X)$ can be given that are conservatively biased in that $E_\theta\hat{\Gamma}(X) \lesssim \Gamma_\theta$, and the frequentist interpretation of the estimated confidence $\hat{\Gamma}(X_i)$ from a large number m of independent experiments is then that $m^{-1}\sum_1^m \hat{\Gamma}(X_i)$ is with high probability exceeded or almost attained by the proportion of correct decisions in those experiments. From a practical point of view, a convenient biased (e.g., ML) estimator may sometimes prove satisfactory, provided its bias (divided by $1 - \Gamma_\omega$ if the latter is near zero) has been shown to be negligible; without such justification, the frequency justification is lost. Other possibilities, such as the use of a lower confidence bound on Γ_θ instead of a point estimator $\hat{\Gamma}$, are discussed in Kiefer (1975).

The use of estimated confidence extends in an obvious way to conditional procedures and to multidecision settings; thus in Example 6(a) it is the Γ of (4.5) that is to be estimated, as we now illustrate. For the sake of brevity, we treat only the ML estimator, despite its bias. As previously mentioned, use of that estimator in an application is not to be recommended unless its bias is regarded by the experimenter to be negligible for the parameter values he believes possible.

Example 6(b): Suppose that we use the fine continuum B procedure of Example 6(a). Then the expression (4.5) has as its ML estimator the same expression with each Δ_i replaced by its ML estimator, b_i:

$$\hat{\Gamma}^{(b_1, b_2)}_{(\Delta_1, \Delta_2)} = [1 + (\exp\{b_1^2\} + \exp\{b_2^2\} + 2\exp\{b_1 b_2\})/(\exp\{b_1^2 + b_2^2\} + \exp(2b_1 b_2))]^{-1}. \quad (5.1)$$

For example, if $X_{[3]} - X_{[1]} = 2.5 = b_1$ and $X_{[3]} - X_{[2]} = 1.0 = b_2$, we obtain $\hat{\Gamma} = .741$.

A computational advantage of estimated conditional confidence for the fine continuum B partition, over estimated unconditional confidence (for the $\mathcal{C}^1$ procedure), is evident in this example. The exact expression for Γ for the $\mathcal{C}^1$ procedure, corresponding to (4.5), is

$$\int_{-\infty}^{\infty} \Phi(x - \Delta_1)d[\Phi(x)\Phi(x - \Delta_1 + \Delta_2)], \quad (5.2)$$

which is the "probability of a correct selection" of Bechhofer. This integral cannot be evaluated in finite terms through Φ and usual elementary functions. Hence the user of estimated unconditional confidence will have to program his computer to evaluate this integral or must use an approximation. Gibbons, Olkin, and Sobel (1977) note the lower bound $P_L = \Phi(2^{-\frac{1}{2}}\Delta_1)\Phi(2^{-\frac{1}{2}}\Delta_2)$ and the upper bound $P_U = \Phi([\Delta_1 + \Delta_2]/2)$, on (5.2); the former is close to (5.2) when Δ_1 and Δ_2 are far apart, while the latter is close when the Δ_i are almost equal. These authors suggest substituting the observed values of $X_{[3]} - X_{[1]}$ and $X_{[3]} - X_{[2]}$ for Δ_1 and Δ_2 to obtain (ML) estimators of P_L and P_U. For the values 2.5 and 1.0 of the previous paragraph, one obtains .729 and .820 for the estimated values for the $\mathcal{C}^1$ procedure, and would assert that the estimated confidence in a correct decision lies between these values. This interval estimator is not of the traditional variety since the difference between P_L and P_U is not due to uncertainty caused by chance, but rather it is the difference between two analytic approximations to the integral (5.2).

For comparing $k > 3$ populations, a more formidable $(k - 2)$-fold integral replaces (5.2), and Gibbons, Olkin, and Sobel suggest a method of clustering populations to obtain closer estimates than the pair (P_L, P_U). Computing the analog of (5.1), however, entails performing roughly $k!$ exponentiations, additions, and multiplications, which a computer can be programmed to carry out quickly and accurately for moderate k, in contrast with the estimate based on *approximations* to the analog of (5.2) for the $\mathcal{C}^1$ procedure. The simplified B of (4.7) produces difficulties like those of the $\mathcal{C}^1$ procedure.

The conditional and estimated confidence approaches are two possible answers to our original goal of a highly variable, data-dependent measure of conclusiveness with frequentist interpretability. (For a confidence *estimator*, the desirable high variability is that of its expectation as a function of ω, not that reflected by its variance for fixed ω.) They are not interchangeable: In the setting of Example 2(a) the estimated confidence for the $\mathcal{C}^1$ partition is the known value .84, which is not at all data-dependent as are the Γ of the conditional partitions of Examples 2(b) and (c). However, in the setting of Example 5 with $\bar{\Delta} = 0$, we have seen that conditional Γ has a useless lower bound, whereas reasonable useful estimates of Γ (even unconditional) can be found. Combining the two approaches, as in Example 6(b), will typically have the desirable effect of yielding the most highly variable measure of conclusiveness, and it will also have the observed computational advantages over the unconditional estimated Γ.

6. CONDITIONAL INTERVAL ESTIMATORS; THE ROLE OF ANCILLARITY

The topic of conditional confidence intervals or sets is too large to give more than brief illustration in this article. When one considers the possible modes of conditioning in various statistical settings, the notion of ancillarity arises naturally (and historically), perhaps mostly in connection with problems of estimation, and hence we include its discussion in the present section.

The papers most closely related to the present section are those of Buehler (1959), Wallace (1959), and Pierce (1973). The emphasis is somewhat different, in that Buehler is interested in a game-theoretic rationale for conditioning, and then primarily in such a possibility as finding a "relevant" set C^1 such that various traditional $\mathcal{C}^1$ interval estimators, of constant confidence γ, have conditional confidence $\Gamma_\omega^1 \geq \gamma + \epsilon > \gamma$ when used in the $\mathcal{C}^2$ procedure with C^1. Variable Γ_ω^b is mentioned for its own interest primarily in the game-theoretic discussion (Buehler 1959, Sec. 9). Pierce, in the same spirit, relates relevant subsets to a non-Bayes structure, treating also finite Ω. Wallace studies existence questions for conditional sets that are relevant (or a variant of that property). Brown (1967) finds relevant subsets in a classical normal setting. Our emphasis, however, is on highly variable Γ_ω^b as a major aim from the outset, and on methods for constructing and comparing procedures with such Γ; the relationship of Γ_ω^b to the Γ_ω of the related $\mathcal{C}^1$ procedure, and the strict constancy of the latter, are not of primary interest to us, although this work of Wallace and Buehler contains conclusions in certain settings on the nonexistence of certain types of $\mathcal{C}^2$ procedures with the desirable property $\inf_\omega \Gamma_\omega^2 > \sup_\omega \Gamma_\omega^1$. Also, interval estimation, while perhaps the most important area of application of these ideas, is only one of many. Nevertheless, we state again that some of the ideas illustrated in the examples of this section have undoubtedly been thought of and used by others in the past; the main novelty, other than the trivial one of fitting interval estimation into our general approach, lies in the variety of possible aims for a conditional interval estimator that we illustrate and also in the use of estimated confidence. We also mention the interesting related work of Pratt (1961b, 1963) on interval estimation and on criticism of classical unconditioned confidence intervals (1961a). Buehler and Fedderson (1963) gave the first examples of relevant subsets in the normal case considered more generally by Brown (1967) and Olshen (1973).

Our discussion of interval estimation here is abbreviated, and does not include the space-consuming discussions of the operating characteristic and optimality considerations introduced in Sections 3.1 and 3.5, as they apply here.

To simplify our discussion, we treat interval estimation problems in the continuous case, with sample space $\mathfrak{X} = R^n$ or a subset thereof and a family $\{f_\omega, \omega \in R^1\}$ of densities on $\mathfrak{X}$. Thus $\Omega = R^1$. (We depart from this in Example 9.) The object is interval estimation of ω. Even without formalizing the consideration of performance characteristics additional to those of Section 2 (and which will be discussed in Section 7.4), we recognize and will deal with the two desirable but conflicting facets of an interval estimator: high conditional confidence coefficient and short length.

In the notation of Section 2, D in set estimation is a collection of subsets of Ω, for simplicity restricted here to be closed intervals. Thus $D_\omega = \{d : \omega \in d\}$. A conditional confidence procedure is again a pair (δ, Z). The conditional confidence coefficient of (2.1) is simply the conditional probability that the interval $\delta(X)$ covers the true value ω. Classical confidence sets are $\mathcal{C}^1$ procedures, often with Γ_ω equal (or almost equal) to some specified value γ.

An important intuitive notion is that of "lucky observation." In the simple hypothesis setting of Section 3, values of X for which $f_1(X)/f_0(X)$ was small or large were typically those associated with large values of Γ (especially in monotone procedures with continuum B). Similarly, in estimating the parameter of a location parameter family, there might be a useful Z such that a reasonable point estimator of ω (that might be thought of as the center of a confidence interval) has large or small conditional dispersion depending on the value of Z. If $X = (X_1, \ldots, X_n)$ with the X_i iid $\mathfrak{N}(\omega, 1)$, there are no lucky observations since the translation group is transitive on the space of the sufficient statistic $\bar{X} = n^{-1} \sum X_i$. In other settings such as the following

Example 1(b), there is a useful Z, which really labels there how lucky X turned out to be in the sense of yielding an accurate estimate of ω.

Sometimes a useful such measure of luck Z can be found in an ancillary statistic. We recall that an ancillary statistic is a rv W whose law does not depend on ω. (In early usage by Fisher and others, the ancillarity was relative to the maximum likelihood estimator $\hat{\omega}$ of ω in contexts where $\hat{\omega}$ was not sufficient but $(\hat{\omega}, W)$ was; we are not limited here to such settings.)

An important and well-known setting where there can be lucky observations of the nature just described and which will be used to illustrate conditional confidence intervals, is that of the uniform density described in Example 1(a). Because of the phenomenon noted in our discussion of that example, Pratt suggests using the procedure of Welch (1939) of conditioning on W_n. Before describing some of the conditional confidence procedures obtainable in this way, we state our view of Pratt's criticism: We agree that this example points up the possible appeal of conditioning but do not feel that the example proves the $\mathcal{C}^1$ procedure is a foolish one *for all customers*. Rather, in the spirit of our feeling that it is impossible to obtain a useful codification of statistics that yields a satisfactory "procedure selector" for all people in all settings, we must leave it up to the customer to choose among various possible conditionings upon reflecting on their operating characteristics. Some, satisfied with fixed confidence and concerned about overall expected interval length (see Pratt 1961b), may stick with the $\mathcal{C}^1$ procedure despite its conditional shortcomings.

We now describe several possibilities that occur if one does use conditioning with Z depending on W_n.

Example 1(b.1): The literature on this example is centered on constructions for which $\Gamma_\omega{}^b$ is constant in ω and b. Since the conditional law of T_n, given $W_n = w$, is uniform from $\omega - \frac{1}{2} + \frac{1}{2}W_n$ to $\omega + \frac{1}{2} - \frac{1}{2}W_n$, the usual conditional prescription amounts to setting $Z = W_n$, $B = (0, 1)$, and $\delta(X) = (T_n - (1 - W_n)\gamma/2,$ $T_n + (1 - W_n)\gamma/2)$. This yields $\Gamma_\omega{}^Z \equiv \gamma$.

Example 1(b.2): An alternative is a procedure yielding specified *fixed length* $d > 0$. The $\mathcal{C}^1$ procedure $\delta = (T_n - d/2, T_n + d/2)$ is of this nature, but it is not the "optimum" procedure described in Example 1(a). Of more interest here is the *conditional* procedure with the same δ and $Z = W_n$ as in Example 1(b.1), for which $\Gamma_\omega{}^Z = \min(d/(1 - Z), 1)$. Both of these procedures are of course sensibly modified by replacing d by $\min(d, 1 - Z)$, without altering $\Gamma_\omega{}^Z$. The conditional procedure reflects the aim mentioned in Section 1, of giving a data-dependent measure of conclusiveness.

Example 1(b.3): Perhaps most satisfactory to many practitioners will be a procedure which guarantees a specified conditional confidence γ but which takes advantage of lucky observations by giving not just a shorter interval as in Example 1(b.1) nor a larger Γ as in 1(b.2), but rather some combination of these (the prescriptions 1(b.1) and 1(b.2) being boundary cases). Any procedure $\delta = (T_n - h(Z)/2,\ T_n + h(Z)/2)$ with $\gamma \leq h(Z)/(1 - Z) \leq 1$ has $\Gamma^Z = h(Z)/(1 - Z) \geq \gamma$, and h can be chosen to achieve the desired balance between short length and high confidence for lucky observations. As an illustration, suppose we want to achieve the following:

$$\text{Find a conditional confidence procedure with } \Gamma_\omega{}^Z \geq .9 \text{ but obtain increased } \Gamma_\omega{}^Z \text{ where possible in preference to trying for length} \leq d\ . \tag{6.1}$$

Here $d < 1$ is specified. It is revealing to condition in terms of the rv $\bar{Z} = h(Z)/(1 - Z)$ rather than Z. A procedure that achieves (6.1) is

$$\begin{aligned}
\bar{Z} &= h(Z)/(1 - Z) \\
&= .9\,, && \text{if } d \leq .9(1 - W_n)\,, \\
&= d/(1 - W_n)\,, && \text{if } .9(1 - W_n) \leq d \leq (1 - W_n)\,, \\
&= 1\,, && \text{if } (1 - W_n) \leq d\,, \\
\delta(X) &= [T_n - (1 - W_n)\bar{Z}/2\,, \\
&\qquad T_n + (1 - W_n)\bar{Z}/2]\,, \\
\Gamma^Z &= \bar{Z}\,.
\end{aligned} \tag{6.2}$$

Each of the discrete parts of $\bar{B}$ here, $\bar{Z} = .9$ and 1, represents a grouping of values of W_n for which the conditional confidence given W_n is the same. For other aims, such mixed procedures will have properties that represent nontrivial averages on the discrete parts of B, as in examples treated in other sections.

There are many other similar examples. In the "problem of the Nile" Fisher (1956) considered the density $\exp\{-(x_1\omega + x_2\omega^{-1})\}$, with $x_1, x_2, \omega > 0$. He treated this density in a fiducial argument to exhibit ancillarity of $Z(x) = (x_1x_2)^{\frac{1}{2}}$, with $\hat{\omega} = (x_1/x_2)^{\frac{1}{2}}$. This density was also mentioned by Buehler (1959), from the conditional confidence viewpoint. Another simple example of this type follows.

Example 7: Suppose that $X = (X_1, X_2)$, the conditional density of X_1 given $X_2 = x_2 > 0$ is $\mathfrak{N}(\omega, x_2)$, and the density of the positive rv X_2 is arbitrary but known. In this case (X_1, X_2) is minimal sufficient and X_2 is ancillary. Thus, for example, a conditional confidence procedure that satisfies the aim (6.1) (with d now any specified positive value) is

$$\begin{aligned}
Z &= \min(1.645X_2{}^{\frac{1}{2}}, d/2)\,, \\
\delta(X_1, X_2) &= [X_1 - h, X_1 + h]\,, \\
\text{where} \\
h &= \max(1.645X_2{}^{\frac{1}{2}}, d/2)\,, \\
\Gamma^Z &= 2\Phi(1.645d/2Z) - 1\,.
\end{aligned} \tag{6.3}$$

Performance characteristics such as C_ω and most reasonable measures of length distribution of course depend on the df of X_2.

In Examples 1(b) and 7 the conditioning variable Z was ancillary, so that the law of Z did not depend on ω. A second feature of these examples was that, for each of the procedures considered, Γ_ω^s was independent of ω because of the invariant structure. The combination of these two features, in general statistical problems, yields such properties as the lack of dependence on ω of G_ω and $P_\omega\{Q_\omega\}$ of Section 2 (and of reasonable measures of loss due to interval length, discussed in Section 7.4). Consequently, conditioning on such an ancillary statistic may be simple and attractive, but it is by no means necessary, nor is it always a fruitful pursuit. Thus it is by now well-known from work starting with that of Buehler (1959) and Wallace (1959) that there are settings, such as that of Example 9 following, where no conditioning of the classical estimator $\bar{\delta}$ can yield a $\mathcal{C}^2$ procedure with constant Γ_ω on each of C^1 and C^2 and with $\Gamma^1 \neq \Gamma^2$. What is perhaps more interesting for our concern is the large class of settings where there is no nontrivial ancillary, and no simple useful one.

Example 8: Suppose $X = (X_1, X_2)$ and that the conditional law of X_1 given $X_2 = x_2$ is $\mathfrak{N}(\omega, x_2)$ as in the previous example, but that the law of X_2 now depends on ω. (For example, think of it as exponential with mean $|\omega|$.) Although X_2 is not ancillary, one can conveniently condition on it. The procedure (6.3) then still achieves the aim set down for it in (6.1), but, of course, G_ω now depends on ω. We have mentioned this conditioning based on X_2 only from the viewpoint of convenience without ancillarity; it may be quite inefficient if X_2 contains much information about ω, since an analog of Theorem 3.1 suggests using a computationally more complex procedure, with conditioning imposed on an unconditionally admissible procedure based on (X_1, X_2).

A brief mention of the use of ancillarity in certain simple symmetric settings is given in Section 1. Once more, Theorem 3.1 implies that many useful (asymmetric) procedures do *not* base conditioning on an ancillary.

In a final comment on ancillarity, we recall some examples of Basu (1964) which are concerned with the possibility of there being several maximal ancillaries. Without discussing his fiducial viewpoint, we can say that these examples are not surprising or upsetting to our approach. The use of different ancillaries, each in the manner that W_n was used in Example 1(b), simply amounts to the consideration of various partitionings; and, as we saw in Section 3, some of these may be less satisfactory than others to our aims. Cox (1971) suggests choosing from the maximal ancillaries one that maximizes the variance of conditional information. It is not known whether the resulting conditioning produces a highly variable Γ compared with other conditionings (ancillary or not).

We turn next to the use in interval estimation of the estimated confidence methodology introduced in Section 5. The inability of conditional confidence alone to satisfy our aims in some settings is typified by problems in which a nuisance parameter substantially affects the accuracy with which one can estimate the parameter of interest. We now illustrate this in the most familiar normal setting.

Example 9: Suppose $X = (X_1, \ldots, X_n)$ with $n \geq 2$ and (departing from $\Omega = R^1$) the X_i are iid $\mathfrak{N}(\omega_1, \omega_2^2)$, with $\Omega = \{(\omega_1, \omega_2): -\infty < \omega_1 < \infty, 0 < \omega_2 < \infty\}$ and with ω_1 to be estimated by an interval of fixed length. As usual, we reduce the problem by sufficiency to procedures that depend on $\bar{X}$ and $S^2 = \sum_1^n (X_i - \bar{X})^2$. The classical constant confidence $\mathcal{C}^1$ procedure $\bar{\delta} = [\bar{X} - cS, \bar{X} + cS]$, of variable length, where c is obtained from Student's distribution, has been a subject of the studies mentioned earlier. In particular, these works have considered the extent to which the Γ_ω^b of a $\mathcal{C}^2$ procedure with the above $\bar{\delta}$ (or any other one with constant unconditional confidence) can vary; thus, for example, it is known that for $\bar{\delta}$ one cannot have Γ_ω^b depending only on b, with $\Gamma^1 \neq \Gamma^2$. Suppose instead that we obey the prescription to use a fixed length interval and thus consider $\delta = [\bar{X} - c, \bar{X} + c]$, for which (as a $\mathcal{C}^1$ procedure) $\Gamma_\omega = 2\Phi(n^{\frac{1}{2}}c/\omega_2) - 1$. Unfortunately, this can be anything between 0 and 1, in analogy with the range of values $\frac{1}{2}$ to 1 for (4.1) or (4.2) when $\bar{\Delta} = 0$, as discussed in Section 5. Intuitively it is appealing to eliminate this variability of Γ_ω by trying to construct a conditional procedure with continuum B and $Z = S$, with the rough idea of using some function of S to estimate the ω_2 in the above expression for Γ_ω, and from our viewpoint it is tempting to regard small values of S as lucky. However, such an attempt achieves nothing in conditional confidence since $\bar{X}$ and S are independent, and thus Γ_ω^s is the same as the above Γ_ω for each value s of S.

A way of implementing our intuitive feeling that the coverage of $[\bar{X} - c, \bar{X} + c]$ is larger when s is small is to use estimated confidence as described in Section 5. Thus the uniformly best (for all convex loss) unbiased estimator of Γ_ω based on S can be shown by well-known methods to be

$$\hat{\Gamma}(S) = B_{\frac{1}{2}, n/2-1}(nc^2/S^2) \ , \quad \text{if} \quad n > 2 \ , \qquad (6.4)$$

where $B_{\frac{1}{2}, n/2-1}$ is the beta df with the given parameters; when $n = 2$, $\hat{\Gamma}(S)$ is the indicator of the event $\{S^2 > 2c^2\}$, of less practical appeal. As in Section 5, we have the frequentist interpretation that an experimenter who uses the $\mathcal{C}^1$ procedure $[\bar{X}_j - c, \bar{X}_j + c]$ with confidence estimator $\hat{\Gamma}(S_j)$ from the jth of a large number N of independent occurrences of this setting, knows that it is very probable that the proportion of correct coverage statements is close to $N^{-1} \sum_1^N \hat{\Gamma}(S_j)$. This assessment can easily be made more quantitative since the $\hat{\Gamma}(S_j)$ are just independent rv's taking on values in $[0, 1]$.

Someone who feels uncomfortable with stating an estimated confidence as a measure of conclusiveness may reflect that, since for fixed known Γ he is willing to estimate σ by S in the $\bar{\delta}$ of variable length, perhaps in the δ of fixed length he should be willing to estimate the unknown Γ_σ!

In parallel with the possibilities that arose in Section 5, there are estimation settings where (as in Example 1(b)) estimated confidence achieves nothing while conditional confidence achieves our aims; in others, the use of the two techniques together yields the most satisfactory variable measure of conclusiveness. As mentioned in Section 5, one may also consider using an interval estimator $\hat{\Gamma}$ of Γ_ω in place of a point estimator. These matters will be discussed again elsewhere.

7. OTHER CONSIDERATIONS

Conditional and estimated confidence procedures can be used in many settings we cannot explore in detail here. Three important statistical considerations will be illustrated briefly in the next three subsections, in some of their conditional and estimated confidence aspects: the use of sequential procedures, the use of asymptotic theory, and nonparametric settings (with a discussion of robustness). Other topics, such as conditional tolerance sets, conditional curve fitting, and conditional experimental designs, and the consequences of restricting to symmetric (invariant) procedures in symmetric settings, must be treated elsewhere, as in Kiefer (1975). The final subsection discusses general loss functions.

7.1 Sequential Procedures

The use of sequential stopping rules presents the possibility, depending on the conditioning, of probabilistic computations as formidable as any encountered in unconditional sequential analysis (for which, however, good simple approximations are often available), or of exceedingly simple ones, simpler than those of unconditional procedures, especially in symmetric settings. We first illustrate this in a symmetric two-hypothesis problem where $X_1, X_2, \ldots$ are iid, each with density f_0 or f_1.

Example 10: We assume $\lambda(X_1) = f_1(X_1)/f_0(X_1)$ to have the same law under f_1 that $1/\lambda(X_1)$ has under f_0. Write $\lambda_n = \prod_1^n \lambda(X_i)$. We will denote by N the random stopping time of a procedure. The use of sequential stopping rules offers the opportunity to specify a set of numbers that can be made to contain the approximate possible values of λ on stopping ("neglecting excess"). For the simplest illustration of this, suppose we specify values $A^{(2)} > A^{(1)} > 1$ and a positive integer M. Let C^1 be the subset $\{N \leq M\}$ of the space of infinite sequences $\{X_i\}$ on which one stops the first time $N = n \leq M$ that λ_n or λ_n^{-1} is $\geq A^{(1)}$, if there is such an n. Let C^2 be the set where $N > M$ and one stops the first time $N = n$ that λ_n or $\lambda_n^{-1} \geq A^{(2)}$. We make decision d_1 if $\lambda_N > 1$ and d_0, otherwise. From the symmetry of laws and procedure, it is easy to see that this $\mathcal{C}^2$ procedure has $\Gamma_i^b \geq A^{(b)}/[1 + A^{(b)}]$. (A better approximation requires study of the distribution of excess, or Wald-type inequalities on it.) The value M can be chosen to yield, at least approximately, the desired apportioning into the two probabilities $P_0\{C_0^b\}$: if ψ is the df of the sample size for the classical sequential probability ratio test (SPRT) with continuation region $1/A^{(1)} < \lambda_n < A^{(1)}$, then $\psi(M) = P_0\{C^1\} \approx (1 + 1/A^{(1)})P_0\{C_0^1\}$. We do not know whether stopping times of N of this form have any optimum properties for $\mathcal{C}^2$ conditioning.

A simple continuum B modification of the preceding conditioning yields exact values of Γ rather than an approximation neglecting excess. If we index B by the values $b \geq A^{(1)}$ and let C^b consist of those values of the sequence $\{X_i\}$ for which $\lambda_N = b$ or b^{-1}, we see at once from the symmetry that $\Gamma_i^b = b/(1+b)$.

In asymmetric settings, only slightly more effort is required to obtain reasonable approximations. The continuum B modification, however, can be unsatisfactory without symmetry, as is illustrated in the extreme by Bernoulli models in which the two possible values of $\log \lambda(X_1)$ are not rationally commensurate; in such cases, (Γ_0^b, Γ_1^b) is either (0, 1) or (1, 0) for each b (just as in Example 2(e)!), and a coarser B is required to obtain a satisfactory conditioning. Nevertheless, a fine conditioning is valuable in other symmetric settings where the $\mathcal{C}^1$ procedure entails difficult computations, as we now illustrate.

Example 11: Suppose the X_i are iid three-vectors (X_{i1}, X_{i2}, X_{i3}), each with the possible laws of the X of Example 6(a). Define $S_{nj} = \sum_{i=1}^n X_{ij}$ and $S_{n[1]} \leq S_{n[2]} \leq S_{n[3]}$ for the ordered S_{nj}'s. One of the stopping rules that has been suggested for this problem (Bechhofer, Kiefer, and Sobel 1968) stops the first time (denoted by n) that the expression on the right side of (4.5), with b_i replaced by $S_{n[3]} - S_{n[i]}$, exceeds a specified value P^*. The population j that yielded $S_{n[3]}$ is again selected as having the largest mean. Even an approximation to the Γ of this $\mathcal{C}^{1\prime}$ procedure is difficult to derive. However, because of the symmetry of the setting and the stopping rule, if we condition on the values (b_1, b_2) of $S_{n[3]} - S_{n[1]}$, $S_{n[3]} - S_{n[2]}$ at the stopping time $N = n$, we again obtain (4.5) as an exact expression for Γ, with (4.6) as a lower bound. Also, as in Section 5, it may be useful not to use (4.6) (especially if $\bar{\Delta} = 0$), but instead to use an estimated $\hat{\Gamma}^{(b_1,b_2)}$ by replacing the Δ_i in (4.5) by estimators of them, such as b_i/N. (Estimated confidence can also be used without conditioning.) An approximation of the law G_ω of Γ, or of the distribution of N, presents the same difficulties as for the $\mathcal{C}^1$ procedure, but at least we have a data-dependent measure of conclusiveness that is simpler and more accurate than the lower bound P^* of the $\mathcal{C}^1$ procedure.

A conditioning procedure analogous to that of the previous paragraph can be employed in any sequential setting in which the model, stopping rule, and decision rule exhibit suitable symmetry.

The comments of Section 1 regarding the relationship of our development to Birnbaum's measure of evidence deserve an addition here. Birnbaum (1962b) remarks that (in the $\mathcal{C}^1$ setting), among all stopping times N that are based on the sequence $\{\lambda_n, n \geq 1\}$ and for which we always have $\lambda_N \geq A$ or $\leq A^{-1}$, the stopping time N^* of the SPRT with bounds A, A^{-1} has the obvious property of being smallest. (Somewhat surprisingly, DeGroot (1962) showed that, even in terms of expectations, N^* can be greater than an N' for which $\lambda'_{N'} \geq A$ or $\leq A^{-1}$, where λ_n' is the likelihood ratio of the rv's $g(X_1), \ldots, g(X_n)$ for a suitable function g. The latter rule cannot have both probabilities of error as small as those of the SPRT, by the Wald-Wolfowitz optimum character of the SPRT, and the existence of N' depends on the excess over the boundary for N^*.) Birnbaum points out the simplicity of this fact compared with the Wald-Wolfowitz theorem on the optimality of the SPRT, but we question the relevance of Birnbaum's remark to any approach other than one that specifies such a minimum value of λ or λ^{-1} (or of Birnbaum's $1 - \alpha(X)$ mentioned in Section 1) as the crucial criterion to be satisfied. A frequentist would not likely be led to this single specification. In any event, in terms of the usual $\mathcal{C}^1$ criteria $P_i\{C_i\}$, competing stopping rules must be considered, and one cannot circumvent the Wald-Wolfowitz optimum character arguments by use of Birnbaum's comments. Hence any corresponding comment in the realm of the sequential conditional procedures of this section cannot yield frequentist optimality results.

7.2 Asymptotic Conditional Confidence Sequences

For simplicity, think of a setting with finite D and Ω, with $X = (X_1, \ldots, X_n)$, the X_i iid, and n large. If we use a $\mathcal{C}^L$ procedure, there is usually no difference in applicability of standard asymptotic statistical theory from the classical $\mathcal{C}^1$ case. Thus the $\Gamma_\omega{}^b$ values we seek are often such that the $P_\omega\{C_d{}^b\}$ can be computed from usual asymptotic normal theory. However, if one or more of the $\Gamma_\omega{}^b$ values is to be very close to one, we may be in the domain of large or intermediate deviation theory.

When we use a procedure with continuum B, not only will such large deviation theory arise, but also we will need local limit theorems to provide results about asymptotic densities rather than probabilities of intervals.

Example 12: In the two-hypothesis setting, suppose the X_i are Bernoulli with mean either .4 or .6. We seek a symmetric ($C_0 = \{\bar{X} < \frac{1}{2}\}$) monotone $\mathcal{C}^2$ procedure with Γ^1 large enough that n must be fairly large, and with Γ^2 extremely close to one, to be achieved by taking $C^2 = \{|\bar{X} - \frac{1}{2}| > c\}$ for some constant c (much larger than $n^{-\frac{1}{2}}$). We can then approximate $P_0\{C_0\}$ by $\Phi([n/.24]^{\frac{1}{2}})$ and $P_0\{C_0{}^2\}$ by the Blackwell-Hodges-Bahadur-Rao large-deviation formula (Blackwell and Hodges 1959; Bahadur and Rao 1960).

The smallness of $P_0\{C_0{}^2\}$ in this example conforms with a result of Bahadur and Raghavachari (1970), that a conditional procedure that is asymptotically optimum in Bahadur's sense of unconditional "slope" must give approximately the same conditional slope, with high probability. However, different asymptotic normalizations may lead to different behavior. For example, if a regular density depending on a location parameter replaces the normal density in Example 6(a), with the Δ_i and $\bar{\Delta}$ of order $n^{-\frac{1}{2}}$, the calculations can be reduced asymptotically to those for the normal model as treated in Section 4.

7.3 Nonparametric Settings; Robustness

Again, only the briefest remarks and illustration can be given here, and they will be given in the simple two-hypothesis setting of Section 3. The typical considerations present in the choice of "robust" $\mathcal{C}^1$ procedures are now magnified by the conditioning process. Thus for a common contamination model (Huber 1964) that assumes the specified f_i is replaced by true density $(1 - \epsilon)f_i + \epsilon g$ where ϵ is small and ϵg is the unknown contamination, the use of a continuum B procedure, or a $\mathcal{C}^L$ procedure with large L or with some nominal $P_i\{C^b\}$ very small, can result in actual conditional probabilities $\Gamma_i{}^b$ that are astronomically different from the nominal ones, even for small ϵ. A rough prescription is thus not to take L greater than two or perhaps three when sample sizes are small and appreciable contamination of the f_i's is expected.

When one treats an application not with the robustness approach of considering contamination from simple specified f_i's as just indicated, but rather in terms of the more classical specifications of nonparametric hypotheses, often the common $\mathcal{C}^1$ methods can be modified to yield satisfactory $\mathcal{C}^L$ procedures. We now give a simple small-sample illustration of a nonparametric version of Example 5(a).

Example 13: Suppose $X = (X_1, X_2, X_3, X_4)$ where the X_i are independent and each of X_1, X_2 has continuous unknown df F_0, while each of X_3, X_4 has df $F_1 = F_0{}^\delta$ for some $\delta > 0$. Such families of Lehmann alternatives were introduced in Lehmann (1953); further detailed study relevant to problems of the present type can be found in Savage (1956). Order the observations X_i from smallest to largest and write $T_i = 0$ if the ith in order is X_1 or X_2, and $T_i = 1$ if it is X_3 or X_4; also, $T = (T_1, T_2, T_3, T_4)$. For any of the six possible values t of T, write $v_i = \sum_{j=1}^{i} t_j$ and $u_i = i - v_i$. The previously cited references show that

$$P_{F_0,F_0{}^\delta}\{T = t\} = 4\delta^2/\prod_{i=1}^{4} (u_i + \delta v_i) . \qquad (7.1)$$

In the ranking formulation corresponding to that of Example 5(a), suppose a value $\bar{\delta} > 1$ is specified. Then $\Omega(d_1) = \{(F_0, \delta): F_0 \text{ continuous}, \delta \geq \bar{\delta}\}$ and $\Omega(d_0) = \{(F_0, \delta): F_0 \text{ continuous}, \delta^{-1} \geq \bar{\delta}\}$. A possible $\mathcal{C}^2$ procedure puts $C^2 = \{X : T = (0, 0, 1, 1) \text{ or } (1, 1, 0, 0)\}$ with C^1 consisting of the other four possibilities. We let C_1 consist of the T values (0, 0, 1, 1), (0, 1, 0, 1), together with a randomization fraction $\frac{1}{2}$ of each of (0, 1, 1, 0) and (1, 0, 0, 1). The associated quantities $P_{F_0,F_0{}^\delta}(C_i{}^b)$ (taking account of the randomization) are easily computed from (7.1) and yield, for $\delta \geq \bar{\delta}$, in an obvious notation,

$\Gamma_\delta{}^b = \Gamma_{1/\delta}{}^b$, and

$$\begin{aligned} \Gamma_\delta{}^1 &= (1 + 5\delta)/6(1 + \delta) \geq \Gamma^1_{\bar\delta}\,, \\ \Gamma_\delta{}^2 &= [1 + (2 + \delta)/\delta^2(1 + 2\delta)]^{-1} \geq \Gamma^2_{\bar\delta}\,. \end{aligned} \tag{7.2}$$

The considerations of Savage (1956) and Savage and Sethuraman (1970) suggest appropriate choices of the $C_i{}^b$ for small sample sizes (larger than those of Example 13). For moderate sample sizes, there is little effect from avoiding the randomization used in this computationally simple example. For large sample sizes, an asymptotic theory is available (Savage 1956, Sethuraman 1970). Corresponding examples with more than two populations and decisions are treated in Kiefer (1975).

7.4 Conditional Expected Loss

Simple (zero-one) loss functions do not always reflect the possible consequence of various decisions, any more than they do in classical $\mathcal{C}^1$ settings. Much of our development requires only minor changes if a loss function W on $\Omega \times D$ is considered in place of the indicator of $\{d \notin D_\omega\}$ which is the loss function of Section 2; we then replace $1 - \Gamma_\omega{}^b$ by the conditional risk

$$R_\omega{}^b = E\{W(\omega, \delta(X)) \,|\, Z = b\}\,. \tag{7.3}$$

The frequentist interpretation in terms of the law of large numbers is analogous to what it was for simple loss. With obvious modifications of the other definitions of Section 2, results corresponding to those of Theorems 3.1 and 3.2 can be proved as before.

In problems with composite hypotheses, such as that of Example 5, ease of calculating an upper bound on $R_\omega{}^b$ (as in (4.1), etc.) depends on growth of $W(\omega, d)$ in ω compared with that of $P_\omega\{C_d{}^b\}$, just as in ordinary $\mathcal{C}^1$ calculations; if W does not grow too rapidly, the bound for analogous procedures will be attained at the same least favorable points as before. Estimated (conditional) expected loss can be used as in Section 5.

Example 14: Suppose the model of Example 5 is altered to make

$$\begin{aligned} W(\theta, d) &= c|\theta|\ , \quad \text{if } \theta < 0,\, d = d_1 \text{ or} \\ &\qquad\qquad\qquad \theta > 0,\, d = d_0\ , \\ &= 0\ , \quad \text{otherwise}\ , \end{aligned} \tag{7.4}$$

where $c > 0$. Thus there is no indifference region, but the loss from selecting the inferior population is proportional to $|\theta_1 - \theta_2|$. Then, for instance, if we use the continuum B of Example 5(b), we obtain

$$R_\theta{}^{|x|} = c|\theta|[1 + e^{2|x\theta|}]^{-1}\,. \tag{7.5}$$

From this we can obtain an upper bound on conditional risk analogous to the lower bound on conditional confidence of (4.2), using the fact that $\max_{u>0} u[1 + e^{2u}]^{-1} = .139$ to obtain $R_\theta{}^{|x|} \leq .139c/|x|$. This last is unbounded as $x \to 0$, and we might instead use estimated conditional risk, where the (biased) ML estimator is $c|X|[1 + e^{2X^2}]^{-1}$. For the symmetric $\mathcal{C}^1$ procedure the corresponding ML estimator of risk is $c|X|\Phi(-|X|)$. An alternative response to the unboundedness of $.139c/|x|$ is to alter the conditioning to a mixed one as in Example 5(c).

In problems of interval estimation, W may now reflect length of interval as well as the event of noncoverage, and dependence on the latter may be refined to grow with the distance by which $\delta(X)$ misses the true ω. Length of interval is of a different character from penalty for noncoverage, in two respects: First, the latter is unknown after the experiment and is assessed in terms of its conditional expectation $1 - \Gamma_\omega{}^b$ or a more general conditional expected loss, while in many problems the loss-from-length contains no such uncertainty after $\delta(X)$ has been computed. Second, the units of the two kinds of loss may be hard to compare. For these reasons, it may seem fitting to some practitioners to list separately the two components of loss. Alternatively, one may prefer to measure the performance of a procedure through such a combination as conditional expected loss from noncoverage and unconditional expected length.

Similarly, in other multidecision problems it may be helpful to consider vector loss functions. For example, when D is finite one can replace $\Gamma_\omega{}^b$ by the vector $\{P_\omega\{\delta(X) = d \,|\, Z = b\},\, d \in D\}$. The cost of experimentation can be adjoined as a component or can be added into a total conditional risk as before. Often that cost will not depend on ω and may consequently be better treated as a separate (or an unconditional) component, as was interval length.

[*Received April 1977.*]

REFERENCES

Anderson, E.B. (1973), *Conditional Inference and Models for Measuring*, Copenhagen: Mentalhygiejnisk Forlag.

Bahadur, R.R., and Lehmann, E.L. (1955), Two comments on "Sufficiency and Statistical Decision Functions," *Annals of Mathematical Statistics* 26, 139–142.

———, and Raghavachari, M. (1970), "Some Asymptotic Properties of Likelihood Ratios on General Sample Spaces," *Proceedings of the Sixth Berkeley Symposium*, Vol. 1, 129–152.

———, and Rao, R.R. (1960), "On Deviations of the Sample Mean, *Annals of Mathematical Statistics*, 31, 1015–1027.

Barnard, G.A. (1949), "Statistical Inference," *Journal of the Royal Statistical Society, Supplement*, 11, 115–139.

Bartholomew, D.J. (1967), "Hypothesis Testing When the Sample Size Is Treated as a Random Variable," *Journal of the Royal Statistical Society*, Ser. B, 29, 53–82.

Basu, D. (1964), "Recovery of Ancillary Information," in *Contributions to Statistics*, ed. C.R. Rao, Oxford: Pergamon Press, 7–20.

Bechhofer, R.E. (1954), "A Single-Sample Multiple Decision Procedure for Ranking Means for Normal Populations With Known Variances, *Annals of Mathematical Statistics*, 25, 16–39.

———, Kiefer, J., and Sobel, M. (1968), *Sequential Identification and Ranking Procedures*, Chicago: University of Chicago Press.

Birnbaum, A. (1961a), "On the Foundations of Statistical Inference: Binary Experiments," *Annals of Mathematical Statistics*, 32, 414–435.

——— (1961b), "Confidence Curves: An Omnibus Technique for Estimation and Testing Statistical Hypotheses," *Journal of the American Statistical Association*, 56, 246–249.

——— (1962a), "On the Foundations of Statistical Inference," *Journal of the American Statistical Association*, 57, 269–306.

——— (1962b), "Intrinsic Confidence Methods," *Bulletin of the International Statistical Institute*, 39, 375–383.

——— (1977), "The Neyman-Pearson Theory as Decision Theory, and as Inference Theory; With a Criticism of the Lindley-Savage Argument for Bayesian Theory," to appear in *Synthèse*.

Blackwell, D., and Hodges, J.L., Jr. (1959), "The Probability in the Extreme Tail of a Convolution, *Annals of Mathematical Statistics*, 30, 1113–1120.

Brown, L.D. (1967), "The Conditional Level of Student's *t*-Test," *Annals of Mathematical Statistics*, 38, 1068–1071.

——— (1977), "An Extension of Kiefer's Theory of Conditional Confidence Procedures," to appear in *The Annals of Statistics*.

Brownie, C., and Kiefer, J. (1977), "The Ideas of Conditional Confidence in the Simplest Setting," *Communications in Statistics* A6(8), 691–751.

Buehler, R.J. (1959), "Some Validity Criteria for Statistical Inference," *Annals of Mathematical Statistics*, 30, 845–863.

———, and Fedderson, A.P. (1963), "Note on a Conditional Property of Student's *t*," *Annals of Mathematical Statistics* 34, 1098–1100.

Cox, D.R. (1958), "Some Problems Connected with Statistical Inference," *Annals of Mathematical Statistics*, 29, 357–372.

——— (1971), "The Choice Between Ancillary Statistics," *Journal of the Royal Statistical Society*, Ser. B, 33, 251–255.

DeGroot, M.H. (1962), Uncertainty, Information, and Sequential Experiments, *Annals of Mathematical Statistics*, 33, 404–419.

Dempster, A.P. (1968), "A Generalization of Bayesian Inference," *Journal of the Royal Statistical Society*, Ser. B, 30, 205–247.

Dvoretzky, A., Wald, A., and Wolfowitz, J. (1951), "Elimination of Randomization in Certain Statistical Decision Procedures and Zero-Sum Two-Person Games, *Annals of Mathematical Statistics*, 22, 1–21.

de Finetti, B. (1974), "Bayesianism," *International Statistical Review*, 42, 117–130.

Fisher, R.A. (1956), *Statistical Methods and Scientific Inference*, Edinburgh: Oliver and Boyd.

Fraser, D.A.S. (1968), *The Structure of Inference*, New York: John Wiley & Sons.

Gibbons, J.D., Olkin, I., Sobel, M. (1977), *Selecting and Ordering Populations, A New Statistical Methodology*. New York: John Wiley & Sons.

Gupta, S.S. (1956), "On a Decision Rule for a Problem in Ranking Means," *Inst. Statistical Mimeo. Series* No. 150, University of North Carolina, Chapel Hill.

Huber, P.J. (1964), "Robust Estimation of a Location Parameter," *Annals of Mathematical Statistics*, 35, 73–102.

Kiefer, J. (1966), "Multivariate Optimality Results," in *Multivariate Analysis*, ed. P.R. Krishnaiah, New York: Academic Press, 255–274.

——— (1975), "Conditional Confidence Approach in Multidecision Problems," *Proceedings of the Fourth Dayton Multivariate Conference*, ed. P.R. Krishnaiah, Amsterdam: North Holland Publishing Co., 143–158.

——— (1976), "Admissibility of Conditional Confidence Procedures," *Annals of Statistics*, 4, 836–865.

Lehmann, E.L. (1953), "The Power of Rank Tests," *Annals of Mathematical Statistics*, 24, 23–43.

——— (1959), "Significance Level and Power," *Annals of Mathematical Statistics* 29, 1167–1176.

——— (1959), *Testing Statistical Hypotheses*, New York: John Wiley & Sons.

Neveu, J. (1965), *Mathematical Foundations of the Calculus of Probabilities*, San Francisco: Holden-Day.

Neyman, J. (1937), "Outline of a Theory of Statistical Estimation Based on the Classical Theory of Probability," *Philosophical Transactions of the Royal Society*, A, 236, 333–380.

Olshen, R.A. (1973), "The Conditional Level of the *F*-Test," *Journal of the American Statistical Association*, 68, 692–698.

Pierce, D.A. (1973), "On Some Difficulties in a Frequency Theory of Inference," *Annals of Statistics*, 1, 241–250.

Pratt, J.W. (1961a), *Review of Lehmann's Testing Statistical Hypotheses, Journal of the American Statistical Association*, 56, 163–166.

——— (1961b), Length of Confidence intervals, *Journal of the American Statistical Association*, 56, 549–567.

——— (1963), "Shorter Confidence Intervals for the Mean of a Normal Distribution with Known Variance, *Annals of Mathematical Statistics*, 34, 574–586.

Quesenberry, C.P., and Gessaman, M.P. (1968), "Nonparametric Discrimination Using Tolerance Regions," *Annals of Mathematical Statistics*, 39, 664–673.

Sandved, E. (1968), "Ancillary Statistics and Estimation of the Loss in Estimation Problems," *Annals of Mathematical Statistics*, 39, 1755–1758.

Savage, I.R. (1956), "Contributions to the Theory of Rank Order Statistics—The Two-Sample Case," *Annals of Mathematical Statistics*, 27, 590–615.

———, and Sethuraman, J. (1970), "Asymptotic Distribution of the Log Likelihood Ratio Based on Ranks in the Two Sample Problem," *Proceedings of the Sixth Berkeley Symposium*, 1, 437–458.

Savage, L.J. (1954), *The Foundations of Statistics*, New York: John Wiley & Sons.

Sethuraman, J. (1970), Stopping Time of a Rank-Order Sequential Probability Ratio Test Based on Lehmann Alternatives—II," *Annals of Mathematical Statistics*, 41, 1322–1333.

Shafer, G. (1976), *A Mathematical Theory of Evidence*, Princeton, N.J.: Princeton University Press.

Wallace, D.L. (1959), "Conditional Confidence Level Properties," *Annals of Mathematical Statistics*, 30, 864–876.

Wald, A. (1950), *Statistical Decision Functions*, New York: John Wiley & Sons.

———, and Wolfowitz, A. (1951), "Two Methods of Randomization in Statistics and the Theory of Games," *Annals of Mathematics*, 53, 581–586.

Weiss, L. (1954), "A Higher Order Complete Class Theorem," *Annals of Mathematical Statistics*, 25, 677–680.

Welch, B.L. (1939), "On Confidence Limits and Sufficiency, with Particular Reference to Parameters of Location," *Annals of Mathematical Statistics*, 10, 58–69.

COMMUN. STATIST.-THEOR. METH., A6(8), 691-751 (1977)

THE IDEAS OF CONDITIONAL CONFIDENCE IN THE SIMPLEST SETTING

C. Brownie and J. Kiefer

Cornell University
Ithaca, New York

Key Words & Phrases: conditioning; confidence coefficients; hypothesis testing.

ABSTRACT

A recently developed framework for comparing the properties of various conditional procedures is studied in detail in the setting of testing between two simple hypotheses, where the ideas are most transparent. In that setting, possible goodness criteria are considered, and illustrations are given. The conditional confidence methodology, unlike the Bayes, fiducial, and likelihood techniques, presents a measure of conclusiveness which has frequentist interpretability; and, unlike traditional Neyman-Pearson procedures, the measure is highly data-dependent.

1. INTRODUCTION

Conditioning is an old and commonly employed tool in statistical inference. However, no comprehensive framework for comparing the properties of different conditional procedures has been studied until recently (Kiefer, 1975, 1976, 1977). Since the other published material concerned with this framework is either expository or of a technical mathematical nature, it seems worthwhile to give a more detailed discussion of some of the considerations that went into the choice of criteria and procedures in the development of this methodology. The present paper does this in Section 3, in the simplest mathematical setting, that of testing between two simple hypotheses, after first defining the general framework in Section 2. Although the setting of simple hypotheses is not the most important for practical applications, many of the phenomena characteristic of the approach occur in it in the clearest form, devoid of the mathematical complications often attending more realistic but complex models; hence, it is expeditious to treat this setup in detail in order to make it easy to understand the current development of the subject and to lay the groundwork for criticism and future modification of it. In fact, this is the setting in which many of the ideas were first studied, and the present paper is an outgrowth of those first considerations. Other settings and proofs are treated in the cited references; these include interval estimation, selection and ranking, nonparametric and sequential techniques, and asymptotic theory.

A principal motivation for this interest in conditional procedures was the desire to have a methodology that gives a highly variable data-dependent measure of conclusiveness in the conclusion inferred from the experiment, with frequentist interpretability of that measure. On the one hand, the classical Neyman-Pearson (NP) development has a frequentist basis in which the practical interpretation of the level and power of a test, or the confidence coefficient of an interval estimator, is found in the long-run frequency of incorrect or correct decisions; however, in such a simple problem as that of testing whether a single normal random variable (rv) X with unit variance has mean -1 or $+1$, using the symmetrical critical region $X \geq 0$ for which each error probability is $.16$, the test does not distinguish, in its statement of error probabilities and decision reached, between the outcomes $X = .4$ and $X = 4$, although intuitively one feels much surer of the decision in the latter case than in the former.

On the other hand, such methodologies as those of subjective Bayesian inference, the likelihood principle, fiducial inference, and the statement of level at which a hypothesis is "just rejected (or accepted)," do give a variable data-dependent measure of conclusiveness, but except in special cases that measure does not have frequentist interpretability in terms of long-run frequencies. (These methodologies are discussed in more detail in Kiefer, 1977, and will not be mentioned further in the present paper, except for a brief illustration of differences among some of these methodologies in Example 3.6.)

The methodology of conditional confidence, which achieves both a data-dependent measure of conclusiveness and frequentist interpretability, is defined in the framework of Section 2, and is discussed and illustrated in the special setting of Section 3. Our approach, being frequentist in nature, can be considered to be in the NP spirit, and the form of conclusion of the latter will be seen to be an extreme case of it. We now explain our departure from the usual NP approach in the simple context of testing H_0: X has density f_0 against H_1: X has density f_1, with possible decisions d_0 and d_1, where d_i is the decision to accept H_i. In this introduction we consider only the simplest form of our methodology, analogous to that where the two error probabilities are equal. Our basic approach is that (i) we partition the sample space $\underline{X}$ into a family of subsets $C = \{C_i^b, b \in B, i = 0,1\}$, where B is a set of labels (one element in the NP case); (ii) when the sample X falls in C_i^b, we state a conclusion (Γ^b, d_i) that "H_i is true, with (conditional) confidence Γ^b"; (iii) the Γ^b (like the level or ordinary confidence coefficient) has a frequency interpretation, but it is now a conditional one: writing $C^b = C_0^b \cup C_1^b$, we shall have, for both $i = 0$ and 1, with P_i denoting probability when H_i is true,

$$P_i\{\text{making } d_i \text{ (the correct decision)} | C^b\} = \Gamma^b \text{ wp1.} \quad (1.1)$$

The frequency interpretation is clear: Suppose for simplicity that B is discrete, so that it is reasonable to assume that

$P_i\{C^b\} > 0$ for all b and i. Then, in repeated use of this approach in a long sequence of independent experiments, if there are many experiments in which the sample falls in C^b, the correct inference will very probably be made in approximately a proportion Γ^b of those experiments (law of large numbers for Bernoulli rv's). Thus, over many experiments in which an experimenter has obtained a value $\Gamma^b = .95$, he will very likely be making a correct assertion in approximately 95% of those experiments. It is clear that the experiments need not be identical, in which case the partition C of course depends on the experiment.

A fundamental idea of the confidence approach is that Γ^b is "attached" to C^b. This can be made more explicit by use of a labeling (not adopted in general because of cases we shall encounter in which the probability of (1.1) depends on i) in which B is taken to be the set of attainable values Γ^b, and the C^b's are labeled so that $\Gamma^b = b$. In that case, if the sample X falls in $C^{.95}$ we are conditionally "95% confident" that our decision is correct. In this case, if $X \in C_0^{.95}$ we would be "95% confident (conditionally) that H_0 is true." Whatever labeling is used, we define the <u>conditional confidence coefficient</u> $\Gamma(X)$ by

$$\Gamma(X) = \Gamma^b \text{ if } X \in C^b . \tag{1.2}$$

The rv Γ then has a conditional frequency interpretation analogous to that of an ordinary confidence coefficient or probability of correct decision.

To see the way in which the NP theory of testing, hereafter called the <u>NP setting</u>, fits into this framework, suppose B has a single element b and that $\underline{X}$ is partitioned into $\underline{X} = C^b = C_0^b \cup C_1^b$ where C_1^b is the critical region, chosen to make both probabilities of error equal $1-\Gamma^b$. We see that (1.1) specializes to express this probability structure in the NP setting. Incidentally, we note that, in this simple case, one does not condition on a single C_j^b, which would lead to

$$P_i\{\text{correct decision} | C_j^b\} = \begin{cases} 1 & \text{if } i = j, \\ 0 & \text{if } i \neq j, \end{cases} \tag{1.3}$$

which is useless. The same type of conditioning would yield (1.3) for general B, and indicates why we instead condition as we do in (1.1). It should be evident why we feel that the present approach is in some sense an analogue of that of NP.

Although "confidence" in earlier literature is usually reserved for the setting of interval estimation, it is used by us in other settings as well; and, as in interval estimation, it can be thought of as approximating the long-run frequency of correct decisions. In the NP setting of testing between simple hypotheses, confidence is thus the complement of the error probability.

Although we discuss (in Section 3) various considerations that might enter into the choice among conditional procedures, we do not make the choice for the practitioner (including the choice of whether to condition at all), but leave it up to him, to be based on the context of his experiment. Thus, if he must take one of two immediate actions, he might choose the NP B of one element. However, if this is an exploratory first stage of a scientific investigation, he might choose the finest possible conditioning partition (discussed in Section 3.5.1). Other experimenters may want to use an in-between partitioning, corresponding to the "strongly prefer H_i, mildly prefer H_i" found in decision theory examples. We do not have any "axiomatics" for the experimenter to use in choosing his partition, but let him choose the B that is intuitively appealing to him and yields satisfactory probabilistic characteristics, whether or not it is easily expressable in terms of simple axioms. In particular, we do not dictate the fineness of his conditioning partition. Our only axiom is the frequentist interpretability of our measure of conclusiveness, its data-dependent variability being a primary additional criterion that leads us away from the unconditional NP framework in many settings.

We feel that any attempt to prescribe a general prescription for choosing a partition, applicable to all settings, will prove unsatisfactory. Similarly, one should not automatically assume the restriction we made for the sake of simplicity in the introduction, to partitions for which the conditional probability of (1.1) is the same for both i. The possible adverse consequences of using such a restriction are familiar in the geometry of decision theory, where in the NP setting one encounters "subminimax" or "subadmissible" examples in which a change of procedure from that dictated by a particular criterion (Bayes, minimax, etc.) can decrease the risk point greatly in some coordinates at the expense of only a slight increase in the others. We do not believe that statistics is so simple a subject that it can be codified in terms of a simple prescription for choice of procedure that yields satisfactory methods in all problems. As a minimum precaution, a practitioner who attacks every problem with the same recipe for constructing a procedure should examine the operating characteristics of nearby competitors to the procedure his rule produces; unfortunately, the followers of many superficially attractive but simplistic approaches to choice of procedure often fail to do this.

2. GENERAL NOTATIONS AND DEFINITIONS

Although the simple specialization of our notation in the two hypothesis setting will be described in Section 3, it may be helpful to introduce our general framework at this point, so that the reader can see how the ideas illustrated in this paper fit into that framework.

To avoid measure-theoretic anomalies, it is convenient to assume the underlying measure space of possible outcomes of the experiment $(\underline{X}, \underline{B}, \nu)$ is of countable type with compactly generated σ-finite ν with respect to which the possible states of nature $\Omega = \{\omega\}$ have densities f_ω. This implies (e.g., Neveu, 1965) existence of regular conditional probabilities in (2.1) below.

(There is no difficulty in extending the considerations to certain other settings such as that where Ω is a nonparametric class of discrete laws on $\underline{X}$.) In the decision space D, we assume for each ω there is specified a nonempty subset D_ω of decisions that are viewed as "correct" when ω is true. (This can be generalized as indicated in Section 3.6 below and in Kiefer (1976, 1977), but we believe the present setup captures the main "confidence" flavor.) Write $\Omega(d) = \{\omega: d \in D_\omega\}$. We often denote the rv describing the experiment's outcome by X.

A <u>decision rule</u> $\delta: \underline{X} \to D$ is required to have $C_\omega \underline{\underline{\text{def}}}\ \delta^{-1}(D_\omega)$ in $\underline{B}$ for each ω. (This is of course easily translated into a more detailed measure-theoretic structure for D.) Thus, we consider nonrandomized rules. The role of randomization, and the reason one can usually neglect it, is discussed in Section 3.4.

Any subfield $\underline{B}_0$ of $\underline{B}$ may be termed a <u>conditioning subfield</u>. In practice it is often convenient to limit consideration (as we shall do) to subfields induced by statistics on $(\underline{X}, \underline{B})$. If Z is such a <u>conditioning statistic</u>, the range of Z is the index set B of Section 1, and (Bahadur and Lehmann, 1951) the corresponding $\underline{B}_0$ is the largest subfield inducing the same partition of $\underline{X}$ as is induced by Z; that partition is $\{C^b, b \in B\}$. We write $C_d^b = C^b \cap \delta^{-1}(d)$ and $K_\omega^b = C^b \cap C_\omega$. The K_ω^b are in general not necessarily disjoint (although they are in Section 3), but the $C_d^b(b \in B, d \in D)$ are, and constitute the partition C of $\underline{X}$; thus, C is the intersection of the partition induced by δ with that induced by Z.

A <u>conditional confidence procedure</u> is a pair $(\delta, \underline{B}_0)$. Its associated <u>conditional confidence function</u> (of $Z(x)$ and ω) is the conditional probability

$$\Gamma_\omega = P_\omega\{C_\omega \mid \underline{B}_0\}\ . \tag{2.1}$$

A <u>conditional confidence statement</u> associated with the procedure $(\delta, \underline{B}_0)$ is the pair (δ, Γ_ω). It is used as follows: if $X = x_0$ and thus $Z(X) = Z(x_0)$, we state that for each ω we have conditional confidence $\Gamma_\omega(x_0) = \Gamma_\omega^{Z(x_0)}$ of being "correct", if ω is true; we make decision $\delta(x_0)$. This has the (conditional) frequentist interpretation described in Section 1. The dependence of this conditional confidence coefficient on ω is no different from the dependence of the power on the true ω in testing between composite hypotheses. However, if there is a function γ on $\underline{X}$ such that $\Gamma_\omega(x) \geq$ (resp., $=$) $\gamma(x)$ for all ω and x, we may simply state more succinctly that "we have confidence at least (resp., exactly) $\gamma(x_0)$ of making a correct decision," when $X = x_0$. The use of such a γ is of less concern in the setting of Section 3 than in such problems as those of testing between composite hypotheses, ranking, or estimation with nuisance parameters; see Kiefer (1975, 1977), where a methodology of estimation of Γ is also discussed for such settings. We note also that Brown (1977) develops an admissibility theory for $\min_\omega \Gamma_\omega$, the smallest "guaranteed confidence".

By C we denote the associated pair $(\delta, \underline{B}_0)$ as well as the equivalent partition $\{C_d^b\}$. A subscript (usually d, i, or ω on C, Γ, etc.) always refers to a decision or state of nature, and a superscript (usually b, y, etc.) refers to the partition $\underline{B}_0$. Different partitions C will be denoted by symbols such as overbars or left superscripts; for example, $\bar{\bar{C}}$ will have $\bar{\bar{\delta}}$, $\bar{\bar{\gamma}}$, $\bar{\bar{\underline{B}}}_0$, and $\bar{\bar{G}}$ (defined below) associated with it, and similarly for ${}^3C$. The class of all partitions will be denoted by $\underline{C}$.

In examples, a particular value of the rv Γ, or a preassigned value for it, is denoted by γ, with appropriate superscripts or subscripts. The rv Γ_ω is constant with probability one (w.p.1) on $\{x: Z(X) = b\}$, and we denote its value there by Γ_ω^b. Whenever Γ_ω^b is independent of ω, we may write it as Γ^b.

It might appear natural to embed our approach in a decision-theoretic framework with space of possible decisions enlarged to $\{e_{d,r};\ d \in D,\ 0 \leq r \leq 1\}$, where r labels the confidence. The fruitlessness of this possibility is discussed in Section 3.2.

From the frequentist point of view in interpreting confidence statements, the probabilistic behavior of Γ_ω is most meaningful when ω is the true state of nature. This behavior is summarized (as discussed in Section 3.1) in the tail probability law

$$G_\omega(t) = P_\omega\{\Gamma_\omega(X) > t;\ \delta(X) \in D_\omega\} . \tag{2.2}$$

One could also examine the law of Γ_ω on $\{\delta \in D_\omega\}$ when the true state is $\omega' \neq \omega$, or even joint laws of $\Gamma_{\omega'}, \Gamma_{\omega'}$. Such "incorrectly asserted confidence statements" will be discussed further in our admissibility considerations of Section 3.1. For now, we mention that, in the simplest setting of Section 3 (or in such problems such as the "k-hypothesis problem"), we shall not require such detailed properties in considering admissibility; rather, it will turn out to be enough to consider $P_\omega\{Q_\omega\}$, where

$$Q_\omega = \{\Gamma_\omega = 0\} . \tag{2.3}$$

This is a maximal $\underline{B}_0$-measurable set among the sets on which we <u>never</u> make a confidence statement correct for ω and with positive Γ_ω.

A procedure will be called <u>nondegenerate</u> if $P_\omega\{Q_\omega\} = 0$ for all ω, and the class of all such procedures will be denoted $\underline{C}^+$. This notion will be discussed in Section 3.1.

From the frequentist viewpoint of making a correct decision, the relevant part of the definition of Γ_ω is on $C_{\omega'}$, where we make a decision in D_ω. This is reflected in (2.2). Thus, $P_\omega\{Q_\omega\}$ is not simply $1\text{-}G_\omega(0)$, and $1\text{-}G_\omega$ is not a proper distribution function (df). We could instead use (2.1) on C_ω but define Γ_ω to be 0 on $\underline{X}\text{-}C_\omega$. This would have the effect of simplifying (2.2) to read simply $P_\omega\{\Gamma_\omega > t\}$, but would make the replacement of (2.3) messier.

The <u>operating characteristic</u> of a procedure in simple cases can be taken to consist of the values of the Γ_ω^b and the probabilities under ω of attaining them, as reflected in G_ω. In more complex settings than that of Section 3, such as that of interval estimation, any sensible operating characteristic for the treatment of optimality or admissibility requires considerations additional to (2.2), for example, reflecting either the law of Γ_ω, when ω is true (which may also arise in simple settings, as in (3.1.3) and Example 3.3), or some measure of the size of d, such as the length of a confidence interval; see Kiefer (1976, 1977). Further notation will be introduced as needed.

For positive integral L we denote by B^L the label set consisting of L elements, and by $\underline{C}^L$ the partitions with $B = B^L$. It is a matter of choice, whether to regard $\underline{C}^L$ as essentially a subset of $\underline{C}^{L+1}$ by allowing the latter to include C^b's with ν-measure 0. (This question does not arise for B a continuum.) It is convenient to allow such sets, in order that each $\underline{C}^L$ have a closed set of operating characteristics $(P_\omega\{Q_\omega\}, G_\omega, \omega \in \Omega)$ in simple problems. We then designate by $\underline{C}^{L*}$ the __proper__ partitions in $\underline{C}^L$, namely, those for which $\nu(C^b) > 0$ for each b; in our simple examples we give this notion more content and avoid trivialities by assuming that every set of positive ν-measure has positive probability under some ω. We also write $\underline{C}^{L+} = \underline{C}^L \cap \underline{C}^+$. Clearly, $\underline{C}^{L+} \subset \underline{C}^{L*}$. It is obvious how to extend these notions to the case L = positive integers.

A case of infinite B of particular interest, termed __continuum B__, is that where B is an interval or a higher dimensional Euclidean analogue and the rv Z has continuous df under each ω with almost every point of B having positive Lebesgue density for some ω. (In the simplest two-hypothesis settings of Section 3, almost every C_i^b is a single point.) It will also be of practical value to consider examples of __mixed B__, where the above definition is altered by allowing some points of B to have positive probability. Partitions in $\underline{C}^1$ are the unconditional ones of classical decision theory (in particular, in Section 3, of the

NP setting), and will often be referred to as such, in the discussion of properties of partitions.

For each partition C, we call $\{C_d\}$ the underlying $\underline{C}^1$ partition of C.

In many of our developments, including the proof of Theorem 3.1, we shall emphasize ideas and minimize measure theory by giving detailed proofs for the case where consideration is restricted to partitions in $\underline{C}^L$ rather than $\underline{C}$. (General proofs, also covering other settings, appear in Kiefer, 1976.) In $\underline{C}^L$ we then interpret the rv Γ_ω of (2.1), as in the introduction, to take on the value $\Gamma_\omega^b = P_\omega\{K_\omega^b\}/P_\omega\{C^b\}$, on the set C_ω^b (only relevant if $P_\omega\{C^b\} > 0$); and Q_ω of (2.3) can be thought of as the union of all C^b's where $P_\omega\{C_\omega^b\} = 0$. We often omit the phrase "with probability one" where no confusion can be caused. The phrase "almost everywhere" always refers to Lebesgue measure on Euclidean sets. Although a "continuum B" often seems of greatest value to us, we will illustrate many ideas in $\underline{C}^2$, where the arithmetic is more transparent.

In examples involving the normal distribution, the law with mean μ and variance σ^2 will be denoted $\underline{N}(\mu,\sigma^2)$; the df of the standard $\underline{N}(0,1)$ will be denoted Φ; its density, ϕ.

3. CHOOSING BETWEEN TWO SIMPLE HYPOTHESES

The setup of Section 2 now specializes to $\Omega = \{0,1\}$. It is easy to transform the original $(\underline{X}, \underline{B}, \nu, f_i)$, after perhaps some randomization to remove atoms (see Section 3.4), into the simple form we often treat, of observing a rv Y taking on values in $(0,1)$, with Lebesgue density either

$$g_0(y) = \begin{cases} 1/a_1, & 0 < y < a_1, \\ 0, & a_1 \le y < 1, \end{cases} \tag{3.0.1}$$

or else

$$g_1(y) = \begin{cases} 0, & 0 < y \leq a_0 \\ \text{positive and nondecreasing}, & a_0 < y < 1; \end{cases} \qquad (3.0.2)$$

here $0 \leq a_0 < 1$ and $0 < a_1 \leq 1$. The set $\{y \leq a_0 \text{ or } y \geq a_1\}$ is that on which it is trivial to choose between g_0 and g_1, and we sometimes simplify our discussion by eliminating such possibilities and considering the case $a_0 = 0$, $a_1 = 1$, in which we have

$$g_0(y) \equiv 1,\ g_1(y) \text{ positive and nondecreasing on } (0,1), \qquad (3.0.3)$$

rather than the general case (3.0.1)-(3.0.2). We shall always make it clear whether we are dealing with the original f_i, or with the general (3.0.1)-(3.0.2), or with (3.0.3); when theoretical results are obtained in the latter setting, it will usually be clear how to extend them to the former.

We denote by P_i the probability measure on $\underline{X}$ (or $(0,1)$) when the true law is f_i (or g_i). We shall try to use X,x when referring to the general setup, and Y,y when referring to (3.0.1)-(3.0.2) or (3.0.3). As in Section 1, the decision space has two elements, $D = \{d_0, d_1\}$, and $\Omega(d_i) = i$. We abbreviate C_{d_i} by C_i, etc. Thus, $C_i^b \equiv C_{d_i}^b = K_i^b$ here. It is often convenient computationally to work in terms of the quantities

$$\rho_i^b = (\Gamma_i^b)^{-1} - 1 = P_i\{C_{1-i}^b\}/P_i\{C_i^b\}, \qquad (3.0.4)$$

the last equality holding when $P_i\{C^b\} > 0$.

Some of our examples will treat <u>symmetric settings</u>, those in which $\lambda(X)$ under P_0 has the same law as $1/\lambda(X)$ under P_1. We often transform such examples to the form $f_1(x) = f_0(-x)$ on $\underline{X} = R^1$. In such settings, $|X|$ is an ancillary statistic; that is, its law is the same under both f_i. In fact, it is "maximal invariant," since $X = |X| \operatorname{sgn} X$ is sufficient. Any

$\underline{B}_0$ which is a subfield of that induced by $|X|$ yields, with the symmetric decision rule δ, a partition with $\Gamma_0^b = \Gamma_1^b$. (The exceptional point 0 has measure zero.)

It is convenient, as indicated in Section 2, to consider conditioning subfields generated by rv's, and it is essentially no restriction to let such rv's Z be real. The sets C_i^b are simply

$$C_i^b = \{y : Z(y) = b,\ \delta(y) = d_i\}\ , \tag{3.0.5}$$

and on C_i^b the rv Γ_i of (2.1) takes on the value Γ_i^b, that can be interpreted as the left side of (1.1).

It may help the reader to understand many of the concepts that arise herein by checking their meaning in the case of the unconditional NP setting $\underline{C}^1$.

3.1. Admissibility Considerations

In the next few paragraphs we shall motivate our admissibility criterion, reserving proofs until the end of this subsection. At first glance it seems natural to regard one partition $\bar{C}$ as superior to another partition $\bar{\bar{C}}$ from the conditional confidence viewpoint if, in the notation of (2.2), the tail probabilities G_ω satisfy

$$\bar{G}_\omega(t) \geq \bar{\bar{G}}_\omega(t) \quad \text{for all } \omega \text{ and } t \geq 0\ , \tag{3.1.1}$$

with strict inequality somewhere. For, the relation (3.1.1) means $\bar{\Gamma}_\omega$ on $\bar{C}_\omega$ is stochastically at least as large as $\bar{\bar{\Gamma}}_\omega$ on $\bar{C}_\omega$ for both ω, and larger under at least one, presumably a situation where most experimenters would prefer to use $\bar{C}$. In fact, if we restrict consideration to the Neyman-Pearson setting $\underline{C}^1$, we see that G_ω has a single jump for $t > 0$, of height $1-\alpha_\omega$ at $1-\alpha_\omega$, where α_ω is the probability of error when ω is true; thus, (3.1.1) reduces to the usual notion of admissibility in that case.

However, (3.1.1) alone is not sensible when B can have more than one element; we shall now show that it leads to a trivial and unsatisfactory conclusion. Suppose we fix the underlying $\underline{C}^1$ partition $\{C_0, C_1\}$ and ask which partitions are admissible, within the class $\underline{C}^{\#}$ consisting of every partition which has the given $\{C_0, C_1\}$ as its underlying $\underline{C}^1$ partition. Every partition in $\underline{C}^{\#}$ has the same value $1-\alpha_i$ of $G_i(0)$. The partition $\bar{C}$ of $\underline{C}^2$ which we now consider, defined by

$$\bar{C}_0^1 = C_0 , \quad \bar{C}_1^1 = \phi ,$$
$$\bar{C}_0^2 = \phi , \quad \bar{C}_1^2 = C_1 , \qquad (3.1.2)$$

has $\bar{G}_i(t)$ with a jump of $1-\alpha_i$ at $t = 1$. Thus, $\bar{C}$ is clearly better than all other partitions in $\underline{C}^{\#}$ that are not essentially identical with it.

This last phenomenon demolishes not the conditional confidence approach, but rather the shortsightedness of considering only (3.1.1). In the NP setting $\underline{C}^1$, it was reasonable to add consideration of the law of $\Gamma_{\omega'}$ when ω is true, but of course it was unnecessary because that was already computable from (3.1.1) and would not alter the notion of admissibility.

Intuitively, when ω is true, if $\omega' \neq \omega$, it should not be very probable that we assert that ω' is true with strong assurance (high $\Gamma_{\omega'}$). Hence, we might amend (3.1.1) to state that $\bar{C}$ is better than $\bar{\bar{C}}$ if (3.1.1) holds and also, with $\omega' \neq \omega$,

$$P_\omega\{\bar{\Gamma}_{\omega'} > t;\ \bar{\delta}(X) \in D_{\omega'}\} \leq P_\omega\{\bar{\bar{\Gamma}}_{\omega'} > t; \bar{\delta}(X) \in D_{\omega'}\}$$
$$\text{for all } t, \omega, \omega' \text{ with } \omega \neq \omega', \qquad (3.1.3)$$

with strict inequality somewhere in either (3.1.1) or (3.1.3). This turns out to be unnecessary in the developments of Section 3.1; we need only the requirement (3.1.3) for $t = 1-$; that is,

$$P_\omega\{\bar{\Gamma}_{\omega'} = 1\} \leq P_\omega\{\bar{\bar{\Gamma}}_{\omega'} = 1\} \text{ for all } \omega , \qquad (3.1.4)$$

which, since $\Gamma_0 + \Gamma_1 = 1$ w.p.1 under each ω, can be rewritten

$$P_\omega\{\bar{Q}_\omega\} \le P_\omega\{\bar{\bar{Q}}_\omega\} \quad \text{for all } \omega . \qquad (3.1.5)$$

Consequently, we define $\bar{C}$ to be better than $\bar{\bar{C}}$ if (3.1.1) and (3.1.5) hold, with strict inequality somewhere, and the notions of admissibility and complete class are derived from this, so that we say "$\bar{\bar{C}}$ is (3.1.1) $\wedge$ (3.1.5) - admissible" if no such better $\bar{C}$ exists. This is a stronger notion of admissibility (yields no more admissible procedures) than would be obtained from using (3.1.3) instead of (3.1.5). Although one can often benefit from examining the functions of (3.1.3) in choosing a procedure (as will be discussed below Theorem 3.2 and in Section 3.3), it does not seem worthwhile to use (3.1.3) in characterizing complete classes, which (Theorem 3.1) are already very large under (3.1.5).

We have written (3.1.1) and (3.1.3) in notation that makes them applicable to other settings such as the k-hypothesis problem, where the conclusions for $k > 2$ are more complex than those of Theorem 3.1. These are treated in Kiefer (1976), as are compactness results that imply that the admissible procedures form a minimal complete class. The fact that $\underline{C}^{L+}$ is not closed causes no difficulty, since for each inadmissible procedure there is an admissible procedure that is better. We remark also that (Section 3.4) the set of operating characteristics of $\underline{C}^{L+}$ procedures is not convex.

A partition C is called a NP partition if $\{C_0, C_1\}$ constitutes an admissible $\underline{C}^1$ NP procedure (test). In that case, $\{C_0, C_1\}$ will be called the underlying NP partition of C.

Our main admissibility result in the simple hypothesis setting characterizes admissible partitions completely in terms of their underlying $\underline{C}^1$ partitions:

Theorem 3.1. A partition C is an admissible partition in $\underline{C}$ in the sense of (3.1.1) $\wedge$ (3.1.5) if and only if it is a Neyman-Pearson partition.

This means that, although very few partitions are admissible in the sense of (3.1.1) alone, there is a very large collection of admissible procedures when (3.1.5) is added. For example, if $\{C_0, C_1\}$ is any NP partition in $\underline{C}^1$, then an admissible partition in $\underline{C}^2$ can be obtained by pairing any measurable subset of C_0 with any measurable subset of C_1 to make up C^1, and letting their complements in the C_i then make up C^2. (The more detailed consequences of such pairings will be discussed in Section 3.3.) Thus, it is doubtful that use of a weaker notion of admissibility than (3.1.1) $\wedge$ (3.1.5) would be fruitful; if anything, one might search for a reasonable but stronger requirement that yields a smaller set of admissible partitions, but not such a trivial set as does (3.1.1) alone.

We now state another rationale that remedies the unsatisfactory result of using (3.1.1) alone. For simplicity we state it when B is finite. A motivation for our approach, as described in Section 1, is the desire to make confidence statements that are sometimes "strong" and sometimes "weak". The intuitive meaning of our label set B is that, in a given conditioning set C^b, we would like the various possible statements $(\delta, \Gamma_\omega^b)$ to have about the same qualitative "strength" (as will be discussed in more detail in Section 3.3). In particular, we do not want one Γ_ω^b to equal 0, if $P_\omega\{C^b\} > 0$; that is, we want

$$P_\omega\{C^b\} > 0 \Longrightarrow P_\omega\{C_\omega^b\} > 0 \text{ for all } \omega \text{ and } b. \qquad (3.1.6)$$

Since (3.1.6) is equivalent to $P_\omega\{Q_\omega\} = 0$ for all ω, we see that (3.1.6) is the restriction to procedures in $\underline{C}^{L+}$. This restriction removes partitions like (3.1.2) from consideration, if $L > 1$. An examination of the proof of Theorem 3.1 will then show that, under this restriction, (3.1.1) alone suffices for admissibility:

Theorem 3.2. A nondegenerate partition C in $\underline{C}^{L+}$ (or $\underline{C}^{+}$) is admissible among partitions in $\underline{C}^{L+}$ (or $\underline{C}^{+}$) in the sense of (3.1.1), if and only if it is a NP partition.

We note that Theorem 3.2 is not completely comforting in eliminating examples such as (3.1.2). For, if we modify the $\bar{C}$ of (3.1.2) by transferring a set of small positive measure from $\bar{C}_0^1$ to $\bar{C}_0^2$ and also from $\bar{C}_1^2$ to $\bar{C}_1^1$, we obtain a partition $\tilde{C}$ whose $\tilde{G}_i$'s are close to the $\bar{G}_i$'s, but which is in $\underline{C}^{2+}$. Thus, if $\bar{\bar{C}}$ is a partition in $\underline{C}^{2+}$ with the same C_0 and C_1 as for $\bar{C}$, but with Γ_0^b and Γ_1^b approximately equal for each b, we do not have $\tilde{C}$ strictly better than $\bar{\bar{C}}$ (as $\bar{C}$ was) in the sense of (3.1.1), but we "almost do": there is a "subadmissibility" phenomenon, in that the $\tilde{G}_i$'s, when altered by very small probability shifts, can be appreciably better than the $\bar{\bar{G}}_i$'s. The same phenomenon will occur for $L > 2$.

There are two possible responses to the criticism of our approach implied in the above phenomenon.

Firstly, although nothing is gained in admissibility considerations (that is, $\bar{\bar{C}}$ above is still admissible in $\underline{C}^{2+}$) if (3.1.5) is replaced by the finer (3.1.3), the function $P_\omega\{\Gamma_{\omega'}>t\}$ is more revealing than is $P_\omega\{Q_\omega\}$ in comparing admissible procedures; this is illustrated in Example 3.3. Thus, it is easily seen that $\bar{C}$ of (3.1.2) assigns appreciable probability to the unfortunate event $\Gamma_{\omega'} = 1$ when the true state is $\omega \neq \omega'$; hence, $\tilde{C}$ assigns such probability to "$\Gamma_{\omega'}$ near 1", which $\bar{\bar{C}}$, better in this respect, does not. There is a trade-off between $\tilde{C}$ (or $\bar{C}$) and $\bar{\bar{C}}$ in the behaviours of the laws of Γ_ω and $\Gamma_{\omega'}$ when ω is true, and if both of these laws are considered there is no subadmissibility phenomenon that arises in the manner described just above.

Secondly, we can take the point of view, alluded to earlier and discussed in detail in Section 3.3, that, within $\underline{C}^{2+}$, an additional criterion for selection of a partition should be imposed, to fulfill our intuitive aim of achieving roughly equal values Γ_0^b and Γ_1^b. The partition $\bar{\bar{C}}$ achieves this goal to a much greater extent than do $\bar{C}$ and $\tilde{C}$.

Each statistician can decide which of these two lines of attack appeals more to him. We feel that the second possibility will probably appeal to more people than the first, and will hence devote further attention to it. This feeling is prompted by the observation that applied workers in problems of interval estimation seem to be much more concerned with the "correct coverage" properties of estimators than with the property of including incorrect parameter values. Of course, in such interval estimation settings (see Kiefer, 1977) it will be appropriate to consider measures of risk other than $P_\omega\{Q_\omega\}$, in addition to G_ω; for example, some notion of expected volume of the conditional confidence set. But, in the traditional unconditional setting, such a notion of volume has received more recent attention than the (sometimes equivalent, as shown by Pratt, 1961) notion of covering incorrect parameter values.

We now turn to the proof of Theorem 3.1. We write in terms of the original f_i, although it may help the reader to think of the setting (3.0.1)-(3.0.2). In the latter setting, if g_1 is strictly increasing, C_0 is to the left of C_1 for a NP partition, and J_0 is to the right of J_1 in part (b) of the proof.

<u>Proof of Theorem 3.1.</u>

<u>(a) Sufficiency of NP structure</u>. We begin by noting two simple facts about the G_i:

$$G_i(t) \text{ has no jump at } t = 0\,; \tag{3.1.7}$$

$$\int_{0+}^{1} t^{-1} d[-G_i(t)] = 1 - P_i\{Q_i\}\,. \tag{3.1.8}$$

To emphasize the ideas and avoid measure theoretic details, we shall conform with the remark of the end of Section 2 and only demonstrate (3.1.7) and (3.1.8) here for partitions C with finite B. For any such C, if $P_i\{C^b\} > 0$ and $\Gamma_i = 0$ on C^b, we have from (2.1) that $P_i\{C_i^b\} = 0$; summing over all b for which $\Gamma_i = 0$ on C^b, we have

$$P_i\{\delta(X) = d_i,\ \Gamma_i(X) = 0\} = 0, \qquad (3.1.9)$$

which is (3.1.7). Next, we note that $P_i\{C_i^b\} > 0$ except when $C_i^b \subset Q_i$; denote the latter set of b's by H_i. The integral on the left side of (3.1.8) is then simply

$$\sum_{b \notin H_i} [P_i\{C_i^b\}/P_i\{C^b\}]^{-1} P_i\{C_i^b\} = \sum_{b \notin H_i} P_i\{C^b\} = 1 - P_i\{Q_i\}, \qquad (3.1.10)$$

as asserted.

Now suppose $\bar{\bar{C}}$ is a NP partition whose underlying NP partition is consequently admissible in the classical $\underline{C}^1$ decision-theoretic sense, but that $\bar{\bar{C}}$ is not admissible in the sense of $(3.1.1) \wedge (3.1.5)$; let $\bar{C}$ be better. Thus (3.1.5) and (3.1.1) hold, with strict inequality somewhere. At $t = 0$, (3.1.1) becomes (by (3.1.7)) $P_i\{\bar{C}_i\} \geq P_i\{\bar{\bar{C}}_i\}$ for both i, and since $\bar{\bar{C}}$ is classically admissible equality must hold; that is, $\bar{G}_i(0) = \bar{\bar{G}}_i(0)$ for both i. We shall hereafter only treat the case where both $\bar{G}_i(0)$ and $\bar{\bar{G}}_i(0)$ are positive; the other case is more trivial.

The functions $\bar{F}_i(t) = 1 - \bar{G}_i(t)/\bar{G}_i(0)$ and $\bar{\bar{F}}_i(t) = 1 - \bar{\bar{G}}_i(t)/\bar{\bar{G}}_i(0)$ are (by (3.1.7)) df's on $(0,1]$, and by (3.1.1) we have

$$\bar{F}_i(t) \leq \bar{\bar{F}}_i(t) \text{ for all } t > 0 \text{ and both } i. \qquad (3.1.11)$$

Since $-t^{-1}$ is strictly increasing on $(0,1]$, it follows that

$$\int_{0+}^{t} -t^{-1}\, d\bar{F}_i(t) \geq \int_{0+}^{t} -t^{-1}\, d\bar{\bar{F}}_i(t), \qquad (3.1.12)$$

if both integrals exist (which they do), and that (3.1.12) holds with strict inequality if there is strict inequality in (3.1.11) for some t . From (3.1.8), this becomes

$$P_i\{\bar{Q}_i\} \geq P_i\{\bar{\bar{Q}}_i\} , \qquad (3.1.13)$$

with strict inequality for any i for which strict inequality holds for some t in (3.1.1). Since we cannot have both (3.1.5) and (3.1.13) with a strict inequality for either i, this contradicts existence of the assumed $\bar{C}$.

<u>(b) Necessity of NP structure</u>. We again emphasize the idea and minimize the measure theory by giving the proof here for partitions $\tilde{C}$ with finite $\tilde{B}$; that is, we shall show that such $\tilde{C}$ which do not have NP structure cannot be admissible in the sense of (3.1.1) $\wedge$ (3.1.5).

If a $\tilde{C}$ with finite $\tilde{B}$ does not have NP structure, then by definition of that structure there are measurable sets $J_i' \subset \tilde{C}_i$ of positive Lebesgue measure and such that $f_1(x)/f_0(x)$ is less throughout J_1' than throughout J_0'. Choose b_0 and b_1 so that $J_i \overset{\text{def}}{=} \tilde{C}^{b_i} \cap J_i'$ has positive Lebesgue measure. (The b_i might be equal or unequal.) The J_i can always be chosen to fulfill the previous conditions and also, for both i (perhaps by reducing the previously chosen J_i), the conditions

$$P_i\{J_i\} > 0 \Longrightarrow P_i\{\tilde{C}_i^{b_i} - J_i\} > 0 ;$$

$$P_{1-i}\{\tilde{C}_i^{b_i} - J_i\} > 0 . \qquad (3.1.14)$$

Now let $J = J_0 \cup J_1$, and consider the class $\underline{C}'$ of partitions which are obtained from $\tilde{C}$ by altering only the $C_i^{b_i}$ on J ; that is, for which $B = \tilde{B}$ and, for $i = 0$ and 1,

$$C_i^b = \tilde{C}_i^b \text{ if } b \neq b_0 \text{ or } b_1 ;$$
$$C_{1-i}^{b_i} = \tilde{C}_{1-i}^{b_i} ;$$
$$C_i^{b_i} \subset \tilde{C}_i^{b_i} \cup J . \qquad (3.1.15)$$

Write $W_i = \tilde{C}_i^{b_i} - J_i$ and $T_i = \tilde{C}_{1-i}^{b_i}$. For C in $\underline{C}'$, write $V_i = C_i^{b_i} \cap J$, so that $C_i^{b_i}$ is the disjoint union of V_i and W_i, and C^{b_i} is the disjoint union of T_i, V_i and W_i. Only the V_i are variable in $\underline{C}'$. The effect of this variability is made explicit by writing out (first assuming $b_0 \neq b_1$)

$$\Gamma_i = \begin{cases} \dfrac{P_i\{V_i\}+P_i\{W_i\}}{P_i\{V_i\}+P_i\{W_i\}+P_i\{T_i\}} & \text{on } C^{b_i}; \\[2ex] \dfrac{P_i\{T_{1-i}\}}{P_i\{T_{1-i}\}+P_i\{W_{1-i}\}+P_i\{V_{1-i}\}} & \text{on } C^{b_{1-i}}. \end{cases} \quad (3.1.16)$$

Thus, the possible values of Γ_i on $C^{b_0} \cup C^{b_1}$ are nondecreasing functions of $P_i\{V_i\}$ and $-P_i\{V_{1-i}\}$. Furthermore, contributions to the law G_i from the two sets C^{b_i} and $C^{b_{1-i}}$ of (3.1.16) are, respectively,

$$P_i\{C_i^{b_i}\} = P_i\{V_i\} + P_i\{W_i\};$$
$$P_i\{C_i^{b_{1-i}}\} = P_i\{T_{1-i}\}. \quad (3.1.17)$$

From the variation of (3.1.17) in $P_i\{V_i\}$, that mentioned above for (3.1.16), and the fact that $-P_i\{V_{1-i}\} = P_i\{V_i\} - P_i\{J\}$ and J is fixed, we conclude that a procedure $\bar{C}$ in $\underline{C}'$ is <u>not</u> dominated by another procedure $\bar{C}$ in $\underline{C}'$ in the sense of (3.1.1), if and only if the pair $(P_0\{V_0\}, P_1\{V_1\})$ is maximal. These maximal pairs (by the NP Lemma on J) are those for which the values of f_1/f_0 on V_0 are almost everywhere no greater than those on V_1. In particular, $\tilde{C}$ is not of this form.

If $b_0 = b_1$, the previous paragraph is modified by simply omitting the second line of (3.1.14) and (3.1.15), and the two appearances of the expression $-P_i\{V_{1-f}\}$.

Finally, it is easy to check from (3.1.14) that all procedures in $\underline{C}'$ have $P_i\{Q_i\} \leq P_i\{\tilde{Q}_i\}$. Thus, $\tilde{C}$ is inadmissible in the sense of (3.1.1) $\wedge$ (3.1.5).

3.2. Relationship to $\underline{C}^1$ Decision-Theoretic Frameworks

In Section 1 we have alluded briefly to differences between our development and those of the likelihood principle, the Bayesian approach, etc. There is a superficial similarity between our framework and certain ones of $\underline{C}^1$ decision theory, which we now discuss. Since the differences are already present under the restriction to procedures in $\underline{C}^{2+}$, we consider only such procedures in the present discussion.

Our procedures then resemble those for a four-decision problem of the type suggested by our motivating comments, with possible decisions $d_{0s}, d_{0w}, d_{1s}, d_{1w}$, where d_{is} means "strong belief that state i is true" and d_{iw} means "weak belief that state i is true." There are several aspects that do not correspond to our treatment. A difference in emphasis, but which may seem superficial mathematically, is the form of our confidence statement as consisting not only of a d_i but also of a value Γ_i between 0 and 1, with its frequency interpretation. It might appear that the decision-theoretic framework could be regarded so as to yield the same interpretation, by including with a procedure δ for the 4-decision setup its operating characteristic $P_i\{\delta(X) = d_{jq}\}$. This correspondence is inadequate, because the probabilities associated with a given partition, and not the labeling (b) of the sets, determine whether a "strong" or "weak" statement is being made. Thus, labeling the sets of our partition as C_i^s and C_i^w is really meaningful to us only if both the Γ_i^s are somewhat larger than both the Γ_i^w. (This is not even a possible labeling for all admissible partitions, as discussed in Sections 3.1 and 3.3.) In the "corresponding" 4-decision problem we would thus have to limit consideration to procedures for which the operating characteristic exhibited such behavior.

In addition, the meaning of (3.1.1) is that d_{is} is preferred to d_{iw} in the sense that $\bar{\delta}$ is at least as good as $\bar{\bar{\delta}}$

if the four quantities $P_i\{\bar{\delta} = d_{is} \text{ or } d_{iw}\}$ and $P_i\{\bar{\delta} = d_{is}\}$ are as large for $\bar{\delta}$ as for $\bar{\bar{\delta}}$. (Thus, if $P_i\{\bar{\bar{C}}_i^b\} \le P_i\{\bar{C}_i^b\}$ and $P_i\{\bar{\bar{C}}_{1-i}^b\} \ge P_i\{\bar{C}_{1-i}^b\}$ for all b and i, which means $\bar{C}$ at least as good as $\bar{\bar{C}}$ in the classical vector sense, it follows that (3.1.1) and even (3.1.3) are satisfied. But (3.1.1) and (3.1.3) do not necessitate these inequalities, because of our ordering of the Γ values.) This, combined with the feature described in the previous paragraph, means that the familiar complete class theorems in terms of operating characteristics for 4-decision problems (e.g., Weiss, 1954) do not yield results for our approach. Indeed, this is indicated by the large class of admissible procedures given by Theorem 3.2, compared with the Bayes-type partitions that are admissible for 4-decision problems, even if admissibility for the latter is defined in terms of the four quantities mentioned earlier in this paragraph.

In Section 3.3(2) we mention the consequences of using $\underline{C}^2$ partitions of a particular ("monotone") structure for the present 4-decision problem rather than for the original conditional confidence setup. We describe there the structure of admissible procedures for certain loss functions in the 4-decision problem.

One could try working with a decision-theoretic setup with an enlarged space of possible decisions $\{d_{ir}, 0 \le r \le 1, i = 0,1\}$, where the label r corresponds to the confidence. One would then have to restrict consideration to procedures for which the operating characteristic yields the frequency interpretation we demand: $P_i\{\delta = d_{ir} | \delta = d_{0r} \text{ or } d_{1r}\} = r$, which is not present in usual decision-theoretic developments; and the ordering in r inherent in (3.1.1) would still have to be imposed.

There are circumstances in which each of the two procedures compared above, the 4-decision $\underline{C}^1$ procedure and 2-decision $\underline{C}^2$ procedure, would appeal more to the practitioner. The former might be appropriate when one of four actions, corresponding in character to the four decisions as we have labeled them, must be taken immediately upon examining the data. The latter might be more fitting when this is the first stage of a scientific investigation whose future stages of experiment and final form of decision may not even be stated at this point; a numerical measure of conclusiveness in the present decision will then be a valuable indicator to the present and future investigators, of what leads are worth pursuing in further experimental stages. In particular, a continuum B conditioning, mentioned earlier and treated in detail in Section 3.5, is perhaps the most natural and appealing conditioning for this last purpose.

3.3. Choice of a Partition

In Section 3.1 we obtained a class of admissible procedures which is too large to be very helpful in the choice of a procedure. We now consider some additional properties that an experimenter might regard as important or desirable in a procedure. This leads to a subclass of procedures with some practical value and intuitive appeal.

The question of how fine a partition to use in conditioning will be ignored here since, as indicated in Section 1, we feel this should be resolved by the experimenter. Thus much of the following discussion concerns partitions where B has a finite predetermined number of elements.

An idea which has been used in different contexts in multiple-decision problems, and which has considerable practical appeal, is that of partitioning the sample space into two regions, one where outcomes (regardless of which state of nature is favoured) are considered "strongly conclusive" in some sense, and the other where outcomes are considered "weakly conclusive" or "inconclusive". (See Quesenberry and Gessaman, 1968, and the discussion of 4-decision procedures of Section 3.2.) As discussed in Section 3.2, these two regions, in our context, are the two C^b's of a partition in $\underline{C}^2$, one where the conditional confidence coefficients Γ_i, attached to decisions d_i, are close to 1, and the other where the Γ_i are smaller. The generalization to more than two regions with different degrees of conclusiveness for procedures in $\underline{C}^L$ with $L > 2$, is obvious.

If the experimenter is deciding between the two possible densities of X, he will often regard f_0 and f_1 as qualitatively the same (thus not assigning f_0 the special role it sometimes has in traditional hypothesis testing, as discussed in Section 3.6(E)); this includes regarding as roughly comparable the rewards from the two types of correct decisions and the penalties from the two incorrect ones. In such a circumstance, an experimenter who chooses to use a classical $\underline{C}^1$ procedure will often find that a partition that minimaxes the error probability (a NP test with $\Gamma_1 = \Gamma_0$), or that comes close to it, is satisfactory, unless a subminimax phenomenon exists for the given f_i's. Similarly, if he chooses to use a procedure in $\underline{C}^L$, he will find it desirable to use a partition with similar properties under the two f_i's. This means requiring that the tail probabilities for a procedure C, given by the functions G_i, are fairly similar. Consequently, the G_i's will have jumps on approximately the same sets of values $\{\Gamma_i^b, b \in B\}$; this, combined with the previously described requirement that each C^b is characterized by a different degree of "conclusiveness", implies the additional requirement that $\Gamma_1^b \approx \Gamma_0^b \ \forall\, b \in B$. (We note here that the constraint $G_0(t) = G_1(t)$, $0 \le t \le 1$, does not necessarily imply $\Gamma_1^b = \Gamma_0^b \ \forall\, b \in B$, as is shown in Example 3.3(ii) below; nor does $\Gamma_1^b = \Gamma_0^b \ \forall\, b \in B$ imply $G_0(t) = G_1(t)$, $0 \le t \le 1$.)

Thus if a procedure $C = \{C_i^b\}$ is to treat f_0 and f_1 comparably, and also have the property that the sets C^b are each characterized by a different degree of conclusiveness, then it should satisfy, at least approximately, the constraints

$$\Gamma_0^b = \Gamma_1^b \ \forall\, b \in B \tag{3.3.1}$$

and

$$G_0(t) = G_1(t), \ 0 \le t \le 1; \tag{3.3.2}$$

also, for each i the values Γ_i^b should be appreciably different as b ranges over B, and the probability masses assigned to these values by G_i should be spread out; for, if the masses are concentrated near one value, one might as well use a $\underline{C}^1$ procedure. (If the Γ_i^b are all close to one, such differences will usually be reflected more appropriately in terms of the ratios $(1-\Gamma_i^b)/(1-\Gamma_i^{b'})$ rather than through the quantities $\Gamma_i^b-\Gamma_i^{b'}$.)

Many formal definitions of a measure of dispersion of G_i, which will reflect this last requirement, could be introduced here, but we shall not use such a precise formalism; instead, we indicate the desired flavor qualitatively in Examples 3.3(i), (ii), (iii) below. The question of what degree of departure from equality in (3.3.1) and (3.3.2) is still satisfactory will not be treated formally either, but the following comments are relevant in this connection.

The imposition of the constraint (3.3.1) is the $\underline{C}^L$ analogue of the conventional $\underline{C}^1$ restriction to confidence intervals with constant confidence coefficient. An obvious practical advantage of doing this in $\underline{C}^L$ as well as $\underline{C}^1$ is that it is simpler to state a single confidence level, than one which depends on the unknown true state of nature. This is to be coupled with the previously described intuitive appeal of having each C^b consist of subsets $\{C_i^b\}$ associated with statements (d_i,Γ_i^b) of comparable "conclusiveness".

For symmetric testing settings, partitions satisfying both (3.3.1) and (3.3.2) exactly are easily constructed in an obvious way, as illustrated in Example 3.3(i) below. However, when symmetry is absent it may be difficult, if not impossible, to obtain a partition C in $\underline{C}^{L+}$, for $L > 1$, which satisfies (3.3.1) and (3.3.2) exactly. In such cases, the considerations described above suggest that it will often be reasonable to choose (3.3.1) as the primary requirement to be satisfied; but the practical advantages of (3.3.1) must be weighed against other considerations, especially when asymmetry is extreme, as is illustrated in a "subadmissibility" anomaly in Example 3.5(i).

Our further discussion of the practical and intuitive considerations involved in the selection of a procedure requires the following definition. In the setting (3.0.3) (where λ is nondecreasing), for arbitrary B we say that a procedure C in $\underline{C}^+$ is a monotone partition if (i) each of the sets C_i^b is an interval, and (ii) C_1^b is to the right of $C_1^{b'} \Longleftrightarrow C_0^b$ is to the left of $C_0^{b'}$, for all b, b' in a subset B^* of B where the C_i^b are nonempty for $b \in B^*$ and $P_i\{Z(X) \in B^*\} = 1$ for both i. Automatically $B = B^*$ for $C \in \underline{C}^{L+}$. (While this definition can be extended to $\underline{C} - \underline{C}^+$, it is less natural and useful there. However, it is natural to extend the definition for $\underline{C}^+$ to the setting (3.0.1) - (3.0.2), even when one but not both of the sets $(0,a_0]$ and $[a_1,1)$ is empty.)

Monotone partitions seem particularly appropriate whenever the testing setting is a symmetric one. In such cases, the most natural procedure with continuum B (mentioned in Section 2) is perhaps that with $C^b = \{x: [1+\lambda(x)]^{-1} = b \text{ or } 1-b\}$ for all $b \in B$, where $B \subset [1/2,1]$. (Thus if $C^b = \{x,x'\}$ where $x \in C_1$, $x' \in C_0$, then $\lambda(x) = 1/\lambda(x') = (1-b)/b \geq 1$.) When symmetry is absent, the Γ of this procedure is no longer functionally related to λ, as one can verify with ease; nevertheless, it still seems intuitively natural to construct sets $C^b = C_1^b \cup C_0^b$ by pairing points in C_1 with relatively high (respectively, low) values of $\lambda(x)$ with points in C_0 with relatively high (respectively, low) values of $1/\lambda(x)$, which for increasing λ leads to the monotone structure.

Some other reasons for using procedures with monotone structure, phrased for simplicity in the restriction to $\underline{C}^2$, are:

(1) Such partitions can be justified in symmetric settings in terms of our desire that Γ^1 and Γ^2 be far apart. For example, suppose $f_0(x) = f_1(-x)$ and $\underline{X} = R^1$, and that $\lambda(x)$ is increasing in x.

Suppose we restrict consideration to "symmetric" procedures, those for which $C_0^b = -C_1^b$ a.e. We may as well consider only admissible procedures: $C_1 = \{x: x \geq 0\}$. The C_i^b need not be intervals. However, among all procedures that yield a prescribed value of $P_0\{C_0^1\}$ (where $\Gamma^1 \leq \Gamma^2$), it is easily seen that Γ^2 is largest, as is $\Gamma^2 - \Gamma^1$, for <u>intervals</u> C_0^2 and (to the right of it) C_0^1, which is monotone structure. Other extremal characterizations of monotone structure can be given similarly. (2) Assuming only that $\lambda(x)$ is increasing, suppose, in terms discussed in Section 3.2, that the experimenter wants also to think of a $\underline{C}^2$ conditional confidence procedure as a procedure for a four-decision problem with decisions d_{is} and d_{iw}. (Such a transformed usage of a procedure C into a procedure $\bar{C}$ for a setup different from that for which C was designed, is of course generally unjustifiable, in that either of C and $\bar{C}$ can be satisfactory while the other is not. Nevertheless, we go through this discussion to obtain some intuitive grasp of considerations that might govern a choice among admissible procedures in $\underline{C}^2$. For special circumstances where, in a different context, such a $\bar{C}$ is "good" if C is, see, e.g., Lehmann (1957).) It is not hard to show that, in the classical decision-theoretic framework, if the losses from making $d_{0s}, d_{0w}, d_{1w}, d_{1s}$ increase under f_0 and decrease under f_1, then every admissible procedure consists of a partition of $\underline{X}$ into four intervals in the same order as these four decisions, and where these respective decisions are made. (Slight additional conditions on the loss are needed if the two central intervals are not always to be empty.) Thus, only those of our conditional confidence procedures that have monotone structure can also be regarded as satisfactory for the classical 4-decision problem. (3) Similarly, suppose that f_0 and f_1 are thought of as members of a monotone likelihood ratio family, and that a secondary worry of the experimenter is that some member of this family other than f_0 or f_1 is the true law. (For the purpose of constructing a procedure, he does not reformulate the problem as one with more than two states in Ω.) In terms of reinterpreting a partition C of $\underline{C}^2$ as in (2) just above, the decisions d_{iw} may seem preferable if the true f is "between" f_0 and f_1, while d_{is} is preferable if the true f is "further out" than f_i. This again leads to procedures with monotone structure. We remark that Brown (1977) has recently obtained further theoretical justification for monotone procedures, in certain settings.

The previous paragraph is not intended as a "rigorous" justification for using monotone procedures. Indeed, the discussion of the three examples that follow describes an anomaly that is present to some extent for all pairs f_0, f_1, in that nonmonotone procedures are not easily eliminated unless one invokes the restriction (3.3.1) together with a criterion relating to the "spread" of the G_i. But the previous discussion may influence some practitioners in their choice of a partition.

It is easy to see that the restriction to monotone procedures deprives us of some operating characteristics that are otherwise achievable, and it will be illustrated in Example 3.5(iv) that this is also the case when the sample size is allowed to vary. Nevertheless, it seems often to be the case that a satisfactory procedure can be found among the monotone partitions.

The following examples are given to illustrate the main features of the above discussion of the considerations involved in choosing a partition which is satisfying to a practitioner. The examples are based on the symmetric setting of choosing between two normal densities f_0 and f_1 with common known variance and different known means θ_0, θ_1. In order to compute actual numerical values easily, let the means be one standard deviation apart with $\theta_0 < \theta_1$, and transform to achieve symmetry about zero; that is, $f_0(x-1/2) = \phi(x) = f_1(x+1/2)$. We shall consider for comparison three procedures contained in $\underline{C}^2$, all with underlying NP structure given by $C_0 = \{x: x < 0\}$, $C_1 = \{x: x \geq 0\}$. We have $\lambda(x) = f_1(x)/f_0(x) = e^{2x}$ strictly increasing in x and logarithmically symmetric about 0, that is, $\lambda(x) = 1/\lambda(-x)$ for all $x > 0$, and $P_0\{C_0\} = P_1\{C_1\} = .691$.

<u>Example 3.3(i). A monotone partition $\bar{\bar{C}}$ with symmetric pairing.</u> This $\bar{\bar{C}}$ is constructed in symmetric fashion with $\bar{\bar{C}}^1 = \{x: |x| < 1.25\}$ and $\bar{\bar{C}}^2 = \{x: |x| > 1.25\}$. This gives

$$\bar{\bar{\Gamma}}_0^1 = \frac{P_0\{\bar{\bar{C}}_0^1\}}{P_0\{\bar{\bar{C}}^1\}} = \frac{\Phi(.50)-\Phi(-.75)}{\Phi(1.75)-\Phi(-.75)} = .63 ,$$

$$\bar{\bar{\Gamma}}_0^2 = \frac{P_0\{\bar{\bar{C}}_0^2\}}{P_0\{\bar{\bar{C}}^2\}} = \frac{\Phi(-.75)}{\Phi(-.75)+1-\Phi(1.75)} = .85 , \tag{3.3.3}$$

and by symmetry $\bar{\bar{\Gamma}}_1^1 = \bar{\bar{\Gamma}}_0^1 = \bar{\bar{\Gamma}}^1$ and $\bar{\bar{\Gamma}}_1^2 = \bar{\bar{\Gamma}}_0^2 = \bar{\bar{\Gamma}}^2$. Also,

$$\bar{\bar{G}}_0(t) = \bar{\bar{G}}_1(t) = \begin{cases} .69, & 0 \le t < .63, \\ .23, & .63 \le t < .85, \\ 0, & .85 \le t \le 1 . \end{cases} \tag{3.3.4}$$

We note that, as described above, constructing the sets $\bar{\bar{C}}_i^b$ in a symmetric fashion, and pairing them in a symmetric way to obtain the sets $\bar{\bar{C}}^b$, gives a procedure $\bar{\bar{C}}$ such that (3.3.1) and (3.3.2) are satisfied. Also note that $\bar{\bar{\Gamma}}^1$ and $\bar{\bar{\Gamma}}^2$ are appreciably different, so that outcomes in the set $\bar{\bar{C}}^2$ can be considered markedly more conclusive than those in $\bar{\bar{C}}^1$.

<u>Example 3.3(ii). A partition $\bar{C}$ which is not montone, but which has sets $\bar{C}_i^b$ with interval structure.</u> This $\bar{C}$ is defined by

$$\bar{C}_0^1 = (-1.0,0),\ \bar{C}_1^1 = (1.0,\infty),$$
$$\bar{C}_0^2 = (-\infty,-1.0),\ \bar{C}_1^2 = (0,1.0).$$

This gives

$$\bar{\Gamma}_0^1 = \frac{P_0\{\bar{C}_0^1\}}{P_0\{\bar{C}^1\}} = \frac{\Phi(0.5)-\Phi(-0.5)}{\Phi(0.5)-\Phi(-0.5)+1-\Phi(1.5)} = .85 \, ,$$

$$\bar{\Gamma}_0^2 = \frac{P_0\{\bar{C}_0^2\}}{P_0\{\bar{C}^2\}} = \frac{\Phi(-0.5)}{\Phi(-0.5)+\Phi(1.5)-\Phi(0.5)} = .56 \, , \qquad (3.3.5)$$

and by symmetry $\bar{\Gamma}_1^1 = \bar{\Gamma}_0^2$ and $\bar{\Gamma}_1^2 = \bar{\Gamma}_0^1$. Note that the "unnatural" asymmetric (nonmonotone) pairing gives $\bar{\Gamma}_0^b$ very different from $\bar{\Gamma}_1^b$ for $b = 1,2$, but

$$\bar{G}_0(t) = \bar{G}_1(t) = \begin{cases} .69, & 0 \le t < .56, \\ .31 & .56 \le t < .85, \\ 0 & .85 \le t \le 1 \, . \end{cases} \qquad (3.3.6)$$

Thus (3.3.2) is satisfied exactly but (3.3.1) is not even approximately satisfied, so that neither of the sets $\bar{C}^1$, $\bar{C}^2$ can be described by a single measure of conclusiveness (even a qualitative one!) that refers to both states of nature.

<u>Example 3.3(iii). A partition $\hat{C}$ for which the sets $\hat{C}^b$ are constructed in a symmetric way, but the sets $\hat{C}_i^b$ do not have interval structure and thus are not monotone.</u> This $\hat{C}$ is defined by $\hat{C}_0^2 = (-1.1,-.6)$, $\hat{C}_1^2 = (.6,1.1)$, $\hat{C}_0^1 = (-\infty,-1.1) \cup (-.6,0)$, $\hat{C}_1^1 = [0,.6) \cup (1.1,\infty)$. Thus the components of $\hat{C}^2$ and $\hat{C}^1$ are symmetric about zero, but $\hat{C}_1^1$ and $\hat{C}_0^1$ do not have interval structure. This gives

$$\hat{\Gamma}^1 = .689, \; \hat{\Gamma}^2 = .697. \qquad (3.3.7)$$

Thus $\hat{C}$ satisfies (3.3.1) and (3.3.2), but $\hat{\Gamma}^1$ and $\hat{\Gamma}^2$ are so similar that the partition of $\underline{X}$ into $\hat{C}^1$ and $\hat{C}^2$ seems pointless, and it is difficult to justify use of $\hat{C}$ over the symmetric $\underline{C}^1$ NP partition.

Comparison of the procedures $\bar{C}$, $\bar{\bar{C}}$, $\hat{C}$ of Examples 3.3(i), (ii), (iii). Assuming that an experimenter wishes to use a procedure in $\underline{C}^2$, then from the viewpoint of G the procedures $\bar{C}$, $\bar{\bar{C}}$ both seem preferable to $\hat{C}$, whose defect has just been described. This preference for $\bar{\bar{C}}$ over $\hat{C}$ illustrates reason (1) for using monotone procedures, given earlier in this subsection.

Comparing $\bar{G}$ with $\bar{\bar{G}}$ shows that both the procedures $\bar{C}$, $\bar{\bar{C}}$ are characterized by a maximum Γ-value of .85, and this value is correctly attained with greater probability under $\bar{C}$; that is, $P_i\{\Gamma_i(X) = .85\}$ is .31 for $\bar{C}$ and .23 for $\bar{\bar{C}}$. Also, the difference between the minimum Γ-values of the two procedures $\bar{C}$, $\bar{\bar{C}}$ is not large, and $\bar{C}$, $\bar{\bar{C}}$ are trivially identical with respect to the criterion (3.1.5) of Section 3.1. Thus, in spite of the asymmetric and intuitively unappealing way in which $\bar{C}^1$ and $\bar{C}^2$ are constructed, comparison of the functions $\bar{G}$ and $\bar{\bar{G}}$ could lead to use of $\bar{C}$ rather than $\bar{\bar{C}}$. In fact, given any desired sufficiently large maximum Γ-value, there is an asymmetric analogue of $\bar{C}$ which will attain this value with greater probability than the symmetric $\bar{\bar{C}}$ analogue. This is because letting the unbounded sets $\bar{C}^1_1$, $\bar{C}^2_0$ move away from the origin, while maintaining the $\bar{C}$ type of structure, results in the maximum Γ-value γ^{max} (say) increasing toward 1 with $P_i\{\Gamma_i = \gamma^{max}, \delta = d_i\}$ approaching its maximum possible value for the given f_i and C_i, the value $P_i\{\delta = d_i\} = .691$; and at the same time the minimum Γ-value decreases toward 0, with the probability of attaining it also approaching 0. The limiting partition obtained in this way is the degenerate partition $\bar{C}$ of (3.1.2) of Section 3.1, which is optimal with respect to criterion (3.1.1) (without (3.1.5)). The discussion for several paragraphs following Theorem 3.2 of Section 3.1 thus becomes relevant, and examination of the probabilities in (3.1.3) may be worthwhile; we find that, as the sets $\bar{C}^1_1$ and $\bar{C}^2_0$ move out and γ^{max} and $P_i\{\Gamma_i = \gamma^{max}, \delta = d_i\}$ approach their upper limits as just described, so also does each of the probabilities of incorrectly stating (d_{1-i}, γ^{max}) (when X has density f_i), given by $P_i\{\Gamma_{1-i} = \gamma^{max}, \delta = d_{1-i}\}$, approach its maximum possible value for the given f_i and C_i, the value $P_i\{\delta = d_{1-i}\} = .309$. In the particular case of the partitions $\bar{\bar{C}}$, $\bar{C}$ of Examples 3.3(i), (ii), we obtain $P_i\{\Gamma_{1-i} = \gamma^{max}, \delta = d_{1-i}\} = .04$ for $\bar{\bar{C}}$ and $.24$ for $\bar{C}$. Thus, from this point of view $\bar{C}$ may no longer seem preferable to $\bar{\bar{C}}$. This, then, is a consideration that opts against asymmetric procedures like $\bar{C}$, additional to the previously emphasized rationale for making $\Gamma_0^b = \Gamma_1^b \; \forall \; b$ so as (1) to give the C^b's meaning as decision sets whose degree of strong or weak conclusiveness is independent of i, and (2) to achieve the practical convenience of being able to state (at least approximately) a conditional confidence coefficient that does not depend on the true state of nature.

3.4. Randomization

In Sections 3.1-3.3 we simplified our development by assuming the possible underlying df's were continuous. This permitted us to restrict consideration to nonrandomized procedures, and to achieve such aims as equality of Γ_0^b and Γ_1^b. We now discuss in more detail the role of randomization, and the reduction to the continuous case.

It is unnecessary to repeat in detail here the familiar practical objections to the use of randomized procedures; perhaps such objections are even more serious in the present framework, where not only the final decision d_i, but also the conditional confidence associated with it, would depend on the outcome of the randomization as well as upon the observation. Nevertheless, for technical reasons of the same sort as in classical decision theory, we must in general allow for randomization in the discrete case. Later in this section we give an example to illustrate the consequences of not permitting randomization; as in classical decision theory, the effect of restricting procedures to be nonrandomized in typical discrete examples is negligible for large sample sizes.

3.4.1. Reduction to Continuous Case

This reduction is familiar. We let $(I,\mathcal{L},\lambda)$ be the unit interval with Lebesgue sets and Lebesgue measure, and replace the original $(\underline{X},\underline{B},P_\omega)$ by its Cartesian product with $(I,\mathcal{L},\lambda)$. In other words, we replace the observed X by $X' = (X,U)$, where U is uniform on $(0,1)$ and independent of X. Then X' has atomless law and the reduction to (3.0.1)-(3.0.2) described in the second paragraph of Section 3 is easily accomplished for X'. (This applies to settings more general than that of two hypotheses; whenever there are densities with respect to a σ-finite ν, then $\nu \times \lambda$ is also σ-finite, and the atomless possible laws of X' have densities relative to $\nu \times \lambda$.)

3.4.2. Randomized Procedures

Next, whether or not we are in (or have reduced to) an atomless setting, we must define "randomized procedure". As in the fundamental classical decision-theoretic work of Wald and Wolfowitz (1951), we consider two methods of randomization:

Method I: We are given a family of nonrandomized partitions $\{{}^tC, t \in T\}$, where the functions ${}^t\delta(x)$ and ${}^tZ(x)$ are measurable on the product of $(\underline{X},\underline{B},\nu)$ and the probability space $(T,\underline{T},\tau)$; we select the nonrandomized ${}^tC$ which will subsequently be used, by performing an experiment, independent of X, according to the probability measure τ on T.

Method II: For a given real index set B (with its Borel sets), for each possible value x of X we are given a probability measure $\pi(\cdot|x)$ (measurable in x) on the space $B \times D$. This is used, when the observed outcome of the experiment is $X = x_0$, by conducting an experiment with possible outcomes in $B \times D$, according to the probability law $\pi(\cdot|x_0)$.

For the sake of simplicity, we once more limit consideration to finite partitions, in discussing these notions. Although it is easy to see that to any Method I procedure there corresponds a Method II procedure equivalent to it in the sense described just below, it might appear that Method II is more general than Method I; however, the results of Wald and Wolfowitz (1951) can be used to show the two methods are "equivalent" in what they can accomplish. This is because the interpretation of a point (b,d) in $B \times D$ in our setting, as a decision in the classical decision-theoretic setting of Wald and Wolfowitz (1951), leads to the following conclusion: For each Method II procedure π, there is a corresponding Method I T, τ, and $\{{}^tC, t \in T\}$, such that, for each b, d, x_0, the conditional (randomization) probability of making decision d, and working with conditioning label b, given that $X = x_0$, is the same for the two methods. Consequently, the quantity

$$P_\omega\{\text{make decision } d, \text{ achieve conditioning label } b\} \tag{3.4.1}$$

(which is $P_\omega\{C_d^b\}$ for a <u>non</u>randomized procedure) is the same for the two randomized procedures, for each ω, b, d. Since the quantities (3.4.1) determine the Γ_ω's and the operating characteristic $(G_\omega, P_\omega\{Q_\omega\})$ of a conditional confidence procedure, the two procedures are indeed equivalent from the (conditional on b) frequentist viewpoint. (The above conclusions also apply if we use the more detailed operating characteristic obtained by replacing $P_\omega\{Q_\omega\}$ by the functions of (3.1.3).)

In the interpretation of these randomized procedures, the conditioning is on the value of the (randomized) $Z(X)$, but not on the outcome of the randomization. For example, a Method I procedure in the setting (3.0.3) might have τ giving probability 1/2 to each of two partitions $\overline{C}$, $\overline{\overline{C}}$ with $\overline{C}_0 = \overline{\overline{C}}_0 = (0,1/2)$ and $B = \{1,2\}$, determined by

$$\overline{C}^1 = (1/5,3/5), \ \overline{\overline{C}}^1 = (2/5,3/5). \tag{3.4.2}$$

(We have replaced the left superscript t by overbars in this simple illustration, to reduce confusion.) We then compute, for example,

$$P_0\{\text{decision } d_0, \text{ conditioning value } b = 1\}$$

$$= \frac{1}{2} P_0\{\bar{C}_0^1\} + \frac{1}{2} P_0\{\bar{\bar{C}}_0^1\} = 1/5, \qquad (3.4.3)$$

and similarly $P_0\{d_1, b = 1\} = 1/10$. Suppose now that $Y = 3/10$. We flip a fair coin to determine whether to use $\bar{C}$ or $\bar{\bar{C}}$. For this value of Y, we then set $Z = 1$ if $\bar{C}$ is chosen but $Z = 2$ if $\bar{\bar{C}}$ is chosen. If $\bar{C}$ is chosen, we then condition not on the outcome $\bar{C}$ (in contrast with $\bar{\bar{C}}$) of the randomization, but rather on the value $b = 1$; thus, we use the probabilities $1/5$ and $1/10$ computed above, to make decision d_0 with $\Gamma = 2/3$. As indicated at the outset of Section 3.4, this interpretation of randomization will be unsatisfactory to many people (see, e.g., Cox (1958) for discussion), who will feel that, once $\bar{C}$ has been chosen, the appropriate interpretation of the outcome of the experiment should be based on $\bar{C}$ alone: the value of Γ would then be $P_0\{\bar{C}_0^1\}/P_0\{\bar{C}^1\} = 3/4$. We respond to this, firstly, with the reiteration that the interpretation we have given is not so much a practical recommendation as a technical consideration which is needed to discuss two methods of randomization; and, secondly, that the interpretation $\Gamma = 3/4$ is included in our framework in the following manner: Relabel the values b so that those for $\bar{C}$ are disjoint from those for $\bar{\bar{C}}$: for example,

$$\bar{C}^1 = (1/5, 3/5), \quad \bar{C}^2 = (0,1) - \bar{C}^1,$$

$$\bar{\bar{C}}^3 = (2/5, 3/5), \quad \bar{\bar{C}}^4 = (0,1) - \bar{\bar{C}}^3 . \qquad (3.4.4)$$

Then, if $Y = 3/10$ and $\bar{C}$ is chosen, the interpretation $\Gamma = 3/4$ is indeed appropriate. But the procedures (3.4.2) and (3.4.4) are, for our interpretation, not the same.

Note that a randomized procedure does not correspond to a fixed partition $\{C_d^b\}$ of the original $\underline{X}$. Instead of C_1^1 being fixed, in the case of (3.4.2) we can think of it as being a random set, being $(1/5, 1/2)$ or $(2/5, 1/2)$ with equal probabilities. The quantity (3.4.1) can be interpreted as $P_\omega\{C_d^b\}$ with this random notion of C_d^b.

3.4.3. Convexity; Elimination of Randomization

The role of randomization in classical decision theory is of course the generation of a convex set of risk functions; the risk functions of the nonrandomized procedures might not be a convex set. The larger collection of randomized procedures might consequently contain a procedure better in the minimax sense, etc.

The Method I randomization between two nonrandomized procedures $\bar{\delta}$ and $\bar{\bar{\delta}}$ in classical $\underline{C}^1$ decision theory, with probabilities ϵ and $1-\epsilon$, yields a randomized procedure whose risk function is linear in ϵ; this produces the convexity just noted. In our conditional confidence framework, the analogous mixing that makes C_ω and $P_\omega\{Q_\omega\}$ (or the functions of (3.1.3)) linear in ϵ is the following: Write $\bar{B}$ and $\bar{\bar{B}}$ for the sets of conditioning labels of $\bar{C}$ and $\bar{\bar{C}}$, respectively, considering $\bar{B}$ and $\bar{\bar{B}}$ to be disjoint. Then choose $\bar{C}$ and $\bar{\bar{C}}$ with respective probabilities ϵ and $1-\epsilon$, according to Method I. This means that, if we are given the two partitions $\bar{C}$ and $\bar{\bar{C}}$ of (3.4.2) which are presented with the same B, we relabel the B of one of the two procedures as in (3.4.4) to make the label sets disjoint, and then mix. In fact, it is not hard to see that mixing in terms of the original label set $\{1,2\}$, as in (3.4.3) when $\epsilon = 1/2$, produces an operating characteristic which is neither linear, nor concave, nor convex, in ϵ. The interpretation (3.4.3) is the appropriate one for the randomization (3.4.2), but the latter is not the mixing that yields linearity in ϵ of the operating characteristics obtained from $\bar{C}$ and $\bar{\bar{C}}$; that of (3.4.4) does yield linearity. An analogous phenomenon occurs for mixing more than two procedures.

Thus, although the set of vectors $\{P_i\{C_i^b\}, 1 \leq b \leq L, i = 0,1\}$ of procedures in $\underline{C}^L$ is convex, the nonlinearity of Γ_i^b in the $P_i\{C_i^b\}$ makes the set of operating characteristics of nonrandomized procedures $\underline{C}^L$ for a fixed L not convex. For, linearly mixing (over t) several ${}^{t}G_{\omega}$'s, each with no more than L jumps, can produce a G_{ω} with many more than L jumps, and that cannot come from a procedure in $\underline{C}^{L}$. (However, mixtures of procedures in $\underline{C}^{+}$ are easily seen to be in $\underline{C}^{+}$ on $\underline{X} \times I$.) Similarly, the randomized procedures (either method) with at most L elements in B, do not have a convex set of operating characteristics. This lack of convexity did not affect the admissibility-complete class conclusions of Theorem 3.2 because those conclusions were obtained by reduction to such results for the underlying NP partitions in the classical $\underline{C}^{1}$ hypothesis testing framework where, in the continuous case, the classical risk points (α,β) (<u>not</u> our G_{ω}!) of all partitions in $\underline{C}^{1}$ <u>are</u> convex.

Now suppose the laws P_{ω} are atomless, and that we are given a randomized procedure (either method) with associated B. It then follows from the work of Dvoretzky, Wald, and Wolfowitz (1951), upon interpretation of the pair (b,d) as a classical decision just as in Section 3.4.2, <u>that there is a nonrandomized procedure C with the same B and with $P_{\omega}\{C_{d}^{b}\}$ equal to the corresponding quantity (3.4.1) of the randomized procedure, for all</u> ω, b, d. Thus, for all of our Section 3 developments we need not consider randomized procedures, if we do not distinguish among procedures with the same operating characteristic. (These conclusions apply to much more general D, and to the use of other operating characteristics, such as the functions of (3.1.3).)

This is why, in Sections 3.1-3.3, we did not consider randomized procedures in the continuous case. It is instructive to relate the random variable U of Section 3.4.4, used there to produce the continuous case from the discrete, to the present randomization. Suppose, in terms of Method I in the discrete case (where randomization is needed to produce convexity), that $(T,\underline{T},\tau) = (I,\mathcal{L},\lambda)$; for any randomization scheme satisfying the mildest regularity assumptions, we can always reduce to this case by relabeling and transforming T. The randomized procedure amounts to a collection of partitions ${}^{t}C$ of $\underline{X}$, one for each t in I, the elements of each partition ${}^tC = \{{}^tC_d^b, b \in B, d \in D\}$ being indexed by the same $B \times D$. (Of course, as in (3.4.4), some of these sets ${}^tC_d^b$ may be empty.) We can then view this procedure, instead, as a nonrandomized partition C of the sample space $\underline{X}' = \underline{X} \times I$, where $C_d^b = \{(x,t): x \in {}^tC_d^b\}$. Thus, procedures on the sample space of the rv X' of Section 3.4.4 can be nonrandomized, since the role of randomization has been taken care of in the construction of X'.

All of the comments of this subsection have referred to our simple setting where there is no choice of experiment (for example, of the sample size, as considered in Example 3.5(v)). If one chooses at random according to a probability measure η on a space H of experiment labels, where the possible probability densities are ${}^ef_i$ on $\underline{X}$ if experiment e is chosen, then (under obvious measurability conditions) the $\underline{X}$ and $P_i\{J\}$ of the original simple setup are replaced here by $\underline{X} \times H$ and $\int_H P_i\{J(e)\}\eta(de)$, where $J(e)$ is the section $\{x: (x,e) \in J\}$ for $J \subset \underline{X} \times H$. This randomization by means of η is like that used with (3.4.4), and cannot be "eliminated". What can be eliminated is the use of a randomized decision-maker for each (x,e). The discussion between (3.4.3) and (3.4.4) now pertains to the much-argued question of whether it is reasonable to condition with respect to anything other than subsets of $J(\bar{e})$, using the ${}^{\bar{e}}P_i$ for computations, if the outcome of the random choice according to η is $\bar{e}$. The brief remarks of this paragraph can be extended to sequential experimentation and sample spaces that vary with e.

3.4.4. Computation of Randomized "Monotone" Procedures in Discrete Cases

We now illustrate, in a binomial example, the possibility of extending the notion of monotone partition to randomized procedures, by replacing $\underline{X} \times I$ by an interval.

Example 3.4. Let ν be counting measure on $\underline{X} = \{0,1,2,\dots,n\}$ and $f_i(x) = \binom{n}{x} p_i^x (1-p_i)^{n-x}$, $x \in \underline{X}$. Suppose $n = 25$, $p_0 = .4$,

$p_1 = .5$. Because of the discreteness and symmetry of the setting, we cannot hope that there is a monotone $\underline{C}^{2+}$ partition with $\Gamma_1^b = \Gamma_0^b$ for $b = 1,2$. In order to present the main ideas as briefly as possible, we shall not search for a nonrandomized procedure that comes as close as possible to satisfying specified goals; instead we choose a particular intuitively natural monotone $\underline{C}^{2+}$ partition and examine its properties. Let C be defined by $C_1 = \{12,13,\ldots,25\}$, $C^1 = \{9,10,\ldots,14\}$. The properties of this partition are summarized as follows:

$$\Gamma_0^1 = .663,\ P_0\{C_0^1\} = .459,\ \Gamma_1^1 = .603,\ P_1\{C_1^1\} = .443,$$

$$\Gamma_0^2 = .889,\ P_0\{C_0^2\} = .274,\ \Gamma_1^2 = .797,\ P_1\{C_1^2\} = .212. \qquad (3.4.5)$$

Thus, as expected, for each b the Γ_i^b's are not equal, but they are reasonably close as examplified by the fact that the differences $|\Gamma_1^b - \Gamma_0^b|$ for $b = 1,2$, are appreciably smaller than the differences $\Gamma_i^2 - \Gamma_i^1$, for $i = 0,1$.

We now apply a method described by Lehmann (1959, p.81) and used for interval estimation by Pratt (1961), to introduce a representation of the randomization scheme that replaces the $\underline{X} \times I$ of our discussion by a real interval $\tilde{\underline{X}}$ containing $\underline{X}$, and on which we can then obtain a monotone $\underline{C}^{2+}$ partition with $\Gamma_1^b = \Gamma_0^b$, $b = 1,2$. Let $\tilde{X} = X + U$ where U is independent of X and uniform on $(0,1)$ under both states of nature. The rv $\tilde{X}$ can be thought of as the sum of an integer-valued observation from the binomial experiment and a fractional component obtained independently using a device such as a spinner. Further, $\tilde{X}$ has a continuous Lebesgue density on the interval $\tilde{\underline{X}} = [0,26)$, with $\tilde{f}_1(\tilde{x})/\tilde{f}_0(\tilde{x})$ nondecreasing and equal to $f_1(x)/f_0(x)$ where x is the integer part of $\tilde{x}$. We can therefore obtain monotone $\tilde{\underline{C}}^{2+}$ partitions with respect to $\tilde{X}$ which satisfy $\Gamma_1^b = \Gamma_0^b$, $b = 1,2$. To make this partition comparable to that of the previous paragraph, we now construct the unique such partition $\tilde{C}$ which in addition has the property that the set $\tilde{C}_1^2$ is equivalent to the set $C_1^2 = \{15,16,\ldots,25\}$ of the partition of $\underline{X}$ above. That is, $\tilde{C}_1^2 = \{x: 15 \leq x < 26\}$, and we need $\tilde{C}_0^2$ such that $\tilde{\Gamma}_0^2 = \tilde{\Gamma}_1^2$, or

$$P_0\{\tilde{C}_1^2\}/P_0\{\tilde{C}_0^2\} = P_1\{\tilde{C}_0^2\}/P_1\{\tilde{C}_1^2\} = (\tilde{\Gamma}^2)^{-1} - 1. \qquad (3.4.6)$$

It is easily seen that $\tilde{C}_0^2 = [0,8+u_0^2)$ for some u_0^2, $0<u_0^2<1$. Substituting in (3.4.6) for $\tilde{C}_1^2$, $\tilde{C}_0^2$ gives the quadratic equation in u_0^2

$$\frac{P_0\{X\in\{15,16,\ldots,25\}\}}{P_0\{X\in\{0,1,\ldots,7\}\}+u_0^2P_0\{X=8\}} = \frac{P_1\{X\in\{0,1,\ldots,7\}\}+u_0^2P_1\{X=8\}}{P_1\{X\in\{15,16,\ldots,25\}\}} \qquad (3.4.$$

from which $u_0^2 = .431$ is obtained as the positive root. This gives $\tilde{\Gamma}^2 = .557$ and $P_0\{\tilde{C}_0^2\} = .205$.

Similarly $\tilde{\Gamma}_0^1 = \tilde{\Gamma}_1^1$ implies $P_0\{\tilde{C}_1^1\}/P_0\{\tilde{C}_0^1\} = P_1\{\tilde{C}_0^1\}/P_1\{\tilde{C}_1^1\} = (\tilde{\Gamma}^1)^{-1} - 1$, which gives $\tilde{C}_0^1 = [8.431, 11.717)$, $\tilde{C}_1^1 = [11.717,15)$, $\tilde{\Gamma}^1 = .639$, $P_0\{\tilde{C}_0^1\} = .456$, $P_1\{\tilde{C}_1^1\} = .480$.

In terms of the more traditional representation of randomization on the original $\underline{X}$, the procedure states

$(d_0,\tilde{\Gamma}^2)$ if $X \leq 7$, and states this with probability .431 if $X = 8$;

$(d_0,\tilde{\Gamma}^1)$ if $X = 9$ or 10, and states this with probability .569 if $X = 8$ and with probability .717 if $X = 11$;

$(d_1,\tilde{\Gamma}^1)$ if $12\leq X\leq 14$, and states this with probability .283 if $X = 11$;

$(d_1,\tilde{\Gamma}^2)$ if $X \geq 15$. (3.4.

Thus to use the procedure $\tilde{C}$, for outcomes of the binomial experiment corresponding to $X = 8$ or $X = 11$, an additional experiment must be performed to determine which of the statements, $(d_0,\tilde{\Gamma}^2)$ or $(d_0,\tilde{\Gamma}^1)$ in the former case, and $(d_0,\tilde{\Gamma}^1)$ or $(d_1,\tilde{\Gamma}^1)$ in the latter case, is made.

The partitions C and $\tilde{C}$ are not very different with respect to the operating characteristics G_i and $\tilde{G}_i$ (e.g., note that

$\Gamma_1^b < \bar{\Gamma}^b < \Gamma_0^b$, $b = 1,2$). The major difference between the two procedures is the equality of $\bar{\Gamma}_1^b$, $\bar{\Gamma}_0^b$ in $\tilde{C}$, which is made possible by the use of randomization. Many practitioners will not find it worth accepting the extra bother and the criticized qualities of randomized procedures, to achieve this equality exactly. Analogous examples with smaller or larger n will of course exhibit the "benefit" of randomization to a greater or lesser degree.

For the same C_1, the finest nonrandomized monotone partition, with two points in each C^b as far as possible, has conditioning sets $C^1 = \{11,12\}$, $C^2 = \{10,13\}, \ldots C^{10} = \{2,21\}$, $C^{11} = \{1,22\}$, $C^{12} = \{0,23,24,25\}$. The corresponding values of (Γ_0^b, Γ_1^b) are (.563,.538), (.680,.614),...,(.981,.977),(.982,.989), (.973,.997). The lack of monotonicity in b of Γ_0^b, which is not due to the lumping of three points in C_1^{12}, but rather to the asymmetry and discreteness, will be discussed in the next paragraph. If this anomaly seems too unattractive, it may be remedied by a further lumping, of C^{11} and C^{12}, with resulting conditional confidence (.981,.990). An analogue of the previous $\tilde{C}$ can be given, but it has little practical attraction, despite the differences between Γ_0^b and Γ_1^b in the nonrandomized partition.

To see the cause of the lack of monotonicity in Γ_0^b, suppose, more generally, that we are testing $p = p_0$ against $p = p_1$ in the Bernoulli case with sample size n, where $p_0 < p_1$. Suppose c is an integer $\leq n/2$ such that $C_0 = \{0,1,\ldots,c\}$ and $C_1 = \{c+1,\ldots,n\}$, and that the C^b are $C^1 = \{c,c+1\}$, $C^2 = \{c-1,c+2\},\ldots,C^c = \{1,2c\}$, $C^{c+1} = \{0,2c+1,\ldots,n\}$ as in the example. Since $(\Gamma_0^b)^{-1} - 1 = f_0(b+c)/f_0(c+1-b)$, we see that Γ_0^b is nondecreasing in b for $1 \leq b \leq c$ if and only if $f_0(b+c)/f_0(c+1-b) \geq f_0(b+c+1)/f_0(c-b)$ for $1 \leq b < c$, which is easily calculated to be equivalent to

$$(p_0^{-1}-1)^2 \frac{(c+1-b)(c+1+b)}{(n-c+b)(n-c-b)} \geq 1 . \qquad (3.4.9)$$

For fixed p_0 and p_1 it is clear that (3.4.9) is often violated for b large (near $c-1$) when n is large, provided $n-2c$ is also large; the latter will always occur, for example, if $p_1 \leq 1/2$, if $P_0\{C_0^b\}$ and $P_1\{C_1^b\}$ are approximately equal. Thus, this lack of monotonicity of Γ_0^b for the finest monotone partition in the discrete case is common under asymmetry for even the nicest distributions. (It also occurs for other common discrete models.) In the illustration, it barely failed to occur between Γ_0^{c-1} and Γ_0^c, but could not be avoided between Γ_0^c and Γ_0^{c+1}.

3.5. Construction of Partitions

The discussion of Section 3.3 was illustrated with examples from a symmetric testing setting, where procedures satisfying $\Gamma_1^b = \Gamma_0^b$ for all b in B were easily obtained by using symmetric pairs of sets (C_1^b, C_0^b). When symmetry is absent in the setting, the actual construction of procedures satisfying a set of restraints on the Γ_i^b or on the sets C_i^b may be difficult or impossible, and this subsection addresses this problem. Monotone partitions only are considered, for the reasons given in Section 3.3. (Because of the convexity of the set of vectors $\{P_i\{C_i^b\}, b\varepsilon B, i = 1,2\}$ for C in $\underline{C}$ or $\underline{C}^I$ mentioned in Section 3.4, which generally fails under the restriction to monotone partitions, the characterization of all achievable sets of $\{\Gamma_j^b\}$'s for <u>arbitrary</u> admissible partitions is in a sense simpler than that of (3.5.12)-(3.5.16) below; but we feel it is of less practical value.) This section is thus concerned with the construction of monotone partitions under the restriction $\Gamma_1^b = \Gamma_0^b$ for all b in B, in Section 3.5.1 when B is a continuum, and in 3.5.2 when B is finite; and in Section 3.6 other types of restrictions on the Γ_i^b and C_i^b, and a "mixed" type of B, are considered.

3.5.1. Continuum B.

As discussed in Section 3.3 (fifth paragraph following that of (3.3.2)), in a symmetric setting when B is a continuum, a monotone partition with $\Gamma_1^b = \Gamma_0^b \; \forall b$ is easily obtained by defining $C^b = \{x: [1+\Lambda(x)]^{-1} = b$ or $1-b\}$, giving $\Gamma^b = b$. We now derive a method for constructing a monotone partition with similar properties in asymmetric settings.

In the context of (3.0.3) we have $g_0(y) = 1$, $0 < y < 1$, and $g_1(y)$ positive and nondecreasing on $(0,1)$. We shall determine a point y^1 in $(0,1)$ and a strictly decreasing continuous function h on $[y^1,1]$ differentiable a.e., and with $h([y^1,1]) = [0,y^1]$, and such that

$$C_1 = [y^1,1), \; C_0 = (0,y^1),$$

$$Z(y) = Z(h(y)) = y \quad \text{on} \quad (y^1,1),$$

$$\Gamma_0^y = \Gamma_1^y = [1 + 1/|h'(y)|]^{-1}$$

$$= [1 + [g_1(h(y))/g_1(y)]^{1/2}]^{-1} \text{ a.e. on } (y^1,1). \quad (3.5.1)$$

Thus, C^y is the pair of points, $\{y, h(y)\}$ for $y^1 < y < 1$, while C^{y^1} (of measure zero) is $\{y^1\}$ and $Z(y^1) = y^1$. The intuitive interpretation of the Γ_i^y for $y^1 < y < 1$ is

$$\Gamma_1^y = P_1\{Y = y \,|\, Y = y \text{ or } h(y)\},$$

$$\Gamma_0^y = P_0\{Y = h(y) \,|\, Y = y \text{ or } h(y)\}. \quad (3.5.2)$$

It is simple to justify the last line of (3.5.1) rigorously in measure-theoretic terms (as holding a.e.), but it will be more instructive to present here only a heuristic derivation of the usual epsilontic variety.

Suppose a y^1 and h of the form stated above (3.5.1) are wanted; we shall see how they can be chosen to satisfy (3.5.1). Let $y > y^1$ be a point in C_1, and let $\varepsilon > 0$ be such that $(y-\varepsilon,y] \subset C_1$; then $h(y) < h(y-\varepsilon)$ and $[h(y),h(y-\varepsilon)) \subset C_0$. Define $C^y(\varepsilon) = \{[h(y),h(y-\varepsilon)) \cup (y-\varepsilon,y]\}$. Then if h is differentiable at y, and g_1 is continuous there, we have

$$P_1\{\delta(Y) = d_1 | Y \in C^y(\varepsilon)\} = P_1\{Y \in (y-\varepsilon,y] | Y \in C^y(\varepsilon)\}$$

$$\approx \frac{\varepsilon g_1(y)}{\varepsilon g_1(y)+\varepsilon g_1(h(y))|h'(y)|}$$

$$= \frac{1}{1+|h'(y)|\, g_1(h(y))/g_1(y)} , \qquad (3.5.3)$$

where the approximation sign denotes that the ratio of the two sides approaches one as $\varepsilon \searrow 0$. Similarly,

$$P_0\{\delta(Y)=d_0 | Y\in C^y(\varepsilon)\} = P_0\{Y\in[h(y),h(y-\varepsilon))| Y\in C^y(\varepsilon)\}$$

$$\approx \frac{\varepsilon|h'(y)|}{\varepsilon|h'(y)|+\varepsilon} = \frac{1}{1+1/|h'(y)|} . \qquad (3.5.4)$$

Note that $C^y(\varepsilon) \to C^y$ as $\varepsilon \searrow 0$. Thus, the conditional probabilities of (3.5.2) are computed as the limits of the left sides of (3.5.3) and (3.5.4), which are just the right sides. These right sides must be equated to yield the last line of (3.5.1). We obtain $[g_1(h(y))]^{1/2}d(h(y)) = -[g_1(y)]^{1/2}dy$, with the boundary condition $h(1) = 0$. Hence, h and y^1 are defined by

$$\int_0^{h(y)} [g_1(t)]^{1/2}dt = \int_y^1 [g_1(t)]^{1/2}dt \text{ for } y^1 < y < 1 ,$$

$$\int_0^{y^1} [g_1(t)]^{1/2}dt = \int_{y^1}^1 [g_1(t)]^{1/2}dt \qquad (3.5.5)$$

We make two brief remarks before turning to an example. Firstly, in the more general case given by (3.0.1)-(3.0.2) the corresponding treatment is obtained by replacing the limits 0,1 in (3.5.5) by a_0, a_1. This defines y^1 in (a_0,a_1) and $h: [y^1,a_1] \to [a_0,y^1]$. We then include an extra point $\{2\}$ (say) in the modified $B = [y^1,a_1) \cup \{2\}$. The last two lines of (3.5.1) now apply on (y^1,a_1), while $Z(y) = 2$ on $(0,a_0] \cup [a_1,1)$ and $\Gamma_i^2 = 1$.

Secondly, it is convenient to have the formula corresponding to (3.5.5) in terms of a pair of densities f_0, f_1 on $(-\infty, \infty)$ as they appear before transformation into (3.0.1)-(3.0.2). If $f_1(x)/f_0(x)$ is nondecreasing, and $(\bar{a}_0, \bar{a}_1)$ is the interval where this ratio is positive and finite, then it is easily seen that the function H on $[x^1, \bar{a}_1)$ corresponding to h on $[y^1, a_1)$ satisfies, in place of (3.5.5) and the last two lines of (3.5.1),

$$\int_{\bar{a}_0}^{H(x)} [f_0(t)f_1(t)]^{1/2} dt = \int_x^{\bar{a}_1} [f_0(t)f_1(t)]^{1/2} dt ,$$

$$\int_{\bar{a}_0}^{x^1} [f_0(t)f_1(t)]^{1/2} dt = \int_{x^1}^{\bar{a}_1} [f_0(t)f_1(t)]^{1/2} dt ,$$

$$\begin{aligned} \Gamma_0^x = \Gamma_1^x &= \{1+|H'(x)|\, f_1(H(x))/f_1(x)\}^{-1} \\ &= \{1+f_0(x)/f_0(H(x))\,|H'(X)|\}^{-1} . \end{aligned} \qquad (3.5.6)$$

The modifications in (3.5.3)-(3.5.4) are obvious. If f_1/f_0 is not monotone, the integrals of (3.5.6) are instead over unions of intervals.

<u>Example 3.5(i)</u>. Suppose $f_i(x) = \theta_i^{-1} e^{-x/\theta_i}$ on $\{x: 0 < x < \infty\}$, where $0 < \theta_0 < \theta_1$. This can be handled by direct reference to (3.5.6), but the transformed setting (3.0.3) will be a convenient one for comparison with other examples. Writing $c = \theta_0/\theta_1$, so that $0 < c < 1$, we obtain $g_1(y) = c(1-y)^{c-1}$ on $(0,1)$. Thus, (3.5.5) yields

$$\begin{aligned} y^1 &= 1 - 2^{-2/(c+1)} , \\ h(y) &= 1 - [1 - (1-y)^{(c+1)/2}]^{2/(c+1)} , \\ \Gamma^y &= \{1 + [(1-y)^{-(c+1)/2} - 1]^{(c-1)/(c+1)}\}^{-1} . \end{aligned} \qquad (3.5.7)$$

It is of interest to look at the structure of the partitions determined by h, y^1 for extreme values of c. As c approaches

1, the density g_1 approaches g_0 and the degree of asymmetry decreases. The NP partition determined by the restriction $\Gamma_0^b = \Gamma_1^b \; \forall b$, approaches the symmetric one (relative to g_0) given by $y^1 = 1/2$. On the other hand, as $c \to 0$ and asymmetry becomes extreme, the underlying NP partition determined by h, y^1 approaches that given by $y^1 = 3/4$. For example, $c = .01$ gives $y^1 = .7465$, with $P_0\{C_0\} = .7465$, and $P_1\{C_1\} = (1-y^1)^c = .9864$. However, if the cut-off point of the underlying NP partition is increased beyond this y^1, a substantial increase in $P_0\{C_0\}$ can be achieved at the expense of a small decrease in $P_1\{C_1\}$. For example, changing C_1 from $(y^1,1)$ to $(.8,1)$ increases $P_0\{C_0\}$ by .0535 while only decreasing $P_1\{C_1\}$ by .0024; and, even in terms of the much less complimentary criterion of relative change in $1-P_i\{C_i\}$, the increase for $i = 0$ is about the same as the decrease for $i = 1$. A choice of a new function h^* (say) can be made for the altered partition so that, for $C^y = \{y, h^*(y)\}$, we still have Γ_0^y and Γ_1^y reasonably close. This example with $c = .01$ (or $\theta_1 = 100\,\theta_0$) is an extreme one, but it serves to illustrate the fact, mentioned elsewhere, that "subadmissibility" anomalies can arise if one blindly insists on the restriction $\Gamma_1^b = \Gamma_0^b \; \forall b \in B$; it is easy to give arithmetically messier examples in which such anomalies are much more striking. The construction of examples of the present type imitates the well known subadmissibility phenomenon for NP $\underline{C}^1$ partitions.

It is of interest to note that the set of C^y's on which the ratio of $\lambda(y)$ to $1/\lambda(h(y))$ is large (respectively, small) has P_1-probability (respectively, P_0-probability) close to one when c is small. Thus, points of quite different likelihood ratio are lumped together in this asymmetric setting. This phenomenon is also present for $\underline{C}^L$ partitions.

Example 3.5(ii). We now illustrate the construction of h in the context of (3.0.2), as described in the paragraph just below (3.5.5). Suppose

$$g_0(y) = \begin{cases} a^{-1}, & 0 < y < a_0 , \\ \quad , & a_0 \le y < 1 ; \end{cases}$$

$$g_1(y) = c(1-y)^{c-1}, \; 0 < y < 1 , \tag{3.5.8}$$

where again $0 < c < 1$. The calculation described below (3.5.5) then yields $h(y) = 1$ and $\Gamma_1^y = 1$ on $[a_0, 1)$; and, corresponding to (3.5.7),

$$y^1 = 1 - \{[1+(1-a_0)^{(c+1)/2}]/2\}^{2/(c+1)} ,$$

$$h(y) = 1 - [1+(1-a_0)^{(c+1)/2} - (1-y)^{(c+1)/2}]^{2/(c+1)} ,$$

$$y_1 < y < a_0 .$$

$$\Gamma^y = \{1+[(1+(1-a_0)^{(c+1)/2})(1-y)^{-(c+1)/2} - 1]^{(c-1)/(c+1)}\}^{-1} ,$$

$$y_1 < y < a_0 . \tag{3.5.9}$$

3.5.2. Finite B

For specified L and g_1 (in the setting (3.0.3)), the explicit analytic representation of the set of achievable vectors $\{P_i\{C_i^b\};\ i = 0,1;\ 1 \le b \le L\}$ for admissible C in $\underline{C}^{L+}$, mentioned in Section 3.3, is not always easy, as we know from classical decision theory. This is also true for the corresponding vectors $\{\Gamma_i^b\}$, and, with increasing difficulty as we add restrictions, for the $\{\Gamma_i^b\}$ of admissible partitions, admissible monotone partitions, and admissible monotone partitions with $\Gamma_0^b = \Gamma_1^b \forall b$. Nevertheless, since the last of these will often be the set of greatest practical use, we shall now indicate how it can be delimited. This is a many-stage description, and in practice the most useful aspect is the explicit description of the one-dimensional coordinate projections, the range of Γ^b (subject to $\Gamma_0^b = \Gamma_1^b$) for each fixed b, given in (3.5.17) and whose use is decribed there.

We now state two general results which are used in the following construction and which apply to the model (3.0.3) (or, appropriately modified, to (3.0.1)-(3.0.2)), regardless of the cardinality of B. (i) If $\Gamma_1^b = \Gamma_0^b \ \forall b \in B$, then monotonicity of the sets C_i^b implies monotonicity of the corresponding Γ^b's. That is, C_0^b to the left of $C_0^{b'}$ and C_1^b to the right of $C_1^{b'}$ implies $\Gamma^b \geq \Gamma^{b'}$. (ii) Using calculations of Section 3.5.1 it is easily shown that

$$\frac{1}{2} \leq \inf_{b \in B} \Gamma^b \leq \sup_{b \in B} \Gamma^b \leq \{1+[g_1(0)/g_1(1)]^{1/2}\}^{-1} \leq 1 \qquad (3.5.10)$$

for every monotone partition with $\Gamma_1^b = \Gamma_0^b = \Gamma^b \ \forall b \in B$; here $g_1(0)$ and $g_1(1)$ are taken as limits of $g_1(y)$ as $y \downarrow 0$ and $y \uparrow 1$.

We also require some additional definitions. Suppose $0 \leq r_0 < r_1 < 1$. Then there is clearly a unique value $s=s(r_0,r_1)$ in (r_0,r_1) such that

$$\int_{r_0}^{s} g_1(y)dy / \int_{s}^{r_1} g_1(y)dy = (r_1-s)/(s-r_0) \quad . \qquad (3.5.11)$$

The two sides of (3.5.11) are simply the values of $(\Gamma_1)^{-1}-1$ and $(\Gamma_0)^{-1}-1$ associated with the pair of sets (r_0,s), $[s,r_1)$. We denote by $\gamma^*(r_0,r_1) = [s(r_0,r_1)-r_0]/(r_1-r_0)$ the associated common value of Γ_0 and Γ_1 when (3.5.11) is satisfied. We also write $\gamma^{**}(r_0,r_1) = \{1+[g_1(r_0)/g_1(r_1)]^{1/2}\}^{-1}$.

Thus, the $\underline{C}^1$ partition with $\Gamma_0 = \Gamma_1$ has $\Gamma=\gamma^*(0,1)= s(0,1)$, the complement of the minimax error probability of NP theory.

We now assume the number of elements $L(\geq 2)$ in $B=\{1,2,\ldots,L$ is specified, and in the context (3.0.3) we restrict consideration to monotone partitions in $\underline{C}^{L+}$ with $\Gamma_0^b = \Gamma_1^b \ \forall b$. We label the C^b so that $\Gamma^1 \leq \Gamma^2 \leq \ldots \leq \Gamma^L$. To avoid space-consuming discussion about when equality can hold in (3.5.14) and (3.5.17), we hereafter assume g_1 continuous and strictly increasing.

Suppose first that a value γ^L is specified, and that we seek a procedure with $\Gamma^L = \gamma^L$. The discussions of (3.5.10) and (3.5.11) show that we must have

$$\gamma*(0,1) < \gamma^L < \gamma**(0,1) \tag{3.5.12}$$

for such a procedure to exist. If (3.5.12) is satisfied C^L is uniquely determined as the pair of sets $(0,y_0^L)$, $[y_1^L,1)$ satisfying

$$\int_0^{y_0^L} g_1(y)dy / \int_{y_1^L}^1 g_1(y)dy = (1-y_1^L)/y_0^L = (\gamma^L)^{-1} - 1 . \tag{3.5.13}$$

It is now clear how to continue. Suppose values $\gamma^1 < \gamma^2 < \ldots < \gamma^L$ are specified and that we have already determined $C^{L-j} = [y_0^{L+1-j}, y_0^{L-j}) \cup [y_1^{L-j}, y_1^{L+1-j})$ (= the previous C^L for $j = 0$) for $0 \leq j \leq J$, where $J < L-1$. Then, as in (3.5.12), a C^{L-J-1} of the form just indicated exists if and only if

$$\gamma*(y_0^{L-J}, y_1^{L-J}) < \gamma^{L-J-1} < \gamma**(y_0^{L-J}, y_1^{L-J}), \tag{3.5.14}$$

and is given by

$$\int_{y_0^{L-j}}^{y_0^{L-J-1}} g_1(y)dy / \int_{y_1^{L-J-1}}^{y_1^{L-J}} g_1(y)dy = (y_1^{L-J} - y_1^{L-J-1})/(y_0^{L-J-1} - y_0^{L-J})$$

$$= (\gamma^{L-J-1})^{-1} - 1. \tag{3.5.15}$$

This procedure continues until $J = L-2$. At that last stage, there is no choice for γ^1, which must satisfy

$$\gamma^1 = \gamma*(y_0^2, y_1^2) \quad , \tag{3.5.16}$$

with $C^1 = [y_0^2, y^1) \cup [y^1, y_1^2)$ given again by (3.5.15) with $y_0^1 = y_1^1 = y^1$. Thus, y^1 is the dividing point of the underlying NP partition $\{C_0, C_1\}$.

Note that we construct $C^L, C^{L-1}, \ldots, C^1$, in that order. One can not hope to guess the appropriate value y^1 and work in the opposite order.

The relations (3.5.12)-(3.5.16) characterize the achievable set $\underline{G}^L$ of vectors $\gamma = (\gamma^1, \gamma^2, \ldots, \gamma^L)$ for monotone partitions in $\underline{C}^L$ with $\Gamma_0^b = \Gamma_1^b = \gamma^b \ \forall b$, but are not very explicit. On the other hand, the projections of $\underline{G}^L$ onto the coordinate axes are simple:

$$\gamma^*(0,1) < \gamma^L < \gamma^{**}(0,1) ;$$
$$1/2 < \gamma^b \leq \gamma^{**}(0,1), \quad 1 < b < L;$$
$$1/2 < \gamma^1 < \gamma^*(0,1) . \qquad (3.5.17)$$

Thus, a reasonable practical approach will often be to start by looking at (3.5.17) and select for γ^1 or γ^L a satisfactory value in the appropriate interval (for example, the smallest acceptable value of γ^1, or a value of γ^L large enough to be indicative of psychological, if not mathematical certainty); the remaining Γ^b may be determined without satisfying precisely assigned values γ^b, as long as the spread of Γ^b's seems satisfactory. In particular, the spread of the pair (Γ^1, Γ^L) seems most important, and its range is the set given by (3.5.12)-(3.5.16) with $L = 2$.

On the other hand, the Γ^b's are not the whole story, and it may be desirable to add to the consideration of the value of a selected Γ^b the examination of associated contributions to the G_i ; that is, of

$$P_0\{C_0^b\} = y_0^b - y_0^{b+1}, \quad P_1\{C_1^b\} = \int_{y_1^b}^{y_1^{b+1}} g_1(y)dy , \qquad (3.5.18)$$

where $(y_0^{L+1}, y_1^{L+1}) = (0,1)$.

This determination of a satisfactory Γ^b and $P_i\{C_i^b\}$'s is most easily implemented for $b = L$, because of the order in which the C^b's must be constructed. Thus, if Γ^1 is specified, it is necessary to work through the whole construction procedure starting with C^L and ending with C^1, and then to modify it if the resulting C^1 is not satisfactory. One possible aid in this case is to use the range of (Γ^1, Γ^L) mentioned just above as a guide to the choice of C^L if $L > 2$. Another possibility is to begin by selecting a C^1 near the center of $(0,1)$ with satisfactory properties, and then to fill out the remaining C^L's; of course, exact equality of Γ_0^b and Γ_1^b for $b > 1$ must then be sacrificed, but one or two shifts of C^1 from the initial trial should usually yield Γ_i^b's that are nearly enough equal, at least for small L. In symmetric settings there is of course no problem: y^1 can be chosen as the center of symmetry, and the C^b's can be chosen in a variety of orders, such as $C^1, C^L, C^2, C^{L-1}, \ldots$.

We omit here the straightforward representation of (3.5.12)-(3.5.16) in terms of original f_i's, analogous to (3.5.6).

Example 3.5(iii). Suppose g_0 and g_1 are as in Example 3.5(i), with $c = .3$. We shall construct a partition in $\underline{C}^{2+}$ with $\Gamma_1^b = \Gamma_0^b$, $b = 1,2$, and satisfying $\Gamma^2 = .9$. We could try to solve (3.5.13) at once, but it is instructive to check (3.5.12) first. Since $g_1(0) = .3$ and $g_1(1) = +\infty$, we obtain $\gamma^{**}(0,1)=1$. Also, from (3.5.11) we obtain $\gamma^*(0,1) = s(0,1) = .698$, so that (3.5.12) is satisfied for $\gamma^2 = .9$. The relation (3.5.13) is

$$\frac{1-(1-y_0^2)^{.3}}{(1-y_1^2)^{.3}} = 1/9 = (1-y_1^2)/y_0^2 \, . \qquad (3.5.19)$$

Substituting for y_1^2 from the second of these equations into the first gives $y_0^2 = .090$, $y_1^2 = .990$, and $P_0\{C_0^2\} = .090$, $P_1\{C_1^2\} = .251$. From (3.5.15) with $J = 0$ we then obtain

$$\frac{(1-.090)^{.3} - (1-y^1)^{.3}}{(1-y^1)^{.3} - (.010)^{.3}} = \frac{.990-y^1}{y^1-.090} \, , \qquad (3.5.20)$$

from which $y^1 = .671$, $P_0\{C_0^1\} = .581$, $P_1\{C_1^1\} = .465$; and, from (3.5.16), $\Gamma^1 = .645$.

Example 3.5(iv). Choice of sample size. We now present some typical computations for the realistic situation in which, as in NP $\underline{C}^1$ settings, requirements are stated on the operating characteristic and we seek a procedure that meets those requirements and which minimizes the sample size. (Alternatively, the costs from sampling and making incorrect decisions can be combined.) Suppose $X_1', X_2', \ldots$ are independent $\underline{N}(\theta,1)$ with mean θ, equal to either $\theta_0 = -.2$ or $\theta_1 = .2$. We shall consider $\underline{C}^{2+}$ procedures which are symmetric and monotone in $X_n = n^{-1/2}\Sigma_1^n X_i'$. Thus, $C_1 = \{X_n \geq 0\}$, $C_0 = \{X_n < 0\}$, and $C^1 = \{|X_n| < c_n\}$. Writing $u_n = .2n^{1/2}$, we obtain

$$\rho^1 = (\Gamma^1)^{-1} - 1 = [\Phi(c_n+u_n)-\Phi(u_n)]/[\Phi(c_n-u_n)-\Phi(-u_n)] ,$$

$$\rho^2 = (\Gamma^2)^{-1} - 1 = [1-\Phi(c_n+u_n)]/[1-\Phi(c_n-u_n)] . \qquad (3.5.21)$$

We now illustrate a few of the types of requirements that can be imposed.

(A) <u>Specification of Γ^1 and Γ^2</u>. In discussing the question of existence of procedures satisfying particular specifications, it is convenient to dispense with the difficulty caused by discreteness of n. This means either thinking of $n^{1/2}X_n$ as a Brownian motion with law $\underline{N}(\theta_i n,n)$ at time n; or to phrase requirements in terms of inequalities or approximate specifications that are more readily satisfied; or to allow randomization among sample sizes (Section 3.4). It is simplest theoretically to think of the first of these in observing that it is not true, as one might think it is at first glance, that for any pair of values (γ^1,γ^2) with $1/2 < \gamma^1 < \gamma^2 < 1$ there is a pair (u_n,c_n) (and hence an (n,c_n)) for which $\Gamma^i = \gamma^i$ for both i. It is interesting that if we allow <u>nonmonotone</u> symmetric procedures, such a (u_n,c_n) <u>does</u> always exist. Roughly speaking, for a monotone procedure Γ^1 and Γ^2 cannot be too close together (how close depending on the value of Γ^1) because C^2 consists of larger values of λ or $1/\lambda$ than does C^1. The exact description of the achievable (γ^1,γ^2), superficially the union over n of the sets characterized by (3.5.12)-(3.5.16) for $L = 2$, will be discussed elsewhere for this and more general settings. The fact that a given (γ^1, γ^2) cannot be achieved exactly is not of much practical concern, since (for example) $\Gamma^1 = \gamma^1$, $\Gamma^2 \geq \gamma^2$ <u>is</u> always achievable.

For an illustration of the determination of n, suppose $\gamma^1 = .8$, $\gamma^2 = .99$, so that we want to achieve $\rho^1 = .25$, $\rho^2 = 1/99$. Using (3.5.21), we proceed by trial and error: pick a value of u_n, find c_n to satisfy $\rho^1 = .25$, compute ρ^2, alter u_n accordingly, etc. We obtain $u_n = 1.42$, $c_n = 1.07$, and $n \approx 51$. Before deciding that this procedure is indeed satisfactory, the practitioner should look at other aspects of the operating characteristic; for, if $P_1\{C_1^2\}$ were very small, the fact that $\Gamma^2 = .99$ might not alone be satisfactory. (In the present example, $P_1\{\Gamma^2 = .99, \delta = d_1\} = P_1\{C_1^2\} = .64$, while $P_1\{C_1^1\} = .29$.) The requirements of the problem could not be stated as equalities to be satisfied by $\Gamma^1, \Gamma^2, P_1\{C_1^2\}$, since these would almost always be contradictory; but a set of appropriate inequalities could obviously be used. Instead of doing this, we consider now satisfying equalities for another pair of the quantities of interest.

<u>(B) Specification of Γ^2 and $P_1\{C_1^2\}$</u>. If values γ^2 and p^2 are given, with $1/2 < \gamma^2 < 1$ and $0 < p^2 < 1$, the specification $\Gamma^2 = \gamma^2$, $P_1\{C_1^2\} = p^2$ becomes (from (3.5.21))

$$1 - \Phi(c_n - u_n) = p^2 ,$$

$$1 - \Phi(c_n + u_n) = p^2[(\gamma^2)^{-1} - 1] , \qquad (3.5.22)$$

which can be solved at once for $c_n \pm u_n$ in terms of Φ^{-1}. For example, if $\gamma^2 = .99$ and $p^2 = .75$, we obtain $c_n = .88$, $u_n = 1.55$, $n \approx 61$. The rest of the operating characteristic would then be checked to make sure the procedure is satisfactory:

$\Gamma^1 = .78$, $P_1\{C_1^1\} = .19$. One might also compare this with the NP $\underline{C}^1$ procedure for the same n, which has $\Gamma = P_1\{C_1\} = .94$.

(C) Specifying $P_1\{C^1\}$. This is an obvious possible beginning. It determines n, and then one can examine the consequences of various choices of c_n.

We reiterate that it is important to keep in mind that not only the Γ^b's, but also their contributions to G, must be examined in choosing a procedure. Thus, for any fixed n, both Γ^b's increase as c_n does; but this does not mean that larger c_n's are automatically "better", since the associated $P_1\{C_1^2\}$ decreases toward 0 as $c_n \to \infty$.

3.6. Modifications of Construction Criteria

We now remark briefly on several possible modifications of the approach of Section 3.5. For simplicity, our discussion will be in terms of the model (3.0.3).

(A) One may decide not simply to construct a procedure in some $\underline{C}^L$ or with continuum B and with Γ_0^b and Γ_1^b equal (or nearly so) for all b, but instead to begin by imposing some other restriction on the procedure to be used. For example, one might fix the sample size and $P_0\{C_0\}$ or $P_0\{C_0\}/P_1\{C_1\}$ or $[1-P_0\{C_0\}]/[1-P_1\{C_1\}]$; or one might set lower bounds on $P_0\{C_0\}$ and $P_1\{C_1\}$ and choose the smallest possible sample size. In any such case, one will usually lose the possibility of the Γ_i being approximately equal over much of $\underline{X}$, and the methods of Section 3.5 must be modified as discussed in (C) below. Lehmann (1958) chooses $(1-\Gamma_0^b)/(1-\Gamma_1^b)$ to be a specified constant.

(B) The impossibility of making the Γ_i approximately equal can also arise because of restrictions on the method of construction For example, an experimenter's intuition may make him insist on separating C_0 from C_1 at $\lambda(x) = 1$. If the setting is asymmetric ($\lambda(x)$ under f_1 has a different law from that of $1/\lambda(x)$ under f_0), he will again usually find it impossible to make Γ_0^b and Γ_1^b equal for all b.

(C) If, for reasons such as those of (A) or (B), one cannot obtain a procedure for which Γ_0^b and Γ_1^b are almost equal for all b, there is a variety of possible constructions of a procedure. If one should insist on a formal criterion, here are two possibilities: (1) Make Γ_0^b and Γ_1^b as close to each other as possible in some definite sense such as $\Gamma_0^b = c_b\Gamma_1^b$ (or perhaps $1-\Gamma_0^b = c_b(1-\Gamma_1^b)$), with the set of constants c_b chosen as close to 1 as possible in some sense. (2) Starting at either the middle or ends of (0,1) depending on whether equality of Γ_i's seems more important for large or small values, construct C_i^b's to give equality over a large extent of (0,1). On the remainder, choose the C_i^b's in some manner such as (1). This last is vacuous if we choose the C_i^b's to give equality of Γ_i's over a maximal set; we are then left with a subset $C_i^{b^*}$ of only one C_i, where the corresponding decision d_i is always made and $\Gamma_i = 1$ while $\Gamma_{1-i} = 0$ (and whose disadvantages have been discussed earlier).

(D) It may be attractive to use a B which is a "mixture" of the discrete and continuous versions. While the continuum B of Section 3.5 has the appeal of yielding a conditional confidence $\Gamma^{Z(x)}$ that varies strictly with x in C_1 if $\lambda(x)$ does, instead of lumping x's with disparate $\lambda(x)$ values into the same C_1^b, it has a feature that may distress some (not all!) practitioners, that $\Gamma^{Z(x)}$ is near 1/2 near the division between C_0 and C_1 if λ is continuous there. A modification that avoids this, while still allowing a continuum of large values of Γ, is to let $B = \{0\} \cup \{x: x \geq x_0\}$, where $h(x_0) < x_0$ in the continuum B construction of Section 3.5. Then C^0 is the interval $(h(x_0), x_0)$ and $C^x = \{h(x), x\}$ for $x \geq x_0$ with Γ^x as in Section 3.5. The point x_0 can be chosen to make $\Gamma_0^0 = \Gamma_1^0$

sufficiently greater than 1/2 to be acceptable. For example, in the setting of Example 3.5(i) with $c = .1$, it can be seen that the choice $C_0^0 = (.424, .717)$, $C_1^0 = [.717, .912)$ yields $\Gamma^0 = .6$ and $\Gamma^y \geq .7$ for $y \geq .912$.

(E) Suppose f_0 is regarded in a special light compared with f_1, as in that traditional testing approach where f_0 represented an old theory or no difference between two populations, and/or in those applications of the NP $\underline{C}^1$ theory in which α is fixed at some traditional value. An analogue for our conditional confidence approach might be to ask that Γ_0^b be set at some preassigned value, possibly depending on b. (One might also try to specify $P_0\{C_0\}$, but this might conflict with the previous restriction.) An extreme case is that in which $\Gamma_0^{Z(x)}$ is specified to be the same preassigned value $1-\alpha_0$ (say) for almost all x. It is clear from (3.1.8) that $P_0\{C_0\} = 1-\alpha_0$. (Here G_0 is like the G_0 for the NP $\underline{C}^1$ setup described below (3.1.1).) A monotone partition that achieves this in our model (3.0.3) is $B = \{y: 1-\alpha_0 \leq y < 1\}$ with $C^y = \{y, \alpha_0^{-1}(1-\alpha_0)(1-y)\}$ for $y > 1-\alpha_0$. We obtain, for $1-\alpha_0 < y < 1$,

$$\Gamma_1^y = \{1+(1-\alpha_0)g_1(\alpha_0^{-1}(1-\alpha_0)(1-y))/\alpha_0 g_1(y)\}^{-1}. \qquad (3.6.$$

One can similarly construct a procedure in $\underline{C}^L$ with corresponding features, or one in the spirit of (D) just above.

An approach resembling (E) may seem appropriate in certain composite hypothesis settings in which one hypothesis is to receive special treatment.

(F) Another modification is that of Brown (1977) of considering the law of Γ_ω on all of $\underline{X}$ rather than only on D_ω as in (2.2). Although the latter reflects our feeling that it is not much comfort to know that Γ_ω is large on the complement of D_ω, Brown gives instances where his formulation is useful.

(G) Although conditional confidence, being associated with probabilities of making correct decisions, is the simplest and intuitively most natural frequentist measure of conclusiveness (just as it is for $\underline{C}^1$ confidence intervals), from a decision-theoretic viewpoint it implies a zero-one loss structure which does not always reflect accurately the relative seriousness of the consequences of various possible decisions, any more than it does in classical $\underline{C}^1$ settings. Much of our development requires only minor changes if a loss function W on $\Omega \times D$ is considered in place of the indicator of $\{d \notin D_\omega\}$ which is the loss function of Section 2; we then replace $1-\Gamma_\omega$ by the conditional risk

$$R_\omega = E\{W(\omega, \delta(X)) | \underline{B}_0\} \; . \tag{3.6.2}$$

The frequentist interpretation in terms of the law of large numbers is analogous to what is was for simple loss. With obvious modifications of the other definitions of Section 2, results corresponding to those of Theorems 3.1 and 3.2 are trivial modifications in the setting of Section 3. Modified loss structure in other settings is discussed in Kiefer (1975, 1976, 1977).

A more complicated structure than that of (3.6.2) would allow W to depend on Γ_ω. Except in the trivial case in which this dependence is linear, the characterization of admissible procedures then appears to be much more difficult. It seems quite reasonable that, for Γ_ω near 1, the scientist's utility with the conclusiveness of the experiment may often be less satisfactorily measured by Γ_ω than by some function such as $-\log(1-\Gamma_\omega)$.

Other modifications of our approach will undoubtedly occur to the reader, and the possibilities multiply when one considers more complex Ω and D.

It will be clear how to alter the calculations of Section 3.5 (in particular, of Example 3.5(iv)) to obtain procedures that satisfy many of these modified criteria. We illustrate only (E) here.

Example 3.6. In the setup of Example 3.5(i), we compute Γ^y from (3.6.1), and for the sake of comparisons that follow we write its complement,

$$1-\Gamma_1^y = \{1+(1-\alpha_0)^{-1}\alpha_0[(1-y)^{-1}-\alpha_0^{-1}(1-\alpha_0)]^{1-c}\}^{-1}. \qquad (3.6.3)$$

Of course, for each α and c this is sometimes (in y) less and sometimes greater than the type II error probability $1-\bar{\Gamma}_1$ of the underlying $\underline{C}^1$ NP procedure $\bar{C}$ (with the same type I error α_0). It is interesting to compare $1-\Gamma_1^y$ with two other quantities that have appeared in the literature. One, related to the likelihood principle, is the "intrinsic significance level" $\alpha(y)$, introduced by Birnbaum (1962), and defined by $\alpha(y) = \min([1+\lambda(y)]^{-1},[1+\lambda(y)^{-1}]^{-1})$. We compare $\alpha(y)$ with $1-\Gamma^y$ for $y \in B \cap \{y:\lambda(y)>1\}$, where both modes of inference make decision d_1. (Here $B = [y^1,1)$, as in Example 3.5(i).) Straightforward analysis shows that for each α_0 we always have $1-\Gamma_1^y < \alpha(y)$ if c is sufficiently small; while for each c we always have the opposite inequality if α_0 is sufficiently small; and there are pairs (α_0,c) for which each inequality holds for some y. Of course, the definition of $\alpha(y)$ has nothing to do with α_0. Finally, when $Y = y_0$ the UMP critical region for testing g_0 against the family of densities $\{cy^{c-1}, 0<c<1\}$, which would "just reject g_0", is $[y_0,1)$, so that the "level $s(y)$ at which $Y = y$ is just significant" is $1-y$. One finds that $s(y) < \Gamma_1^y$ on B for all α_0 and c, and that $s(y) < \alpha(y)$ on $\{y:\lambda(y) > 1\}$ for all c (a triviality since g_1 is increasing).

For settings with complicated densities, s, being an integral, is much more difficult to compute than is Γ. The analogous difference is even greater in multiple decision problems; see Kiefer (1975, 1977).

ACKNOWLEDGEMENT

This research was sponsored by the National Science Foundation, MCS 75-22481.

BIBLIOGRAPHY

Bahadur, R.R. and Lehmann, E.L. (1955). Two comments on "Sufficiency and statistical decision functions." Ann. Math. Statist. 26, 139-42.

Birnbaum, A. (1962). Intrinsic confidence methods. Bull. Int. Statist. Inst. 39, 375-83.

Brown, L.D. (1977). An extension of Kiefer's theory of conditional confidence procedures. To appear in Ann. Statist. 5.

Cox, D.R. (1958). Some problems connected with statistical inference. Ann. Math. Statist. 29, 357-72.

Dvoretzky, A., Wald, A., and Wolfowitz, J. (1951). Elimination of randomization in certain statistical decision procedures and zero-sum two-person games. Ann. Math. Statist. 22, 1-21.

Kiefer, J. (1975). Conditional confidence approach in multidecision problems. Proc. 4th Dayton Multivariate Conf. New York: Academic Press.

Kiefer, J. (1976). Admissibility of conditional confidence procedures. Ann. Math. Statist. 4, 836-65.

Kiefer, J. (1977). Conditional confidence statements and confidence estimators. (With comments.) To appear in J. Amer. Statist. Assoc. 72.

Lehmann, E. (1957a,b). A theory of some multiple decision problems, I and II. Ann. Math. Statist. 28, 1-25 and 547-72.

Lehmann, E.L. (1959). Testing Statistical Hypotheses. New York: John Wiley and Sons, Inc.

Neveu, J. (1965). Mathematical Foundations of the Calculus of Probabilities. San Francisco: Holden-Day.

Pratt, J.W. (1961). Length of Confidence intervals. J. Amer. Statist. Assoc. 56, 549-67.

Quesenberry, C.P. and Gessaman, M.P. (1968). Nonparametric discrimination using tolerance regions. *Ann. Math. Statist.* 39, 664-73.

Wald, A. and Wolfowitz, A. (1951). Two methods of randomization in statistics and the theory of games. *Ann. Math.* 53, 581-86.

Weiss, L. (1954). A higher order complete class theorem. *Ann. Math. Statist.* 25, 677-80.

Received April, 1977.

Recommended by Walter T. Federer, Cornell University, Ithaca, NY.

P.R. Krishnaiah, ed., *Multivariate Analysis–IV*
© North-Holland Publishing Company (1977) 143–158

CONDITIONAL CONFIDENCE AND ESTIMATED CONFIDENCE IN MULTIDECISION PROBLEMS (WITH APPLICATIONS TO SELECTION AND RANKING)*

J. KIEFER

Cornell University, Ithaca, N.Y., U.S.A.

The framework introduced earlier by the author is applied to certain multidecision models. Basic is the principle of exhibiting a measure of conclusiveness Γ_ω of the decision (when ω is true). Moreover, this Γ_ω should be highly "data-dependent", taking advantage of "lucky" observed values and thus perhaps giving a frequentist response to criticisms of Neyman–Pearson methodology raised by Bayesians. As is the case for unconditional confidence coefficients, it is also convenient to have Γ_ω almost independent of ω. The construction of admissible procedures that meet these requirements, characterized earlier, is implemented in various ranking and selection problems, including nonparametric ones, of both the Gupta and Bechhofer forms.

1. Introduction

This is one of a series of papers in a study of conditional confidence procedures and estimated confidence coefficients. There is a considerable literature of conditioning in statistical inference, but no previous methodical presentation of a frequentist non-Bayesian framework that considers possible criteria of goodness of such procedures, and methods for constructing them, in general statistical settings.

The present development was motivated by the desire to find an answer in Neyman–Pearson–Wald (NPW) frequentist terms, to a criticism that has been raised frequently against the NPW approach by authors supporting other foundational approaches. These critics are often disturbed at the prospect of making a decision that is not accompanied by some data-dependent measure of "conclusiveness" of the experimental outcome. For example, in the ranking problem for deciding which of two normal rv's X_i with unit variance has mean $+1$ and which has mean -1, where the standard symmetric NP test makes a decision accompanied only by the

* Research under NSF Grant MPS 72–04998 A02.

assessment of error probabilities $\Phi(-2^{1/2})$, some statisticians may be disturbed by an intuitive feeling that they are much surer of the conclusion when $X_1 - X_2 = 10$ than when $X_1 - X_2 = 0.5$. Authors adhering to Bayesian, likelihood, fiducial, or evidential foundations have criticized other aspects of the NP approach as well; but this notion of wanting a measure of conclusiveness that depends on the experimental outcome has also appealed to many practitioners, as is evident in the old and continued practice of stating the level at which a significance test "just rejects" a null hypothesis. Intuitively, even for testing between two simple hypotheses one tries to make use of "lucky outcomes" to make stronger assertions.

The principle adopted herein, in developing a methodology that may satisfy the objection mentioned in the previous paragraph for some workers, is that our data-dependent measure of conclusiveness should have a frequentist (law of large numbers) meaning similar to that emphasized by Neyman for classical NP tests and confidence intervals. As is discussed in [3] and [6], this implies that, except in rare cases of symmetry, our approach and form of conclusion cannot agree with those of the critics mentioned above. The conditional confidence procedures are suggested as possibilities that may sometimes have appeal to practitioners. No attempt is made to give a prescription for deciding whether to use any of them, or which one to use. In particular, the fineness of the conditioning partition is to be chosen by the practitioner; one possibility is the degenerate unconditional conditioning, which is the classical NPW framework; if nondegenerate conditioning is used, it should yield a highly variable conditional confidence, or one might as well use an unconditional procedure, instead.

We assume the underlying measure space of possible outcomes of the experiment, $(\mathscr{X}, \mathscr{B}, \nu)$, to be of countable type with compactly generated σ-finite ν with respect to which the possible states of nature $\Omega = \{\omega\}$ have densities f_ω. This implies existence of regular conditional probabilities in the sequel. (See, e.g., [8].) We write X for the rv representing the experiment with outcome in $\mathscr{X}$.

In the decision space D, we assume that for each ω there is specified a nonempty subset D_ω of decisions that are "correct" when ω is true. We write $\Omega_d = \{\omega : d \in D_\omega\}$. The confidence "flavor" is best exhibited in terms of the simple loss function implied by consideration in these terms; other possible treatments will be described in Section 7.

A (nonrandomized) decision rule $\delta : \mathscr{X} \to D$ is required to have $\delta^{-1}(D_\omega) \in \mathscr{B}$ for each ω. The apparent neglect of randomization is intended to aid in clarity of exposition. It is not a genuine neglect, since $\mathscr{X}$

can be regarded as the product of a more primitive sample space and a randomization space.

Any subfield $\mathscr{B}_0$ of $\mathscr{B}$ may be called a *conditioning subfield*, but it is convenient to consider only those subfields generated by statistics on $(\mathscr{X}, \mathscr{B})$. If Z is such a statistic, $\mathscr{B}_0$ is the largest subfield inducing the same partition of $\mathscr{X}$ that Z induces [1]. Writing $Z(\mathscr{X}) = B$, we denote that partition by $\{C^b, b \in B\}$. We also write, for a given δ and $\mathscr{B}_0$,

$$C_d^b = C^b \cap \delta^{-1}(d), \qquad C_\omega = \delta^{-1}(D_\omega). \tag{1.1}$$

The C_ω are not necessarily disjoint, but the C_d^b $(b \in B,\ d \in D)$ are, and constitute the *partition* C of $\mathscr{X}$. Subscripts always index D or Ω; superscripts index B; symbols such as overbars distinguish different partitions and their operating characteristics.

A *conditional confidence procedure* is a pair $(\delta, \mathscr{B}_0)$ or (δ, Z) or, equivalently, the corresponding partition C. Its associated *conditional confidence function* is the set $\Gamma = \{\Gamma_\omega, \omega \in \Omega\}$ of conditional probabilities

$$\Gamma_\omega = P_\omega\{C_\omega \mid \mathscr{B}_0\}, \qquad \omega \in \Omega. \tag{1.2}$$

For each ω, this $\mathscr{B}_0$-measurable function on $\mathscr{X}$ can be regarded as a function of Z, and we write Γ_ω^b for its value on the set $Z = b$. A *conditional confidence statement* associated with $(\delta, \mathscr{B}_0)$ is a pair (δ, Γ). It is used as follows: if $X = x_0$, and thus $Z(x_0) = z_0$, we state that "for each ω, we have conditional confidence $\Gamma_\omega(x_0) = \Gamma_\omega^{z_0}$ of being correct if ω is true". We make decision $\delta(x_0)$.

Such statements have a conditional frequentist interpretation analogous to that of the NPW setting $\mathscr{B}_0 = \{\mathscr{X}, \phi\}$, where Γ_ω is simply the probability of a correct decision when ω is true. In a sequence of n independent experiments, if ω_i is true and X_i is observed in the ith experiment, then the proportion of correct decisions made will be close to $n^{-1}\Sigma_1^n \Gamma_{\omega_i}(X_i)$ with probability near one when n is large. This is true even if different $(\delta, \mathscr{B}_0)$'s are used in the experiments, but the frequentist meaning is of most intuitive value in cases where there is no dependence of Γ_ω on ω, just as in the familiar NP treatment of confidence intervals or composite hypotheses with specified minimum power. Thus, if there is a function ψ on $\mathscr{X}$ such that $\Gamma_\omega(x) \geq$ (resp. $=$) $\psi(x)$ for all ω and all x, we may make the more succinct statement that "we have conditional confidence at least (resp., exactly) $\psi(x_0)$ that $\delta(x_0)$ is a correct decision;" if $X = x_0$. The frequentist interpretation in terms of the law of large numbers is then simpler, and its practical meaning is made clearer if we look only at those experiments in which $\Gamma(X) \geq$ (resp. $=$) 0.95 (for example): a correct decision is very likely made in almost 95% or more (resp. almost 95%) of such experiments, if their number is large.

Whenever $\Gamma_\omega^Z(x)$ can be chosen not to depend on ω, we write Γ^b in place of Γ_ω^b. The examples that arise often enjoy a symmetry that permits construction of invariant C's with Γ_ω independent of ω throughout $\mathscr{X}$. We shall give examples in which the conditioning partition $\{C^b\}$ is "as fine as possible" while permitting this construction, and also in which it consists of the coarsest non-trivial partitioning, of two elements. Intermediate examples will also be given.

The operating characteristic of C can be thought of as the collection consisting of each sub-df of Γ_ω on C_ω, with corresponding sub-df's that refer to incorrect confidence statements. This is treated in detail in [6], and its use is indicated in (2), just below.

For finite (or denumerably infinite) B, we interpret (1.2) through the simple conditional probability formula: if $P_\omega\{C^b\} > 0$,

$$\Gamma_\omega^b = P_\omega\{C^b \cap C_\omega\}/P_\omega\{C^b\}. \tag{1.3}$$

The normal law with mean μ and variance σ^2 is denoted $\mathcal{N}(\mu, \sigma^2)$. The standard $\mathcal{N}(0,1)$ df and density are denoted Φ and ϕ.

To summarize briefly the main ideas behind our conditioning approach and the selection of a conditional confidence procedure, we want:

(1) a data-dependent measure of conclusiveness with frequentist interpretability;

(2) admissibility of the procedure as derived from some notion that a "good" procedure has a high (resp., low) probability of yielding a high value of the measure Γ_ω when ω is true (resp., false);

(3) high variability of that measure (of Γ_ω^b, as a function of b), to achieve a wide range of possible strengths of conclusiveness in the inferential statement that is lacking in an unconditional procedure (and without which we might as well use the latter);

(4) small variability of Γ_ω^b in ω, for the practical reason of making simpler confidence statements, and to avoid certain theoretical anomalies (alluded to later).

Discussion of the relationship with other approaches, shortcomings of various goodness criteria, general theoretical developments, etc., are contained in [3] and [6]. In particular, it is shown there that, for various admissibility criteria of the type (2) above, any unconditionally admissible δ, when conditioned aribtrarily, yields an admissible conditional procedure (but that, except when Ω has only 2 elements, there are other admissible conditional procedures, as well). In the present paper we omit theoretical details and give illustrations of the approach in fixed sample size ranking and selection settings. In addition to considering conditional procedures, we discuss the technique of *estimating* a conditional or unconditional Γ_ω that depends strongly on ω, and without which one may be forced to use an unsatisfactory lower bound ψ. This is also considered from a frequentist justification.

2. Bechhofer's indifference zone approach

For definiteness, we consider the problem of selecting exactly one of three normal populations, it being desired to select the one with the largest mean, the three variances and sample sizes being assumed equal. The simpler case of two populations is studied extensively in [3], and more than three populations offer only added computational and notational complexity, but no new conceptual difficulties. (A three-population example with perhaps less realistic D_ω's is treated in [3].)

After transformation, the model reduces to $X = (X_1, X_2, X_3)$ with $X_i \sim \mathcal{N}(\theta_i, 1)$; the X_i are independent, and d_i is the decision that "θ_i is largest." We write $\theta_{\max} = \max_i \theta_i$. There are a number of different possibilities for specifying parameter regions where only d_1 (for example) is correct, and we choose one of the simplest: $\theta_1 > \theta_{\max} - \bar{\Delta}$, where $\bar{\Delta} > 0$ is specified. Thus, we still view the selection as correct if the chosen population has mean "not too far" ($< \bar{\Delta}$) from the largest mean. This means that, if $\theta_1 > \theta_2 > \theta_3$,

$$D_\omega = \begin{cases} d_1 & \text{if } \theta_1 - \bar{\Delta} > \theta_2, \theta_3, \\ \{d_1, d_2\} & \text{if } \theta_1 > \theta_2 > \theta_1 - \bar{\Delta} \geq \theta_3, \\ \{d_1, d_2, d_3\} & \text{if } \theta_1 > \theta_2 > \theta_3 \geq \theta_1 - \bar{\Delta}, \end{cases} \tag{2.1}$$

and similarly for other possible orderings of the θ_i. Bechhofer's formulation is geared to guaranteeing a specified probability P^* of being correct when one d_i is correct (first line of (2.1)). Alternatively, one can think of this formulation for the above "correctness" structure when Ω is restricted to the set where one θ_i exceeds the others by at least $\bar{\Delta}$.

We hereafter denote by $\Omega(j)$ the subset of Ω where j of the d_i's are correct.

The Bechhofer unconditional procedure [2] selects the d_i for which X_i is largest. We shall use that δ in what follows. Write $X_{[1]} \leq X_{[2]} \leq X_{[3]}$ for the ordered X_i's, and similarly for the θ_i's. For $i = 1$ or 2, write $b_i = x_{[3]} - x_{[i]}$ and $\Delta_i = \theta_{[3]} - \theta_{[i]}$.

Example 2.1. *Two-element B.* We first consider the simplest possible conditioning, that given by a partition $\{C^1, C^2\}$, where C^2 consists of the more conclusive, "lucky", observations. A simple such possibility is $C^2 = \{b_2 \geq c\}$, where $c > 0$. (Questions of admissibility and other goodness properties of such partitions are discussed in [3] and [6].) This yields, in $\Omega(1)$, in an obvious notation,

$$\Gamma^2_{(\Delta_1,\Delta_2)} \geq \Gamma^2_{(\bar{\Delta},\bar{\Delta})}$$

$$= \frac{\mathbf{P}\{Y_1 > -\bar{\Delta} + \max(Y_2, Y_3)\}}{\mathbf{P}\{Y_1 > c - \bar{\Delta} + \max(Y_2, Y_3)\} + 2\mathbf{P}\{Y_1 > c + \max(Y_2, Y_3 + \bar{\Delta})\}}, \quad (2.2)$$

where the Y_i are independent and $\mathcal{N}(0, 1)$. To reduce repetitiveness, we omit discussion of the inequality of (2.2) and of the expression for Γ in $\Omega(2)$, since similar calculations arise in other examples we shall treat. $\Gamma^1_{(\Delta_1,\Delta_2)}$ can be computed similarly, and tables of the $\Gamma^b_{(\bar{\Delta},\bar{\Delta})}$ as a function of c and $\bar{\Delta}$ can be obtained from bivariate normal tables or from tables that appear in the work of Bechhofer and his school. (In more complex settings one cannot expect that the lower bound on Γ^b_ω will be obtained at the same configuration for each b, as it is in the present example.) Since Bechhofer's lower bound P^* on the probability of a correct selection for his unconditional procedure on the set $\Delta_2 \geq \bar{\Delta}$ satisfies

$$P^* = \Gamma^1_{(\bar{\Delta},\bar{\Delta})} P_{(\bar{\Delta},\bar{\Delta})}\{C^1\} + \Gamma^2_{(\bar{\Delta},\bar{\Delta})} P_{(\bar{\Delta},\bar{\Delta})}\{C^2\}, \quad (2.3)$$

we see the way in which the Γ^b's must balance under the "least favorable (LF) configuration", $(\Delta_1, \Delta_2) = (\bar{\Delta}, \bar{\Delta})$. For example, if $P^* = 0.9$ and we choose to use a conditional procedure to attain the higher value $\Gamma^2_{(\bar{\Delta},\bar{\Delta})} = 0.99$ for the more conclusive set C^2 of (b_1, b_2)-values, and if C^2 occurs with probability 0.3 under the LF configuration (and hence $P_{(\bar{\Delta},\bar{\Delta})}$ {correct selection, $\Gamma^2_{(\bar{\Delta},\bar{\Delta})} = 0.99\} = 0.3(0.99)$), then $\Gamma^1_{(\bar{\Delta},\bar{\Delta})} = [0.9 - 0.3(0.99)]/0.7 = 0.86$. Practitioners may encounter settings in which they are willing to have their confidence in a correct selection reduced from 0.9 to 0.86 with probability 0.7 (in the least favorable case) in order to achieve the more highly conclusive value 0.99 in the case of a "lucky sample," which occurs with probability 0.3.

Those who feel uncomfortable with conditional inference may prefer to use the 6-decision unconditional procedure with decision sets C_i^b, where in C_i^1 the decision is "weakly conclusive choice of population i" and C_i^2 stands for a strongly conclusive choice, and where the properties of the procedure are assessed entirely in terms of unconditional probabilities $P_\omega\{C_i^b\}$. (The author would certainly use that procedure in many circumstances.) The theoretical relationship of this procedure to the corresponding conditional one, in terms of admissibility, is treated in [6]. From a foundational viewpoint, such a procedure is in the NPW frequentist mold; but its frequency interpretation is not in terms of a data-dependent proportion of correct decisions with the interpretation Γ_ω^b had, since $P_\omega\{C_i^b\}$ for a single b does not have the "confidence" flavor and (summing over b) $P_\omega\{C_\omega\}$ is the confidence of an unconditional 3-decision procedure.

Example 2.2. *Continuum B.* We now turn to a possible "continuum" conditioning. The "finest" invariant B is $\{(b_1, b_2)\colon 0 \le b_1 \le b_2\}$. Each point in B corresponds to six possible orderings of the X_i's. Writing out the densities of these six orderings (e.g., when $\theta_1 = 0$, $\theta_2 = -\Delta_1$, $\theta_3 = -\Delta_2$, since the results are translation and permutation invariant), we obtain

$$\Gamma_{(\Delta_1,\Delta_2)}^{(b_1,b_2)} = \begin{cases} [1 + (e^{b_1\Delta_1} + e^{b_1\Delta_2} + e^{b_2\Delta_1} + e^{b_2\Delta_2})/(e^{b_1\Delta_1+b_2\Delta_2} + e^{b_1\Delta_2+b_2\Delta_1})]^{-1} & \text{in } \Omega(1), \\ [1 + (e^{b_1\Delta_1} + e^{b_2\Delta_2})/(e^{b_1\Delta_2} + e^{b_2\Delta_1} + e^{b_1\Delta_1+b_2\Delta_2} + e^{b_1\Delta_2+b_2\Delta_1})]^{-1} & \text{in } \Omega(2), \\ 1 & \text{in } \Omega(3). \end{cases} \tag{2.4}$$

It is not hard to see that each of these expressions is nondecreasing in $\Delta_1 \ge \Delta_2$ (fixed); putting $\Delta_1 = \Delta_2$, the resulting expressions are nondecreasing in Δ_2. The infimum of (2.4) on $\Omega(1)$ is thus attained at $\Delta_1 = \Delta_2 = \bar{\Delta}$, and this is the limit of a sequence of points (Δ_1, Δ_2) at which the infimum on $\Omega(2)$ is approached: Thus,

$$\Gamma_{(\Delta_1,\Delta_2)}^{(b_1,b_2)} \begin{cases} \ge \Gamma_{(\bar{\Delta},\bar{\Delta})}^{(b_1,b_2)} = [1 + (e^{b_1\bar{\Delta}} + e^{b_2\bar{\Delta}})/e^{b_1\bar{\Delta}+b_2\bar{\Delta}}]^{-1} & \text{on } \Omega(1), \\ \ge \tfrac{1}{2} + \tfrac{1}{2}(1 + e^{-b_1\bar{\Delta}} + e^{-b_2\bar{\Delta}})^{-1} & \text{on } \Omega(2), \\ = 1 & \text{on } \Omega(3). \end{cases} \tag{2.5}$$

The first line of the right-hand side of (2.5) is the smallest of the three, and thus can be used as the function ψ of Section 1.

In some respects, this finest symmetric partition is the most natural one, and its use obviates the choice of c in Example 2.1, or the choice among B's of various sizes.

Example 2.3. *A mixed procedure.* The procedure of Example 2.2 may appeal to some practitioners who want their conditional confidence statement to reflect the fact that, for b_1 and b_2 near 0, they do not feel very conclusive about their selection. Others may want to achieve the benefits of the procedure of Example 2.2 for large b_1 and b_2 without completely sacrificing the value of $\Gamma^1_{(\bar{\Delta},\bar{\Delta})}$ of the procedure of Example 2.1 when b_1 and b_2 are small. One possible procedure, with the same δ, is given by

$$Z(x_1, x_2, x_3) = \begin{cases} 0 & \text{if } b_2 < c, \\ (b_1, b_2) & \text{if } b_2 \geq c, \end{cases} \tag{2.6}$$

for which Γ^0_ω is the Γ^1_ω of Example 2.1 (corresponding to (2.2)) and $\Gamma^{(b_1,b_2)}_\omega$ is given by (2.5) for $b_2 \geq c$. Other possibilities with similar properties will occur to the reader.

Other B's. It may be desirable to have a finite number of C^b's but more than 2. On the other hand, in place of the procedure of Example 2.2 one may find it desirable to use a continuum B procedure with a simpler B (especially if the number of populations is large). An example of the latter is $B = \{b_2\colon 0 \leq b_2\}$. The result, derived in [3], is, in $\Omega(1)$,

$$\Gamma^{b_2}_{(\bar{\Delta},\bar{\Delta})} = \left\{1 + \frac{\Phi(-6^{-1/2}b_2)\phi(2^{-1/2}(b_2+\bar{\Delta})) + \Phi(-6^{-1/2}(b_2+\bar{\Delta}))\phi(2^{-1/2}b_2)}{\Phi(6^{-1/2}(\bar{\Delta}-b_2))\phi(2^{-1/2}(\bar{\Delta}-b_2))}\right\}^{-1} \tag{2.7}$$

Ranking problems. We shall not take the space to give the straightforward details of the parallels of the previous procedures for other "goals". Briefly, suppose, for example, that we want a complete ranking of the three θ_i's, and condition as in Example 2.2 for Bechhofer's procedure of ranking the θ_i's in the same order as the X_i's, and compute Γ^b_ω on the set where $\theta_{[3]} - \theta_{[2]} \geq \Delta^*_2$, $\theta_{[2]} - \theta_{[1]} \geq \Delta^*_1$, where the Δ^*_i are specified; on this set, only the fully correct ranking is a "correct decision". The expression for Γ is then obtained by moving the last exponential in the first line of (2.4) into the numerator; it is more convenient to represent the result with the substitution $\Delta_1 = \Delta_2 + \Delta'_1$ where $\Delta'_1 = \theta_{[2]} - \theta_{[1]}$. The LF configuration is then seen to be $\Delta_2 = \Delta^*_2$, $\Delta'_1 = \Delta^*_1$, yielding an appropriate lower bound as in (2.5). This must be considered in combination with the evaluation of Γ on other parameter sets analogous to those of (2.1), for whatever schedule of "correct decisions" is adopted.

3. Gupta's subset selection approach

In this approach of Gupta (1956), the unconditional format allowed for seven possible decisions, the nonempty subsets of $\{1, 2, 3\}$; the subscript of the largest θ_i is asserted to be in that subset. We consider the normal probability model stated at the outset of Section 2. One of the rules δ studied by Gupta, and which we consider here, is

$$\delta(x_1, x_2, x_3) = \{i : x_{\max} - x_i \leq c\}, \tag{3.1}$$

where $c > 0$.

We discuss only the analogue of Example 2.2, again with $B = \{(b_1, b_2): 0 \leq b_2 \leq b_1 < \infty\}$. The formulas for Γ over the several regions of interest are given by (2.4), whether "correctness" corresponds to (2.1) (with, for example, $\{d_1, d_2\}$ there being replaced by $\{1, 2\}$ here) with $\bar{\Delta} > 0$, or to the original Gupta formulation with $\bar{\Delta} = 0$ in (2.1). But there is an important difference in the form of assertion possible here, from that in Example 2.2.

In Example 2.2, we would never *know* that $\omega \in \Omega(3)$, and are led to use the $\Gamma^{(b_1,b_2)}_{(\bar{\Delta},\bar{\Delta})}$ of $\Omega(1)$ as lower bound ψ, as described there. In the present subset selection formulation, though, if $\delta(x)$ contains two elements we know $\omega \in \Omega(2) \cup \Omega(3)$ and can use the smaller $\Gamma^{(b_1,b_2)}_{(\bar{\Delta},\bar{\Delta})}$ of $\Omega(2)$ as ψ; and if $\delta(x) = \{1, 2, 3\}$, we know $\omega \in \Omega(3)$ and can take $\psi(x) = 1$.

Some practitioners may feel uncomfortable about asserting "conditional confidence 1" in the correctness of $\delta(x)$ whenever $b_1 \leq c$. The author does not feel that the value $\Gamma = 1$ here is defective in the way it is in certain examples discussed in [3] and [6], which in the present context of Section 2 would amount to taking $B = \{1, 2, 3\}$ and $C^b = \{x_b = x_{\max}\}$ a.e. For the latter ill-advised conditioning, one can make a "$\Gamma = 1$" statement about a *decision which might be wrong*; this anomaly is reflected in the great variation of Γ^b_ω as a function of ω, for that conditioning. In the present example of Gupta's procedure δ, when $\delta(x) = \{1, 2, 3\}$ so that $\Gamma(x) = 1$, *the decision must be correct.* Hence, many practitioners may feel more comfortable about saying $\Gamma = 1$ in this event, than in using the unconditional procedure that yields a constant value $\psi < 1$ that refers also to this case of a decision that *must be correct*; in fact, the latter type of behavior is one of the subjects of criticism of classical unconditional confidence statements by authors such as Pratt (1961). Thus, it does not seem unreasonable to the author to make the assertion $\Gamma = 1$ when $\delta(x) = \{1, 2, 3\}$ in the present example. (In the above "ill-advised conditioning", $\psi = 0$.)

Although the other types of conditioning partitions mentioned in Section

2 can be used here, one motivation for the procedure of Example 2.3 is now absent: the ψ of the conditional subset selection procedure discussed here does not approach $\frac{1}{3}$ as $b_1 \to 0$, as does that given by the first line of (2.5); the form of (3.1) guarantees a more satisfactory confidence statement.

4. Conditioning in other selection and ranking problems

The conditioning partitions $\{C^b\}$ in the previous examples were adopted without the justification, discussed in [3], [6], in terms of operating characteristics (the law of Γ_ω on C_ω when ω is true, with analogues for incorrect decisions). We mention here only some intuitive ideas on the choice of $\{C^b\}$.

Aş mentioned in Section 1, we try to choose C to make Γ_ω^b highly variable in b, since the desire for a data-dependent measure of conclusiveness motivated the development. Some optimum properties, in terms of such variability, are discussed in [3] in simpler examples with conditionings analogous to those chosen here. On the other hand, lack of strong dependence of Γ_ω^b on ω, for each fixed b, or at least existence of a useful lower bound ψ (since such independence of ω is impossible for nontrivial procedures in the examples we have treated), is very convenient; it also eliminates anomalies related to that mentioned in the next to last paragraph of Section 3, in the possibility of incorrectly asserting a Γ_ω value close to 1 with appreciable probability under an $\omega'' \neq \omega'$.

In symmetric examples in which there is a group leaving the problem invariant that is transitive on Ω, the maximal invariant (or a function of it) will be an ancillary statistic that can often be used for a convenient conditioning. In the setting of Section 2, (b_1, b_2) is such a statistic if the original parameter space is replaced by $\{(\theta_1, \theta_2, \theta_3): \Delta_1 = \bar{\Delta}_1, \Delta_2 = \bar{\Delta}_2\}$ for fixed $\bar{\Delta}_i$. For the actual Ω of Section 2, the use of conditioning based on functions of (b_1, b_2) achieves constant Γ_ω^b on orbits given by fixing the Δ_i, and thus yields the convenient ψ from the least favorable configuration.

In other settings no such invariant structure may be present, and the guessing of a good $\{C^b\}$, and calculation of useful ψ, may be difficult. As an example, suppose the X_i of Sections 2 and 3 are now independent binomial (n, θ_i) rv's. The only symmetry is that associated with population labels (symmetric group) and the transformation $X_i \to n - X_i$ (reflection); even if only two populations are present, a parameter set such as $|\theta_1 - \theta_2| = \bar{\Delta}$ admits no transitive action as it did in the normal case.

A conditioning used for decades in a variety of problems involving binomial populations is that based on $b = \Sigma X_i$; generally this was used in a spirit opposite to ours, in conditional calculations that made Γ_ω^b (or an analogue) as independent of b as possible, in achieving some unconditional aim. Gupta, Huang and Huang (1974) have used this conditioning in subset selection (and other) problems, where, however, a specified minimum unconditional probability of correct selection is again the goal, and the decision rule is based in part on ΣX_i so as to yield the desired unconditional results through convenient conditional calculations.

By way of illustration of conditional confidence, we now describe, for two binomial populations, what can be achieved by such a conditioning. We treat the Bechhofer goal of selecting the population with largest θ_i. The conditional law of $Y = X_2 - X_1$, given $X_1 + X_2 = b$, is well known, and it follows from calculations in [5] that the Bechhofer rule (with equal randomization probabilities if $Y = 0$) achieves

$$\Gamma_\omega^b = [1 + \rho_n(\lambda, b)]^{-1} \tag{4.1}$$

where $\lambda = \theta_{[2]}(1-\theta_{[1]})/\theta_{[1]}(1-\theta_{[2]})$ and

$$\rho_n(\lambda, b) = \frac{\sum_{r<\frac{1}{2}t} \binom{n}{t-r}\binom{n}{r}\lambda^r + h_n(\lambda, b)}{\sum_{r>\frac{1}{2}t} \binom{n}{t-r}\binom{n}{r}\lambda^r + h_n(\lambda, b)};$$

$$h_n(\lambda, b) = \begin{cases} 0 & \text{if } b \text{ is odd}, \\ \frac{1}{2}\binom{n}{b/2}^2 \lambda^{b/2} & \text{if } b \text{ is even}. \end{cases} \tag{4.2}$$

This Γ_ω^b, for fixed λ, is *roughly* decreasing in $|b - \frac{1}{2}n|$, as one might expect; there is a fine-structure oscillation, even values of b generally producing smaller values of Γ than both neighboring odd values because of the randomization term h_n. (Thus, one might conveniently lump together *at least* the four values $b = 2j,\ 2j+1,\ n-2j,\ n-2j-1$, into a coarser conditioning partition.)

As a numerical example, when $n = 9$ one calculates that, if $\lambda = 2$, Γ_ω^b ranges from $\frac{2}{3}$ when $b = 1$ (or $\frac{1}{2}$ for the trivial $b = 0$) to 0.78 when $b = 9$. This is not much variability, and it is not hard to see that a conditioning based on values of $|X_1 - X_2|$ can achieve much more in this respect. (An example is treated in [3].) We have gone through this example to illustrate that a conditioning that is attractive for calculations, and which has proved useful in calculating procedures for various unconditional aims, may also yield simple calculations for our framework, but may often produce procedures whose variability of Γ is less satisfactory than for other conditionings.

For large sample sizes, approximate computations of conditional confidence in examples like this last one can often be considerably simplified by the use of the central limit theorem (local version for fine B) or large deviation theory. An example is treated in [3].

5. Nonparametric models

It is possible to apply the conditional confidence approach to a variety of nonparametric settings. As in parametric models, a good conditional procedure can generally be constructed by imposing an appropriate conditioning partition $\{C^b\}$ on a satisfactory unconditional procedure δ. In many problems it is possible thereby to obtain a useful ψ for fairly large Ω with the use of traditional tools such as sample quantiles and simple rank statistics. An example is the decision as to whether iid rv's X_i with continuous df have median >0 or ≤ 0, based on $\Sigma_i \operatorname{sgn} X_i$. Obviously, coarse conditionings will have better "robustness" properties than fine ones. We now illustrate the more detailed analysis that can accompany a stronger structural assumption on Ω.

Suppose X_{ij} are independent rv's, $1 \leq j \leq M_i$, $1 \leq i \leq k$, where the X_{ij} have law F_i. Write $N = \Sigma_i M_i$. The stronger structure we assume is that the F_i belong to the same family of continuous Lehmann alternatives [7]; that is, the elements of Ω can be labeled $\omega = (F, \theta_1, \theta_2, \ldots, \theta_k)$ where F is a continuous df on the reals, $\theta_i > 0$ for all i, and $F_i = F^{\theta_i}$. (It is trivial but unnecessary to specify F further so as to make this representation unique.) We consider the formulation of Section 2, of selecting the population for which θ_i is largest. When $k = 3$, for specified $\bar{\Delta} > 1$ we adopt a correct decision structure corresponding to (2.1):

$$D_\omega = \begin{cases} d_1 & \text{if } \min(\theta_1/\theta_2, \theta_1/\theta_3) > \bar{\Delta}, \\ \{d_1, d_2\} & \text{if } \theta_1/\theta_3 \geq \bar{\Delta} > \theta_1/\theta_2 > 1, \\ \{d_1, d_2, d_3\} & \text{if } \bar{\Delta} > \theta_1/\theta_3 \geq \theta_1/\theta_2 > 1, \end{cases} \tag{5.1}$$

and similarly for other orderings. Order the X_{ij}, N in number, and let $Z_{it} = 0$ if the tth largest observation is an X_{ij}, and $Z_{it} = 1$ otherwise, for $1 \leq t \leq N$. Also write $V_{it} = \Sigma_{s=1}^{t} Z_{is}$. Finally, write $T_i = \Sigma_{t=1}^{N} V_{it}/t$. (These definitions are employed in order to be able to use tables of [11] easily, below.)

The use of rank order statistics for inference in models where Ω consists of Lehmann alternatives has been discussed extensively. In particular, Savage (1956) gives tables when $k = 2$ and N is small, of the probabilities of various rank orders, which depend only on θ_1/θ_2; and he discusses the use of T_1 for testing $\theta_1 = \theta_2$.

For $k > 2$, analogous formulas are known, but tables like Savage's can be used for the present ranking problem, as we now illustrate. Suppose $k = 3$ and all $M_i = 2$, so that $N = 6$. A possible unconditional δ selects the population for which T_i is smallest. The possible values of T_i are listed in Table 1 of [11], p. 607, case "$N = 6$, $m = 2$, $n = 4$" there. We consider a B of two elements, as in Example 2.2, and let C^2 be the event that $\min_i T_i$ is its smallest possible value, 2.1, this being the event that the selected population yielded the two largest of the 6 observed values. It is not hard to verify that, for this C, the LF configuration for both Γ^b's is again that in which one θ_i exceeds the other two by exactly $\bar{\Delta}$. Moreover, because of the small value of N in the present example, δ reduces to choosing the population that yielded the largest observation. Using the formula for the probability of a rank order when $k = 2$, one can show that, in the present case $k = 3$,

$$\begin{aligned} \Gamma^2_{\mathrm{LF}} &= [1 + 2(2 + 3\bar{\Delta})/\bar{\Delta}^4(2\bar{\Delta} + 3)]^{-1}, \\ P_{\mathrm{LF}}\{C^2, \text{correct decision}\} &= 6\bar{\Delta}^4/(2 + \bar{\Delta})(1 + \bar{\Delta})(2 + 3\bar{\Delta})(1 + 2\bar{\Delta}). \end{aligned} \tag{5.2}$$

Thus, for example, if $\bar{\Delta} = 3.62$ (chosen for use of a column of Savage's tables) the unconditional δ has confidence ≥ 0.805 (LF probability of correct selection). The conditional procedure just described has $\Gamma^2_{\mathrm{LF}} = 0.986$, $P_{\mathrm{LF}}\{C^2, \text{correct}\} = 0.374$, and similarly $\Gamma^1_{\mathrm{LF}} = 0.69$, $P_{\mathrm{LF}}\{C^1, \text{correct}\} = 0.431$. The practitioner can compare this operating characteristic with that of the unconditional procedure based on the same δ, to choose whether it is worth lowering the conditional confidence from 0.805 to 0.69 on a set where a correct selection is made with LF probability 0.431, in order to increase it to 0.986 on a set of corresponding probability 0.374.

For larger k and n_i the computations are lengthier; for sufficiently large M asymptotic approximations can be made, as indicated earlier.

6. Estimating Γ

In some selection problems of the type considered in Section 2 it may be appropriate to consider the format (2.1) with $\bar{\Delta} = 0$. In that case the lower bound ψ is $\frac{1}{3}$, a rather useless fact. Intuitively, one may feel that large b_2

makes one feel very confident in the decision, and it should be possible to quantify this. Thus, for either a conditional or unconditional procedure we may want to estimate the Γ_ω which is unknown and for which we have no useful lower bound.

This is the spirit of estimating the power or risk function, a subject that has received attention (e.g., in [10]) for unconditional procedures. For arithmetical simplicity, we summarize some of the features in the simpler case $k = 2$. In that case the problem, in terms of $Y = 2^{-1/2}(X_1 - X_2)$, becomes that of deciding whether a $\mathcal{N}(\theta, 1)$ rv has $\theta > 0$ or $\theta < 0$, using the symmetric δ based on sgn Y. For the unconditional procedure, $\Gamma_\theta = \Phi(|\theta|)$. Intuitively, one may want to estimate this by $\Phi(|Y|)$. This is not an unbiased estimator, and a simple analyticity argument shows that none exists. The estimator $\Phi(|Y|)$ overestimates $\Phi(|\theta|)$ with probability $\Phi(-2|\theta|)+\frac{1}{2}$ and is thus not median unbiased, either. However, other simple estimators of Γ_θ can be constructed which are conservative from either the bias or median bias viewpoint; e.g., for which the expectation is $\leq \Gamma_\theta$. (For example, $\Phi(|Y| - c)$, for c sufficiently large, is of this nature.) What may seem more satisfactory to some practitioners is the estimation of Γ_θ in terms of what may be termed a *secondary confidence statement*, for example in terms of a one-sided lower confidence limit with associated confidence function (possibly also depending on θ!) As an illustration, let τ be nondecreasing on the non-negative reals, with $0 < \tau < 1$. When θ is the true parameter value, Y^2 has df $G_{|\theta|}$ (say), the non-central chi-square df with noncentrality parameter $E_\theta Y^2 - 1 = \theta^2$ and one degree of freedom. Let $h(|\theta|) = G_{|\theta|}^{-1}(\tau(|\theta|))$ for all $|\theta| \geq 0$. Also, denote by h^{-1} the inverse of h on $[h(0), \infty)$ with $h^{-1} = 0$ otherwise. Then the one-sided lower confidence limit $\Phi(h^{-1}(|Y|))$ on $\Gamma_\theta = \Phi(|\theta|)$ is calculated, in standard fashion, to have confidence function (correctness probability)

$$\begin{aligned} \mathbf{P}_\theta\{\Phi(h^{-1}(|Y|)) \leq \Phi(|\theta|)\} &= \mathbf{P}_\theta\{|Y| \leq h(|\theta|)\} \\ &= G_{|\theta|}(h(|\theta|)) = \tau(|\theta|). \end{aligned} \tag{6.1}$$

Thus, if τ is chosen to be a constant γ_0, we obtain a lower confidence bound on Γ_θ with confidence coefficient γ_0 (median unbiased estimator if $\gamma_0 = \frac{1}{2}$); since it may seem unsatisfactory for $1 - \Phi(h^{-1}(|Y|))$ often to be much smaller than $1 - \gamma_0$, one may want to let τ vary accordingly, e.g., $\tau = \Phi^{1/2}$.

Although the exposition and illustration of the previous paragraph were given in terms of unconditional procedures, exactly the same considerations occur for estimating Γ_ω^b for conditional procedures. We remark only that one can examine such properties as expectations or covering probabilities as computed either conditionally or unconditionally for estimators of Γ^b_ω.

The frequentist interpretation, conditional or unconditional, of such estimators is clear. For example, if $E_\omega\{\hat{\Gamma}(X)|Z(X)=b\}\leq\Gamma^b_\omega$ for all ω, b, then the average of the values $\hat{\Gamma}(X_i)$, over a large number of experiments in which $Z(X_i)=b_0$, is with high probability exceeded or almost attained by the proportion of correct decisions in those experiments.

There are certainly many multidecision problems for which an unconditional procedure with estimated confidence will seem to some workers to answer the desire for a data-dependent highly variable measure of conclusiveness (with frequentist interpretation) as satisfactorily as does any conditional procedure. The two-decision normal problem just considered is such a setting, and it is worthwhile to point out the differences in what the two ideas — conditional confidence and estimated confidence — achieve. Although a useful conservatively biased estimator $\hat{\Gamma}(|y|)$ of the confidence $\Phi(|\theta|)$ of the unconditional procedure will approach 1 as $|y|\rightarrow\infty$, such an estimator of $\Gamma^{|y|}_\theta$ for our *conditional* procedure will do so more rapidly. Still, some of those who are uneasy with the conditioning framework will prefer the unconditional procedure with estimated Γ in that case. However, in settings where unconditional Γ_θ is bounded away from 1 but where likelihood ratios are unbounded (as in the example of the second paragraph of Section 1, or in the example just above but with Ω restricted to configurations where $|\theta_1-\theta_2|=\bar{\Delta}$), the conditional procedure can take advantage of lucky observations to state an estimated confidence close to 1, which the unconditional procedure cannot (the probability of a correct selection being *known* to be $\Phi(\bar{\Delta})$, in the example). Th greater conceptual simplicity of estimating Γ for an unconditional procedure, over both conditioning and also estimating Γ, is of course an attraction of the former. A weighing of these features will indicate, I believe, different settings where it is attractive to use one of the two techniques, or both, or neither.

7. Other variations

We have discussed conditional inference using the simple zero-one loss function that is probably of greatest intuitive value in developing concepts of conditional as well as unconditional inference. Other loss structures are considered in [3], [6]. In particular, admissibility results are obtained for problems in which, as in the formulation of Section 3, the geometric size of the confidence set is variable, and in which a penalty is assessed that depends on that size. Roughly, it is again the case that a procedure which was unconditionally admissible for such a model is still admissible after conditioning in any way, and there are also other admissible conditional procedures. The sequential framework, discussed briefly in [3], leads to more difficult probabilistic calculations, because the sets C^b are now obtained from a stopping rule. This will be treated in detail in another paper.

References

[1] Bahadur, R.R. and Lehmann, E.L. (1955). Two comments on "Sufficiency and statistical decision functions". *Ann. Math. Statist.* **26**, 139–142.

[2] Bechhofer, R.E. (1954). A single-sample multiple decision procedure for ranking means of normal populations with known variances. *Ann. Math. Statist.* **25**, 16–39.

[3] Brownie, C. and Kiefer, J. (1975). Conditional confidence statements. [To appear.]

[4] Gupta, S.S. (1956). On a decision rule for a problem in ranking means. *Inst. Stat. Mimeo Ser.* No. 150, Univ. of N.C., Chapel Hill.

[5] Gupta, S.S., Huang, D.-Y., and Huang, W.-T. (1974). On ranking and selection procedures and tests of homogeneity for binomial populations. Purdue Mimeo Series # 375.

[6] Kiefer, J. (1975). Admissibility of conditional confidence procedures. *Ann. Statist.*, to appear.

[7] Lehmann, E.L. (1953). The power of rank tests. *Ann. Math. Statist.* **24**, 23–43.

[8] Neveu, J. (1965). *Mathematical Foundations of the Calculus of Probabilities.* Holden-Day (San Francisco).

[9] Pratt, J.W. (1961). *Review of Lehmann's Testing Statistical Hypotheses. J.A.S.A.* **56**, 163–166.

[10] Sandved, E. (1968). Ancillary statistics and estimation of the loss in estimation problems. *Ann. Math. Statist.* **39**, 1755–1758.

[11] Savage, I.R. (1956). Contributions to the theory of rank order statistics — the two-sample case. *Ann. Math. Statist.* **27**, 590–615.

J. KIEFER

THE FOUNDATIONS OF STATISTICS— ARE THERE ANY?

1. BEGINNINGS

The fact that an author asks such a question as that of the title no doubt makes his answer easy for most readers to guess. Still, some explanation of the answer may be called for. Perhaps it is best to begin by explaining the question, and that will require some definitions. Many contributors to this volume, experienced in philosophical and foundational writing, may find my interpretation of some words and ideas shallow, biased, or incomplete. (Also, the paper, like almost all on foundations, will no doubt be viewed as somewhat polemical!) These remarks will nevertheless outline some of the misgivings I share with a number of statisticians, at the seriousness with which certain foundational developments are viewed by their proponents, as the salvation of statistics.

To narrow the discussion, my meaning of *foundations* will be restricted to the context of a *predictive science*, and in such a science a *successful foundational scheme* will be a set of undefined objects and axioms from which useful predictive rules can be derived. (This need not mean that the statistical problems referred to later are necessarily phrased as problems of prediction of the value of a random variable to be observed in the future.) We are not concerned merely with a collection of observations about some aspect of the real world, but about a collection of rules of sufficient simplicity and generality in their applicability that they can be useful in predicting what will be observed in other settings involving the same type of phenomena. The obvious example is physics, and I note only that we allow for more than one successful foundational scheme and need not even assume there is any God-given 'correct' one; on the other hand, we are not interested here in most of the foundational schemes that appear as answers to the proverbial final examination question in

Synthese **36** (1977) 161–176.

cosmology, 'define 'universe' and give three examples'; if such foundational schemes are included, the question of the title has a positive but valueless answer.

Thus, also, we are not talking about areas of knowledge such as political history, and we may argue about whether we include economics. There are some economic principles that are usefully applied, but we have all seen the limited applicability of much of macro-economics in a real world that rarely satisfies the necessary conditions for proposed theories to be used. Humans, and collections of them, are less regular than inanimate objects and plants, at least in terms of our current ability to predict their behavior.

How do statistics and foundational systems for statistics fit into this picture? First, some more definitions.

A *probability structure* here means a classical (Kolmogorov) one, in terms of which the *frequentist* viewpoint of statistics, as emphasized by Neyman, is based; *subjective probability* will be labeled as such, and *Bayesian* will mean *subjective Bayesian* here unless the existence of a physical prior law is mentioned. (Thus, notwithstanding the opinion of a colleague of mine that, for statistics, the important ideas of the meaning of 'probability' are those *not* to be found in Feller or Loève, the term will be used here as it is by such authors as these.) Probability theory studies the behavior of complex chance systems made up of simpler chance entities whose laws are given. It provides a useful foundational system in a number of scientific disciplines, in which its conclusions have been seen repeatedly to correspond to what is observed in many real world situations; such settings exhibit outcomes that conform closely to what we would guess from such results as the law of large numbers. If insurance companies and gamblers couldn't make long run predictions, they wouldn't be in business.

A *setting for statistical inference* is a collection of *possible* probability structures, exactly one of which (called the *true state of nature*) is the mechanism actually operating in the setting at hand; which one, is unknown. *Statistical inference* in the broad sense is the collection of procedures for reaching a conclusion about which possible state of nature is the true one in such a setting, together with the probabilistic properties (operating characteristics) of such procedures.

This last definition is purposely quite broad, and the controversial possible choices of narrower definitions will be discussed in the next section.

2. WHAT IS STATISTICAL INFERENCE?

Which real world settings, where data are collected, should be regarded as settings of statistical inference? We shall take a broad view here, not for the sake of generality, but because the distinctions that are often made seem to me to be either illusory or else due to practical difficulties in defining the setting precisely. Thus, a number of authors have distinguished between problems of *decision* and of *inference* or reaching *conclusions* (e.g., Tukey, 1960); and Birnbaum (for example, in the present volume) has made a similar distinction between the *behavioral* and *evidential* interpretations of decision. Examples are often given of the difference between typical statistical problems of technology and of science: a manufacturer decides (with resulting *action*) whether a lot is defective or not; but a scientist (let us call him 'A') might only state that a particular hypothesis (= probability structure) is to be rejected without asserting which state is guessed to be true, or that a particular subset of the possible states of nature seems to him to contain the true one (in which case, if the set is too large to be useful, or if it carries little confidence, we shall call it a *weak decision*, and, otherwise, a *strong decision* – a simplistic dichotomy for the sake of brevity), or that a likelihood function (or perhaps a fiducial distribution or subjective posterior law) he presents reflects the plausibility of various possible states being true. Compare this scientist with scientist B who analyzes a series of observations and decides that they verify (as a useful approximation to truth) the law of gravity for rocks thrown near the earth's surface, including the value of the gravitational constant. Certainly B has used statistics as a tool of predictive science, the gravitational law being useful for predicting what will occur in some future settings; moreover, his results may serve as building blocks for further development of the subject. His results are not 'actions', but his decision implies future actions by him and others, and might be made with greater certainty than that of the manufacturer (who *must* act, whether or not confident).

Almost every A would like to be a B. But A's setting is generally 'dirtier' in one of several respects: (i) he does not have enough observations to feel confident about making a precise assertion about what is true (a strong decision); (ii) he could not, before taking (or analyzing) the observations, make a complete list of the possible true laws and assertions (or decisions) for his setting. Because of (ii), the framework of formal decision theory may be useless to A; and, even if he were not faced with (ii), and hence could list the possible states of nature and corresponding assertions, the presence of (i) would deter him from making anything but a weak decision. Still, he would like to discover (approximate) truth and to build upon it, or have others do so, in further investigations. One may regard his experiment as a first stage in a complex sequential undertaking in which *the possible decisions* (other than, perhaps, weak ones) *are themselves not delimited at the start.* While A cannot now list all possible outcomes of the various stages of this undertaking, or the subsequent stages they will suggest, he would not be carrying out the present stage without the conviction that those subsequent stages will develop in such a way as to lead to the discovery and/or verification of a truth about nature (or to the strong decision that he should quit). It seems more fruitful, in assessing the foundations of statistics, to regard A and B as having the same ultimate goal in the use of statistics, even though the settings they encounter may differ in complexity by orders of magnitude, rather than to regard their problems (decision or inference, etc.) as being fundamentally different because this *first stage* is so different.

From this point of view, the presentation of the likelihood function as illustrative of inference rather than decision is unsatisfactory. For, a next stage of experimentation, and perhaps a more precise listing of possible future terminal decisions, must be *decided upon* from that function. The same is true for the listing of all data and details of the experiment, that some of us would prefer to give in poorly defined settings with inconclusive observations. The choice may be more complex and less well-based than in simple technological settings; but it is dodging the issue and ignoring what happens next, to view the likelihood function (with perhaps different modifications corresponding to the prior laws of different Bayesians) as a different *ending* to a statistical problem, to be contrasted with a decision. (With this in mind, we shall aim our criticism in most subsequent discussion, at Bayesian decision theory rather than at axioms that supposedly lead to inference rather than decision. We will not take the space here to discuss other criticisms that have been raised about the likelihood principle, regarding identical interpretation of two identical likelihood functions in settings in which the two experiments and sets of possible states of nature differ.)

The separation of descriptive statistics from a more narrowly defined inference or decision theory often seems artificial, in this light. The informative summarization and presentation of data is frequently not only for the purpose of recording history, but for the purpose of assessing the possible mechanisms and deciding on the next stage of experimentation or a future action. (Comprehensive studies of pollution, such as that of Box and Tiao, or of income and wealth distribution, are best viewed in this way.) Even when the presentation is primarily of historical interest, the choice of statistical method (width of histogram intervals, method of smoothing, etc.) is presumably motivated by the same desire to approximate the truth well as is the choice of the physical or biological scientist who is trying to answer a more specific question. In another methodological domain, Tukey's extensive collection of methods of exploratory data analysis is aimed at treating settings in which the possible states of nature and decisions are typically not easy to formulate in advance of the experiment; but the methods are surely also to be viewed as being used in one stage of a larger undertaking in which there lurks the aim of an ultimate strong decision or an accurate assertion about nature.

The *foundations* of statistics should not consider decision and inferential measure such as likelihood as *final steps* of two kinds of problems and distinguish between them in that form. A more sensible distinction would be one that accepted the similar ultimate purpose of statistics in complex problems of varying completeness in their formulation, and which would regard decision theory (as the term is now understood) as a simple but useful case of a more general incompletely formulated setting of the nature mentioned above. (The latter is what most scientific investigators are faced with, but an appropriate foundational system that yields useful consequences about such complex settings, has not yet appeared.) It would be pretentious to view these as analogues of, say, Newtonian and relativistic mechanics, but there is at least a partial analogy.

In summary, the distinction between decision and inference (as final products of the axioms) seems questionable at the foundational level;

however, even if the distinction is accepted, in settings of inference we are left with the unanswered question of what to do next on the basis of something like the likelihood function; we now look further at the question of the title in settings of decision.

3. FOUNDATIONS OF WHAT?

Successful theories in the physical or biological sciences are able to predict how entities (or collections of them) behave under various circumstances; questions of how they *should* behave (e.g., occasionally violating gravity or the laws of thermodynamics, to aid mankind during the energy shortage; or to achieve some more heavenly purpose) do not any longer arise in the minds of many scientists. Social sciences often do not stop short of such questions; economists recommend some action to their government (or perhaps revolution against it) to achieve the 'better' of several possible economic (or economic-political) states. This is a concern with how people and other economic entities *should* behave. Genetic engineering, as distinguished from an understanding of the DNA mechanisms, may be viewed similarly.

Are the foundations of statistics concerned with how statisticians behave or how they *should* behave? To start with, we accept the unfortunate reality that human behavior is necessarily involved in planning and analyzing experiments. Scientific knowledge and technological invention are human perceptions and developments, and they have proceeded from the knowledge at previous stages only with the intervention of human ingenuity. No computerized knowledge producer is available to avoid this; moreover, the programming of the rules by which such a computer would make choices would itself ultimately be a human choice. These obvious comments are made because many foundational developments are evidently attempts to find a single simple prescription which is meant to provide a satisfactory 'program' for a large collection of settings of widely differing character. A step in these developments is the confusion between the consequences of systems of axioms of rational (or coherent, etc.) behavior, which presumably tell us how (rational) statisticians behave, and the question of how one *should* behave (choose a terminal decision or the next stage of experimentation) in an actual setting.

These are not the same because the axioms, which lead to 'acting like a Bayesian', are not complete in the sense of mathematical logic. (We are discussing here the axioms for decision making; as indicated earlier, axioms that lead to the presentation of, say, the likelihood function, do not come to grips with the experimental or terminal decision that must still be made.) That is, anyone who acts relative to any subjective prior law whatsoever is 'rational'. My own observation, on questioning a number of practicing Bayesians, is that many do not appreciate that there is no absolute (as distinguished from subjective and individualistic) basis for asserting that this *should* be their behavior; nor even that, if a physical prior law exists, their rational behavior might not minimize the actual expected loss. In a setting where we have a precise listing of all components of a suitably simple and regular decision-theoretic model, Wald's work shows that this rational subjective Bayesian behavior is equivalent to picking an admissible procedure. (*Variation* in settings is discussed in Section 5.) Although the picking out of an admissible procedure is equivalent to acting like a Bayesian relative to *some* prior law, many of us do not find it helpful in practice to know this conclusion of the mathematics of behavior; we are not comforted by the unthinking self-satisfaction each of a number of subjectivists may attain upon using his own prior law, all in the same setting. As Le Cam (1968) put it, "Bayesian statistics is intended for personal views, it does not provide any way of scientific discourse."

To summarize, these foundations for a theory of how rational individuals *do* act, at best cannot provide the desired program for automatic decision making, and at worst have deluded some practitioners into believing that the foundations provide such a program for how they *should* act.

It may be objected that (in the simple, well-defined model discussed shortly above) the non-Bayesian decision theorist's presentation of the minimum complete class of all admissible procedures is also not a satisfactory end. I agree. One must still pick the admissible procedure actually to be used. In a given setting, two non-Bayesians might pick different procedures. A propos of this, a theoretical Bayesian who is among the deeper thinkers about foundations has written me, in criticizing some non-Bayesian, frequentist conditional confidence procedures, that "telling [the experimenter] that his confidence should depend on the procedure he was using is lame, lame, lame." To me it seems just as lame that a scientist announces a conclusion and his measure of belief in it, both dependent upon his subjective prior law.

This degree of arbitrariness non-Bayesians perceive in the choice of procedure is an inescapable feature of most statistical settings. Moreover, the subject matter of statistics – the collection of settings of statistical inference – is so complex that many of us believe it impossible to codify it in terms of a simple program that will pick a suitable procedure in each setting. Extant systems of statistical foundations are aimed at achieving such simplistic codification.

What remains of our original question is, ignoring the objections thus far stated, and being pragmatic, has such a codification proved *useful* in predictive sciences, in practice?

4. DOESN'T IT WORK?

Since subjective Bayesian analysis is used by so many more people than it was twenty years ago, must its foundational system not be a useful one? In answering this, we look at fairly strict adherents to the principle of making decisions in terms of one's own subjective prior law. We do this because the choice of a decision is not primarily a consequence of the subjective Bayesian foundations if, for example one uses a number of possible subjective prior laws to obtain operating characteristics of Bayes procedures that are compared further before choosing a procedure (which therefore, from our previous discussion, amounts in a fixed setting to choosing among that number of admissible procedures); or if the prior law used is obtained as a good approximation to a genuine physical prior law (which, if it exists and is known, is what any of us would use, with some precautions as described in connection with the procedures T_0 and T_1, below).

Some Bayesians have let it be known that, by this stage of development of the methodology, there is more than one Bayesian school. The present discussion cannot be trusted to reflect understanding of all nuances of difference, but one belief that seems no longer to be as widely held as it once was in some Bayesian quarters, is the view that a scientist or business man can always find a prior law that has a strong *physical* basis, by sufficiently thorough examination of the background of knowledge related to his present setting. There is a philosophical question here, of whether it is meaningful that such a physical prior law exists (does Nature throw dice in choosing the true natural law?), let alone whether we can come close to knowing it if it does. When I once asked a leader of Bayesian thought what prior law he would use for the distance to a new astronomical body the like of which no one had seen before (e.g., the first pulsar), he responded with a collection of betting assertions such as 'surely you wouldn't bet more than a million to one that the body is more than 5×10^9 light years away.' He then went on to point out that, for sufficiently large sample sizes, the exact nature of the prior law tends to disappear in the posterior law and in the numerical value of our estimate of the true distance. This line of argument indicates no belief in the availability of a good approximation to a physical prior law, and the reliance on asymptotic theory (mathematically astronomical sample sizes rather than actual ones of astronomy!) does not support the usefulness of Bayesian foundations in this setting. A more satisfactory answer would have been, 'in such settings where we are dealing with new phenomena and have few observations, the Bayesian approach will not accomplish any more than others.' Of course some problems are hard, why not admit it? Incidentally, this asymptotic theory that Bayesians rely upon in taking comfort in the use of diffuse prior laws, was initiated by v. Mises (1919) in the Bernoulli case almost sixty years ago, was later extended by Kolmogorov, and in more recent years has been treated in great generality by Wolfowitz, Le Cam, and others. A principal conclusion is that, for sufficiently large sample sizes, *many* procedures have good properties; but how large a sample is needed for a Bayes procedure to have such properties *depends on the prior law employed*, so that we can take little comfort in what we would get as an estimate from using a prior law that reflects the mentioned million-to-one odds, if the true distance were in fact near to or less than 5×10^9 light years.

Why is the Bayesian approach so prevalent in business schools (at least in the U.S.)? Businessmen must often make immediate terminal decisions (actions). Sometimes they do encounter settings that fit into a background that produces a sensible physical prior law (or approximation thereof). Where they don't, they must still act, and the Bayesian approach provides quick and simple recipes for action. It is an old political proverb, that American businessmen mind an indecisive government more than they mind one whose policies are harsh but known. Presumably this attitude is reflected inwardly in the importance businessmen give to decisiveness even at the expense of making some incorrect decisions. We, and they, *will never know how much more or less they lose* by being subjective Bayesians than they would by using another approach. Also, whatever success the introduction of the Bayesian approach has led to in the business world has in part been due to the coincident introduction of quantitative methods of assessing various possible prospects, that would also benefit other methodologies.

Another occasional contributor to Bayesian success is actually a departure from the strict Bayesian scheme. It is astounding how many practicing Bayesians who argue articulately about the wisdom of incorporating into a prior law one's feelings about the relative importance of different parts of the risk function (perhaps not an outrageous *first step* in trying to select a procedure), then act blindly in using the Bayes procedure that results, without ever examining the risk function (operating characteristic) of that procedure and comparing it with those of other procedures. It is easy to give examples in which the Bayes procedure T_0 relative to a given prior law P_0 has large (even unbounded) risk on a sizable set B_0 of small P_0 probability, whereas another procedure T_1 has much smaller risk than T_0 on B_0 and only slightly larger risk on the complement of B_0; T_1 is not Bayes, having slightly larger integrated risk with respect to P_0 than T_0 does. (This, somewhat loosely, is described as a *subadmissibility* phenomenon; the precise definition is given in terms of sequences of procedures and settings.) You may feel that I don't know the right practicing Bayesians, but the overwhelming majority of those I have questioned had never thought of this; many are thereafter quick to realize that P_0 was only an approximate input and T_1 is more sensible to use than T_0. I am told by John Pratt that sensitivity analysis is the name of the game some Bayesians (largely complementary to my sample) go through in concluding that they shouldn't use T_0. It is comforting that this occurs sometimes; but it is disturbing to learn, continually, of much-heralded Bayesian interactive computing packages that query and reexamine the customer's prior law, but that typically then grind out the posterior law or some form of decision without ever having looked at the operating characteristic of the Bayes procedure. This last is but one shortcoming of routine, increasingly mechanized Bayesian analysis; at least as serious is the fact that there are so many customers who do not understand fully the consequences of what they are doing in using such routines, to which they have been attracted by the oversimplified promise of such ease and comfort in reaching a quick decision to which such beautiful names are attached. Such mechanical routines also discourage the possibility of deserved distrust in the model and of insight the data might yield in the form of a possible new natural law to be verified in subsequent experiments.

The modification of strict Bayesian analysis, that compares the risk function of T_0 with that of other procedures, is not in conformity with Bayesian foundations. The scheme of picking a single subjective P_0 and looking only at the risk functions of nearby competitors of T_0 to see whether one of them is more sensible, also seems too rigid; but, in any event, it bears as much relationship to a routine that might be followed by a non-Bayesian decision theorist, as it does to the strict dictates of the axiom systems for rational behavior. (The spirit of this modification of the Bayes routine has been formalized in various criteria; for example, one of these chooses a minimax procedure among the procedures that are Bayes relative to some law in a specified set of prior laws that replaces P_0. This also seems too rigid; for example, there is still the possibility of occurrence of the subadmissibility phenomenon. Again, it is a matter of statistics being too complex a subject for any simple prescription – even one that requires the space of a book or two – to yield a satisfactory choice of procedure in all settings.)

Thus, strict subjective Bayesianism does not seem useful, and the sensible modifications some Bayesians have suggested remove the resulting methodology from being an illustration of the usefulness of Bayes *foundations*.

5. OTHER COMMENTS

5.1. *More on Bayes*

The subjective Bayesian approach has other difficulties than those mentioned earlier. For example, as Wolfowitz (1962) pointed out, the rational behavior axiom that assumes comparability of all pairs of choices is more severe than one that is much more straightforward (although phrased in terms of less primitive choices), that the statistician must be able to select a procedure he prefers from the class of all those available. Moreover, the crucial Savage-Lindley conclusion that rational statisticians behave like Bayesians is of limited impact for several reasons. On the one hand, if we restrict consideration to a fixed setting, the conclusion amounts only to that mentioned earlier, that the admissible and Bayes procedures coincide in suitably regular settings – not very startling, once understood. On the other hand, as Birnbaum (1977) discusses elsewhere in this volume, the basic *assumption* of preservation of indifference sets under probabilistic mixing is highly debatable, as is therefore the conclusion that, as the experiment varies but the states of nature remain fixed, a rational person should use the same prior law. This means that, theoretically, it is not as conclusive as we have been told by Bayesian theorists, that we should be able to select our procedure on the basis only of a subjective prior law and without looking at the set of available risk functions in the setting at hand except through their integrated risk relative to that prior law; and, practically, an examination of the variety of possible geometric configurations of risk functions in different settings, which may exhibit such phenomena as that of subadmissibility alluded to earlier, adds to one's misgivings about the practical virtues of the Bayesian prescription.

In assessing the practical impact of the foundations of rational behavior, we must also face the reality that the set of possible states of nature a statistician encounters generally *does* change as he goes from setting to setting involving different phenomena, so that, even if we accept the axioms and ignore Birnbaum's criticism, the conclusion of rational behavior can then only be that we act like Bayesians in each setting *with a different subjective prior law in each setting*. Except when a relationship among those prior laws on different sets in different settings can be established, the rational behavior axioms imply only that (under regularity conditions) the statistician uses *some* admissible procedure in each setting.

Among the Bayesians I know there are some who are followers of Fisher in using randomization of the experiment but followers of Lindley in conditioning (after the randomization is performed) on the experiment actually used. This is inconsistent with the axioms, as many Bayesian theorists may agree: a Bayesian can only think of randomizing among the *best* (Bayes) experiments, but then has no reason to randomize at all. Perhaps this is a reflection on my sampling of Bayesians. Perhaps, too, it is another illustration of the tendency mentioned earlier, of the Bayesian rationale, to invite an unwarranted satisfaction with a simplistic approach that yields quick answers, often without any understanding of the consequences in terms of actual probabilities of error or expected loss. The increasing use of the approach in applications by unthinking practitioners is a more serious worry than any axiomatic inadequacy *per se*.

5.2. *Probability and Betting*

A Bayesian colleague told me he cannot understand how I can look at risk-functions and choose among them without thinking in terms of 'probabilities' of the various states of nature. This reflects not only his understandable tendency to see the choices of others through his own choice mechanism, but also a logical fallacy in his perception of psychology: even assuming Bayesian axioms, the fact that one can be viewed as acting *just as though he is a Bayesian* relative to some prior law (a description of mathematical psychology), does not mean he *does* choose a decision in that way with a conscious injection of his prior law.

A related misunderstanding occurs in connection with the odds and bets that are often a hallmark of Bayesian foundational discussions. These discussions arise, for example, in the Bayesian dissatisfaction with the Neyman-Pearson confidence coefficient, that non-Bayesian frequentists refuse to regard (after the experiment) as a 'probability' of correctly covering the true parameter value, instead interpreting its meaning in terms of the expectation viewpoint before the experiment is conducted, or in terms of the actual frequency of successes in many (not necessarily identical) experiments as justified by the law of large numbers. To make the controversy clearer, let me describe it in termsof the single flip of a fair coin – not a problem of inference about the true probability model, which is known, but, rather, one of guessing the outcome of the flip. Before the coin is flipped we would all speak of probability $\frac{1}{2}$ of the coin coming up 'heads'. Some subjectivists have asked, 'if the coin is flipped but the outcome is hidden from you, (a) wouldn't you still bet as though heads is a 50–50 chance, and, if so, (b) would you still insist on using the familiar (non-Bayesian frequentist) language that, once the coin is flipped, the probability is no longer $\frac{1}{2}$ that the coin will come up "heads", but is either 0 or 1 (you know not which) that it now *is* heads?' My answer to both questions is yes. My betting habits may be the same as they were *before* the coin was flipped, but that does not mean I must think the situations are the same or use the same language for them. There is no chance mechanism still to be used after the coin has been flipped, so it is not a (classical) probability situation. The fact that di Finetti has integrated personal and classical physical probabilities into one system, the fact that Bayesians therefore use 'probability' in the resulting broader sense, does not force me to do so, or even to believe in personal probability, even though my betting can be described by Bayesians in such terms. (The language of probability is also used in other contexts where there is no probability mechanism, such as in deterministic settings of mechanics with many particles; but there, in statistical mechanics, the language and probability calculus are being used, knowingly, to describe a probability model that turns out to yield good approximations of certain properties of the deterministic setting which are much more difficult to compute exactly. This is quite unlike the spirit of the Bayesian discussion.)

Incidentally, another common misunderstanding is present in the interpretation of frequentist probability (e.g., that a confidence interval will by chance cover the true parameter value) and posterior probability relative to a physical prior law and given that the observation $X = 2.9$. The former refers to a chance experiment still to be conducted, the latter does not, and this is not altered by one's similarity of betting behavior in the two instances. For a frequentist, the posterior law relative to the yet-to-be-observed X is the random quantity described in terms of an experiment yet to be conducted, and whose behavior in terms of the law of large numbers justifies the way we use it when $X = 2.9$. Of course, subjectivists require no such distinction.

We shall not enter here into the question of whether it is reasonable to develop a theory of inference about scientific truth that rests on willingness to make certain bets; faced with such objections, some Bayesians have attempted to rework some of their developments in other terms. See, e.g., Buehler (1976).

5.3. *What to do instead*

My only 'foundations', other than my reliance on Neyman's frequentist point of view, are qualitative: that, as stated more than once earlier, statistics is too complex to be codified in terms of a simple prescription that is a panacea for all settings, and that one must look as carefully as possible at a variety of possible procedures, both in the difficult settings in which one cannot easily exhibit all operating characteristics, and also in ones that are simple superficially, in which there is an approximate physical prior law with which to make a preliminary selection. Few non-Bayesian frequentists (if any) now have agreement with the straw men exhibited by the Bayesians, and who supposedly use tests of level 0.05 or 0.01 on every possible occasion without further examination of the possible power functions. (No doubt these Bayesians have met some such flesh-and-blood statisticians, as have I, and just as I have met the unthinking Bayesians they will regard as *my* straw men!)

There are a few simple settings, no doubt assumed to be valid more often than is warranted, in which some of the 'standard' procedures of statistics have been justified theoretically (for example, as being almost minimax, close to Bayes relative to a variety of prior laws, and without a sizable subadmissibility defect). Although theoretical statisticians work on such simple, often unrealistic models, it is a mistake to regard their conclusions, as some data analysts have, as practical recommendations; rather, they are guides to understanding and clues, obtained in tractable models, as to what one might investigate as possible good procedures in the more complex real-world settings where exact calculations are impossible. Whatever may be the eventual value of the current expanding trend to make simulation robustness studies of various procedures, it is doubtful that this development, initiated by Tukey and other data analysts, would have taken its present direction without the theoretical studies of Huber and others in simple cases; and it is difficult to imagine that it could have come out of elegant Bayesian computations with convenient conjugate prior laws.

Perhaps some day someone will prove me wrong and provide the satisfactory codification we do not now have. Until that new Copernicus arrives, I can only say to the personal probability advocates that the world of scientific inference is no more egocentric than it is geocentric.

Cornell University

ACKNOWLEDGMENT

Alan Birnbaum suggested the writing of this paper and contributed stimulation to its views in his criticism of Kiefer (1977), as did the criticism of George Barnard, Bob Buehler, and

John Pratt. Thanks are also due to Bob Hogg for inviting the talk that produced an earlier version, and to Mel Novick, who, in trying to convince me that my views were wrong, had the opposite effect.

REFERENCES

Birnbaum, A.: 1977, 'The Neyman-Pearson Theory as Decision Theory, and as Inference Theory; with a Criticism of the Lindley-Savage Argument for Bayesian Theory', this volume, pp. 19–49.

Buehler, R.: 1976, 'Coherent Preferences', *Ann. Math. Statist.* 4, 1051–1064.

Kiefer, J.: 1977 (discussion), 'Conditional Confidence Statements and Estimated Confidence Coefficients', *J. Amer. Statist. Assn.*

LeCam, L.: 1968, in D. G. Watts (ed.), *The Future of Statistics*, Academic Press, New York, p. 143.

Savage, L. J.: 1954, *The Foundations of Statistics*, John Wiley, New York.

Tukey, J.: 1960, 'Conclusions versus Decision', *Technometrics* 2, 423–433.

v. Mises, R.: 1919, 'Fundamentalsätze der Wahrscheinlichkeitsrechnung', *Math. Zeit.* 4. 1–97.

Wolfowitz, J.: 1962, 'Bayesian Inference and Axioms of Consistent Decision', *Econometrica* 30, 470–479.

SEQUENTIAL STATISTICAL METHODS

*J. Kiefer**

Article reprinted from:

MAA Studies in Mathematics,
Vol. 18, "Studies in Probability
Theory," M. Rosenblatt, editor,
M.A.A., 1978

1. INTRODUCTION

Most of the standard techniques of statistical analysis are based on experiments in which the amount of experimentation to be conducted is decided before the experiment begins. For example, in trying to select the best of several drugs to cure a disease, the medical research worker might decide in advance how many patients will be treated with each drug. An alternative occurs naturally to the experimenter who finds the resulting data inconclusive: perhaps he should try the drugs on some additional patients. On the other hand, if he does not have the facilities to treat all the patients at once and finds conclusive enough evidence to select the best drugs after observing only a fraction of the intended total number, perhaps he should save time and money by terminating the experiment early. Thus, he decides to lengthen or shorten the experiment, based on the

* Supported by NSF Grant MCS 75-22481.

1

outcome of the part of the experiment conducted up to the moment of decision. Such a technique is called a *sequential procedure.*

With such natural impetus, experimenters undoubtedly conducted sequential experiments long before the establishment of probability theory as a mathematical tool with which precise properties of experiments could be studied. Even to this day, whether out of ignorance, sloppiness, or deception, some experimenters publish results obtained from sequential experiments, as though the experiments were not of that nature. As we shall see, the meaning of a given set of results on two drugs, each applied to ten patients, is quite different, if the sample size of twenty was chosen in advance and adhered to, from what it is if the experimenter had not specified his sample size in advance but simply continued treating patients until the results looked conclusive, which turned out to occur after twenty patients had been observed. Such subtleties, as well as the definition and calculation of meaningful characteristics of sequential procedures, require the calculus of probability and in particular the study of random walks. This article illustrates the way in which various parts of probability theory arise as the tools for studying sequential procedures.

The formal mathematical study of sequential procedures was initiated during World War II by the Statistical Research Group at Columbia University, under the leadership of Abraham Wald. His work *Sequential Analysis* (1947) has served as the source book for many practical applications of the techniques and as the starting point for a great deal of additional theoretical development in the succeeding thirty years.

The motivation for developing sequential techniques during the war was that those techniques saved considerably over the classical nonsequential methods in the number of observations needed in sampling inspection of various mass-produced items in order to achieve specified standards of quality. This idea of saving observations will be illustrated in a common statistical setting in Section 3, as will be a second motivation: there are some statistical problems that cannot be handled at all, let alone efficiently, by nonsequential methods. For now, we illustrate both of these notions in an example chosen for its elementary probability structure rather than for practicality.

Suppose the value of an integer-valued physical parameter is to be guessed at from independent measurements $X_1, X_2, \ldots$, each of which takes on the value $\theta - 1$ or $\theta + 1$ with probability 1/2 each if the actual parameter value is θ. This true value of the parameter is unknown to the experimenter. If N observations $X_1, X_2, \ldots, X_N$ are taken, an integer-valued function $t_N(X_1, \ldots, X_N)$ is used to estimate the true value θ of the parameter, and we suppose the experimenter views the outcome as satisfactory only if t_N takes on exactly the value θ. By chance the X_i may turn out such that $t_N = \theta$ or $t_N \neq \theta$. In order to make clear our computation of the chance that an estimator performs unsatisfactorily, we must adopt a notation that makes precise *which* of the infinite family of probability mechanisms (corresponding to different values of the true θ) actually governs the X_i in terms of which the event under consideration is written. In statistical models, where a class of probability laws is under consideration, we denote by $P_\theta\{A\}$ the probability of an event A when θ is the true parameter value, using E_θ for the corresponding expectation operation. Then $P_\theta\{t_N(X_1, X_2, \ldots, X_N) \neq \theta\}$ is a measure of the inaccuracy of the procedure t_N in the present example, often called the *risk function* of the procedure, as a function of θ.

Suppose B_n is the experiment which, for a positive integer n chosen in advance of observing the X_i, takes exactly n measurements $X_1, \ldots, X_n$. This is the *nonsequential* or *fixed sample size* experiment with sample size $N = n$. Define $U_n = \min(X_1, \ldots, X_n)$ and $V_n = \max(X_1, \ldots, X_n)$. The estimator $t'_n(X_1, \ldots, X_n) = U_n + 1$ can be shown to yield a maximum (over θ) risk that is as small as that of any other estimator, and for it we have

$$P_\theta\{t'_n \neq \theta\} = P_\theta\{X_1 = X_2 = \cdots = X_n = \theta + 1\} = 2^{-n}. \qquad (1.1)$$

This can be made as small a positive value as the experimenter specifies, by taking n sufficiently large; but of course he must pay the expense of taking n observations. If the cost of experimentation is proportional to the number of observations taken (often but not always the case), the cost of this experiment is then cn where c is the cost per observation.

For a sequential procedure, the number of observations N is a random variable. (Here and in the sequel, the fixed sample size procedure based on n observations is viewed as a degenerate member of the class of all sequential procedures, with $P_\theta\{N = n\} = 1$ for all θ.) If each observation again costs c, the *expected* cost is $cE_\theta N$ when θ is true. Let N^* be the first n for which $V_n - U_n = 2$, and let B^* be the procedure which stops after N^* observations and estimates θ by $t'_{N^*} = U_{N^*} + 1$. Since $V_{N^*} - U_{N^*} = 2$, we obtain $P_\theta\{t'_{N^*} \neq \theta\} = 0$. Furthermore, for $m \geqslant 2$ we have

$$P_\theta\{N^* = m\} = P_\theta\{X_1 = X_2 = \cdots = X_{m-1} \neq X_m\} = 2^{1-m},$$

so that

$$E_\theta N^* = \sum_{m=2}^{\infty} m2^{1-m} = 3. \tag{1.2}$$

(This series may be summed as

$$\left[\frac{d}{dz}\sum_{2}^{\infty} z^m\right]\Bigg|_{z=1/2} = \left[\frac{d}{dz} z^2(1-z)^{-1}\right]\Bigg|_{z=1/2}.)$$

Thus, if $n > 3$, the procedure B_n is inferior to B^* in both respects, that the former results in a higher probability of yielding an incorrect estimate of θ, and also entails on the average taking more observations. In addition, we see that, if we required a procedure for which the probability of an incorrect estimate is 0, for no fixed sample size n would B_n suffice, but B^* would.

This example, extreme for the sake of simplicity, nevertheless illustrates a phenomenon that often enables sequential procedures to be superior to nonsequential ones: there can be "lucky observations." If $V_2 - U_2 = 2$, the procedure B^* takes advantage of the resulting perfect knowledge of θ to stop earlier than B_n does for $n \geqslant 3$; and if $U_n = V_n$, an unlucky circumstance in which B_n stops, B^* goes on to collect more information. In more realistic examples, the possible states of knowledge will not be so extreme and transparent, but the idea of lucky observation sequences remains.

Incidentally, the example also illustrates that the much used estimator $t''_n = n^{-1}\sum_1^n X_i$, the "sample mean," is not appropriate for all models; it performs much worse than t'_n in the above setting.

Next, to illustrate the way in which the interpretation of an outcome depends on the sampling rule used to achieve it, we consider another model with independent, identically distributed random variables (i.i.d. r.v.'s) $X_1, X_2, \ldots$, this time each X_i being "uniformly distributed" from 0 to an unknown parameter value $\theta > 0$, so that each X_i has continuous distribution function (df) given by

$$P_\theta\{X_i \leqslant x\} = \begin{cases} 0 & \text{if } x < 0, \\ x/\theta & \text{if } 0 \leqslant x \leqslant \theta, \\ 1 & \text{if } x > \theta; \end{cases} \tag{1.3}$$

the "uniform" density refers to the constant value θ^{-1} of (d/dx) $P_\theta\{X_i \leqslant x\}$ on $0 < x < \theta$. Thus, θ is an unknown upper bound on the possible value of an observable X_i, and the statistician wants to make some inference about θ. A type of inference somewhat different from estimation of a parameter as considered earlier is *hypothesis testing*, exemplified here as the process of guessing, when it is known only that $\theta \geqslant 1$, whether the true θ is 1 or is >1 (these being the two hypotheses). A statistical procedure t_n based on $X_1, \ldots, X_n$ now makes one of two assertions g_0 and g_1, where these denote, respectively, the guesses that "the true θ is 1" and that "the true θ is >1." By chance the X_i's can yield a correct or incorrect guess for a given procedure (just as in the previous example), and we see that there are two types of incorrect inference that are possible, guessing (from observations on the X_i's) g_0 when the true θ is >1, or guessing g_1 when $\theta = 1$. We would like both types of error to be unlikely; but for any specified number of observations, it is easy to see that a procedure can be chosen to make one error probability very small only at the expense of making the other large, at least for values θ near 1. Classically, a value α between 0 and 1 was often specified, reflecting the maximum error probability that the experimenter was willing to tolerate under that one of the two hypotheses that was singled out as representing (for example) a more traditional or conservative scientific theory. Subject to this restriction, one would select a procedure that would perform as well as possible, in some well-defined sense, under the other hypothesis.

In the present example of uniformly distributed X_i, let us consider nonsequential procedures t_n based on $X_1, X_2, \ldots, X_n$, and suppose $\theta = 1$ is the hypothesis under which the probability of error is to be $\leqslant \alpha$. If we again write $V_n = \max_{1 \leqslant i \leqslant n} X_i$, it can be shown that, among all procedures t_n for which the probability of error is $\leqslant \alpha$ when $\theta = 1$, the procedure t_n^* defined by

$$t_n^*(X_1, \ldots, X_n) = \begin{cases} g_0 & \text{if } V_n \leqslant (1-\alpha)^{1/n}, \\ g_1 & \text{if } V_n > (1-\alpha)^{1/n}, \end{cases} \tag{1.4}$$

yields the maximum probability of a correct guess when $\theta > 1$. Since, for $0 < c < \theta$,

$$P_\theta\{V_n \leqslant c\} = P_\theta\{X_1 \leqslant c, X_2 \leqslant c, \ldots, X_n \leqslant c\} = (c/\theta)^n, \tag{1.5}$$

the probability of an incorrect guess under t_n^* is

$$P_\theta\{t_n^* \text{ is incorrect}\} = \begin{cases} \alpha & \text{if } \theta = 1, \\ (1-\alpha)/\theta^n & \text{if } \theta > 1. \end{cases} \tag{1.6}$$

(In this elementary setting, other procedures than that of (1.4) also yield (1.6); the one given by (1.4) is often used because it also performs optimally for true values $\theta < 1$ if such values are possible, where g_0 is also viewed as correct when $\theta \leqslant 1$.)

If α is a small number, such as 10^{-3}, scientists would often publish the fact that $t_n^* = g_1$ as fairly conclusive evidence that the true θ is not 1; for the small probability ($\leqslant .001$) of having the X_i unluckily turn out to yield g_1 when in fact $\theta = 1$ leads us to turn to belief in the alternative that $\theta > 1$.

Suppose a dishonest scientist (perhaps a consultant to the manufacturer of a stultifying aerospray the public will buy if it can be shown $\theta > 1$) wants to be sure to publish the conclusion g_1 using t_n^*, with a small α such as .001, *even if* $\theta = 1$. He can achieve this by using a sequential procedure while pretending he is not doing so, as follows: Let $N\#$ be the first integer n for which $V_n > (1-\alpha)^{1/n}$, if such an n exists; if no such n exists with positive probability for some θ value, we shall say $N\#$ is *not well defined* for that θ value. If $N\# = n$, the scientist pretends he had decided on the sample size n and procedure t_n^* from the outset, and according to (1.6) can thus publish the statement indicated at the end of the previous paragraph. But actually he is using a sequential procedure based on $N\#$, and as long as

$$P_\theta\{N\# \text{ is well defined}\} = 1 \quad \text{for } \theta \geqslant 1 \tag{1.7}$$

the resulting procedure $t^*_{N\#}$ achieves

$$P_\theta\{t^*_{N\#} \text{ is incorrect}\} = \begin{cases} 1 & \text{if } \theta = 1, \\ 0 & \text{if } \theta > 1, \end{cases} \tag{1.8}$$

in contrast to the claim (1.6) he is making.

We shall verify (1.7) when $\theta = 1$, it being more trivial when $\theta > 1$. For $i \geqslant 1$, let A_i be the event that $N\# \leqslant 2^i$, and let A_i^c be its complement. Clearly A_{i+1}^c entails

$$\max_{2^i < j \leq 2^{i+1}} X_j \leqslant (1 - \alpha)^{1/2^{i+1}},$$

and since this last event is independent of $A_1, \ldots, A_i$, we have

$$P_{\theta=1}\{A_{i+1}^c \mid A_1^c \cap A_2^c \cap \cdots \cap A_i^c\} \leqslant P_{\theta=1}\{\max_{2^i < j \leq 2^{i+1}} X_j \leqslant (1-\alpha)^{1/2^{i+1}}\} = (1 - \alpha)^{1/2}. \tag{1.9}$$

Since the sets A_i^c are decreasing, we obtain

$$P_{\theta=1}\{A_k^c\} = P\left\{\bigcap_{i=1}^{k} A_i^c\right\}$$
$$= P\{A_1^c\} \prod_{i=1}^{k-1} P\left\{A_{i+1}^c \,\middle|\, \bigcap_{j=1}^{i} A_j^c\right\} \leqslant (1 - \alpha)^{(k+1)/2}, \tag{1.10}$$

from which $P_{\theta=1}\{A_k\} \to 1$ as $k \to \infty$, so that (1.7) is satisfied.

2. RANDOM WALKS AND MARTINGALES

Throughout this section we shall restrict consideration to discrete random variables X_i, Y_i, Z_i, in order to dispense with the measure-theoretic details that are inessential to the concepts of interest here.

The analysis of sequential procedures for many models that are less simple than those of Section 1, can be made in terms of certain chance processes we now discuss. Let $Y_1, Y_2, \ldots$, be i.i.d. r.v.'s. Define $S_0 = 0$ and $S_n = \sum_1^n Y_i$ for $n \geqslant 1$. The process $S_0, S_1, S_2, \ldots$, is called a *random walk* (more precisely, a one-dimensional random walk with discrete time, and which is homogeneous in space and time: the law of the motion $S_{n+1} - S_n$, from "time" n to $n + 1$, is independent of the position S_n and of the time n). The possible values of the sequence $\omega = (S_0, S_1, S_2, \ldots)$ can be thought of as points in the probability space of the random walk, and it is often convenient to join successive pairs (n, s_n), $n \geqslant 0$, to make a piecewise-linear graph that can be viewed as the "sample path" graph of the random function ω. In the next section we will consider such random walks with "absorbing sets" defined by two sequences $\{a_n, n > 0\}$ and $\{b_n, n > 0\}$ with $b_n \leqslant a_n$; the stopping variable N is defined as $N = \min\{n : S_n \leqslant b_n \text{ or } S_n \geqslant a_n\}$, and in physical applications one thinks of the "particle" whose path is given by ω as being "absorbed" by the set $[a_N, +\infty)$ or $(-\infty, b_N]$ into which it falls when the process is stopped (Fig. 1).

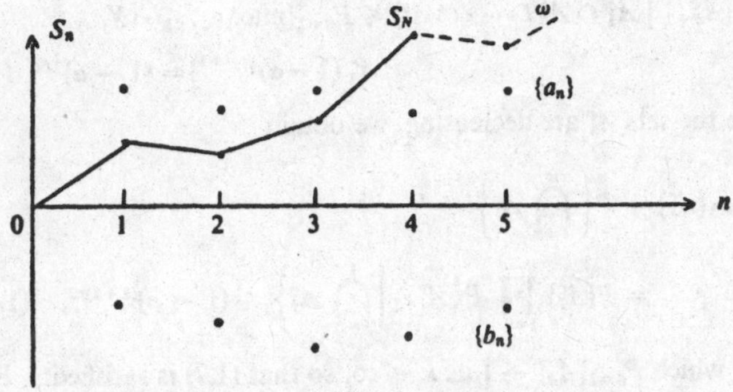

FIG. 1

If $EY_i = 0$, we see that, for any sequence of possible values $(0, s_1, \ldots, s_n)$ of $(S_0, S_1, \ldots, S_n)$, we have

$$E\{S_{n+1} \mid S_0 = 0, S_1 = s_1, \ldots, S_n = s_n\} = s_n$$

for the conditional expectation of the position S_{n+1} given the "past history." This means that the process $\{S_n\}$ is a special case of a 0-expectation *martingale*, a chance process $\{Z_n, n \geqslant 0\}$ with the properties $Z_0 = 0$, $E|Z_n| < \infty \forall n$, and

$$E\{Z_{n+1} \mid Z_0 = 0, Z_1 = z_1, \ldots, Z_n = z_n\} = z_n \tag{2.1}$$

for all n and any possible realization $(0, z_1, \ldots, z_n)$ of the past. This is more general than the 0-expectation random walk because the

$Z_{n+1} - Z_n$ do not have to be i.i.d. It is an immediate consequence of (2.1) that $EZ_n = 0 \forall n$.

Martingales are a mathematical model of "fair games of chance." If you play a game repeatedly and Z_n is your total "profit" (negative profit being loss) after n plays of the game, the game may be thought to be fair if, no matter what the history of your successes through n plays of the game, the conditional expectation of the amount $Z_{n+1} - Z_n$ you win on the $(n + 1)$st play is 0. That is precisely (2.1).

Everyone wants a strategy that will let him quit while ahead. A *stopping time* N based on $\{Z_i\}$ is a positive integer-valued random variable with the property that, for each positive integer n, the event $\{N = n\}$ is a set in the space of $Z_1, Z_2, \ldots, Z_n$. In the gambling context, this formalizes the notion that you can not in practice decide whether to stop gambling at time 10 on the basis of knowledge of whether or not the 11th play is to be favorable for you.

Can we obtain a favorable stopping strategy N, one for which $EZ_N > 0$, for any simple games? If N is bounded above, it is not hard to use (2.1) and to work backward from that upper bound (as $n + 1$), step by step, to show $EZ_N = 0$. On the other hand, in the simple game of repeatedly flipping a fair coin (with independent flips) while betting an adversary one unit on each flip that you will guess the outcome correctly, so that the amount you win on the nth flip (regardless of how you guess) is $Z_{n+1} - Z_n = Y_n$ where the Y_i are independent and $P\{Y_i = 1\} = P\{Y_i = -1\} = 1/2$, a well-known strategy is to "continue betting until you're 1 unit ahead, and then stop." In other words, $N = \min\{n : Z_n = 1\}$, in which case $Z_N = 1$ with probability one, and hence $EZ_N = 1$. The only feature of this development that is not obvious is that N is well defined with probability one, and this can be proved by an analogue of the proof of (1.7): defining $n_1 = 2$ and $n_{i+1} - n_i = n_i^2$, it can be shown, in analogy with (1.9), that $P\{Z_{n_{i+1}} \geqslant 1 \mid Z_{n_i} = z\} > \epsilon$ for all $i > 1$ and all $z \geqslant -n_i$ (the minimum possible value), for some $\epsilon > 0$.

However, a "practical" difficulty of this strategy is that it requires infinite capital to use it, since Z_n can be arbitrarily large and negative for some n prior to N. A theoretical difficulty of the strategy is that, although $P\{N < \infty\} = 1$, it can be shown that the probability law of N has such large "tails" that $EN = \sum_{n=1}^{\infty} nP\{N = n\} = +\infty$. Thus, although the gambler with infinite capital would eventually win with this strategy, the expected length of time it would take him to do it is infinite.

This last difficulty can be eliminated by using a different betting strategy, if your opponent will permit it: instead of betting one unit on each flip of the coin, you double your bet until you win. This means $Z_{n+1} - Z_n = \pm 2^n$, each possibility with probability 1/2, for $n \geqslant 0$, and again that $N = \min\{n : Z_n = 1\}$ and hence $Z_N = 1$ with probability 1. In this case $N = n$ if you lose the first $n - 1$ matches and win the nth, an event of probability 2^{-n}. This decreases rapidly, and $EN = \sum_{n=1}^{\infty} n2^{-n} = 2$ (see (1.2)), so the shortcoming of the previous strategy has been overcome. Alas, it has been replaced by another: the *expected amount you bet* on the final play is $E|Z_N - Z_{N-1}| = E2^{N-1} = \sum_{n=1}^{\infty} 2^{n-1}2^{-n} = +\infty$. Thus, the "practical" difficulty of the previous paragraph is even more evident here.

The peculiarities of these two strategies typify what one must resort to in order to obtain a positive expected gain from imposing a stopping time on a fair game. Doob's *optional stopping theorem* for martingales makes precise the impossibility of designing a well-behaved winning strategy, by asserting that, if the laws of N and $Z_{n+1} - Z_n$ are suitably regular, $EZ_N = 0$. The result has an especially simple statement in the case of the random walk $Z_n = S_n$:

$$EY_i = 0 \quad \text{and} \quad EN < \infty \quad \text{imply} \quad ES_N = 0. \tag{2.2}$$

This last is a version of what is known as *Wald's equation*, and it can be proved by using the validity of (2.2) for the previously mentioned case of a bounded stopping variable $N_m = \min(N, m)$ and letting $m \to \infty$. We shall use (2.2) in the next section.

3. WALD'S SEQUENTIAL PROBABILITY RATIO TEST

Suppose $X_1, X_2, \ldots,$ are i.i.d. r.v.'s, it being known that the common probability function of the X_i is one of two different specified laws f_0 and f_1, where $f_j(x) = P_j\{X_i = x\}$; it is unknown whether f_0 or f_1 is the true law, and the problem, in the language of Section 1, is to decide between these two hypotheses. Again we illustrate the ideas in the discrete case, and for convenience we assume f_0 and f_1 positive on the same domain.

We define the r.v. $\lambda_n = \prod_{i=1}^{n} [f_1(X_i)/f_0(X_i)]$, the "probability ratio" of the observations under the two possible laws. Intuitively, if $X_1, \ldots, X_n$ turn out such as to make λ_n large, one might feel secure in making the guess g_1 that f_1 is the true law, and if λ_n is small one would vote for f_0. This notion is made precise in the Neyman-Pearson Lemma, according to which the nonsequential procedure

$$t_n^*(X_1, \ldots, X_n) = \begin{cases} g_0 & \text{if } \lambda_n < c, \\ g_1 & \text{if } \lambda_n \geqslant c, \end{cases} \tag{3.1}$$

where c is a nonnegative constant, has the property that, if t_n is any other procedure, and

$$P_0\{t_n \neq g_0\} \leqslant P_0\{t_n^* \neq g_0\}, \tag{3.2}$$

then

$$P_1\{t_n \neq g_1\} \geqslant P_1\{t_n^* \neq g_1\}. \tag{3.3}$$

Thus, as indicated in connection with (1.4) (which conforms with (3.1) in an extreme example), no procedure can simultaneously minimize both probabilities of error $P_i\{t_n \neq g_i\}$, but for each value of c the corresponding t_n^* has the property that no t_n can be better than it in terms of both of these error probabilities. Alternatively, in terms of the language used in Section 1, if c has been chosen in such a way as to make $P_0\{t_n^* \neq g_0\}$ equal a specified value α, then among all t_n with $P_0\{t_n \neq g_0\} \leqslant \alpha$, the procedure t_n^* minimizes $P_1\{t_n \neq g_1\}$. The Neyman-Pearson Lemma is not difficult to prove, but we shall not take the space to prove it here.

Intuitively, large or small values of λ_n seem more conclusive than values near 1. This suggests the possible wisdom of using a sequential procedure that continues to observe the X_i until the first time $N = n$ that λ_n is "suitably large or small." This last means choosing two sequences A_n and B_n with $0 < B_n < A_n < \infty$ and stopping at $N = \min\{n : \lambda_n \leqslant B_n \text{ or } \lambda_n \geqslant A_n\}$. The simplest such sequence is obtained by taking A_n and B_n to be constants A and B, independent of n, and which satisfy $0 < B < 1 < A < \infty$. This is the form of Wald's *sequential probability ratio test:*

$$\bar{N} = \min\{n : \lambda_n \leqslant B \text{ or } \lambda_n \geqslant A\},$$
$$\bar{t}_{\bar{N}} = \begin{cases} g_0 & \text{if } \lambda_{\bar{N}} \leqslant B, \\ g_1 & \text{if } \lambda_{\bar{N}} \geqslant A. \end{cases} \tag{3.4}$$

This procedure takes advantage of "lucky observation sequences" as described in Section 1, to stop at an early time n if λ_n is far from 1 then.

The simple form of this test enabled Wald to compute good approximations to the performance characteristics $P_i\{\bar{t}_{\bar{N}} \neq g_i\}$ and $E_i\bar{N}$ discussed in the introduction, and also suggested an optimum property of this test among all sequential (and nonsequential) tests. We shall discuss these in this section and will also give a simple numerical example. First we note that, if we write $Y_i = \log[f_1(X_i)/f_0(X_i)]$, the Y_i are i.i.d. r.v.'s under either f_0 or f_1 (as the law of X_i), and in the notation of the random walk of Section 2 we can rewrite (3.4) as

$$\bar{N} = \min\{n : S_n \leqslant \log B \text{ or } S_n \geqslant \log A\}$$
$$\bar{t}_{\bar{N}} = \begin{cases} g_0 & \text{if } S_{\bar{N}} \leqslant \log B, \\ g_1 & \text{if } S_{\bar{N}} \geqslant \log A. \end{cases} \tag{3.5}$$

We can thus think of Wald's test in terms of a random walk with one of two laws and with the absorbing sets of Figure 1 defined by $a_n = \log A$ and $b_n = \log B$. That, in accordance with our intuition, the random walk tends to drift toward the upper barrier $\log A$ if f_1 is true and toward the lower barrier $\log B$ if f_0 is true, is indicated by the fact that

$$E_0 Y_1 < 0 < E_1 Y_1, \tag{3.6}$$

which is a consequence of Jensen's inequality: since $\log y$ is concave, the average of its value at several points is less than its value at the average point. Hence,

$$\begin{aligned} E_0 Y_1 = E_0 \log[f_1(X_1)/f_0(X_1)] &< \log E_0[f_1(X_1)/f_0(X_1)] \\ &= \log \sum_x [f_1(x)/f_0(x)] f_0(x) = \log 1 = 0, \end{aligned} \tag{3.7}$$

the strictness of the inequality following from the fact that there is more than one possible value of Y_1 since the f_i are different. The other half of (3.6) is proved similarly.

For any well-defined N and associated guessing procedure t_N, we hereafter abbreviate $P_i\{t_N \neq g_i\}$ by $\pi_i(t_N)$.

It is not hard to verify that $\bar{N}$ is well defined and that $E_i\bar{N} < \infty$. Simple inequalities on the probabilities of error for $\bar{t}_{\bar{N}}$ are

$$\pi_0(\bar{t}_{\bar{N}})/[1 - \pi_1(\bar{t}_{\bar{N}})] \leqslant A^{-1}; \qquad \pi_1(\bar{t}_{\bar{N}})/[1 - \pi_0(\bar{t}_{\bar{N}})] \leqslant B. \quad (3.8)$$

The first of these, for example, is proved by noting that, if $\bar{Q}_n = \{(x_1, \ldots, x_n) : \bar{N} = n, \bar{t}_{\bar{N}} = g_1\}$, we have $\lambda_n \geqslant A$ on $\bar{Q}_n$, and thus

$$\begin{aligned}\pi_0(\bar{t}_{\bar{N}}) &= \sum_{n=1}^{\infty} P_0\{\bar{Q}_n\} = \sum_{n=1}^{\infty} \sum_{\bar{Q}_n} \prod_{i=1}^{n} f_0(x_i) \\ &\leqslant \sum_{n=1}^{\infty} \sum_{\bar{Q}_n} A^{-1} \prod_{i=1}^{n} f_1(x_i) = A^{-1} \sum_{n=1}^{\infty} P_1\{\bar{Q}_n\} \\ &= A^{-1}[1 - \pi_1(\bar{t}_{\bar{N}})]. \end{aligned} \quad (3.9)$$

The inequality in (3.9) arises because $S_{\bar{N}}$ might be $> \log A$ rather than exactly $\log A$, on $\bar{Q}_n$. If $S_{\bar{N}}$ can only take on the values $\log A$ and $\log B$, the inequalities (3.8) become equalities, and we obtain $\pi_0(\bar{t}_N) = (1 - B)/(A - B)$ and $\pi_1(\bar{t}_N) = B(A - 1)/(A - B)$. In practice, even when $S_{\bar{N}}$ is not of this special character, these formulas usually give good approximations.

We now turn to a computation of a lower bound on E_iN for any well-defined stopping time N for which $E_iN < \infty$ and $\pi_i(t_N) < 1$ for both i. First, writing $Q_n = \{(x_1, \ldots, x_n) : N = n, t_N = g_1\}$, we compute in a manner similar to (3.9),

$$\begin{aligned}E_1\{1/\lambda_N \mid t_N = g_1\} &= \sum_n \sum_{Q_n} \left[\prod_1^n (f_0(x_i)/f_1(x_i))\right] \prod_1^n f_1(x_i)/P_1\{t_N = g_1\} \\ &= \pi_0(t_N)/[1 - \pi_1(t_N)]. \end{aligned} \quad (3.10)$$

Next, using Jensen's inequality as in (3.7), with (3.10), we obtain

$$\begin{aligned}E_1\{S_N \mid t_N = g_1\} &= -E_1\{\log(1/\lambda_N) \mid t_N = g_1\} \geqslant -\log E_1\{1/\lambda_N \mid t_N = g_1\} \\ &= \log\{[1 - \pi_1(t_N)]/\pi_0(t_N)\}. \end{aligned} \quad (3.11)$$

A similar computation yields

$$E_1\{S_N \mid t_N = g_0\} \geqslant \log\{\pi_1(t_N)/[1 - \pi_0(t_N)]\}.$$

Together these give

$$\begin{aligned} E_1 S_N &= \sum_{i=0}^{1} E_1\{S_N \mid t_N = g_i\} P_1\{t_N = g_i\} \\ &\geqslant \pi_1(t_N) \log\{\pi_1(t_N)/[1 - \pi_0(t_N)]\} \\ &\quad + [1 - \pi_1(t_N)] \log\{[1 - \pi_1(t_N)]/\pi_0(t_N)\} \\ &= q(\pi_0(t_N), \pi_1(t_N)) \text{ (say).} \end{aligned} \tag{3.12}$$

Defining Z_n as the sum $\sum_1^n (Y_i - E_1 Y_i)$, still with

$$Y_i = \log[f_1(X_i)/f_0(X_i)],$$

we may apply Wald's equation (2.2) to obtain $E_1 S_N = (E_1 Y_1) E_1 N$ and hence, from (3.12) and (3.6),

$$E_1 N \geqslant (E_1 Y_1)^{-1} q(\pi_0(t_N), \pi_1(t_N)). \tag{3.13}$$

Similarly, one can prove that

$$E_0 N \geqslant (-E_0 Y_1)^{-1} q(\pi_1(t_N), \pi_0(t_N)). \tag{3.14}$$

We note that the inequality sign in (3.12), and thus in (3.13) or (3.14), becomes equality if S_N can take on only one value when $t_N = g_1$ and only one when $t_N = g_0$. For Wald's $\bar{N}$, this means that if $S_{\bar{N}}$ can only take on the values $\log B$ and $\log A$, the right sides of (3.13) and (3.14) give exact expressions for $E_i\bar{N}$. As in the case of (3.8), even when $S_{\bar{N}}$ is not of this special character, the expressions of (3.13) and (3.14) often give excellent approximations for Wald's stopping rule, in practice.

Wald conjectured that the simple sequential rule (3.4) possesses a very strong optimum character, compared with all others. Suppose we specify positive values α_0 and α_1, α_j being the maximum probability of an incorrect guess that we are willing to tolerate when f_j is the true law of the X_i. Subject to the two restrictions $\pi_i(t_N) \leqslant \alpha_i$ $(i = 0, 1)$, we might seek the procedure (N, t_N) that minimizes $E_0 N$; or we might try to find the procedure that minimizes $E_1 N$. In mathematics one may expect two different minimization problems to have different solutions; the problems are concerned with achieving quick stopping time on the average under two quite different drifts, as indicated in (3.6). However, remarkably, if there is a Wald procedure (3.4) for which $\pi_i(\bar{t}_{\bar{N}}) = \alpha_i$ for $i = 0, 1$ (which there always is if $\alpha_0 + \alpha_1 < 1$ and Y_1 has a continuous law under both f_i), then that procedure is simultaneously the solution to both problems! This is stated as:

Optimum character of Wald's test. *If* $(\bar{N}, \bar{t}_{\bar{N}})$ *is a procedure of the form* (3.4) *and* (N, t_N) *is any other procedure for which*

$$\pi_i(t_N) \leqslant \pi_i(\bar{t}_{\bar{N}}), \qquad i = 0, 1, \tag{3.15}$$

then

$$E_i N \geqslant E_i \bar{N}, \qquad i = 0, 1. \tag{3.16}$$

Wald proved this in his book assuming $E_i N < \infty$ and that $S_{\bar{N}}$ can only equal $\log A$ or $\log B$. This last assumption is very restrictive, and the optimum character in the general setting without this assumption was first proved by Wald and Wolfowitz (1948). The proof assuming that $S_{\bar{N}}$ can only equal $\log A$ or $\log B$ and that the $E_i N$ are finite is quite easy: Differentiation shows that $q(\pi_0, \pi_1)$ is decreasing in its arguments provided $\pi_0 + \pi_1 < 1$, a condition always satisfied for a Wald procedure. Hence, from (3.15), the expressions on the right sides of (3.13) and (3.14) are always at least equal to the corresponding expressions with the $\pi_i(\bar{t}_{\bar{N}})$ in place of the $\pi_i(t_N)$. On the other hand, our earlier remarks indicated that these last expressions are equal to the $E_i \bar{N}$ for $S_{\bar{N}}$ of the assumed special character. While this proof does not carry over to the general case, it makes the conclusion there plausible.

To illustrate the savings in $E_i N$ that are possible from using a Wald sequential test rather than a nonsequential one, we consider the problem of deciding whether a coin has probability .4 or .6 of coming up "heads" on a single toss, assuming those are known to be the only possible values of that probability. If $X_i = 1$ or -1 corresponding to a head or tail, this means that

$$f_0(-1) = .6 = 1 - f_0(1), \qquad f_1(-1) = .4 = 1 - f_1(1), \tag{3.17}$$

with the $f_i(x) = 0$ otherwise. Consequently, Y_i can only take on the values $\pm \log 1.5$, with

$$P_1\{Y_i = \log 1.5\} = .6, \qquad P_0\{Y_i = \log 1.5\} = .4. \tag{3.18}$$

If, for example, we specify $\alpha_0 = \alpha_1 = 1/25$ as the maximum probability of error we will tolerate under either f_0 or f_1, binomial probability function tables show that 75 observations are needed for a nonsequential test to achieve this, the optimum (Neyman-Pearson) procedure being $t_{75} = g_1 \Leftrightarrow S_{75} > 0$ (which yields $\pi_i(t_{25}) = .0396$). For a sequential Wald test, we take $\log A = -\log B$ to be of the form $m \log 1.5$ where m is an integer. Then $S_{\bar{N}}$ can only equal $\log A$ or $\log B$, and the discussion below (3.9) shows that, in order to achieve $\pi_0(\bar{t}_{\bar{N}}) = \pi_1(\bar{t}_{\bar{N}}) \leqslant 1/25$, we need only solve $1/25 \geqslant (1 - B)/(A - B) = 1/(A + 1)$, for which we only need $m = 8$, the smallest integer $\geqslant \log 24/\log 1.5$ (which yields $\pi_i(\bar{t}_{\bar{N}}) = .0376$). The expressions (3.13) and (3.14) are equalities for the Wald test, and (since (3.18) yields $E_1 Y_1 = .2 \log 1.5$) they yield $E_i\bar{N} = 37.0$. Thus, the expected number of observations required by the Wald test is slightly *less than half* that required by the best nonsequential test. This is typical of what occurs in more complex settings.

4. SEQUENTIAL PROCEDURES FOR OTHER PROBLEMS

The striking success of sequential analysis in Section 3 was presented there for the simple hypothesis testing problem in which there was only one possible probability law of the X_i under each hypothesis. In this section we shall mention briefly a few of the many more complicated statistical settings which have been handled with sequential procedures. For simplicity, for the most part we consider once more flips of a coin, but no longer with the probability of a head being restricted to two possible values as in the example of Section 3. Thus, we consider i.i.d. "Bernoulli r.v.'s" X_i with

$$P_p\{X_i = 1\} = 1 - P_p\{X_i = -1\} = p \tag{4.1}$$

for $0 < p < 1$.

(A) First suppose, in extension of the hypothesis testing problem of Section 3, that we want to guess whether the true but unknown value of p that governs this model satisfies $0 < p < 1/2$ or $1/2 < p < 1$—do we guess that the coin is biased in favor of tails (g_0) or in favor of heads (g_1)? A moment's reflection shows that, since the law of X_i is almost the same when $p = 1/2 - \epsilon$ as when $p = 1/2 + \epsilon$ when ϵ is small, no nonsequential procedure can achieve

$$\sup_{p \neq 1/2} P_p\{t_N \text{ is incorrect}\} \leqslant \alpha_0, \tag{4.2}$$

no matter how large the fixed sample size $N = n$ is, if $\alpha_0 < 1/2$. Similarly, the requirement (4.2) cannot be achieved by the sequential procedure that stops the first time $\sum_1^n X_i$ is $\leqslant C$ or $\geqslant D$ (this being the form that (3.5) takes for testing whether $p = 1/2 - \epsilon_0$ or $p = 1/2 + \epsilon_0$, for a *fixed* ϵ_0). The simplest type of procedure that will achieve (4.2) is based on a stopping rule $N' = \min\{n : |\sum_1^n X_i| \geqslant a_n\}$, where $\{a_n\}$ is a sequence of positive constants for which

$$\sup_{p<1/2} P_p\left\{\sup_{n>0}\left(\sum_1^n X_i - a_n\right) \geqslant 0\right\} \leqslant \alpha_0 \tag{4.3}$$

and

$$P_p\left\{\inf_{n>0}\left(\sum_1^n X_i + a_n\right) < 0\right\} = 1, \qquad 0 < p < 1/2. \tag{4.4}$$

If we use such an N' with $t_{N'} = g_0$ or g_1 according to whether $\sum_1^{N'} X_i < 0$ or > 0, equation (4.4) insures that N' is well defined, since, for $p < 1/2$, the random walk $\sum_1^n X_i$ eventually hits or crosses the lower barrier $\{-a_n, n > 0\}$, if it does not cross the upper barrier first. On the other hand, (4.3) insures that, when $p < 1/2$, the probability that the random walk *ever* hits the upper barrier $\{a_n, n > 0\}$, which is greater than the probability that it results in the incorrect guess g_1 due to hitting the upper barrier before the lower one, is $\leqslant \alpha_0$. The case $p > 1/2$ is covered similarly.

The condition (4.4) follows if the a_n satisfy

$$\lim_{n\to\infty} a_n/n = 0, \tag{4.5}$$

since the well-known law of large numbers asserts for these X_i with $E_pX_i = 2p - 1$, that

$$\lim_{n\to\infty} P_p\left\{\left|n^{-1}\sum_1^n X_i - (2p-1)\right| < \epsilon\right\} = 1 \tag{4.6}$$

for every $\epsilon > 0$; choosing ϵ to be $<1 - 2p$, we obtain (4.4). In order to study (4.3), we first note the intuitively plausible statement that the probability of (4.3) is nondecreasing in p. This statement can be proved by constructing r.v.'s X_i' and X_i'' as functions of the same r.v. H_i uniformly distributed on (0, 1), with $X_i' = 1$ or $= -1$ depending on whether $H_i \leqslant p'$ or $>p'$, and with $X_i'' = 1$ or -1 depending on whether $H_i \leqslant p''$ or $>p''$; then X_i' and X_i'' have the law (4.1) with $p = p'$ and p'', and if $p' < p''$ we have $X_i'' \geqslant X_i'$ and thus $\sum_1^n X_i'' \geqslant \sum_1^n X_i'$, from which the monotonicity of (4.3) follows. As a consequence of this monotonicity, (4.3) follows from

$$P_{1/2}\left\{\sup_{n>0}\left(\sum_1^n X_i - a_n\right) \geqslant 0\right\} \leqslant \alpha_0. \tag{4.7}$$

(We use the law (4.1) with $p = 1/2$ in (4.7), even though it is not considered to be a possible law for the actual coin at hand.) There are many possible sequences $\{a_n\}$ that satisfy (4.5) and (4.7). From our earlier discussion we know that a_n cannot be chosen constant, and we seek unbounded a_n. Elementary probability inequalities on the tails of the binomial probability law (e.g., in Feller (1950)) show that $a_n = Cn^\beta$, with $1/2 < \beta < 1$ and C sufficiently large, suffices; this is reflected in the fact that Cn^β is $Cn^{\beta-1/2}$ "standard deviation units" of the law of $\sum_1^n X_i$. It is thus not hard to show that

$$\sum_{n=1}^{\infty} P_{1/2}\left\{\sum_1^n X_i \geqslant a_n - 1\right\} \leqslant \alpha_0$$

for C sufficiently large, and this sum dominates the probability of (4.7). In order that E_pN' not be too large when p is near 1/2 (the most difficult values of p to judge correctly), we want the a_n to increase as slowly as possible with n subject to (4.7), and this suggests letting β be near 1/2. In fact, it can be shown that $Cn^{1/2}\log n$ also works for appropriate C. A more subtle argument is involved in showing that, for each $\delta > 0$, the sequence

$$C + [(2 + \delta)n \log\log(n + 3)]^{1/2}$$

works for sufficiently large C. In a sense this is the smallest type of regular sequence that works, in that for $\delta < 0$ it can be shown that the probability of (4.7) is 1, and that the left side of (4.2) equals 1/2 for that sequence. The properties of these borderline sequences are the subject of the celebrated "Law of the Iterated Logarithm" of probability theory. (See Feller (1950).)

The property (4.7) implies that, when $p = 1/2$, the r.v. N' is not well defined, since with probability at least $1 - 2\alpha_0$ the rule never tells us to stop! This is not merely a property of rules constructed through the use of (4.3)–(4.4), and it can be shown that every stopping rule that satisfies (4.2) has this property. We excluded $p = 1/2$ from our set of possible values governing the law of the X_i, but in our next example we use this property of the value $p = 1/2$ (and a corresponding one for $p > 1/2$).

(B) Our second problem is to find a procedure that still achieves (4.2) for $p \geqslant 1/2$, but that demands *no error* if $p < 1/2$; that is, we require

$$P_p\{t_N \text{ incorrect}\}\begin{cases} \leqslant \alpha_0 & \text{if } p \geqslant 1/2, \\ = 0 & \text{if } p < 1/2. \end{cases} \tag{4.8a}$$

The idea of considering such a requirement is due to Robbins. It is impossible to achieve (4.8a) if N is well defined for all p, and the essence of the procedure is that it makes use of the fact that $P_p\{\text{the procedure never stops}\} \geqslant 1 - \alpha_0$ for $p \geqslant 1/2$ to assert the first half of (4.8a). The second half of (4.8a) can be written in the stronger form $P_p\{N \text{ well defined and } t_N \text{ correct}\} = 1$ for $p < 1/2$. A procedure that achieves these results is obtained by using only the lower boundary $\{-a_n\}$ of the previous problem: now $N = \min\{n : \sum_1^n X_i \leqslant -a_n\}$, and otherwise N is undefined. We guess g_0 if we stop. Then (4.4) yields the second half of (4.8a), and the first half follows from the analogue for $p > 1/2$ of (4.3) and (4.7):

$$\sup_{p \geq 1/2} P_p\left\{\inf_{n>0}\left(\sum_1^n X_i + a_n\right) \leqslant 0\right\} = P_{1/2}\left\{\inf_{n>0}\left(\sum_1^n X_i + a_n\right) \leqslant 0\right\} \leqslant \alpha_0. \tag{4.8b}$$

One may imagine applications of such a test: a physical process of some kind is "in control" if $p \geqslant 1/2$, and out of control (hence, requiring action) if $p < 1/2$. In the latter case, we want definitely to stop and "guess g_0." In the former case, we keep observing except for the probability α_0 that we incorrectly stop and go to the trouble of attempting to fix the process when it is actually under control. Obviously, the sense in using such a procedure will depend on the cost of taking many observations, perhaps of never stopping, compared with the possible cost of not detecting that the process is out of control when $p < 1/2$.

In certain extreme (non-Bernoulli) models for the law of the X_i's (replacing (4.1)) it is possible to find a sequential test that achieves $P_\theta\{N \text{ well defined and } t_N \text{ correct}\} = 0$ for all true values θ of the parameter value, even though the two sets of values $\{\theta : g_i \text{ is a correct guess}\}$ represent collections of laws with a common limit point (as with $p = 1/2$ in the example just above). For example, if X_i has uniform density from $\theta - 1/2$ to $\theta + 1/2$, for deciding whether $\theta < 0$ or $\theta > 0$ we need only stop at $N = \min\{n : |X_n| > 1/2\}$ and make the obvious guess.

(C) For a final problem requiring the use of sequential analysis, we consider estimation of the parameter p of the law (4.1) of the X_i where $0 < p < 1$ and p is otherwise unknown. If, for some specified value c with $0 < c < 1$, we regard a guess (estimate) of the true value of p as correct if it differs by at most c units from p, and want an estimator t_n that is incorrect with probability at most α_0 (specified), there is no difficulty in finding an n for which the nonsequential procedure based on n observations, and which estimates p by

$$t_n(X_1, \ldots, X_n) = (n^{-1} \sum_1^n X_i + 1)/2$$
$$= n^{-1} \text{ (number of heads in } n \text{ tosses)}, \qquad (4.9)$$

achieves this goal. However, suppose instead of wanting the *absolute error* $|t_n - p| \leqslant c$ as above that we are interested in making the *relative error* small, so that we want, for specified positive $c' < 1$ and α_0' (with $0 < \alpha_0' < 1$),

$$\sup_{0<p<1} P_p\{|t_n(X_1, \ldots, X_n) - p|/p \geqslant c'\} \leqslant \alpha_0'. \qquad (4.10)$$

We now show that no fixed sample size n suffices. For, whatever value n is chosen, it is clear from (4.1) that

$$\lim_{p \downarrow 0} P_p\{X_1 = X_2 = \cdots = X_n = -1\} = 1.$$

Hence, whatever value $t_n(-1, -1, \ldots, -1) = h$ (say) may be, the probability approaches 1 as $p \downarrow 0$, that $|t_n - p|/p = |h/p - 1|$, which is either unbounded (if $h \neq 0$) or 1 (if $h = 0$ is permitted as a guess, even though it is not a possible value of p). In either case, (4.10) is violated.

The difficulty here occurs for p near 0. If there were a known value $r > 0$ and we *knew* $r \leqslant p < 1$, it would be easy to solve the problem with a nonsequential procedure. For example, suppose with $c = c'r$ that $\nu(r)$ is an integer such that taking $n = \nu(r)$ observations solves the original "absolute error" problem of the previous paragraph. Then, for $p \geqslant r$, we have

$$\alpha_0 \geqslant P_p\{|t_{\nu(r)} - p| \geqslant c'r\} \geqslant P_p\{|t_{\nu(r)} - p|/p \geqslant c'\}, \quad (4.11)$$

in accordance with (4.10) for $\alpha_0 = \alpha_0'$. The trouble, of course, is that we do not know such an r. Suppose we could find a stopping time N_1 (well defined for $0 < p < 1$) and a positive r.v. $R_{N_1}(X_1, \ldots, X_{N_1})$, which is "usually right" as a guess of such an r, in the sense that

$$P_p\{R_{N_1} > p\} \leqslant \alpha_0'/2 \qquad \text{for } 0 < p < 1. \quad (4.12)$$

We could then proceed as follows: (1) Observe $X_1, X_2, \ldots, X_{N_1}$, and compute R_{N_1}; (2) take $\nu^*(R_{N_1})$ *additional* observations, where $\nu^*(r)$ is the function ν that yields (4.11) for $p \geqslant r$ when $\alpha_0 = \alpha_0'/2$; (3) Use $t_{\nu^*(R_{N_1})}(X_{N_1+1}, \ldots, X_{N_1+\nu^*(R_{N_1})}) = t^*$ (say) to estimate p as in (4.9), but using only the additional observations of (2). The total number of observations is $N_1 + \nu^*(R_{N_1})$, the final stopping time. We then have, writing G_p for the d.f. of R_{N_1} when p is true,

$$\begin{aligned} P_p\{|t^* - p|/p \geqslant c'\} &\leqslant P_p\{R_{N_1} > p\} \\ &\quad + \int_0^p P_p\{|t^* - p|/p \geqslant c' \,|\, R_{N_1} = r \leqslant p\} dG_p(r) \\ &\leqslant \alpha_0'/2 + \alpha_0'/2 = \alpha_0', \end{aligned} \quad (4.13)$$

thus achieving (4.10).

It remains to define an N_1 and R_{N_1} that yield (4.12). Intuitively, the smaller p, the longer it will tend to take before any X_i is 1, so we might try to use $N_1 = \min\{n : X_n = 1\}$ as a stopping time, and also make R_{N_1} small when N_1 is large. For this N_1, we have $P_p\{N_1 > m\} = P_p\{X_1 = X_2 = \cdots = X_m = -1\} = (1-p)^m$ when m is a nonnegative integer (from which it follows that N_1 is well defined for $0 < p < 1$), and consequently

$$P_p\{N_1 > u\} \geqslant (1-p)^u \tag{4.14}$$

for all nonnegative u. Define

$$R_{N_1} = 1 - (1 - \alpha_0'/2)^{1/N_1}. \tag{4.15}$$

Then, for $0 < p < 1$,

$$\begin{aligned} P_p\{R_{N_1} \leqslant p\} &= P_p\{1 - (1-\alpha_0'/2)^{1/N_1} \leqslant p\} \\ &= P_p\left\{N_1 \leqslant \frac{\log(1-\alpha_0'/2)}{\log(1-p)}\right\} \\ &= 1 - P_p\left\{N_1 > \frac{\log(1-\alpha_0'/2)}{\log(1-p)}\right\} \\ &\leqslant 1 - (1-p)^{\log(1-\alpha_0'/2)/\log(1-p)} = \alpha_0'/2, \end{aligned} \tag{4.16}$$

where the inequality follows from (4.14). Thus, this N_1 and the R_{N_1} defined by (4.15) achieve (4.12).

The method used to solve this problem has many other applications. It amounts to first finding a (random) domain which is very likely to contain the true parameter value, and which is such that, *assuming* the true parameter value lies in that domain, we know a fixed sample size procedure (with sample size depending on the domain) that solves the problem. The idea of using such a method is originally due to Charles Stein.

There are many other statistical problems than those illustrated in this paper, in which only sequential procedures can yield solutions, or in which some sequential procedure is superior to any fixed sample size procedure. This is not always the case. For example, in the problem of estimating the unknown mean θ of i.i.d. normal r.v.'s X_i with known variance, it can be shown that, among all procedures with $\sup_{-\infty<\theta<\infty} E_\theta N \leq m_0$ (= specified integer), the maximum risk $\sup_{-\infty<\theta<\infty} P_\theta\{|t_N - \theta| \geqslant c\}$ is minimized by the fixed sample size procedure $N = m_0$, $t_{m_0} = m_0^{-1} \sum_1^{m_0} X_i$. (It can be made precise that there are no "lucky" observation sequences here since this t_n contains all the information about θ based on $X_1, \ldots, X_n$, and no value of t_n is more informative than another.) But such settings are really the rare ones, and sequential analysis often provides a valuable practical device as well as an interesting implementation of various probabilistic ideas.

SELECTED REFERENCES

W. Feller, *An Introduction to Probability Theory and Its Applications*, Wiley, New York, 1950 and thereafter.

Z. Govindarajulu, *Sequential Statistical Procedures*, Academic Press, New York, 1975.

J. Kiefer, "Invariance, sequential estimation, and continuous time processes," *Ann. Math. Statist.*, **28** (1957), 573–601.

H. Robbins, "Some aspects of the sequential design of experiments," *BAMS*, **58** (1952), 527–535.

———, "Statistical methods related to the law of the iterated logarithm," *Ann. Math. Statist.*, **41** (1970), 1397–1409.

C. Stein, "A two-sample test for a linear hypothesis whose power is independent of the variance," *Ann. Math. Statist.*, **16** (1945), 243–258.

A. Wald, *Sequential Analysis*, Wiley, New York, 1947.

A. Wald and J. Wolfowitz, "Optimum character of the sequential probability ratio test," *Ann. Math. Statist.*, **19** (1948), 326–339.

J. Wolfowitz, "Minimax estimates of the mean of a normal distribution with known variance," *Ann. Math. Statist.*, **21** (1950), 218–230.

Reprinted from
Studies in Probability Theory (M. Rosenblatt, ed.)
Washington, DC: Math. Assoc. Amer., 1978, pp. 1-23

Reprinted from MYCOLOGIA, Vol. LXXI, No. 2, pp. 343–378, March–April, 1979

COMMENTS ON TAXONOMY, INDEPENDENCE, AND MATHEMATICAL MODELS (WITH REFERENCE TO A METHODOLOGY OF MACHOL AND SINGER)

J. KIEFER[1,2]

Department of Mathematics, Cornell University, Ithaca, New York 14853

SUMMARY

Some principles of statistics, and of science more generally, are presented in a taxonomic context. Some previously promised criticism of the approach of Machol and Singer, additional to that of Kiefer, is given in the framework of a broader discussion that may serve to alert mycologists to some of the dangers that attend the use of unjustified mathematical models, and to suggest bases for future numerical taxonomy.

1. INTRODUCTION

Machol and Singer (1971) presented a numerical technique for assigning a lower-level taxon to one of several higher-level taxa. A review of this work, Kiefer (1975), concentrated on what I viewed as the most evident and important failure in its basis: the assumption of statistical independence of characters. In Machol and Singer (1977) a further application of the technique appeared. These three papers will be referred to, respectively, as MS-I, K and MS-II, herein.

A paper presenting further criticism of the technique of MS-I was promised in K, and some readers may by now have remarked on its absence as have MS-II, p. 1163. The paper has not appeared until now because, instead of submitting it earlier, I have exchanged letters with Machol and Singer for several years, in an attempt to explain the major difficulties I perceive in their approach, with the hope that they would understand and correct or resolve them. It would be inappropriate for me to quote the correspondence extensively here, but it will be no revelation to the reader of MS-II that the authors believe that they have obtained sensible taxonomic conclusions (I agree), that nobody else's approach is better (I disagree), and that (MS-II, 1165 bottom-1168, line 5, 1171, lines 15–18) they admittedly have no justification for their assumption of independence of characters. Lines 1–7, p. 1164, of MS-II, show that the authors continue

[1] Research supported by the National Science Foundation.

[2] Present address: Department of Statistics, University of California, Berkeley, California 94720.

343

to miss my point that their probability ratios are meaningless, even as magnitudes. Thus, although my file contains over 100 pages of correspondence with them, including my unsuccessful prepublication comments on MS-II that were intended to clarify the misinterpretations there of K, my efforts in correspondence have failed almost completely.

It may be unclear to many that the issue deserves further publication—the overcrowded journals cannot afford every such critique of a new methodology—however, as mentioned in K, the stamp of authority given the method by Singer's eminence in mycology presents a special danger of widespread misuse of the disastrously misleading arithmetic of this approach, which is unjustified on biological-statistical grounds and also full of errors in its execution. The present article is motivated by these considerations, by the misinterpretation my views have been given in MS-II, and by the hope that other mycologists may at least be prompted by my comments to feel uneasy enough about the MS methodology not to adopt it without at least approaching their own friendly statisticians or biostatisticians for further discussion. Such discussion may result in a better explanation than mine of the important issue of statistical independence of characters. Even more important (since it is concerned with fundamental principles of scientific method) would be a resulting understanding of whether it is preferable to use an inappropriate mathematical model and accompanying mechanical routine that yield precise numerical odds, rather than to use intelligent but less formal data analysis in settings that cannot even approximately support such a mathematical structure. Perhaps the present critique may serve, more broadly, as a warning to biologists of some of the pitfalls that are present in yielding to the temptation to use superficially attractive mathematics.

My criticisms of MS-I, II are many, beginning with the authors' opening phrase "we have developed . . ." (since the technique was in fact first used long ago) and extending through every page. A selection is made herein, based on importance of the issue, and with a choice of representative occurrences in MS. At the same time, the correspondence mentioned above, as well as the conversations I have had with a number of mycologists, made it evident that an understandable presentation of the issues must include definitions and discussion of concepts from basic probability and statistical theory. This accounts for the paper's length; I hope it can serve as a useful survey of some of the principles that must be faced by a user of mathematics in taxonomy or elsewhere in biology.

Some readers may believe it unnecessary, or even repugnant, that the definitions and discussion, especially in Sections 2 and 3, are given in such formal mathematical terms. I do not believe the present state of the art in taxonomy of higher fungi requires or benefits from such treatment, as this paper should make evident. But Machol and Singer have couched their methodology in the language of probability theory, and have used its arithmetic to obtain the striking odds ratios that are used to justify the technique and its conclusions. One can only examine the approach of MS meaningfully in terms of the mathematical structure they use. Although the relevant notions will be introduced with brief illustrations of their meaning, it may aid the reader unfamiliar with the criticism of K, or with any such probabilistic concepts, if I now give an informal, unmathematical exposition of the concept of statistical independence that is central to one of my main criticisms of MS.

To this end, suppose we are interested in two character traits of the species in a particular taxon. We select one of the species at random. If our ability to predict the second trait in the chosen species is unaffected by whether or not we know the first trait for that species, the two traits are *statistically independent*. But if knowledge of the first trait affects, even slightly, our prediction of the second trait, the traits are not statistically independent, and are said to be *statistically dependent*. The cause of this statistical dependence is not under discussion here; unfortunately, the terms "independence" and "correlation" are used in a variety of ways (discussed further in Section 4), and such a usage as that of independence referring to lack of regulation of two character traits by the same gene must be separated from the concept of statistical independence. Two traits may fail to be statistically independent for any of a number of reasons, perhaps having nothing to do with regulation by common genes.

Think of species as being cards in a deck (taxon), the two characters being denomination and suit. If the deck is an ordinary deck of 52 cards, there is a chance of $\frac{1}{4}$ that a chosen card will be a spade. If I tell you that the chosen card is a king, it does not change your predictive ability regarding the suit of that card, because exactly one out of the four kings is a spade. The two characters are thus statistically independent. But suppose we instead regard a second deck, obtained from an ordinary deck by removing the black jacks and queens and the red kings and aces. It has 44 cards, of which 11 are spades, so again the chance is $\frac{1}{4}$ that a chosen card will be a spade. However, now if I tell you the chosen card is a king, you know it can only be a club or spade, so the chance of it being a spade is $\frac{1}{2}$, no longer the value $\frac{1}{4}$ that was relevant if you did not know the denomination.

The two characters for this second deck are not statistically independent. This last is also true for a third deck obtained from an ordinary deck by removing the red jacks and queens and the black kings and aces.

Suppose you did not know which deck was at hand, and made predictions and probabilistic calculations based on the assumption that the characters are independent. These will be correct if you are lucky enough to have the first deck, but will be incorrect for either of the other two decks. Moreover, the direction of your error is unpredictable: given that the chosen card is a king, the value $\frac{1}{4}$ for the chance of it being a spade is lower than the true value ($\frac{1}{2}$) for the second deck, but higher than the true value (zero) for the third. Other decks could exhibit different amounts of error.

In the biological world, there are many documented examples of taxa in which characters are not statistically independent, a few references being given in Section 5. Indeed, as I will discuss in Section 6C, the very process of grouping similar lower taxa into higher ones will often increase the departure from independence. With a large enough number of species, one can infer whether or not two characters are approximately independent by counting the number of species with each possible pair of character-trait values. In terms of the cards, if enough cards have been chosen we can judge whether our deck exhibits the independence of denomination and suit that are present in an ordinary deck.

Coming to the methodology of MS, I assert that they are making calculations assuming independence of characters (as in an ordinary deck) without ever having justified this assumption. As I have illustrated above, they cannot know the direction or magnitude of error in the resulting calculations of probabilities. There are other errors in their calculations, but this important one, by itself, vitiates the mathematical assertions that are at the heart of the technique.

Section 2 contains definitions and development of the "classification" problem treated in MS. Section 3 discusses further the question of independence of characters raised in K and described above. Section 4 is concerned with the broader issue of mathematical modeling in science. A review of some related literature appears only in Section 5, in order that reference can be made to the ideas presented in the earlier sections. Section 6 contains additional comments that refer to the MS methodology in its relationship to Bayesian inference, to sphericity of taxa, to additional considerations of modeling, to various statistical principles, and to the notion of care in handling data.

2. DEFINITIONS AND BACKGROUND

The establishment of a useful taxonomy of species is a very complicated undertaking. The hierarchical structure with taxa of different levels is the subject of continual dispute, rearrangement, and even abandonment (by some taxonomists who prefer simply to list the values of similarity coefficients between any two species). The complexity of the problem invites consideration of various easier problems whose solution may contribute to that of the larger problem. Thus, *cluster analysis* might be used to group species (or possibly collections of them) into disjoint clusters each of which is to be regarded as a taxon of the same level; for examples, the species of Agaricales could be clustered into genera or into families, or the genera (previously established) could be clustered into families. A somewhat simpler problem is that of *classification*. (This is usual statistical terminology, which I shall use; some authors, e.g., Cormack (**1971**) call this *identification* or *assignment*, and use "classification" for the larger taxonomic problem. Similarly, some of the other nomenclature varies in the literature.) In this problem we have two populations (taxa) A and B of individuals (species), and a third group C (lower taxon) of individuals which we are to *classify* as all belonging to A or else all belonging to B. We are treating the model in which it is assumed that one of these two possible classifying decisions is appropriate; for example, we know that C is not to be split into some species to be assigned to A and some to B, although this and other variants could be treated.

The form of classification problem just described is that considered in the model used in MS-I, II. The treatments given in the statistical literature of this problem, and that used by MS, are based on a precise mathematical structure. Before describing that model, we mention the phenomena with which it deals. We observe some number n_A of individuals from population A, n_B from B, and n_C from C. There is a set of m characters that we observe for each individual. The jth character may be in any one of several different "states" (or can be thought of as taking on one of several possible "values"), depending on the individual. For example, the 17th character in the list of Table I of MS-II, lamellae width, can be in any one of the three states (1) about $\frac{1}{10}$ pileus diam, (2) narrower, or (3) broader; many characters refer to presence or absence of a feature and are hence dichotomous. For any individual, we can list the state d_j taken on by the jth character, and we abbreviate a possible "list" or m-vector $(d_1, d_2, \ldots, d_m)$ of states of all characters by **d**. Thus, if an individual had lamellae width $> \frac{1}{10}$ pileus diam, its vector **d** would have 17th component $d_{17} = 3$.

Each of the possible *combinations* of character states is represented by a different vector **d**, and there are many possible **d**'s (2^m if all characters are only dichotomous). In practice we will observe individuals with only a small fraction of the possible different d's, unless m is small. The number of characters m used in the classification problems of MS-I, II varies between 33 and 56.

The model assumes population A is characterized by a probability function P_A defined on the collection D of all possible vectors **d**. Thus, for each possible vector **d**, the value $P_A(\mathbf{d})$ is nonnegative, and the sum $\Sigma_{\text{all } d} P_A(\mathbf{d})$ of all such values is unity. For any of the possible values of **d**, $P_A(\mathbf{d})$ is the probability that an individual selected at random from population A will turn out to have its list of character states be exactly those listed in **d**. Thus, if $m = 3$, an individual with $\mathbf{d} = (d_1, d_2, d_3) = (2, 2, 3)$ in terms of the vocabulary of Table I of MS-I would have spores that are inamyloid, subglobose, over 10 μm long. If $\mathbf{A}_r$ is the vector of character states we shall observe for the rth individual selected at random from population A, the probability that that individual will turn out by chance to have exactly these character states is then

$$\text{Prob}\{\mathbf{A}_r = (2, 2, 3)\} = P_A((2, 2, 3)) \tag{1}$$

Here $1 \leq r \leq n_A$. We adopt a similar notation, P_B and P_C for the probability functions associated with populations B and C, and $\mathbf{B}_s$ and $\mathbf{C}_t$ for the chance m-vectors of characters we will observe from randomly selected members of those populations.

A common assumption is that the n_C character lists $\mathbf{C}_1, \mathbf{C}_2, \ldots, \mathbf{C}_{n_C}$ of the sample chosen from population C are *statistically independent*. Coupled with the fact that each $\mathbf{C}_s$ is governed by probability function P_C, this means that, for any collection $\mathbf{d}_1, \mathbf{d}_2, \ldots, \mathbf{d}_{n_C}$ of n_C (not necessarily distinct) m-vectors in D,

$$\text{Prob}\{\mathbf{C}_1 = \mathbf{d}_1, \mathbf{C}_2 = \mathbf{d}_2, \ldots, \mathbf{C}_{n_C} = \mathbf{d}_{n_C}\} = \prod_{t=1}^{n_C} P_C(\mathbf{d}_t) \tag{2}$$

(where $\prod_{t=1}^{n_C} q_t = q_1 q_2, \ldots, q_{n_C}$). This use of an independence assumption, which will be discussed in Section 6C, refers to the absence of statistical dependence among *species* rather than the absence of such relationship among *characters* (the chief concern of K, of Section 3, and of the illustration of Section 1). In the card selection illustration of Section 1, the assumption (2) refers to prediction of the second card chosen from knowledge of the first, rather than to the prediction of the second character from the first. Thus, if the chance mechanism that is operative prescribes that the first card chosen is returned to the deck and the deck is well shuffled before the second card is chosen, knowledge of the first card cannot help us to predict the second, and the cards are independent. However, if the mechanism is such that the first card is *not returned* and you are told it is the queen of hearts, you can with certainty predict that the second card is not the queen of hearts, an extreme illustration of the fact that the cards are not independent. In mycological terms, assumption (2) means, for example, that if we restrict attention to the first character and first two species chosen from C,

$$\begin{aligned}\text{Prob}\{\text{species 1 has amyloid spores, species 2 has inamyloid spores}\}\\ = \text{Prob}\{\text{species 1 has amyloid spores}\}\\ \times \text{Prob}\{\text{species 2 has inamyloid spores}\};\end{aligned}$$

phrased another way, whether the first individual selected from population C has inamyloid spores (first component of $\mathbf{C}_1$ turned out to be 2) or amyloid spores, the "conditional" probability that the second individual selected from C will have inamyloid spores (first component of $\mathbf{C}_2$ will be 2) is the same. Relation (2) implies other similar but more complex statements, all of which mean that knowing the character states taken on for any set of individuals selected from C does not change the odds we assign to any character states being taken on by the other individuals. It is often reasonable to make this assumption in classification studies in which n_C is small compared with the size of the population C from which the n_C individuals are chosen at random; for example, where 100 individuals are selected at random out of several million, or where the size of C can be regarded as "infinite," as in studies of genetically identical yeast cells which can be supplied without end. For now, we shall not discuss the reasonableness of (2) in the MS study (see Section 6C); we only mention that their model assumes it.

For the sake of exposition in understandable steps, it is convenient, temporarily, to make one other assumption in addition to (2): that the functions P_A and P_B are known exactly. With this assumption, deciding whether C is actually the same population as A or the same as B reduces to the well known problem of "testing between two simple hypotheses," and the famous Neyman-Pearson Lemma asserts that any reasonable classification procedure is of the form that, for some constant $k \geq 0$, it decides C is the same as A if

$$\prod_{t=1}^{n_C} [P_A(\mathbf{C}_t)/P_B(\mathbf{C}_t)] > k, \tag{3}$$

and decides C is the same as B otherwise. The product of expressions in square brackets in (3) is called a "likelihood ratio" in statistics. The choice of k depends on additional specification of features one wants the procedure to have; for example, if it is deemed a much more serious error to misclassify C as A when C is really B, than to suffer the opposite misclassification, then k is chosen to be large. In the Bayesian approach, it is assumed that, before observing the n_C individuals from C, we know "prior probabilities" π_A and π_B that C is actually A or B, respectively (with $\pi_A + \pi_B = 1$). In that case, it is correct to choose $k = \pi_B/\pi_A$ (the "prior odds ratio that C is B, against C being A") in (3); thus, if the two possibilities for C are equally likely, one chooses k = 1, which is what is done in MS-I, II. With this choice of π_B/π_A, the probability ratio on the left side of (3), with the observed values of the $\mathbf{C}_t$ substituted, is identical with the "posterior probability (odds) ratio that C is A, against C being B." This ratio thus expresses to what value the data $\mathbf{C}_1, \mathbf{C}_2, \ldots, \mathbf{C}_{n_C}$ have changed the prior assessment of odds ratio $\pi_A/\pi_B = 1$. Each of the expressions that is called a "probability ratio" in the conclusion of MS is *not* simply such a posterior value, but rather its n_Cth root,

$$\left\{ \prod_{t=1}^{n_C} [P_A(\mathbf{C}_t)/P_B(\mathbf{C}_t)] \right\}^{1/n_C}. \tag{4}$$

While it is made clear by MS that (4) is used, it is an unfortunate confusion that they still use the term "likelihood ratio" for (4); in fact, "probability ratio" (MS-I, pp. 760, 763, 766) is also used. (It cannot help the reader to be told, in MS-II, p. 1165, that if a ratio of 1.64 were obtained, "we did not thereby state or even imply that, literally, it was 1.64 times as probable in the taxonomic context that the genus came from the second family." Machol and Singer never tell us just what the 1.64, based on (2) and on (5) below, *is* supposed to mean. Further discussion of this and of the "subjective Bayesian" viewpoint are contained in Sections 4 and 6A. The term Bayesian is now often used in reference to subjective Bayesians, who employ subjective prior probabilities in settings where π_A and π_B are unknown. The development described in the present paragraph would also be usable by one who is not a subjective Bayesian, but for whom previous long experience yields frequencies from which π_A and π_B can be estimated reliably.)

Often the functions P_A and P_B are not known, and in that case they are estimated from the values assumed by $\mathbf{A}_1, \mathbf{A}_2, \ldots, \mathbf{A}_{n_A}$ and $\mathbf{B}_1, \mathbf{B}_2, \ldots, \mathbf{B}_{n_B}$. (In the classification literature, these are often called the "training samples," while the C_t constitute the "testing sample.") This is the situation encountered by Machol and Singer, who discuss (MS-I, expecially pp. 761–763) why they believe their method of estimation of P_A and P_B is preferable to others. Further comments on this will be found in Section 6A. For now, let us remark that caution must obviously be exercised in ascribing to (4) the same meaning when P_A and P_B are estimated from data, as when they are known exactly. We hereafter denote an estimate of P_A by $\hat{P}_A$, and

$$\{ \prod_{t=1}^{nc} [\hat{P}_A(C_t)/\hat{P}_B(C_t)]\}^{1/nc} \qquad (4')$$

will denote the expression obtained from (4) when the estimates $\hat{P}_A$ and $\hat{P}_B$ are substituted for P_A and P_B.

3. INDEPENDENCE OF CHARACTERS

The additional assumption that is made in the methodology of MS-I, II is that of *statistical independence of the m characters*. This states that, for population A (with an analogue for B), there are probability functions $P_{A,1}, P_{A,2}, \ldots, P_{A,m}$, where $P_{A,j}$ is defined on the possible states of character j, and such that, for each possible m-vector $\mathbf{d} = (d_1, d_2, \ldots, d_m)$ in D,

$$P_A((d_1, d_2, \ldots, d_m)) = \prod_{j=1}^{m} P_{A,j}(d_j). \qquad (5)$$

I have given examples in Section 1, as well as in K, p. 204, of the meaning of this assumption. In terms of the discussion of the previous section, it means, for example, when C is the same as A, that knowing whether or not the spore is amyloid (first component of C_1 = 1 or 2) does not alter the probability $P_{A,2}(2)$ we assign to its being subglobose. (Thus, relation (5) refers to independence of different characters for the *same* species, while (2) referred to independence of *different* species.) Independence again implies various more complex statements, which in summary say that the (conditional) chance that a random individual from A has any subset of its characters in specified states, given the states of the remaining characters, does not depend on the latter. In terms of the ball-and-urn model described in MS-I (especially p. 759, bottom), it says that knowning whether the ball is red or yellow does not change the odds we assign to its being rough or smooth: obvious nonsense for an urn in which all red balls are rough and all yellow balls are smooth, and also nonsense for less extreme cases of lack of independence, as indicated in the example of K and that below. Statistical independence of characters is a precise mathematical assumption, and calculation of expression (4) assuming each factor $P_A(C_t)$ can be computed as a product of m character probabilities as in formula (5), will lead to a meaningless conclusion if the independence assumption (5) is violated.

How meaningless can the odds-ratios computed in MS-II be? One of the authors has written me his hope that, even if assumption (5) is violated, a meaningful value of (4) would not be off by being less than the square root of the value of (4) computed under assumption (5), so that a value of 10^4 obtained for (4) using (5) would still be interpretable as at least 10^2 without independence of characters. I believe that this is wishful thinking, justified neither by (a) theoretical principle nor (b) empirical evidence that (5) is close to being correct.

As for (a), it is easy to give examples to show that, *in the absence of* (5), the data of MS-I, II are consistent with values of the estimated functions $\hat{P}_A$ and $\hat{P}_B$ that lead to the ***opposite conclusions*** from those obtained in MS. (Notice: I am ***not*** stating disagreement with the actual taxonomic conclusions, but only that the huge "probability ratios" obtained in MS-I, II from (4) and (5) do not carry any of the weight given to them there as support for those conclusions.)

It would be best to illustrate (a) and (b) with data of character combinations from the studies of MS; unfortunately, the data for counts of combinations are not available. Hence, I will use marginal (individual) character counts from MS, but the illustrative frequencies of combinations I have chosen for examples may of course not be factual. (The reader may find actual examples in several of the references of Section 5.) Consider characters 4 (cheilocystidia differentiated vs. undifferentiated or absent) and 32 (tropical or subtropical vs. temperate) in MS-II. I have chosen this pair because the characters are dichotomous, leading to simplicity of arithmetic, and the observed counts can be used to illustrate the effect of violation of assumption (5) without the violation being extreme; the same phenomenon can be illustrated with other pairs (and also in the studies of MS-I), and with even more extreme anomalies when one considers not merely *pairs* of characters but *combinations of 33 characters* as used in MS-II. Using the estimates $\hat{P}$ described in MS-I, p. 762 (to be discussed in Section 6A), one obtains, with A = *Hydropus* without *Floccipedes* and B = *Gerronema*,

$$\hat{P}_{A,4}(1) = 1 - \hat{P}_{A,4}(2) = 0.523, \qquad \hat{P}_{B,4}(1) = 1 - \hat{P}_{B,4}(2) = 0.168,$$
$$\hat{P}_{A,32}(1) = 1 - \hat{P}_{A,32}(2) = 0.574; \quad \hat{P}_{B,32}(1) = 1 - \hat{P}_{B,32}(2) = 0.778; \tag{6}$$

thus, for example, 0.778, the estimated probability of a species in *Gerronema* being tropical or subtropical, is obtained as [26(0.799) + 1]/[26 + 2] according to this method, since there are 26 species in *Gerronema* and an observed proportion 0.799 are tropical or subtropical. The corresponding observed proportions for C = *Floccipedes* are 0.6 with cheilocystidia differentiated and 0.2 tropical or subtropical. From this Machol and Singer compute

$$\prod_t [\hat{P}_{A,j}(C_t)/\hat{P}_{B,j}(C_t)]^{1/5} = \begin{cases} 1.283 & \text{for } j = 4, \\ 1.583 & \text{for } j = 32, \end{cases} \tag{7}$$

and under the assumption (5) these two factors contribute 1.283 × 1.583 = 2.501 to (4′), which with the contributions of the other 31 characters makes (4′) equal 8380. Thus, $8380^{1/33} = 1.315$ is the "average" (geometric mean) advantage per character of A over B in conforming with the data from C, so that the two characters of (7) are fairly typical. (The article MS-II never lists the value of n_C for *Floccipedes*, but the reader is evidently presumed to have read the footnote on p. 762 of MS-I, where we are told that such numbers are counted from the lists in Singer (1962); see Section 6E with regard to the general quality of "data recording" in MS.)

We may represent the above data for the two characters in terms of three 2 × 2 tables. (All calculations below were carried to four significant figures and then rounded off.) The observed frequencies in table C should obviously be multiples of $1/n_C = 0.1$, as they are in (9), but only the marginal totals are used in the computation from (8). I have not bothered with such details in the entries of tables B and C, since there is some question about the appropriateness of using the MS estimates, as is described in Section 6A, and since slight changes in the numbers will not affect the conclusions reached from this illustration.)

A

0.300	0.274	0.574
0.223	0.203	0.426
0.523	0.477	

B

0.131	0.647	0.778
0.037	0.185	0.222
0.168	0.832	

C

0.12	0.08	0.2
0.48	0.32	0.8
0.6	0.4	

(8)

The numbers below each square refer to character 4, those to the right refer to character 32, and these marginal frequencies are the *only* data listed in MS-II concerning these characters. The numbers in the cells of the squares are obtained using the independence assumption (5); for example, $0.168 \times 0.222 = 0.037$ is the estimated probability $\hat{P}_{B,4}(1) \times \hat{P}_{B,32}(2)$ that a randomly chosen species of *Gerronema* has differentiated cheilocystidia and comes from a temperate climate. Of course, two numbers in any row or column of a square add to yield the given marginal total, and the pair of marginal totals at the bottom or side of each square totals one. Now, if (5) is not satisfied, there are many sets of numbers that could appear in the squares and that *yield the same marginal totals as* (8), so that they are consistent with all the data given in MS-II. One such example, not the most extreme, obtained by modifying the numbers of (8) only slightly, is

A

0.423	0.151	0.574
0.100	0.326	0.426
0.523	0.477	

B

0.068	0.710	0.778
0.100	0.122	0.222
0.168	0.832	

C

0	0.2	0.2
0.6	0.2	0.8
0.6	0.4	

(9)

In this table, the probability that a randomly chosen species of *Gerronema* has differentiated cheilocystidia and comes from a temperate climate is 0.100. The values of (9) yield, for the contribution of characters 4 and 32 to the MS ratio (4′), *not* the value 2.501 obtained from (8), but rather 0.893, reversing the effect of these two characters by now *slightly favoring B! And this value* 0.893 *is consistent with the given marginal data.* Other examples consistent with those data can be given in place of (9), that favor A either by more than 2.501 or by less than 0.893.

I am not asserting that (9) is more correct than (8); but only that the authors have given no justification for assuming (8) is more accurate. When one looks at combinations of 3, 4 or 33 characters in place of only 2, the possible departures from the conclusion obtained from (4′) and (5) are much more striking, and there is no theoretical justification for thinking that a meaningful value of (4′) in the absence of (5) would come out >1, let alone 8,380.

It is up to Machol and Singer to justify (5). My inquiries to them, for records of data counts of combinations of at least *pairs* of characters (to yield either a verification of the analogues of (8) for such pairs in the studies of MS-I, II, or else an indication of accurate substitutes for (8) and (9)), have received the response that such data are not available. Evidently the data were not recorded in the form one might expect, of the character list $\mathbf{A}_r$, $\mathbf{B}_s$, or $\mathbf{C}_t$ for each species (e.g., on an IBM card for each species), but only in the form of the marginal totals that the authors compiled by counting, for each character separately, how many species in the taxon (A, B, or C) were in each state. (I invite the collaboration of other mycologists who have suitable data they would like to investigate in the terms I have mentioned.) Independence of pairs of characters does not imply independence of larger sets of characters as assumed in formula (5), but without such a verification there is no basis at all for using (5). As we shall see in Section 5, other biologists have studied the question of independence for many years, and other scientists have used (4) without assuming (5).

It is in order to discuss the various uses of the words "independence" and "correlation" among taxonomists, since confusion about that usage appears (in our correspondence as well as in MS-I, II) to have helped cloud the issues. Thus, despite our many exchanges of letters, MS-II, p. 1165 asserts that "taxonomic correlation" as defined by Jardine and Sibson (**1971**) seems "closer to us to what Kiefer had in mind" than "statistical correlation." *This explanation is totally incorrect.* With regard to correlations, my criticism in K of the use of assumption (5) in MS-I, as well as the lengthy subsequent correspondence, is concerned with statistical correlation among character states within a single population such as A. (For 2×2 tables such as those of expressions (8) and (9), statistical correlation and statistical independence are equivalent notations.) Taxonomic correlation of two characters (say j_1 and j_2) is used by Jardine and Sibson (**1971**, pp. 26–28) to refer to the tendency of two characters (which they call "attributes") to "discriminate the OTU's in similar ways;" this means only that the *marginal* probability functions P_{A,j_1} and P_{A,j_2} are similar, *and has nothing to do with combinations of characters* as considered in expressions (5), (8) (9). In fact, Jardine and Sibson discuss the use of taxonomic and statistical correlation, and consequent redundancy of attributes, as being of interest primarily for the purpose of "selection rather than weighing of attributes." They beg the issue of what to do if statistically correlated attributes have been included (except in passing, on pp. **33–34**, and then in the context of normally distributed attributes rather than the discrete ones we are considering). In my opinion they consider correlation and redundancy in terms that are too black-or-white (pp. **33–34**, **171–172**), as though one need usually only consider the possibility of selecting almost totally redundant or totally nonredundant characters. In any event, the effectiveness of their methodology appears to rest on the assumption, without much comment, that the characters satisfy assumption (5). This is disappointing in a book reputed to be one of the most mathematical treatments of numerical taxonomy.

In correspondence, Singer has also explained the frequent use by biologists of the phrase "independence of characters" to refer to lack of regulation by the same gene or combination of genes. In view of the uncertain knowledge of gene linkages, as well as the evolutionary advantage of various combinations of characters, it is clear that this explanation cannot justify the assumption of (5). (See also Section 6C.)

Whatever the unfortunate ambiguity of language, there is no ambiguity in the way that (5) is used by Machol and Singer to compute their odds ratios—they have assumed statistical independence of characters. In attacking the "kind of precise use of 'dependence' and independence' which Kiefer would like to have as the basis for a taxonomic model," MS-II (p. 1168) explains that one "would have to start with a precise specification of, for example, the probability that fungi of a certain genus have a certain character, e.g., amyloid spores." To the contrary, whatever model I would need if enough data were present, it would certainly not give a "precise specification" of the type just indicated, but would leave the precise value as an *unknown parameter*, along with the probabilities of various *combinations of characters*, to be estimated from the data. More to the point, the authors' condescending further explanation to the statistically naive of the facts of life as they perceive them, "alas, in the biological world with which we deal, no such probability can be measured or defined, at least with the biological methods now at our disposal," should have been taken by them as a caveat against their assumption of the rather delicate, precise assumption (5), unchecked by any data, and which is the basis for their numerical assessments. For it is not I, but MS, who have promulgated the precise probability model (5)!

I am unable to comprehend how the use of (5) is consistent with the statement in MS-II (p. 1164) that the authors are *not* seeking characters that are independent (in K's sense) but are seeking "precisely a measure of that dependence." It is certainly a goal to elucidate such dependence (although MS in fact quantify no such correlations). In an extreme case in which all m characters are in state 1 for almost all individuals in taxon A and in state 2 for almost all those in B, an experienced biologist (certain that he has observed enough characters) might be able to reach clearcut taxonomic conclusions, and perhaps evolutionary conclusions. But with such statistical dependence of the characters, it would be meaningless for him to use probability ratios assuming (5) for A and B, to classify C, as Machol and Singer have done.

Further comments bearing on the question of independence are those of Section 6B (relating to sphericity of taxa) and Section 6C. The description of MS-I on the contorted, nonspherical shape of actual higher taxa is in fact a direct contradiction of their independence assumption (5).

A final remark on independence: despite the botanical literature that illustrates dependence of characters (Section 5), I assert neither the extensive presence of such dependence here, not its effect on the odds ratios. But I have given examples of what *could* happen. It is the authors' obligation to justify the use of (5) or the efficacy of their procedure if (5) is violated; my attempts to help by obtaining counts of character combinations from them have failed.

4. SCIENTIFIC METHOD AND THE USE OF MATHEMATICAL MODELS

Perhaps more disturbing than the use of assumption (5) by Machol and Singer is the motivation they give for their use of a mathematical model. In our correspondence, as well as in MS-II (p. 1165, lines 7–12), they repeatedly view the use of such a mathematical model, with its attractive precise conclusions (strings of formulas, final probability ratios given to three significant figures) as automatically progressive, and the use of less mathematical techniques (exploratory data analysis, selection of a few of the most important characters that can lead to meaningful results while taking account of dependence, conclusions given without the assertion of such precise odds ratios) as scientifically backward, in their words a "fallacy!"

This is frightening.

Good science has never been, can never be, based on the use of algebraically attractive but biologically unjustified formulas that are used for the sake of giving a mechanical, routine, numerically oriented methodology. If one knows as little about the *actual* underlying probability structure as Machol and Singer do here (with *not a single data check* that (5) is correct, and with the possible consequence of meaningless odds ratios that we have seen in the discussion of expressions (8) and (9)), it is simply bad science to use such a structure in the way they have. As was stated in K, an important further tragedy is the fact that appearance of such a methodology in the work of an author of great stature gives it a stamp of authority that will lead the statistically inexperienced mycologist to use it without realizing its lack of meaning. On reading the odds ratios of MS-I, one such mycologist wrote to me "WOW!" with such enthusiasm that it motivated the writing of K. And yet, of the several dozen biologists or biostatisticians with whom I have discussed MS, and who understood the notion of statistical independence of characters, all have recoiled in horror on being told of the MS methodology based on (5).

Even the more theoretically skilled and model-minded of experienced contemporary statisticians work in constant consciousness of the philosophy repeatedly emphasized by J. W. Tukey (**1962**, **1977**) and which warns in statistical terms against the use of a methodology induced by a precise mathematical model such as (5), in settings in which the data cannot support such a model; that is, in which previous or present data cannot confirm (even roughly, in the present instance) the model's validity, and in which departures of the true state of nature from the assumed model can make the methodology yield misleading conclusions. Thus, leading statisticians who have worked in taxonomy, would agree with MS-I, p. 757 that "one does not learn about biology by sitting in an armchair . . .; one learns about biology by observing organisms;" but, having not observed independence, they would not assume (5). And they would not publish their work in the form Machol and Singer use; it is not that they have not *thought* of such techniques (which in fact are rather old, as we shall see in Section 5), but rather that they have rejected them. As much as they would enjoy the reassurance such large odds-ratios yield, the best cluster analysts and workers with discrete biological data that I know of have told me their strong preference for using the approach I have consequently described in K, p. 205 (and which is criticized in MS-II, p. 1165), of selecting a few important characters *and checking or taking account of the correlations between them* (which one cannot do for the full set of characters). When classification is replaced by the larger problem of clustering into taxa, these workers sometimes cluster species and characters simultaneously. Of course, they cannot then think of odds ratios, calculated from the same data that were used to select the characters, as meaningful in the way Machol and Singer do; such odds ratios would appear, if at all, only as a guide with an appropriate warning; still, if those ratios were obtained with consideration of the question of dependence, they would be of more value than those of MS. Such scientists would refrain from using simplifying mathematical models for which they have no justification. The implication in MS-II, p. 1165 (and, more strongly, in our correspondence) that such an approach is a step backward in contrast to their more mathematical approach, is thus based on a naive view that more mathematics makes better science.

It is not my position to defend all the "conventional numerical taxonomy" published prior to MS. Much of it is bad (although it does not all deserve such a criticism as the categorical assertion of MS-I, p. 759 that it has "no way of dealing with two matches or, worse, one match"). I agree with Machol and Singer that little of it incorporates consideration of statistical dependence (except for continuous-valued normally distributed characters, which, as described in Section 5, are more easily treated than discrete-valued characters), so that such a standard work as that of Sneath and Sokal (**1973**) barely considers the subject. Still, almost all of those authors have a virtue over Machol and Singer: *they do not state meaningless probability ratios.* Data analysts often use a variety of quick tests and estimates in the complicated analysis of data such as that of typical taxonomic studies; but they realize that such data do not support, even approximately, any useful probabilistic assertions concerning the final conclusions.

That does not, of course, mean that such studies cannot be important, or that we should give up taxonomy until we have sufficient data to use precise mathematical models. One can gain insights from those studies, including those whose "information-theoretic" approach (assuming (5)) is close to that of MS; hence, also, one can learn from MS-I, II. One can learn much more conclusive information by the selection of characters and consideration of dependence I have suggested. The assertion in MS-I, p. 753, and MS-II, p. 1168, that the MS method is "a quantification of what competent taxonomists have always done" is flawed by the drastic shortcomings of the quantifying model!

MS-II, p. 1164, contains a defense against the criticism in K of the precise statement of these odds ratios in MS-I. I believe the setting forth of these meaningless numbers, to three significant figures (not just in the Tables, but in the text discussion, as on p. 766 of MS-I or 1169 of MS-II), can easily induce a reader's belief (even a "WOW") in them that not only is unwarranted (because of the use of (5)), but that also is not adequately disposed of by the comments of MS-I, II. If indeed (MS-I, p. 761) "we are only interested in whether it [the ratio (4')] is greater or less than unity" (a limitation of interest that would be made by no scientist interested in the conclusiveness of his results, if he really believed the model), why repeatedly quote the gigantic odds ratios to three significant figures? Where is the evidence that the authors have heeded their warning that "judgement must be exercised" (MS-I, p. 764, p. 765), when they use (5) without the slightest evidence that it is valid, or that the phenomenon illustrated in comparing expressions (8) and (9) could not occur? (The statement on MS-I, p. 766 that "we are inclined to utilize these magnitudes only to estimate significance," quoted in MS-II in response to K's criticism of the lack of such warnings, was actually used in MS-I in a context different from the present one of correctly classifying a lower taxon; rather, it was used in a discussion about not concluding that a ratio magnitude measures "the size of the hiatus between families.") In MS-I, p. 766, we are in fact told that "the smallest of our final probability ratios was 17,200, which seems adequately significant;" it is then explained what a probability ratio *would* mean if it came out closer to unity, for example between 0.01 and 100; *nowhere are we warned that the probability ratios should not be taken so seriously because the assumed model may not be adequate!* The exposition of pp. 759–761 of MS-I, using the phrase "probability ratio" for the estimate (4′) of (4) under assumption (5), evidences intended probabilistic interpretation of the resulting ratio.

It was mentioned in the discussion of expression (4) that the ratios given in MS are *not* really likelihood ratios or probability ratios (although they are referred to as such repeatedly, as in a quotation in the previous paragraph), but are the n_Cth roots of such numbers. Thus, for example, using the fact that there are five species of *Ripartites*, five of *Floccipides*, and 21 of *Squamanita-Cystoderma*, the estimates (4′) of the so-called "probability ratios," listed as 1.35×10^6, 8,380, and 1.06×10^8, respectively, in MS-I, are translated into estimates of *properly* called estimated probability ratios of, respectively, 4×10^{30}, 4×10^{19}, and 3×10^{168}. Realizing the numbers of observations in these data, the casual way in which they were recorded (see Section 6E), and the unjustified use of (5), it seems to me that such huge ratios should engender not a "WOW," but rather total skepticism that the data can support any such model or any confidence in the methodology.

In connection with the definition of (4) as a "likelihood ratio" or "probability ratio," MS-I (p. 761) indicates appreciation of the enormity of these *actual* ratios before reduction to the n_Cth root (which reduction, whatever its meaning, may make the ratios seem less absurd to a reader.) The statement that "the result is the same whether we had drawn 100 balls or some other number," while true of expression (4), may well invite an incorrect inference: the *conclusiveness* of a value of (4) is *not* the same regardless of n_C; if the model given by (2) and (5) is correct, then it is (4) *without* the root being taken that is meaningful. But my main point is that the authors have not shown that (4′) as it stands, or with any other root instead, is "adequately significant" provided it does not "come out close to unity, for example in the range 0.01–100."

Unfortunately, the history of science is littered with elegant but empty mathematical models. One may aptly quote Machol (**1975**) in his discussion of others' use of independence: "Explicitly or implicitly, people multiply together probabilities of events of low probability and come out with impossibly small numbers when in fact the events are dependent."

5. SOME HISTORY

It is not my purpose to give a comprehensive history of classification problems, let alone of numerical taxonomy. A selection of references, relevant to the work of MS-I, II and indicative of its relation to earlier literature in the light of the previous discussion, will be given. Much of this cited literature, and many other similar papers, were present in obvious sources long before the MS study was begun, as the titles show. Without being a biologist, taxonomist, or expert on classification, I found it easy to uncover this material; presumably Machol and Singer could have done the same.

At the outset, one may note that Bayesian classification methodology is not new at all. A treatment of this approach that is important enough (because of its basic scientific quality) that it has appeared in at least three places, is that of Birnbaum and Maxwell (**1961**). These authors study mental patients in terms of $m = 9$ dichotomous characters such as depression, anxiety, etc. They use the data to obtain a method for classifying such patients into four categories (neurotic, schizophrenic, etc.), which correspond to our populations or taxa, A and B, and the method can then be used to classify future patients, the analogue of our C. This setting is perhaps not as large or complex as the taxonomic one of MS, *but it is of the same basic classification structure.* Notable is the awareness of Birnbaum and Maxwell that they have no justification for making the independence assumption (5), *and so they do not make it.* Instead, they look at frequencies of various *combinations* of characters, not just the marginal individual character frequencies of Machol and Singer. (They also do not assume equality of the prior probabilities π_A, π_B, etc. described in Section 2, as MS do; this is less important than their avoidance of (5), but again illustrates the care evident in good scientific method that does not assume away difficulties for the convenience of using an elegant formula!) It is true that it is easier to work with 9 characters than with 30 (that is, with 2^9 possible combinations rather than 2^{30}); that is why selection of characters and less attempt to make precise probabilistic assessments may be necessary in studies such as those of MS. Also, it is true that Birnbaum and Maxwell's study was not aimed at the particular context of classifying a genus into one of two families. Nevertheless, this reference alone should indicate to the reader that Machol and Singer, who refer to no earlier methodology of the form (3) (with or without (5)), have neither developed a new technique as they indicate, nor used the well-known technique it turns out to be with the care it requires.

An interesting earlier classification (and, to some extent, clustering) study with 13 dichotomous characters of six (tentative) populations of mental patients is that of Rao and Slater (**1949**). The development is not carried out in the manner of Birnbaum and Maxwell, no doubt largely because the number of characters (13) is larger. The characters are consequently grouped into three coarser psychological attributes, and the total count of characters in each coarser attribute recorded. *The resulting 3-vectors are then studied for dependence.* It is found that one linear function of the counts essentially accounts for all "significant" classifiability of the patients, and that only three population groups (each containing two populations) can be distinguished. Much of this study is carried out using a perhaps slightly optimistic normal approximation for the sums of counts in each coarser attribute. The multivariate normal classification theory, pioneered (and in a taxonomic setting!) in the famous paper of Fisher (**1936**) and by now the subject of an extensive probabilistic theory, yields a methodology for which it is much simpler to obtain conclusive probabilistic assessments than with dichotomous data, because there are only $m(m-1)/2$ correlation coefficients to be estimated to take account of dependence in departures from (5), rather than the almost 2^m parameters (that must be estimated from the data) in a model with dichotomous characters in which neither (5) nor any other structure is assumed. (A subsequent paragragraph discusses this, further.) I have described the work of Rao and Slater more than in passing terms because it is an early treatment of dichotomous characters which deals with the problems both of grouping characters (of which selection can be viewed as a special case), and also of dependence and a thoughtful unmechanical further use of the data. It may not offer exactly the optimum sequence of steps for the MS setting, but it illustrates, almost 30 years ago, a far better feeling for sound scientific analysis in classification based on dichotomous characters.

Another landmark paper on classification with discrete characters is that of Cochran and Hopkins (**1961**). Taking a Bayesian approach, they use combinations of characters rather than to assume (5). The plug-in rule they consider is a standard one MS-I (p. 762, top) criticizes; without defending it, I note that Cochran and Hopkins investigate the actual performance of the resulting rule and how far the misclassification probabilities are from the naive numbers one would obtain by *ignoring* the fact that P_A and P_B have been estimated in computing (4'). Machol and Singer have presented no corresponding investigation of the performance of their method.

There is an extensive development of statistical theory for dichotomous (or other discrete) characters. There are models that assume neither the naive independence of (5) (which requires only m parameters to be estimated) nor totally unrestricted dependence (which requires $2^m - 1$). For example, models have appeared that parallel the multivariate normal theory by explaining all dependence in terms of probabilities $P_{A,jj'}$ associated with combinations of only *pairs* (j,j') of characters, or in related terms. The need to estimate roughly $m^2/2$ parameters in such models makes them unusable for the values of $m (> 30)$ considered in MS, since n_A and n_B are always smaller than $m^2/2$. Nevertheless, it is clear that the use of such models would be possible for a selection of the important characters. The validity of the model should of course be tested by looking at combinations of more than two characters. Some of the pioneering papers on such models for discrete data are those of Bahadur (**1961a**, **b**), Goodman (**1964**, **1970**, **1971**, **1973**), Martin and Bradley (**1972**). Some of these papers illustrate selection of a model appropriate for accounting for dependence in the data at hand. A good illustration of the investigation of dependence among selected characters also appears in Fienberg (**1970**). The recent book by Bishop, Fienberg, and Holland (**1975**) gives a broad coverage of some of these models and applications of them. Other good early illustrations of the study of dependence, combinations of characters, and selection of a few characters, may be found in Solomon (**1961**) and Raiffa (**1961**); the latter is a leader of the Bayesian school.

The literature of models and methodologies for classification problems is reviewed comprehensively (as of five years ago) by Das Gupta (**1973**). Estimation of $P_{A,i}$ and $P_{B,i}$ in the MS methodology, as described in Section 2 above and in Section 5A, is an example of a "plug-in rule." The MS rule itself may be viewed not as we have earlier in this section but, instead, in the form of the logarithm of the product in expression (3); it is then in the family of methodologies that are referred to as "information-theoretic" in the literature. The many papers on such methods include those by Rescigno and Maccacato (**1961**) (who discuss the danger of not taking account of dependence), and Estabrook (**1967**), among the earliest. Information-theoretic methods for the broader (than classification) context of cluster analysis and taxonomy are considered, e.g., in Orlocci (**1968**) and Hartigan (**1975**), the latter an excellent compendium of various numerical techniques and a good source of references; both of these refer to the problem of dependence. Hills (**1967**) discusses likelihood ratio (as in (4)) and other classification methods, with other "plug-in rules" than that of MS; his stepwise method studies *combinations* of characters. A recent volume of contributions is edited by van Ryzin (**1977**). The earliest history of classification has been treated by Hodges (**1950**), who quotes the warning of Pearson (**1926**) on the great attention that need be paid in selecting characters as little correlated as possible (which Pearson would presumably have checked by examining the data), and mentions an 80-year-old work of Heincke (**1898**) which includes a *Bayesian classification procedure for herring, based primarily on discrete characters, assuming* (5) *and equality of the prior probabilities.* Heincke does at least show awareness of the role and tentative nature of the assumption of independence "was noch nicht bewiesen ist," although he then justifies its use in an unconvincing, unquantitative discussion of biologists' observations that, he believes, confirm the assumption. Hodges lists subsequent criticism of this work. Those who cannot remember the past are indeed condemned to repeat it!

Finally, we mention that there is a large botanical literature in which dependence of discrete-valued (often dichotomous) characters is investigated in connection with taxonomic-evolutionary studies. An early reference is Sinnott and Bailey (**1914**), where (p. **446**) counts of character *combinations*, of the type lacking MS, are listed for various families. Of course, this and other early papers cannot be expected to be highly mathematical; more important, this paper shows an awareness that (5) is not to be assumed automatically. The classic work of Sporne (**1949**), which considers 12 floral and vegetative characters of 259 families of Dicotyledons, lists (p. 269–270) many highly significant correlations; in particular (p. 270), there are 10 pairs of characters (j, j′) for which $P_{A,jj'}/P_{A,j}P_{A,j'}$ (assumed = 1 in (5)) is as large as 2 or small as $\frac{1}{2}$ (and frequently more extreme), and 10 more pairs in which the departure from independence is almost this large. Although the purpose of Sporne's work was an assessment of the "advancement" of the families rather than a classification, we note that correlations of the magnitude Sporne found would, for many pairs, produce changes in 2 × 2 tables from the ones obtained assuming independence, that are greater than those observed in our hypothetical excursion from (8) to (9). In another context, studies of procedures based on a number of dichotomous characters for identifying microbial cultures, such as that of Niemelä, Hopkins, and Quadling (**1968**), show a consciousness of the role of dependence, of patterns of *combinations* of characters, and of selection of a subset of the most meaningful characters. Medical diagnoses provide another area of classification problems; as is aptly stated (p. 71) in a paper on this subject by Croft and Machol (**1974**), "most previous studies have made the assumption of independence of symptoms, an assumption which is always suspect and frequently wrong."

6. OTHER COMMENTS

The discussion in MS-I, II raises a number of other taxonomic and statistical issues, the most important of which I will group into five topics here.

A. Bayesian features.—The first paragraph of MS-II, in its reference to the "controversial implications" of the word "Bayesian," may lead the reader to feel that the Bayesian aspect of the MS methodology is a major reason for my criticism. I hope that the emphasis of Sections 3 and 4 has shown that my main objections are more fundamental, and that my reference to Birnbaum and Maxwell (**1961**) in Section 5 has indicated approval of properly implemented Bayesian classification where the setting warrants it. It certainly *is* my belief that the unthinking and mechanical use of Bayesian statistics, in the assumption of values for the prior probabilities (π_A and π_B or their analogues in other settings) for which there is little scientific basis, or in the frequent use, instead, of the scientist's "subjective prior probabilities," can have disastrous consequences. The use of this attractive-appearing approach, which yields posterior probability ratios so easily, contrasts with the complexity many of us perceive in the settings scientists commonly study, that we feel cannot be codified in terms of such a simple recipe. Thorough statisticians find it essential to study the "performance characteristics" of their method as did Cochran and Hopkins (**1961**), mentioned in Section 5. This performance can be assessed in terms such as the probability that the method yields an erroneous conclusion, as a function of the unknown true state of nature (the actual P_C, and the actual P_A and P_B if they are unknown and hence estimated, in the setting of Section 2). Statisticians also find it important to study the "robustness" of the method, its performance under departures from the assumed model such as departures from (5). (In MS-I, p. 766, the authors acknowledge not having studied the performance characteristics of their method, but erroneously refer to what is needed as a study of 'confidence intervals.') Some of the better Bayesians also make such studies for their Bayes rules (analogue of (3) with $k = \pi_B/\pi_A$), but less sophisticated or careful Bayesians simply give the posterior odds ratio for the data at hand, as MS have done. In Kiefer (**1977**), I have commented on the delusion of rational, optimal statistical analysis that such mechanical use of the Bayes approach has yielded. It is not useful to reply to the Bayesian polemics of MS-I (e.g., footnote, p. 766), but some technical omissions and inaccuracies in MS deserve comment.

The use of (3) with $k = 1$ assumes $\pi_A = \pi_B$. The statement on p. 759 of MS-I, that the Bayes rule is this likelihood ratio rule in their explanatory ball-and-urn model, assumes this, which is *not* a part of the Bayes model. While the effect of this assumption on the results is negligible in the present context (and is microscopic compared with the effect of (5)!), it is important in Bayesian analysis to make clear such assumptions and their basis or extent of effect on the results. Similarly, the formula illustrated above (7), and discussed on p. 762 of MS-I, for computing the Bayesian estimates $\hat{P}_{A,j}(1)$ in MS, is based on the particular assumptions (a) that the unknown *true* value $P_{A,j}(1)$ is, for each taxon A and each character j, *uniformly distributed* between 0 and 1 according to the prior law (for example, that there is exactly 0.34 prior probability that $P_{B,32}(1)$ lies between 0.12 and 0.46, in the *Gerronema* illustration); and (b) that the "penalty" from misestimating the actual $P_{A,j}(1)$ by a value $\hat{P}_{A,j}(1)$ is proportional to $(P_{A,j}(1) - \hat{P}_{A,j}(1))^2$. In this case MS-I (p. 762) tells us these assumptions; but neither of them is an *automatic* assumption of Bayesian methodology; an indication of, or literature reference to, the performance characteristics of the resulting estimation rule, should be given. It is not sufficient simply to say that "other formulas are possible but they would not affect the argument." The statement on p. 761 of MS-I that "Bayesian statistics does not have this difficulty," of estimating $P_{A,j}(1)$ to be 0 if there are no observed species in state 1 for character j, is misleading from two viewpoints: on the one hand, Bayesian statistics, which does not consist of a single rule, might reach *exactly* the disfavored conclusion for a different prior law and penalty function; on the other hand, it is not the case, as MS imply, that every classical (= non-Bayesian) statistician would automatically be inclined to estimate $P_{A,j}(1)$ to be 0 in this circumstance (as is known, for example, to users of the chi-square test with correction for small cell counts).

Actually, the MS method is *not* properly constructed from a Bayesian viewpoint. Even assuming $\pi_A = \pi_B$ and that the $P_{A,j}(1)$'s are uniformly distributed, a Bayesian would *average* each of the numerator and denominator of the left side of (3) with respect to those prior probabilities, not "plug in" Bayesian *estimators* of the $P_{A,j}(1)$'s, which are used in answering a different question. This averaging involves knowing not merely that the marginal prior probability laws of the individual $P_{A;j}(1)$'s are uniform, but also what the prior law states about *combinations* of the $P_{A,j}(1)$'s, for different j's. In particular, we again encounter the question of whether it is reasonable to assume, at this level of prior knowledge of the parameters $P_{A,j}(1)$ of the model, the kind of independence that was assumed in (5) for the characters, once the parameters $P_{A,j}(1)$ were treated as known. Estimating the $P_{A,j}(1)$'s and then substituting the estimates into (3) avoids worrying about prior dependence of the parameters, but it gives an incorrect answer even for a Bayesian!

On p. 762 bottom of MS-I, the estimation technique is modified to take account of what it otherwise yields as a likelihood ratio MS consider anomalous in certain situations: if families of sizes $n_A = 20$ and $n_B = 180$ both yield *zero* frequency of state 1 for some dichotomous character for which *all* members of C occurred in state 1, their method described above would yield 182/22 for the contribution of that character to (4′). The authors consider this "bizarre," and consequently modify the ratio to be unity in such cases. Actually, the result is not so bizarre at all from a Bayesian viewpoint: if the character probability $P_{A,j}(1)$ is indeed uniformly distributed, the posterior odds ratio for such a single character in this circumstance, if treated in true Bayesian fashion as described in the previous paragraph, would be 181/21. This extreme case is stated by MS not to occur, but their modification is used in nearby less extreme cases. Ironically, the modification amounts to use of the kind of rule they view as defective in the hands of the classical statistician, as described earlier!

Even assuming one is going to use a likelihood ratio method with Bayesian plug-in estimators, this being what MS really employ, the general estimation formula given on p. 762 (middle) of MS-I is inappropriate if the character j in question is not dichotomous; as given, the estimated probabilities of the states *do not sum to unity* if the number of states $L > 2$, and the formula $(Nf_{ijk} + 1)/(N + L)$ would be more appropriate, where N is the number of members of the taxon (our n_A) and f_{ijk} is the proportion of species in the taxon k under consideration, for which character j is in state i. It is *impossible* that all $P_{A,j}(i)$, $1 \leq i \leq L$, be uniformly distributed if $L > 2$, which accounts for this error. My check of the numerical results listed by MS shows that they indeed used their erroneous formula (with $N + L$ above replaced by $N + 2$); for example, the ratio value 1.338 listed for character 17 in Table II of MS-II (where $L = 3$) is obtained from the incorrect formula, the correct value being 1.02 times larger; the final ratio 8380 should be 1.54 times larger in terms of the correct calculation (which does not, obviously, make it any more meaningful to me!) In MS-II, $n_A > n_B$. When $n_A < n_B$, as in all computations of MS-I, the final listed ratio should, similarly, be reduced appropriately. This will not affect the order of magnitude of the listed odds ratios, but is indicative of careless mathematics; and even fervent believers in the MS approach (or users of the FORTRAN program they offer) will wonder when the estimated probabilities of states for a given character in taxon A sum to $(n_A + L)/(n_A + 2)$ rather than to unity!

B. Sphericity of taxa.—The discussion of MS-I rightly criticizes the tacit assumption that taxa are spherical (more correctly, ellipsoidal) in the shape of the collection of points (m-vectors) $\mathbf{A}_t$ for each taxon A. But I believe MS give a false impression of the rarity of consideration by others of nonspherical taxa, as the reader will see upon glancing at the cluster analysis computer routines for step-by-step build-up of taxa, often yielding chain-like portions, that have appeared frequently in such journals as *Systematic Zoology* and, more recently, in Hartigan (**1975**). But MS do not seem to realize what their model, particularly (5), assumes about sphericity. It should be evident to the reader that the description in MS-I, p. 757 of the families in Agaricales as "far from spherical . . . rather, highly convoluted chains, anastomosed, and folded in many dimensions" implies that the *characters in those families cannot come close to satisfying* (5)! Such taxa are not well represented by their centers of gravity, but those that satisfy (5) often are. In fact, it is a well-known feature of statistical theory that, especially when n_A is moderately large, under assumption (5) the character m-vectors $\mathbf{A}_t$ of n_A randomly chosen members of A will exhibit an ellipsoidal pattern almost as they would for the better known case of continuous, normally distributed, character values. Thus, despite the assertion (p. 757) that insistance upon spheroidal taxa would be "contrary to everything useful discovered in science in recent centuries," the assumption (5) assumes away the chains MS would like to allow and accepts by assumption essentially the spherical or ellipsoidal taxa of which they are correctly skeptical! It is not accurate that (MS-I, p. 758) "when one describes the entire family *Strophariaceae* quantitatively, one is in fact describing the center of gravity of the chain;" this is true of the MS method of description based on (5), but one might not use that inappropriate method! In defense of MS it must be said that their use of smaller intrafamilial taxa (p. 758) has possibly reduced this effect, but cannot be expected to have eliminated it. (However, they do not hesitate, in Tables 3 and 4 of MS-I, to combine dissimilar genera into new, hypothetical taxa they cannot entirely accept. See Hartigan (**1975**) for more systematic methods of approaching this problem of combining clusters.) Indeed, since they do not diagram frequencies of *combinations* of characters, there is no way that MS can judge the extent of actual nonsphericity of their taxa, or can even contemplate the possible wisdom of constructing taxa with chain-like portions, whose consideration they have rightly praised.

C. More on the model.—There are several reasons why characters of species within a given taxon may exhibit the kind of statistical correlation mentioned in some of the references cited in Section 5, and which is exhibited in the nonspherical taxa shapes described in MS-II and quoted in Section 6B. There is the obvious possibility of genetic linkage, and there is the correlation induced by the mutual advantage contributed by favorable states of two or more characters in evolutionary adaptation; it would be presumptuous of me to discuss either, but many biologists have done so. There is also the fact that taxa are *not* random samples from some God-given ball-and-urn scheme (the explanatory model used by MS), constructed to yield independent characters; as Gould (**1974**) aptly puts it in his perceptive review of Sneath and Sokal (**1973**), "a classification is a human decision, constrained by a bevy of facts, about how best to order nature." The very act of *selecting* the similar species that have been put into a genus, or the genera in a family, introduces correlations. This is so even in an ideal (for the believer in independence) scheme in which, in a larger universe (higher taxon), characters are independent. This induced dependence is a well-known statistical phenomenon, which the interested reader can check by generating (say) a hundred random vectors with 10 or so components, the components being chosen independently, each with equally likely possible states 1 and 2; when the resulting character vectors are clustered into "taxa" in almost any reasonable way (for example, by using a scheme based on number of matches), there is a great chance that, *within* the separate taxa, substantial character correlations will appear. For, chain-like clusters or parts of clusters will arise by chance, and such chains usually exhibit considerable dependence. Almost nobody would believe that this "ideal" model is a believable one in nature; I mention it because, if selection of the species that are put into a taxon introduces dependence in a scheme in which characters were independent before that selection, it may make it easier to appreciate the dependence introduced in the selection of the species in the higher taxa A and B in the actual classification problems encountered by taxonomists. Of course, insistance on almost-spherical taxa (recommended by neither MS nor me) would yield smaller correlation of characters; but when there are 30 characters, it is doubtful that the set of a few hundred species selected in a typical genus or family can truly be close to spherical, even if taxonomists use arithmetic based on *assuming* sphericity.

We note that the processes by which taxa are formed also make the independence exhibited in formula (2) (as distinct from that of (5)) questionable; hence this also affects the meaning of the MS plug-in estimators $\hat{P}_{A,j}$, especially if n_A is small.

Realization of the process by which genera are joined to make families, be it by nonmathematical taxonomy or by a scheme like that of MS, reinforces one's understanding of the presence of dependence: chains and convolutions of the type described on p. 758 of MS-II are created. Indeed, the lumping together of *Squamanita* and *Cystoderma* for some of the comparisons in MS-I (pp. 774–5), with the dissimilarities of these genera MS have noted, throws additional question on the use of (5) for such a group.

In MS-I, p. 763, it is argued that whenever the frequency of state 1 of a dichotomous character j is 0 in one subgenus of genus C and is 1 in another subgenus, the frequency for genus C should be obtained as (a) an unweighted average of the frequencies in the two subgenera, rather than as (b) the frequency of all species in the genus; the two computations could of course lead to very different results if the two subgenera are of quite different sizes. Without taking the space to criticize all the arguments of MS in detail, I will mention two points. Firstly, although their "if one counts species, why not subspecies and varieties" indicates an arbitrariness in (b) (be it at the expense of not maintaining the central concept of species), it ignores the perhaps comparable degree of arbitrariness in the establishment of subgenera, present in (a). Secondly, although (b) is indeed dependent on unequal levels of floristic exploration as MS point out, one should realize that the whole MS model outlined in Section 2 undergoes a drastic conceptual change upon the introduction and use of subgenera in this fashion. Whatever the meaningfulness of that model as explained by the MS ball-and-urn analogy, we must now think of the hypothetical probabilities $P_{A,j}$ (relevant if C comes from family A) as referring to balls that are subgenera of C for characters of the indicated type, but to balls that are subsections or lower levels (perhaps species, if the prescription just below (f) on p. 763 of MS-II is followed and we never reach a zero-one frequency split at a higher level) for other characters. One may wonder what kind of ball-and-urn mechanism produces such balls! As a consequence, relevance of the model and the calculations that proceed from it seem all the more dubious to me.

D. Other statistical principles.—I will list here some of the other assertions in MS that are misleading.

On p. 765 of MS-I, the methodology is termed "robust, in the sense that a few errors of judgment on marginal questions are not likely to affect the overall results." It is true that, assuming (5), a few character likelihood ratio factors of 2 or even 10 will not alter the conclusion yielded by a value of (4) that is in the millions. But the term "robust" is employed in statistics for a procedure whose high probability of yielding a correct inference is not much reduced when the model upon which it is based is somewhat incorrect. For the MS method, we have seen in the example of (8) and (9) how moderate departures from the assumed (5) can alter the true value of (4) from what MS calculate for (4) assuming (5). Moreover, no investigation by MS or (to my knowledge) anyone else has delimited those departures from (5) under which the MS technique will probably perform well. In a remarkable statement (MS-I, p. 763) about their method, the authors assert that "unlike other models in numerical taxonomy, it is superfluous to test it against accepted taxonomies." Such a test would not have been conclusive evidence for using the methodology, but it would have been more than MS have given us. The reader may decide why the MS model, alone, should be exempt from testing.

The notion (MS-I, pp. 754, 764), that the use *without weighing* of all available characters "in which there is any measurable difference between the two major taxa being compared" avoids judgmental weighing and the controversy often attending selection of characters, is misleading because of the fallacious assumption of independence. If proper account were taken of dependence among characters, that would automatically induce appropriate "weighing" when (4) is used, which has been obliterated by the assumption of (5); several highly correlated characters would correctly have less effect on the final odds ratio than they do in MS. Also the paragraph on p. 764 of MS-I that refers to "independence of characters" (which addresses the question raised in Section 3 and in K only in the obvious and extreme case of correlation near unity) shows in its illustration that MS, too, must exercise some judgement in their list of characters and states. They acknowledge this judgmental factor (p. 765), and I certainly do not believe one can entirely avoid it (although taking account of lack of statistical independence would help to do so). But the consequence is that the MS scheme is not so free as p. 764 implies from that factor whose presence in other numerical taxonomy work they criticize. I do not share their belief that their judgment in this respect has been superior to that of all conventional numerical taxonomists.

It is well known that, when a data set suggests the method of analysis to be used on that data set, one obtains a different *actual* precision of the procedure (usually much worse, in practice) from what one calculates pretending the method had been selected before seeing the data. Hence, a routine of selecting a few important characters upon seeing the counts of character vectors, and of analyzing them taking account of dependence of characters (as suggested in Section 4) yields results that should not be accompanied by a measure of conclusiveness (such as an odds ratio) computed as though those characters had been selected before seeing the data. Indeed, as I have stated earlier, the better of the workers who use such methods refrain from stating such meaningless measures, notwithstanding which I regard their methods as superior to that of MS for the reasons stated in Section 4. In the present setting, "seeing the data" means having even rough perceptions about the families in question (or even the Agaricales in general, of which these families are a part), which yielded such a decision as that of not listing "presence or absence of an annulus" and "presence or absence of a cortina" as separate characters, in MS-I, pp. 764–765, instead listing a three-state character "annulus, cortina, neither;" or of adopting only the two states of lamellar attachment "concurrent, other" in the study of *Phaeomarasmius*; or of selecting as A and B the subgroups of two families that appear closest to C, in MS-I, p. 758; or, in MS-II, p. 1169, of deciding "we found only 33 characters useful for our purpose;" or of combining *Squamanita* and *Cystoderma* for further comparisons, because of what the data have shown (MS-I, pp. 774–775). I do not argue here with any of the choices of characters or states or even with the step of combining taxa for at least some further comparisons; but the fact that such selection or combination was based on the present data, even if only in the qualitative terms of some recollection of the contents of Singer (1962), makes the probability ratios additionally suspect, be it in a slightly more subtle way than would be the case if one had selected the five most important characters. An unfortunate feature of the application of a probabilistic model in taxonomy, such as that of (2) and (5), is that we do not have the luxury of the experimental scientist who can let a data set suggest a model and then legitimately compute probabilities from that model for a subsequent data set obtained under the same experimental conditions. In taxonomy, an "observation" is a character vector of a species, and one cannot treat that same species as one can a new observation in a second experiment. One could make legitimate probabilistic calculations based on the character and state selections that came from previously observed species, on only the species collected in the future; but to do so would mean throwing out so much information that it is preferable to use the past data with a sensible selection but without the company of a meaningless odds ratio. (In some statistical studies, a random subsample of the data is used to suggest a procedure to be used on the remaining data. In the present context, I believe this approach to be of doubtful usefulness.)

An obvious scientific principle, exercised in any careful statistical study, is that one should not assert bold inferences that are based on a particular model whose validity is unknown, without also testing the adequacy of that model. I have mentioned that the crucial independence assumption (5) is never examined by MS in terms of actual data. Moreover, even if one could sensibly assume (5), there is the further fact that the approach allows us to decide that C is overwhelmingly favored to belong to A rather than B, or vice versa, without entertaining the possibility that C belongs to neither. It is perfectly possible for C to be very distinct from *both* A and B, but to be somewhat closer to A and to have this manifested in a huge odds ratio. (The consideration of two or three different B's in each of the four studies of MS-I does not alter this, the question still being whether the data support the model's implication that C belongs to A.) Some quick "goodness-of-fit" tests I conducted indicated that, *assuming the statistical independence of* (5) as MS do, the data of MS-II would barely support the "C belongs to A" assertion (significant difference at about 0.01 level), and that the results are somewhat worse in the case of Tables II and III of MS-I, very much worse for Table I, astronomically worse for Table IV, the study for which MS certainly indicate least support. (I will gladly supply details for those who are interested.) These results are not unexpected: typically there had been some previous doubt about putting genus C into higher taxon A precisely because of the differences between the frequencies of some of the character states in C and A. So we recognize that our purpose is really to adjoin to A (or B) a collection of character vectors $\mathbf{C}_t$ that are at best on the edge of the collections of character vectors obtained from A and B, rather than being typical of either. That is the way clustering in numerical taxonomy often proceeds. But realization of these circumstances throws additional doubt on the meaningfulness of the odds ratios computed from the MS model. For example, does *Phaeomarasmius* really belong in the Pholiotoideae (Table I of MS-I) when a chi-square-type test finds the former significantly different from the latter at any reasonable level ($P \ll 10^{-10}$)? As I stated in K, I believe this *taxonomic* conclusion, and am certainly not even qualified to argue about most of the biological aspects; but that belief is bolstered by the nonmathematical factor of Singer's skill, and cannot be supported by the MS mathematics. The numerical taxonomy MS criticize, superficially less quantitative in its restraint from presenting a numerical measure of conclusiveness, is less misleading than the arithmetic of MS, because it recognizes that it is not tenable that C is a sample from A, does not compute odds ratios based on that untenable hypothesis, but is willing to adjoin C to A as a useful taxonomic step. I cannot emphasize enough the difference between making such a reasonable taxonomic conclusion and accompanying it by a startling and meaningless odds ratio.

The discussion on the middle of p. 766 of MS-I, that one is tempted to say A is relatively close to B if the odds ratio (4) is near unity, may be settled by a similar comment; and one should note that although MS raise this temptation, they refrain from succumbing to it. It can easily be that A and B are quite distant from each other, and that C, roughly equidistant from them, yields an odds ratio near unity; but A′ and B′, much closer than A was to B, can yield a huge odds ratio because C′ is so much closer to A′ than to B′. Thus, the odds ratio may tell us nothing conclusive about the hiatus between families A and B.

E. Care in handling data.—I have mentioned the presence in MS of such careless errors, aside from the major ones that evoked my main criticism, as those of terming "likelihood ratio" or "probability ratio" what is actually the n_Cth root of this, and of giving estimates of probabilities of states of a character that sum to more than unity if there are more than two states. Other shortcomings that make the method confusing and difficult to use include the MS treatment of frequencies discussed in Section 6A, and the method of recording the data, mentioned in Section 3, that eliminates the possibility of studying statistical dependence among characters. Additionally, and despite the reassurance in MS-II, p. 1171, of the "careful work required for the present study," the basic recording of data in MS appears to me to be rather sloppy, and I doubt whether anyone, given the same list of character vectors, could reproduce the tables. For example, there certainly is a problem of dealing with character vectors for which some components are missing, but it is not insurmountable, and there is a vast literature of methods for dealing with such "censored" observations, as they are termed. However, MS-I, p. 765, deals with the problem by making estimates "if we believed that the possible error resulting from such estimates was unlikely to be significant. For example, we have estimated that 50% of *Galerina* are lignicolous . . .. We have not gone through the monograph counting species." Which of the tabled character state frequencies involve such estimates? We are not told. Another taxonomist cannot hope to check the results, let alone know how reliable the 50% estimate is. Evidently he must take on faith the reliability of the MS numbers and the correctness of their belief that their likelihood ratio (4′) is, nevertheless, substantially correct in magnitude.

The tables also contain an unbelievably large number of entries that are simply impossible, not explainable by any operation of the MS conventions of frequency computation discussed in Section 6A. For example, in MS-II, Table II, the smallest possible nonzero frequency for *Gerronema* is $1/n_B = 1/26 = 0.0385$, but there are entries of 0.011 and 0.022; these are not printers' errors, as one can check from the frequency sum over the states of each character being unity and from the likelihood ratio contribution listed for each of those characters. Again, *Floccipedes* contains five species, but two cases of frequency 0.100 are listed. *Ripartites*, in Table II of MS-I, with five species, produces frequencies of 0.150 and 0.167. There are other anomalies that appear in the tables in higher frequency values, unexplainable by the MS prescription on frequencies described in Section 6A, and which could arise only out of arithmetical carelessness or by use of some estimate obtained without the authors having made actual frequency counts, in the manner quoted in the previous paragraph or by some other scheme of which we have not been told. Every table of frequencies in MS-I, II contains a large number of these mysterious entries. Does the entry 0.011 for state 3 of character 30 of *Gerronema* mean that no species had a sweetish or fruity odor but that one species had an odor that was regarded as 29% (= 0.011/0.0385) sweetish or fruity? Or does it mean that 29% of observers regard the species as having such an odor? We are never told, nor are we told that this type of assessment was ever used, or how the 0.011 was otherwise obtained.

I am not even arguing that such an assessment was necessarily inappropriate, only that the best mycologists could not hope to reproduce these figures, which reflect a cavalier disregard either for accuracy in computation or for the principle of telling other scientists what one has done, so that they can at least hope to check the results. As to whether this sloppiness affects the final odds ratios appreciably, you must evidently trust MS that it does not; although I have seen no numerical basis (as distinct from the authors' intuition) for their faith in the magnitude of the resulting ratios, my argument in this paragraph has not been directed at that faith or at the larger reasons discussed earlier for distrusting the odds ratio. Rather, it is simply a plea for reliable arithmetic and recording of data.

7. CONCLUSION

The Machol-Singer methodology assesses the conclusiveness of its findings in terms of probability ratios which are as meaningless as they are strikingly large. It is based on questionable assumptions the authors have made no effort to justify. The reliability of its performance has never been shown when the assumptions are violated, while examples show it can easily lead to incorrect conclusions in such cases. This is not to say that the *taxonomic* conclusions reached by the authors are incorrect; or even that one might not use their scheme as a preliminary step, the tables at least being useful for ascertaining which characters might be most important. Indeed, a more careful methodology, of then selecting the most important characters but taking account of their statistical dependence, could often reach the same taxonomic conclusions. Such a technique, although suggested by the best statisticians who work on such problems, is nevertheless deprecated by Machol and Singer, who prefer to assume an unjustified but more mathematical model and record its consequences. I view this last as bad science; and the prominent presentation of those probability ratios, despite the authors' occasional assertion that only the closeness to unity of those ratios is of interest, encourages a belief in the conclusiveness of the results which is totally unwarranted.

Taxonomy is a difficult subject, aiming at evolutionary insight, to be explored with a careful hunt for chains and correlations and with verification that any mathematical models employed are supported by the data or past experience. The authors have engaged in none of these steps. However, I repeat my earlier comment: don't believe me; ask your own friendly statistician.

8. ACKNOWLEDGMENTS

Over the years since MS-I appeared, I have received helpful comments from S. E. Fienberg, L. A. Goodman, J. A. Hartigan,

J. L. Hodges, Jr., R. P. Korf, F. Mosteller, A. Rosenberg, H. Solomon, and A. H. Smith. In their many letters to me, R. E. Machol and R. Singer contributed considerably to the choice of form and emphasis of the present paper. I am especially grateful to Jim and Karen Reeds for their detailed criticism of my writing; and, above all, to William Kruskal for continual advice on this undertaking and the correspondence that led to it.

LITERATURE CITED

Bahadur, R. R. 1961a. A representation of the joint distribution of responses to n dichotomous items. Pp. 158–168. *In: Studies in items analysis and prediction.* Ed. H. Solomon. Stanford University Press.

——. 1961. On classification based upon responses to n dichotomous items. Pp. 169–176. *In: Studies in item analysis and prediction.* Ed. H. Solomon. Stanford University Press.

Birnbaum, A., and A. E. Maxwell. 1961. Classification procedures based on Bayes's formula. *Appl. Statist.* **9**: 152–169.

Bishop, Y., S. E. Fienberg, and P. Holland. 1975. *Discrete multivariate analysis.* MIT Press, Cambridge, Mass. 557 p.

Cochran, W. G., and C. E. Hopkins. 1961. Some classification problems with multivariate qualitative data. *Biometrics* **7**: 10–32.

Cormack, R. M. 1971. A review of classification. *J. Roy. Statist. Soc.*, Series A, **134**: 321–353.

Croft, D. J., and R. M. Machol. 1974. Mathematical methods in medical diagnosis. *Ann. Biomed. Engin.* **2**: 69–89.

Das Gupta, S. 1973. Theories and methods in classification. Pp. 77–138. *In: Discrimination: a review, analysis and applications.* Ed. T. Cacoullos. Academic Press, New York.

Estabrook, G. F. 1967. An information theory model for characteristic analysis. *Taxon* **16**: 86–97.

Fienberg, S. E. 1970. The analysis of multidimensional contingency tables. *Ecology* **51**: 419–433.

Fisher, R. A. 1936. The use of multiple measurements in taxonomic problems. *Ann. Eugen.* **7**: 179–188.

Goodman, L. A. 1964. Interactions in multidimensional contingency tables. *Ann. Math. Statist.* **35**: 632–646.

——. 1970. The multivariate analysis of qualitative data, interactions among multiple cross-classifications. *J. Amer. Statist. Assoc.* **65**: 226–256.

——. 1971. The analysis of multidimensional contingency tables: stepwise procedures for direct estimation methods for building models for multiple classifications. *Technometrics* **13**: 33–62.

——. 1973. Guided and unguided methods for the selection of models for a set of T multidimensional contingency tables. *J. Amer. Statist. Assoc.* **68**: 165–175.

Gould, S. J. 1974. Review of: *Numerical taxonomy,* by Sneath and Sokal. *Science,* **183**: 739–740.

Hartigan, J. A. 1975. *Clustering algorithms.* John Wiley, New York. 351 p.

Heincke, F. 1898. Naturgeschichte des Herings. *Abh. Deutsch. Seef.* **2**: (2 vol.): 1–128 and 1–223.

Hills, M. 1967. Discrimination and allocation with discrete data. *Appl. Statist.* **16**: 237–250.

Hodges, J. L., Jr. 1950. *Discrimatory analysis. I. Survey.* USAF School of Aviation Medicine, Randolph Field, Texas. 115 p.

Jardine, N., and R. Sibson. 1971. *Mathematical taxonomy.* John Wiley, London. 286 p.

Kiefer, J. 1975. Review of Bayesian analysis of generic relations in Agaricales, by R. E. Machol and R. Singer. *Mycologia* **67**: 203–205.

——. 1977. The foundations of statistics—are there any? *Synthese* **36**: 161–176.

Machol, R. E. 1975. The titanic coincidence. *Interfaces* (of T.I.M.S.) **5**: 53–54.

——, **and R. Singer.** 1971. Bayesian analysis of generic relations in Agaricales. *Nova Hedwigia* **21**: 753–787.

——, **and** —. 1977. Taxonomic position of *Hydropus floccipes* and allied species—a quantitative approach. *Mycologia* **69**: 1162–1172.

Martin, D. C., and R. A. Bradley. 1972. Probability models, estimation, and classification for multivariate dichotomous populations. *Biometrics* **28**: 203–321.

Niemelä, S. I., J. W. Hopkins, and C. Quadling. 1968. Selecting an economical binary test battery for a set of microbial cultures. *Canad. J. Microbiol.* **14**: 271–279.

Orlocci, L. 1969. Information theory models for hierarchic and nonhierarchic classifications. Pp. 148–164. *In: Numerical taxonomy.* Ed. A. J. Cole. Academic Press, New York.

Pearson, K. 1926. On the coefficient of racial likeliness. *Biometrika* **18**: 105–117.

Raiffa, H. 1961. Statistical decision theory approach to item selection for dichotomous test and criterion variables. Pp. 187–220. *In: Studies in items analysis and prediction.* Ed. H. Solomon. Stanford University Press.

Rao, C. R., and P. Slater. 1949. Multivariate analysis applied to differences between neurotic groups. *Brit. Jnl. Psych. Stat. Sec.* **2**: 17–29.

Rescigno, A., and G. A. Maccacato. 1961. The information content of biological classifications. Pp. 437–446. *In: Information theory.* Ed. C. Cherry. Butterworths, London.

Singer, R. 1962. *The Agaricales in modern taxonomy,* 2nd ed. Cramer, Weinheim. 915 p.

Sinnott, E. W., and I. W. Bailey. 1914. Investigations of the phylogeny of the angiosperms. *Amer. J. Bot.* **1**: 441–453.

Sneath, P. H., and R. R. Sokal. 1973. *Numerical taxonomy.* W. H. Freeman, San Francisco. 547 p.

Solomon, H. 1961. Classification procedures based on dichotomous response vectors. Pp. 177–186. *In: Studies in items analysis and prediction.* Ed. H. Solomon. Stanford University Press.

Sporne, K. R. 1949. A new approach to the problem of the primitive flower. *New Phytologist* **48**: 259–275.

Tukey, J. W. 1962. The future of data analysis. *Ann. Math. Statist.* **33**: 1–67.

——. 1977. *Exploratory data analysis.* Addison-Wesley, Redding, Mass. 688 p.

van Ryzin, J., Ed. 1977. *Classification and clustering.* Academic Press, New York. 488 p.

Accepted for publication September 14, 1978

G. Kallianpur, P.R. Krishnaiah, J.K. Ghosh, eds., *Statistics and Probability: Essays in Honor of C.R. Rao*
© North-Holland Publishing Company (1982) 419–428

OPTIMUM RATES FOR NON-PARAMETRIC DENSITY AND REGRESSION ESTIMATES, UNDER ORDER RESTRICTIONS*

J. KIEFER (deceased 1981)
University of California, Berkeley, CA, U.S.A.

1. Introduction

Among the many contributions for which Professor Rao will be remembered is his work on the information inequality for estimators in regular parametric (finite-dimensional space of distributions) problems. The lower bound on squared error obtained by that method is useful in certain non-parametric problems as well, by reduction to an appropriate finite-dimensional problem. Čenčov (1962) was the first to announce a lower bound result for a case of the global density estimation problem in which the "best rate" is obtained, and he mentioned the Cramér–Rao inequality as part of the method. Farrell (1967, 1980) pioneered the use of the inequality in the local density estimation problem, and also (1972) the use of the Bayes method to obtain general results on rates in these problems. [The latter method has not always been termed such, but is the principal method used in the lower bound results mentioned herein, except for Farrell (1967, 1980), who alone makes essential use of the inequality in his calculations. Čenčov, Hasminskii, and Samarov make use of a Hajek-type asymptotic efficiency result, which can be proved by asymptotic Bayesian methods; Bretagnolle and Huber (1979) use Kullback–Leibler information in an inequality on risk, but then use the device of Farrell (1972) of estimating a maximum risk from an average.] It is to be emphasized that these lower bounds apply to *all* estimators, not merely usual ones of kernel types, etc. Thus, results on the properties of good kernel estimators are referred to later in this section not because they yield lower bounds for all estimators, but rather because they justify that such bounds give an attainable rate.

Some problems which have been attacked by one or both of these methods are those of density estimation, failure rate estimation, spectral density estimation, and regression function estimation. Moreover, both the local and global performance of estimators have been considered. In principle, at least for squared error loss, there are thus 16 possible combinations of method, problem, performance, not all of which have been carried out. The Bayes method has seemed easiest to apply, both in terms of details of the proof and also in the greater generality of the loss functions it treats; but the information inequality method has thus far yielded the most explicit results in the form of a close-to-best constant [C in (1.2) below, or its analogue for local estimation] that multiplies the power of the sample size, n, shown to be

*Research sponsored by NSF Grant MCS 78-25301 and ONR Grant N00014-75-C-0444.

optimum in these investigations, in Farrell (1980); and Farrell has used Wolfowitz's extension of the information inequality to consider also sequential procedures. These advantages of the two methods may change as more precise (and perhaps sequential) Bayes estimates are used, or as the extensions of Barankin and Rao of the information inequality for other loss functions, are invoked. In any event, it seems worth while to give both developments here, even though the result of Section 4 can also be obtained by the method of Section 2 if one keeps track of the constants.

The various investigations differ in the classes of functions (regularity conditions) imposed, but typically are of the following nature, described here for the problem of local density estimation. The rv's $X_1, X_2, \ldots, X_n$ are iid with common density, f on R^d, assumed to belong to some class $\mathfrak{F}$. For global performance, one estimates all of f; for local performance, one estimates $f(b)$ for some specified point b, usually represented as 0. (More generally, estimation of Tf has been considered, where T is a differential operator.) General loss functions have been considered, but simplest are zero–one and squared error. For example, for zero–one loss and local performance, one shows for an appropriate p (depending on d and $\mathfrak{F}$) that

$$\lim_{n\to\infty} \sup_{f\in\mathfrak{F}} P_f\{|t_n - f(0)| > cn^{-p}\}$$

approaches 1 as $c \to 0$, for every sequence of estimators $t_n = t_n(X_1, \ldots, X_n)$; particular t_n's are then shown to yield 0 as the limit of (1.1) when $c \to \infty$; thus, n^{-p} is the best "rate in probability" possible for t_n, a rough asymptotic minimax result. Similarly, for squared integrated error and global performance, with $t_n(x) = t_n(x; X_1, \ldots, X_n)$ estimating $f(x)$, one shows that

$$\lim_{n\to\infty} n^{2p} \sup_{f\in\mathfrak{F}} E_f \int_B [t_n(x) - f(x)]^2 \, dx \geq C, \tag{1.2}$$

where B has positive measure and C is a positive constant depending on B and $\mathfrak{F}$; particular t_n's are shown to make the left side of (1.2) finite. [Usually $B = R^d$ in the literature, but we must modify this in Section 3. For the local analogue of (1.2), at the point x, omit the integration in (1.2).] While the assumptions of different authors vary slightly, a typical formulation for local density estimation is that, if k^{th} derivatives of f are assumed bounded uniformly in $\mathfrak{F}$ [this can be weakened to a Lipschitz condition on the $(k-1)$st derivative], then $p = k/(2k+d)$ in (1.1) and (1.2). (If Tf is estimated where T is a differential operator of order $s < k$, the corresponding result is $p = (k-s)/(2k+d)$.) The analogue of (1.2) for the maximum deviation of t_n from f on an interval B is, for the same p,

$$\lim_{n\to\infty} (n/\log n)^{2p} \sup_{f\in\mathfrak{F}} E_f \sup_{x\in B} [t_n(x) - f(x)]^2 \geq C. \tag{1.3}$$

In some non-parametric settings, particularly ones arising in life testing or reliability theory, f is restricted by the context by an *ordering condition* of a different nature from the *regularity condition* indicated above. Examples of such orderings are f decreasing, (and with $d = 1$) f with decreasing failure rate (DFR), f "star-shaped", or f with increasing failure rate on average (IFRA). The problem of estimating the df F corresponding to f, or f itself, or the failure rate $r_f = f/(1-F)$, has been considered frequently; see Barlow et al. (1972). One may wonder, if one imposes such an additional restriction, can one increase the value of p in (1.1) and (1.2)? For example, if $\mathfrak{F}$ consists of those f that are decreasing on $(-1,\infty)$ and for which $-f' \leqslant M < \infty$ (as in the $k=1$ condition above), is there a sequence t_n that makes use of the monotonicity of f to achieve some $p > \frac{1}{3}$ in (1.1), or (1.2) for B a compact subinterval of $(-1,\infty)$? The present paper answers "no" to this question, although this does not prove that the best constant C in (1.2) might not be improved by imposing the additional restriction.

In view of the variety of problems, losses, performance characteristics, and regularity and ordering conditions of $\mathfrak{F}$ that have been (or can be) considered, no attempt is made herein to treat all cases. Rather, three examples are treated in detail to indicate the type of argument needed to modify existing proofs when an ordering condition is imposed. These are the treatments of Farrell (1967), Stone (1980) and Samarov (1978) for the density estimation problem. Modifications, imposing an order restriction on the class of functions to be estimated, can be made in such works as those of Farrell (1972), Meyer (1977), Bretagnolle and Huber (1979), Hasminskii (1979) and Wahba (1975) for this problem, of those of Stone (1980, 1981) and Ibragimov and Hasminskii (1980) for the regression problem, and of those of Samarov (1977) and Farrell (1980) for the spectral density problem (perhaps the least natural of these on which to impose an order restriction), with varying amounts of difficulty; for example, the choice of $\mathfrak{F}$ in Bretagnolle and Huber (1979) (in the explicit form of f_0 of the next paragraph) is such that more alterations are required in the proof to handle (say) decreasing densities than in the presentation of Samarov (1978), although one eventually obtains a more explicit C in (1.2) from the former; the regularity restrictions of $\mathfrak{F}$ differ in the two works. The lower bound rate for estimating f, where $F < 1-\varepsilon$, yields the same rate for estimating r_f. We remark that, for some of the classes $\mathfrak{F}$ considered by Wahba and by Meyer, the order obtained in the lower bound is not quite attained by any known sequence t_n. In the three treatments considered herein, the order given by the bound is attainable.

The idea of the modification of the Bayes method needed to handle order restrictions, is fairly obvious. The lower bound is typically obtained in the treatments without order restrictions by considering a subfamily of densities of the form $f_0 + q_n$, where f_0 is fixed and q_n is restricted to a sufficiently large finite-dimensional family. By considering $f_0^* + \varepsilon q_n$ where ε is sufficiently small and f_0^* satisfies the order property *strictly* enough (e.g., for decreasing densities, f_0^* and q_n on R^1 with compact support and $(f_0^*)' < -L < 0$) one obtains a family satisfying the order restrictions for which the derivation of the lower bound can be carried through with appropriate modifications. This can be formalized in general in terms of $\mathfrak{F}$ (with order restrictions) containing an appropriate ball, but the resulting condition must still be verified in each case in the manner indicated above (and which is used in Sections 2 and 3). This is more difficult for some order restrictions than for others.

The information inequality method (Section 4) requires more effort for modification, since one must go through all the information inequality calculations for the parametric family, satisfying the order restriction, that replaces the family used by Farrell.

We do not discuss in detail here the existence of sequences of estimators satisfying the order restriction and which achieve the optimum rate. (Of course, the rate is achieved by estimators in the literature that take on values f not necessarily satisfying the order restriction, but it seems unsatisfactory to use such estimators.) Some estimators satisfying order restrictions are in the literature (e.g., Barlow et al. (1972), Section 5.4 for local estimation of r_f when $k=2$); it can be verified that others can be obtained by isotonizing standard estimators for problems without restrictions. Some of the many papers that study estimators for the latter problems are listed in the references to the present paper. The papers of Rosenblatt (1971), Wahba (1975), Walter and Blum (1979) and Wegman (1972a, 1972b) contain additional references and summaries of many results. We omit reference to the work of the author and Wolfowitz, and of Millar, on the basically different problem of asymptotically optimum estimation of a df under order restrictions.

A word is in order regarding estimation of f at or near an endpoint of the support of f, say the left end point x_f in the DFR case. Most of the literature does not handle this case, but Sacks and Ylvisaker (1978) give asymmetric kernels that attain the desired rate for estimation at known x_f or at $x_f+n^{-2/(2k+d)}t$ for $t>0$. Since x_f can be estimated to within order n^{-1} in probability (with corresponding results on the moments of the discrepancy), the S–Y estimator can be used to estimate f at known or unknown x_f in the setting of estimation at a point (Section 2). The family of estimators at $x_f+n^{-2/(2k+d)}t$ seem suitable for the global estimation problem of Section 3 where (1.2) is used with $B=R^1$. When $B=R^1$ (rather than $B=I_1$ of Section 3), (1.3) is no longer valid with x_f unknown, since an error of order 1 is made in estimating f, just to the right or left of x_f.

2. Local density estimation*

For brevity we treat only estimation of $f(0)$, although estimation of $(Tf)(0)$ (see Section 1) is handled similarly; also, we describe regularity on f in terms of derivatives, although this can be replaced by Stone's Lipschitz condition. Although the desired conclusion for decreasing densities is a consequence of that for DFR considered in the next two paragraphs, we first consider the almost trivial application of our method to the former case, in order to illustrate the method when the calculations are simplest. Thus, we first consider the problem of estimating $f(0)$ for suitably regular densities f for which f is non-increasing in all d coordinate variables, on its support D_f; let I be a fixed rectangle of R^d with $0\in\operatorname{int} I$, and for k a positive integer let $\mathfrak{F}_{D,k,M}$ consist of all such f for which $I\subset\operatorname{int} D_f$ and for which all kth derivatives are bounded in absolute value on int D_f by $2M<\infty$. For convenience rescale so that $I\subset(-1,1)^d$. Let $f_0(x_1,\ldots,x_d)=h[1-d^{-1}\Sigma_1^d x_i]$ on $D_{f_0}=[-1,1]^d$ where $h=\min(1,M)$ (in case $k=1$). Let $x_0=(\frac{1}{2},\ldots,\frac{1}{2})$. We let g_n be the functions

*Stone (1980).

defined by Stone (1980, Section 3). These are infinitely differentiable, have compact support that shrinks to $\{0, x_0\}$ as $n \to \infty$, have kth derivatives bounded in absolute value by M, have $\lim_n \max_x g_n(x) = 0$, and have $\int f_0 g_n \, dx = 0$. Clearly $f_0(1+\varepsilon g_n)$ has negative first derivatives for n large and ε sufficiently small and positive, and is in $\mathcal{F}_{D,k,M}$. Stone's Bayes argument applies to f_0, $f_0(1+\varepsilon g_n)$ and thus yields $p = k/(2k+d)$ in (1.1) and in the local analogue of (1.2). [Stone's lower bound argument is valid although f_0 here is not continuous. The reason for assuming $D_f \supset I$ is that Stone's or other standard estimators, or modifications of them to make them non-increasing, will then achieve the rate p uniformly on $\mathcal{F}$; if we only assume $0 \in \operatorname{int} D_f$, the proof of uniformity fails because f can have a discontinuity arbitrarily near 0.]

Now suppose $d=1$. The failure rate of f with df F is defined to be $r_f = f/(1-F)$ on int D_f. If f is DFR (r_f non-decreasing), f is decreasing on int D_f and $D_f = [b, +\infty)$ for some finite b. Let $\mathcal{F}_{\mathrm{DFR},k,M}$ be the subset of $\mathcal{F}_{D,k,M}$ with DFR. Now let $f_0(x) = L(x+2)^{-(L+1)}$ on $D_{f_0} = [-1, \infty)$ where $L > 0$ is small enough that $|f_0^{(k)}| \leqslant M$ on $(-1, \infty)$. Then $r_{f_0}(x) = L(x+2)^{-1}$ and $r'_{f_0}(x) = -L(x+2)^{-2}$. For g_n as above, defined w.r.t. this f_0 and with support taken to be in $[-\frac{1}{2}, 1]$, the failure rate of $f_0(1+\varepsilon g_n)$ is $f_0(1+\varepsilon g_n)/[1 - F_0 - \varepsilon \int_{-\infty}^{x} f_0 g_n] = f_0(1+\varepsilon g_n)/U_n$ (say), with derivative

$$U_n^{-2}\{U_n[f_0' + \varepsilon(f_0 g_n)'] + [f_0(1+\varepsilon g_n)]^2\}. \tag{2.1}$$

For $x > 1$, $f_0 g_n$ vanishes; and for ε small, (2.1) on $(-1, 1]$ is uniformly within $L/10$ of r'_{f_0}, while $r'_{f_0} \leqslant -L/9$ on $(-1, 1]$. Thus, for ε sufficiently small, $f_0(1+\varepsilon g_n)$ is in $\mathcal{F}_{\mathrm{DRF},k,M}$. Again applying Stone's result, we conclude that $p = k/(2k+1)$ for $\mathcal{F}_{\mathrm{DFR},k,M}$.

A natural generalization of DFR to $d > 1$ is obtained by defining $G_f(x) = \int_{H_x} f(u) \, du$ where $H_x = \{u : u_i \geqslant x_i \forall i\}$, $r_f = f/G_f$, and f to be DFR if r_f is non-increasing on int D_f. Taking $f_0(x) = L^d \prod_1^d (x_i + 2)^{-(L+1)}$ on $D_{f_0} = [-1, \infty)^d$, one sees that a development like that above yields $p = k/(2k+d)$ for $\mathcal{F}_{\mathrm{DFR},k,M}$, the subset of DFR densities of $\mathcal{F}_{D,k,M}$.

Such classes as densities increasing on D_f, or with increasing failure rate (IFR) or IFR on average or star-shapedness can be treated by the same method. The same derivation of optimum rate holds for estimation of f at an endpoint (known or unknown) of D_f; see also the end of Section 1.

3. Global density estimation*

For $d=1$, let I_1 and I_2 be fixed intervals with $I_1 \subset \operatorname{int} I_2$ and I_1 of positive (perhaps infinite) length. Let $\mathcal{F}_{\mathrm{DFR},k,M}$ consist of densities f for which $D_f \supset I_2$, f has DFR on int D_f, and the kth derivative satisfies $|f^{(k)}| \leqslant M < \infty$ on I_2. (Samarov considers a more general Lipschitz condition, as well as an alternative integral-Lipschitz condition; for brevity we omit consideration of these, but the treatment that follows works for these regularity conditions.) The reason for the formulation involving I_1 and I_2 is that we want to show the optimum rate has nothing to do with the discontinuity at

*Samarov (1978).

the left end of D_f, and conventional estimators suffice for estimation on I_1. Thus, to obtain a rate attainable by a known method of estimation (suitable kernel with compact support), or by a suitable DFR modification thereof, we consider (1.2) with $B=I_1$. This causes only obvious modifications in Samarov's proof for $B=R^1$, which we do not discuss. The lower bound of course also holds with $B=R^1$; see the end of Section 1 regarding attainment of the rate in that case.

We verify that $p=k/(2k+1)$ in (1.2) for $\mathfrak{F}_{\mathrm{DFR},k,M}$, and this order is consequently also optimum for non-increasing densities satisfying the same regularity restrictions. Take $I_2=[0,\infty)$, $I_1\subset\operatorname{int} I_2$, and $f_0(x)=L(x+2)^{-(L+1)}$ on $D_f=[-1,\infty)$ as in Section 2. Let ϕ be a bounded function with support $[-1,1]$, bounded kth derivative, and $\int_{-1}^{1}\phi(x)dx=0$. Let $[a,b]$ be a non-degenerate interval interior to I_1, and, with m the integral part of $n^{1/(2k+1)}(b-a)/2$, let $x_{n,j}=a+n^{-1/(2k+1)}(2j-1)$ and $\phi_{n,j}(x)=n^{-k/(2k+1)}\phi(n^{1/(2k+1)}(x-x_{n,j}))$ for $1\leqslant j\leqslant m$. Following Samarov, we let $\theta=(\theta_1,\ldots,\theta_m)$ and $f_\theta=f_0+\varepsilon\sum_{j=1}^{m}\theta_j\phi_{n,j}$ where $\theta\in\Omega_n=\{\theta:\max_{1\leqslant i\leqslant m}|\theta_i|\leqslant 1\}$. The functions $\sum_j\theta_j\phi_{n,j}$ have the role of the $f_0 g_n$ of Section 2, and it is easily seen that the $\phi_{n,j}$ have disjoint supports in $[a,b]$ and that for ε sufficiently small (by an argument like that of Section 2) $f_\theta\in\mathfrak{F}_{\mathrm{DFR},k,M}$ $\forall\theta\in\Omega_n$. Samarov's argument then yields the desired result.

We remark that for $d>1$ Samarov's argument can be extended to yield $p=k/(2k+d)$ and the f_0 of Section 2 yields the desired result in this case. Indeed, Bretagnolle and Huber have sketched the modification to the case $d>1$ of their development for $d=1$, which shares some elements (a similar class of f_θ's, but with each $\theta_i=\pm 1$) with Samarov's development.

The development for DFR above applies also to the proof of (1.3) by Hasminskii (1979b). In fact, the latter works with essentially the above f_θ's but with a simpler subset of the above Ω_n.

4. Information inequality development*

The author is indebted to Roger Farrell for pointing out that differentiation under the integral sign is not justified for the particular family considered in his 1967 paper. However, his basic outline of *method* is correct and can be applied to more regular families to obtain his main conclusion on lack of uniform consistency. The DFR family $\{g_{\gamma,\theta}\}$ defined below is such a family, and it yields the desired conclusion (of our being unable to estimate $g_{\gamma,\theta}(0)$ with uniformly small risk, for *any* sample size). This family has support depending on γ (*not* the differentiation parameter); we spend additional time to define $\{f_{\gamma,\theta}\}$ with support $[-\frac{1}{4},\infty)$ for all members of the family because the conclusion is then more satisfactory.

In all that follows, $\Omega=\{(\gamma,\theta):0<\gamma<\frac{1}{8},\ 3\gamma<\theta<4\gamma\}$. The development will take γ as known but arbitrary in the interval $0<\gamma<\frac{1}{8}$, and the information inequality will be used w.r.t. θ. We first define densities $g_{\gamma,\theta}$ with support $[-\gamma,\infty)$ and then densities $h_{\gamma,\theta}$ with support $[-\frac{1}{4},-\gamma]$. These are defined so that $g_{\gamma,\theta}(-\gamma)=h_{\gamma,\theta}(-\gamma)$.

*Farrell (1967, 1980).

Let $h^*=h$ except that $h^*(-\gamma)=0$. Since the densities g and h are each continuous on their support, are bounded uniformly on Ω, and are piecewise continuously differentiable, it follows that $f_{\gamma,\theta}=\frac{1}{2}(g_{\gamma,\theta}+h^*_{\gamma,\theta})$ is also bounded uniformly on Ω, and is continuous and piecewise continuously differentiable on $[-\frac{1}{4},\infty)$. As Farrell (1967, p. 472) points out, the result for $\{f_{\gamma,\theta}\}$ then implies it for continuously differentiable f. Making a scale transformation (to alter the bound on f), we obtain the

Theorem. *For $M>0$, there is a value $b<0$ such that, if $\mathfrak{F}$ consists of the DFR densities bounded by M and with continuous derivative on (b,∞), then for some $c>0$*

$$\inf_{t_n}\sup_{f\in\mathfrak{F}} E_f(t_n-f(0))^2>c, \tag{4.1}$$

for every $n>0$.

In fact, it is not difficult to modify the definition of $f_{\gamma,\theta}$ to allow b, in the theorem, to be specified, $-\infty<b<0$; we omit the extra arithmetic. Also, the corresponding sequential statement (below (4.1), "for every stopping variable n with $\sup_f En<M'$ where $M'<\infty$" can be verified as in Farrell's development. We have not bothered herein to obtain a sharp constant for $\{f_{\gamma,\theta}\}$, with which to replace the right side of (4.1).

Proof of theorem. We define

$$g_{\gamma,\theta}(x)=\gamma(\theta-\gamma)^{\gamma}(x+\theta)^{-\gamma-1}, \quad \text{on } D_g=[-\gamma,\infty). \tag{4.2}$$

Then $r_g(x)=\gamma(x+\theta)^{-1}$ on D_g, so g is DFR. A direct calculation yields the Fisher information

$$I_g=-E_{\gamma,\theta}\partial^2\log g_{\gamma,\theta}(X)/\partial\theta^2=\gamma/(\gamma+2)(\theta-\gamma)^2. \tag{4.3}$$

We write $J=g_{\gamma,\theta}(-\gamma)=\gamma/(\theta-\gamma)$.

We define

$$h_{\gamma,\theta}(x)=\begin{cases}\alpha(x+\beta)^{-2}, & \text{if } -\frac{1}{4}\leqslant x\leqslant -2\gamma,\\ A-B(x/\gamma)+C(x/\gamma)^2, & \text{if } -2\gamma\leqslant x\leqslant -\gamma,\end{cases} \tag{4.4}$$

where

$$\beta=\tfrac{1}{4}+\frac{(\frac{1}{4}-2\gamma)^2}{\frac{1}{4}+\frac{17}{12}\gamma}, \qquad \alpha=2(\beta-2\gamma)^2;$$
$$A=-4+8J, \qquad B=5-10J, \qquad C=3J-1. \tag{4.5}$$

It can be verified that the definitions (4.5) make (4.4) a density, make the two lines of (4.4) have the common value 2 at $x=-2\gamma$, and make $h_{\gamma,\theta}(-\gamma)=J=g_{\gamma,\theta}(-\gamma)$. Note that $h_{\gamma,\theta}(x)$ does not depend on θ for $-\frac{1}{4}\leqslant x\leqslant -2\gamma$. On $-2\gamma\leqslant x\leqslant -\gamma$, $h_{\gamma,\theta}(x)=P_1(x/\gamma)+P_2(x/\gamma)J$, where the P_i's are quadratic polynomials with constant coefficients. Thus, on $(-2\gamma,-\gamma)$, we have $\partial h_{\gamma,\theta}(x)/\partial\theta=-P_2(x/\gamma)\gamma/(\theta-\gamma)^2$. Since $\frac{1}{3}<J<\frac{1}{2}$ on Ω, we have $B>0, C>0$, and hence $h_{\gamma,\theta}$ is decreasing on $[-2\gamma,-\gamma]$ where it has minimum $h_{\gamma,\theta}(-\gamma)=J>\frac{1}{3}$. Consequently, if $\max_{1\leq u\leq 2}|P_2(u)|=K$, we have

$$I_h=\int_{-2\gamma}^{-\gamma}\left\{\left[\partial h_{\gamma,\theta}(x)/\partial\theta\right]^2/h_{\gamma,\theta}(x)\right\}dx\leq 3\gamma K^2\left[\gamma/(\theta-\gamma)^2\right]^2<K^2/\gamma. \quad (4.6)$$

Defining $f=(g+h^*)/2$ as described earlier, we have $I_f=(I_g+I_h)/2$ since the supports of g and h^* are disjoint. From (4.3) and (4.6), $I_f<K'/\gamma$ on Ω where $K'=\frac{1}{2}(K^2+1)$. We define

$$Q=\partial f_{\gamma,\theta}(0)/\partial\theta=\gamma(\theta-\gamma)^{\gamma-1}\theta^{-\gamma-2}[\gamma\theta-(\gamma+1)(\theta-\gamma)]$$
$$<-K''/\gamma, \quad (4.7)$$

where $K''>0$. The information inequality for an estimator t_n of $f(0)$ with bias $b(\gamma,\theta)=E_{\theta,\gamma}t_n-f_{\gamma,\theta}(0)$ is

$$E_{\theta,\gamma}\left(t_n-f_{\gamma,\theta}(0)\right)^2\geq b^2+\left(Q+\partial b/\partial\theta\right)^2/nI_f. \quad (4.8)$$

Following Farrell, if there is a θ in $(3\gamma,4\gamma)$ for which $\partial b/\partial\theta\leq -Q/2$ (positive), we see that the right side of (4.1) is $\geq Q^2/4nI_f$ at that θ. If there is no such θ, then b is increasing in θ on $(3\gamma,4\gamma)$, with derivative $>-Q/2$, and hence at some θ in that interval we have $|b|>(\gamma/2)(-Q/2)$ and the right side of (4.1) is $\geq Q^2\gamma^2/16$ at that θ. Thus, since $I_f<K'/\gamma$ and $Q<-K''/\gamma$, we obtain

$$\max_{3\gamma<\theta<4\gamma} E_{\theta,\gamma}(t_n-f_{\gamma,\theta}(0))^2\geq$$
$$\geq\min(Q^2/4nI_f,\,Q^2\gamma^2/16)$$
$$>(K'')^2\min(1/4K'n\gamma,\,\tfrac{1}{16}). \quad (4.9)$$

Taking the supremum over $0<\gamma<\frac{1}{8}$ yields the right side of (4.1), but we must still show f is DFR.

We write F, H for the df's of f, h. Since $f=g/2$ on $[-\gamma,\infty)$, $r_f=r_g$ there and thus f has DFR on $[-\gamma,\infty)$. On $[-\frac{1}{4},-\gamma]$, $r_f=h/2[1-H/2]$, so $(2-H)^2r_f'=(2-H)h'+h^2\leq h'+h^2$ since $h'<0$. We shall show $h'+h^2<0$ on $[-\frac{1}{4},-\gamma]$.

On $[-\frac{1}{4},-2\gamma]$ we have, for $0<\gamma<\frac{1}{8}$,

$$\alpha^{-1}(x+\beta)^4\left[h'(x)+h^2(x)\right]=-2(x+\beta)+\alpha$$
$$\leq\tfrac{1}{2}-2\beta+\alpha$$
$$=2\left(\tfrac{1}{4}-2\gamma\right)^2\left(\tfrac{1}{2}+\tfrac{17}{6}\gamma\right)^{-2}\left(-8\gamma+\tfrac{49}{36}\gamma^2\right)<0. \quad (4.10)$$

On $[-2\gamma,-\gamma]$ we have

$$h'+h^2=-B/\gamma+2Cx/\gamma^2+\left[A-B(x/\gamma)+C(x/\gamma)^2\right]^2$$
$$<-(B+2C)/\gamma+(A+2B+4C)^2 \quad (4.11)$$
$$\leq-(3-4J)/\gamma+4<-1/\gamma+4<0.$$

This completes the proof.

Farrell (1980) uses the information inequality method to obtain the lower bound [local analogue of (1.2)] when $k=2$, with a determination of a C that is close to best possible. One can define a family of DFR densities to obtain the right rate in that case using the present method, but without invoking Farrell's technique to obtain a good constant. The derivation is tedious compared to that of Section 2. However, for decreasing (rather than DFR) densities there may be hope that a reasonable C can be obtained by Farrell's method.

Acknowledgements

The author is grateful to Jerry Sacks for pointing out the applicability of the S–Y estimators in Section 1.

References

[1] Alekseev, V. (1974). On uniform convergence of spectral density estimates of a Gaussian stationary random process. *Teor. Veroyatnost. i Primenen.* **19**, 198–206.

[2] Barlow, R.E., Bartholomew, D.J., Bremner, J.M. and Brunk, H.D. (1972). *Statistical Inference Under Order Restrictions*. Wiley, London.

[3] Bickel, R.J. and Rosenblatt, M. (1973). On some global measures of the deviations of density function estimates. *Ann. Statist.* **1**, 1071–1095.

[4] Bretagnolle, J. and Huber, C. (1979). Estimation des densités: risque minimax. *Z. Wahrsch.* **47**, 119–137.

[5] Cencov, N.N. (1962). Evaluation of an unknown distribution density from observations. (Translation) *Soviet Math.* **3**, 1559–1562.

[6] Farrell, R.H. (1967). On the lack of a uniformly consistent sequence of estimators of a density function in certain cases. *Ann. Math. Statist.* **38**, 471–474.

[7] Farrell, R.H. (1972). On best obtainable asymptotic rates of convergence in estimation of a density function at a point. *Ann. Math. Statist.* **43**, 170–180.

[8] Farrell, R.H. (1979). Asymptotic lower bounds for the risk of estimators of the value of a spectral density function. *Z. Wahrsch.* **49**, 221–234.

[9] Farrell, R.H. (1980). On the efficiency of density function estimators. (To appear.)

[10] Hasminskii, R.Z. (1979a). Lower bound for the risk of nonparametric estimates of the mode. In *Contribution to Statistics* (Hajek Memorial Volume) Jurečkova, J. Ed., pp. 91–97. Academia, Prague.

[11] Hasminskii, R.Z. (1979b). A lower bound for risks of nonparametric estimates of density in the uniform metric. *Teor. Veroyathost. i Primenen.* **24**, 824–828.

[12] Hasminskii, R.Z. and Samarov, A. (1978). On the quality of some non-parametric estimates in uniform metric. (To appear.)

[12a] Ibragimov, I.A. and Hasminskii, R.Z. (1980). On nonparametric estimation of regression. *Soviet Math. Dokl.* **21**, 810–814.

[13] Leadbetter, M.R. and Watson, G.S. (1963). On the estimation of a probability density, I. *Ann. Math. Statist.* **34**, 480–491.

[14] Meyer, T.G. (1977). Bounds for estimation of density functions and their derivatives. *Ann. Statist.* **5**, 136–142.

[15] Nadaraya, E.A. (1965). On nonparametric estimates of regression curves. *Theor. Prob. Applns.* **10**, 186–190.

[16] Nadaraya, E.A. (1970). Remark on nonparametric estimates of density functions and regression curves. *Theor. Prob. Applns.* **15**, 134–37.

[17] Nadaraya, E.A. (1974). On the integral mean square error of some nonparametric estimates for the density function. *Theor. Prob. Applns.* **19**, 133–141.

[18] Parzen, E. (1962). On the estimation of a probability density function and the mode. *Ann. Math. Statist.* **33**, 1065–1076.

[19] Pickands. J. (1969). Efficient estimation of a probability density. *Ann. Math. Statist.* **40**, 854–864.

[20] Revesz, P. (1972). On empirical density function. *Periodica Math. Hungar.* **2**, 85–110.

[21] Rosenblatt, M. (1956). Remarks on some nonparametric estimates of a density function. *Ann. Math. Statist.* **27**, 832–837.

[22] Rosenblatt, M. (1971). Curve estimates. *Ann. Math. Statist.* **42**, 1801–1823.

[23] Sacks, J. and Ylvisaker, D. (1978). Asymptotically optimum kernels for density estimation at a point. (To appear.)

[24] Samarov, A. (1977). Lower bound for risk of spectral density estimates. *Problems of Info. Theor. Transm.* **13**, 48–51.

[25] Samarov, A. (1978). Lower bound for integral risk of density estimates. In *Problems of Construction of Systems for Information Transmission*. Bloch, E. Ed., (To appear.)

[26] Schuster, E.F. (1969). Estimation of a probability density function and its derivatives. *Ann. Math. Statist.* **40**, 1187–1195.

[27] Schwartz, S.C. (1967). Estimation of a probability density by an orthogonal series. *Ann. Math. Statist.* **38**, 1261–1265.

[28] Stone, C.J. (1980). Optimal rates of convergence for nonparametric estimators. (To appear in *Ann. Math. Statist.*)

[28a] Stone, C.J. (1981). Optimal global rates of convergence for nonparametric regression. Preprint.

[29] Van Ryzin, J. (1973). A histogram method of density estimation. *Comm. Statist.* **2**, 493–506.

[30] Wahba, G. (1975). Optimal convergence properties of variable knot, kernel, and orthogonal series methods for density estimation. *Ann. Math. Statist.* **3**, 15–29.

[31] Walter, G. and Blum, J. (1979). Probability density estimation using delta sequences. *Ann. Statist.* **7**, 328–340.

[32] Wegman, E.J. (1972a). Nonparametric probability density estimation: I. A summary of available methods. *Technometrics.* **14**, 533–546.

[33] Wegman, E.J. (1972b). Nonparametric probability density estimation: II. A comparison of density estimation methods. *J. Statist. Comput. Simul.* **1**, 225–245.

[34] Weiss, L. and Wolfowitz, J. (1967). Estimation of a density at a point. *Z. Wahrsch.* **7**, 327–335.

[35] Winter, B.B. (1975). Rate of strong consistency of two nonparametric density estimators. *Ann. Statist.* **3**, 759–766.

[36] Woodroofe, M.B. (1970a). On the maximum deviation of the sample density. *Ann. Math. Statist.* **38**, 475–481.

[37] Woodroofe, M.B. (1967b). The maximum deviation of sample spectral densities. *Ann. Math. Statist.* **38**, 1558–1569.

[38] Woodroofe, M.B. (1970). On choosing a delta-sequence. *Ann. Math. Statist.* **41**, 1665–1671.

[39] Woodroofe, M.B. and Van Ness, J.W. (1967). The maximum deviation of sample spectral densities. *Ann. Math. Statist.* **38**, 1558–1569.

Eight Lectures on Mathematical Statistics

J. KIEFER

Preface. These are notes of Kiefer's lectures at Peking University (Spring, 1980) about recent developments in certain branches of mathematical statistics. The notes were prepared by Cheng Ping (first lecture), Chen Jiading (second lecture), Chen Xiru (third and fourth lectures), and Zhang Yao-ting (lectures 5–8). Kiefer reviewed these notes and made certain additions.

Lecture 1. Multivariate Estimation

We use the following notation:

X—observation, usually a random variable or vector;
S—sample space, domain of X;
$\Omega = \{P_\theta\} \approx \{\theta\}$ = set of possible distributions on S,
Q = set of states of nature;
$D = \{d\}$ = action (decision) space;
$L(\theta, d)$ = loss incurred by decision d when θ is the time state;
t: $S \to D$ = statistical decision rule that makes decision $t(x_0)$ if $X = x_0$;
$R_t(\theta) = E_\theta L(\theta, t(X))$ = risk function of t.

We say that t is better than t' if $R_t(\theta) \le R_{t'}(\theta)$ for all θ and strict inequality holds for some θ. We say that t' is inadmissible if there exists t which is better than t'; otherwise t' is admissible (cf. Figure 1).

Additional notation includes the following:

Π: prior distribution on Ω;
t_π: Bayes rule with respect to π, i.e., t_π minimizes $\int_\Omega R_t(\theta)\, d\Pi(\theta)$;
$\bar{t}$: minimax solution, i.e., $\sup_\theta R_{\bar{t}}(\theta) = \min_t \sup_\theta R_t(\theta)$.

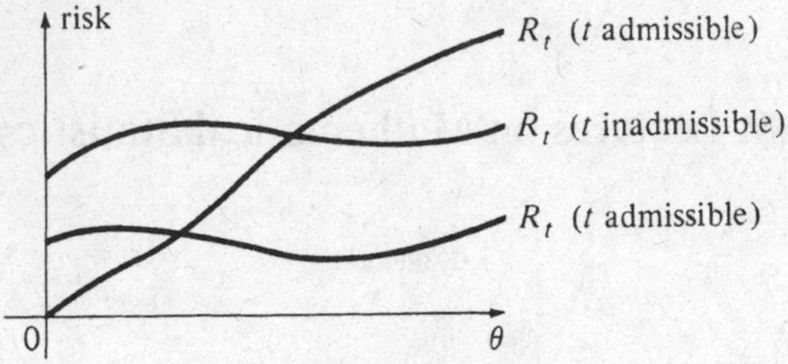

Figure 1.

We now consider an example. Let S be the p-dimensional Euclidean space R^p and let $X \sim N(\theta, I_p)$, where $\theta = (\theta_1, \ldots, \theta_p)^T$. For the problem of estimating θ, we take $D = R^p$. (For the problem of estimating only θ_1, take $D = R^1$.) When θ is the true parameter and the estimator d is used, the loss is

$$L(\theta, d) = \sum_1^p (\theta_i - d_i)^2 = \|\theta - d\|^2.$$

The classical estimator $t^c(x) = x$ and the ridge estimator $t^r(x) = (k_1 x_1, \ldots, k_p x_p)$ $(0 < k_i < 1)$ have risk functions

$$R_{t^c}(\theta) = p,$$

$$R_{t^r}(\theta) = \sum_1^p E_\theta[k_i X_i - \theta_i]^2 = \sum_1^p \left[k_i^2 + (1 - k_i)^2 \theta_i^2\right].$$

If we assume the prior distribution $\Pi(\theta) = N(0, bI_p)$ and minimize $\int_\Omega E_\theta[t_i(X) - \theta_i]^2 \Pi(d\theta)$, then we obtain the Bayes estimator

$$t_i(X) = E_\Pi(\theta | X_i) = \frac{bX_i}{1+b},$$

noting that

$$\begin{pmatrix} \theta_i \\ x_i \end{pmatrix} \sim N\left(\begin{pmatrix} 0 \\ 0 \end{pmatrix}, \begin{pmatrix} b & b \\ b & b+1 \end{pmatrix}\right)$$

and that

$$k_i = \frac{b}{b+1} = 1 - \frac{1}{b+1}.$$

It is easy to prove that if t_Π is the unique Bayes estimator with respect to Π then t_Π is admissible. For example, the Bayes estimator t^r of the p-dimensional normal mean θ with respect to a normal prior Π is admissible. However, since the risk function of t^r is unbounded, it is very risky to use this estimator unless θ is known to lie in a bounded set or Π is completely known (cf. Hoerl and Kennard, *Technometrics* (1970) for ridge estimators). We now discuss other kinds of admissible estimators.

To begin with, consider Stein's ideas of combining estimators from p separate problems. First note that when the covariance matrix of X is $\sigma^2 I_p$, we have p independent observations $x_1,\ldots,x_p$. Since they are independent, how can they be combined together in estimating their individual parameters? The "Stein effect" in combining separate problems is indeed a very surprising phenomenon. We give here two viewpoints to explain this phenomenon, which basically follows from the form of the loss function that incorporates various unrelated parameters into a single total loss.

(1) Empirical Bayes (Robbins, 1951, *Proc. 2nd Berkeley Symposium*, 131–148). Suppose we know that the prior distribution Π is $N(0, bI_p)$ but that b is unknown. Since $E_\theta\|X\|^2 = \|\theta\|^2 + p$, we can use $(\Sigma_{i=1}^p x_i^2 - p)/p$ to estimate b and use instead of t_Π the estimator

$$t^*(X) = \left(1 - \frac{p}{\|X\|^2}\right)X.$$

This is an "empirical Bayes" estimator, which may also be regarded as a "self-adaptive" ridge estimator. While this intuitive idea does not necessarily mean that the estimator must have a satisfactory risk function, it suggests that we should "shrink" X.

(2) Since $E_\theta\|X\|^2 = \|\theta\|^2 + p$, $\|X\|$ tends to be larger than $\|\theta\|$. This also suggests the shrinkage of X in estimating θ.

While the above arguments are merely heuristic, such heuristic reasoning helps us to find estimators that are better than the classical maximum likelihood estimator t^c, at least when p is large. It also leads to general methods to study this class of problems.

To prove that an estimator t' is inadmissible, the usual method is to find an estimator t that is better than t'. Proofs of admissibility, however, are usually very difficult. Blyth (1951, *A.M.S.*, 22–42) proved that t^c is admissible when $p=1$, and his proof proceeds as follows. If t^c should be inadmissible, then

$$\int_{-\infty}^{\infty} [R_{t^c}(\theta) - R_t(\theta)]\,\Pi_b(d\theta) \ge \frac{\varepsilon}{\sqrt{b}} \qquad (*)$$

for some $\varepsilon > 0$ and for some estimator t, where Π_b is the $N(0, b)$ distribution. Since

$$t_{\Pi_b}(X) = \frac{bX}{b+1}, \qquad R_{t_{\Pi_b}}(\theta) = \left(1 - \frac{1}{b+1}\right)^2 + \frac{\theta^2}{(b+1)^2},$$

it then follows that

$$\int_{-\infty}^{\infty} \left[1 - R_{t_{\Pi_b}}(\theta)\right]\Pi_b(d\theta) = 0\left(\frac{1}{b}\right) \quad \text{as } b \to \infty,$$

which contradicts $(*)$. The proof of admissibility using this approach in more general situations involves complicated computations. We refer to the following papers on the admissibility of t^c in estimating a normal mean when $p=2$:

James and Stein, 1960, *4th Berkeley Symp.*, Vol. 1, 361–379;
Brown, *A.M.S.* (1966), 1087–1136;
Brown and Fox, *A.M.S.* (1974), 248–266;
Stein, *A.M.S.* (1955), 518–522.

Concerning the admissibility of AX as an estimator of θ, where A is a $p\times p$ matrix, Cohen (1966, *A.M.S.*) obtained a necessary and sufficient condition for AX to be admissible. The condition is that A be symmetric with eigenvalues ≤ 1, where equality holds for no more than two eigenvalues.

Inadmissibility of t^c when $p\geq 3$ was established by Stein (*3rd Berkeley Symp.*, Vol. 1, 197–206) and James and Stein (*4th Berkeley Symp.*). From the preceding heuristic argument, we should try shrinking X to obtain an estimator that is better than t^c. The original method of Stein is to evaluate the risk functions of estimators of the form

$$t^s(X)=\left(1-\frac{c}{\|X\|^2}\right)X$$

and then to choose c appropriately thereby.

We now consider another method (Stein, *Prague Symp.*, 1973, 345–381; Hudson, *A.M.S.*, 1978, 473–484). Let $p=1$, and let $t(X)=X+g(X)$. Then

$$\begin{aligned}R_t(\theta)-R_{t^c}(\theta)&=E_\theta\left[(X+g(X)-\theta)^2-(X-\theta)^2\right]\\&=2E_\theta\{(X-\theta)g(X)\}+E_\theta[g(X)]^2\\&=E_\theta\{2g'(X)+g^2(X)\}.\end{aligned}$$

The last equality above can be obtained using integration by parts: Letting ϕ denote the standard normal density, we have

$$\begin{aligned}E_\theta g'(X)&=\int_{-\infty}^{\infty}g'(y)\phi(y-\theta)\,dy\\&=-\int_{-\infty}^{\infty}g(y)\,d\phi(y-\theta)=E_\theta[(X-\theta)g(X)].\end{aligned}$$

Note that since $2g'(x)+g^2(x)$ does not involve θ, if we can find g such that $2g'(x)+g^2(x)<0$ for all x, then t is better than t^c. This is not possible for the case $p=1$.

For the case $p>1$, let

$$t_i^s(X)=x_i-\frac{cx_i}{\sum_1^p x_i^2}\triangleq x_i+g_i(X).$$

Then

$$\begin{aligned}\sum_{i=1}^{p}\left[2\frac{\partial g_i}{\partial x_i}+g_i^2\right]&=\sum_{i=1}^{p}2c\frac{1}{\|X\|^4}(2x_i^2-\|X\|^2)+\frac{c^2}{\|X\|^2}\\&=\frac{1}{\|X\|^2}[c^2-2c(p-2)]<0\text{ when }c^2<2c(p-2).\end{aligned}$$

In fact, $c=p-2$ gives the best of these estimators. This approach is very simple and useful.

Note that although t^s is better than t^c, it is inadmissible since $(1-(c/\|X\|^2))^+X$ is better than $t^s(X)$. However, since $h(X)=(1-(c/\|X\|^2))^+X$ is not an analytic function, $h(X)$ is still inadmissible.

The problem of constructing admissible estimators that are better than t^c when $p \geq 5$ has been studied by several authors. In particular, Strawderman and Cohen (*A.M.S.*, 1970) and Bock (*A.S.*, 1975) obtained estimators that are Bayes, admissible and minimax.

Minimax estimation of θ has also been studied by many authors, for example, Baranchik (*A.M.S.*, 1970), Strawderman (*A.M.S.*, 1970), Berger (*A.S.*, 1975, 1976), and Brown (*A.S.*, 1979). Note that t^c is a minimax estimator and therefore $\inf_t \sup_\theta R_t(\theta) = p$.

The aforementioned results can be extended to other loss functions (Brown (*A.M.S.*, 1966), etc.), and to the case of unknown common variance (James-Stein, Berger, etc.). However, in the latter case, admissible minimax estimators have not yet been found and admissible minimax estimation of the variance even in the 1-dimensional case remains an open problem (Stein, *A.M.S.*, 1964).

Similar results have been found for other parametric families of distributions. For example, there are better minimax estimators than the classical ones for the Poisson family and the Γ family when $p \geq 2$ (Clevenson and Zidek, *A.S.* 1977; Peng, *A.S.* 1978; Berger, *A.S.* 1978).

Is the use of a linear combination of $(\theta_i - d_i)^2$, to describe the loss function, of any practical interest? Efron and Morris (*J. Amer. Stat. Assoc.* 1973, 379–421) and Brown (*J. Amer. Stat. Assoc.* 1975, 417–427) have considered examples to illustrate the practical relevance of such loss functions. While it may be difficult in practice to completely specify the c_i in the loss function $L = \Sigma c_i(\theta_i - d_i)^2$, we have the following results when $p \geq 3$:

(A) There does not exist t that is better than t^c for all loss functions L.
(B) If $\max_{i,j} c_i/c_j < M < \infty$, then there exists t that is better than t^c for all loss functions $L = \Sigma c_i(\theta_i - d_i)^2$.

Different loss functions of the form $\Sigma_{i=1}^p h(\theta_i)(\theta_i - d_i)^2/\Sigma_{i=1}^p h(\theta_i)$ may lead to different conclusions. Berger (*A.S.*, 1978) studied this problem for p independent $\Gamma(\theta_i, \alpha_i)$ distributions with $h(\theta_i) = \theta_i^m$ for different values of m. For $p = 2$, the usual minimax estimator $x_i/(1 + x_i)$ can be admissible for $m = 2$ but cannot be admissible for $m \neq 2$. Similar results have also been found for normal distributions (Brown, *A.S.* 1979).

Finally, it should be pointed out that any admissible estimator is Bayes or limit of Bayes estimators (Wald, *A.M.S.* 1950). For $p = 1$, Sacks (*A.M.S.* 1963) obtained the following result. Let ν be the limit of normalized Π_i. For example, if $\Pi_b = N(0, b)$, then $\sqrt{2b\pi}\,\Pi_b \to$ Lebesgue measure as $b \to \infty$. Generalized Bayes estimators of the form

$$t_\nu(X) = \frac{\int_{-\infty}^{\infty} \theta\phi(X-\theta)\nu(d\theta)}{\int_{-\infty}^{\infty} \phi(X-\theta)\nu(d\theta)}$$

may be inadmissible, but they include all admissible solutions. The Baranchik–Strawderman estimators are generalized Bayes solutions for $p = 3, 4$ (where there do not exist spherically symmetric Bayes solutions) but are (proper) Bayes solutions when $p \geq 5$.

A very important question is when t_ν is admissible. In the normal case with $p=1$, if ν is absolutely continuous and $d\nu/dx = f$ is smooth, then

$$t_\nu \text{ is admissible } \Leftrightarrow \int_{-\infty}^{\infty} \frac{1}{f(x)}\,dx = \infty$$

(cf. Farrell (*A.M.S.* 1964), Zidek (*A.M.S.* 1970), Brown (*A.M.S.* 1971), and Stein (1975, *Bernoulli Anniversary Volume*, 217–241)). The case $p>1$ is much less clear. Brown (*A.M.S.* 1971, *A.S.* 1974) has provided some results. Other important open problems include: If $X \sim N(\theta, \Sigma)$, where θ and Σ are unknown parameters, how should we estimate θ? How should we estimate Σ? (The minimax estimator of James and Stein is not very satisfactory because of its complicated form and because it is possibly inadmissible.)

Lecture 2. Sequential Methods

Suppose that without fixing the sample size in advance we observe $x_1, x_2, \ldots$ sequentially until the information accumulated is adequate to make a satisfactory decision. We define the "stopping time" N to be a random variable that satisfies the following condition: For all $n \geq 1$, $\{\mathbf{x}\colon N(\mathbf{x}) = n\}$ depends only on $(x_1, \ldots, x_n)$. A "sequential decision rule" $\delta = (N_\delta, t_\delta)$ consists of the stopping time (or stopping rule) N_δ and the terminal decision rule $t_\delta = t_\delta(x_1, \ldots, x_N)$. We will discuss statistical problems for which:

(A) sequential methods are more efficient than nonsequential methods; or
(B) sequential methods provide the only feasible solutions.

(A) Sequential methods were first systematically studied by Wald, who developed the Sequential Probability Ratio Test (SPRT): Let $x_1, x_2, \ldots$ be i.i.d. random variables. Let $\Omega = \{f_0, f_1\}$ (= set of two possible identities), $D = \{d_0, d_1\} = ($ = set of possible actions). The loss function is

$$L(f_i, d_j) = \begin{cases} 0, & i = j \\ W_i, & i \neq j. \end{cases}$$

Define the stopping rule

$$N = \min\left\{ n\colon \sum_1^n \log \frac{f_1(x_i)}{f_0(x_i)} \geq A \text{ or } \leq B \right\}.$$

For any sequential decision rule $\delta = (N_\delta, t_\delta)$ the error probabilities are $\alpha_{\delta,i} = P_{f_i}\{t_\delta \neq \delta_i\}$ and the expected sample size is $E_i N_\delta$. Optimality of a sequential decision rule can be formulated in two ways: (1) minimization (in the Bayes or minimax sense) of $r_\delta(i) = W_i \alpha_{\delta,i} + cE_i N_\delta$, where c denotes the cost per observation; (2) minimization of $b_0 E_0 N_\delta + b_1 E_1 N_\delta$ subject to error constraints $\alpha_{\delta,i} \leq q_i$, $i = 0, 1$, where q_0, q_1, b_0, b_1 are prescribed constants.

Wald and Wolfowitz (1948, *A.M.S.*, 326) established the following results: (1) Given any prior probabilities for f_0 and f_1, there exist A, B such that the SPRT with

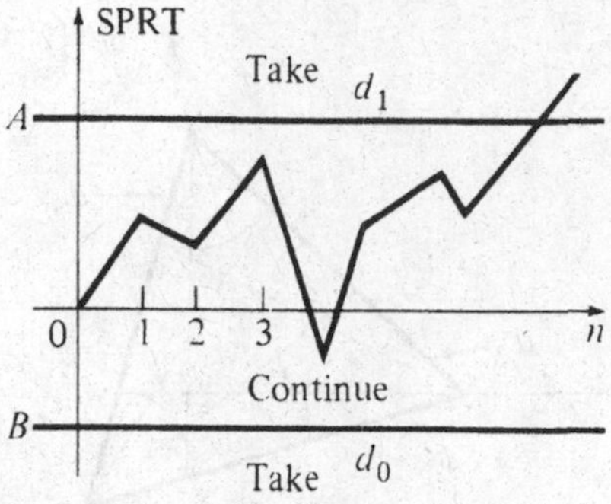

Figure 2.

stopping bounds A, B is Bayes. (2) Suppose there exist A, B for which $\alpha_{\mathrm{SPRT},i} = q_i$ $(i = 0, 1)$. Then the SPRT with stopping bounds A, B minimizes $E_0 N_\delta$ and $E_1 N_\delta$ among all δ for which $\alpha_{\delta,i} \leq q_i$ $(i = 0, 1)$.

K-decision problems. We want to decide which of $f_1, \ldots, f_k$ is the time density (cf. Wald, 1950, *Statistical Decision Function*, Wiley). Here admissible (Bayes) rules δ are difficult to find. For example, when $k = 3$, the stopping region consists of three sector-shaped convex sets; however, to find the exact boundary is very difficult.

Kiefer and Sacks (1963, *A.M.S.*, 705) showed that as $c \to 0$ (or all $q_i \to 0$) an appropriate combination of SPRTs is asymptotically efficient. (In Figure 4 every pair of dotted lines passing through a common vertex corresponds to an SPRT.)

Lorden (1977, *A.S.*, 1–21) proposed a nearly optimal test (denoted by $L = (N_L, t_L)$) satisfying

$$\alpha_{L,i} \leq \alpha_{\mathrm{Bayes},i}, \qquad E_i(N_L - N_{\mathrm{Bayes}}) = 0(1) \quad \text{as } c \to 0.$$

Lucky Observations. A reason why the SPRT is better than fixed sample size tests is that it waits for the likelihood ratio to be large or small (near 0). Note in this connection that the likelihood ratio gives us information about which hypothesis is true only when it is large or small but not when it is close to 1. This idea of waiting

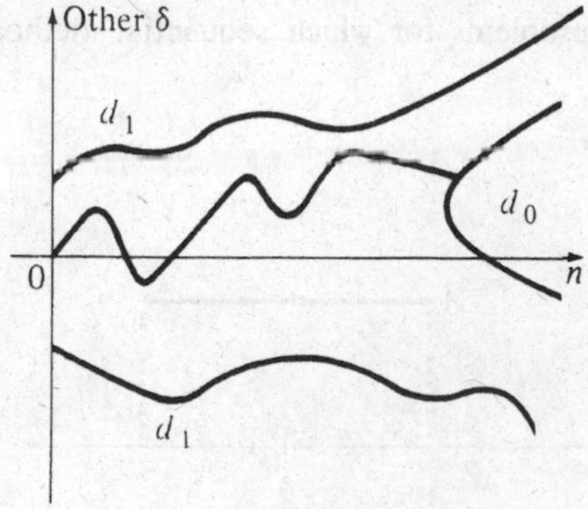

Figure 3.

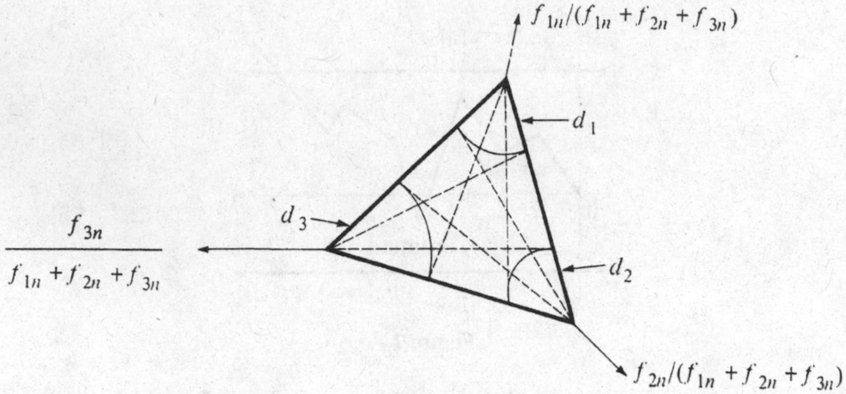

Figure 4.

for "lucky observations" appears in other problems, and we give here the following example on estimation.

Let $\Omega = R^1$, and let $x_1, x_2, \ldots$ be i.i.d. with common density $f_0(x-\theta)$. The action space D is R^1, and the loss function L is $|\theta - d|$.

(a) First consider the uniform density

$$f_0(x-\theta) = \begin{cases} 1 & \text{if } |x-\theta| \leq \frac{1}{2} \\ 0 & \text{otherwise.} \end{cases}$$

A sufficient statistic from n observations is $(\min_{1\leq i\leq n} x_i, \max_{1\leq i\leq n} x_i)$. If the difference $\max_{1\leq i\leq n} x_i - \min_{1\leq i\leq n} x_i$ is close to 0, we do not get much information about θ, so we should continue sampling until $\max_{1\leq i\leq n} x_i - \min_{1\leq i\leq n} x_i$ is close to 1, in which case we can determine θ with high confidence. In this problem, sequential methods are helpful (cf. Wald, 1950).

(b) Now consider the case where $f_0(x-\theta)$ is the $N(\theta,1)$ density. Here $\bar{x}_n = 1/n\Sigma_1^n x_i$ is a sufficient statistic and has the $N(\theta, 1/n)$ distribution. In this problem, "lucky observations" do not arise, and sequential methods do not help (cf. Wolfowitz, 1950, *A.M.S.*, 218).

(B) We now turn to problems for which sequential methods provide the only feasible solutions.

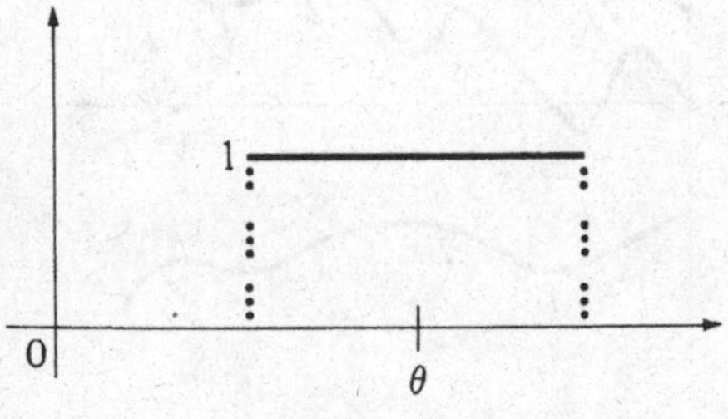

Figure 5.

Example (Stein, 1945, *A.M.S.*, 243). Let $x_1, x_2, \ldots$ be i.i.d. $N(\mu, \sigma^2)$ random variables. Here $\Omega = \{(\mu, \sigma^2)\colon \mu \in R^1, \sigma > 0\}$. We want to find an estimator t_δ of μ such that

(a) $\quad E_{\mu,\sigma}(t_\delta - \mu)^2 \leq B < \infty,$

(b) $\quad P_{\mu,\sigma}\{|t_\delta - \mu| > d\} \leq \gamma < 1$ (leading to the confidence interval $[t_\delta - d, t_\delta + d]$).

For any fixed n, when σ is large, the variance of $\bar{x}_n$ would be large. This suggests that there does not exist an estimator t_δ satisfying (a) or (b) if the sample size n is fixed in advance. Following is Stein's two-stage solution of (b). (A similar result holds for (a) and is omitted here.)

If σ is known, we should take n^* observations, where

$$n^*(\sigma, \gamma) = \min\left\{n\colon n \geq 1, \Phi\left(-\frac{d\sqrt{n}}{\sigma}\right) \leq \frac{\gamma}{2}\right\},$$

and Φ is the standard normal distribution function. Setting $t_\delta = \bar{x}_{n^*}$, the problem is solved. When σ is unknown, we use the following two-stage procedure:

(1) Take an initial sample $x_1, \ldots, x_{n_0}$ $(n_0 \geq 2)$ to find an upper confidence bound

$$\hat{\sigma} = \hat{\sigma}_{n_0} = c_{n_0}\sqrt{\sum_1^{n_0}(x_i - \bar{x}_{n_0})^2}$$

for σ such that

$$P_{\mu,\sigma}\{\sigma \leq \hat{\sigma}\} \geq 1 - \frac{\gamma}{2} \quad \text{for all } \mu, \sigma.$$

(2) Then continue sampling until we have a sample size of $n^*(\hat{\sigma}, \gamma/2)$. Hence letting $t_\delta = \bar{x}_{n^*}$ we have

$$\begin{aligned} P_{\mu,\sigma}\{|t_\delta - \mu| > d\} &\leq P_{\mu,\sigma}\{\sigma > \hat{\sigma}\} + P_{\mu,\sigma}\{|t_\delta - \mu| > d \mid \sigma \leq \hat{\sigma}\} \\ &\leq \frac{\gamma}{2} + \frac{\gamma}{2} = \gamma. \end{aligned}$$

This kind of idea is useful for many other problems; sometimes more than two stages are needed.

In the above example, it seems better to use the following approach. Using the sequential observations, we can compute $\hat{\sigma}_n$ for every n with $c_n = 1/\sqrt{n-1}$. Then $P_{\mu,\sigma}\{\hat{\sigma}_n \to \sigma\} = 1$. Replacing σ by $\hat{\sigma}_n$ in the formula for $n^*(\sigma, \gamma)$ leads to the stopping rule.

$$N = \min\left\{n\colon n \geq 2, \Phi\left(-\frac{d\sqrt{n}}{\hat{\sigma}_n}\right) \leq \frac{\gamma}{2}\right\}.$$

Let $t_\delta = \bar{x}_N$. This idea was first proposed by Anscombe (1953, *J.R.S.S.*, 1–21). Without adjusting $\hat{\sigma}_n$ for small n, we cannot guarantee that $P_{\mu,\sigma}\{|\bar{x}_N - \mu| > d\} \leq \gamma$.

However, it is not difficult to show that

$$P_{\mu,\sigma}\{|\bar{x}_N-\mu|>d\}\to\gamma \quad \text{as } \sigma\to\infty. \tag{1}$$

In addition, Chow and Robbins (1965, *A.M.S.*, 457) proved that

$$\frac{E_{\mu,\sigma}N}{n^*(\sigma,\gamma)}\to 1 \quad \text{as } \sigma\to\infty. \tag{2}$$

We now consider two related problems:

(i) Can we modify the above method to obtain a stopping time N' such that for all μ and σ,

$$P_{\mu,\sigma}\{|\bar{x}_{N'}-\mu|>d\}\le\gamma? \tag{3}$$

The answer is affirmative, as shown by Starr (1966, *A.M.S.*, 1137). Simons (1968, *A.M.S.*, 1946) further found that we need only take $N'=N+K$, where N is the same as above and K is a sufficiently large positive integer.

(ii) Is there any better stopping rule than N in the sense of achieving a smaller order of magnitude than $E_{\mu,\sigma}N$ in (2) while preserving either (1) or (3)? The answer is essentially negative, and the above stopping rule N is nearly optimal for large σ.

Vardi (1979, *A.S.*, 1034) proved this point for a related problem. He considered (a) instead of (b), so we define for a sequential rule (t_δ, N_δ) its risk

$$r_\delta(\mu,\sigma)=E_{\mu,\sigma}(t_\delta-\mu)^2+cE_{\mu,\sigma}N_\delta.$$

Let

$$N=\min\left\{n\colon \frac{\hat{\sigma}_n}{\sqrt{c}}\le 1\right\}, \qquad t_N=\bar{x}_N.$$

The sequential rule (N,t_N) is asymptotically optimal as $\sigma\to\infty$. This rule is motivated by the observation that $\sigma^2/n+cn$ reaches its minimum when $n=\sigma/\sqrt{c}$. It has been studied by Robbins (1959, *Cramér Volume*, 235) and Starr and Woodroofe (1969), *Proc. Nat. Acad. Sci. U.S.A.*, 285).

SOME RESEARCH PROBLEMS

(1) Many results in sequential analysis are of an asymptotic nature (e.g., the formula (1)) because exact computations are prohibitively difficult. It is therefore desirable to develop small-sample approximations (as in Lorden's work for the K-decision problem) that are applicable even to small EN.

(2) In many nonparametric problems (cf. Lectures 3 and 4) there are asymptotic results concerning various sequences of statistics $\{t_n\}$, but there are no results on stopping rules N, for example, of the type (3) for parametric problems. For instance, if $x_1, x_2, \ldots$ are i.i.d. with common unknown density f, how should we estimate f sequentially? This is a worthwhile research problem.

(3) Selection problems: An example is the selection of the best among K treatments. The experimental protocol requires the use of only one treatment per

patient, and we would like to minimize the number of times when the inferior treatments are used (cf. Flehinger, Louis, Robbins, and Singer, *Proc. Nat. Acad. Sci. U.S.A.* 1972, 2993).

Lecture 3. Nonparametric Estimation

Let $x_1, \ldots, x_n, \ldots$ be a sequence of i.i.d. random variables such that the distribution function F of x_1 belongs to some family Ω. In parametric estimation, Ω is a set whose elements can be identified by real numbers, e.g., $\Omega = \{N(a, \sigma^2): -\infty < a < \infty, \sigma > 0\}$. In nonparametric estimation, however, Ω cannot be identified in this way, and we usually take Ω as the family of all distribution functions that are absolutely continuous with respect to Lebesgue measure, or as the family of all distribution functions having finite α^{th} absolute moments ($\alpha > 0$), etc.

The usual formulation of nonparametric estimation is as follows. Let $\theta(F)$ be a known transformation taking values in some space $\mathscr{D}$, and the objective is to estimate $\theta(F)$. For example, letting Ω be the family of distributions having finite second moments, we define

$$\theta(F) = \text{mean of } F = \int x\,dF(x)$$

or

$$\theta(F) = \text{variance of } F = \int x^2\,dF(x) - \left[\int x\,dF(x)\right]^2,$$

etc. It should be noted that $\theta(F)$ need not be real valued (or vector valued). For example, if F itself is what is to be estimated, then $\mathscr{D} = \Omega$ and $\theta(F) = F$.

In the past few decades, there have accumulated a number of results in nonparametric estimation, but on the whole these results are not as systematic and complete as in parametric estimation. Here we consider three problems in nonparametric estimation.

1. Estimation of the Distribution Function

On the basis of sample observations $x_1, \ldots, x_n$, define the empirical distribution function

$$F_n(x) = \frac{1}{n}\#\{i: 1 \le i \le n,\ x_i \le x\} \qquad (-\infty < x < \infty).$$

Here $\#A$ denotes the number of elements of a set A. $F_n(x)$ is a step function, with jump points at $x_1, \ldots, x_n$ and with values that are multiples of $1/n$.

Without any assumptions about F, F_n is a natural estimator of F since the use of $F_n(x)$ to estimate $F(x)$ for fixed x corresponds to using the observed relative frequency to estimate a probability. By Borel's law of large numbers, $F_n(x) \to F(x)$

a.s. as $n \to \infty$ for every fixed x. On a global level, we have

Glivenko–Cantelli Theorem.

$$P\left\{ \sup_x |F_n(x) - F(x)| \to 0 \right\} = 1.$$

Kolmogorov's Theorem.

$$\lim_{n \to \infty} P\left\{ \sqrt{n} \sup_x |F_n(x) - F(x)| \le y \right\} = \sum_{k=-\infty}^{\infty} (-1)^k e^{-2k^2 y^2}, \qquad y > 0.$$

Sometimes we may want to modify F_n somewhat, for example, by smoothing F_n to obtain a continuous function as an estimator.

Are there any optimal properties of F_n as a global estimator of F? General answers to this question have not yet been found. However, in 1956, Dvoretzky, Kiefer, and Wolfowitz proved the following asymptotically optimal property of F_n: Let T_n be the set of all estimators $t_n = t_n(x) = t_n(x; x_1, \dots, x_n)$ of F (based on $x_1, \dots, x_n$). For $t_n \in T_n$, define

$$\gamma_{t_n}(F) = P_F\left\{ \sup_x \sqrt{n} |t_n(x) - F(x)| > c \right\}, \tag{1}$$

where $c > 0$ is a fixed number. Let Ω be the set of all distribution functions on the real line. Then we have

Theorem (asymptotically minimax character of F_n).

$$\lim_{n \to \infty} \left\{ \frac{\sup_{F \in \Omega} \gamma_{F_n}(F)}{\inf_{t_n \in T_n} \sup_{F \in \Omega} \gamma_{t_n}(F)} \right\} = 1. \tag{2}$$

The meaning of this theorem is that for the risk function (1), F_n is approximately the minimax estimator of F for large n. In fact, this theorem can be extended to other loss functions. The proof of the theorem is rather long. We refer the details to the paper of Dvoretzky, Kiefer and Wolfowitz (1956, *A.M.S.*), and only outline the proof here.

1) First show that we can restrict to continuous distribution functions and therefore take Ω in (2) to be the set of all continuous distribution functions.

2) If F is continuous, we can replace it by the uniform distribution function U on $(0,1)$, where

$$U(x) = \begin{cases} 0, & x \le 0, \\ x, & 0 < x < 1 \\ 1, & x \ge 1. \end{cases}$$

In fact, it is not hard to show that

$$\gamma_{F_n}(F) = \gamma_{U_n}(U), \tag{3}$$

where U_n is the empirical distribution function based on $F(x_1),\ldots, F(x_n)$ i.e.,

$$U_n(x) = \frac{1}{n}\#\{i\colon 1 \le i \le n,\ F(x_i) \le x\}.$$

3) For every fixed positive integer h, define a family $\mathscr{F}_h$ of distribution functions by

$$\mathscr{F}_h = \{F\colon F(0) = 0,\ F(1) = 1 \text{ and } F'(x) \text{ is constant on } I_{h,j} \text{ for } j = 1,\ldots, h\},$$

when

$$I_{h,j} = \left[\frac{j-1}{h}, \frac{j}{h}\right).$$

Any distribution $F \in \mathscr{F}_h$ can be characterized by $p = (p_1,\ldots, p_n)$, where p_j is the probability of $I_{h,j}$ under F. The parameter space can therefore be identified as

$$\Pi = \left\{ p = (p_1,\ldots, p_h)\colon p_1 \ge 0,\ldots, p_h \ge 0, \sum_1^h p_j = 1\right\}.$$

This family admits an h-dimensional sufficient statistic

$$S_n = \left(F_n\left(\frac{1}{h}\right) - F_n\left(\frac{0}{h}\right),\ldots, F_n\left(\frac{h}{h}\right) - F_n\left(\frac{h-1}{h}\right)\right),$$

and nS_n has the multinomial distribution $M(n; p_1,\ldots, p_h)$. Let

$$H_h = \left(0, \frac{1}{h},\ldots,1\right).$$

Noting the one-to-one correspondence between F and p, we define for any F in $\mathscr{F}_h$ and its estimator $t_n = t_n(x)$ the risk

$$\tilde{\gamma}_{t_n}(p) = \tilde{\gamma}_{t_n}(F) = P_F\left\{ \sup_{x \in H_h} \sqrt{n}\,|t_n(x) - F(x)| > c\right\}.$$

Let $p^* = (1/h,\ldots,1/h)$.

4) For any (continuous) F in Ω, we obtain by (3) that $\gamma_{F_n}(F) = \gamma_{U_n}(U)$. Moreover, by analyzing the stochastic process $\{\sqrt{n}\,(U_n(x) - x)\colon 0 \le x \le 1\}$ (which is approximately a Brownian bridge), it can be shown that

$$\gamma_{U_n}(U) \approx \tilde{\gamma}_{U_n}(p^*) \quad \text{as } n \to \infty.$$

5) We now consider the minimax problem for $\mathscr{F}_h$. By using the fact that S_n is a sufficient statistic with respect to $\mathscr{F}_h$ together with the minimax solution of a related multinomial estimation problem, it can be shown that

$$\tilde{\gamma}_{U_n}(p^*) \approx \min_{t_n} \max_{p \in \Pi} \tilde{\gamma}_{t_n}(p) \quad \text{as } n \to \infty.$$

Since the right-hand side cannot exceed $\min_{t_n} \max_F \gamma_{t_n}(F)$, we have shown the desired conclusion.

For multi-dimensional distribution functions F (on R^p), $\sup_x |F_n(x) - F(x)|$ is no longer "distribution-free". However, the preceding theorem can still be generalized to this setting, and the empirical distribution function F_n is still asymptotically minimax under certain assumptions on the loss function (cf. Kiefer and Wolfowitz (1959, *A.M.S.*)).

Another noteworthy point is that the theorem can be extended to situations where there are certain restrictions on the one-dimensional distribution functions (cf. Kiefer and Wolfowitz (*ZfW*, 1976, Band 43), Miller (*ZfW*, 1976, Band 48), etc.). An example is

$$\Omega = \{F\colon F([0,\infty)) = 1 \text{ and } F \text{ is concave on } [0,\infty)\}.$$

Here, although F_n is still asymptotically minimax, there are better estimators than F_n.

2. Estimation of the Density Function

Let $\Omega_m = \{F\colon F$ is a one-dimensional distribution function with density function f such that the mth derivative $f^{(m)}$ exists and is bounded on $R^1\}$. The problem is to estimate $f(x)$ at a given x or to estimate f globally. We consider here the former case.

1. *The Histogram Estimator*. This is the earliest and still one of the most commonly used density estimators. The method is to split R^1 into sub-intervals $I_j = [jh_n, (j+1)h_n)$, $j = 0, \pm 1, \pm 2, \ldots$, where $h_n > 0$ is suitably chosen. For given x, find j such that $x \in I_j$ and determine the number n_j of sample observations that lie in I_j. Define

$$\hat{f}_n(x) = \frac{n_j}{nh_n} = \frac{\hat{p}_j}{h_n} \qquad \left(\text{where } \hat{p}_j = \frac{n_j}{n}\right) \tag{4}$$

as an estimator of $f(x)$. Note that $\hat{f}_n$ is constant on I_j and therefore $\hat{f}_n$ is a step function. This is why we call it the "histogram estimator." The mean squared error of the estimator is given by

$$\begin{aligned}\text{MSE}(\hat{f}_n(x)) &= E[\hat{f}_n(x) - f(x)]^2 \\ &= \text{Var}(\hat{f}_n(x)) + [E(\hat{f}_n(x)) - f(x)]^2.\end{aligned}$$

Since n_j has the binomial $B(n, p_j)$ distribution with

$$p_j = \int_{I_j} f(x)\,dx,$$

it follows easily that

$$\text{Var}(\hat{f}_n(x)) = \frac{p_j(1-p_j)}{nh_n^2}.$$

If f is bounded in some neighborhood of x, then $p_j = 0(h_n)$, and therefore $\text{Var}(\hat{f}_n(x)) \le c^*/nh_n$ (as $h_n \to 0$). On the other hand,

$$E\hat{f}_n(x) - f(x) = \frac{1}{h_n}\int_{I_j}[f(t) - f(x)]\,dt,$$

and therefore if f has a third derivative at x, then

$$|E\hat{f}_n(x) - f(x)| \le ch_n \quad (\text{as } h_n \to 0).$$

Hence $\mathrm{MSE}(\hat{f}_n(x)) \le c^*/nh_n + c^2h_n^2$. Letting $h_n = Kn^{-1/3}$ (for some positive constant K), we then obtain that $\mathrm{MSE}(\hat{f}_n(x)) = 0(n^{-2/3})$, which is of a larger order of magnitude than $0(n^{-1})$ commonly encountered in the estimation of parameters or of distribution functions.

Can we improve this order of magnitude? To obtain any improvement we need to use higher order derivatives of f and apply smoother "windows" to exploit the smoothness properties of f.

2. *The Kernel Estimator*. This method has been thoroughly studied in the literature. One of the shortcomings of the histogram estimator is that the point x may be near the boundary of I_j. To overcome this, Rosenblatt (1959, *A.M.S.*, 832) introduced the following modification. For given x, let $I_{n,x} = [x - h_n, x + h_n]$ and let n_x denote the number of sample observations lying in $I_{n,x}$. Using $\tilde{f}_n(x) = n_x/(2nh_n)$ as an estimator of $f(x)$, Rosenblatt established the following result: Suppose that $f(x) \neq 0$ and $f''(x) \neq 0$. Then as $h_n \to 0$,

$$\mathrm{MSE}[\tilde{f}_n(x)] = \frac{f(x)}{2nh_n} + \frac{[f''(x)]^2}{36} h_n^4 + o\left(\frac{1}{nh_n} + h_n^4\right),$$

and therefore for $h_n = Kn^{-1/5}$ (where $K > 0$ is a constant), $\mathrm{MSE}[\tilde{f}_n(x)] = 0(n^{-4/5})$.

Rosenblatt's estimator can be expressed as

$$\tilde{f}_n(x) = \frac{1}{nh_n} \sum_{i=1}^{n} k\left(\frac{x - x_i}{h_n}\right), \tag{5}$$

where

$$k(x) = \begin{cases} \frac{1}{2} & \text{if } |x| \le 1, \\ 0 & \text{if } |x| > 1, \end{cases} \tag{6}$$

is a symmetric probability density function. Different functions k can also be used instead. We call these density estimators "kernel estimators." Parzen (1962, *A.M.S.*, 1068) gave a systematic theory of kernel estimators, and subsequently Wahba and several other authors continued the study of such estimators.

Since in practice we require (5) to be nonnegative, it is natural to choose k to be nonnegative. In this case, we cannot do better than the order of $n^{-4/5}$ for $\mathrm{MSE}[\tilde{f}_n(x)]$. However, if we do not require k to be nonnegative, then we can improve the convergence rate. For example, suppose that f is m times continuously differentiable in some neighborhood of x and suppose that we choose the kernel k such that

$$k \text{ is bounded}, \qquad \lim_{|y| \to \infty} yk(y) = 0, \qquad \int_{-\infty}^{\infty} k(y) = 1,$$

$$\int_{-\infty}^{\infty} y^i k(y)\, dy = 0 \quad (i = 1, \ldots, m-1), \qquad \int_{-\infty}^{\infty} |y|^m k(y) < \infty.$$

Then it can be shown that

$$\mathrm{MSE}(\tilde{f}_n(x)) \le \frac{c^*}{nh_n} + c^2 h_n^{2m} \quad (\text{where } c^*,\ c \text{ are constants}).$$

Therefore if we choose $h_n = Kn^{-1/(2m+1)}$, then $\mathrm{MSE}(\tilde{f}_n(x)) = O(n^{-2m/(2m+1)})$.

There are many papers giving similar results when polynomials, splines, or orthogonal series are used as density estimators, and concerning adaptive choices of the optimum constant K in h_n (cf. Nadaraya, *Th. Prob.* 1974, 133). For $f \in \Omega_m$, the preceding result is probably the best possible. For example, if $f \in \Omega_1$, then all density estimators $\hat{f}_n$ (not only the kernel estimators) have $\text{Risk}(\hat{f}_n) \geq cn^{-2/3}$. Such lower bounds for the MSE have been developed by Farrell (1967, 1972, *A.M.S.*) and others.

3. Nonparametric Regression

Nonparametric regression is regression analysis under no assumptions about the form of the regression function. Stone (1977, *A.S.* 596) proposed a method of nonparametric regression that has attracted a lot of recent attention. The key idea of Stone's method is as follows. Consider the random vector (X, Y), where X, Y may be multidimensional. The basic problem of regression analysis is to find the conditional distribution of Y given $X = x$, or to estimate $E[g(Y)|X = x]$ for some given function $g(Y)$. For example, the case $g(Y) = Y$ reduces to the regression function $E[Y|X = x]$. Suppose that we have i.i.d. sample observations (X_i, Y_i), $i = 1, \ldots, n$. Choose appropriate "weight functions" $W_{ni}(x) = W_{ni}(x; X_1, \ldots, X_n)$, $i = 1, \ldots, n$, to reflect the importance of the sample value (X_i, Y_i) in estimating the conditional distribution of Y given $X = x$. Let

$$I_A(Y) = \begin{cases} 1 & \text{if } Y \in A, \\ 0 & \text{if } Y \notin A. \end{cases}$$

Then use $\hat{P}_n^Y(A|x) = \sum_{i=1}^n W_{ni}(x) I_A(Y_i)$ as an estimator of $P^Y(A|x) = P\{Y \in A | X = x\}$. Likewise use $\hat{E}_n(g(Y)|x) = \sum_{i=1}^n W_{ni}(x) g(Y_i)$ as an estimator of $E[g(Y)|X = x]$.

Stone introduced a notion of consistency for these regression estimators, using convergence in the rth mean, and obtained some important general results for such consistency. Recently there have been several investigations into the convergence rates in nonparametric regression, similar to those in density estimation. Another worthwhile research problem is about the performance of these estimators when n is not very large.

Lecture 4. Robustness and Efficiency of Nonparametric Methods

In this lecture, Ω is the same as in Lecture 3 while D is somewhat simpler. We will study the loss functions γ_t more carefully.

A statistical procedure is said to be "robust" if its properties do not change much when the underlying assumptions are slightly violated. It resembles a robust person who can maintain balance after being lightly pushed. For example, although $\bar{X}$ is a very good estimator of the mean a of the normal $N(a, \sigma^2)$ distribution, it may perform rather poorly when the underlying distribution deviates from the normal distribution. Therefore $\bar{X}$ is not a robust estimator. For many years, robustness has been associated with insensitivity to "outliers," which are the relatively few atypical values in the data, possibly due to some systematic error. A "good" statistical procedure should not be much influenced by these atypical outliers. For example, the sample median $\tilde{X}_n$ is more robust than the sample mean $\overline{X}_n$ in estimating the center of a symmetric distribution.

A commonly discussed problem in the literature is the robust estimation of the center of a symmetric distribution. Let $\mathscr{F}_s$ denote the family of one-dimensional probability density functions symmetric about the origin, and let

$$\mathscr{F} = \{ f(x-\theta) \colon f \in \mathscr{F}_s, \theta \in R^1 \}.$$

We want to estimate θ under squared loss

$$L(\theta, d) = (\theta - d)^2.$$

If f is known, then we can use parametric methods to get good estimators, e.g., $t_n = \overline{X}_n$ (sample mean) when f is the standard normal density, and $t_n = \tilde{X}_n$ (sample median) for the Laplace density $f(x) = \frac{1}{2}e^{-|x|}$. The question is how these estimators work for other densities. Also, are there "universally good" estimators? These problems have attracted much recent interest. Generally speaking, $\tilde{X}_n$ appears to be a universally better estimator than $\overline{X}_n$. Noting that $\tilde{X}_n$ is approximately $N(\theta, 1/4\pi f^2(0))$, $\tilde{X}_n$ should perform reasonably well if f is unimodal. Although $\mathrm{Var}(\tilde{X}_n)/\mathrm{Var}(\overline{X}_n)$ is approximately $\pi/2$ for normal f, $\tilde{X}_n$ can perform well for those f which cause trouble to $\overline{X}_n$.

From a historical perspective, the normal distribution has been the most important and most widely used distribution that appears almost everywhere. The use of $\overline{X}_n$ has therefore been a customary practice. This common practice of using $\overline{X}_n$ has been criticized by Box and Anderson (1955, *J.R.S.S.*), Tukey (1962, *A.M.S.*), and others, who pointed out that $\overline{X}_n$ may be very poor when f is not normal and especially when f has "fat" tails. The notion of "fat tails" is closely related to that of "outliers." It is therefore important to look for estimators which have good properties for all f in $\mathscr{F}_s$ and which are also relatively easy to compute. This problem has attracted growing interest from both applied and theoretical statisticians.

A method to construct such robust estimators was proposed by Huber, who introduced the concept of "quantitative robustness." The idea is to find some kind of optimal solution by restricting to a certain subset (of $\mathscr{F}$) that has both practical relevance and analytical tractability. For example, Huber (1964, *A.M.S.*) examined the "contaminated normal family"

$$\mathscr{F}_{N,\varepsilon} = \{ f \colon f = (1-\varepsilon)N(0,1) + \varepsilon g, \ g \in \mathscr{F}_s \},$$

and considered the minimax solution with respect to $\mathscr{F}_{N,\varepsilon}$ and squared loss. Specifically, he found t_n^* that approximately minimizes over all t_n the expression

$$\sup_{F \in \mathscr{F}_{N,\varepsilon} \times R^1} E_F(t_n - \theta_F)^2,$$

where θ_F is the center of symmetry of the symmetric distribution F and

$$\mathscr{F}_{N,\varepsilon} \times R^1 = \{ f(x-\theta) \colon f \in \mathscr{F}_{N,\varepsilon}, \theta \in R^1 \}.$$

Another approach to robustness is the so-called "qualitative robustness," which is based on some sort of consistency ideas due to Fisher. Let F be the population distribution. We regard θ_F, the quantity to be estimated, as a functional $T(F)$ of F. From sample observations $X_1,\ldots,X_n$, construct the empirical distribution function F_n and use $T(F_n)$ to estimate θ_F. Examples are

$$T_1(F)=\int x\,dF(x),\qquad T_1(F_n)=\bar{X}_n;$$

$$T_2(F)=F^{-1}(\tfrac{1}{2}),\qquad T_2(F_n)=\tilde{X}_n;$$

$$T_3(F)=\frac{1}{1-2\alpha}\int_{F^{-1}(\alpha)}^{F^{-1}(1-\alpha)}x\,dF(x);$$

$$T_3(F_n)=\frac{1}{n(1-2\alpha)}\sum_{i=[n\alpha]}^{[n(1-\alpha)]}X_{(i,n)}.$$

Here $X_{(1,n)}\le X_{(2,n)}\le\cdots\le X_{(n,n)}$ denote the order statistics of $X_1,\ldots,X_n$. $T_3(F_n)=\bar{X}_{\alpha,n}$ is called the "α-trimmed mean" and is an important robust estimator with a long history.

The performance of $T(F_n)$ as an estimator of θ_F depends on the properties of the functional T, so it is helpful to study these properties. In this connection, two concepts have proved useful. The first is due to Von Mises (1947, *A.M.S.*) who used the Gateaux derivative of T at F in the direction of G:

$$\lim_{\varepsilon\to 0}\frac{T[(1-\varepsilon)F+\varepsilon G]-T(F)}{\varepsilon}.$$

The other is the "influence curve" of Hampel (1971, *A.M.S.*), who used the derivative $IC(T,F;x)$ of T at F in the direction of δ_x, where δ_x is the degenerate distribution putting point mass at x. For example, if $F'\in\mathscr{F}_s$, then

$$IC(T_1,F;x)=x,\qquad IC(T_2,F;x)=\frac{1}{2F'(0)}\operatorname{sgn}x.$$

The concept of IC is particularly useful in comparing different estimators that correspond to different functionals.

Breakdown Point. Suppose that for a given estimator t_n of θ, its maximal risk in $\mathscr{F}_{N,\varepsilon}$ becomes worst possible (for example, reaches ∞) when the contamination proportion ε exceeds some number α. The smallest such α is called the "breakdown point." For instance, the breakdown point of $\bar{X}_{\alpha,n}$ is α. All other things being equal, we would like the breakdown point to be largest possible, as it reflects the resistance of the estimator to contamination. Usually the concept of breakdown point is applicable only to large sample sizes.

The concept of efficiency or asymptotic efficiency of an estimator is well known and is based on the Cramer–Rao inequality and the Fisher information number. When $f=F'$ is known, under certain regularity assumptions, the Fisher information number is defined by $V^*(f)=\int(f')^2\,dx/f$, and the asymptotic efficiency of an estimator t_n is defined by

$$e_{(t_n)}(f) = \lim_{n \to \infty} \frac{1}{n} V^*(f) / \gamma_{t_n}(f).$$

The asymptotic relative efficiency (A.R.E.) of an estimator t'_n to another estimator t''_n is defined as $\lim_{n \to \infty}[\gamma_{t''_n}(f)/\gamma_{t'_n}(f)]$.

Efficiency and robustness are usually conflicting goals. It is not hard to see that to achieve robustness an estimator has to sacrifice some efficiency. As a robust estimator represents some sort of compromise between various distribution forms, it is not comparable to an efficient estimator specifically designed for a particular distribution form. The issue is that when the assumptions for the particular distribution family do not hold the robust estimator can still perform reasonably well while the efficient estimator may become very poor. For instance, increasing the α in the trimmed mean $\overline{X}_{\alpha, n}$ will increase its robustness but decrease its efficiency. The primary objective in the theory of robustness is robustness itself; efficiency only plays a secondary role.

We now discuss several kinds of robust estimators.

1. *L-estimator:* This is a linear combination of the order statistics $X_{(1,n)} \leq \cdots \leq X_{(n,n)}$ of the form

$$t_n = \sum_{i=1}^{n} a_{in} X_{(i,n)},$$

where the a_{in} are constants. Commonly used estimators such as the sample mean $\overline{X}_n$, the sample median $\tilde{X}_n$, and trimmed means are examples of L-estimators. If f is known, then there exists a fully efficient L-estimator $t_{f,n}$.

2. *R-Estimator:* Let $X_1, \ldots, X_n$ denote the sample observations and $X_{(1,n)} \leq \cdots \leq X_{(n,n)}$ the order statistics. If $X_i = X_{(R_i, n)}$, then we say that the rank of X_i is R_i. An estimator based on the ranks R_i is called a "rank estimator." In the past two decades there have appeared many such estimators in the literature. In particular, Hodges and Lehmann (1963, *A.M.S.*, 598) proposed the estimator

$$t_{\text{HL}, n} = \text{median}\left\{ \frac{X_i + X_j}{2}, 1 \leq i \leq j \leq n \right\},$$

which is commonly known as the "Hodges–Lehmann estimator."

3. *M-Estimator:* This is a "maximum-likelihood-type" estimator proposed by Huber in 1964. The idea is to find a suitable function ρ such that $\sum_{i=1}^{n} \rho(X_i - \theta)$ reaches its minimum at $\theta = t_{\rho, n}$. Under certain regularity assumptions, minimization can be carried out by solving the equation $\sum_{i=1}^{n} \rho'(X_i - \theta) = 0$. The underlying motivation is that when f is known the maximum likelihood estimator of θ is obtained by minimizing

$$\sum_{i=1}^{n} [-\ln f(X_i - \theta)],$$

which corresponds to taking $\rho(x) = -\ln f(x)$. For example, when f is normal, we take $\rho(x) = x^2$; when f is Laplace ($f(x) = \frac{1}{2} e^{-|x|}$), we take $\rho(x) = |x|$.

In 1964, Huber established the asymptotic normality of such estimators, and obtained the following result related to robustness: There exists least favorable f^* in $\mathscr{F}_{N,\varepsilon}$ such that the maximum likelihood estimator $\tilde{t}_n^*$ with respect to f^* (i.e., M.L.E. of θ in $f^*(x-\theta)$) satisfies

$$\lim_{n\to\infty}\left[n\gamma_{\tilde{t}_n^*}(f^*)\right] = \sup_{f\in\mathscr{F}_{N,\varepsilon}}\left[\lim_{n\to\infty}\left(n\gamma_{\tilde{t}_n^*}(f)\right)\right].$$

The f^* is given by

$$f^*(x)=\begin{cases}(1-\varepsilon)\varphi(x), & \text{if } |x|<c\\ Ke^{-c|x|}, & \text{if } |x|\ge c,\end{cases}$$

where $\varphi(x)=(1/\sqrt{2\pi})e^{-x^2/2}$ and $K>0$ is some constant. In particular, this leads to

$$\rho(x)=\begin{cases}x^2 & \text{if } |x|<c,\\ 2c|x|-c^2 & \text{if } |x|\ge c.\end{cases}$$

In practice, since there is no sample closed form for $\tilde{t}_n^*$, it has to be found by complicated numerical computations. Moreover, the solutions differ for different ε and also for other distributions than the normal in the "central" distribution in $\mathscr{F}_{N,\varepsilon}$.

From an asymptotic efficiency point of view, it seems that R-estimators are the best among the L-, R-, and M-estimators. In particular, if f is bimodal, M-estimators may not perform well. To give an example,

$$\lim_{n\to\infty}\gamma_{\bar{X}_n}(f)/\gamma_{\bar{X}_{\alpha,n}}(f)\ge 1/(1-2\alpha)^2 \text{ and may be } \infty,$$

$$\text{A.R.E. of } t_{\mathrm{HL}} \text{ to } \bar{X}_{\alpha,n}\ge 0.9 \text{ and may be } \infty,$$

$$\text{A.R.E. of } t_{\mathrm{HL}} \text{ to } \tilde{t}_n\ge 0.86 \text{ and may be } \infty.$$

These results show that t_{HL} performs very well.

We can also carry out simulations to study the risks of various estimators and to compare their performance (cf. Andrews et al., Princeton University Press, 1972).

Another related idea is adaptive estimation. (1) *Crude method*: Hogg (1974, *JASA*, 909) suggested using the sample observations to estimate the size of the tail of the population distribution and thereby deciding which robust estimator to use. (2) *Sophisticated method*: First use $\hat{f}_n$ (see Lecture 3) to estimate $f\in\mathscr{F}_s$. Letting $t_{f,n}$ denote the fully efficient L-estimator of θ when f is known, we modify $t_{\hat{f},n}$ suitably to construct our estimator (so that it has the same asymptotic behavior as $t_{f,n}$). This idea dates back to Stein (*3rd Berk. Symp.* 1956). Other contributions are due to Van Eeden (1970, *A.M.S.*, 1972), Beran (1974, *AS*, 63), Stone (1975, *AS*, 267), Sacks (1975, *AS*, 285). The difficulty with these methods is that for the asymptotic results to be applicable, we need a very large sample size n which may also vary with f.

OTHER RESEARCH DIRECTIONS AND PROBLEMS

(1) Presence of unknown scale parameter (Huber 1964, 1972, *A.M.S.*, etc.); regression models (Huber 1973, *AS*, 199; Bickel 1973, *AS*, 597; 1975, *JASA* 428); dependent X_i (Portnoy 1977, *AS*, 22).

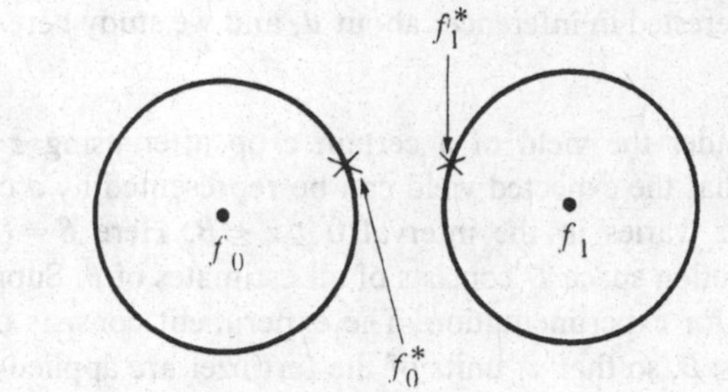

Figure 6.

(2) f not symmetric: What is the parameter to be estimated? Median? (See Bickel–Lehmann 1975, *AS*, 1038).

(3) Many other models in estimation remain to be settled; there are only partial results to date.

(4) Testing hypotheses (Huber 1965, *A.M.S.*, 1753): For the simple vs. simple problem, f_0 consists only of a single f_0 and H_1 consists only of a single H_1, and the Neyman–Pearson test involves the likelihood ratio $L(\mathbf{x}) = f_1(\mathbf{x})/f_0(\mathbf{x})$. However, if there is contamination, how should we perform the test? A method can roughly be described as follows: In H_0 (containing other densities than f_0) and H_1 (containing other densities than f_1) find the least favorable pair of densities f_0^* and f_1^*, and apply the Neyman–Pearson test to test f_0^* versus f_1^*. In a certain sense, this pair represents the "shortest distance" between H_0 and H_1 (see Figure 6). Let $L^* = f_1^*/f_0^*$. Then $\prod_{i=1}^n L^*(x_i)$ is the test statistic based on sample observations $x_1, \ldots, x_n$. Moreover, L^* represents a two-sided truncation of L, i.e., there exist c, c' such that

$$L^* = \begin{cases} L & \text{if } c \le L \le c', \\ c & \text{if } L \le c, \\ c' & \text{if } L \ge c'. \end{cases}$$

To extend this idea to composite hypotheses, we ask when there exist f_i^* like those discussed above (cf. Huber–Strassen 1973, *AS*, 251). Extensions to multiple decision problems are even more complicated but represent a worthwhile research direction.

(5) Sequential analysis: The sequential testing problem of f_0 versus f_1 can be treated by the aforementioned approach (cf. Huber's paper). Other problems that remain to be settled include robustness properties of sequential procedures, such as the Anscombe–Robbins estimator (see Lecture 2).

(6) Nonparametric problems: How should we do density estimation for $f \in \mathscr{F}_{N,\varepsilon}$ (see Lecture 3)?

(7) Robustness in other branches of statistics: experimental design (see Lecture 8).

Lecture 5. Experimental Design—Basic Concepts

Let S denote the sample space, and let $\{P_{\bar{\theta}}\}$ denote the set of all possible distributions for the sample $\mathbf{x}$. Available to us in designing an experiment are the different choices of $\varepsilon = \{S, \{P_{\bar{\theta}}\}\}$. In simple situations, we can represent $\{P_{\bar{\theta}}\}$ by

$\Omega = \{\bar{\theta}\}$. We are interested in inferences about $\bar{\theta}$, and we study here how to choose ε for this purpose.

Example 1. Consider the yield of a certain crop after using z units of a new fertilizer. Suppose that the expected yield can be represented by a cubic polynomial $\Sigma_{j=1}^4 \theta_j z^{j-1}$, where z varies in the interval $0 \le z \le B$. Here $\theta = (\theta_1, \theta_2, \theta_3, \theta_4)^T$ is unknown, and the action space D consists of all estimates of θ. Suppose that n such crops are available for experimentation. The experiment consists of choosing $Z = (z_1, \dots, z_n)^T$, $0 \le z_i \le B$, so that z_i units of the fertilizer are applied to the i-th crop. Let x_i denote the yield of the i-th crop, and assume that $x_1, \dots, x_n$ are independent random variables with a common variance σ^2 and means $Ex_i = \Sigma_{j=1}^4 \theta_j z_i^{j-1}$. Thus, $\mathbf{x} = (x_1, \dots, x_n)^T$ represents the sample, $\bar{\theta} = (\theta, \sigma)^T$, and $\bar{\theta}$ and Z determine the distribution $P_{\bar{\theta}}$.

Some possible objectives of the experiment are:

(a) to estimate θ_2 (which represents approximately the rate of increase of yield per unit increase of fertilizer when the amount of fertilizer used is small); here $D = R^1$ (or some subset of R^1);
(b) to test $\theta_2 \le 1.1$ versus $\theta^2 > 1.1$ (where 1.1 represents some standard value for conventional fertilizers); here $D = \{d_0, d_1\}$;
(c) to estimate $\theta = (\theta_1, \theta_2, \theta_3, \theta_4)^T$; here $D = R^4$;
(d) to estimate $\Sigma\theta_j z^{j-1}$ for $0 \le z \le B$;
(e) to predict the "response" value $\Sigma\theta_j b^{j-1}$ when it is decided to apply b units of the fertilizer for each plot in the future; when $b > B$, this corresponds to "extrapolation," assuming that the same cubic curve $\Sigma\theta_j z^{j-1}$ holds $z = b$.

In this lecture we will discuss primarily (a), (c), (d), (e). The loss functions will be described below.

A General Model

Let $Z^* = \{Z\}$ = set of all possible experiments. (In general, as in Example 1, Z^* is the Cartesian product Q^n of some set Q. $Q = [0, B]$ in Example 1.) Once the

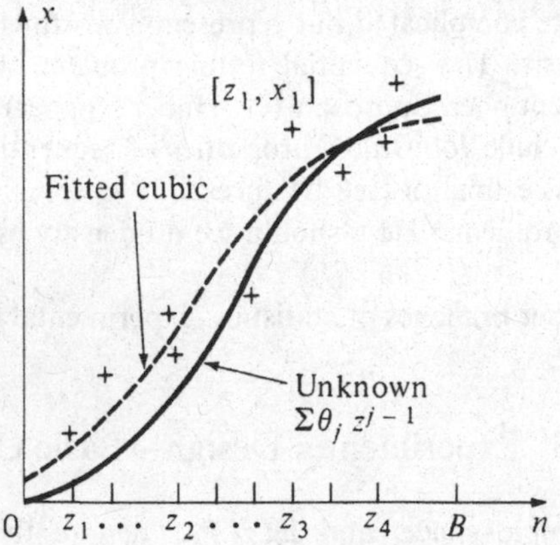

Figure 7.

experiment Z has been chosen, the possible probability distributions depend only on $\bar{\theta} = (\theta, \sigma)$, where θ is the parameter of interest and σ is a nuisance parameter. The observation vector $\mathbf{x} = (x_1, \ldots, x_n)^T$ satisfies $E_{\theta,\sigma}\mathbf{x} = F(\theta, Z)$, where F is a known function. The model is said to be *linear* (with respect to θ) if θ is p-dimensional and there exists an $n \times p$ matrix B_Z such that

$$E_{\theta,\sigma}\mathbf{x} = F(\theta, Z) = B_Z\theta.$$

In Example 1, $p = 4$ and the model is linear, although $E_{\theta,\sigma}x_i$ is a cubic polynomial in z_i; here,

$$B_Z = \begin{pmatrix} 1 & z_1 & z_1^2 & z_1^3 \\ 1 & z_2 & z_2^2 & z_2^3 \\ \vdots & \vdots & \vdots & \vdots \\ 1 & z_n & z_n^2 & z_n^3 \end{pmatrix}.$$

If $Z^* = Q^n$ and the observations x_i are independently distributed with the distribution of x_i depending only on z_i and $\bar{\theta}$, then we call it a "repeatedly observable model," as we are allowed to repeat as many measurements as we want at the same design level z_i to estimate the common distribution of the x_i at this design level. In particular, Example 1 is a repeatedly observable model.

If a repeatedly observable model is linear, then there exists a function $f\colon Q \to R^p$ such that

$$E_{\theta,\sigma}X_i = \theta^T f(z_i) = \sum_{j=1}^{p} \theta_j f_j(z_i),$$

where $f^T(z_i)$ is the i-th rows vectors of B_Z (e.g., in Example 1, $f^T(z) = (1, z, z^2, z^3)$).

Note that not all statistical models are repeatedly observable, as illustrated by

Example 2. Suppose that we want to compare V different types of crops. The yield of each crop depends on its type and on the field conditions. If the field is too large, then conditions may vary over different plots of the field, and there may be a "plot effect" on the yield. This suggests dividing the field into small plots so that the field conditions may be regarded as uniform within each plot. Suppose that we can plant k crops in each plot and that $k < V$. Let b be the number of plots, and let $n = bk$. Assume that if the r-th crop is planted in the j-th plot then the expected yield is $\alpha_r + \beta_j$, where α_r is the type effect and β_j is the plot effect. This is therefore a linear model with $p = V + b$ parameters. An experimental design is an assignment of n crops of V different types to b plots of the field, as illustrated by Figure 8. The model is not repeatedly observable, since there can at most be k (not n) x_i's that have the same expectation.

In Example 2 our primary interest is in the differences among the α_r and not in the β_j. The design theme is an example of "incomplete block designs." The attribute "incomplete" means that at least one block does not contain all the types

1 2 3	4 4 1
2 3 4	1 1 2
3 4 1	2 2 1

$V = 4, k = 3, b = 6$

Figure 8.

Measure of the Efficiency of a Design. Suppose we design an experiment to estimate the parameters of a linear model. Assume that the sample observations x_i are independent random variables with a common variance σ^2. The usual least squares estimator of the parameter vector θ is given by

$$\bar{t} = (t_1, \ldots, t_p)^T = (B_Z^T B_Z)^- B_Z^T \mathbf{x}.$$

If $c^T\theta$ is estimable, then $c^T\bar{t}$ is the best linear unbiased estimator of $c^T\theta$. $M_Z = B_Z^T B_Z$ is called the "information matrix" of the experiment Z. If M_Z is nonsingular, then

$$\operatorname{cov}(\bar{t}) = \sigma^2 M_Z^{-1}.$$

How should we compare the two covariance matrices

$$M_Z^{-1} = \begin{pmatrix} 2 & 0 \\ 0 & 6 \end{pmatrix}, \qquad M_Z^{-1} = \begin{pmatrix} 4 & 0 \\ 0 & 3 \end{pmatrix}?$$

One of them gives a better estimate of θ_1, while the other gives a better estimate of θ_2. In view of this, we introduce certain "optimality functionals" to evaluate and compare designs. If a design Z minimizes some functional Φ of M_Z, then it is said to be optimal (with respect to Φ). Examples are:

$\Phi(M_Z) = (M_Z^{-1})_{11}$ —related to estimation of θ_1,

$\Phi_A(M_Z) = \operatorname{tr}(M_Z^{-1})$ —related to the so-called "A-optimality,"

$\Phi_D(M_Z) = \operatorname{Det}(M_Z^{-1})$ —related to "D-optimality,"

$\Phi_E(M_Z) =$ largest eigenvalue of M_Z^{-1} —related to "E-optimality."

Geometric Meaning of the Optimality Criteria. Figure 9 illustrates the case $p = 2$. Different optimality functionals Φ represent different ways of assessing the size of the "concentration ellipsoids" (or "confidence ellipsoids").

The minimization of $\Phi(M_Z)$ is typically a large discrete optimization problem; there are no simple computational algorithms and one has to deal with a separate problem for each n. However, we can develop a simple approximate solution that holds for all n, provided that the model is repeatedly observable.

Approximate Theory of Optimal Designs. Define the "design measure"

$$\xi_Z(z) = (\text{number of } z_i\text{'s equal to } z)/n.$$

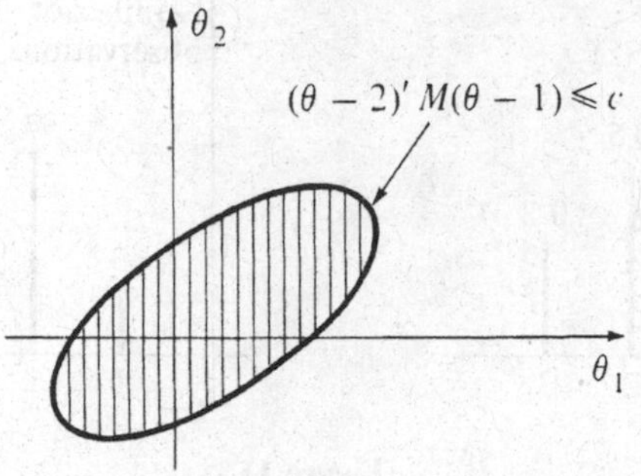

Figure 9.

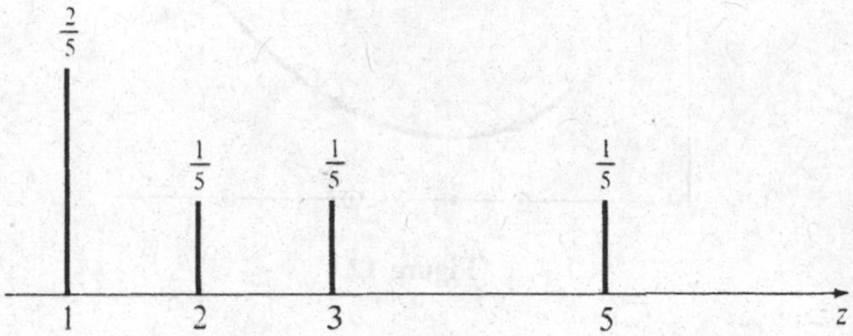

Figure 10.

For example, when $n=5$ and the z_i assume the values 1, 1, 2, 3, 5, the design measure ξ_Z is given by Figure 10.

Let

$$M_Z = B_Z^T B_Z = \sum_{i=1}^{n} \underbrace{f(z_i)}_{p\times 1} f^T(z_i) = n\int_Q f(z) f^T(z) \xi_Z(dz).$$

We call the last term above, after dividing by n, the "information matrix" $M(\xi_Z)$ of the design measure ξ_Z. Therefore, when Φ is homogeneous (i.e., there exists a constant k such that $\Phi(aM)=a^k\Phi(M)$ for all a), the minimization of $\Phi(M_Z)$ reduces to the minimization of $\Phi(M(\xi))$ over the set (ξ: for every z there exists $k=0,\ldots,n$ such that $\xi(z)=k/n\}$.

The approximate theory involves minimization of $\Phi(M(\xi))$ over the set

$$\Xi = \{\xi : \xi \text{ is a discrete probability distribution on } Q\}.$$

This problem admits a practical solution for many Φ's. Once we have found the minimizing ξ^*, we can use it to construct a design for every n by simply rounding off the $n\xi^*(z)$ into integers. This idea is illustrated in Figure 11, where the design is approximately optimal although not exactly optimal for the given n. The notion of "approximately optimal" is illustrated in Figure 12, where the points $\odot$ represent approximations to the true minimum $\cdot$ such that these approximations belong to a prescribed subset of the real line.

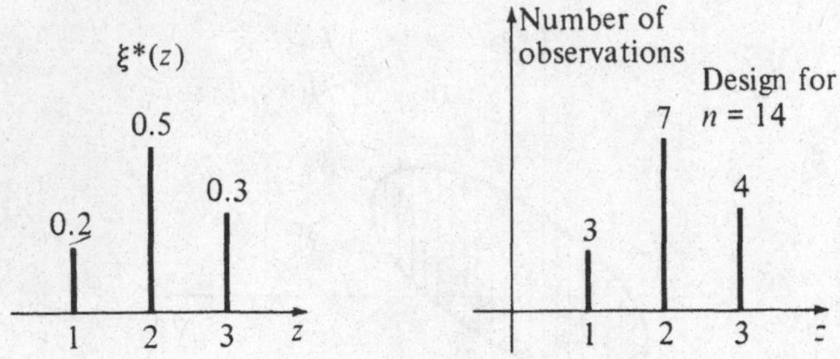

Figure 11.

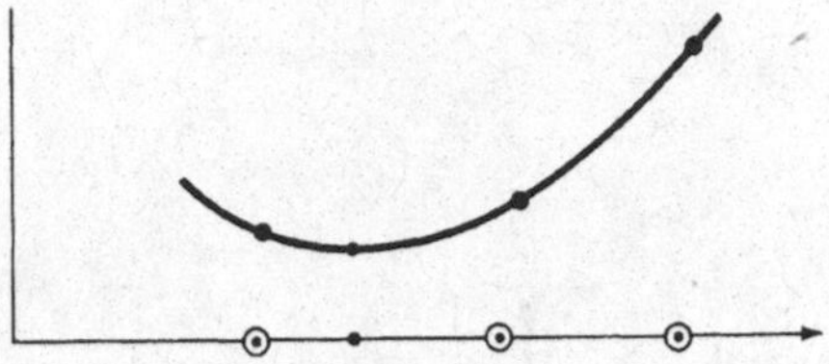

Figure 12.

Historical Note. K. Smith (1918, *Biometrika*) considered the use of $f^T(z)\bar{t}$ to estimate the mean response $f^T(z)\theta$. Noting that

$$E_{\theta,\sigma}\left[f^T(z)\bar{t} - f^T(z)\theta\right]^2 = \sigma^2 f^T(z) M_Z^{-1} f(z),$$

we take

$$\Phi_G(M(\xi)) = \max_{z\in Q} V_\xi(z),$$

where $V_\xi(z) = f^T(z)M^{-1}(\xi)f(z)$. The design measure ξ^* that minimizes $\Phi_G(M(\xi))$ is called G-optimal. Smith discussed optimal designs for polynomial models of degree $d \le 6$. P. G. Guest (1958, *A.M.S.*, 294) extended the result to all d. P. G. Hoel (1958, *A.M.S.*, 1134) discussed D-optimal designs for all d. In general, a design that is optimal with respect to one criterion may not be optimal with respect to another criterion, but here we have

$$D\text{-optimality} \Leftrightarrow G\text{-optimality}$$

(assuming that Q is compact and f is continuous on Q, as in this example). The equivalence of G-optimality and D-optimality holds generally in all other design problems, and this will be discussed in the next lecture.

Lecture 6. Complete Class and Designs in Regression Models

(A) *Complete Class.* Before continuing with the subject of the previous lecture, we first digress and introduce a complete class of designs. In the previous lecture, when we were introducing the concept of "optimality functionals," we used two covariance matrices to illustrate that neither can be regarded as the better one without specifying Φ. We now say that a design measure ξ' is inadmissible if there exists ξ such that

$$M^{-1}(\xi') - M^{-1}(\xi) \tag{6.1}$$

is nonnegative definite and is not zero. This definition makes sense since when $c \in R^p$ the variance of the estimator $c^T\bar{t}$ of $c^T\theta$ is $\sigma^2 c^T M_z^{-1} c$. Therefore, from the viewpoint of the approximate theory, (6.1) shows that ξ is as good as ξ' for estimating all linear forms $c^T\theta$ and is better than ξ' for estimating some $c^T\theta$. This in turn suggests that $\Phi(M(\xi')) \geq \Phi(M(\xi))$ for all reasonable Φ.

If a design measure is not inadmissible, then we call it admissible. The class $\mathscr{C}$ of all admissible design measures is called a complete class. Thus, for every $\xi' \notin \mathscr{C}$, there exists $\xi \in \mathscr{C}$ such that ξ is better than ξ' in the sense that (6.1) holds.

In Example 1 (cubic polynomial) of Lecture 5, the admissible designs take no more than four points. The situation is similar for higher degree polynomial models (cf. Kiefer, *J. Roy. Stat. Soc. Ser. B* (1959), 272–319). However, why do we often take more than four design levels in practice for cubic models? The reason is that we usually do not believe that it is really cubic and would therefore like to have further information to test this hypothesis.

An area we suggest for further research is the case of polynomials of several variables, for which almost nothing is known about $\mathscr{C}$.

(B) *Φ-optimal Designs for Response Curves and Surfaces.* We consider linear, repeatedly observable models with independent errors that have the same variance. With the same notation as in Lecture 5,

$$E_{\theta,\sigma} x_i = \underbrace{f^T(z_i)}_{1\times p} \underbrace{\theta}_{p\times 1}, \qquad z_i \in Q,$$

$\xi(z)$ = proportion of design levels that assume the level z (for the approximate theory),

$$M(\xi) = \int_Q f(z) f^T(z) \xi(dz) \quad \text{(information matrix)},$$

$$V_\xi(z) = f^T(z) M^{-1}(\xi) f(z),$$

$\sigma^2 n^{-1} V_\xi(z)$ = variance of fitted curve at z.

Let $\Phi_D(M(\xi)) = \operatorname{Det} M^{-1}(\xi)$, $\Phi_G(M(\xi)) = \max_{z\in Q} V_\xi(z)$. Kiefer and Wolfowitz (*Canad. J. Math.* (1960), 363) proved the following equivalence theorem: If f is continuous and Q is compact, then

$$\begin{aligned} &\xi^* \text{ is } D\text{-optimal},\\ \Leftrightarrow\ &\xi^* \text{ is } G\text{-optimal},\\ \Leftrightarrow\ &\max_{z\in Q} V_{\xi^*}(z) = p\,(= \text{number of parameters}),\\ \Rightarrow\ &\xi^*(\{z: V_{\xi^*}(z) = p\}) = 1. \end{aligned}$$

Outline of the proof:

(1) $\Phi_D(M)$ is convex.
(2) Note that $M((1-\alpha)\xi^* + \alpha\xi) = (1-\alpha)M(\xi^*) + \alpha M(\xi)$.
(3) It is easy to show that for

$$M = \|m_{ij}\|, \qquad \left\| \frac{\partial \operatorname{Det} M^{-1}}{\partial m_{ij}} \right\| = -(\operatorname{Det} M^{-1}) M^{-1}.$$

(4) For every ξ we have

$$\begin{aligned} \max_{Z \in Q} V_\xi(z) &\geq \int_Q V_\xi(z)\xi(dz) \\ &= \int_Q \operatorname{tr}\left[M^{-1}(\xi) f(z) f^T(z)\right]\xi(dz) \\ &= \operatorname{tr} I_p = p. \end{aligned}$$

Therefore

$$\begin{aligned} \xi^* \text{ is } D\text{-optimal} &\Leftrightarrow 0 \leq \frac{\partial}{\partial \alpha} \Phi_D[(1-\alpha)M(\xi^*) + \alpha M(\xi)]\bigg|_{\alpha = 0^+} \quad \forall \xi, \quad \text{(by (1) and (2))} \\ &\Leftrightarrow 0 \leq \operatorname{tr}\left\{ [M(\xi) - M(\xi^*)] \left\| \frac{\partial \Phi_D(M)}{\partial m_{ij}} \right\|_{M = M(\xi^*)} \right\} \quad \forall \xi, \\ &\Leftrightarrow 0 \leq \operatorname{tr}\left\{ \left[f(z) f^T(z) - M(\xi^*) \right]\left(-M^{-1}(\xi^*)\right) \right\} \quad \forall_z, \quad \text{by (3)}, \\ &\Leftrightarrow V_{\xi^*}(z) \leq \operatorname{tr} I_p = p \ \forall_z, \\ &\Rightarrow \xi^* \text{ is } G\text{-optimal}, \quad \text{by (4)}. \end{aligned}$$

This therefore shows that D-optimality $\Rightarrow$ G-optimality. The converse implicated that G-optimality $\Rightarrow$ D-optimality is proved by using a result in game theory. The remaining parts of the theorem can be proved by developing the preceding arguments further.

From the above discussion we can also obtain some other useful results:

Suppose we find that $\operatorname{Det} M^{-1}(\xi')$ is small but know nothing about $\operatorname{Det} M^{-1}(\xi')/\min_\xi \operatorname{Det} M^{-1}(\xi)$. However, if it is known that $\max_z V_{\xi'}(z) \leq p(1+\varepsilon)$, then we can conclude that

$$\max_z V_{\xi'}(z) / \min_\xi \max_z V_\xi(z) \leq 1 + \varepsilon,$$

and therefore

$$\operatorname{Det} M^{-1}(\xi') / \min_\xi \operatorname{Det} M^{-1}(\xi) \leq 1 + \varepsilon.$$

Another useful fact is that since Φ_D is strictly convex all D-optimal designs ξ^* have the same values of M and V.

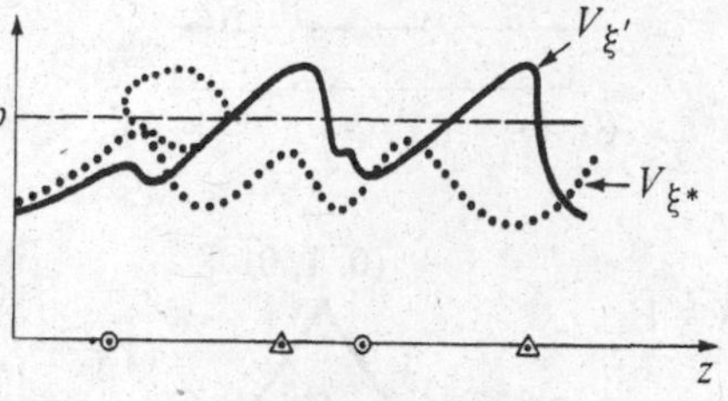

Figure 13.

Using an iterative method for solving two-person games, Wynn (1970, *A.M.S.*, 1655) showed how a suboptimal design ξ' can be improved by transferring the measure on those points of which $V_{\xi'}$ is small to those points at which $V_{\xi'}$ is large. This is illustrated in Figure 13 where we transfer measure from the ⊙ points to the ▲ points. Figure 14 shows the ξ^* in Example 1 of Lecture 5, and V_{ξ^*} is shown in Figure 13.

There are also "equivalence theorems" for other criteria. However, criteria that are equivalent to Φ_G may have much less statistical meaning than Φ_G itself (cf. Kiefer 1974, *AS*, 849).

As to E-optimal designs, the function $\Phi_E(M)$(= largest eigenvalue of M^{-1}) is not a differentiable function of the elements of M. Systematic methods for finding E-optimal designs have not yet been developed and are worthwhile research topics. Another research topic is about the use of various Φ to compare designs, as the experimenter is usually not convinced by results on the basis of only one such Φ.

Reasons why D-Optimality is often used.

(1) D-optimality is equivalent to G-optimality.
(2) The solution can be found by relatively direct methods without computing the inverse matrix (thus avoiding many numerical problems).
(3) It is invariant under linear transformations $\theta \to L\theta$ on the parameter space. Sometimes it is also invariant under linear transformations $z \to Lz$ on Q. For example, the interval $[0, B]$ can be mapped into the interval $[-1,1]$ by the linear transformation $z \to az + b$. Thus, D-optimal designs for $[0, B]$ can be transformed into D-optimal designs for $[-1,1]$.

We now demonstrate how the preceding results and some further considerations can greatly simplify the computations for D-optimal designs. In classical problems, the multivariate versions of Q are spheres, cubes, or simplices.

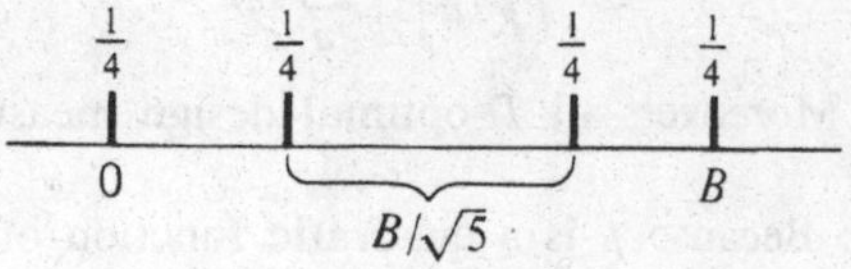

Figure 14.

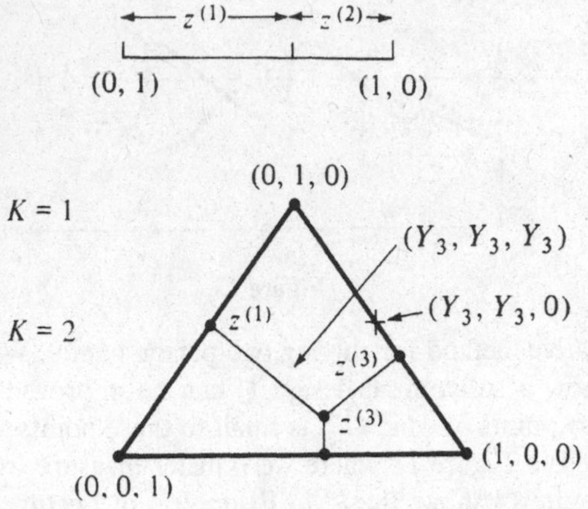

Figure 15.

Example 1. (Experimentation with mixtures; cf. Scheffe, 1958, *J.R.S.S.*, 344; Kiefer, 1961, *A.M.S.*, 298)

$$Q = K\text{-dimensional simplex}$$
$$= \left\{ Z = (z^{(1)}, \ldots, z^{(K+1)}) : z^{(j)} \geq 0\ \forall_j,\ \sum_i z^{(j)} = 1 \right\}.$$

Suppose that an alloy consists of $K+1$ ingredients and that $z^{(j)}$ is the proportion of the j-th ingredient. Assume that the "hardness" of the alloy can be well approximated by a quadratic function of Z, i.e., for the composition Z the expected hardness of the alloy in an experiment is

$$\sum_{j=1}^{K+1} \theta_j z^{(j)} + \sum_{1 \leq i < j \leq K+1} \theta_{ij} z^{(i)} z^{(j)},$$

(noting that this form includes also the constant term and the square terms since $\Sigma_j z^{(j)} = 1$). Therefore,

$$\theta = \{\theta_i, \theta_{ij}\}, \qquad p = (K+1)(K+2)/2,$$

and we want to find a D-optimal ξ.

(1) Symmetry: Suppose that we have the D optimal design in Figure 16 where the support is given by the $\cdot$ points. Because of symmetry in this problem and in Φ_D, ξ^* that is obtained from ξ by symmetry transformations (see Figure 17) is also optimal. Because of the convexity of Φ_D, averaging all such ξ_d^* over the group of symmetry transformations gives again an optimal

$$\bar{\xi} = \frac{1}{(K+1)!} \sum_d \xi_d^*,$$

with a symmetric $V_{\bar{\xi}}$. Moreover, all D-optimal design measures ξ have the same such V.

(2) Properties of V: Because f is a quadratic function of Z, V (variance) is a quadratic function. Moreover, M^{-1} is positive definite. It is therefore easy to prove

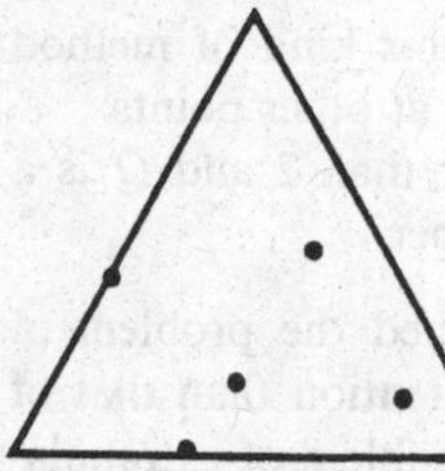

Figure 16.

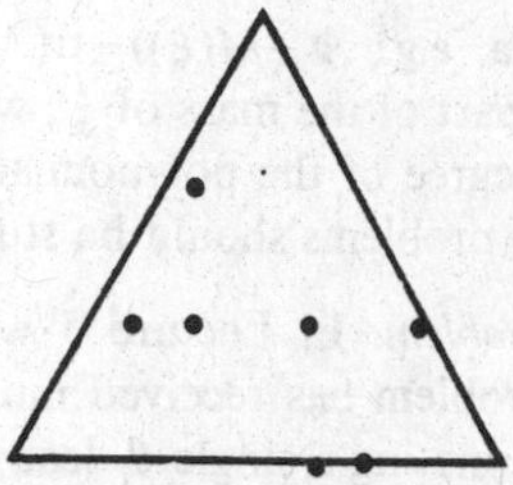

Figure 17.

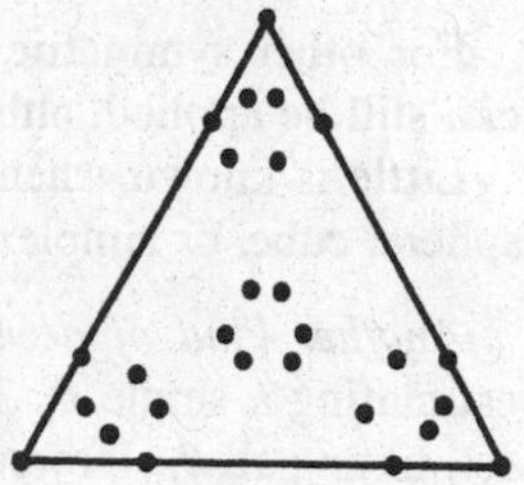

Figure 18.

the following property of V as a function on $\{\mathbf{z}: \sum_j z^{(j)} = 1\}$:

$$V(\mathbf{z}) \to \infty \quad \text{as } |\mathbf{z}| \to \infty.$$

(3) From (1) and (2) we can see that the maximum of V in Q can only be attained at the points $(1,0,0,\ldots,0)$, $(\frac{1}{2},\frac{1}{2},0,\ldots,0),\ldots,(1/k+1,1/k+1,\ldots,1/k+1)$, and points obtained by interchanging the coordinates thereof.

(4) From the equivalence theorem of this lecture, the support of any D-optimal ξ^* is a subset of the set of points (3). This therefore greatly simplifies the ξ to be considered. Two techniques can then be used to determine ξ^*.

(a) Consider the K values

$$q_j = \xi\left(\frac{1}{j},\ldots,\frac{1}{j},0,\ldots,0\right), \qquad 1 \le j \le K,$$

$(q_{K+1} = 1 - \sum_{j=1}^{K} q_j)$. Maximize Det $M(\xi)$ as a function of the q_j's.

(b) Try to guess a symmetric ξ^* that may be D-optimal. Then check whether $V_{\xi^*}(\mathbf{z}) \le p$ for all $\mathbf{z} \in Q$.

This is a very convenient method for the example under discussion. Since M is nonsingular, the support of ξ^* must have at least p points. Here are exactly $p = (K+1)(K+2)/2$ such points: $(1,0,\ldots,0)$, $(\frac{1}{2},\frac{1}{2},0,\ldots,0)$, and points obtained from them by interchanging the coordinates. Let ξ^* put mass $1/p$ at each of these points. It is not hard to compute $M^{-1}(\xi^*)$ and $V_{\xi^*}(Z)$. We then find that $V_{\xi^*} = p$ at each of the points in the support of ξ^*. In view of symmetry and (2), it remains only to check that $V_{\xi^*}(Z) \le p$ for the $K-1$ points $(1/j,\ldots,1/j,0,\ldots,0)$, $3 \le j \le K+1$, and this is not hard to do.

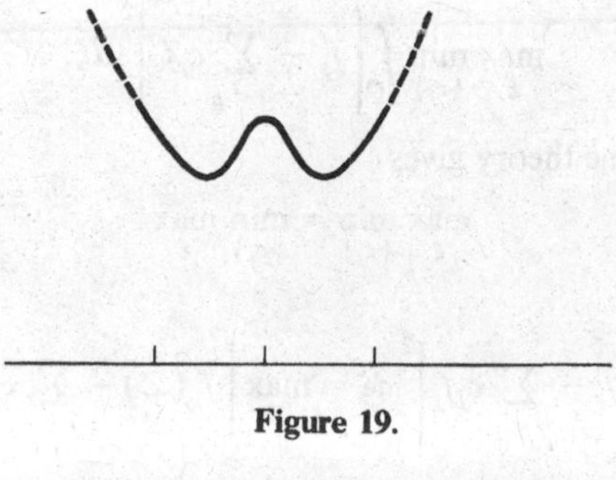

Figure 19.

For other symmetric criteria, e.g., $\Phi_A(M(\xi)) - \mathrm{tr}(M^{-1}(\xi))$, this kind of method can still be applied, although part of the mass of ξ^* will be put at other points.

Little is known when the degree of the polynomial is higher than 2 and Q is a sphere, cube, or simplex. Such problems should be studied further.

Another kind of decision problem. In Lecture 5 we mentioned the problem of estimating a single θ_i. This problem has received much less attention than that of estimating all θ_i. However, estimating a single θ_i leads to the problem of extrapolation which has received much attention. We therefore will consider it here.

Estimating θ_p (Kiefer and Wolfowitz, *A.M.S.* (1959), 271): We want to minimize $[M^{-1}(\xi)]_{p,p}$ (i.e., the $(p, p)^{\text{th}}$ element of the matrix $M^{-1}(\xi)$). In order to transform this problem into another form, consider the $p \times p$ matrix

$$C = \begin{bmatrix} 1 & 0 & 0 & \cdots & 0 & C_1 \\ & 1 & 0 & \cdots & 0 & C_2 \\ & & & \ddots & \vdots & \vdots \\ & & & & 1 & C_{p-1} \\ & & & & & 1 \end{bmatrix}$$

and note that $[M^{-1}]_{pp} = [CM^{-1}C^T]_{pp} = [\{(C^T)^{-1}MC^{-1}\}]_{pp}$ and

$$(C^T)^{-1} = \begin{bmatrix} 1 & & & & \\ 0 & 1 & & & \\ \vdots & \vdots & \ddots & & \\ 0 & 0 & \cdots & 1 & \\ -C_1 & -C_2 & \cdots & -C_{p-1} & 1 \end{bmatrix}$$

Let $H = (C^T)^{-1}MC^{-1}$. Then $[M^{-1}]_{pp} = 1/H_{pp}$ if we choose $\{c_j\}$ such that

$$H_{ip} = \int_Q \Big[f_p - \sum_{j<p} c_j f_j \Big] f_i \, d\xi = 0, \qquad 1 \le i \le p-1,$$

which is equivalent to choosing $\{c_j\}$ to minimize

$$\int_Q \Big[f_p - \sum_{j<p} c_j f_j \Big]^2 d\xi.$$

Therefore, in order to find ξ that minimizes $1/H_{pp}$, we need only solve

$$\max_{\xi} \min_{\{c_j\}} \int_Q \Big[f_p - \sum_{j<p} c_j f_j \Big]^2 d\xi. \tag{6.2}$$

A simple result of game theory gives

$$\max_{\xi} \min_{\{c_j\}} = \min_{\{c_j\}} \max_{\xi}.$$

Since

$$\max_{\xi} \int_Q \Big[f_p - \sum_{j<p} c_j f_j \Big]^2 d\xi = \max_{Z} \Big[f_p(Z) - \sum_{j<p} c_j f_j(Z) \Big]^2,$$

it then follows that

$$\sqrt{(6.2)} = \min_{\{c_j\}} \max_{Z} \left| f_p(Z) - \sum_{j<p} c_j f_j(Z) \right|.$$

The least quantity above is the "best Chebyshev L_∞-approximation" of f_p by $\Sigma_{j<p} c_j f_j$, and there is a large literature on this kind of approximation. If we find the best approximation $\Sigma_{j<p} c_j^* f_j$, we can use it to determine the subset Q_0 (of Q) at which $|f_p - \Sigma_{j<p} c_j^* f_j|$ attains its maximum. From the preceding discussion, it then follows that the ξ^* such that

$$\int_{Q_0} \left[f_p - \sum_{j<p} c_j^* f_j \right] f_i \, d\xi^* = 0, \qquad 1 \le i \le p-1, \tag{6.3}$$

gives the optimal design for estimating θ_p. Since Q_0 is usually a finite set, (6.3) usually reduces to a set of linear equations.

Example 2. Consider a polynomial model in one dimension with $Q = [-1,1]$ and $f_j(z) = z^{j-1}$. The difference $f_p - \Sigma_{j<p} c_j^* f_j$ between f_p and its best approximation $\Sigma_{j<p} c_j^* f_j$ is a Chebyshev polynomial of the first kind, and

$$Q_0 = \{\lambda_j : 0 \le j \le p-1\},$$

where

$$\lambda_j = \cos\left(\frac{j\pi}{p-1} \right).$$

The $(p-1)$ equations in (6.3) together with the equation $\Sigma_{j=0}^{p-1} \xi^*(\lambda_j) = 1$ uniquely determine $\xi^*(\lambda_0), \ldots, \xi^*(\lambda_{p-1})$, and give

$$\xi^*(\lambda_0) = \xi^*(\lambda_{p-1}) = \frac{1}{2(p-1)},$$

$$\xi^*(\lambda_j) = \frac{1}{p-1}, \qquad 1 \le j \le p-2.$$

Extrapolation. Suppose that at $Z \in Q^* \supset Q$ the value of the regression function remains $\theta^T f(Z)$. While we only have observations at design levels in Q, we want to estimate $\theta^T f(Z_0)$, where $Z_0 \in Q^* - Q$. This is the "extrapolation" problem of estimating the response at Z_0. For many f, the problem of minimizing $\mathrm{Var}(\hat{t}^T f(Z_0))$ can be transformed to a Chebyshev approximation problem which yields the same Q_0 as before. Although the analogue of (6.3) may lead to different $\xi^*(\lambda_j)$, the most

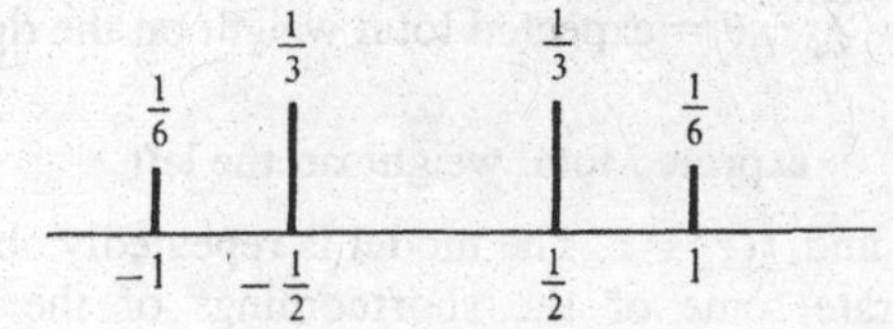

Figure 20. Optimal design ξ^* for cubic ($p = 4$).

difficult part of the computation lies in the determination of Q_0, which has been solved before.

A surprising fact in the problem of estimating $\theta^T f(z_0)$, where Q is one-dimensional and f is a polynomial, is that the support of the optimal design remains the same for all $|z_0|>1$. This was first noticed by Hoel and Levine (1964, *A.M.S.*, 1533). For sufficiently large z_0, the optimal design measure ξ also minimizes $\max_{-1\le z\le z_0} V_\xi(z)$.

There are, however, few satisfactory results to date if the dimension of Q exceeds 1.

Lecture 7. Combinatorial Designs

In statistical applications one often uses balanced incomplete block designs (B.I.B.D.) and encounters Latin squares and Hadamard matrices. The intuitive meaning of these designs is that M_Z is always "completely symmetric"; its diagonal elements are all equal and its off-diagonal elements are also all equal. This implies that

(1) a similar form holds for M_Z^- and therefore the least squares estimates of θ_i have the same variance for all i;
(2) it is easy to compute $\bar{t} = M_Z^- B_Z^T \mathbf{x}$.

The extensive use of such designs is partly due to the pervasive influence of R. A. Fisher.

Wald (1943, *A.M.S.*, 134) was the first to demonstrate certain optimum properties of Latin squares. However, under certain experimental conditions, designs with the largest possible symmetry may not be optimal (cf. Kiefer, *Int. Congress Math.* (1970), Vol. 3; Ruiz and Seiden, *AS* (1973)).

The following two examples illustrate the ideas underlying these designs. The different combinatorial designs therein are all optimal.

Example 1 (Weighing designs). Here $Q=\{-1,0,1\}^p$ and there are p objects to be weighed in n weighing experiments. After the i-th weighing, we have $z_i = (z_{i1},\ldots,z_{ip})^T$, where

$$\begin{aligned} z_{ij} &= 1 \quad \text{if object } j \text{ is placed on the right side,}\\ &= -1 \quad \text{if object } j \text{ is placed on the left side,}\\ &= 0 \quad \text{if object } j \text{ is not involved in the weighing,} \end{aligned}$$

(cf. Figure 21). The problem is to estimate the weight θ_j of object j, $j=1,\ldots,p$. Let $\theta^T=(\theta_1,\ldots,\theta_p)$. The reading x_i of the i-th weighing has the expected value

$$\begin{aligned} z_i^T\theta = \sum_j z_{ij}\theta_j &= \text{expected total weight on the right}\\ &\quad - \text{expected total weight on the left.} \end{aligned}$$

This is a linear model and $f(z)=z$. The model is repeatedly observable. It provides an example to illustrate some of the shortcomings of the approximate theory discussed in Lecture 6. Like the simplex example in Lecture 6, we can multiply any coordinate of z by -1 and obtain by symmetry that the optimal design measure ξ^*

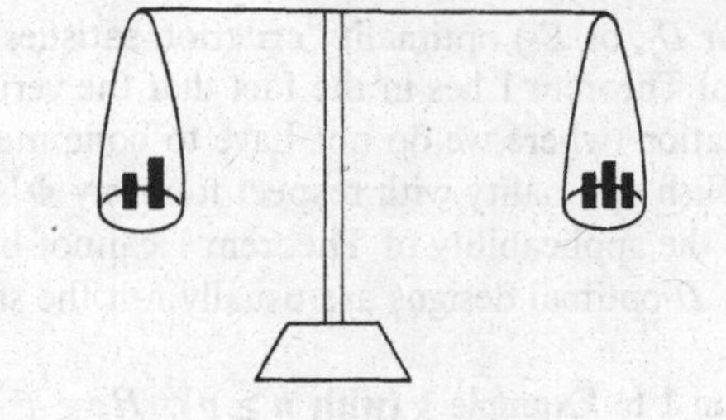

Figure 21.

satisfies

$$\xi^*(z) = 1/2^p$$

for every z without any 0 component. When $p = 10$, since $2^{10} = 1024$, we need $n \approx 1000$ weighings before the approximate theory can become accurate. This suggests the need of using the "exact theory" instead of the approximate theory.

The following elementary result is sometimes useful in proving optimality with respect to all commonly used symmetric Φ.

Theorem 1. *Given a linear model (not necessarily that of Example 1), suppose that Z^* is a design satisfying*

(1) $M_{Z^*} = cI_p$ *for some* $c > 0$,
(2) $\operatorname{tr} M_{Z^*} = \max_Z \operatorname{tr} M_Z$.

Then Z^* is Φ-optimal for every Φ satisfying the following conditions:

(a) *Φ is convex;*
(b) *Φ is invariant under orthogonal transformations, i.e., $\Phi(LML^T) = \Phi(M)$ for every orthogonal matrix L;*
(c) $\Phi(cI_p) \searrow$ *as* $c \nearrow$.

Proof. Let μ denote the invariant probability measure on the orthogonal group. Let Z^* satisfy (1) and (2), and let Z' be another design. Let

$$pc'(= \operatorname{tr} c'I_p) = \operatorname{tr} M_{Z'}.$$

Then

$$\int (LM_{Z'}L^T)\mu(dL) = c'I_p. \tag{7.1}$$

Therefore

$$\begin{aligned}
\Phi(M_{Z'}) &= \int \Phi(LM_{Z'}L^T)\mu(dL), && \text{by (b)},\\
&\geq \Phi\left(\int (LM_{Z'}L^T)\mu(dL)\right), && \text{by (a)},\\
&= \Phi(c'I_p), && \text{by (7.1)},\\
&\geq \Phi(cI_p), && \text{by (1), (2) and (c)},\\
&= \Phi(M_{Z^*}), && \text{by (1)}.
\end{aligned}$$

Notes. (i) The A- (or D-, or E-) optimality criterion satisfies (a), (b) and (c).

(ii) The usefulness of Theorem 1 lies in the fact that the verification of (1) and (2) involves little computation (where we do not have to compute quantities like M_Z^{-1}) and that we can establish optimality with respect to many Φ's.

On the other hand, the applicability of Theorem 1 cannot be very extensive since we know that A- and D-optimal designs are usually not the same.

Application of Theorem 1 to Example 1 (with $n \geq p$). *Here*

$$Ex = \begin{pmatrix} z_1' \\ z_2' \\ \vdots \\ z_n' \end{pmatrix} \theta, \qquad B_Z = \begin{pmatrix} z_1' \\ \vdots \\ z_n' \end{pmatrix},$$

$\operatorname{tr} M_Z (= \Sigma_{i,j} z_{ij}^2)$ *attains its maximum value* $np \Leftrightarrow |z_{ij}| = 1 \ \forall_{i,j}$.

For every Z satisfying the above property, (2) clearly holds. When $p = 1$, (1) also holds. When $p = 2$, (1) holds $\Leftrightarrow n$ is even. The interesting case is $p \geq 3$. Here we obtain from (1) that $n \equiv 0 \pmod 4$. For all the (n, p) used "in practice," Z^* can be obtained by choosing any p columns of the $n \times n$ Hadamard matrix H_n, which has all entries equal to ± 1 and which satisfies $H_n H_n^T = nI_n$. In this way we obtain the design Z^* that is optimal in the sense of Theorem 1. Such designs are clearly very appealing. Note that if we were interested only in the weight of object 1 then we would have weighed it n times on the right side of the scale and would have omitted all the other objects (i.e., $z_{ij} = 0$ if $j > 1$). Thus, the estimate of θ_1 would have variance σ^2/n where $\operatorname{Var} x_i = \sigma^2$. On the other hand, we would have no information to estimate the other θ_j. However, the Z^* obtained from the Hadamard matrix H_n provides an estimate for every θ_j with the same variance σ^2/n.

There is only a small literature on the case $n \not\equiv 0 \pmod 4$. Ehlich (*Math. Z.* (1964), Vol. 83, p. 183) did pioneering work on this subject. The only problems treated in this case are almost exclusively on D-optimality, and there is still a long way to go to completely solve these problems.

Example 2 (Block designs, cf. Lecture 5). Suppose there are v types and b blocks, each block having size k, so that $n = bk$. If the i-th observation is of type h and in block j, then $Ex_i = \alpha_h + \beta_j$. Thus, $\theta^T = (\alpha_1, \ldots, \alpha_v, \beta_1, \ldots, \beta_k)$, $p = v + b$, and a design Z represents the allocation of types to the blocks.

For example, let $v = 7$, $k = 4$. Let $x_1, \ldots, x_4$ be the observations in the first block, $x_5, \ldots, x_8$ be the observations in the second block, etc. Suppose that the types in the first block are

$$\boxed{2,\ 1,\ 1,\ 6}\,.$$

Then the first row of B_Z is

$$\big(\underbrace{0,1,0,\ldots,0}_{v},\underbrace{1,0,\ldots,0}_{b}\big) = u_1,$$

since $u_1 \theta = \alpha_2 + \beta_1$ is just what we want.

The approach here differs from that used for the previous examples in the following two respects: (A) Here we are interested only in the α_h and not in the β_j. (B) We cannot estimate these α_h's since they are unidentifiable in this model. From our observations the most we can expect to estimate is $\gamma_{hj} = \alpha_h + \beta_j$, but the $\gamma_{hj} = \alpha_h + \beta_j$ cannot determine the α_h and the β_j, since $\gamma_{hj} = (\alpha_h + c) + (\beta_j - c)$ for every c. However, for suitably chosen Z, we can indeed estimate all the differences $\alpha_{h_1} - \alpha_{h_2}$.

To deal with (A) and (B) in general models (not necessarily block designs), let $\theta^T = (\alpha^T, \beta^T)$ and partition $\bar{t}$ and M_Z accordingly as

$$\bar{t}^T = (\bar{t}_1^T, \bar{t}_2^T), \qquad M_Z = \begin{pmatrix} M_{Z_1} & M_{Z_2} \\ M_{Z_2}^T & M_{Z_3} \end{pmatrix}.$$

Let $C_Z = M_{Z_1} - M_{Z_2} M_{Z_3}^- M_{Z_2}^T$. This is the information matrix for α. If $c^T\alpha$ is estimable, then the variance of $c^T\bar{t}_1$ is $c^T C_Z^- c$. In many designs (e.g., Example 2), since each row sum of C_Z is 0, we have

$$\text{every } \alpha_{h_1} - \alpha_{h_2} \text{ is estimable} \Leftrightarrow \operatorname{rank}(C_Z) = v - 1.$$

The analogue of Theorem 1 is

Theorem 2. *Theorem* 1 *still holds if we replace* M_Z *by* C_Z *and* (1) *by*

(1′) $\qquad C_{Z^*}$ is completely symmetric.

Here we regard Φ *as a function of* C_Z *(e.g.,* Φ_A *is a constant multiple of* Σ_{h_1,h_2}Var *(least squares estimate of* $\alpha_{h_1} - \alpha_{h_2}$)).

Application of Theorem 2 to Example 2. *If* $v > b$, *then* (1′) *and* (2) *are satisfied when* Z^* *is a B.I.B.D. For example, for* $v = 7 = b$ *and* $k = 4$,

$$Z^* = \begin{array}{|c|c|c|c|c|c|c|} 1 & 2 & 3 & 4 & 5 & 6 & 7 \\ 2 & 3 & 4 & 5 & 6 & 7 & 1 \\ 3 & 4 & 5 & 6 & 7 & 1 & 2 \\ 5 & 6 & 7 & 1 & 2 & 3 & 4 \end{array}.$$

This is obtained by applying cycles (mod 7) *to the first block.*

In problems where B.I.B. designs do not exist (for the given v, b, k), we encounter a much more difficult situation. Consider for example the case $n \not\equiv 0$ (mod 4) in Example 1. C. S. Cheng (1978, *A.M.S.*, 1239) has given a thorough study of such situations. However, there are still many unsolved problems.

Suppose that in each block neighbors are positively correlated and have a common correlation coefficient. Then in a good B.I.B.D. every pair of different types should be adjacent the same number of times. The Z^* above, however, does not have this property, since 1,2 are adjacent twice but 1,4 are never adjacent. If we modify the first block as | 3 1 2 5 |, six adjacent pairs are repeated symmetrically and we can thereby construct a desired design from this first block. Under what conditions can this be done in general? We do not yet know the answer.

The "most symmetric" combinatorial design need not always be optimal. For example, consider Latin squares in a two-way classification. For a two-way table with R rows and C columns, suppose that there are ν types, so that the expected value of an observation of type i in row r and column c is $\alpha_i+\beta_r+\gamma_c$. The problem is to estimate pairwise differences of the α_i's. *A generalized Youden design* (GYD) is a balanced block design similar to a B.I.B.D. but allowing $k>\nu$ (cf. Kiefer, 1958).

Example. $\nu=4,\ R=C=6$. Let

$Z^* =$

1	4	4	2	2	3
2	1	2	4	3	3
3	3	1	4	2	4
1	4	3	1	4	2
4	1	2	3	1	2
3	2	1	3	4	1

.

If we change the lower left-hand corner as

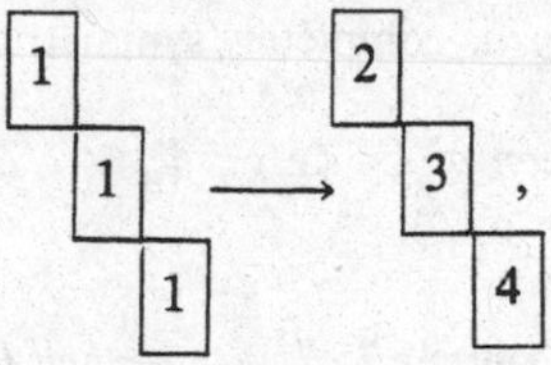

,

the design becomes better than Z^* under the D-optimality criteria. Although Z^* has a completely symmetric 4×4 C_{Z^*}, $\operatorname{tr} C_{Z^*} < \max_Z \operatorname{tr} C_Z$.

Theorem. GYD *is A-optimal and E-optimal. When* $\nu\neq 4$, *it is also D-optimal.*

The proof requires heavier machinery than Theorem 2.

Theorem 3. *Let g be a nonincreasing convex function and let* C_{Z^*} *be completely symmetric. Suppose that* Z^* *minimizes*

$$\sum_1^{\nu} g\left(\frac{\nu}{\nu-1}[C_Z]_{ii}\right).$$

Then Z^* *also minimizes* $\Sigma g(\lambda_j(Z))$, *where the* $\lambda_j(Z)$ *are the nonzero eigenvalues of* C_Z.

Lecture 8. Nonlinear Models, Sequential Designs, and Robust Designs

(A) Nonlinear Models

First recall that for a linear, repeatedly observable model with $p=1$ (single parameter) such that $E_{\theta,\sigma}x_i=\theta f(z_i)$, $z_i\in Q$, we have

$$M_Z=\sum_{i=1}^{n}f^2(z_i),$$

so the optimal design which minimizes Var(E) should take all observations at the point that maximizes $f(z)$. We now consider a simple illustrative example for nonlinear models.

Example 1. Suppose that the observations x_i are independent random variables with a common variance σ^2 and

$$E_{\theta,\sigma}x_i=e^{-\theta z_i},\qquad z_i\in Q=[0,B],$$

and that the design levels are repeatable. If we know that the time value of θ lies in a small neighborhood of θ_0, then we can write

$$e^{-\theta z}\doteq e^{-\theta_0 z}-ze^{-\theta_0 z}(\theta-\theta_0),$$

and therefore

$$E_{\theta,\sigma}\left[x_i-(z_i\theta_0+1)e^{-\theta_0 z_i}\right]\doteq z_i e^{-\theta_0 z_i}\theta.$$

Treating $x_i-(z_i\theta_0+1)e^{-\theta_0 z_i}$ as the observations in the linear model $\theta f(z)$ with $f(z)=ze^{-\theta_0 z}$, we therefore choose all design levels at the value of z that maximizes $ze^{-\theta_0 z}$ in Q, viz.,

$$z_i=\begin{cases}1/\theta_0 & \text{when } \theta_0\geq 1/B,\\ B & \text{when } \theta_0<1/B.\end{cases}$$

Note that this "optimal" design (assuming the linear approximation) depends on the true value of θ, unlike the case of linear models. We should also note the following point. Suppose that n is large and $z_i=z$ for all i. Since we use $\bar{x}_n=1/n\Sigma_{i=1}^n x_i$ to estimate $e^{-\theta z}$, we may as well use $t_n=-1/z\log\bar{x}_n$ to estimate θ. By the central limit theorem, $\sqrt{n}(t_n-\theta)\to N(0,\sigma^2/z^2e^{2z\theta})$ in distribution as $n\to\infty$. For fixed values of z, the functions $z^{-2}e^{2z\theta}$ (as functions of θ) that appear in the variances of the asymptotic distributions are shown in Figure 22. If $B\geq 2$, then for $\theta=\frac{1}{2}$ (or 1, 2), the lowest curve in Figure 22 corresponds to $z=2$ (or 1, $\frac{1}{2}$ respectively). Unlike linear models, there does not exist a "uniformly best" curve. The design which takes all z_i equal to $\frac{1}{2}$ is called "locally asymptotically optimal" at $\theta=2$.

Without knowing in advance that θ is near θ_0, we can use a minimax or Bayes approach to select the design.

Example 1 (continued). Suppose that the observations are available one at a time. Then the choice of z_i can depend on (z_j,x_j), $1\leq j\leq i-1$, and we can choose z_i adaptively and sequentially by one of the following methods.

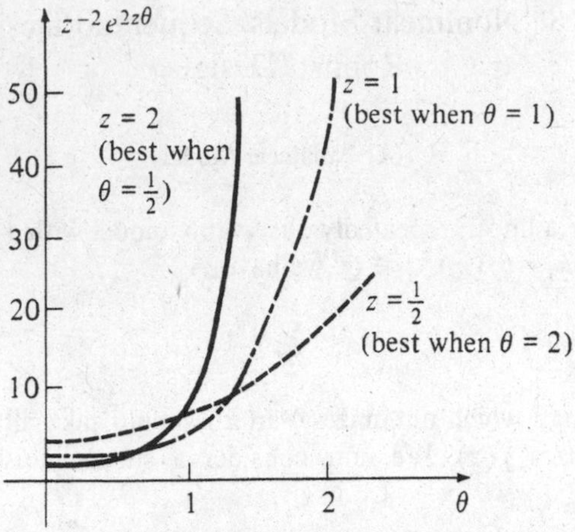

Figure 22.

(1) *Two-stage procedure:* Take n_0 observations during the first stage, where the design levels z_i should be widely dispersed. These sample data provide an initial estimate of θ. We then choose the remaining $n - n_0$ design points at the level that is optimum for the estimated value of θ.
(2) *Fully sequential procedure:* Each level z_i is chosen as the optimum level that corresponds to the estimated value of θ based on all the previous observations.

Chernoff (1959, *A.M.S.*, 755), Kiefer and Sacks (1963, *A.M.S.*, 705), and Atkinson and Federov (1975, *Biometrika*, 57) considered several sequential designs in detail. When θ is known to lie in a given bounded set, it is not hard to construct sequential designs that are asymptotically optimal, i.e.,

$$\sqrt{n}\,(t_n - \theta) \to N\left(0, \sigma^2 \min_{z \in Q} z^{-2}e^{2z\theta}\right)$$

as $n \to \infty$ (for fully sequential designs (2)), or as $n \to \infty$ and $n_0 \to \infty$ such that $n_0/n \to 0$ (for two-stage designs (1)). However, there are no satisfactory results for small n, and much more research is needed in this area.

(B) Robustness

Consider robustness in the following four kinds of situations:

(1) heavy tails and other difficulties in the distribution of $x_i - Ex_i$;
(2) discrepancy between $f(x)$ and the assumed model;
(3) errors in the design levels z_i;
(4) missing observations.

Although these issues should be considered together, such a treatment would be very difficult. Therefore we discuss them separately.

(1) has been partly discussed in Lecture 4. In particular, Huber (1973, *AS*), Bickel (1973 *AS*, 1975 *J.A.S.A.*) considered robust regression methods. For certain models, robust estimators t do not depend on the design Z. A more useful problem is to consider models in which robust estimators t_Z are associated with designs Z and to optimize both t_Z and Z.

In discussing (3), let us assume that although z_i is the design level that we want to choose, what is actually chosen is $z_i + e_i$, where e_i represents error such that $Ee_i = 0$. How should we choose Z and modify the least squares estimator $\bar{t}$ in this case? Some work has been done, but the results are not very satisfactory. This is an important and worthwhile research direction (cf. Box and Draper, 1975, *Biometrika*, 62).

In considering (4), suppose that we may have at most m missing observations and we want to avoid problems when that occurs. One approach is to look for designs Z such that M_Z still preserves its rank if we get m fewer observations, where m can be very large. Some theoretical results concerning B.I.B.D. and several illustrative examples have been given by Andrews and Herzberg (1976, *J.R.S.S.*, 284). Another approach is to consider $\tilde{\Phi}(Z) = \max_{M(Z)} \Phi(M(Z))$, where $M(Z)$ is a matrix that is obtained by deleting m rows from B_Z (the model being $E_{\theta,\sigma}x = B_z\theta$). We then try to find Z that minimizes $\tilde{\Phi}(Z)$. Little has been done in this area, and the work also only discusses B.I.B. designs in which the missing observations all come from one single type. This is an important area for future research.

We now discuss (2). Suppose that the assumed model is $E_{\theta,\sigma}x_i = f^T(z_i)\theta$ while the true model is $E_{\theta,\sigma}x_i = f^T(z_i)\theta + g(z_i)$, g being a "contamination" function. Let G denote the set of all possible contamination functions. Then, as in Lecture 4, to be able to resist contamination we need first to understand the properties of $\bar{t}$ under all $g \in G$.

Box and Draper (1959 *J.A.S.A.*, 622) studied the following situation. Let h be a given s-dimensional vector function defined on Q, and let $G = \{g\colon g = \rho^T h,\ p \in R^s\}$. For example, consider a quadratic function on an interval Q (i.e., $f^T(z) = (1, z, z^2)$). If we are worried that the true function may in fact also involve third and fourth powers, then we take $s = 2$ and $h^T(z) = (z^3, z^4)$. Let $C = M_Z^{-1}B_Z^T$, $E = Cx$. Define the $n \times S$ matrix

$$A_Z = \begin{pmatrix} h^T(z_1) \\ h^T(z_2) \\ \vdots \\ h^T(z_n) \end{pmatrix}.$$

Therefore, when there is contamination $\rho^T h$, we have

$$E\bar{t} = CEx = C[B_Z \vdots A_Z]\begin{pmatrix} \theta \\ \rho \end{pmatrix}.$$

Let $D^T = C[B_Z \vdots A_Z]$. The mean squared error of the fitted response at z is

$$E\left[\bar{t}^T f(z) - (\theta^T, \rho^T)\begin{pmatrix} f(z) \\ h(z) \end{pmatrix}\right]^2 \tag{8.1}$$

$$= \sigma^2 f^T(z) M_Z^{-1} f(z) + (\theta^T, \rho^T)\left[Df(z) - \begin{pmatrix} f(z) \\ h(z) \end{pmatrix}\right]\left[Df(z) - \begin{pmatrix} f(z) \\ h(z) \end{pmatrix}\right]^T \begin{pmatrix} \theta \\ \rho \end{pmatrix}$$

$$= \text{variance} + \text{square of bias.}$$

Imitating Φ_G, we want to choose Z to minimize $\max_{z \in Q}(8.1)$. This is a very difficult problem. Box and Draper therefore replaced $\max_{z \in Q}(8.1)$ by the integral $\int_Q (8.1)\nu(dz)$, where ν is a prescribed probability measure. Let

$$\int_Q \begin{pmatrix} f(z) \\ h(z) \end{pmatrix}\begin{pmatrix} f(z) \\ h(z) \end{pmatrix}^T \nu(dz) = \begin{pmatrix} \Gamma_{11} & \Gamma_{12} \\ \Gamma_{21} & \Gamma_{22} \end{pmatrix}.$$

Then

$$\sigma^{-2}\int_Q (8.1)\nu(dz) = \operatorname{tr}(\Gamma_{11} M_Z^{-1})$$

$$+ \sigma^{-2}\begin{pmatrix} \theta \\ \rho \end{pmatrix}^T\left[D\Gamma_{11}D^T - \begin{pmatrix} \Gamma_{11} \\ \Gamma_{21} \end{pmatrix} D^T - D\begin{pmatrix} \Gamma_{11} \\ \Gamma_{21} \end{pmatrix}^T + \Gamma\right]\begin{pmatrix} \theta \\ \rho \end{pmatrix}$$

$$\overset{\text{Def}}{=} V + B.$$

We cannot minimize $V + B$ as we do not know $\sigma^{-1}\binom{\theta}{\rho}$. Box and Draper therefore tried to minimize B instead and claimed that the minimizing z should also have high efficiency with respect to $V + B$, provided that $\sigma^{-1}\binom{\theta}{\rho}$ lies within some bounded (but wide) interval.

Note that $B = \sigma^2\binom{\theta}{\rho}^T \Lambda \binom{\theta}{\rho}$, where

$$\Lambda = D\Gamma_{11}D^T - \begin{pmatrix} \Gamma_{11} \\ \Gamma_{21} \end{pmatrix} D^T - D\begin{pmatrix} \Gamma_{11} \\ \Gamma_{21} \end{pmatrix}^T + \Gamma$$

$$= \left[D - \begin{pmatrix} I \\ \Gamma_{21}\Gamma_{11}^{-1} \end{pmatrix}\right]\Gamma_{11}\left[D - \begin{pmatrix} I \\ \Gamma_{21}\Gamma_{11}^{-1} \end{pmatrix}\right]^T + \begin{bmatrix} 0 & 0 \\ 0 & \Gamma_{22} - \Gamma_{21}\Gamma_{11}^{-1}\Gamma_{12} \end{bmatrix}.$$

The last summand above does not involve Z. If we let

$$C[B_Z \vdots A_Z](= D^T) = [I \vdots \Gamma_{11}^{-1}\Gamma_{12}], \tag{8.2}$$

then we have

$$D - \begin{pmatrix} I \\ \Gamma_{21}\Gamma_{11}^{-1} \end{pmatrix} = 0,$$

making Λ as small as possible. Since $C = M_Z^{-1}B_Z^T$, (8.2) is equivalent to

$$[I \vdots M_Z^{-1}B_Z^T A_Z][I \vdots \Gamma_{11}^{-1}\Gamma_{12}],$$

which is satisfied if

$$M_Z = k\Gamma_{11} \quad \text{and } B_Z^T A_Z = k\Gamma_{12} \quad \text{for some } k > 0.$$

Box and Draper derived these results in the context of polynomials with $f^T(z) = (1, z, \ldots, z^{p-1})$, $h^T(z) - (z^p, \ldots, z^{p+s-1})$ and concluded that the moments of ξ_Z are equal to those of ν up to order $2p + s - 2$. These moments are in fact the elements of $[n^{-1}M_Z \vdots n^{-1}B_Z^T A_Z]$ and of $[\Gamma_{11} \vdots \Gamma_{12}]$. In particular, if ν is Lebesgue measure and s is large, then we can take Z to be n equally spaced points that partition Q.

Note that V decreases with n while B does not. Moreover, the Box–Draper design has a relatively large $\max_{z \in Q}(8.1)$ when compared to the D-optimal design under the assumed model (cf. Galil and Kiefer, *Technometrics*, 1977). The Box–Draper method has been extended by Karson, Manson, and Hader (*Technometrics*, 1969) to other linear estimators Cx. The deficiency just described, however, still persists in the extension.

Other models (involving different classes G) have been considered by Huber (1975, *Statistical Design and Linear Models*, 287), Marcus and Sacks (1976, *Statistical Decision Theory II*, 245), and Li and Notz (unpublished). Although there have been several papers about this subject, much remains to be done. A related problem is which order we should use in (polynomial) curve fitting. For given designs Z there is a large literature on this subject (subset selection in regression). C. J. Stone's forthcoming paper provides a decision-theoretic treatment of subset selection.

CONDITIONAL INFERENCE

Conditional probability* appears in a number of roles in statistical inference*. As a useful tool of probability theory, it is in particular a device used in computing distributions of many statistics used in inference. This article is not concerned with such purely probabilistic calculations but with the way in which conditioning arises in the construction of statistical methods and the assessment of their properties. Throughout this article, essentially all measurability considerations are ignored for the sake of emphasizing important concepts and for brevity. Random variables (rv's) may be thought of as having discrete case or absolutely continuous (with respect to Lebesgue measure*) case densities. Appropriate references may be consulted for general considerations.

SUFFICIENCY*

Suppose that $\{P_\theta, \theta \in \Omega\}$ is the family of possible probability laws on the sample space S (with an associated σ-field $\mathcal{C}$). Think of the rv X under observation as the identity function on S, and $\{f_\theta, \theta \in \Omega\}$ as the corresponding family of densities of X. The usual definition of a *sufficient statistic** T on S is in terms of conditional probabilities: for all A in $\mathcal{C}$, $P_\theta\{X \in A \mid T(X)\}$ is independent of θ. Often it is convenient to think of a sufficient partition of S, each of whose elements is a set where T is constant. The two concepts are equivalent in most common settings.

Using the definition of conditional probability and writing in the discrete case for simplicity, we have, if T is sufficient,

$$\begin{aligned} f_\theta(x) &= P_\theta\{X = x\} \\ &= P_\theta\{X = x, T(X) = T(x)\} \\ &= P_\theta\{X = x \mid T(X) = T(x)\} \\ &\quad \times P_\theta\{T(X) = T(x)\} \\ &= h(x) g(\theta, T(x)), \end{aligned} \qquad (1)$$

where $h(x) = P\{X = x \mid T(X) = T(x)\}$, independent of θ. In usual cases this development can be reversed and one has the *Fisher–Neyman decomposition* $f_\theta(x) = h(x) g(\theta, T(x))$ as necessary and sufficient for sufficiency of T. In graphic terms, Fisher's assertion that T contains all the information in X about θ is evidenced by the fact that, given that $T(X) = t$, one can conduct an experiment with outcomes in S and not depending on θ, according to the law $h(x \mid t) = P\{X' = x \mid T(x) = t\}$, thereby producing a rv X' with the same unconditional law $\sum_t h(x \mid t) P_\theta\{T = t\}$ as X, for all θ; we can recover the whole sample X, probabilistically, by this randomization that yields X' from T.

This excursion into sufficiency is made both because of its relation to ancillarity discussed below (*see also* ANCILLARY STATISTICS) and also because some common developments of statistical decision theory that use the development amount formally to a conditional inference, although the usual emphasis about them is not in such terms. One such development rephrases the meaning of sufficiency by saying that, for any statistical procedure δ, there is a procedure δ^* depending only on the sufficient statistic T that has the same operating characteristic. Indeed, if δ denotes a randomized decision* function, with $\delta(\Delta \mid x)$ the probability assigned to the set Δ of decisions (a subset of the set D of all possible decisions) when $X = x$, then

$$\delta^*(\Delta \mid t) = E\{\delta(\Delta \mid X) \mid T(X) = t\} \qquad (2)$$

defines a procedure on $T(S)$ with the desired property; δ^* and δ have the same risk function for every loss function L on $\Omega \times D$ for which the risk of δ is defined. The procedure δ^* is defined in terms of a conditioning, although the emphasis is on its unconditional properties. In particular, if $L(\theta, d)$ is convex in d on D, now assumed a convex Euclidean set, then the nonrandomized procedure d^{**} on $T(S)$ [for which $\delta^{**}(d^{**}(t) \mid t) = 1$] defined by

$$d^{**}(t) = E\left\{\int_d r\delta(dr \mid X) \mid T(X) = t\right\} \qquad (3)$$

has risk at least as small as δ: d^{**} is the *conditional expected decision* of δ, given that $T(X) = t$, and the stated improvement from δ to d^{**} is the Rao–Blackwell theorem. Thus, in unconditional decision theory, use is made of procedures defined conditionally; the emphasis in *conditional inference*, though, is usually on conditional rather than unconditional risk.

Many treatments of conditional inference use extensions of the sufficiency concept, often to settings where nuisance parameters are present. For example, if $\theta = (\phi, \tau)$ and the desired inference concerns ϕ, the statistic T is *partially sufficient* for ϕ if the law of T depends only on ϕ and if, for each τ_0, T is sufficient for ϕ in the reduced model $\Omega = \{(\phi, \tau_0)\}$. This and related concepts are discussed by Basu [5]. This topic and many other matters such as conditional asymptotics for maximum likelihood (ML) estimators* are treated by Anderson [1]; a detailed study for exponential families* is given in Barndorff-Nielsen [3]. Hájek [17] discusses some of these concepts in general terms.

ANCILLARY AND OTHER CONDITIONING STATISTICS

Fisher [14, 15] in his emphasis on ML, defined an ancillary statistic U as one that (a) has a law independent of θ and (b) together with an ML estimator $\hat{d}$ forms a sufficient statistic. Currently in the literature, and herein, we take (a) without (b) as the definition. However, whether or not we are concerned with ML estimation, Fisher's rationale for considering ancillarity is useful: U by itself contains no information about θ, and Fisher would not modify $\hat{d}$ in terms of U; however, the value of U tells us something about the precision of $\hat{d}$, e.g., in that $\text{var}_\theta(\hat{d} \mid U = u)$ might depend on u. If we flip a fair coin to decide whether to take $n = 10$ or $n = 100$ independent, identically distributed (i.i.d.) observations ($X = (X_1, X_2, \ldots, X_n)$ above), normally distributed on R^1 with mean θ and variance 1, and denote the sample mean by $\bar{X}_n$, then $\hat{d} = \bar{X}_n$ is ML but not sufficient, $U = n$ is ancillary, and $(\hat{d}, U)$ is minimal sufficient. The unconditional variance of $\hat{d}$ is $\frac{11}{200}$. Fisher pointed out that, knowing that the experiment with 10 observations was conducted, one would use the conditional variance $\frac{1}{10}$ as a more meaningful assessment of precision of $\hat{d}$ than $\frac{11}{200}$, and would act similarly if $n = 100$.

Much of the argumentative literature attacking or defending unconditional Neyman–Wald assessment of a procedure's behavior in terms of examples such as this last one is perhaps due to an unclear statement of the aim of the analysis of procedures. If, before an experiment, procedures are compared in terms of some measure of their performance, that comparison must be unconditional, since there is nothing upon which to condition; even procedures whose usefulness is judged in terms of some conditional property once X is observed can only be compared before the experiment in an unconditional expectation of this conditional property. At the same time, if that conditional property is of such importance, account of its value should be taken in the unconditional comparison. An example often cited in criticism of unconditional inference is that of Welch [27], the model being that $X_1, X_2, \ldots, X_n$ are i.i.d. with uniform law* on $[\theta - \frac{1}{2}, \theta + \frac{1}{2}]$. If $W_n = \min_i X_i$, $V_n = \max_i X_i$, $Z_n = (V_n + W_n)/2$, and $U_n = V_n - W_n$, a confidence interval on θ with various classical *unconditional* optimality properties is of the form

$$\left[\max(W_n + q, V_n) - \tfrac{1}{2}, \min(W_n, V_n - q) + \tfrac{1}{2}\right]$$

for an appropriate q designed to give the desired confidence coefficient γ. Pratt [23], in a criticism from a Bayesian perspective, points out various unappealing features of this procedure; e.g., it *must* contain θ if $U_n > q$, and yet the confidence coefficient* is only γ. One may indeed find it more satisfactory to give an interval and confidence assessment *conditional* on U_n, as Welch suggests. The classical interval is what one would use if its optimum properties were criteria of chief concern, but many practitioners will not find those unconditional properties as important as conditional assessment of precision based on the value of U_n.

The last example illustrates an intuitive idea about the usefulness of conditioning. If U_n is near 1, X has been "lucky" and θ can be estimated very accurately, whereas the opposite is true if U_n is near 0. A conditional assessment is an expression of how lucky, by chance, X was in the sense of accuracy of the inference; unconditional risk or confidence averages over all possible values of X.

Many other examples of ancillarity, exhibiting various phenomena associated with the concept, occur in the literature. A famous example is that of Fisher [16] in which $X = ((Y_1, Z_1), (Y_2, Z_2), \ldots, (Y_n, Z_n))$, the vectors (Y_i, Z_i) being i.i.d. with common Lebesgue density $e^{-\theta y - \theta^{-1} z}$ for $y, z > 0$, with $\Omega = \{\theta : \theta > 0\}$. In this case

$$\hat{d} = \left(\sum_i Z_i \Big/ \sum_i Y_i\right)^{1/2}, \quad U = \left[\left(\sum_i Z_i\right)\left(\sum_i Y_i\right)\right]^{1/2},$$

and the conditional variance of $\hat{d}$ given that $U = u$ depends on u; $(\hat{d}, U)$ is minimal sufficient.

An instructive example is that of i.i.d. rv's X_i with Cauchy density $1/\{\pi[1 + (x - \theta)^2]\}$, for which the ML estimator (or other invariant estimator, such as the Pitman best invariant estimator* for quadratic loss if n is large enough) has conditional distribution depending on $U = (X_2 - X_1, X_3 - X_1, \ldots, X_n - X_1)$. For example, when $n = 2$, $\hat{d} = \bar{X}_2$ and the conditional density of $Z = \hat{d} - \theta$ given that $U = u$ is

$$2[1 + (u/2)^2]/\{\pi[1 + (u/2 + z)^2][1 + (u/2 - z)^2]\},$$

and a rough view of the spread of this density can be seen from its value $2/\{\pi[1 + (u/2)^2]\}$ at $z = 0$: large values of $|U|$ give less precise conditional accuracy of $\hat{d}$.

It is often convenient to replace (S, X) by $(T(S), T)$ for some minimal sufficient T, in these considerations. In the Cauchy example U becomes the set of *order statistic** differences. When $X_1, \ldots, X_n$ are $\mathfrak{N}(\theta, 1)$,

$$U = (X_1 - X_2, X_1 - X_3, \ldots, X_1 - X_n)$$

is ancillary on S, but in terms of $(T(S), T)$ with $T = \bar{X}_n$ we have no nontrivial ancillary: we cannot obtain a better conditional assessment of the accuracy of $\hat{d} = \bar{X}_n$ by conditioning on an ancillary.

In all of the foregoing examples, U is a *maximal ancillary*; no ancillary U^* induces a partition of S that is a refinement of the partition induced by U. Moreover, in these examples the maximal ancillary is unique. When that is the case, a further argument along Fisherian lines would tell us that, since a maximal ancillary gives the most detailed information regarding the (conditional) accuracy of $\hat{d}$, we should condition on such a maximal ancillary. Unfortunately, ancillary partitions do not parallel sufficient partitions in the existence of a unique finest such partition in all cases. Basu, in a number of publications (e.g., refs. 4 and 5), has considered several illustrations of this phenomenon. A simple one is a $X_1, \ldots, X_n$ i.i.d. 4-nomial with cell probabilities $(1 - \theta)/6$, $(1 + \theta)/6$, $(2 - \theta)/6$, and $(2 + \theta)/6$; the vector $T = (Y_1, Y_2, Y_3, Y_4)$ of the four observed cell frequencies is minimal sufficient, and each of $U_1 = Y_1 + Y_2$ and $U_2 = Y_1 + Y_4$ is maximal ancillary. If one adopts the *conditioning principle*, of conditioning on a maximal ancillary in assessing the accuracy of $\hat{d}$, the question arises whether to condition on U_1 or U_2.

Among the attempts to answer this are those by Cox [12] and Basu [4]. The former suggests conditioning on the ancillary U (if there is a unique one) that maximizes the variance of the conditional information. Roughly, this will give a large spread of the conditional accuracies obtained for different values of U, reflecting as much as possible the "luckiness" of X that we have mentioned; it was variability of the conditional accuracy that made conditioning worthwhile. Basu suggests that the difficulty of nonuniqueness of maximal U may lie in the difference between a real or performable experiment, such as creation of the sample size n(10 or 100) in the first example, and a conceptual or nonperformable experiment such as one from which U_1 or U_2 would result in the last example above. Basu implies that one should condition in the former case but not necessarily in the latter, and that in practice the nonuniqueness problem will not arise in terms of any ancillary representable as the result of a real experiment.

The problem of which maximal ancillary to use attracts attention in large part because of insistence on the use of an ancillary for conditioning. One may consider conditional inference based on an arbitrary conditioning variable V, and (a) require that some conditional measure of accuracy of, or confidence in, the decision, is approximately constant, given the value of V. At the same time (b) one would try, in the spirit of our comments about lucky observations and Cox's suggestion, to choose V and the decision procedure to make the variability of that conditional accuracy or confidence as large as possible. A development of Kiefer [18, 19] gives a framework in terms of which such conditional procedures can be compared. In this framework the statistician's goals are considered to be flexible so that, in an example such as that above of X_i uniformly distributed from $\theta - \frac{1}{2}$ to $\theta + \frac{1}{2}$, the length of the confidence interval* and the conditional confidence given U_n may both vary with U_n. A modification of the theory by Brown [8] includes a precise prescription for conditionings that produce most variable conditional confidence coefficients, in some settings.

The use of conditioning other than in terms of an ancillary is not new. For example, a common test of independence in 2×2 tables* conditions on the marginal totals, which are not ancillary. Similarly, inference about the difference between two Bernoulli parameters, each governing n observations, is often based on conditioning on the sum of successes in the $2n$ observations, also not an ancillary. Both of these are useful tools for which tables have been constructed.

BAYESIAN INFERENCE* AND OTHER AXIOMATICS

We have alluded to the *conditioning principle*. Various systems of foundational axioms considered by Birnbaum [6], Barnard, and others imply that inference should be based on a sufficient statistic, on the likelihood function, or conditionally on an ancillary statistic. A detailed discussion here would wander too far from the main subject. A popular axiomatic system related to conditioning is that of the Bayesian approach [25]. It is impossible to list and discuss here usual axioms of "rational behavior"* that lead to the use of Bayesian inference based on a subjective (or, infrequently, physical) prior law. Only the result of using such an approach will be described here.

If π is a prior probability law* on Ω, the element of the posterior law* of θ given that $X = x$ is given by Bayes' theorem as

$$\pi(d\theta \mid x) = f_\theta(x)\pi(d\theta) \Big/ \int_\Omega f_\theta(x)\pi(d\theta). \tag{4}$$

This may be thought of as the earliest basis for "conditional inference." Whatever the meaning of π (subjective or physical), $\pi(d\theta \mid x)$ updates $\pi(d\theta)$ to give probabilistic assessments in the light of the information $X = x$. Bayes procedures of statistical decision theory*, or informal "credibility intervals" that contain θ with state posterior probability, flow from (4).

The conditioning framework of Bayesian inference is conceptually quite different from that of classical conditioning in the frequentist framework (such as, in the uniform example, the assessment of a conditional confidence coefficient for an interval estimator, given that $U_n = u$). In the Bayesian context the conditioning is on the entire observation X or a sufficient $T(X)$, and the rv whose conditional law is ascertained is θ; in the conditioning of the preceding section, the conditioning was most often on a (by-itself-uninformative) ancillary U, and conditioning on X or $T(X)$ would yield nothing useful because the conditional probability assesses the accuracy of $\hat{d}$ or the coverage probability of an interval, *both functions of* X, and θ is not a rv. Thus direct comparison of the achievements of the two approaches in producing "conditional procedures" is not obvious.

Bayesians list, among the advantages of their approach, the lack of any arbitrariness in choice of point estimator or confidence interval method, or of the conditioning partition; which of two maximal ancillaries to use simply does not arise. Of course, a Bayesian credibility interval depends on the choice of π, as does the probabilistic meaning of that interval. Non-Bayesians regard the credibility intervals resulting from the use of subjective π's as appropriate for the expression of a Bayesian's subjective views, but not as meaningful for scientific discourse about θ as evidenced in X. These comments do not apply to (a) large-sample considerations in which Bayesian methods and certain frequentist methods yield essentially the same results; (b) settings in which the problem is transitive under a group on Ω that leaves the problem invariant and the Bayesian uses a (possibly improper) invariant π, yielding a procedure of frequentist invariance theory*; or (c) use of π to select a procedure which is then analyzed on a frequentist basis.

RELEVANT SUBSETS

A considerable literature, beginning with the work of Buehler [9] and Wallace [26], is concerned with questions such as the following: Given a confidence procedure with confidence coefficient γ, is there a conditioning partition $\{B, S - B\}$ such that, for some $\epsilon > 0$, the conditional confidence is $> \gamma + \epsilon$ for all θ, given that $X \in B$, and is $< \gamma - \epsilon$ for all θ, given that $X \notin B$. The set B is then called *relevant*. (The considerations have been simplified here, and a number of variants of the stated property are treated in the references.) Thus the set $B = \{U_n > c\}$, c a constant, is relevant for the classical confidence interval mentioned in the uniform example. In the example of $X_1, \ldots, X_n$ i.i.d. and $\mathcal{N}(\mu, \sigma^2)$, the usual confidence interval $[\bar{X}_n - cS_n, \bar{X}_n + cS_n]$ on μ, with $S_n^2 = \sum_i (X_i - \bar{X}_n)^2/(n-1)$, was proved by Buehler and Fedderson [11] to have conditional confidence (probability of containing μ) $> \gamma + \epsilon$ for some $\epsilon > 0$, given that $X \in B$, $= \{|\bar{X}_n|/S_n < c'\}$ for some $c' > 0$. Intuitively, $E\bar{X}_n^2 = \mu^2 + \sigma^2 n^{-1}$ while $ES_n^2 = \sigma^2$, so that if $\bar{X}_n^2/S_n^2 < n^{-1}$, there is evidence that S_n overestimates σ. This work has been extended by others, such as Brown [7] and Olshen [21]. Related work is due to Robinson [24].

Pierce [22], Buehler [10], and others have constructed a theory of "coherence"* of statistical procedures, based on the concept of relevant conditioning. If Peter takes γ confidence coefficient to mean that he will give or take $\gamma : 1 - \gamma$ odds on his interval containing θ, then Paul with a relevant B can beat him by betting for or against coverage depending on whether or not $X \in B$. The proponents of this theory regard such an "incoherent" procedure as unacceptable. Under certain assumptions they show that the only coherent procedures are obtained by using the Bayesian approach for some proper π.

These developments are interesting mathematically, and the existence of relevant sets is sometimes surprising. But a non-Bayesian response is that the confidence coefficient of the incoherent procedure is being compared unfavorably at a task for which it was not designed. All that was claimed for it was meaning as an unconditional probability γ, and the resulting frequentist interpretability of γ in terms of repeated experiments and the law of large numbers. That certain conditional probabilities differ from γ may seem startling because of our being used to unthinking unconditional employment and interpretation of such intervals, but if a finer conditional assessment is more important, the frequentist can use such an assessment with the same interval. The chance that the third toss of a fair coin is heads, given that there is one head in the first four tosses, is $\frac{1}{4}$; this does not shake one's belief in the meaning of $\frac{1}{2}$ as the unconditional probability that the third toss is a head. Which is the more useful number depends on what one is after.

OTHER CONSIDERATIONS

Among the many other topics related to conditional inference, we mention three.

By now it is well known that a test of specified level α with classical optimum properties is not necessarily obtained by using a family of conditional tests, each of conditional level α. An early paper giving a possible prescription for construction of conditional tests is Lehmann [20].

Bahadur and Raghavachari [2], in an asymptotic study of conditional tests, showed that a conditional procedure that is asymptotically optimum in Bahadur's sense of unconditional "slope" must give approximately constant conditional slope, with probability near 1.

Efron and Hinkley [13] showed that, in assessing the precision of the ML estimator $\hat{d}$ from i.i.d. $X_1, \ldots, X_n$ with density f_θ, a useful approximation to the conditional variance of $\hat{d}$, given an appropriate ancillary, is $1/I_n(X,\hat{d})$, where $I_n(X,\theta) = -\sum_i \partial^2 \log f_\theta(X_i)/\partial\theta^2$. This is Fisher's "observed information" as contrasted with the ML estimator $1/I_n(\hat{d})$ of unconditional asymptotic variance, where $I_n(\theta)$ is Fisher's information $E_\theta I_n(X,\theta)$. The observed information seems often to provide a more accurate picture.

References

[1] Anderson, E. B. (1973). *Conditional Inference and Models for Measurement*. Mentalhygiejnisk Forlag, Copenhagen.

[2] Bahadur, R. R. and Raghavachari, M. (1970). *Proc. 6th Berkeley Symp. Math. Statist. Prob.*, Vol. 1. University of California Press, Berkeley, Calif., pp. 129–152.

[3] Barndorff-Nielsen, O. (1978). *Information and Exponential Families in Statistical Theory*. Wiley, New York.

[4] Basu, D. (1964). In *Contributions to Statistics*, C. R. Rao, ed. Pergamon Press, Oxford, pp. 7–20.

[5] Basu, D. (1977). *J. Amer. Statist. Ass.*, **72**, 355–366.

[6] Birnbaum, A. (1962). *J. Amer. Statist. Ass.*, **57**, 269–326.

[7] Brown, L. D. (1967). *Ann. Math. Statist.*, **38**, 1068–1075.

[8] Brown, L. D. (1978). *Ann. Statist.*, **6**, 59–71.

[9] Buehler, R. J. (1959). *Ann. Math. Statist.*, **30**, 845–863.

[10] Buehler, R. J. (1976). *Ann. Statist.*, **4**, 1051–1064.

[11] Buehler, R. J. and Fedderson, A. P. (1963). *Ann. Math. Statist.*, **34**, 1098–1100.

[12] Cox, D. R. (1971). *J. R. Statist. Soc. B*, **33**, 251–255.

[13] Efron, B. and Hinkley, D. V. (1978). *Biometrika*, **65**, 457–488.

[14] Fisher, R. A. (1935). *J. R. Statist. Soc. A*, **98**, 39.

[15] Fisher, R. A. (1936). *Proc. Amer. Acad. Arts Sci.*, **71**, 245.

[16] Fisher, R. A. (1956). *Statistical Methods and Scientific Inference*. Oliver & Boyd, London.

[17] Hájek, J. (1965). *Proc. 5th Berkeley Symp. Math. Statist. Prob.*, Vol. 1. University of California Press, Berkeley, Calif., pp. 139–162.

[18] Kiefer, J. (1976). *Ann Statist.*, **4**, 836–865.

[19] Kiefer, J. (1977). *J. Amer. Statist. Ass.*, **72**, 789–827.

[20] Lehmann, E. L. (1958). *Ann. Math. Statist.*, **29**, 1167–1176.

[21] Olshen, R. A. (1973). *J. Amer. Statist. Ass.*, **68**, 692–698.

[22] Pierce, D. A. (1973). *Ann. Statist.*, **1**, 241–250.

[23] Pratt, J. W. (1961). *J. Amer. Statist. Ass.*, **56**, 163–166.

[24] Robinson, G. K. (1975). *Biometrika*, **62**, 155–162.

[25] Savage, L. J. (1954). *The Foundations of Statistics*. Wiley, New York.

[26] Wallace, D. L. (1959). *Ann. Math. Statist.*, **30**, 864–876.

[27] Welch, B. L. (1939). *Ann. Math. Statist.*, **10**, 58–69.

(ANCILLARY STATISTICS
BAYESIAN INFERENCE
COHERENCE
CONDITIONAL PROBABILITY
CONFIDENCE INTERVALS
CREDIBILITY THEORY
DECISION THEORY)

J. KIEFER

CONDITIONAL INVERSE (OF A MATRIX) *See* GENERALIZED INVERSE

Reprinted from
Encyclopedia of Statistical Sciences, Volume 2,
Wiley-Interscience, (1982) 103-109

BOOK REVIEWS

MAURICE G. KENDALL AND ALAN STUART, *The Advanced Theory of Statistics, Volume 2, "Inference and Relationship."* Hafner Publishing Company, New York, 1961. \$21.00, 132 shillings, x + 676 pp.

Review by J. KIEFER

Cornell University

The preface to this second volume of the three-volume work (hereafter referred to as K-S) states that the volume bears little resemblance to the original Volume 2 of Kendall (1946) (hereafter referred to as K), and that it was "planned and written practically *ab initio*, owing to the rapid development of the subject over the past fifteen years." Even superficial page counting shows that the three new volumes will contain much more material than did the old two, and the list of references indicates the extent to which the authors have updated the work.

A chief asset of the work is this large content, which will probably not be surpassed by any other reference work in statistics for many years. Also on the positive side is the successful presentation of the material of some of the more classical and unmathematical chapters. My main criticism of the book is that it has a very high density of errors in statements and proofs. Another negative aspect is the exclusion, in a work of this encyclopaedic nature, of much of the content and almost all of the spirit of modern mathematical statistics (as typified, for example, by the papers in these *Annals* which have elicited the most interest in the last two decades). I shall expand on these assessments in the next few paragraphs, and shall then list more detailed comments. My attempt will be to criticize this book in terms of what its aims appear to be, although this attempt may not be successful, since the aims are not stated.

One notices almost immediately that the book is written on an unfortunate mixture of mathematical levels, especially in view of the promise of its title. Unquestionably, one could produce a valuable reference work which lists statistical topics, models, procedures, etc., without giving proofs. One could also write a book which proves precisely stated results, but which keeps the level down by doing this only under restricted but stated conditions. But it seems uneven to make mention of sets of measure zero on one page and then, repeatedly, to give proofs which are only stated to hold "under regularity conditions" which are not specified. (A footnote on page 8 is evidently meant to justify this loose approach, but it is unsatisfactory both because the necessary regularity conditions vary, and also because they are sometimes assumed to hold where they do not.) Reading is also made difficult by the authors' deriving incorrect equalities or giving unsatisfactory proofs, and only *afterward* stating that the equality is only approximate or the proof only heuristic! The large number of errors, some quite fundamental, makes matters worse. Thus, I believe that it would be difficult for a student to use this book as a text in a first graduate course if he really wanted to understand the proofs and many of the deeper ideas. Referring to Hoeffding's (1962) comprehensive review of the recent book of Wilks (1962), I would say that, despite Hoeffding's well-founded criticisms, that book has a uniformity of level and a mathematical preciseness which the present volume does not touch, and Wilks' book therefore seems to me more appropriate as a text or reference for a first course for students with advanced calculus. The comparison with Lehmann (1959), on its own special topic (which occupies at least four chapters and a hundred pages of K-S, not counting preliminaries and applications in other sections), is therefore even more severe on K-S. As a reference work, K-S is much broader than Wilks, and in some areas it contains modern material not mentioned in the latter. I do not think, however, that K-S has succeeded in capturing the spirit of major parts of recent statistical theory, so that the present owner of K who buys K-S will gain more in the updating of the more classical statistical developments than in the addition of important modern developments. This is not a book on decision theory, and it is not and should not necessarily be written in that framework; but lack of even mention of risk functions, admissibility, complete classes, multiple decision procedures, etc., is a noticeable omission in a book of this encyclopaedic nature. What is more, as should become apparent in this review, the book could gain enormously in terms of presentation from the clarity and preciseness associated with modern statistics, without ever adopting its general framework and abstract results or mentioning its applications.

Some of the chapters on the more classical and less mathematical topics, such as the first two of the four chapters on statistical relations (26 and 27) and the chapter on categorized data (33), contain comprehensive and well-presented treatments. A few chapters (especially 33) are admirable in their inclusion of many practical examples. Some of the chapters on topics which have undergone considerable advancement in recent years, such as those on tests of fit (30) and robustness and distribution-free procedures (31), are notably up-to-date; but the deficient mathematical developments are all the more noticeable in such chapters.

Thus, the book's main usefulness will be as a reference work (which, although incomplete, is by far the most complete of its kind) and as a source-book of problems for a student who is able to rectify the errors.

It would require an undue amount of space to list all the book's errors. As a selection, I shall therefore go into more (although not complete) detail in the first chapters (17 and 18) and two others (29 and 30) than in other chapters.

The first three chapters (17, 18, 19) cover point estimation in about one hundred pages. It is disappointing that a work aiming for such completeness should omit mention here of Bayes procedures, completeness, invariance, and the minimax principle, all of which are treated in more elementary books. (The first three of these are given inadequate coverage in other contexts in later sections.) One notes at once an unfortunate choice of notation which makes reading more difficult than necessary throughout the book: the writing of E or P instead of something which evidences the underlying probability law, such as E_θ or P_θ. The terminology "consistent estimator" instead of "consistent family (or sequence) of estimators" seems unfortunate. (The terminology for tests on page 240 is similar.) On page 5 it is incorrectly stated that "a consistent estimator with finite mean value must tend to be unbiased in large samples." This is repeated on pages 42, 55, and on page 19 in a further discussion which is made precise only in the discussion of super-efficiency on page 44. (The corresponding discussion of consistency and unbiasedness in testing on page 445 is also incorrect.) The limiting variance and variance of the limiting d.f. are not distinguished. Although there is no discussion of the *local* character of best unbiased estimators, the discussion of (globally) best unbiased estimators is reasonable up to page 16, at which point a function of an unbiased estimator of a parameter is discussed as though it is an unbiased estimator of the function of the parameter. In Example 17.14 the "estimator" obtained depends on the unknown parameter. Sufficiency is unfortunately introduced (pages 22 and 27) in terms of estimators. The entire discussion of sufficiency, "single sufficient statistics", etc., would benefit greatly from a precise definition of "statistic", which from its usage seems usually (e.g., on p. 25) but not always (e.g., p. 29) to mean "continuously differentiable real function." The discussion of Equation (17.73) et seq. treats the likelihood function for a rectangular family as though the range $0 < x < \theta$ in which the functional form θ^{-n} holds is not part of the functional definition. Regarding what they consider ambiguities in choosing a sufficient statistic, the authors state that "we simply choose a function t which is a consistent estimator, and usually also an unbiased estimator, of θ." Exercise 17.13 is incorrect as stated; for example, if $X_1, X_2, \cdots X_n$ are independent with common Lebesgue density $c(\theta, \phi) \exp\{-x\theta - x^2\phi - x^3\theta\phi\}$ for $x > 0$ where $\theta, \phi > 0$, then there is a "single sufficient statistic" for θ if ϕ is known and for ϕ if θ is known, but no sufficient pair when both are unknown.

The appraisal of ML on pages 38 and 61–62 is good as far as it goes. The proof of consistency of ML which is presented as "a simplified form of Wald's proof," misses the point of Wald's compactness argument by incorrectly concluding that $\hat{\theta}_n \to \theta_0$ (the true value) in probability from the fact that $P_{\theta_0}\{\hat{\theta}_n \to \theta^*\} = 0$ for each $\theta^* \neq \theta_0$. (The "proof" is made all the more difficult by its incorrect conclusion that $n^{-1} \log [L(x \mid \theta_0)/L(x \mid \theta^*)] > 0$ "for large n with probability unity" from the corresponding conclusion about expectations, and by the writing of the event of convergence as "$\hat{\theta} = \theta_0$".) The discussion of Hurzubazar's work does not consider the possibility of inconsistent solutions of the likelihood equation. Example 18.4 reflects lack of identifiability rather than lack of uniqueness of ML estimation itself. On page 45 differentiation under an integral sign is incorrectly justified by the fact that the derivative of the integrand fails to exist at only one point. The "efficiency" of ML estimation for several parameters is presented in terms of generalized variance; more satisfactory is the fact that the limiting covariance matrix of any sufficiently regular sequence of estimators, minus that of the ML sequence, is nonnegative definite; this implies, for example, the (one-dimensional) efficiency of ML estimation of any regular real function of the parameters. (On page 81, LS estimators receive a similar treatment.) Exercise 18.33 is incorrect: if the two variances are $(\log n)^{-1}$ and $(\log n - \log 2)^{-1}$, they are of the same order $(\log n)^{-1}$, but the relative efficiency, as defined on the top of page 20, is $\frac{1}{2}$. The ML discussion admirably includes references to such recent work as LeCam's and Bahadur's but regretably omits the v. Mises-Kolmogorov-LeCam-Wolfowitz asymptotic Bayes result.

A more geometric approach to LS estimation would have enhanced some of the discussion; identifiability is discussed on page 87, but the linear subspace of identifiable linear parametric functions is unfortunately not described in settings where some parameters are unidentifiable. On page 91 it is incorrectly stated that "the addition to $\hat{\theta}$ of an arbitrary function of the observations, which tends to zero in probability, will make no difference to its asymptotic properties." On the positive side, such topics as BAN estimators are discussed.

Chapters 20 and 21 discuss interval estimation. The presentation, like that of K, gives one chapter to confidence intervals and one to fiducial intervals (plus Bayesian intervals), but includes somewhat more criticism than did K. The treatment would be enhanced if confidence *sets* were considered from the outset. On page 115 shortness (referring to length) is discussed as though it is a global rather than local property. Neyman shortness is also considered, but the relation between the two concepts is not. Asymptotic, but no small sample, optimality results are given (in which the super-efficiency mentioned two sections earlier is forgotten). In the discussion of randomization, the use of the actual sample values to aid in randomization based on the sufficient statistic, which is of practical importance, is omitted. Tolerance intervals are discussed briefly.

The chapter on fiducial intervals contains several additions from that in K, and is quite lucid in some of its criticism. The problem of two normal means is given extensive treatment, including work of Scheffé, Wald, and Welch. The Bayesian approach is based on infinite measures (following Jeffreys to some extent) without justification; moreover, when the normal mean and variance are both unknown, $d\mu\, d\sigma/\sigma$ rather than $d\mu\, d\sigma/\sigma^2$ is used. Invariance of either the Jeffreys or non-Bayesian variety is not mentioned. On page 153 there is a confusing passage regarding "imaginary" confidence intervals and the "need for sufficiency," and a discussion of Bayesian inference which is too brief and which unfortunately does not discuss the question of where the prior distribution comes from. An example on that page is misleading; after computing the a posteriori law of a normal mean μ which is assumed to be uniformly distributed on $(0, 1)$, the authors criticize the confidence interval approach by ignoring the ways in which it can properly make use of such information limiting the domain of

μ; this last appears to be at least partly due to an overemphasis on "exactness" (corresponding to similarity for tests).

Chapters 22 through 25 are concerned with hypothesis testing. The definitions of "parameters" and "simple" on page 162 are rather loose. In the discussion of simple hypotheses on page 166, Equation (22.5) supposes $L(x \mid H_0) > 0$ whenever $L(x \mid H_1) > 0$. The conclusion (22.7) that the ratio can equal a constant only on a set of measure zero is false. The result "proved" on page 173 regarding noncxistence of two-sided UMP tests for regular distributions with range independent of parameter value is false, as the density $[1 + \theta^6 x^{2+\text{sgn}\,\theta}]/2$, $-1 < x < 1$, $-1 < \theta < 1$, shows for $\theta_0 = 0$; an error in the proof is that θ^*, θ^{**} depend on x, θ_1, so that $\theta_1 - \theta_0$ need not change sign as claimed. The reference on page 174 to the rectangular distribution is irrelevant in view of the nondifferentiability in that case. Example 22.9, which is used in Example 22.10 and Exercises 22.10 and 22.11, is incorrect; the critical region should also include $\{\min_i x_i \leqq \theta_0\}$, and thus does not depend only on $\bar{x}$. Thus, Example 22.10 is not an example of what it claims to be (and, if the critical region *were* $\{\bar{x} \leqq c_\alpha\}$ as incorrectly claimed, Condition (2) of Section 22.21 *would* be satisfied, which it is correctly stated not to be). The criticism of conventional significance levels is good. The reason given for sufficiency in Excerise 22.9 is a wrong one.

In Chapter 23, page 187, it is incorrect to conclude that there is a critical *region*, rather than a randomized test, depending only on a sufficient statistic. (Several parts of the book would receive easier treatment in terms of randomized test functions.) Feller's example on page 188 is not exactly relevant to the discussion there, since the number of parameters increases with n. Example 23.2 is misleading, since the *minimal* sufficient statistic is complete. The statement on page 196 about H_1 being reduced to a simple alternative is false. The proof on page 208 of a generalized NP Lemma (e.g., Lemma 8.2 of Lehmann and Scheffé's work on completeness) is somewhat garbled. In the geometric interpretation on page 216, the condition is sufficient but not necessary as stated; a function can maximize without its derivative maximizing. In the discussion of ancillary statistics, the phrase "$T_r \mid T_s$ is sufficient for θ_k" sets the tone for an unprecise presentation which later mentions "the two statistics $(T_r \mid T_s)$ and T_s." The function h does not necessarily disappear as stated when $r + s = n$. In Exercise 23.1, the value $\alpha = \frac{1}{2}$ should be excluded. Exercise 23.4 incorrectly states that $\text{cov}\,(Z, Y) = 0$ implies $E(Y \mid Z) = 0$ (in a context $Y = \partial \log L(x \mid \theta)/\partial\theta$ where a counterexample is provided by $Z = X$ normal with mean 0, variance θ^2) and incorrectly ascribes the statement to Neyman; fortunately, Exercises 23.6–23.9 do not depend on this statement as suggested. Read $(x_1, \cdots, x_n)$ for θ at the end of Exercise 23.27.

Chapter 24 deals with LR tests and the general linear hypothesis. The sloppy "proof" of the asymptotic LR statistic distribution in Section 24.7 could well have been omitted. The "asymptotic sufficiency" proof at the beginning of that section is also inadequate. The study of the nonregular case is an interesting nonstandard inclusion, but there are errors in it; for example, in Section 24.15, the LR statistic is 0 with probability approaching 1 as $n_1 \to \infty$ when $\theta^* > \theta_0$, and does not have the stated chi-square distribution; the statement about bias is also wrong, as Example 22.6 shows. On page 243 there is another confusing development concerning conditional statistics; for example, a conditional statistic $t \mid L$ is integrated with respect to L to obtain t. The LR approach receives fair criticism. A total of about one page is spent in the book's main effort on invariance (there being slight reference later in conjunction with rank-order tests). The Hunt-Stein theorem is never mentioned, nor is any of Stein's recent work.

It is a pity that part of the general discussion of Chapter 25, "The Comparson of Tests," did not begin the treatment of hypothesis testing, to present earlier developments in a better-directed framework. Most of the development here is, however, asymptotic. On page 265 it is concluded from the fact that often power $\to$ 1 at a fixed alternative as $n \to \infty$, that one "must" consider a sequence of alternative approaching the null hypothesis θ_0; this ignores another well known approach, which is based on log (1 − Power). Mention is made on pages 267 and 270 of the fact that "no case where $m > 2$ seems to be known," where $(\theta - \theta_0)^m$ is the first nonconstant term of the power function; reparametrization in standard examples yields such an m, and families of competing tests with any two values of m can be obtained easily. The discussion of asymptotic power of tests at the point of greatest difference seems quite useful.

Chapters 26–29 cover statistical relationships, and the coverage is more extensive than that of earlier chapters, relative to the existing literature. Chapter 26 deals with linear regression and correlation. The subject receives a good classical treatment, as do partial and multiple correlation in the next chapter. (The notation convention introduced on page 323 regarding secondary subscripts is somewhat confusing.) The development is continued in Chapter 28 ("The General Theory of Regression"). The demonstration of Section 28.5 is incorrect; the authors speak of "completeness" of a single density g and use it in the manner of Chapter 23; what is needed is completeness of the family $\{e^{it_1x}g(x)\}$, and the factor e^{it_1x} should be deleted from (28.29) (which, as it stands, could never follow from (28.28)). In Section 28.15, the question, "how should the elements of $\mathbf{X}$ be chosen so that the estimators $\hat{\beta}_i$ are uncorrelated?" and the subsequent discussion could be misleading, since such a choice is not always possible. The precise quantitative basis for deciding which degree polynomials "evidently . . . are good fits" on page 361 is unfortunately not given. The development on page 368 of confidence regions for a regression line is unconvincing in its statement that the functional I cannot be effectively minimized without further restriction on the form of g; there is no mention of why or whether the form chosen loses nothing. In Exercise 28.18 the matrix should be replaced by its inverse. It is disappointing to see exercises such as 28.22 in which the testing theory developed earlier is completely forgotten, and tests which could have been derived from that theory are instead devised entirely on intuitive grounds.

Chapter 29, "Functional and Structural Relationship," is the least successful of these four chapters. It begins with a statement of the need for a "clear terminology and notation"; the reader can decide for himself whether Section 29.2 achieves this. (See also footnote, page 383.) Since (Section 29.3) errors are assumed uncorrelated and not independent, the conclusion of Section 29.9 regarding joint normality is incorrect. In the second paragraph on page 380 are some more misleading comments on identifiability; only one of the six given parameters is identifiable, but many combinations are. The idea of treating the four cases of the structural relationship under normality with various combinations of parameters known is a good one, but the treatment of the "overidentified" case as "a somewhat embarrassing position" which is resolved by changing the model, rather than merely by using ML or some other method of estimation on the original model (with the given underlying restrictions on the parameter space), is very bad. The pages which follow again indicate how much more satisfactory it would have been to discuss identifiability per se at the outset of the chapter, rather than to drag it in where it arises because the ML method breaks down. (The alternatives presented in Section 29.27 reflect the same spirit.) The properties incorrectly described for various procedures later in the chapter would be more clearly understood if the lack of identifiability for normally distributed incidental parameter (Reiersøl) were kept in mind. It seems unwise to say (page 388) that an estimator has "performed rather well," on the evidence of one experiment based on 9 observations. "Imaginary confidence regions" arise again on page 391. The properties stated for Geary's method are not very appealing; in fact, except in trivial parametric examples, no satisfactory estimation procedure is described for the linear structural relation in the whole chapter; Wolfowitz's minimum distance method and the methods of Neyman and Scott and Stein are not described. Since normality of errors has been assumed in the development leading up to page 399, one may question the logarithmic transformation discussed there (as do the authors, but on page 413). The discussion of Wald's procedure on page 400 is inadequate; the use of the instrumental variables is not the same as in (29.94), since they are no longer independent, and the procedure (since Neyman's critique) is well-known not to have the indicated validity; the treatment through page 405 suffers accordingly. Section 29.42 deserves similar comment; the subject is much more delicate than indicated in this chapter, and the assumption that "the values x are so far spread out compared with error variances that the series of observed ξ's is in the same order as the series of unobserved x's" is not a very meaningful one. Criticism similar to the above applies to the material on controlled variables, curvilinear regression, etc.

Chapters 30–32, and to some extent 33, are concerned with nonparametric problems. The first of these chapters, entitled "Tests of Fit," is introduced by a commentary on the point of formulating the problem in a certain manner, which omits mention of the most important aspect, the infinite dimensional character of the alternatives. In all of the asymptotic developments here and elsewhere, $O(n^{-\frac{1}{2}})$ is used without explanation to mean also $O_p(n^{-\frac{1}{2}})$. The choice of the usual chi-square one-tail critical region receives an interesting discussion on page

422, but the asymptotic optimum power properties are not mentioned as reasons. The developments of this chapter are otherwise admirably up to date in the inclusion of work of Chernoff and Lehmann, Watson, etc., although the execution is sloppy. The discussion of Section 30.22 (see also 30.44) is vague about where the power should be maximized and the sense in which it is maximized by Mann and Wald; the derivation on page 438 of the results of these authors is given in terms of a different criterion than theirs (which can be, but is not, shown to be equivalent); in this derivation (30.66) is false for large k, and when $k \to \infty$ (as it does) the asymptotic normality results cannot be introduced to obtain (30.70) as stated. (Some such oversights, which have been investigated by T. Taylor in a Cornell master's essay, appear in the original, but they are not justified by the irrelevant reason on page 440 that "an upper limit to k is provided by the fact that the multinormal approximation to the multinomial distribution cannot be expected to be satisfactory if the np_{0i} are very small.") As an example of the style and attitude toward proofs which is sometimes infuriating, we cite the fact that, in the *following* numbered section, it is stated that Mann and Wald use "a much more sophisticated and rigorous argument . . . our own heuristic argument . . ." (the last referring also to a generalization to composite hypotheses, which is not as "clear" as it is stated to be). Pity again the poor student who spends hours trying to understand the "proof" which precedes, and *then* reads this! The avoidance of the noncentral chi-square distribution in the asymptotic discussion of Section 30.24 is mysterious. The argument on the top of page 436 is not quite as obvious as stated; for example, (30.59) requires that $(E_1 - E_0)(x^2 - x)I_R > 0$ where I_R is the indicator of the critical region. The top of page 440 makes a conclusion, "clearly," on the basis of a single sampling experiment. The discussion of the "more conservative" equal probability case on page 440 seems incorrect, since the expected minimum frequency is greater there than for unequal probabilities. Section 30.34 and Exercise 30.7 assume H_0. Section 30.36 and Exercise 30.10 are incorrect unless t_1 and t_2 are assumed appropriately invariant. The comparison of the p_k^2 tests and chi-square test on page 448 is not very meaningful: the alternatives differ. On page 451, the normalization $n\omega^2$ is cited as different from the $n^{\frac{1}{2}}$ of the CLT; of course, they are really the same normalization. Feller's derivation of the Kolmogorov-Smirnov law is given essentially in toto except for justifying taking the limit of the generating function; an equation on page 456 reads $p_n(0) \to (2\pi n)^{-\frac{1}{2}}$(as $n \to \infty$). In Exercise 30.9 the χ^2-distribution is only asymptotic.

Chapter 31, on robust and distribution-free procedures, is one of the relatively better chapters in converage of what it sets out to cover, although point estimation (Winsorisation, etc.) receives little attention in this or the next chapter. The relationship of various problems given in Section 31.14 is somewhat misleading. The inclusion of much material on rank order tests, including, for example, the Chernoff-Savage results on the Fisher-Yates statistic, is to be commended. Section 31.69 seems slightly overenthusiastic about the Wilcoxon and Fisher-Yates two-sample tests relative to all others; the statement that "although little is known of the power of these tests" (such as the Kolmogorov-

Smirnov and other two-sample tests), "it is clear that they are less efficient" certainly requires at least a limitation of alternatives from the general (31.111) that had been discussed. The reduction from (3.1.160) to (3.1.161) enlarges H_0, since symmetry about 0 of the d.f. of $x_1 - x_2$ does not imply that $F(x_1, x_2) = F(x_2, x_1)$. On page 509, it is incorrectly stated that the method of breaking ties affects limiting distributions in the *normal* case (where ties have probability zero); it is not clear what the authors meant to refer to. Chapter 32, which covers uses of order statistics, is brief but fairly adequate, except again on point estimation. (The parametric use of linear combinations of order statistics was treated briefly in Chapter 19.)

Chapter 33 covers categorized data comprehensively. This chapter, in its many numerical examples, seemed different in spirit from, and more successful than, most other chapters. A more critical development is needed in some places. For example, Yule's approach is developed without comment by using a questionable "invariance principle" on page 546. Three possible degrees of "fixing" in 2×2 tables are explained well; the principle of conditioning could have received longer comment. On page 556, the sense in which Lancaster's pooling prescription is "best" is unclear. The canonical analysis of page 569 receives a good start, but (Sections 33.46 and 33.50) sample and population properties are not sufficiently distinguished. The analysis of Section 33.57 does not mention the extent to which the Poisson dispersion test is really a test of smallness of variance/mean. Formula (33.127) is complete nonsense.

The final Chapter 34 covers sequential methods. In describing excess over the boundaries, we read (page 597), "there is no exact probability of reaching the boundary at M—and, in fact, this point is inaccessible." The consideration of the OC and ASN function for other states than the two simple hypotheses which have been under consideration, should be motivated. The explanation of why optimum properties of SPRT's "will not, then, come as a surprise" does not make it clear why nonconstant bounds should not yield such properties. The "proof" of Wald's equation, ascribed to Johnson (who later published a correction), is a sloppy version of Wolfowitz's proof; it incorrectly does not use the finiteness of En, but justifies the interchanges E and $\sum_1^n$ (with n a random variable) as "being legitimate since $E|z_i|$ exists and is finite." Similar remarks apply to Section 34.31. On page 611, the authors do not specify the sense in which SPRT's for composite hypotheses "can be used in the ordinary way" when weighting functions are used and the tests therefore are not based simply on the product of independent factors. This is one of the least precisely written chapters, and page 613 exemplifies this: the passage to the limit in using nonintegrable weight functions is never justified; a statement about the distribution of s^2 and $(\bar{x} - \mu_0)/s$ should instead refer to the joint distribution over all stages; the reason given for monotonicity oversimplifies the situation greatly; "optimality" is used in a sense which has no visible relationship to the ASN; "excess" is neglected everywhere (as it was announced earlier that it would be); the restriction here and on page 612 to tests with given bounds A and B and the weight function-LR structure is never given meaning. The statement on page 616 that "it cannot be said that, in general, sequential procedures are very satisfactory for estimation" is never explained. In Exercise 34.4, replace g by $-g$ in the last two sentences. Exercises 34.5, 34.6, and 34.7 require unstated conditions. The differentiation in Exercise 34.9 is never justified. The results of Exercises 34.11, 34.12, and 34.13 only hold asymptotically. In Exercise 34.17, an upper bound which depends on the unknown parameter is given on the choice of sample size.

Finally, on the last page of text, one-half page is spent on "decision functions." I feel obliged to repeat that, while I do not feel that the ideal three-volume work on the advanced theory of statistics is a three-volume work on decision theory, I question the relegation of this topic to one-half page of a 676 page volume which is part of a three-volume work of such supposedly broad coverage. Why include such detail on some theoretical aspects of hypothesis testing and measures of association (for example), and never mention admissibility, multiple decision procedures, etc.? The general results of abstract decision theory are probably not so important for inclusion in a work of this sort. But, even if the general theory and interesting recent applications are omitted from the book, the sampling of criticisms included in this review certainly shows that K-S would benefit greatly from absorption of some of the spirit of modern statistics, in the careful way in which the latter looks at problems and compares procedures. Clarity and precise statements in the motivation and definitions, and in the development of solutions, does not entail mathematical abstraction or the rigid use of oversimplified decision-theoretic models. Nor does the complex nature of many practical statistics problems mean that one should not try to be as clear and accurate as possible in formulating them; if anything, the importance of such care is greatest in such settings which are farthest from the simplest decision-theoretic models.

In summary, this useful reference would be much improved if it contained fewer invalid proofs and misleading motivations, but greater precision and clarity.

REFERENCES

[1] Hoeffding, W. (1962). Book Reviews. *Ann. Math. Statist.* **33** 1467.
[2] Kendall, M. G. (1946). *Advanced Theory of Statistics*, **2**. Griffin, London.
[3] Lehmann, E. L. (1959). *Testing Statistical Hypotheses*. Wiley, New York.
[4] Wilks, S. S. (1962). *Mathematical Statistics*. Wiley, New York.

Reprinted from The Annals of Mathematical Statistics
Vol. 35, No. 3, September, 1964

THE SAVORY WILD MUSHROOM *Second Edition.*

By Margaret McKenny; revised and enlarged by Daniel E. Stuntz. University of Washington Press, Seattle. $8.95 (cloth); $4.95 (paper). xxii + 242 p. + 32 p. pl.; ill.; index. 1971.

This book (hereafter abbreviated M2) is derived from the late Margaret McKenny's 1962 work of the same title (M1). Both invite comparison with Alexander H. Smith's, *Mushroom Hunter's Field Guide*, to the 1964 printing of which reference will be made; all concentrate on macroscopic field characteristics for the amateur, with M1 and M2 being more frankly biased toward use in the Pacific Northwest.

The "edible-poisonous-to be avoided" trichotomy of M1 has been replaced by traditional and more useful groupings in M2. Unlike Smith, M2 has no keys or glossary of terms. The introductory description is somewhat sparse; for example, someone whose first reading about mushrooms in M2 will begin the section on *Boletes* without having read a definition of pores.

Choice of species undoubtedly depends on the author's personal experiences. Smith is remarkable in its usefulness to the novice, in including what one is likely to find which is also of interest. M2 has such omissions as *Agaricus edulis*, *Lepiota procera*, *Lactarius corrugis*, *Craterellus*, *Clitocybe illudens*, the last a dangerous one. The descriptions in M2, sometimes slightly altered from those in M1, seem less useful to me than those of Smith. The latter does not always give the full "cap-gills-stem" list of features that M2 does, but concentrates on those features which are most distinctive and useful in identification. Some descriptions in M2 can be misleading; thus, that of *Cantherellus cibarius*, unlike Smith's, would make one expect always to find gills with interlacing veins. Moreover, useful contrasts among confusable species are more prevalent in Smith; for example, the difference between the stem cross-section in *Gyromitra gigas* and *G. esculenta* is emphasized there.

The black-and-white photographs are satisfactory, the colored ones are inferior to those in M1 or Smith. The views are often more informative in M2 than in M1, but most colors are inaccurately dingy yellow or brown. Even *A. muscaria* appears to be made of chocolate. Color reproduction is frequently a problem, but it is difficult to explain this deterioration from the brighter and usually more accurate colors in the same publisher's M1.

An entire chapter of recipes (but fewer than in M1!) exceeds Smith's occasional line on the subject. The same is true of Professor Tyler's chapter on toxins. The latter is updated from M1 and is perhaps more detailed than needed for likely readers of M2; moreover, one questions whether "European specimens of *Gyromitra esculenta* are almost uniformly toxic" and whether *Paxillus involutus* should be placed in the same danger category as *Rhodophyllus lividus*.

A delightful bonus is Angelo Pellegrini's instructive initiation to mushroom hunting and eating, including an amusing lesson in restraint from broadcasting the location of one's major finds. But these three chapters do not make up for the greater usefulness of Smith in almost every respect except for its shape, both for identifying what one has found and for guessing where to look. Thus, the reader of M2 would not know the most productive habitat for morels in the East; unfortunately, the reviewer must heed Mr. Pellegrini's advice and not be more specific!

J. KIEFER, *Mathematics, Cornell University*

Reprinted from *Quarterly Review Biology* **47** (1972), 342–343.

A FIELD GUIDE TO WESTERN MUSHROOMS.
By Alexander H. Smith, University of Michigan Press, Ann Arbor. $16.50. vi + 280 p.; ill.; index. 1975.

Here is the master again, with a volume whose coverage differs greatly from that of his classic, *The Mushroom Hunter's Field Guide* (*MH*). Only a quarter of the 201 illustrated species were in MH. The format is similar, reflecting Smith's singular ability to pinpoint the few most critical identifying features of a species rather than give a complete botanical description that often overwhelms and confuses beginners. Microscopic features are now also listed.

Perhaps the style is less chatty. I confess missing the frequent opportunity MH afforded to sympathize with this great expert who suffered his "usual allergic reaction" to a morsel his students cherished. Maybe, too, additional experience has made his attitude on edibility just slightly less conservative. Thus, now omitted is "I discourage eating species of *Russula*;" and to the warning that the host of *Hypomyces lactifluorum* may be poisonous is added, "Fortunately, this does not happen very often" (p. 45). Nomenclature and taxonomy have been updated. The photographs are usually excellent, with only an occasional false tone (p. 224).

My few months' use of the book in the West were too short for this Easterner to make the most informed judgment. The book's appeal will probably parallel that of MH, still the first book I suggest to inquiring mushroom hunters in my own area. Smith can write the guide for *anywhere*, and

Should I not end up in the hunter's hell
Devoid of Agaricus or Morel
Heaven will be to hunt mushrooms with
The Ultimate Guide by A. H. Smith.

J. KIEFER, *Mathematics*, *Cornell University*

Reprinted from
Quarterly Review Biology **52**, 91 (1977)

NOTE ON ASYMPTOTIC EFFICIENCY OF M.L. ESTIMATORS IN NONPARAMETRIC PROBLEMS

by J. Kiefer[1)] and J. Wolfowitz[2)]

Cornell University

Let $X_1, X_2, \ldots$ be independent and identically distributed random variables taking on values in a space S, each with probability measure P_θ on an appropriate σ-field of S. Here θ is an unknown element of a set Ω. Let $\hat{t}_n$ be a maximum likelihood (M.L.) estimator of θ based on $X_1, \ldots, X_n$, assuming one exists. Let ξ be an a priori probability measure on Ω.

It is well known [4], [5], [6], [7] that, when θ is finite dimensional and P_θ is suitably regular, then, no matter which "suitably smooth" ξ is assumed, the a posteriori distribution of θ based on $X_1, \ldots, X_n$ is approximately normal about $\hat{t}_n$ with probability near unity when n is large, in any of several precise senses: (1) If J_θ is the covariance matrix of $n^{1/2}(\hat{t}_n - \theta)$ under P_θ and $H_\theta' H_\theta = J_\theta^{-1}$ (with H_θ suitably smooth), then, for any $\epsilon > 0$, with probability approaching one as $n \to \infty$ <u>when the</u> X_i's <u>are actually distributed according to</u> P_{θ_0}, the a posteriori distribution function (d.f.) of $n^{1/2}H_{\theta_0}(\theta - \hat{t}_n)$ differs everywhere by less than ϵ from the normal d.f. with means 0 and identity covariance matrix. This statement also holds if $n^{1/2}H_{\theta_0}(\theta - \hat{t}_n)$ is replaced by $n^{1/2}H_{\hat{t}_n}(\theta - \hat{t}_n)$. (2) As a consequence of (1), with probability approaching one as $n \to \infty$ <u>when θ is distributed according to</u> ξ (or, for that matter, according to <u>any</u> suitably smooth ξ', not necessarily the ξ used in calculating the a posteriori law), the conclusion of (1) regarding $n^{1/2}H_{\hat{t}}(\Theta - \hat{t}_n)$ is valid. In view of such phenomena as superefficiency, (1) and (2) are perhaps the most satisfactory statements of asymptotic optimality of M.L. estimators.

The purpose of the present note is to extend the above results to cases where Θ is infinite-dimensional. These new results seem to be of both theoretical and practical interest. Certain difficulties arise which are not present in the finite-dimensional case. The results in the latter case are usually proved by expanding the logarithm of the likelihood function of Θ about $\hat{t}_n$, and an attempt to duplicate this procedure in even the simplest infinite-dimensional examples is fruitless. In fact, the first derivatives of the likelihood function will no longer generally all be zero at $\Theta = \hat{t}_n$, as can be seen in the simplest possible (multinomial) example, Example 1 below. There $x_1 \log(1 - \Sigma_2^\infty \Theta_i) + \Sigma_2^\infty x_i \log \Theta_i$ has a negative derivative with respect to Θ_i at $\Theta = \hat{t}_n$ whenever $x_1 > 0$ and $x_i = 0$ (the latter holding, for each n, for all but a finite number of values i).

Our approach in certain simple examples of the infinite-dimensional case is to use the structure of P_Θ to reduce the problem to one of considering an infinite collection of finite-dimensional situations which can be treated by the method mentioned above. This approach has limited applicability, and cannot be applied to many of the simple but important infinite-dimensional cases for which consistency of the M.L. estimator was proved in[2]. However, such common practical problems as those where the sample d.f. is used to estimate the d.f. P_θ of X_1 are included (Examples 2 and 3). In fact, the results of [1] and [3] on the asymptotic minimax character of the sample d.f. for a wide variety of loss functions suggests some sort of "asymptotic efficiency" result such as that obtained here.

Suppose, then, that $\phi = (\phi_1, \phi_2, \ldots)$ is a one-to-one function of θ, where the ϕ_i's are real. It will be no loss of generality to assume that the range of ϕ is the infinite-dimensional unit cube. We write $\phi^{(k)} = (\phi_1, \ldots, \phi_k)$ and $\phi^{[k]} = (\phi_{k+1}, \phi_{k+2}, \ldots)$. Similarly, write $\hat{t}_n = (\hat{t}_n^{(k)}, \hat{t}_n^{[k]})$ for the m.l. estimator of ϕ and $x^{(n)} = (x_1, \ldots, x_n)$. We now assume the following:

(a) ξ assigns measure 1 to a subset Ω_ξ of Ω such that all P_θ's, $\theta \in \Omega_\xi$, have densities f_ϕ relative to a common σ-finite measure μ;

(b) for each k and n, there are functions $g_{k,n}$ and $h_{k,n}$ such that

$$\prod_{i=1}^{n} f_\phi(x_i) = g_{k,n}(\hat{t}_n^{(k)}, \phi^{(k)})\, h_{k,n}(x^{(n)}, \phi^{[k]}) ,$$

where $g_{k,n}$ is the induced density (with respect to, say, μ') of $\hat{t}_n^{(k)}$;

(c) for each k, $d\xi$ can be written on Ω_ξ as

$$q(\phi^{(k)})\, d\phi^{(k)}\, r(\phi^{[k]}|\phi^{(k)})\, d\eta_k(\phi^{[k]})$$

where q is the continuous positive Lebesgue density of $\phi^{(k)}$ according to ξ and r is the conditional density (according to ξ), with respect to a σ- finite measure η_k, of $\phi^{[k]}$ given $\phi^{(k)}$; furthermore, q and r are assumed to be bounded and bounded away from zero and are uniformly continuous in $\phi^{(k)}$ (with the same modulus of continuity for each value of $\phi^{[k]}$, in the case of r);

(d) (classical result for the finite-dimensional case) for each k and each set of positive numbers ϵ, c_1 c_2, c_3, if $\hat{t}_n^{(k)}$ has density $g_{k,n}(\cdot, \phi^{(k)})$ with respect to μ' and $\phi_0^{(k)}$ is the true parameter value, there is a number $N = N(k, \phi_0^{(k)}, \epsilon, c_1, c_2, c_3)$ such that $n > N$ implies that there is a set A_n of values $x^{(n)}$ with $P_{\phi_0^{(k)}}\{A_n\} > 1 - \epsilon$ and such that, if $x^{(n)} \epsilon A_n$ and if the assumed a priori Lebesgue density λ of $\phi^{(k)}$ has modulus of continuity c_1 and satisfies $0 < c_2 < \lambda < c_3 < \infty$, then the a posteriori d.f. of $n^{1/2} H_{\hat{t}_n}^{(k)} (\phi_n^{(k)} - \hat{t}_n^{(k)})$ differs by at most ϵ from the k-variate normal d.f. with zero means and covariance matrix the identity (Here $H^{(k)}$ has an obvious interpretation.) This assumption, which implies (1) and (2) in the finite-dimensional case, is an assumption only on finite-dimensional densities.

Under these assumptions, it is easy to show that the conclusions of (1) and (2) above hold also in our infinite-dimensional case in the sense of describing the asymptotic a posteriori density of $n^{1/2} H_{\hat{t}_n}^{(k)} (\phi_n^{(k)} \hat{t}_n^{(k)})$ for each k.

(In assumption (d), λ was a density on a finite-dimensional space, so some argument is needed here.) In fact, one has only to note that the function

$$\psi_k(n, x^{(n)}, \phi^{(k)}) = \frac{\int h_{k,n}(x^{(n)}, \phi^{[k]})\, r(\phi^{[k]}|\phi^{(k)})d\eta_k(\phi^{[k]})}{\int h_{k,n}(x^{(n)}, \phi^{[k]})d\eta_k(\phi^{[k]})},$$

being a probabilistic average of functions r (of $\phi^{(k)}$) satisfying (c) (say with modulus c_1 and with $0 < c_2 < r < c_3 < \infty$), also satisfies (c); that is, ψ_k is bounded and bounded away from zero and is uniformly continuous in $\theta^{(k)}$ with the same modulus c_1 and bounds c_2 and c_3 as r, independent of n and $x^{(n)}$ (the integrals clearly exist except on a set which can be ignored). Hence

$$\bar{\psi}_k(n,x^{(n)},\phi^{(k)}) = q(\phi^{(k)})\, \psi_k(n,x^{(n)},\phi^{(k)}) / \int q(\phi^{(k)})\psi_k(n,x^{(n)},\phi^{(k)})d\phi^{(k)}$$

is a Lebesgue probability density of $\phi^{(k)}$ with the same properties (the c_i's being replaced by obvious functions thereof), and (even though $\bar{\psi}_k$ depends on $x^{(n)}$) the desired result now follows from (d) and the fact that the a posteriori Lebesgue density of $\phi^{(k)}$ is $g\bar{\psi} / \int g\bar{\psi}\, d\phi^{(k)}$.

The above assumptions can obviously be relaxed, but we omit details for brevity. In particular, conclusion (2) holds without such stringent continuity assumptions on f_ϕ as those of (b) and (d) holding, since the validity of such conditions on a set of $\phi^{(k)}$-values (projection of Ω_ξ) to which ξ assigns probability one suffices for the proof of (2). (μ and thus the form of f_ϕ may depend on ξ.) The boundedness and continuity conditions of (c) can also be weakened.

We now consider several examples.

Example 1. Multinomial with infinitely many classes.

Let $\theta = (\theta_1, \theta_2, \ldots)$ with $\Omega = \{\theta: \text{all } \theta_i \geq 0, \Sigma_1^\infty \theta_i = 1\}$. Let $X_i = (X_{i1}, X_{i2}, \ldots)$; let μ give measure one to each vector $x = (x_1, x_2, \ldots)$ with exactly one x_i unity and all others zero, and zero measure elsewhere, and let

$$f_\theta(x) = \prod_1^\infty \theta_i^{x_i}.$$

There are many ways of describing ξ's for which our result holds, and we present only one simple class of such ξ's. Let $(i_1, i_2, \ldots)$ be a permutation of $(1, 2, \ldots)$, and let $\phi_1 = \theta_{i_1}$ and $\phi_j = \theta_{i_j} / (1 - \Sigma_{s=1}^{j-1} \theta_{i_s})$ for $j > 1$. (We shall discuss only ξ's for which $\Sigma_{s=1}^{j-1} \theta_s < 1$ for all j.) If the ϕ_i's are independently distributed according to ξ, each with continuous bounded and boundedly positive Lebesgue density on $[0, 1)$, our assumptions hold. More generally, the ϕ_i's can be dependent if ξ satisfies (c) or the weaker conditions mentioned above.

Example 2. Estimating a univariate d.f.

Without essential loss of generality we take X_1 to have the interval $[0, 1]$ as its range; the case where the range of X_1 is the whole line is treated similarly. Let Ω be the class of d.f.'s θ on $[0, 1]$, or a suitably rich subclass. (The symbol θ is used here to denote the d.f. itself, not just a label.)

The sample d.f. S_n can be regarded as the M.L. estimator of θ (see [2], page 893). There are again many ways of representing ξ's, for example, as measures on spaces Ω_ξ of nondecreasing random functions θ. One simple representation for which the assumptions are easily verified is the following: Let $t_1, t_2, \ldots$ be any dense sequence of distinct points of (0, 1). Let $\phi_1 = \theta(t_1)$ and, for $j > 1$, let $\phi_j = [\theta(t_j) - \theta(t_j)] / [\theta(t_j'') - \theta(t_j')]$ where t_j' (resp., t_j'') is the largest t_i less than (resp., smallest t_i greater than) t_j among $t_1, \ldots, t_{j-1}$, and where $\theta(t_j')$ (resp., $\theta(t_j'')$) is replaced by 0 (resp., 1) if no such t_i exists. This allows the range of each ϕ_j to be the unit interval, and $\phi = (\phi_1, \phi_2, \ldots)$ determines θ. If now ξ satisfies the conditions of the previous example regarding ϕ, the desired conclusion again holds. One can similarly construct ξ's on Ω_ξ's which consist of absolutely continuous or more restricted d.f.'s θ. It would be interesting to obtain conditions which yield asymptotic a posteriori normality of $\theta(t)$ for t different from all t_i, as well as the more general normality of linear functionals of θ (this applies also to our other examples), but this seems to invoke a finer analysis.

The confirmed Bayesian can evidently conclude from the previous example that he should base much of his large sample nonparametric inference on the values of the sample d.f. corresponding to a few appropriate abcissa values, rather than on sample moments.

Example 3. Estimating a multivariate d.f.

Now let Ω consist of d.f.'s θ on the unit k-cube I^k. We proceed exactly as in Example 2, except that now the ϕ_j's represent a successive breakup of I^k into rectangles of decreasing diameter. For example, for $k = 2$ one might take $\phi_1 = \theta(1/2, 1)$, $\phi_2 = \theta(1/2, 1/2) / \theta(1/2, 1)$, $\phi_3 = [\theta(1,1/2)-\theta(1/2,1/2)]/[1-\theta(1/2,1)]$, $\phi_4=\theta(1/4,1/2)/\theta(1/2,1/2)$, etc.

Example 4. Generalized Poisson process.

Let X_i be a random function, and suppose that, for $i = 1, 2, \ldots$, we observe $X_i(t)$ for $0 \leq t \leq 1$. (The X_i are separable, and in fact we need only "observe" them for rational t.) The X_i are again independently and identically distributed, each being the sample function of a generalized Poisson process; that is, X_i has independent Poisson increments and $EX_i(t) = \theta(t)$ where θ is continuous and nondecreasing and $\theta(0) = 0$. We now let $\phi = (\phi_0, \phi_1, \ldots)$ where $\phi_0 = \theta(1)$ and the other ϕ_i's are obtained as in Example 3. The conclusion follows as before.

Example 5. Wiener process with varying drift.

In a manner similar to that of Example 4 we can treat the problem where $X_i(t) = Y_i(t) + \theta(t)$, the Y_i's being independent Wiener processes with $EY_i(t) = 0$, $EY_i^2(t) = t$, and where $\theta(t)$ is an unknown continuous function which is to be estimated. We omit the details.

Footnotes

1) Research sponsored by the Office of Naval Research.

2) The research of this author was supported by the U.S Air Force under Contract No. AF 18(600)-685, monitored by the Office of Scientific Research.

References

[1] A. Dvoretzky, J. Kiefer, and J. Wolfowitz, "Asymptotic minimax character of the sample distribution function and of the classical multinomial estimator," Ann. Math.Stat.27(1956), pp.642-669.

[2] J. Kiefer and J. Wolfowitz, "Consistency of the maximum likelihood estimator in the presence of infinitely many incidental parameters," Ann. Math. Stat.27(1956) pp. 887-906.

[3] J. Kiefer and J. Wolfowitz, "Asymptotic minimax character of the sample distribution function for vector chance variables " Ann. Math. Stat.30(1959), pp. 463-489.

[4] A.N. Kolmogoroff, "Determination of the center of dispersion and of the accuracy, on the basis of a finite number of observations," Izv. Akad. Nauk SSSR, Math. Series 6(1942),pp.3-32.

[5] L. LeCam, "On some asymptotic properties of maximum likelihood estimates and related Bayes' estimates," U. Of Cal. Pub. in Stat. 1(1953), pp. 277-330.

[6] R. von Mises, Wahrscheinlichkeitsrechnung und ihre Anwendungen, M.S. Rosenberg, N.Y., 1945 (see also Math.Zeit.4(1919)pp.1-97

[7] J. Wolfowitz, "The method of maximum likelihood and the Wald theory of decision functions," Konikl.Ned. Akad. van Wet.56(1953), pp. 114-119.

Math 371 Final Exam

10:45-1:15, 15 Dec., 1972.

NOTE: Problems have been designed so that they may use information given in other problems, but each problem can be worked without your having worked the others.

The problems or their parts are assigned points in the margin, totaling 115. You may select any collection totaling 100 points. (When a problem has several parts, you may select whichever parts you wish.) Write the numbers of the problems (and parts) on which you wish to be graded, on the cover of your exam book.

The examination is in the form of a story. It is suggested that you read it once quickly before selecting the problems you decide to work.

Only the names have been changed to protect the innocent.

-1-

ONCE UPON A TIME there was a rumble at the bottom of the ocean. With probability $\frac{1}{3}$, one volcano would arise, while with probability $\frac{2}{3}$ two volcanos would arise. Given that one arises, there is probability $\frac{3}{4}$ that enough lava spouts up to make an island (and probability $\frac{1}{4}$ that no island forms). But, if two volcanos arise, the lava is split between them, and the conditional probability that i islands form is $(3-i)/6$ for $i = 0, 1, 2$.

(5) (a) Determine whether or not the event that a single volcano forms is independent of the event that a single island forms.

(10) (b) Find the probability function of the rv N = number of islands formed.

-2-

Actually, a single island formed. Its shape was a square with sides parallel to the compass directions, 200 meters on a side.

Over the years, simple plants washed ashore and grew. The barren rock became soil. A lovely and edible flower eventually covered much of the island.

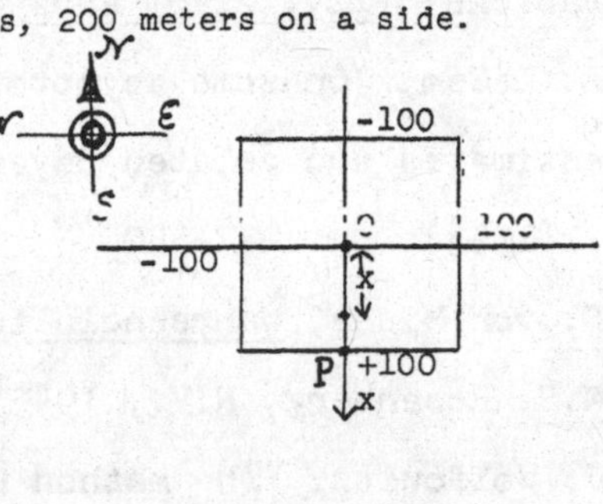

The island had a peak at its center (O in diagram), and the slope was such that little sun hit the north half of the island, while the noonday temperature (°c) of the soil a distance x meters south of the center was $20 + \frac{x^2}{500}$ for $0 \leq x \leq 100$.

A pair of insects land at random (uniform distribution) on the segment OP between the center and the point on the shore due south of it, and their eggs are deposited there.

10 (a) Show that the probability density function of the noonday temperature T where they land is

$$f_T(t) = \begin{cases} \dfrac{5}{2\sqrt{500t-10,000}} & \text{if } 20 \leq t \leq 40, \\ 0 & \text{otherwise.} \end{cases}$$

5 (b) Show that the expected noonday temperature where they land is $26\frac{2}{3}$. [Hint: do <u>NOT</u> use (a)!]

5 (c) The eggs only hatch if they land where the noonday temperature is between 25 and 27.2. What is the probability that this happens?

10

One night a great storm washes a cherry ashore. The cherry lands exactly at the center O of the island, and its seed produces a tree. The number of full years Y that such a tree lives has a geometric law,

$$P\{Y=y\} = (1-p)p^y \quad \text{for} \quad y = 0,1,2,\ldots$$

Here p can depend on various environmental factors, as we shall see later.

A tree can only bear fruit and thus have progeny if it lives at least 10 years.

How large must p be in order that the tree has a 50-50 chance of living at least 10 years? (Leave answer in terms of logarithms, etc.

10 -4-

The insect eggs of problem 2 hatch, and generations of insects follow.

A man is shipwrecked on the island.

No food floats ashore with him, but he can survive on the lovely and edible flower mentioned earlier.

What <u>does</u> float ashore with him is a crate of MIRA-KILL, the greatest insect-destroying marvel of modern technology.

The man is a great scientist.

He notices that the insects have started to devour the leaves of the cherry tree, which would drastically reduce the quantity p mentioned in problem 3.

This year's insects die, but their eggs linger on.

In order to make sure p is large enough to give the tree the 50-50 chance mentioned in problem 3, the scientist will spread his MIRA-KILL over the island so that it has a good chance of destroying the insect eggs and saving the tree. If he can destroy all the eggs except perhaps one, this insect species will die out.

There are 10^6 eggs on the island, and MIRA-KILL is so potent that each egg has only a chance of 2×10^{-6} of surviving, with survival being independent from egg to egg.

What (approximately) is the chance that at most one egg survives?

-5-

The scientist once took a probability course.

To relieve his boredom on the island, he tries to recall and apply some of the ideas he learned.

He notices that the actual amount of MIRA-KILL which lands on each egg, in milligrams, is uniformly distributed between 0 and 4, and the amounts on different eggs are independently distributed. (Perhaps you realize that a chemical balance also floated ashore!)

(10) (a) He shows that the mean and standard deviation of the amount of MIRA-KILL on a random egg are 2 and $2/\sqrt{3}$, respectively. Do the same.

(10) (b) He notices a pair of eggs next to each other, and wonders about the total amount Z of poison on the two eggs. He remembers something called the MGF of Z, and computes that it is

$$M_Z(t) = \begin{cases} 1 & \text{if } t = 0, \\ [e^{8t} - 2e^{4t} + 1]/16t^2 & \text{if } t \neq 0. \end{cases}$$

Show that he is correct.

(10) (c) He suspects that the total amount of poison in the two eggs is likely to be between 2 and 6 milligrams. But, having computed M_Z, he finds he doesn't know how to use it to find the probability of this event $2 \leq Z \leq 6$. Compute this probability <u>without</u> making use of M_Z.

(5) (d) He remembers, from his probability course, that he can use the result of (a) to show that $EZ = 4$. But he doubts whether he could have reached that conclusion if he knew only that the amount of MIRA-KILL on each egg had marginal distribution as given above, but that the amounts on different eggs might not be independent. Explain briefly whether he is right to have such doubts.

-6-

Enough insect eggs are killed (as described in problem 4) that p is large enough, and the tree is lucky enough, that it lives exactly 10 years, so it produces one crop of fruit.

The seeds would ordinarily be spread by a migrating flock of birds which pass through the island.

Unfortunately, the poisoned insect eggs are eaten by the birds before the cherries are ripe. Each bird eats 75 insects, and dies if it gets more than 120 milligrams of MIRA-KILL.

(5) (a) Use the result of Problem 5(a) to find the mean and variance of the total amount of poison a bird eats.

(10) (b) Use the result of (a) to show that the probability that a given bird survives is approximately

$$\frac{1}{\sqrt{2\pi}} \int_{-\infty}^{-3.0} e^{-x^2/2}\, dx.$$

[This is about .0013, which you needn't show.]

(10) -7-

All the birds die.

Moreover, they never had a chance to eat the cherries, which therefore drop directly under the tree, where by now there are too many leaves for the seeds to reach the soil and germinate.

The tree dies, childless.

-7-
(cont)

The lovely and edible flowers, which had relied on the insects for pollenization, gradually die out now that the insects are gone.

The man has nothing to eat and gradually starves. He finally manages to catch a small fish. If only he had 4 fish, he could survive for a week. He notices that these fish attract each other, and hits upon a scheme for catching more:

He holds his fish in the water. With probability 1/2 another fish will swim near slowly enough that the man can capture it with his other hand. But the other possibility of probability 1/2 is that this other fish will dart at his hand very rapidly and bite it, and he will lose his first fish, too, and have nothing. However, if he gets a second fish in this way, he can use one of his two fish to try to attract a third fish (or lose the second one) in the same way, and so on. He stops when his total gets to 0 or 4 fish, whichever happens sooner.

What is the <u>expected</u> number of fish he ends up with? (A one-line explanation should accompany your answer.)

(0)

EPILOGUE

(FOR ENTERTAINMENT ONLY; 0 POINTS; WORK ONLY IF YOU HAVE TIME, BUT PLEASE DO SO IF YOU HAVE IT, TO SATISFY THE INSTRUCTOR'S AND TA'S CURIOSITY!)

The fish escaped.

The man died of hunger.

The island, devoid of vegetation, was eroded by the rain and the sea, until it disappeared.

Many ages later, the ocean floor rumbles again. A single square island, 200 meters on side, arises, its center uniformly distributed throughout the ocean.

(a) Is it correct that the probability is 0 that the second island's center appears exactly where the first island's center was?

(b) If your answer to (a) is that this is indeed an event of probability 0, does this mean that it's <u>impossible</u> for history exactly to repeat itself?

MORAL: Performing well in a probability course may not be as earth shaking a consideration as it seems to you at this moment ...

Commentary on Papers [36], [38], [39]

INGRAM OLKIN
Stanford University

The three papers [36], [38], [39] deal with questions of minimaxity in a multivariate setting. These problems have remained elusive even in the simplest case of $p = 2$ or 3 dimensions.

In the case of Hotelling's T^2 test for testing $H: \xi = 0$ versus $A: \xi'\Sigma^{-1}\xi > 0$, it was known long ago (Simaika [1940]) that T^2 is uniformly most powerful among all level α tests whose power function depends on $\xi'\Sigma^{-1}\xi$. Further, Stein (1956) showed that T^2 is admissible. Paper [36] attacks the problem whether T^2 maximizes, among all level α tests, the minimum power under the hypothesis $H_1: N\xi'\Sigma^{-1}\xi = \delta > 0$. The results are positive for the very special case $p = 2$, $N = 3$. In this special case the problem reduces to solving an integral transform. Unfortunately, the method does not generalize to higher dimensions in p or to other values of N.

In a subsequent paper, Linnik, Pliss, and Šalaevskiĭ (1966) extended the result to $N = 4$ (for $p = 2$ dimensions). However, the method of proof was different so perhaps could be extended. This was accomplished by Šalaevskiĭ (1969) for all N and $p = 2$. To date, results for $p > 2$ are unavailable.

Because the multiple correlation test displays a similarity to Hotelling's T^2, paper [39] obtains a parallel result for the multiple correlation test in the simplest case $p = 3$ and $N = 4$. Thus was extended to $N > 4$ by Šalaevskiĭ (1969). (Note that when ξ is known, this is equivalent to the case $N = 3$.)

Again, a similar result for the complex analogue to Hotelling's T^2 was obtained by Halfina (1967) for $p = 2$ and $N = 3$, and for $N > 3$ by Šalaevkiĭ (1969).

Paper [38] is concerned with local and asymptotic minimaxity. In particular, it is shown that for every level α, p, and N, Hotelling's T^2 test is locally and asymptotically minimax for testing $\delta = 0$ versus $\delta = \lambda$ as $\lambda \to 0$ and $\lambda \to \infty$, respectively. A parallel result for local minimaxity of the multiple correlation test is obtained, however, the counterpart for asymptotic minimaxity does not carry over.

REFERENCES

Halfina, N. M. (1967). The minimax properties of the complex analogue of the T^2 test, (Russian) *Mat. Zametki* **2**, 635–644

Linnik, Ju. V., Pliss, V. A., and Šalaevskiĭ, O. V. (1966). On the theory of the Hotelling test, (Russian) *Dokl. Akad. Nauk SSSR* **168**, 743–746, [English translation: *Soviet Math. Dokl.* **7** (1966) 719–722]

Šalaevskiĭ, O. V. (1968). The minimax character of Hotelling's T^2 test, (Russian) *Dokl. Akad. Nauk SSSR* **180**, 1048–1050, [English translation: *Soviet Math. Dokl.* **9** (1968) 733–736]

Šalaevskiĭ, O. V. (1969). Minimax character of Hotelling's T^2-test I, (Russian) *Zap. Naučn. Sem. Leningrad. Otdel. Mat. Inst. Steklov (LOMI)* **13**, 138–182

Šalaevskiĭ, O. V. (1969). Minimax character of the R^2 test of a relation, (Russian) *Zap. Naučn. Sem. Leningrad. Otdel. Mat. Inst. Steklov (LOMI)* **13**, 183–248

Commentary on Paper [45]

INGRAM OLKIN
Stanford University

Paper [45] is one of my favorites in that a very simple device yields a surprising number of important results. The goal is to show that certain procedures in a multivariate context are admissible Bayes procedures. To illustrate the idea consider the canonical form for the general linear hypothesis: $X = (\dot{X}, \ddot{X}, \dddot{X})$ where $\dot{X}: p \times r$, $\ddot{X}: p \times n$, $\dddot{X}: p \times h$, $EX = (\xi, 0, \nu)$, and the columns of X are independent with common variance matrix Σ. The problem is to test $H: \xi = 0$. The key point is that under both H and the alternative we assign measures to Σ of the form $\Sigma^{-1} = I + \eta\eta'$, for some $p \times r$ matrix η, and under the alternative we assign measures to ξ of the form $\xi = \Sigma\eta$.

In this framework, it is shown that the critical region $\operatorname{tr}(\dot{X}\dot{X}' + \ddot{X}\ddot{X}')^{-1}\dot{X}\dot{X}' > c$ is an admissible Bayes procedure. Also, the likelihood ratio test: $\det(\dot{X}\dot{X}' + \ddot{X}\ddot{X}')^{-1}\ddot{X}\ddot{X}' < c$ is admissible Bayes (for $n > p + r - 1$).
Variations on this theme lead to results for testing for the independence of sets of variates, the Behrens–Fisher problem, classification, equality of covariance matrices, and others.

To carry out the various computations, the authors use a prior (under H) of the form

$$d\pi_0(\eta)/d\eta = |I + \eta\eta'|^{-m/2}.$$

The use of this prior led to some restrictions on the dimensions n, p, r. In trying to weaken these restrictions I wrote to Jack on February 16, 1965 suggesting an alternative prior. Jack's reply of February 26, 1965 follows:

"Many thanks for your helpful comments on our papers. You are indeed right that the stupid divisibility restriction in Sec. 7 can be largely avoided (actually, not by taking c near 1 as you wrote, I think, but rather with $c = \varepsilon + (p-1)/\min_i n_i$). The integrability condition is now $\min_i n_i > 2(p-1)$, which is slightly worse than before, reflecting the reason why η was always a vector where possible, in order to get as close as possible to settling the min sample size case. Thus, in Sec. 5 your remark yields a wider spectrum of Π's than before, but only for larger minimum n. This apriori dist'n on Σ^{-1} is that of Sec. 4(iv)(b), p. 25, but we never listed all the modifications in other sections as someone with more energy and competence would have; thus, with $\eta\ p \times r$, $r \geq p$,

$$|V|^{(r-p-1+t)/2}/|I+V|^{n/2} \sim |\eta\eta'|^{t/2}/|I+\eta\eta'|^{n/2};$$

one gets a different test from that with $t = 0$ in Sec. 4, so it is not surprising that you didn't yet see how to do this case in terms of V, although perhaps a somewhat different function of V and ξ would work. It is interesting that for $p < r$ in Sec. 7 if $|\eta\eta'|^{t/2}$ is replaced by $|\eta\eta'|^{t/2}$ one gets a different test, but you're probably familiar with this sort of thing. We're very much obliged to you for your findings, and hope it's o.k. with you if we stick in a brief remark stating and crediting them.

This whole process of inserting a sufficiently large power of $|\Sigma^{-1}|$ and getting the LR test (in Secs. 5 and 7) is actually a special case of part of Schwartz's thesis, wherein he shows in fair generality that, for certain problems with groups acting, some modification of a natural test like the LR test (actually, based on such a test statistic for a slightly different sample size, which statistic often doesn't depend much on " ") is Bayes. He'll send you a copy when the typing is finished.

In going over our correspondence on the Giri reference in the paper, I conclude that the ref to you, Shrik, Coch, Bliss is appropriate in 6(ii), NOT in 6(iv) with Giri as you stated. If $\xi = (\xi_1, \xi_2)$, you consider hypotheses involving ξ_1, while Giri considers η_1 where $(\eta_1, \eta_2)' = \Sigma^{-1}\xi'$. [Aren't you proud of my now writing *NOT column* vectors, but instead, *row* vectors as you insist? I actually just wrote up a design paper in this way, so maybe I learned something!] This is quite different from the point of view of which tests might be useful for either problem. The η formulation still seems a little unnatural to me from a practical viewpoint. I believe Rao considered the η formulation before Giri, so the section heading should be changed as you suggest, in any case. Let me know, please, if you think the above is in error, so we can make appropriate galley changes if necessary.

Again, thanks for all your help, and best wishes to all O's."

Jack
Feb. 26, 1965

Since you say it's limericktime:

A student of Ted one day chose
To take orals in "land of the Mose";
But Red almost wrecked hers:
She wrote out her vectors
In columns instead of in rows.

[*Editorial note*: The names Ted, Mose, Red refer to T. W. Anderson, Lincoln Moses, Ingram Olkin.]

Commentary on Paper [4]

David Siegmund

Stanford University

Paper [4] is concerned with recursive estimation of the maximum (or minimum) of a function which can be determined experimentally up to some additive stochastic error. Kiefer, in his masters' thesis which later appeared as [8], had earlier considered the problem when there is no error. The method estimates the location of the maximum after $2n$ observations by a linear combination (depending on n) of the estimate after $2(n-1)$ observations and an estimate of the slope of the function at this previous estimate. The method was stimulated by that of Robbins and Monro (1951), who proposed a similar recursive scheme for estimation of the zero of a function observed with error.

Since 1952 more than a hundred papers and at least three books have been written on the subject of stochastic approximation. One outstanding contribution is Dvoretzky's theorem (1956), which showed that the results of both Kiefer–Wolfowitz and Robbins–Monro are special cases of an abstract stochastic approximation scheme. Other noteworthy theoretical contributions of the subsequent twenty years are those of Sacks (1958), Venter (1967), and a series of papers by Fabian (e.g. Fabian, 1971).

During the past decade problems of on line identification and adaptive control of stochastic systems have stimulated renewed interest in generalized stochastic approximation schemes and their relation to recursive least squares estimation. Although recursive least squares estimators seem to converge more rapidly and hence to be more useful in practice, the simpler stochastic approximation algorithms are more easily analyzed and often provide important insights for dealing with complex control problems. For important contributions to the mathematical theory see Lai and Robbins (1979) and Lai and Wei (1982). A discussion of applications in engineering is given by Ljung (1977).

REFERENCES

Dvoretzky, A. (1956). On stochastic approximation, *Proc. Third Berkeley Symp.*, Vol. 1, 39–56, University of California Press

Fabian, V. (1971). Stochastic approximation, in *Optimizing Methods in Statistics*, (J. S. Rustagi, ed.), Academic Press, New York, 439–470

Lai, T. L. and Robbins, H. (1979). Adaptive design and stochastic approximation, *Ann. Statist.* **7**, 1196–1221

Lai, T. L. and Wei, C. Z. (1982). Least squares estimates in stochastic regression models with applications to identification and control of dynamic systems, *Ann. Statist.* **10**, 154–166

Ljung, L. (1977). Analysis of recursive stochastic algorithms, *IEEE Trans. Aut. Control*, **AC-22**, 551–555

Robbins, H. and Monro, S. (1951). A stochastic approximation method, *Ann. Math. Statist.*, **22**, 400–407

Sacks, J. (1958). Asymptotic distribution of stochastic approximation procedures, *Ann. Math. Statist.* **29**, 373–405

Venter, J. (1967). An extension of the Robbins-Monro procedure, *Ann. Math.*

Permissions

Springer-Verlag would like to thank the original publishers of the papers of Jack Kiefer for granting permission to reprint specific papers in this collection:

[1] Reprinted from *Ann. Math. Statist.* **22,** © 1951 by The Institute of Mathematical Statistics.
[2] Reprinted from *Econometrica* **20,** ©1952 by The Econometric Society.
[3] Reprinted from *Econometrica* **20,** © 1952 by The Econometric Society.
[4] Reprinted from *Ann. Math. Statist.* **23,** © 1952 by The Institute of Mathematical Statistics.
[5] Reprinted from *Ann. Math. Statist.* **23,** © 1952 by The Institute of Mathematical Statistics.
[6] Reprinted from *Ann. Math. Statist.* **23,** © 1952 by The Institute of Mathematical Statistics.
[7] Reprinted from *Ann. Math. Statist.* **24,** © 1953 by The Institute of Mathematical Statistics.
[8] Reprinted from *Proc. Amer. Math. Soc.* **4,** © 1953 by The American Mathematical Society.
[9] Reprinted from *Ann. Math. Statist.* **24, 30,** © 1953, 1954 by The Institute of Mathematical Statistics.
[10] Reprinted from *Ann. Math. Statist.* **24,** © 1953 by The Institute of Mathematical Statistics.
[11] Reprinted from *Econometrica* **21,** © 1953 by The Econometric Society.
[12] Reprinted from *Trans. Amer. Math. Soc.* **78,** © 1955 by The American Mathematical Society.
[13] Reprinted from *Ann. Math. Statist.* **26,** © 1955 by The Institute of Mathematical Statistics.
[14] Reprinted from *Ann. Math. Statist.* **27,** © 1956 by The Institute of Mathematical Statistics.
[15] Reprinted from *Ann. Math. Statist.* **27,** © 1956 by The Institute of Mathematical Statistics.
[16] Reprinted from *Ann. Math. Statist.* **27,** © 1956 by The Institute of Mathematical Statistics.
[17] Reprinted from *Naval Res. Logist. Quart.* **3,** © 1956 by The Office of Naval Research.
[18] Reprinted from *Ann. Math. Statist.* **28,** © 1957 by The Institute of Mathematical Statistics.
[19] Reprinted from *Ann. Math. Statist.* **28,** © 1957 by The Institute of Mathematical Statistics.
[20] Reprinted from *J. Soc. Indust. Appl. Math.* **5,** © 1957 by The Society for Industrial and Applied Mathematics.
[21] Reprinted from *Trans. Amer. Math. Soc.* **87,** © 1958 by The American Mathematical Society.
[24] Reprinted from *Ann. Math. Statist.* **30,** © 1959 by The Institute of Mathematical Statistics.
[25] Reprinted from *Ann. Math. Statist.* **30,** © 1959 by The Institute of Mathematical Statistics.
[27] Reprinted from *Proc. Cambridge Philos. Soc.* **55,** © 1959 by The Cambridge Philosophical Society.